Light-Cone Gauge: $\quad X^+ = \beta\alpha' p^+ \tau, \quad p^+$

$$\mathcal{P}^{\tau\mu} = \frac{1}{2\pi\alpha'}\,\dot{X}^\mu\,, \quad \mathcal{P}^{\sigma\mu} = -\frac{1}{2\pi\alpha'}\,X^{\mu\prime}\,, \quad 0 = X^\mu - X^{\mu\prime\prime}$$

$$\dot{X}^- \pm X^{-\prime} = \frac{1}{2\beta\alpha' p^+}\left(\dot{X}^I \pm X^{I\prime}\right)^2$$

Open String: $\quad X^\mu(\tau,\sigma) = x_0^\mu + \sqrt{2\alpha'}\,\alpha_0^\mu\,\tau + i\sqrt{2\alpha'}\sum_{n\neq0}\frac{1}{n}\,\alpha_n^\mu\,e^{-in\tau}\,\cos n\sigma$

$$\alpha_0^\mu = \sqrt{2\alpha'}\,p^\mu\,, \quad \left[\alpha_m^I,\alpha_n^J\right] = m\,\delta_{m+n,0}\,\delta^{IJ}$$

$$M^2 = \frac{1}{\alpha'}\left(N^\perp - 1\right)\,, \quad N^\perp = \sum_{p=1}^{\infty}\alpha_{-p}^I\,\alpha_p^I$$

$$\left[L_m^\perp, L_n^\perp\right] = (m-n)L_{m+n}^\perp + \frac{D-2}{12}\,(m^3-m)\delta_{m+n,0}$$

$$L_n^\perp = \frac{1}{2}\sum_{p=-\infty}^{\infty}\alpha_{n-p}^I\,\alpha_p^I\;(n\neq0)\,, \quad L_0^\perp = \alpha'\,p^I\,p^I + N^\perp$$

$$\alpha_n^- = \frac{1}{\sqrt{2\alpha'}\,p^+}L_n^\perp\,, \quad \left[L_m^\perp,\alpha_n^J\right] = -n\,\alpha_{m+n}^J$$

Closed String: $\quad X^\mu(\tau,\sigma) = x_0^\mu + \sqrt{2\alpha'}\,\alpha_0^\mu\,\tau + i\sqrt{\frac{\alpha'}{2}}\sum_{n\neq0}\frac{e^{-in\tau}}{n}\left(\alpha_n^\mu\,e^{in\sigma} + \bar{\alpha}_n^\mu\,e^{-in\sigma}\right)$

$$\alpha_0^\mu = \sqrt{\frac{\alpha'}{2}}\,p^\mu\,, \quad L_0^\perp = \frac{\alpha'}{4}\,p^I\,p^I + N^\perp\,, \quad \bar{L}_0^\perp = \frac{\alpha'}{4}\,p^I\,p^I + \bar{N}^\perp$$

$$N^\perp = \sum_{p=1}^{\infty}\alpha_{-p}^I\,\alpha_p^I\,, \quad \bar{N}^\perp = \sum_{p=1}^{\infty}\bar{\alpha}_{-p}^I\,\bar{\alpha}_p^I$$

$$M^2 = \frac{2}{\alpha'}\left(N^\perp + \bar{N}^\perp - 2\right)\,, \quad \bar{N}^\perp = N^\perp$$

NS sector: $\quad (-1)^F = -1$ on ground state $|NS\rangle\,,$

$$a_{NS} = -\frac{1}{48}\,, \quad \alpha'M^2 = -\frac{1}{2} + N^\perp\,, \quad NS+ : (-1)^F = +1$$

R-sector: 8 fermion zero modes $\rightarrow\; 2^{8/2} = 16$ ground states

$$(-1)^F = -1 \text{ on } |R_a\rangle,\; a = 1,\dots8\,, \quad (-1)^F = +1 \text{ on } |R_{\bar{a}}\rangle,\; \bar{a} = \bar{1},\dots\bar{8}\,.$$

$$a_R = \frac{1}{24}\,, \quad \alpha'M^2 = N^\perp\,.$$

A First Course in String Theory
Second Edition

Barton Zwiebach is once again faithful to his goal of making string theory accessible to undergraduates. He presents the main concepts of string theory in a concrete and physical way to develop intuition before formalism, often through simplified and illustrative examples. Complete and thorough in its coverage, this new edition now includes the AdS/CFT correspondence and introduces superstrings. It is perfectly suited to introductory courses in string theory for students with a background in mathematics and physics.

This new edition contains completely new chapters on the AdS/CFT correspondence, an introduction to superstrings, and new sections covering strings on orbifolds, cosmic strings, moduli stabilization, and the string theory landscape. There are almost 300 problems and exercises, with password protected solutions available to instructors at www.cambridge.org/zwiebach.

Barton Zwiebach is Professor of Physics at the Massachusetts Institute of Technology. His central contributions have been in the area of string field theory, where he did the early work on the construction of the field theory of open strings and then developed the field theory of closed strings. He has also made important contributions to the subjects of D-branes with exceptional symmetry and tachyon condensation.

From the first edition
'A refreshingly different approach to string theory that requires remarkably little previous knowledge of quantum theory or relativity. This highlights fundamental features of the theory that make it so radically different from theories based on point-like particles. This book makes the subject amenable to undergraduates but it will also appeal greatly to beginning researchers who may be overwhelmed by the standard textbooks.'

Professor Michael Green, University of Cambridge

'Barton Zwiebach has written a careful and thorough introduction to string theory that is suitable for a full-year course at the advanced undergraduate level. There has been much demand for a book about string theory at this level, and this one should go a long way towards meeting that demand.'

Professor John Schwarz, California Institute of Technology

'There is a great curiosity about string theory, not only among physics undergraduates but also among professional scientists outside of the field. This audience needs a text that goes much further than the popular accounts but without the full technical detail of a graduate text. Zwiebach's book meets this need in a clear and accessible manner. It is well-grounded in familiar physical concepts, and proceeds through some of the most timely and exciting aspects of the subject.'

Professor Joseph Polchinski, University of California, Santa Barbara

'Zwiebach, a respected researcher in the field and a much beloved teacher at MIT, is truly faithful to his goal of making string theory accessible to advanced undergraduates – the test develops intuition before formalism, usually through simplified and illustrative examples . . . Zwiebach avoids the temptation of including topics that would weigh the book down and make many students rush it back to the shelf and quit the course.'

Marcelo Gleiser, Physics Today

'. . . well-written . . . takes us through the hottest topics in string theory research, requiring only a solid background in mechanics and some basic quantum mechanics . . . This is not just one more text in the ever-growing canon of popular books on string theory . . .'

Andreas Karch, Times Higher Education Supplement

'. . . the book provides an excellent basis for an introductory course on string theory and is well-suited for self-study by graduate students or any physicist who wants to learn the basics of string theory'.

Zentralblatt MATH

'. . . excellent introduction by Zwiebach . . . aimed at advanced undergraduates who have some background in quantum mechanics and special relativity, but have not necessarily mastered quantum field theory and general relativity yet . . . the book . . . is a very thorough introduction to the subject . . . Equipped with this background, the reader can safely start to tackle the books by Green, Schwarz and Witten and by Polchinski.'

Marcel L. Vonk, Mathematical Reviews Clippings

Cover illustration: a composite illustrating open string motion as we vary the strength of an electric field that points along the rotational axis of symmetry. There are three surfaces, each composed of two lobes joined at the origin and shown with the same color. Each surface is traced by a rotating open string that, at various times, appears as a line stretching from the boundary of a lobe down to the origin and then out to the boundary of the opposite lobe. The inner, middle, and elongated lobes arise as the magnitude of the electric field is increased. For further details, see Problem 19.2.

A First Course in String Theory

Second Edition

Barton Zwiebach

Massachusetts Institute of Technology

CAMBRIDGE
UNIVERSITY PRESS

Shaftesbury Road, Cambridge CB2 8EA, United Kingdom

One Liberty Plaza, 20th Floor, New York, NY 10006, USA

477 Williamstown Road, Port Melbourne, VIC 3207, Australia

314–321, 3rd Floor, Plot 3, Splendor Forum, Jasola District Centre, New Delhi – 110025, India

103 Penang Road, #05–06/07, Visioncrest Commercial, Singapore 238467

Cambridge University Press is part of Cambridge University Press & Assessment,
a department of the University of Cambridge.

We share the University's mission to contribute to society through the pursuit of
education, learning and research at the highest international levels of excellence.

www.cambridge.org
Information on this title: www.cambridge.org/9780521880329

First published 2004
Reprinted 2005, 2007
Second edition 2009 (version 12, November 2022)

Printed in the United Kingdom by TJ Books Limited, Padstow Cornwall

A catalogue record for this publication is available from the British Library

ISBN 978-0-521-88032-9 Hardback

To my parents, Oscar and Betty Zwiebach, with gratitude

Contents

Foreword

String theory is one of the most exciting fields in theoretical physics. This ambitious and speculative theory offers the potential of unifying gravity and all the other forces of nature and all forms of matter into one unified conceptual structure.

String theory has the unfortunate reputation of being impossibly difficult to understand. To some extent this is because, even to its practitioners, the theory is so new and so ill understood. However, the basic concepts of string theory are quite simple and should be accessible to students of physics with only advanced undergraduate training.

I have often been asked by students and by fellow physicists to recommend an introduction to the basics of string theory. Until now all I could do was point them either to popular science accounts or to advanced textbooks. But now I can recommend to them Barton Zwiebach's excellent book.

Zwiebach is an accomplished string theorist, who has made many important contributions to the theory, especially to the development of string field theory. In this book he presents a remarkably comprehensive description of string theory that starts at the beginning, assumes only minimal knowledge of advanced physics, and proceeds to the current frontiers of physics. Already tested in the form of a very successful undergraduate course at MIT, Zwiebach's exposition proves that string theory can be understood and appreciated by a wide audience.

I strongly recommend this book to anyone who wants to learn the basics of string theory.

David Gross
Director, Kavli Institute For Theoretical Physics
University of California, Santa Barbara

From the Preface to the First Edition

The idea of having a serious string theory course for undergraduates was first suggested to me by a group of MIT sophomores sometime in May of 2001. I was teaching Statistical Physics, and I had spent an hour-long recitation explaining how a relativistic string at high energies appears to approach a constant temperature (the Hagedorn temperature). I was intrigued by the idea of a basic string theory course, but it was not immediately clear to me that a useful one could be devised at this level.

A few months later, I had a conversation with Marc Kastner, the Physics Department Head. In passing, I told him about the sophomores' request for a string theory course. Kastner's instantaneous and enthusiastic reaction made me consider seriously the idea for the first time. At the end of 2001, a new course was added to the undergraduate physics curriculum at MIT. In the spring term of 2002 I taught *String Theory for Undergraduates* for the first time. This book grew out of the lecture notes for that course.

When we think about teaching string theory at the undergraduate level the main question is, "Can the material really be explained at this level?". After teaching the subject two times, I am convinced that the answer to the question is a definite yes. Although a complete mastery of string theory requires a graduate-level physics education, the basics of string theory can be well understood with the limited tools acquired in the first two or three years of an undergraduate education.

What is the value of learning string theory, for an undergraduate? By exposing the students to cutting-edge ideas, a course in string theory can help nurture the excitement and enthusiasm that led them to choose physics as a major. Moreover, students will find in string theory an opportunity to sharpen and refine their understanding of most of the undergraduate physics curriculum. This is valuable even for students who do not plan to specialize in theoretical physics.

This book was tailored to be understandable to an advanced undergraduate. Therefore, I believe it will be a readable introduction to string theory for any graduate student or, in fact, for any physicist who wants to learn the basics of string theory.

Acknowledgements

I would like to thank Marc Kastner, Physics Department Head, for his enthusiastic support and his interest. I am also grateful to Thomas Greytak, Associate Head for Education, and to Robert Jaffe, Director of the Center for Theoretical Physics, both of whom kindly supported this project.

Teaching string theory to a class composed largely of bright undergraduates was both a stimulating and a rewarding experience. I am grateful to the group of students that composed the first class:

Jeffrey Brock	Adam Granich	Trisha Montalbo
Zilong Chen	Markéta Havlíčková	Eugene Motoyama
Blair Connely	Kenneth Jensen	Megha Padi
Ivailo Dimov	Michael Krypel	Ian Parrish
Peter Eckley	Francis Lam	James Pate
Qudsia Ejaz	Philippe Larochelle	Timothy Richards
Kasey Ensslin	Gabrielle Magro	James Smith
Teresa Fazio	Sourav Mandal	Morgan Sonderegger
Caglar Girit	Stefanos Marnerides	David Starr
Donglai Gong		

They were enthusiastic, funny, and lively. My lectures were voice-recorded and three of the students, Gabrielle Magro, Megha Padi, and David Starr, turned the tapes and the blackboard equations into LaTeX files. I am grateful to the three of them for their dedication and for the care they took in creating accurate files. They provided the impetus to start the process of writing a book. I edited the files to produce lecture notes.

Additional files for a set of summer lectures were created by Gabrielle and Megha. In the next six months the lecture notes became the draft for a book. After teaching the course for a second time in the spring term of 2003 and a long summer of edits and revisions, the book was completed in October 2003.

By the time the lecture notes had become a book draft, David Starr offered to read it critically. He basically marked every paragraph, suggesting improvements in the exposition and demonstrating an uncanny ability to spot weak points. His criticism forced me to go through major rewriting. His input was tremendous. Whatever degree of clarity has been achieved, it is in no small measure thanks to his effort.

I am delighted to acknowledge help and advice from my friend and colleague Jeffrey Goldstone. He shared generously his understanding of string theory, and several sections in this book literally grew out of his comments. He helped me teach the course the second time that it was offered. While doing so, he offered perceptive criticism of the whole text. He also helped improve many of the problems, for which he wrote elegant solutions.

The input of my friend and collaborator Ashoke Sen was critical. He believed that string theory could be taught at a basic level and encouraged me to try to do it. I consulted repeatedly with him about the topics to be covered and about the strategies to present them. He kindly read the first full set of lecture notes and gave invaluable advice that helped shape the form of this book.

The help and interest of many people made writing this book a very pleasant task. For detailed comments on all of its content I am indebted to Chien-Hao Liu and to James Stasheff. Alan Dunn and Blake Stacey helped test the problems that could not be assigned in class. Jan Troost was a sounding board and provided advice and criticism. I've relied on the knowledge of my string theory colleagues – Amihay Hanany, Daniel Freedman, and Washington Taylor. I'd like to thank Philip Argyres, Andreas Karch, and Frieder Lenz for testing the lecture notes with their students. Juan Maldacena and Samir Mathur provided helpful input on the subject of string thermodynamics and black holes. Boris Körs, Fernando Quevedo, and Angel Uranga helped and advised on the subject of string phenomenology. Thanks are also due to Tamsin van Essen, editor at Cambridge, for her advice and her careful work during the entire publishing process.

Finally, I would like to thank my wife Gaby and my children Cecile, Evy, Margaret, and Aaron. At every step of the way I was showered with their love and support. Cecile and Evy read parts of the manuscript and advised on language. Questions on string theory from Gaby and Margaret tested my ability to explain. Young Aaron insisted that a ghost sitting on a string would make a perfect cover page, but we settled for strings moving in electric fields.

Barton Zwiebach
Cambridge, Massachusetts, 2003

Preface to the Second Edition

It has been almost five years since I finished writing the first edition of *A First Course in String Theory*. I have since taught the undergraduate string theory course at MIT three times, and I have received comments and suggestions from colleagues all over the world. I have learned what parts of the book are most challenging for the students, and I have heard requests for extra material.

As in the first edition, the book is broadly divided into Part I (Basics) and Part II (Developments). In this second edition I have improved the clarity of many arguments and the general readability of Part I. This part is studied by the largest number of readers, many of them independently and outside of the classroom setting. The changes should make study easier. There are more figures and the number of problems has been increased to better cover the range of ideas developed in the text. Part I has five new sections and one new chapter. The new sections discuss the classical motion of closed strings, cosmic strings, and orbifolds. The new chapter, Chapter 14, is the last one of Part I. It explains the basics of superstring theory.

Part II has changed as well. The ordering of chapters has been altered to bring T-duality earlier into the book. The material relevant to particle physics has been collected in Chapter 21 and includes a new section on moduli stabilization and the landscape. Chapter 23 is new and is entirely devoted to strong interactions and the AdS/CFT correspondence. I aim to give there a gentle introduction to this lively area of research. The number of chapters in the book has gone from twenty-three to twenty-six, a nice number to end a book on string theory!

I want to thank Hong Liu and Juan Maldacena for helpful input on the subject of AdS/CFT. Many thanks are also due to Alan Guth, who helped me teach the string theory course in the spring term of 2007. He tested many of the new problems and offered very valuable criticism of the text.

About this book

A First Course in String Theory should be accessible to anyone who has been exposed to special relativity, basic quantum mechanics, electromagnetism, and introductory statistical physics. Some familiarity with Lagrangian mechanics is useful but not indispensable.

Except for the introduction, all chapters contain exercises and problems. The exercises, called *Quick calculations*, are inserted at various points throughout the text. They are control calculations that are expected to be straightforward. Undue difficulty in carrying them out may indicate problems understanding the material. The problems at the end of the chapters are more challenging and sometimes develop new ideas. A problem marked with a dagger[†] is one whose results are cited later in the text. A mastery of the material requires solving all the exercises and many of the problems. All the problems should be read, at least.

Throughout most of the book the material is developed in a self-contained way, and very little must be taken on faith. Chapters 14, 21, 22, and 23 contain a few sections that address subjects of much interest for which a full explanation cannot be provided at the level of this book. The reader will be asked to accept some reasonable facts at face value, but otherwise the material is developed logically and should be *fully* understandable. These sections are *not* addressed to experts.

This book has two parts. Part I is called "Basics," and Part II is called "Developments." Part I begins with Chapter 1 and concludes with Chapter 14. Part II comprises the rest of the book: it begins with Chapter 15 and it ends with Chapter 26.

Chapter 1 serves as an introduction. Chapter 2 reviews special relativity, but it also introduces concepts that are likely to be new: light-cone coordinates, light-cone energy, compact extra dimensions, and orbifolds. In Chapter 3 we review electrodynamics and its manifestly relativistic formulation. We make some comments on general relativity and study the effect of compact dimensions on the Planck length. We are able at this point to examine the exciting possibility that large extra dimensions may exist. Chapter 4 uses nonrelativistic strings to develop some intuition, to review the Lagrangian formulation of mechanics, and to introduce terminology. Chapter 5 uses the relativistic point particle to prepare the ground for the study of the relativistic string. The power and elegance of the Lagrangian formulation become evident at this point. The first encounter with string theory happens in Chapter 6, which deals with the classical dynamics of the relativistic string. This is a very important chapter, and it must be understood thoroughly. Chapter 7 solidifies the understanding of string dynamics through the detailed study of string motion, both for open and for closed strings. It includes a section on cosmic strings, a topic of potential experimental relevance. Chapters 1 through 7 could comprise a mini-course in string theory.

Chapters 8 through 11 prepare the ground for the quantization of relativistic strings. In Chapter 8, one learns how to calculate conserved quantities, such as the momentum and the angular momentum of free strings. Chapter 9 gives the light-cone gauge solution of the string equations of motion and introduces the terminology that is used in the quantum theory. Chapter 10 explains the basics of quantum fields and particle states, with emphasis on the counting of the parameters that characterize scalar field states, photon states, and graviton states. In Chapter 11 we perform the light-cone gauge quantization of the relativistic particle. It all comes together in Chapter 12, another important chapter that should be understood thoroughly. This chapter presents the light-cone gauge quantization of the open relativistic string. The critical dimension is obtained and photon states are shown to emerge. Chapter 12 contains a section on the subject of tachyon condensation. Chapter 13 discusses the quantization of closed strings and the emergence of graviton states. It also contains two sections that deal with quantum closed strings on the simplest orbifold, the half-line. Chapter 14 is the last chapter of Part I. It introduces the subject of superstrings. The Ramond and Neveu–Schwarz sectors of open strings are presented and combined to obtain a supersymmetric theory. The chapter concludes with a brief discussion of type II closed string theories.

The first part of this book can be characterized as an uphill road that leads to the quantization of the string at the summit. In the second part of this book the climb is over. The pace slows down a little, and the material elaborates upon previously introduced ideas. In Part II one reaps many rewards for the effort exerted in Part I.

The first chapter in Part II, Chapter 15, deals with the important subject of open strings on various D-brane configurations. The discussion of orientifolds has been relegated to the problems at the end of the chapter. Chapter 16 introduces the concept of string charge and demonstrates that the endpoints of open strings carry Maxwell charge. The next four chapters are organized around the fascinating subject of T-duality. Chapters 17 and 18 present the T-duality properties of closed and open strings, respectively. Chapter 19 studies D-branes with electromagnetic fields, using T-duality as the main tool. Chapter 20 introduces the general framework of nonlinear electrodynamics. It demonstrates that electromagnetic fields in string theory are governed by Born–Infeld theory, a nonlinear theory in which the self-energy of point charges is finite.

String models of particle physics are considered in Chapter 21. This chapter explains in detail the particle content of the Standard Model and discusses one approach, based on intersecting D6-branes, to the construction of a realistic string model. The chapter concludes with some material on moduli stabilization and the landscape.

Chapter 22 begins with string thermodynamics, followed by the subject of black hole entropy. It presents string theory attempts to derive the entropy of Schwarzschild black holes and the successful derivation of the entropy for a supersymmetric black hole. The applications of string theory to strong interactions are studied in Chapter 23. After a discussion of Regge trajectories and the quark–antiquark potential, the subject turns to the AdS/CFT correspondence. The correspondence is discussed in some detail, with emphasis on the geometry of AdS spaces. A section on the quark–gluon plasma is included.

Chapter 24 gives an introduction to the Lorentz covariant quantization of strings. It also introduces the Polyakov string action. The last two chapters in the book, Chapters 25 and 26, examine string interactions. We learn that the string diagrams which represent the processes of string interactions are Riemann surfaces. These two chapters assume a little familiarity with complex variables and have a mathematical flavor. One important goal here is to provide insight into the absence of ultraviolet divergences in string theory, the fact that made string theory the first candidate for a theory of quantum gravity.

In this book I have tried to emphasize the connections with ideas that students have learned before. The quantization of strings is described as the quantization of an infinite number of oscillators. String charge is visualized as a Maxwell current. The effects of Wilson lines on circles are compared with the Bohm–Aharonov effect. The modulus of an annulus is related to the capacitance of a cylindrical conductor, and so forth and so on. The treatment of topics is generally explicit and detailed, with formalism kept to a minimum.

The choice was made to use the light-cone gauge to quantize the strings. This approach to quantization can be understood in full detail by students with some prior exposure to quantum mechanics. The same is *not* true for the Lorentz covariant quantization of strings, where states of negative norms must be dealt with, the Hamiltonian vanishes, and there is no conventional looking Schrödinger equation. The light-cone approach suffices for most physical problems and, in fact, simplifies the treatment of several questions.

This book as a textbook

Part I of the book is structured tightly. Little can be omitted without hampering the understanding of string quantization. The first chapter in Part II (on D-branes) is important for much of the later material. Many choices among the remaining chapters are possible. Different readers/instructors may take different routes.

My experience suggests that the complete book can be covered in a full-year course at the undergraduate level. In a school with an academic year composed of three quarters, Part I and four chapters from Part II may be covered in two quarters. In a school with an academic year composed of two semesters, Part I and two chapters from Part II may be covered in one semester. In either case, the choice of chapters from Part II is a matter of taste. Chapters 21, 22, and 23 give an appreciation for current research in string theory. Lecturers who prefer to focus on T-duality and its implications will cover as much as possible from Chapters 17–20. If this book is used to teach exclusively to graduate students, the pace can be quickened considerably.

An updated list of corrections can be found at http://xserver.lns.mit.edu/~zwiebach/firstcourse.html. Solutions to the problems in the book are available to lecturers via solutions@cambridge.org.

BASICS

A brief introduction

Here we meet string theory for the first time. We see how it fits into the historical development of physics, and how it aims to provide a unified description of all fundamental interactions.

1.1 The road to unification

Over the course of time, the development of physics has been marked by unifications: events when different phenomena were recognized to be related and theories were adjusted to reflect such recognition. One of the most significant of these unifications occurred in the nineteenth century.

For a while, electricity and magnetism had appeared to be unrelated physical phenomena. Electricity was studied first. The remarkable experiments of Henry Cavendish were performed in the period from 1771 to 1773. They were followed by the investigations of Charles Augustin de Coulomb, which were completed in 1785. These works provided a theory of static electricity, or electrostatics. Subsequent research into magnetism, however, began to reveal connections with electricity. In 1819 Hans Christian Oersted discovered that the electric current on a wire can deflect the needle of a compass placed nearby. Shortly thereafter, Jean-Baptiste Biot and Felix Savart (1820) and André-Marie Ampère (1820–1825) established the rules by which electric currents produce magnetic fields. A crucial step was taken by Michael Faraday (1831), who showed that changing magnetic fields generate electric fields. Equations that described all of these results became available, but they were, in fact, inconsistent. It was James Clerk Maxwell (1865) who constructed a consistent set of equations by adding a new term to one of the equations. Not only did this term remove the inconsistencies, but it also resulted in the prediction of electromagnetic waves. For this great insight, the equations of *electromagnetism* (or electrodynamics) are now called "Maxwell's equations." These equations unify electricity and magnetism into a consistent whole. This elegant and aesthetically pleasing unification was not optional. Separate theories of electricity and magnetism would be inconsistent.

Another fundamental unification of two types of phenomena occurred in the late 1960s, about one-hundred years after the work of Maxwell. This unification revealed the deep relationship between electromagnetic forces and the forces responsible for weak interactions. To appreciate the significance of this unification it is necessary first to review the main developments that occurred in physics since the time of Maxwell.

An important change of paradigm was triggered by Albert Einstein's special theory of relativity. In this theory one finds a striking conceptual unification of the separate notions of space and time. Different from a unification of forces, the merging of space and time into a spacetime continuum represented a new recognition of the nature of the *arena* where physical phenomena take place. Newtonian mechanics was replaced by relativistic mechanics, and older ideas of absolute time were abandoned. Mass and energy were shown to be interchangeable.

Another change of paradigm, perhaps an even more dramatic one, was brought forth by the discovery of quantum mechanics. Developed by Erwin Schrödinger, Werner Heisenberg, Paul Dirac and others, quantum theory was verified to be the correct framework to describe microscopic phenomena. In quantum mechanics classical observables become operators. If two operators fail to commute, the corresponding observables cannot be measured simultaneously. Quantum mechanics is a framework, more than a theory. It gives the rules by which theories must be used to extract physical predictions.

In addition to these developments, four fundamental forces had been recognized to exist in nature. Let us have a brief look at them.

One of them is the force of gravity. This force has been known since antiquity, but it was first described accurately by Isaac Newton. Gravity underwent a profound reformulation in Albert Einstein's theory of general relativity. In this theory, the spacetime arena of special relativity acquires a life of its own, and gravitational forces arise from the curvature of this dynamical spacetime. Einstein's general relativity is a classical theory of gravitation. It is not formulated as a quantum theory.

The second fundamental force is the electromagnetic force. As we discussed above, the electromagnetic force is well described by Maxwell's equations. Electromagnetism, or Maxwell theory, is formulated as a classical theory of electromagnetic fields. As opposed to Newtonian mechanics, which was modified by special relativity, Maxwell theory is fully consistent with special relativity.

The third fundamental force is the weak force. This force is responsible for the process of nuclear beta decay, in which a neutron decays into a proton, an electron, and an antineutrino. In general, processes that involve neutrinos are mediated by weak forces. While nuclear beta decay had been known since the end of the nineteenth century, the recognition that a new force was at play did not take hold until the middle of the twentieth century. The strength of this force is measured by the Fermi constant. Weak interactions are much weaker than electromagnetic interactions.

Finally, the fourth force is the strong force, nowadays called the color force. This force is at play in holding together the constituents of the neutron, the proton, the pions, and many other subnuclear particles. These constituents, called quarks, are held so tightly by the color force that they cannot be seen in isolation.

We are now in a position to return to the subject of unification. In the late 1960s the Weinberg–Salam model of *electroweak* interactions put together electromagnetism and the weak force into a unified framework. This unified model was neither dictated nor justified only by considerations of simplicity or elegance. It was necessary for a predictive and consistent theory of the weak interactions. The theory is initially formulated with four massless particles that carry the forces. A process of symmetry breaking gives mass to three of these particles: the W^+, the W^-, and the Z^0. These particles are the carriers of the weak force. The particle that remains massless is the photon, which is the carrier of the electromagnetic force.

Maxwell's equations, as we discussed before, are equations of classical electromagnetism. They do not provide a quantum theory. Physicists have discovered quantization methods, which can be used to turn a classical theory into a quantum theory – a theory that can be calculated using the principles of quantum mechanics. While classical electrodynamics can be used confidently to calculate the transmission of energy in power lines and the radiation patterns of radio antennas, it is neither an accurate nor a correct theory for microscopic phenomena. Quantum electrodynamics (QED), the quantum version of classical electrodynamics, is required for correct computations in this arena. In QED, the photon appears as the quantum of the electromagnetic field. The theory of weak interactions is also a quantum theory of particles, so the correct, unified theory is the quantum electroweak theory.

The quantization procedure is also successful in the case of the strong color force, and the resulting theory has been called quantum chromodynamics (QCD). The carriers of the color force are eight massless particles. These are colored gluons, and just like the quarks, they cannot be observed in isolation. The quarks respond to the gluons because they carry color. Quarks can come in three colors.

The electroweak theory together with QCD form the Standard Model of particle physics. In the Standard Model there is some interplay between the electroweak sector and the QCD sector because some particles feel both types of forces. But there is no real and deep unification of the weak force and the color force. The Standard Model summarizes completely the present knowledge of particle physics. So, in fact, we are not certain about any possible further unification.

In the Standard Model there are twelve force carriers: the eight gluons, the W^+, the W^-, the Z^0, and the photon. All of these are bosons. There are also many matter particles, all of which are fermions. The matter particles are of two types: leptons and quarks. The leptons include the electron e^-, the muon μ^-, the tau τ^-, and the associated neutrinos ν_e, ν_μ, and ν_τ. We can list them as

$$\text{leptons}: \quad e^-, \mu^-, \tau^-, \nu_e, \nu_\mu, \nu_\tau.$$

Since we must include their antiparticles, this adds up to a total of twelve leptons. The quarks carry color charge, electric charge, and can respond to the weak force as well. There are six different types of quarks. Poetically called flavors, these types are: up (u), down (d), charm (c), strange (s), top (t), and bottom (b). We can list them as

$$\text{quarks}: \quad u, d, c, s, t, b.$$

The u and d quarks, for example, carry different electric charges and respond differently to the weak force. Each of the six quark flavors listed above comes in three colors, so this gives $6 \times 3 = 18$ particles. Including the antiparticles, we get a total of 36 quarks. Adding leptons and quarks together we have a grand total of 48 matter particles. Adding matter particles and force carriers together we have a total of 60 particles in the Standard Model.

Despite the large number of particles it describes, the Standard Model is reasonably elegant and very powerful. As a complete theory of physics, however, it has two significant shortcomings. The first one is that it does not include gravity. The second one is that it has about twenty parameters that cannot be calculated within its framework. Perhaps the simplest example of such a parameter is the dimensionless (or unit-less) ratio of the mass of the muon to the mass of the electron. The value of this ratio is about 207, and it must be put into the model by hand.

Most physicists believe that the Standard Model is only a step towards the formulation of a complete theory of physics. A large number of physicists also suspect that some unification of the electroweak and strong forces into a Grand Unified Theory (GUT) will prove to be correct. At present, however, the unification of these two forces appears to be optional.

Another attractive possibility is that a more complete version of the Standard Model includes supersymmetry. Supersymmetry is a symmetry that relates bosons to fermions. Since all matter particles are fermions and all force carriers are bosons, this remarkable symmetry unifies matter and forces. In a theory with supersymmetry, bosons and fermions appear in pairs of equal mass. The particles of the Standard Model do not have this property, so supersymmetry, if it exists in nature, must be spontaneously broken. Supersymmetry is such an appealing symmetry that many physicists believe that it will eventually be discovered.

While the above extensions of the Standard Model may or may not occur, it is clear that the inclusion of gravity into the particle physics framework is not optional. Gravity must be included, with or without unification, if one is to have a complete theory. The effects of the gravitational force are presently quite negligible at the microscopic level, but they are crucial in studies of cosmology of the early universe.

There is, however, a major problem when one attempts to incorporate gravitational physics into the Standard Model. The Standard Model is a quantum theory, while Einstein's general relativity is a classical theory. It seems very difficult, if not altogether impossible, to have a consistent theory that is partly quantum and partly classical. Given the successes of quantum theory, it is widely believed that gravity must be turned into a quantum theory. The procedures of quantization, however, encounter profound difficulties in the case of gravity. The resulting theory of quantum gravity appears to be ill-defined. As a practical matter, in many circumstances one can work confidently with classical gravity coupled to the Standard Model. For example, this is done routinely in present-day descriptions of the universe. A theory of quantum gravity is necessary, however, to study physics at times very near to the Big Bang, and to study certain properties of black holes. Formulating a quantum theory that includes both gravity and the other forces seems fundamentally necessary. A *unification* of gravity with the other forces might be required to construct this complete theory.

1.2 String theory as a unified theory of physics

String theory is an excellent candidate for a unified theory of all forces in nature. It is also a rather impressive prototype of a complete theory of physics. In string theory all forces are truly unified in a deep and significant way. In fact, all the particles are unified. String

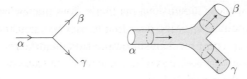

Fig. 1.1   The decay $\alpha \rightarrow \beta + \gamma$ as a particle process (left) and as a string process (right).

theory is a quantum theory, and, because it includes gravitation, it is a quantum theory of gravity. Viewed from this perspective, and recalling the failure of Einstein's gravity to yield a quantum theory, one may conclude that in string theory all other interactions are necessary for the consistency of the quantum gravitational sector! While it may be difficult to measure the effects of quantum gravity directly, a theory of quantum gravity such as string theory may have testable predictions concerning the other interactions.

Why is string theory a truly unified theory? The reason is simple and goes to the heart of the theory. In string theory, each particle is identified as a particular vibrational mode of an elementary microscopic string. A musical analogy is very apt. Just as a violin string can vibrate in different modes and each mode corresponds to a different sound, the modes of vibration of a fundamental string can be recognized as the different particles we know. One of the vibrational states of strings is the graviton, the quantum of the gravitational field. Since there is just one type of string, and all particles arise from string vibrations, all particles are naturally incorporated into a single theory. When we think in string theory of a decay process $\alpha \rightarrow \beta + \gamma$, where an elementary particle α decays into particles β and γ, we imagine a single string vibrating in such a way that it is identified as particle α that breaks into two strings that vibrate in ways that identify them as particles β and γ (Figure 1.1). Since strings may turn out to be extremely tiny, it may be difficult to observe directly the string-like nature of particles.

Are we sure that string theory is a good quantum theory of gravity? There is no complete certainty yet, but the evidence is very good. Indeed, the problems that occur when one tries to quantize Einstein's theory do not seem to appear in string theory.

For a theory as ambitious as string theory, a certain degree of uniqueness is clearly desirable. It would be somewhat disappointing to have several consistent candidates for a theory of all interactions. The first sign that string theory is rather unique is that it does not have adjustable dimensionless parameters. As we mentioned before, the Standard Model of particle physics has about twenty parameters that must be adjusted to some precise values. A theory with adjustable dimensionless parameters is not really unique. When the parameters are set to different values one obtains different theories with potentially different predictions. String theory has one dimensionful parameter, the string length ℓ_s. Its value can be roughly imagined as the typical size of strings.

Another intriguing sign of the uniqueness of string theory is the fact that the dimensionality of spacetime is fixed. Our physical spacetime is four-dimensional, with one time dimension and three space dimensions. In the Standard Model this information is used to build the theory, it is not derived. In string theory, on the other hand, the number of spacetime dimensions emerges from a calculation. The answer is not four, but rather ten.

Some of these dimensions may hide from plain view if they curl up into a space that is small enough to escape detection in experiments done with low energies. If string theory is correct, some mechanism must ensure that the observable dimensionality of spacetime is four.

The lack of adjustable dimensionless parameters is a sign of the uniqueness of string theory: it means that the theory cannot be deformed or changed continuously by changing these parameters. But there could be other theories that cannot be reached by continuous deformations. So how many string theories are there?

Let us begin by noting two broad subdivisions. There are open strings and there are closed strings. Open strings have two endpoints, while closed strings have no endpoints. One can consider theories with only closed strings and theories with both open and closed strings. Since open strings generally can close to form closed strings, we do not consider theories with only open strings. The second subdivision is between bosonic string theories and superstring theories. Bosonic strings live in 26 dimensions, and all of their vibrations represent bosons. Since they lack fermions, bosonic string theories are not realistic. They are, however, much simpler than the superstrings, and most of the important concepts in string theory can be explained in the context of bosonic strings. The superstrings live in ten-dimensional spacetime, and their spectrum of states includes bosons *and* fermions. In fact, these two sets of particles are related by supersymmetry. Supersymmetry is therefore an important ingredient in string theory. All realistic models of string theory are built from superstrings. In all string theories the graviton appears as a vibrational mode of closed strings. In string theory gravity is unavoidable.

By the mid 1980s five ten-dimensional superstring theories were known to exist. In the years that followed, many interrelations between these theories were found. Moreover, another theory was discovered by taking a certain strong coupling limit of one of the superstrings. This theory is eleven-dimensional and has been dubbed M-theory, for lack of a better name. It has now become clear that the five superstrings and M-theory are only facets or different limits of a *single* unique theory! At present, this unique theory remains fairly mysterious. It is not yet clear whether or not the set of bosonic string theories is connected to the web of superstring theories.

All in all, we see that string theory is a truly unified and possibly unique theory. It is a candidate for a unified theory of physics, a theory Albert Einstein tried to find ever since his discovery of general relativity. Einstein would have been surprised, or perhaps disturbed, by the prominent role that quantum mechanics plays in string theory. But string theory appears to be a worthy successor of general relativity. It is almost certain that string theory will give rise to a new conception of spacetime. The prominence of quantum mechanics in string theory would not have surprised Paul Dirac. His writings on quantization suggest that he felt that deep quantum theories arise from the quantization of classical physics. This is precisely what happens in string theory. This book will explain in detail how string theory, at least in its simplest form, is nothing but the quantum mechanics of classical relativistic strings.

1.3 String theory and its verification

It should be said at the outset that, as of yet, there has been no experimental verification of string theory. In order to have experimental verification one needs a sharp prediction. It has been difficult to obtain such a prediction. String theory is still at an early stage of development, and it is not so easy to make predictions with a theory that is not well understood. Still, some interesting possibilities have emerged.

As we mentioned earlier, superstring theory requires a ten-dimensional spacetime: one dimension of time and nine of space. If string theory is correct, extra spatial dimensions must exist, even if we have not seen them yet. Can we test the existence of these extra dimensions? If the extra dimensions are the size of the Planck length ℓ_P (the length scale associated with four-dimensional gravity), they will remain beyond direct detection, perhaps forever. Indeed, $\ell_P \sim 10^{-33}$ cm, and this distance is many orders of magnitude smaller than 10^{-16} cm, which is roughly the smallest distance that has been explored with particle accelerators. This scenario was deemed to be most likely. It was assumed that in string theory the length scale ℓ_s coincides with the Planck length, in which case extra dimensions would be of Planck length, as well.

It turns out, however, that string theory allows extra dimensions that are as large as a tenth of a millimeter! Surprisingly, extra dimensions that large may have gone undetected. To make this work out, the string length ℓ_s is taken to be of the order of 10^{-18} cm. Moreover, our three-dimensional space emerges as a hypersurface embedded inside the nine-dimensional space. The hypersurface, or higher-dimensional membrane, is called a D-brane. D-branes are real, physical objects in string theory. In this setup, the presence of large extra dimensions is tested by gravitational experiments. Extra dimensions much larger than ℓ_P but still very small may be detected with particle accelerators. If extra dimensions are detected, this would be strong evidence for string theory. We discuss the subject of large extra dimensions in Chapter 3.

A striking confirmation of string theory may result from the discovery of a cosmic string. Left-over from early universe processes, a cosmic string can stretch across the observable universe and may be detected via gravitational lensing or, more indirectly, through the detection of gravitational waves. No cosmic strings have been detected to date, but the searches have not been exhaustive and they continue. If found, a cosmic string must be studied in detail to confirm that it is a string from string theory and not the kind of string that can arise from conventional theories of particle physics. We discuss the subject of cosmic strings in Chapter 7.

Another interesting possibility has to do with supersymmetry. If we start with a ten-dimensional superstring theory and compactify the six extra dimensions, the resulting four-dimensional theory is, in many cases, supersymmetric. No unique predictions have emerged for the specific details of the four-dimensional theory, but supersymmetry may be a rather generic feature. An experimental discovery of supersymmetry in future accelerators would suggest very strongly that string theory is on the right track.

Leaving aside predictions of new phenomena, we must ask whether the Standard Model emerges from string theory. It should, since string theory is supposed to be a unified theory of all interactions, and it must therefore reduce to the Standard Model for sufficiently low energies. While string theory certainly has room to include all known particles and interactions, and this is very good news indeed, no one has yet been able to show that they actually emerge in fine detail. In Chapter 21 we will study some models which use D-branes and have an uncanny resemblance to the world as we know it. In these models the particle content is in fact *precisely* that of the Standard Model (the particles are obtained with zero mass, however, and it is not clear whether the process that gives them mass can work out correctly). Our four-dimensional world is part of the D-branes, but these D-branes happen to have more than three spatial directions. The additional D-brane dimensions are wrapped on the compact space (we will learn how to imagine such configurations!). The gauge bosons and the matter particles in the model arise from vibrations of open strings that stretch between D-branes. As we will learn, the endpoints of open strings must remain attached to the D-branes. If you wish, the musical analogy for strings is improved. Just as the strings of a violin are held stretched by pegs, the D-branes hold fixed the endpoints of the open strings whose lowest vibrational modes could represent the particles of the Standard Model!

String theory shares with Einstein's gravity a problematic feature. Einstein's equations of gravitation admit many cosmological solutions. Each solution represents a consistent universe, but only one of them represents *our* observable universe. It is not easy to explain what selects the physical solution, but in cosmology this is done using arguments based on initial conditions, symmetry, and simplicity. The smaller the number of solutions a theory has, the more predictive it is. If the set of solutions is characterized by continuous parameters, selecting a solution is equivalent to adjusting the values of the parameters. In this way, a theory whose formulation requires no adjustable parameters may generate adjustable parameters through its solutions! It seems clear that in string theory the set of solutions (string models) is characterized by both discrete and continuous parameters.

In order to reproduce the Standard Model it seems clear that the string model must not have continuous parameters; such parameters imply the existence of massless fields that have not been observed. It was not easy to find models without continuous parameters, but that became possible recently in the context of flux compactifications; models in which the extra dimensions are threaded by analogs of electric and magnetic fields. There is an extraordinary large number of such models, certainly more than 10^{500} of them. There may be even more models that manage to avoid continuous parameters by other means. Physicists speak of a vast *landscape* of string solutions or models.

In this light we can wonder what are the possible outcomes of the search for a realistic string model. One possible outcome (the worst one) is that no string model in the landscape reproduces the Standard Model. This would rule out string theory. Another possible outcome (the best one) is that one string model reproduces the Standard Model. Moreover, the model represents a well-isolated point in the landscape. The parameters of the Standard Model are thus predicted. The landscape may be so large that a strange possibility emerges: many string models with almost identical properties all of which are consistent with the Standard Model to the accuracy that it is presently known. In this possibility there is a loss of predictive power. Other outcomes may be possible.

String theorists sometimes say that string theory has already made at least one successful *prediction*: it predicted gravity! (I heard this from John Schwarz.) There is a bit of jest in saying so – after all, gravity is the oldest known force in nature. I believe, however, that there is a very substantial point to be made here. String theory is the quantum mechanics of a relativistic string. In no sense whatsoever is gravity put into string theory by hand. It is a complete surprise that gravity emerges in string theory. Indeed, none of the vibrations of the *classical* relativistic string correspond to the particle of gravity. It is a truly remarkable fact that we find the particle of gravity among the *quantum* vibrations of the relativistic string. You will see in detail how this happens as you progress through this book. The striking quantum emergence of gravitation in string theory has the full flavor of a prediction.

1.4 Developments and outlook

String theory has been a very stimulating and active area of research ever since Michael Green and John Schwarz showed in 1984 that superstrings are not afflicted with fatal inconsistencies that threaten similar particle theories in ten dimensions. Much progress has been made since then.

String theory has provided new and powerful tools for the understanding of conventional particle physics theories, gauge theories in particular. These are the kinds of theories that are used to formulate the Standard Model. Close cousins of these gauge theories arise on string theory D-branes. We examine D-branes and the theories that arise on them in detail beginning in Chapter 15. A remarkable physical equivalence between a certain four-dimensional gauge theory and a closed superstring theory (the AdS/CFT correspondence) is discussed in Chapter 23. As we will explain, the correspondence has been used to understand hydrodynamical properties of the quark-gluon plasma created in the collision of gold nuclei at heavy ion colliders.

String theory has also made good strides towards a statistical mechanics interpretation of black hole entropy. We know from the pioneering work of Jacob Bekenstein and Stephen Hawking that black holes have both entropy and temperature. In statistical mechanics these properties arise if a system can be constructed in many degenerate ways using its basic constituents. Such an interpretation is not available in Einstein's gravitation, where black holes seem to have few, if any, constituents. In string theory, however, certain black holes can be built by assembling together various types of D-branes and strings in a controlled manner. For such black holes, the predicted Bekenstein entropy is obtained by counting the ways in which they can be built with their constituent D-branes and strings. In fact, the class of black holes amenable to string theory analysis continues to grow. We discuss this important development in Chapter 22.

String theory will be needed to study cosmology of the Very Early Universe. String theory may provide a concrete model for the realization of inflation – a period of dramatic exponential expansion that the universe is likely to have experienced at the earliest times. The theory of inflation suggests that our universe is a growing bubble or region inside a space that continues to inflate for eternity. Bubbles continue to emerge forever and some have remarked that every model in the landscape may be physically realized in some bubble. Inflation does not appear to be eternal in the past, so some kind of beginning seems necessary. The deepest mysteries of the universe seem to lie hidden in a regime where classical general relativity surely breaks down. String theory should allow us to peer into this unknown realm. Some day we may be able to understand how the universe comes into being, if it does, or how the universe could have existed forever in the past, if it did.

Most likely, answering such questions will require a mastery of string theory that goes beyond our present abilities. String theory is in fact an unfinished theory. Much has been learned about it, but in reality we have no complete formulation of the theory. A comparison with Einstein's theory is illuminating. Einstein's equations for general relativity are elegant and geometrical. They embody the conceptual foundation of the theory and feel completely up to the task of describing gravitation. No similar equations are known for string theory, and the conceptual foundation of the theory remains largely unknown. String theory is an exciting research area because the central ideas remain to be found.

Describing nature and formulating the theory – those remain the present-day challenges of string theory. If surmounted, we will have a theory of all interactions, allowing us to understand the fate of spacetime and the mysteries of a quantum mechanical universe. With such high stakes, physicists are likely to investigate string theory until definite answers are found.

2 Special relativity and extra dimensions

The word relativistic, as used in the term "relativistic strings," indicates consistency with Einstein's theory of special relativity. We review special relativity and introduce the light-cone frame, light-cone coordinates, and light-cone energy. We then turn to the idea of additional, compact space dimensions and show with an example from quantum mechanics that, if small, these dimensions have little effect at low energies.

2.1 Units and parameters

Units are nothing other than fixed quantities that we use for purposes of reference. A measurement involves finding the unit-free ratio of an observable quantity to the appropriate unit. Consider, for example, the definition of a second in the international system of units (SI system). The SI second (s) is defined to be the duration of 9 192 631 770 periods of the radiation emitted in the transition between the two hyperfine levels of the cesium-133 atom. When we measure the time elapsed between two events, we are really counting a unit-free, or dimensionless, number: the number that tells us how many seconds fit between the two events or, alternatively, how many periods of the cesium radiation fit between the two events. The same goes for length. The unit called the meter (m) is nowadays defined as the distance traveled by light in a certain fraction of a second (1/299 792 458 of a second, to be precise). Mass introduces a third unit, the prototype kilogram (kg), kept safely in Sèvres, France.

When doing dimensional analysis, we denote the units of length, time, and mass by L, T, and M, respectively. These are called the three basic units. A force, for example, has units

$$[F] = MLT^{-2}, \tag{2.1}$$

where $[X]$ denotes the units of the quantity X. Equation (2.1) follows from Newton's law that equates the force on an object to the product of its mass and its acceleration. The newton (N) is the SI unit of force, and it equals kg·m/s^2.

It is interesting that no additional basic units are needed to describe other quantities. Consider, for example, electric charge. Do we need a new unit to describe charge? Not really. This is easy to see in Gaussian units. In these units, Coulomb's law for the force $|\vec{F}|$ between two charges q_1 and q_2 separated by a distance r reads

$$|\vec{F}| = \frac{|q_1 q_2|}{r^2}. \tag{2.2}$$

The units of charge are fixed in terms of other units because we have a force law where charges appear and all other quantities have known units. The esu is the Gaussian unit of charge, and it is defined by stating that two charges of one esu each, placed at a distance of one centimeter apart, repel each other with a force of one dyne (the Gaussian unit of force, 10^{-5} N). Thus

$$\text{esu}^2 = \text{dyne} \cdot \text{cm}^2 = 10^{-5} \text{ N} \cdot (10^{-2}\text{m})^2 = 10^{-9} \text{ N} \cdot \text{m}^2. \tag{2.3}$$

It follows from this equation that

$$[\text{esu}^2] = [\text{ N} \cdot \text{m}^2], \tag{2.4}$$

and, using (2.1), finally we get

$$[\text{esu}] = M^{1/2}L^{3/2}T^{-1}. \tag{2.5}$$

This expresses the esu in terms of the three basic units.

In SI units, charge is measured in coulombs (C). The situation in SI units is a little more intricate, but the essential point is the same. A coulomb is defined in SI units as the amount of charge carried by a current of one ampere (A) in one second. The ampere itself is defined as the amount of current that, when carried by two wires separated by a distance of one meter, produces a force of 2×10^{-7} N/m. The coulomb, as opposed to the esu, is not expressed in terms of meters, kilograms, and seconds. Coulomb's law in SI units is

$$|\vec{F}| = \frac{1}{4\pi\epsilon_0}\frac{|q_1 q_2|}{r^2}, \quad \text{with} \quad \frac{1}{4\pi\epsilon_0} = 8.99 \times 10^9 \frac{\text{N} \cdot \text{m}^2}{\text{C}^2}. \tag{2.6}$$

Note the presence of C^{-2} in the definition of the constant prefactor. Since each charge carries one factor of C, all the factors of C cancel in the calculation of the force. Two charges of one coulomb each, placed one meter apart, will each experience a force of 8.99×10^9 N. This fact allows you to deduce (Problem 2.1) how many esus there are in a coulomb. Even though we do not write coulombs in terms of other units, this is just a matter of convenience. Coulombs and esus are related, and esus are written in terms of the three basic units.

When we speak of parameters in a theory, it is convenient to distinguish between dimensionful parameters and dimensionless parameters. Consider, for example, a theory in which there are three types of particles with masses m_1, m_2, and m_3. We can think of the theory as having one dimensionful parameter, the mass m_1 of the first particle, say, and two dimensionless parameters, the mass ratios m_2/m_1 and m_3/m_1.

String theory is said to have no adjustable parameters. By this it is meant that no dimensionless parameter is needed to formulate string theory. String theory does, however, have one dimensionful parameter. That parameter is the string length ℓ_s. This length sets the scale in which the theory operates. In the early 1970s, when string theory was first being formulated, the theory was thought to be a theory of hadrons. Back then, the string length was taken to be comparable to the nuclear scale. Nowadays, we think that string theory is a theory of fundamental forces and interactions. Accordingly, we set the string length to be much smaller than the nuclear scale.

2.2 Intervals and Lorentz transformations

Special relativity is based on the experimental fact that the speed of light ($c \simeq 3 \times 10^8$ m/s) is the same for all inertial observers. This fact leads to some rather surprising conclusions. Newtonian intuition about the absolute nature of time, the concept of simultaneity, and other familiar ideas must be revised. In comparing the coordinates of events, two inertial observers, henceforth called Lorentz observers, find that the appropriate coordinate transformations mix space and time.

In special relativity, events are characterized by the values of four coordinates: a time coordinate t and three spatial coordinates x, y, and z. It is convenient to collect these four numbers in the form (ct, x, y, z), where the time coordinate is scaled by the speed of light so that all coordinates have units of length. To make the notation more uniform, we use indices to relabel the space and time coordinates as follows:

$$x^\mu = (x^0, x^1, x^2, x^3) \equiv (ct, x, y, z). \tag{2.7}$$

Here the superscript μ takes the four values $0, 1, 2$, and 3. The x^μ are spacetime coordinates.

Consider a Lorentz frame S in which two events are represented by the coordinates x^μ and $x^\mu + \Delta x^\mu$. Consider now a second Lorentz frame S', in which the same two events are described by the coordinates x'^μ and $x'^\mu + \Delta x'^\mu$, respectively. In general, not only are the coordinates x^μ and x'^μ different, so too are the coordinate differences Δx^μ and $\Delta x'^\mu$. On the other hand, both observers will agree on the value of the invariant interval Δs^2. This interval is defined by

$$-\Delta s^2 \equiv -(\Delta x^0)^2 + (\Delta x^1)^2 + (\Delta x^2)^2 + (\Delta x^3)^2. \tag{2.8}$$

Note the minus sign in front of $(\Delta x^0)^2$, as opposed to the plus sign appearing before the spacelike differences $(\Delta x^i)^2$ ($i = 1, 2, 3$). This sign encodes the fundamental difference between time and space coordinates. The agreement on the value of the intervals is expressed as

$$-(\Delta x^0)^2 + (\Delta x^1)^2 + (\Delta x^2)^2 + (\Delta x^3)^2 = -(\Delta x'^0)^2 + (\Delta x'^1)^2 + (\Delta x'^2)^2 + (\Delta x'^3)^2, \tag{2.9}$$

or, in brief:

$$\Delta s^2 = \Delta s'^2. \tag{2.10}$$

The minus sign on the left-hand side of (2.8) implies that $\Delta s^2 > 0$ for events that are *timelike separated*. Timelike separated events are events for which

$$(\Delta x^0)^2 > (\Delta x^1)^2 + (\Delta x^2)^2 + (\Delta x^3)^2. \tag{2.11}$$

The history of a particle is represented in spacetime as a curve, the *world-line* of the particle. Any two events on the world-line of a particle are timelike separated, because no particle can move faster than light and therefore the distance light would have traveled in the time interval that separates the events must be larger than the space separation between the events. This is the content of (2.11). You at the time you were born and you at this moment are timelike separated events: a long time has passed and you have not gone that far. Events connected by the world-line of a photon are said to be *lightlike separated*. For such a pair of events, we have $\Delta s^2 = 0$, because in this case the two sides of (2.11) are identical: the spatial separation between the events coincides with the distance that light would have traveled in the time that separates the events. Two events for which $\Delta s^2 < 0$ are said to be *spacelike separated*. Events that are simultaneous in a Lorentz frame but occur at different positions in that same frame are spacelike separated. It is because Δs^2 can be negative that it is not written as $(\Delta s)^2$. For timelike separated events, however, we define

$$\Delta s \equiv \sqrt{\Delta s^2} \quad \text{if} \quad \Delta s^2 > 0 \quad \text{(timelike interval)}. \tag{2.12}$$

Many times it is useful to consider events that are infinitesimally close to each other. Small coordinate differences are needed to define velocities and are also useful in general relativity. Infinitesimal coordinate differences are written as dx^μ, and the associated invariant interval is written as ds^2. Following (2.8), we have

$$-ds^2 = -(dx^0)^2 + (dx^1)^2 + (dx^2)^2 + (dx^3)^2. \tag{2.13}$$

The equality of intervals is the statement

$$ds^2 = ds'^2. \tag{2.14}$$

A very useful notation can be motivated by trying to simplify the expression for the invariant ds^2. To do this, we introduce symbols that carry subscripts instead of superscripts. Let us define

$$dx_0 \equiv -dx^0, \quad dx_1 \equiv dx^1, \quad dx_2 \equiv dx^2, \quad dx_3 \equiv dx^3. \tag{2.15}$$

The only significant change is the inclusion of a minus sign for the zeroth component. All together, we write

$$dx_\mu = (dx_0, dx_1, dx_2, dx_3) \equiv (-dx^0, dx^1, dx^2, dx^3). \tag{2.16}$$

Now we can rewrite ds^2 in terms of dx^μ and dx_μ:

$$\begin{aligned}
-ds^2 &= -(dx^0)^2 + (dx^1)^2 + (dx^2)^2 + (dx^3)^2 \\
&= dx_0 dx^0 + dx_1 dx^1 + dx_2 dx^2 + dx_3 dx^3,
\end{aligned} \tag{2.17}$$

and we see that the minus sign in (2.13) is gone. The invariant interval has become

$$-ds^2 = \sum_{\mu=0}^{3} dx_\mu dx^\mu. \tag{2.18}$$

Throughout the rest of the text we will use Einstein's summation convention. In this convention, indices repeated in a single term are to be summed over the appropriate set of values. We do not consider indices to be repeated when they appear in different terms. For example, there are no sums implied by $a^\mu + b^\mu$ or $a^\mu = b^\mu$, but there is an implicit sum in $a^\mu b_\mu$. A repeated index must appear once as a subscript and once as a superscript and should not appear more than twice in any one term. The letter chosen for the repeated index is not important; thus $a^\mu b_\mu$ is the same as $a^\nu b_\nu$. Because of this, repeated indices are sometimes called dummy indices! Using the summation convention, we can rewrite (2.18) as

$$-ds^2 = dx_\mu dx^\mu. \tag{2.19}$$

Just as we did for finite coordinate differences in (2.12), for infinitesimal timelike intervals we define the quantity

$$ds \equiv \sqrt{ds^2} \quad \text{if} \quad ds^2 > 0 \quad \text{(timelike interval)} . \tag{2.20}$$

We can also express the interval ds^2 using the Minkowski metric $\eta_{\mu\nu}$. This is done by writing

$$-ds^2 = \eta_{\mu\nu} \, dx^\mu dx^\nu . \tag{2.21}$$

Equation (2.21), by itself, does not determine the metric $\eta_{\mu\nu}$. If, in addition, we require $\eta_{\mu\nu}$ to be symmetric under an exchange in the order of its indices,

$$\eta_{\mu\nu} = \eta_{\nu\mu} , \tag{2.22}$$

then this together with (2.21) completely determines a metric called the Minkowski metric. It is reasonable to declare that $\eta_{\mu\nu}$ be symmetric for, as we shall now see, any antisymmetric component would be irrelevant.

Any two-index object $M_{\mu\nu}$ can be decomposed into a symmetric part and an antisymmetric part:

$$M_{\mu\nu} = \frac{1}{2}(M_{\mu\nu} + M_{\nu\mu}) + \frac{1}{2}(M_{\mu\nu} - M_{\nu\mu}) . \tag{2.23}$$

The first term on the right-hand side, the symmetric part of M, is invariant under exchange of the indices μ and ν. The second term on the right-hand side, the antisymmetric part of M, changes sign under exchange of the indices μ and ν. If $\eta_{\mu\nu}$ had an antisymmetric part $\xi_{\mu\nu}$ $(= -\xi_{\nu\mu})$, then its contribution would drop out of the right-hand side of (2.21). We can see this as follows:

$$\xi_{\mu\nu} \, dx^\mu dx^\nu = (-\xi_{\nu\mu}) \, dx^\mu dx^\nu = -\xi_{\mu\nu} \, dx^\nu dx^\mu = -\xi_{\mu\nu} \, dx^\mu dx^\nu. \tag{2.24}$$

In the first step, we used the antisymmetry of $\xi_{\mu\nu}$. In the second step, we relabeled the dummy indices: the μ were changed into ν and vice versa. In the third step, we switched the order of the dx^μ and dx^ν factors. The result is that $\xi_{\mu\nu}dx^\mu dx^\nu$ is identical to minus itself, and therefore it vanishes. It would be useless for $\eta_{\mu\nu}$ to have an antisymmetric part, so we simply declare that it has none.

Since repeated indices are summed over, equation (2.21) means that

$$-ds^2 = \eta_{00}dx^0 dx^0 + \eta_{01}dx^0 dx^1 + \eta_{10}dx^1 dx^0 + \eta_{11}dx^1 dx^1 + \cdots. \qquad (2.25)$$

Comparing with (2.17) and using (2.22), we see that $\eta_{00} = -1$, $\eta_{11} = \eta_{22} = \eta_{33} = 1$, and all other components vanish. We collect these values in matrix form:

$$\eta_{\mu\nu} = \begin{pmatrix} -1 & 0 & 0 & 0 \\ 0 & 1 & 0 & 0 \\ 0 & 0 & 1 & 0 \\ 0 & 0 & 0 & 1 \end{pmatrix}. \qquad (2.26)$$

In this equation, which follows a common identification of two-index objects with matrices, we think of μ, the first index in η, as the row index, and ν, the second index in η, as the column index. The Minkowski metric can be used to "lower indices." Indeed, equation (2.15) can be rewritten as

$$dx_\mu = \eta_{\mu\nu}\, dx^\nu. \qquad (2.27)$$

If we are handed a set of quantities b^μ, we always define

$$b_\mu \equiv \eta_{\mu\nu}b^\nu. \qquad (2.28)$$

Given objects a^μ and b^μ, the relativistic *scalar product* $a \cdot b$ is defined as

$$a \cdot b \equiv a^\mu b_\mu = \eta_{\mu\nu}\, a^\mu b^\nu = -a^0 b^0 + a^1 b^1 + a^2 b^2 + a^3 b^3. \qquad (2.29)$$

Applied to (2.19), this gives $-ds^2 = dx \cdot dx$. Note that $a^\mu b_\mu = a_\mu b^\mu$.

It is convenient to introduce the matrix inverse for $\eta_{\mu\nu}$. Written conventionally as $\eta^{\mu\nu}$, it is given by

$$\eta^{\mu\nu} = \begin{pmatrix} -1 & 0 & 0 & 0 \\ 0 & 1 & 0 & 0 \\ 0 & 0 & 1 & 0 \\ 0 & 0 & 0 & 1 \end{pmatrix}. \qquad (2.30)$$

You can see by inspection that this matrix is indeed the inverse of the matrix in (2.26). When thinking of $\eta^{\mu\nu}$ as a matrix, as with $\eta_{\mu\nu}$, the first index is a row index and the second index is a column index. In index notation the inverse property is written as

$$\eta^{\nu\rho}\eta_{\rho\mu} = \delta^\nu_\mu, \qquad (2.31)$$

where the Kronecker delta δ^ν_μ is defined by

$$\delta^\nu_\mu = \begin{cases} 1 & \text{if } \mu = \nu \\ 0 & \text{if } \mu \neq \nu. \end{cases} \qquad (2.32)$$

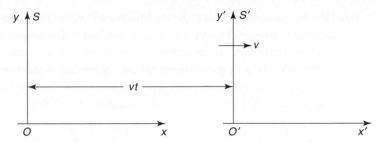

Fig. 2.1 **Two Lorentz frames connected by a boost. S' is boosted along the $+x$ direction of S with boost parameter $\beta = v/c$.**

Note that the repeated index ρ in (2.31) produces the desired matrix multiplication. The Kronecker delta can be thought of as the index representation of the identity matrix. The metric with upper indices can be used to "raise indices." Using (2.28) and (2.31), we get

$$\eta^{\rho\mu} b_\mu = \eta^{\rho\mu} (\eta_{\mu\nu} b^\nu) = (\eta^{\rho\mu} \eta_{\mu\nu}) b^\nu = \delta^\rho_\nu b^\nu = b^\rho. \tag{2.33}$$

The lower μ index of b_μ was raised by $\eta^{\rho\mu}$ to become an upper ρ index. The last step in the above calculation needs a little explanation: $\delta^\rho_\nu b^\nu = b^\rho$ because as we sum over ν, δ^ρ_ν vanishes unless $\nu = \rho$, in which case it equals one.

Lorentz transformations are the relations between coordinates in two different inertial frames. Consider a frame S and a frame S', that is moving along the positive x direction of the S frame with a velocity v, as shown in Figure 2.1. Assume the coordinate axes for both systems are parallel, and that the origins coincide at the common time $t = t' = 0$. We say that S' is boosted along the x direction with velocity parameter $\beta \equiv v/c$. The Lorentz transformations in this case read

$$
\begin{aligned}
x' &= \gamma (x - \beta ct), \\
ct' &= \gamma (ct - \beta x), \\
y' &= y, \\
z' &= z,
\end{aligned}
\tag{2.34}
$$

where the Lorentz factor γ is given by

$$\gamma \equiv \frac{1}{\sqrt{1 - \beta^2}} = \frac{1}{\sqrt{1 - \frac{v^2}{c^2}}}. \tag{2.35}$$

Using indices, and changing the order of the first two equations, we arrive at

$$
\begin{aligned}
x'^0 &= \gamma (x^0 - \beta x^1), \\
x'^1 &= \gamma (-\beta x^0 + x^1), \\
x'^2 &= x^2, \\
x'^3 &= x^3.
\end{aligned}
\tag{2.36}
$$

In the above transformations, the coordinates x^2 and x^3 remain unchanged. These are the coordinates orthogonal to the direction of the boost. The inverse Lorentz transformations give the values of the x coordinates in terms of the x' coordinates. They are readily found by solving for the x in the above equations. The result is the same set of transformations with x and x' exchanged and with β replaced by $(-\beta)$, as required by symmetry.

The coordinates in the above equations satisfy the relation

$$(x^0)^2 - (x^1)^2 - (x^2)^2 - (x^3)^2 = (x'^0)^2 - (\dot{x}'^1)^2 - (x'^2)^2 - (x'^3)^2 , \qquad (2.37)$$

as you can show by direct computation. This is just the statement of invariance of the interval Δs^2 between two events: the first event is represented by $(0, 0, 0, 0)$ in both S and S', and the second event is represented by coordinates x^μ in S and x'^μ in S'. By definition, *Lorentz transformations are the linear transformations of coordinates that respect the equality* (2.37).

In general, we write a Lorentz transformation as the linear relation

$$x'^\mu = L^\mu_{\ \nu} x^\nu , \qquad (2.38)$$

where the entries $L^\mu_{\ \nu}$ are constants that define the linear transformation. For the boost in (2.36), we have

$$[L] = L^\mu_{\ \nu} = \begin{pmatrix} \gamma & -\gamma\beta & 0 & 0 \\ -\gamma\beta & \gamma & 0 & 0 \\ 0 & 0 & 1 & 0 \\ 0 & 0 & 0 & 1 \end{pmatrix} . \qquad (2.39)$$

In defining the matrix L as $[L] = L^\mu_{\ \nu}$, we are following the convention that the first index is a row index and the second index is a column index. This is why the lower index in $L^\mu_{\ \nu}$ is written to the right of the upper index.

The coefficients $L^\mu_{\ \nu}$ are constrained by equation (2.37). In index notation, this equation requires

$$\eta_{\alpha\beta} x^\alpha x^\beta = \eta_{\mu\nu} x'^\mu x'^\nu . \qquad (2.40)$$

Using (2.38) twice on the right-hand side above gives

$$\eta_{\alpha\beta} x^\alpha x^\beta = \eta_{\mu\nu} (L^\mu_{\ \alpha} x^\alpha)(L^\nu_{\ \beta} x^\beta) = \eta_{\mu\nu} L^\mu_{\ \alpha} L^\nu_{\ \beta} x^\alpha x^\beta . \qquad (2.41)$$

Equivalently, we have the equation

$$k_{\alpha\beta} x^\alpha x^\beta = 0 , \quad \text{with} \quad k_{\alpha\beta} \equiv \eta_{\mu\nu} L^\mu_{\ \alpha} L^\nu_{\ \beta} - \eta_{\alpha\beta} . \qquad (2.42)$$

Since $k_{\alpha\beta} x^\alpha x^\beta = 0$ must hold for all values of the coordinates x, we find that

$$k_{\alpha\beta} + k_{\beta\alpha} = 0 , \qquad (2.43)$$

as you should convince yourself by writing out the sums over α and β. Since $k_{\alpha\beta}$ is in fact symmetric under the exchange in the order of its indices, (2.43) implies $k_{\alpha\beta} = 0$. This means that

$$\eta_{\mu\nu} L^\mu_{\ \alpha} L^\nu_{\ \beta} = \eta_{\alpha\beta} . \qquad (2.44)$$

Rewriting (2.44) to make it look more like matrix multiplication, we have

$$L^{\mu}{}_{\alpha} \, \eta_{\mu\nu} \, L^{\nu}{}_{\beta} = \eta_{\alpha\beta} \,. \tag{2.45}$$

The sum over the ν index works well: it is a column index in $\eta_{\mu\nu}$ and a row index in $L^{\nu}{}_{\beta}$. The μ index, however, is a row index in $L^{\mu}{}_{\alpha}$, while it should be a column index to match the row index in $\eta_{\mu\nu}$. Moreover, the α index in $L^{\mu}{}_{\alpha}$ is a column index, while it is a row index in $\eta_{\alpha\beta}$. This means that we should exchange the columns and rows of $L^{\mu}{}_{\alpha}$, which is the matrix operation of transposition. Therefore equation (2.45) can be rewritten as the matrix equation

$$L^{\mathrm{T}} \eta \, L = \eta \,. \tag{2.46}$$

Here η is the matrix whose entries are $\eta_{\mu\nu}$. This neat equation is the constraint that L must satisfy to be a Lorentz transformation.

An important property of Lorentz transformations can be deduced by taking the determinant of each side of equation (2.46). Since the determinant of a product is the product of the determinants, we get

$$(\det L^{\mathrm{T}})(\det \eta)(\det L) = \det \eta \,. \tag{2.47}$$

Cancelling the common factor of $\det \eta$ and recalling that the operation of transposition does not change a determinant, we find

$$(\det L)^2 = 1 \; \longrightarrow \; \det L = \pm 1 \,. \tag{2.48}$$

You can check that $\det L = 1$ for the boost in (2.39). Since $\det L$ never vanishes, the matrix L is always invertible and, consequently, all Lorentz transformations are invertible linear transformations.

The set of Lorentz transformations includes boosts along each of the spatial coordinates. It also includes rotations of the spatial coordinates. Under a spatial rotation, the coordinates (x^0, x^1, x^2, x^3) of a point transform into coordinates (x'^0, x'^1, x'^2, x'^3), for which $x^0 = x'^0$, because time is unaffected. Since the spatial distance from a point to the origin is preserved under a rotation, we have

$$(x^1)^2 + (x^2)^2 + (x^3)^2 = (x'^1)^2 + (x'^2)^2 + (x'^3)^2 \,. \tag{2.49}$$

This, together with $x^0 = x'^0$, implies that (2.37) holds. Therefore spatial rotations are Lorentz transformations.

Any set of four quantities which transforms under Lorentz transformations in the same way as the x^{μ} do is said to be a four-vector, or Lorentz vector. When we use index notation and write b^{μ}, we mean that b^{μ} is a four-vector. Taking differentials of the linear equations (2.36), we see that the linear transformations that relate x' to x also relate dx' to dx. Therefore the differentials dx^{μ} define a Lorentz vector. In the spirit of index notation, a quantity with no free indices must be invariant under Lorentz transformations. A quantity has no free indices if it carries no index or if it contains only repeated indices, such as $a^{\mu} b_{\mu}$.

A four-vector a^{μ} is said to be timelike if $a^2 = a \cdot a < 0$, spacelike if $a^2 > 0$, and null if $a^2 = 0$. Recalling our discussion below (2.11), we see that the coordinate differences between timelike-separated events define a timelike vector. Similarly, the coordinate differences between spacelike-separated events define a spacelike vector, and the coordinate differences between lightlike-separated events define a null vector.

Quick calculation 2.1 Verify that the invariant ds^2 is indeed preserved under the Lorentz transformations (2.36).

Quick calculation 2.2 Consider two Lorentz vectors a^μ and b^μ. Write the Lorentz transformations $a^\mu \to a'^\mu$ and $b^\mu \to b'^\mu$ analogous to (2.36). Verify that $a^\mu b_\mu$ is invariant under these transformations.

2.3 Light-cone coordinates

We now discuss a coordinate system that will be extremely useful in our study of string theory, the light-cone coordinate system. The quantization of the relativistic string can be worked out most directly using light-cone coordinates. There is a different approach to the quantization of the relativistic string, in which no special coordinates are used. This approach, called Lorentz covariant quantization, is discussed briefly in Chapter 24. Lorentz quantization is very elegant, but a full discussion requires a great deal of background material. We will use light-cone coordinates to quantize strings in this book.

We define the two light-cone coordinates x^+ and x^- as two independent linear combinations of the time coordinate and a chosen spatial coordinate, conventionally taken to be x^1. This is done by writing

$$x^+ \equiv \frac{1}{\sqrt{2}}(x^0 + x^1),$$

$$x^- \equiv \frac{1}{\sqrt{2}}(x^0 - x^1). \tag{2.50}$$

The coordinates x^2 and x^3 play no role in this definition. In the light-cone coordinate system, (x^0, x^1) is traded for (x^+, x^-), but the other two coordinates x^2, x^3 are kept. Thus, the complete set of light-cone coordinates is (x^+, x^-, x^2, x^3).

The new coordinates x^+ and x^- are called light-cone coordinates because the associated coordinate axes are the world-lines of beams of light emitted from the origin along the x^1 axis. For a beam of light moving in the positive x^1 direction, we have $x^1 = ct = x^0$, and thus $x^- = 0$. The line $x^- = 0$ is, by definition, the x^+ axis (Figure 2.2). For a beam of light moving in the negative x^1 direction, we have $x^1 = -ct = -x^0$, and thus $x^+ = 0$. This corresponds to the x^- axis. The $x^\pm$ axes are lines at 45° with respect to the x^0, x^1 axes.

Can we think of x^+, or perhaps x^-, as a new time coordinate? Yes. In fact, both have equal right to be called a time coordinate, although neither one is a time coordinate in the standard sense of the word. Light-cone time is not quite the same as ordinary time.

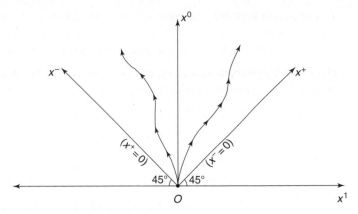

Fig. 2.2 A spacetime diagram with x^1 and x^0 represented as orthogonal axes. Shown are the light-cone axes $x^\pm = 0$. The curves with arrows are possible world-lines of physical particles.

Perhaps the most familiar property of time is that it goes forward for any physical motion of a particle. Physical motion starting at the origin is represented in Figure 2.2 as curves that remain within the light-cone and whose slopes never go below 45°. For all these curves, both x^+ and x^- increase as we follow the arrows. The only subtlety is that, for special light rays, light-cone time will freeze! As we saw above, x^+ remains constant for a light ray in the negative x^1 direction, while x^- remains constant for a light ray in the positive x^1 direction.

For definiteness, we will take x^+ to be the *light-cone time* coordinate. Accordingly, we will think of x^- as a spatial coordinate. Of course, these light-cone time and space coordinates will be somewhat strange.

Taking differentials of (2.50), we readily find that

$$2\, dx^+ dx^- = (dx^0 + dx^1)\,(dx^0 - dx^1) = (dx^0)^2 - (dx^1)^2 . \qquad (2.51)$$

It follows that the invariant interval (2.13), expressed in terms of the light-cone coordinates (2.50), takes the form

$$-ds^2 = -2\, dx^+ dx^- + (dx^2)^2 + (dx^3)^2 . \qquad (2.52)$$

The symmetry in the definitions of x^+ and x^- is evident here. Notice that, if we are given ds^2, solving for dx^- or for dx^+ does not require us to take a square root. This is a very important feature of light-cone coordinates, as we will see in Chapter 9.

How do we represent (2.52) with index notation? We still need indices that run over four values, but this time the values will be called

$$+, -, 2, 3. \qquad (2.53)$$

Just as we did in (2.21), we write

$$-ds^2 = \hat{\eta}_{\mu\nu}dx^\mu dx^\nu. \tag{2.54}$$

Here we have introduced a light-cone metric $\hat{\eta}$ which, like the Minkowski metric, is also defined to be symmetric under the exchange of its indices. Expanding this equation, and comparing with (2.52), we find

$$\hat{\eta}_{+-} = \hat{\eta}_{-+} = -1, \quad \hat{\eta}_{++} = \hat{\eta}_{--} = 0. \tag{2.55}$$

In the $(+, -)$ subspace, the diagonal elements of the light-cone metric vanish, but the off-diagonal elements do not. We also find that $\hat{\eta}$ does not couple the $(+, -)$ subspace to the $(2, 3)$ subspace:

$$\hat{\eta}_{+I} = \hat{\eta}_{-I} = 0, \quad I = 2, 3. \tag{2.56}$$

The matrix representation of the light-cone metric is

$$\hat{\eta}_{\mu\nu} = \begin{pmatrix} 0 & -1 & 0 & 0 \\ -1 & 0 & 0 & 0 \\ 0 & 0 & 1 & 0 \\ 0 & 0 & 0 & 1 \end{pmatrix}. \tag{2.57}$$

The light-cone components of any Lorentz vector a^μ are defined in analogy with (2.50):

$$a^+ \equiv \frac{1}{\sqrt{2}}(a^0 + a^1),$$

$$a^- \equiv \frac{1}{\sqrt{2}}(a^0 - a^1). \tag{2.58}$$

The scalar product between vectors, shown in (2.29), can be written using light-cone components. This time we have

$$a \cdot b = -a^- b^+ - a^+ b^- + a^2 b^2 + a^3 b^3 = \hat{\eta}_{\mu\nu} a^\mu b^\nu. \tag{2.59}$$

The last equality follows immediately from summing over the repeated indices and using (2.57). The first equality needs a small computation. In fact, it suffices to check that

$$-a^- b^+ - a^+ b^- = -a^0 b^0 + a^1 b^1. \tag{2.60}$$

This is quickly done using (2.58) and the analogous equations for $b^\pm$. We can also introduce lower light-cone indices. Consider the expression $a \cdot b = a_\mu b^\mu$, and expand the sum over the index μ using the light-cone labels:

$$a \cdot b = a_+ b^+ + a_- b^- + a_2 b^2 + a_3 b^3. \tag{2.61}$$

Comparing with (2.59), we find that

$$a_+ = -a^-, \quad a_- = -a^+. \tag{2.62}$$

When we lower or raise the zeroth index in a Lorentz frame, we get an extra sign. In light-cone coordinates, the indices of the first two coordinates switch and we get an extra sign.

Since physics described using light-cone coordinates looks unusual, we must develop an intuition for it. To do this, we will look at an example where the calculations are simple but the results are surprising.

Consider a particle moving along the x^1 axis with speed parameter $\beta = v/c$. At time $t = 0$, the positions x^1, x^2, and x^3 are all zero. Motion is nicely represented when the positions are expressed in terms of time:

$$x^1(t) = vt = \beta x^0, \quad x^2(t) = x^3(t) = 0. \tag{2.63}$$

How does this look in light-cone coordinates? Since x^+ is time and $x^2 = x^3 = 0$, we must simply express x^- in terms of x^+. Using (2.63), we find

$$x^+ = \frac{x^0 + x^1}{\sqrt{2}} = \frac{1 + \beta}{\sqrt{2}} x^0. \tag{2.64}$$

As a result,

$$x^- = \frac{x^0 - x^1}{\sqrt{2}} = \frac{(1 - \beta)}{\sqrt{2}} x^0 = \frac{1 - \beta}{1 + \beta} x^+. \tag{2.65}$$

Since it relates light-cone position to light-cone time, we identify the ratio

$$\frac{dx^-}{dx^+} = \frac{1 - \beta}{1 + \beta} \tag{2.66}$$

as the light-cone velocity. How strange is this light-cone velocity? For light moving to the right ($\beta = 1$) it equals zero. Indeed, light moving to the right has zero light-cone velocity because x^- does not change at all. This is shown as line 1 in Figure 2.3. Suppose you have a particle moving to the right with high conventional velocity, so that $\beta \simeq 1$ (line 2 in the figure). Its light-cone velocity is then very small. A long light-cone time must pass for this particle to move a little in the x^- direction. Perhaps more interestingly, a static particle in standard coordinates (line 3) is moving quite fast in light-cone coordinates. When $\beta = 0$ the particle has unit light-cone speed. This light-cone speed increases as β grows negative: the numerator in (2.66) is larger than one and increasing, while the denominator is smaller than one and decreasing. For $\beta = -1$ (line 5), the light-cone velocity is infinite! While this seems odd, there is no clash with relativity. Light-cone velocities are just unusual. The light-cone is a frame in which kinematics has a nonrelativistic flavor and infinite velocities are possible. Note that light-cone coordinates were introduced as a change of coordinates, not as a Lorentz transformation. There is no Lorentz transformation that takes the coordinates (x^0, x^1, x^2, x^3) into coordinates $(x'^0, x'^1, x'^2, x'^3) = (x^+, x^-, x^2, x^3)$.

Quick calculation 2.3 Convince yourself that the last statement above is correct.

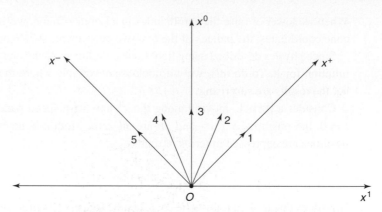

World-lines of particles with various light-cone velocities. Particle 1 has zero light-cone velocity. The velocities increase through that of particle 5, which is infinite.

2.4 Relativistic energy and momentum

In special relativity there is a basic relationship between the rest mass m of a point particle, its relativistic energy E, and its relativistic momentum $\vec{p}$. This relationship is given by

$$\frac{E^2}{c^2} - \vec{p} \cdot \vec{p} = m^2 c^2. \tag{2.67}$$

The relativistic energy and momentum are given in terms of the rest mass and velocity by the following familiar relations:

$$E = \gamma m c^2, \quad \vec{p} = \gamma m \vec{v}. \tag{2.68}$$

Quick calculation 2.4 Verify that the above E and $\vec{p}$ satisfy (2.67).

Energy and momentum can be used to define a momentum four-vector, as we will prove shortly. This four-vector is

$$p^\mu = (p^0, p^1, p^2, p^3) \equiv \left(\frac{E}{c}, p_x, p_y, p_z \right). \tag{2.69}$$

Using the last two equations, we have

$$p^\mu = \left(\frac{E}{c}, \vec{p} \right) = m\gamma (c, \vec{v}). \tag{2.70}$$

We use (2.28) to lower the index in p^μ:

$$p_\mu = (p_0, p_1, p_2, p_3) = \eta_{\mu\nu} p^\nu = \left(-\frac{E}{c}, p_x, p_y, p_z \right). \tag{2.71}$$

The above expressions for p^μ and p_μ give

$$p^\mu p_\mu = -(p^0)^2 + \vec{p} \cdot \vec{p} = -\frac{E^2}{c^2} + \vec{p} \cdot \vec{p}, \tag{2.72}$$

and, making use of (2.67), we have

$$p^{\mu} p_{\mu} = -m^2 c^2.$$ (2.73)

Since $p^{\mu} p_{\mu}$ has no free index it must be a Lorentz scalar. Indeed, all Lorentz observers agree on the value of the rest mass of a particle. Using the relativistic scalar product notation, condition (2.73) reads

$$p^2 \equiv p \cdot p = -m^2 c^2.$$ (2.74)

A central concept in special relativity is that of *proper time*. Proper time is a Lorentz invariant measure of time. Consider a moving particle and two events along its trajectory. Different Lorentz observers record different values for the time interval between the two events. But now imagine that the moving particle is carrying a clock. The proper time elapsed is the time elapsed between the two events *on that clock*. By definition, it is an invariant: all observers of a particular clock must agree on the time elapsed on that clock!

Proper time enters naturally into the calculation of invariant intervals. Consider an invariant interval for the motion of a particle along the x axis:

$$-ds^2 = -c^2 dt^2 + dx^2 = -c^2 dt^2 (1 - \beta^2).$$ (2.75)

Now evaluate the interval using a Lorentz frame attached to the particle. This is a frame in which the particle does not move and time is recorded by the clock that is moving with the particle. In this frame $dx = 0$ and $dt = dt_p$ is the proper time elapsed. As a result,

$$-ds^2 = -c^2 dt_p{}^2.$$ (2.76)

We cancel the minus signs and take the square root (using (2.20)) to find

$$ds = c \, dt_p.$$ (2.77)

This shows that, for timelike intervals, ds/c is the proper time interval. Similarly, cancelling minus signs and taking the square root of (2.75) gives

$$ds = cdt\sqrt{1 - \beta^2} \longrightarrow \frac{dt}{ds} = \frac{\gamma}{c}.$$ (2.78)

Being a Lorentz invariant, ds can be used to construct new Lorentz vectors from old Lorentz vectors. For example, a velocity four-vector u^{μ} is obtained by taking the ratio of dx^{μ} and ds. Since dx^{μ} is a Lorentz vector and ds is a Lorentz scalar, the ratio is also a Lorentz vector:

$$u^{\mu} = c \frac{dx^{\mu}}{ds} = c \left(\frac{d(ct)}{ds}, \frac{dx}{ds}, \frac{dy}{ds}, \frac{dz}{ds} \right).$$ (2.79)

The factor of c is included to give u^{μ} the units of velocity. The components of u^{μ} can be simplified using the chain rule and (2.78). For example,

$$\frac{dx}{ds} = \frac{dx}{dt} \frac{dt}{ds} = \frac{v_x \gamma}{c}.$$ (2.80)

Back in (2.79), we find

$$u^\mu = \gamma(c, v_x, v_y, v_z) = \gamma(c, \vec{v}).$$ (2.81)

Comparing with (2.70), we see that the momentum four-vector is just mass times the velocity four-vector:

$$p^\mu = mu^\mu.$$ (2.82)

This confirms our earlier assertion that the components of p^μ form a four-vector. Since any four-vector transforms under Lorentz transformations as the x^μ do, we can use (2.36) to find that under a boost in the x-direction the p^μ transform as

$$\frac{E'}{c} = \gamma\left(\frac{E}{c} - \beta\, p_x\right),$$
$$p_x' = \gamma\left(-\beta\frac{E}{c} + p_x\right).$$ (2.83)

2.5 Light-cone energy and momentum

The light-cone components p^+ and p^- of the momentum Lorentz vector are obtained using the rule (2.58):

$$p^+ = \frac{1}{\sqrt{2}}(p^0 + p^1) = -p_-,$$
$$p^- = \frac{1}{\sqrt{2}}(p^0 - p^1) = -p_+.$$ (2.84)

Which component should be identified with light-cone energy? The naive answer would be p^+. In any Lorentz frame, both the time and energy are the zeroth components of their respective four-vectors. Since light-cone time was chosen to be x^+, we might conclude that light-cone energy should be taken to be p^+. This is not appropriate, however. Light-cone coordinates do not transform as Lorentz ones do, so we should be careful and examine this question in detail. Both $p^\pm$ are energy-like, since both are positive for physical particles. Indeed, from (2.67), and with $m \neq 0$, we have

$$p^0 = \frac{E}{c} = \sqrt{\vec{p}\cdot\vec{p} + m^2c^2} > |\vec{p}| \geq |p^1|.$$ (2.85)

As a result, $p^0 \pm p^1 > 0$, and thus $p^\pm > 0$. While both are plausible candidates for energy, the physically motivated choice turns out to be $-p_+$, which happens to coincide with p^-.

Before we explain this choice, let us first evaluate $p_\mu x^\mu$. In standard coordinates,

$$p \cdot x = p_0 x^0 + p_1 x^1 + p_2 x^2 + p_3 x^3.$$ (2.86)

In light-cone coordinates, using (2.61),

$$p \cdot x = p_+ x^+ + p_- x^- + p_2 x^2 + p_3 x^3.$$ (2.87)

In standard coordinates, $p_0 = -E/c$ appears together with the time x^0. In light-cone coordinates, p_+ appears together with the light-cone time x^+. We would therefore expect p_+ to be minus the light-cone energy.

Why is this pairing significant? Energy and time are conjugate variables. As you learned in quantum mechanics, the Hamiltonian operator measures energy and generates time evolution. The wavefunction of a point particle with energy E and momentum $\vec{p}$ is given by

$$\psi(t, \vec{x}) = \exp\left(-\frac{i}{\hbar}(Et - \vec{p} \cdot \vec{x})\right). \tag{2.88}$$

Indeed, this wavefunction satisfies the Schrödinger equation

$$i\hbar \frac{\partial \psi}{\partial x^0} = \frac{E}{c} \psi. \tag{2.89}$$

Similarly, light-cone time evolution and light-cone energy E_{lc} should be related by

$$i\hbar \frac{\partial \psi}{\partial x^+} = \frac{E_{\mathrm{lc}}}{c} \psi. \tag{2.90}$$

To find the x^+ dependence of the wavefunction, we recognize that

$$\psi(t, \vec{x}) = \exp\left(\frac{i}{\hbar}(p_0 x^0 + \vec{p} \cdot \vec{x})\right) = \exp\left(\frac{i}{\hbar} p \cdot x\right), \tag{2.91}$$

and, using (2.87), we have

$$\psi(x) = \exp\left(\frac{i}{\hbar}(p_+ x^+ + p_- x^- + p_2 x^2 + p_3 x^3)\right). \tag{2.92}$$

We can now return to (2.90) and evaluate:

$$i\hbar \frac{\partial \psi}{\partial x^+} = -p_+ \psi \quad \longrightarrow \quad -p_+ = \frac{E_{\mathrm{lc}}}{c}. \tag{2.93}$$

This confirms our identification of $(-p_+)$ with light-cone energy. Since, presently, $-p_+ = p^-$, it is convenient to use p^- as the light-cone energy in order to eliminate the sign in the above equation:

$$p^- = \frac{E_{\mathrm{lc}}}{c}. \tag{2.94}$$

Some physicists like to raise and lower $+$ and $-$ indices to simplify expressions involving light-cone quantities. While this is sometimes convenient, it can easily lead to errors. If you talk with a friend over the phone, and she says ". . . p-plus times . . .," you will have to ask, "plus up, or plus down?" In the rest of this book we will not lower the $+$ or $-$ indices. They will always be up, and the energy will always be p^-.

We can check that the identification of p^- as light-cone energy fits together nicely with the intuition that we have developed for light-cone velocity. To this end, we confirm that a particle with small light-cone velocity also has small light-cone energy. Suppose we have a particle moving very fast in the $+x^1$ direction. As discussed below (2.66), its light-cone velocity is very small. Since p^1 is very large, equation (2.67) gives

$$p^0 = \sqrt{(p^1)^2 + m^2 c^2} = p^1 \sqrt{1 + \frac{m^2 c^2}{(p^1)^2}} \simeq p^1 + \frac{m^2 c^2}{2 p^1}. \qquad (2.95)$$

The light-cone energy of the particle is therefore

$$p^- = \frac{1}{\sqrt{2}} (p^0 - p^1) \simeq \frac{m^2 c^2}{2\sqrt{2}\, p^1}. \qquad (2.96)$$

As anticipated, both the light-cone velocity and the light-cone energy decrease as p^1 increases.

2.6 Lorentz invariance with extra dimensions

If string theory is correct, we must entertain the possibility that spacetime has more than four dimensions. The number of time dimensions must be kept equal to one – it seems very difficult, if not altogether impossible, to construct a consistent theory with more than one time dimension. The extra dimensions must therefore be spatial. Can we have Lorentz invariance in worlds with more than three spatial dimensions? Yes. Lorentz invariance is a concept that admits a very natural generalization to spacetimes with additional dimensions.

We first extend the definition of the invariant interval ds^2 to incorporate the additional space dimensions. In a world with five spatial dimensions, for example, we would write

$$-ds^2 = -c^2 dt^2 + (dx^1)^2 + (dx^2)^2 + (dx^3)^2 + (dx^4)^2 + (dx^5)^2. \qquad (2.97)$$

Lorentz transformations are then defined as the linear changes of coordinates that leave ds^2 invariant. This ensures that every inertial observer in the six-dimensional spacetime will agree on the value of the speed of light. With more dimensions, come more Lorentz transformations. While in four-dimensional spacetime we have boosts in the x^1, x^2, and x^3 directions, in this new world we have boosts along each of the five spatial dimensions. With three spatial coordinates, there are three basic spatial rotations: rotations that mix x^1 and x^2, those that mix x^1 and x^3, and finally those that mix x^2 and x^3. The equality of the number of boosts and the number of rotations is a special feature of four-dimensional spacetime. With five spatial coordinates, we have ten rotations, which is twice the number of boosts.

The higher-dimensional Lorentz invariance includes the lower-dimensional one: if nothing happens along the extra dimensions, then the restrictions of lower-dimensional Lorentz invariance apply. This is clear from (2.97). For motion that does not involve the extra dimensions, $dx^4 = dx^5 = 0$, and the expression for ds^2 reduces to that used in four dimensions.

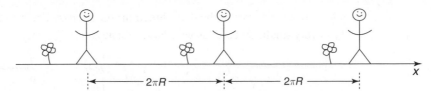

Fig. 2.4   A one-dimensional world that repeats each $2\pi R$. Several copies of Phil are shown.

2.7 Compact extra dimensions

It is possible for additional spatial dimensions to be undetected by low energy experiments if the dimensions are curled up into a compact space of small volume. In this section, we will try to understand what a compact dimension is. We will focus mainly on the case of one dimension. In Section 2.10 we will explain why small compact dimensions are hard to detect.

Consider a one-dimensional world, an infinite line, say, and let x be a coordinate along this line. For each point P along the line, there is a unique real number $x(P)$ called the x-coordinate of the point P. A good coordinate on this infinite line satisfies two conditions.

- Any two distinct points $P_1 \neq P_2$ have different coordinates: $x(P_1) \neq x(P_2)$.
- The assignment of coordinates to points is continuous: nearby points have nearly equal coordinates.

If a choice of origin is made for this infinite line, then we can use distance from the origin to define a good coordinate. The coordinate assigned to each point is the distance from that point to the origin, with a sign depending upon which side of the origin the point lies.

Imagine that you live in a world with one spatial dimension. Suppose you are walking along and notice a strange pattern: the scenery repeats each time you move a distance $2\pi R$, for some value of R. If you meet your friend Phil, you see that there are Phil clones at distances $2\pi R, 4\pi R, 6\pi R, \ldots$ down the line (see Figure 2.4). In fact, there are clones up the line, as well, with the same spacing.

There is no way to distinguish an infinite line with such a strange property from a circle with circumference $2\pi R$. Indeed, saying that this strange line is a circle *explains* the peculiar property – there really are no Phil clones; you meet the same Phil again and again as you go around the circle!

How do we express this mathematically? We can think of the circle as the open line with an *identification*. That is, we declare that points with coordinates that differ by $2\pi R$ are the same point. More precisely, two points are declared to be the same point if their coordinates differ by an integer number of $2\pi R$:

$$P_1 \sim P_2 \quad \longleftrightarrow \quad x(P_1) = x(P_2) + 2\pi R n, \quad n \in \mathbb{Z}. \qquad (2.98)$$

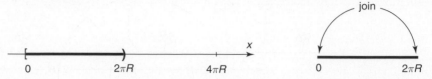

Fig. 2.5 The interval $0 \le x < 2\pi R$ is a fundamental domain for the line with the identification
(2.99). The identified space is a circle of radius R.

This is precise, but somewhat cumbersome, notation. With no risk of confusion, we can
simply write

$$x \sim x + 2\pi R, \qquad (2.99)$$

which should be read as "identify any two points whose coordinates differ by $2\pi R$." With
such an identification, the open line becomes a circle. The identification has turned a non-
compact dimension into a compact one. It may seem to you that a line with identifications
is only a complicated way to think about a circle. We will see, however, that many physical
problems become clearer when we view a compact dimension as an extended one with
identifications.

The interval $0 \le x < 2\pi R$ is a *fundamental domain* for the identification (2.99) (see
Figure 2.5). A fundamental domain is a subset of the entire space that satisfies two
conditions.

1. No two points in the fundamental domain are identified.
2. Any point in the entire space is in the fundamental domain or is related by the
   identification to some point in the fundamental domain.

Whenever possible, as we did here, the fundamental domain is chosen to be a connected
region. To build the space implied by the identification, we take the fundamental domain
together with its boundary, and implement the identifications on the boundary. In our case,
the fundamental domain together with its boundary is the segment $0 \le x \le 2\pi R$. In this
segment we identify the point $x = 0$ with the point $x = 2\pi R$. The result is the circle.

A circle of radius R can be represented in a two-dimensional plane as the set of points that
are at a distance R from a point called the center of the circle. Note that the circle obtained
above has been constructed directly, without the help of an embedding two-dimensional
space. For our circle, there is no point, anywhere, that represents the center of the circle.
We can still speak, figuratively, of the radius R of the circle, but in our case, the radius is
simply the quantity which multiplied by 2π gives the total length of the circle.

On the circle, the coordinate x is no longer a good coordinate. The coordinate x is now
either multivalued or discontinuous. This is a problem with any coordinate on a circle.
Consider using angles to assign coordinates on the unit circle (Figure 2.6). Fix a reference
point Q on the circle, and let O denote the center of the circle. To any point P on the circle
we assign as a coordinate the angle $\theta(P) = \angle POQ$. This angle is naturally multivalued.
The reference point Q, for example, has $\theta(Q) = 0°$ and $\theta(Q) = 360°$. If we force angles

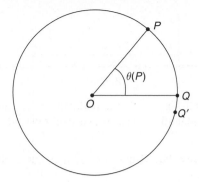

Fig. 2.6   Using the angle θ to define a coordinate on a circle. The reference point Q is assigned zero angle: $\theta(Q) = 0$. The coordinate θ is naturally multivalued.

to be single valued by restricting $0° \leq \theta < 360°$, for example, then they become discontinuous. Indeed, two nearby points, Q and Q', then have very different angles: $\theta(Q) = 0$, while $\theta(Q') \sim 360°$. It is easier to work with multivalued coordinates than it is to work with discontinuous ones.

If we have a world with several open dimensions, then we can apply the identification (2.99) to one of the dimensions, while doing nothing to the others. The dimension described by x turns into a circle, and the other dimensions remain open. It is possible, of course, to make more than one dimension compact. Consider, for example, the (x, y) plane, subject to *two* identifications:

$$x \sim x + 2\pi R, \quad y \sim y + 2\pi R. \tag{2.100}$$

It is perhaps clearer to show both coordinates simultaneously while writing the identifications. In that fashion, the two identifications are written as

$$(x, y) \sim (x + 2\pi R, \ y), \tag{2.101}$$

$$(x, y) \sim (x, \ y + 2\pi R). \tag{2.102}$$

The first identification implies that we can restrict our attention to $0 \leq x < 2\pi R$, and the second identification implies that we can restrict our attention to $0 \leq y < 2\pi R$. Thus the fundamental domain can be taken to be the square region $0 \leq x, y < 2\pi R$, as shown in Figure 2.7. To build the space implied by the identifications, we take the fundamental domain together with its boundary, forming the full square $0 \leq x, y \leq 2\pi R$, and implement the identifications on the boundary. The vertical edges are identified because they correspond to points of the form $(0, y)$ and $(2\pi R, y)$, which are identified by (2.101). The horizontal edges are identified because they correspond to points of the form $(x, 0)$ and $(x, 2\pi R)$, which are identified by (2.102). The resulting space is called a two-dimensional torus. One can visualize the torus by taking the fundamental domain (with its boundary)

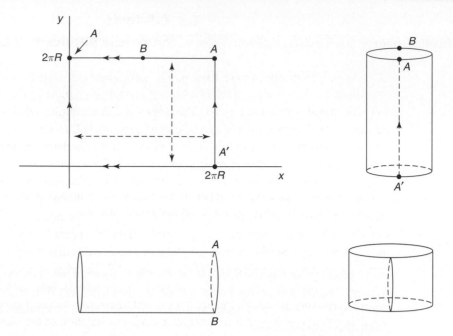

Fig. 2.7 **A square region in the plane with identifications indicated by the dashed lines and arrowheads. The resulting surface is a torus. The identification of the vertical lines gives a cylinder, shown to the right of the square region. The cylinder, shown horizontally and flattened in the bottom left, must have its edges glued to form the torus.**

and gluing the vertical edges as their identification demands. The result is a cylinder, as shown in the top right corner of Figure 2.7 (with the gluing seam dashed). In this cylinder, however, the bottom circle and the top circle must also be glued, since they are nothing other than the horizontal edges of the fundamental domain. To do this with paper, you must flatten the cylinder and then roll it up to glue the circles. The result looks like a flattened doughnut. With a flexible piece of garden hose, you could simply identify the two ends to obtain the familiar picture of a torus.

We have seen how to compactify coordinates using identifications. Some compact spaces are constructed in other ways. In string theory, however, compact spaces that arise from identifications are particularly easy to work with. We shall focus on such spaces throughout this book.

Quick calculation 2.5 Consider the plane (x, y) with the identification

$$(x, y) \sim (x + 2\pi R, y + 2\pi R). \tag{2.103}$$

What is the resulting space? Hint: the space is most clearly exhibited using a fundamental domain for which the line $x + y = 0$ is a boundary.

2.8 Orbifolds

Sometimes identifications have fixed points, points that are related to themselves by the identification. For example, consider the real line parameterized by the coordinate x and subject to the identification $x \sim -x$. The point $x = 0$ is the unique fixed point of the identification. A fundamental domain can be chosen to be the half-line $x \geq 0$ (Figure 2.8). Note that the boundary point $x = 0$ must be included in the fundamental domain. The space obtained by the above identification is in fact the fundamental domain $x \geq 0$. This is the simplest example of an *orbifold*, a space obtained by identifications that have fixed points. An orbifold is singular at the fixed points. While the half-line $x \geq 0$ is a conventional one-dimensional manifold for $x > 0$, neighborhoods of the point $x = 0$ fail to be typical. This orbifold is called an $\mathbb{R}^1/\mathbb{Z}_2$ orbifold. Here $\mathbb{R}^1$ stands for the (one-dimensional) real line, and $\mathbb{Z}_2$ describes a basic property of the identification when it is viewed as the transformation $x \rightarrow -x$: if applied twice, it gives back the original coordinate.

Certain two-dimensional cones can be obtained as orbifolds. Begin with the (x, y) plane and identify every point with the image obtained by rotation around the origin through the angle $2\pi/N$, where $N \geq 2$ is an integer. A simple description of the identification makes use of the complex coordinate $z = x + iy$:

$$z \sim e^{\frac{2\pi i}{N}} z . \tag{2.104}$$

This identification does as expected: multiplication of any complex number by a phase $e^{i\alpha}$ (α real) rotates the complex number by the angle α. The identification is of $\mathbb{Z}_N$ type: viewed as the transformation $z \rightarrow e^{\frac{2\pi i}{N}} z$, if applied N times, it gives back the original coordinate. A point must be identified with all the $N - 1$ images obtained by repeated action of the transformation. The only fixed point of the $\mathbb{Z}_N$ transformation is the origin $z = 0$. A fundamental domain for (2.104), as we will explain below, is provided by the points z that satisfy the constraint

$$0 \leq \arg(z) < \frac{2\pi}{N} . \tag{2.105}$$

Here we recall that for $z = re^{i\theta}$, with r and θ real, we have $\arg(z) = \theta$. The fundamental domain is shown in Figure 2.9. To the right one sees the cone, obtained by gluing the rays $\arg(z) = 0$ and $\arg(z) = 2\pi/N$ using the identification (2.104). The resulting cone is called the $\mathbb{C}/\mathbb{Z}_N$ orbifold, where $\mathbb{C}$ denotes the complex plane, namely, the original two-dimensional plane equipped with a complex coordinate. The cone is singular at the apex $z = 0$ in the sense that it has concentrated curvature.

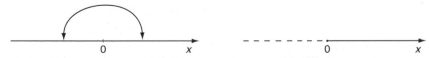

Fig. 2.8   The identification $x \sim -x$ on the real line yields the half-line. This is the $\mathbb{R}^1/\mathbb{Z}_2$ orbifold.

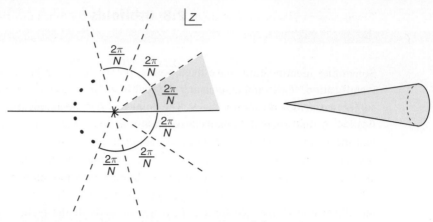

Fig. 2.9 The fundamental domain for the identification $z \sim e^{\frac{2\pi i}{N}} z$ is shown shaded. After identification we obtain a cone – the $\mathbb{C}/\mathbb{Z}_N$ orbifold.

Let us explain why the region defined by (2.105) is a fundamental domain. Acting repeatedly on the region with the transformation $z \to e^{\frac{2\pi i}{N}} z$, we find $N - 1$ images that together with the region cover seamlessly the full complex plane. Since each point in the region has exactly $N - 1$ copies, all outside the region, no two points in the region are identified. Any point in $\mathbb{C}$ that is not in the region must lie in one of its $N - 1$ copies and therefore it has an image in the region. Note that our argument used the fact that N is an integer; for N irrational, for example, any point would have an infinite number of images. Our construction gives only the cones whose total angle at the apex is 2π divided by an integer. Cones with other angles exist, but they are not obtained as orbifolds. The case when N is replaced by a rational number is examined in Problem 2.7. Additional examples of orbifolds are considered in Problems 2.5, 2.6, and 2.10.

Physics on spaces with generic singularities is typically complicated and sometimes even inconsistent. Orbifolds are spaces with tractable singularities, at least as far as strings are concerned. The physics of quantum strings on an orbifold, as we will see in Chapter 13, is completely regular because the orbifold arises from identifications applied to a non-singular space where the quantum string is simple. The strings on the orbifold inherit that simplicity.

2.9 Quantum mechanics and the square well

Planck's constant $\hbar$ appears as the constant of proportionality relating the energy E and the angular frequency ω of a photon:

$$E = \hbar\omega. \tag{2.106}$$

Since ω has units of inverse time, $\hbar$ has units of energy times time. Energy has units of ML^2T^{-2} and, therefore,

$$[\hbar] = [\text{Energy}] \times [\text{Time}] = ML^2T^{-1}. \tag{2.107}$$

The value of Planck's constant is $\hbar \simeq 1.055 \times 10^{-27}$ erg $\cdot$ s.

The constant $\hbar$ appears in the basic commutation relations of quantum mechanics. The Schrödinger position and momentum operators satisfy

$$[x, p] = i\hbar. \tag{2.108}$$

If we have several spatial dimensions, then the commutation relations are

$$\left[x^i, \, p_j \right] = i\hbar \, \delta^i_j, \tag{2.109}$$

where the Kronecker delta is defined as in (2.32):

$$\delta^i_j = \begin{cases} 1 & \text{if } i = j \\ 0 & \text{if } i \neq j. \end{cases} \tag{2.110}$$

In three spatial dimensions, the indices i and j run from 1 to 3. The generalization of quantum mechanics to higher dimensions is straightforward. With d spatial dimensions, the indices in (2.109) simply run over the d possible values.

To set the stage for the analysis of a small extra dimension, we review a standard quantum mechanics problem. Consider the time-independent Schrödinger equation

$$-\frac{\hbar^2}{2m}\nabla^2\psi(x) + V(x)\psi(x) = E\,\psi(x) \tag{2.111}$$

applied to the case of a one-dimensional square-well potential of infinite height:

$$V(x) = \begin{cases} 0 & \text{if } x \in (0, a) \\ \infty & \text{if } x \notin (0, a). \end{cases} \tag{2.112}$$

For $x \notin (0, a)$, the infinite potential implies $\psi(x) = 0$. In particular, $\psi(0) = \psi(a) = 0$. This is just the quantum mechanics of a particle living on a *segment*, as shown in Figure 2.10.

When $x \in (0, a)$, the Schrödinger equation becomes

$$-\frac{\hbar^2}{2m}\frac{d^2\psi}{dx^2} = E\psi. \tag{2.113}$$

The solutions of (2.113), consistent with the boundary conditions, are

$$\psi_k(x) = \sqrt{\frac{2}{a}} \sin\left(\frac{k\pi x}{a}\right), \quad k = 1, 2, \ldots, \infty. \tag{2.114}$$

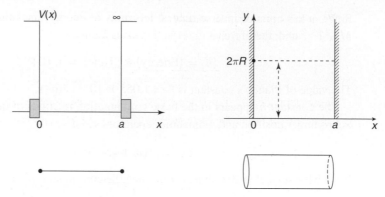

Fig. 2.10 Left: the square-well potential in one dimension. Here the particle lives on a segment. Right: in the (x, y) plane the particle must remain within $0 < x < a$. The direction y is identified as $y \sim y + 2\pi R$. The particle lives on a cylinder.

The value $k = 0$ is not allowed since it would make the wavefunction vanish everywhere. By performing the differentiation indicated in (2.113), we see that the energy E_k associated with the wavefunction ψ_k is

$$E_k = \frac{\hbar^2}{2m} \left(\frac{k\pi}{a} \right)^2 . \tag{2.115}$$

2.10 Square well with an extra dimension

We now add an extra dimension to the square-well problem (no pun intended!). In addition to x, we include a dimension y that is curled up into a small circle of radius R. In other words, we make the identification

$$(x, y) \sim (x, \; y + 2\pi R). \tag{2.116}$$

The original dimension x has not been changed (see Figure 2.10). Since the y direction has been turned into a circle of circumference $2\pi R$, the space where the particle moves is now a *cylinder*. The cylinder has length a and circumference $2\pi R$. The potential $V(x, y)$ will remain given by (2.112) and is y-independent.

We will see that, as long as R is small, and as long as we look only at low energies, the quantum mechanics of the particle on the segment is very similar to the quantum mechanics of the particle on the cylinder. The only length scale in the original problem is the size a of the segment, so small R means $R \ll a$.

In two dimensions, the Schrödinger equation (2.111) becomes

$$-\frac{\hbar^2}{2m} \left(\frac{\partial^2 \psi}{\partial x^2} + \frac{\partial^2 \psi}{\partial y^2} \right) = E\psi. \tag{2.117}$$

We use separation of variables in order to solve this equation. We let $\psi(x, y) = \psi(x)\phi(y)$ and find that the equation takes the form

$$-\frac{\hbar^2}{2m}\frac{1}{\psi(x)}\frac{d^2\psi(x)}{dx^2} - \frac{\hbar^2}{2m}\frac{1}{\phi(y)}\frac{d^2\phi(y)}{dy^2} = E. \tag{2.118}$$

The x-dependent and y-dependent terms of this equation must separately be constant, and the solutions are of the form $\psi_{k,l}(x, y) = \psi_k(x)\phi_l(y)$, where

$$\psi_k(x) = c_k \sin\left(\frac{k\pi x}{a}\right), \tag{2.119}$$

$$\phi_l(y) = a_l \sin\left(\frac{ly}{R}\right) + b_l \cos\left(\frac{ly}{R}\right). \tag{2.120}$$

The physics along the x dimension is unchanged, since the wavefunction must still vanish at the ends of the segment. Therefore (2.119) takes the same form as (2.114) and $k = 1, 2, \ldots$. The boundary condition for $\phi_l(y)$ arises from the identification $y \sim y + 2\pi R$. Since y and $y + 2\pi R$ are coordinates that represent the same point, the wavefunction must take the same value at these two arguments:

$$\phi_l(y) = \phi_l(y + 2\pi R). \tag{2.121}$$

As opposed to $\psi_k(x)$, the function $\phi_l(y)$ need not vanish for any y. As a result, the general periodic solution, recorded in (2.120), includes both sines and cosines, with $l = 0, 1, 2, \ldots$. The presence of cosines allows a nonvanishing *constant* solution: for $l = 0$, we get $\phi_0(y) = b_0$. This solution is the key to understanding why a small extra dimension does not change the low energy physics very much.

The energy eigenvalues of the $\psi_{k,l}$ are

$$E_{k,l} = \frac{\hbar^2}{2m}\left[\left(\frac{k\pi}{a}\right)^2 + \left(\frac{l}{R}\right)^2\right]. \tag{2.122}$$

These energies correspond to doubly degenerate states when $l \neq 0$, because in this case (2.120) contains two linearly independent solutions. The extra dimension has changed the spectrum dramatically. We will see, however, that if $R \ll a$, then the *low-lying* part of the spectrum is unchanged. The rest of the spectrum changes, but these changes are not accessible *at low energies*.

Since $l = 0$ is permitted, the energy levels $E_{k,0}$ coincide with the old energy levels E_k! The new system contains all the energy levels of the old system. But it also includes additional energy levels. What is the lowest *new* energy level? To minimize the energy, each of the terms in (2.122) must be as low as possible. The minimum occurs when $k = 1$, since $k = 0$ is not allowed, and $l = 1$, since $l = 0$ gives us the old levels. The lowest new energy level is

$$E_{1,1} = \frac{\hbar^2}{2m}\left[\left(\frac{\pi}{a}\right)^2 + \left(\frac{1}{R}\right)^2\right]. \tag{2.123}$$

When $R \ll a$, the second term is much larger than the first, and

$$E_{1,1} \sim \frac{\hbar^2}{2m}\left(\frac{1}{R}\right)^2. \tag{2.124}$$

This energy is comparable to that of the level k eigenstate of the original problem (see (2.115)) when

$$\frac{k\pi}{a} \sim \frac{1}{R} \rightarrow k \sim \frac{1}{\pi}\frac{a}{R}. \tag{2.125}$$

Since R is much smaller than a, k is a very large number. So the first new energy level appears at an energy far above that of the low-lying original states. We therefore conclude that an extra dimension can remain hidden from experiments at a particular energy level as long as the dimension is small enough. Once the probing energies become sufficiently high, the effects of an extra dimension can be observed.

Curiously, the quantum mechanics of a string introduces new features. For an extra dimension *much* smaller than the already small string length ℓ_s, new low-lying states can appear! These correspond to strings that wrap around the extra dimension. They have no analog in the quantum mechanics of a point particle, and we will study them in detail in Chapter 17. In string theory, the conclusion remains true that no new low energy states arise from a small extra dimension, but there is a small qualification: the dimension must not be significantly smaller than ℓ_s. We will learn in Chapter 17 that, in string theory, the effects of a compact dimension of radius smaller than ℓ_s cannot be distinguished from those of another compact dimension with a radius larger than ℓ_s.

Problems

Problem 2.1 Exercises with units.

(a) Find the relation between coulombs (C) and esus.

(b) Explain the meaning of the unit K (degree kelvin) used for measuring temperatures, and explain its relation to the basic length, mass, and time units.

(c) Construct and evaluate a dimensionless number using the charge e of the electron (as defined in the Gaussian system of units), $\hbar$, and c. (In Heaviside–Lorentz units, the Gaussian e^2 is replaced by $\frac{e^2}{4\pi}$.)

Problem 2.2 Lorentz transformations for light-cone coordinates.

Consider coordinates $x^\mu = (x^0, x^1, x^2, x^3)$ and the associated light-cone coordinates (x^+, x^-, x^2, x^3). Write the following Lorentz transformations in terms of the light-cone coordinates.

(a) A boost with velocity parameter β in the x^1 direction.

(b) A rotation with angle θ in the x^1, x^2 plane.

(c) A boost with velocity parameter β in the x^3 direction.

Problem 2.3 Lorentz transformations, derivatives, and quantum operators.

(a) Give the Lorentz transformations for the components a_μ of a vector under a boost along the x^1 axis.

(b) Show that the objects $\frac{\partial}{\partial x^\mu}$ transform under a boost along the x^1 axis in the same way as the a_μ considered in (a) do. This checks, in a particular case, that partial derivatives with respect to upper-index coordinates x^μ behave as a four-vector with lower indices, which is why they are written as ∂_μ.

(c) Show that, in quantum mechanics, the expressions for the energy and momentum in terms of derivatives can be written compactly as $p_\mu = \frac{\hbar}{i} \frac{\partial}{\partial x^\mu}$.

Problem 2.4 Lorentz transformations as matrices.

A matrix L that satisfies (2.46) is a Lorentz transformation. Show the following.

(a) If L_1 and L_2 are Lorentz transformations so is the product $L_1 L_2$.
(b) If L is a Lorentz transformation so is the inverse matrix L^{-1}.
(c) If L is a Lorentz transformation so is the transpose matrix L^T.

Problem 2.5 Constructing simple orbifolds.

(a) Consider a circle S^1, presented as the real line with the identification $x \sim x + 2$. Choose $-1 < x \le 1$ as the fundamental domain. The circle is the space $-1 \le x \le 1$ with the points $x = \pm 1$ identified. The orbifold $S^1/\mathbb{Z}_2$ is defined by imposing the (so-called) $\mathbb{Z}_2$ identification $x \sim -x$. Describe the action of this identification on the circle. Show that there are two points on the circle that are left fixed by the $\mathbb{Z}_2$ action. Find a fundamental domain for the two identifications. Describe the orbifold $S^1/\mathbb{Z}_2$ in simple terms.

(b) Consider a torus T^2, presented as the (x, y) plane with the identifications $x \sim x + 2$ and $y \sim y + 2$. Choose $-1 < x, y \le 1$ as the fundamental domain. The orbifold $T^2/\mathbb{Z}_2$ is defined by imposing the $\mathbb{Z}_2$ identification $(x, y) \sim (-x, -y)$. Prove that there are four points on the torus that are left fixed by the $\mathbb{Z}_2$ transformation. Show that the orbifold $T^2/\mathbb{Z}_2$ is topologically a two-dimensional sphere, naturally presented as a square pillowcase with seamed edges.

Problem 2.6 Constructing the $T^2/\mathbb{Z}_3$ orbifold.

Consider the complex plane $z = x + iy$ subject to the following two identifications

$$z \sim T_1(z) = z + 1, \quad \text{and} \quad z \sim T_2(z) = z + e^{i\pi/3} .$$

(a) A fundamental domain, with its boundary, is the parallelogram with corners at $z = 0, 1$, and $e^{i\pi/3}$. Where is the fourth corner? Make a sketch and indicate the identifications on the boundary. The resulting space is an oblique torus.

(b) Consider now the *additional* $\mathbb{Z}_3$ identification

$$z \sim R(z) = e^{2\pi i/3} z .$$

To understand how this identification acts on the oblique torus, draw the short diagonal that divides the torus into two equilateral triangles. Describe carefully the $\mathbb{Z}_3$ action on each of the two triangles (recall that the action of R can be followed by arbitrary action with T_1, T_2, and their inverses).

(c) Determine the three fixed points of the $\mathbb{Z}_3$ action on the torus. Show that the orbifold $T^2/\mathbb{Z}_3$ is topologically a two-dimensional sphere, naturally presented as a triangular pillowcase with seamed edges and corners at the fixed points.

Problem 2.7 A more general construction for cones?

Consider the (x, y) plane and the complex coordinate $z = x + iy$. We have seen that the identification $z \sim e^{\frac{2\pi i}{N}} z$, with N an integer greater than two, can be used to construct a cone.

Examine now the identification

$$z \sim e^{2\pi i \frac{M}{N}} z, \quad N > M \geq 2,$$

where M and N are relatively prime integers (their greatest common divisor is one). One may guess that a fundamental domain is provided by the points z that satisfy $0 \leq \arg(z) < 2\pi \frac{M}{N}$. Play with low values of M and N to convince yourself that this is *not* true. Determine a fundamental domain for the identification. [Hint: use the following result. Given two relatively prime integers a and b, there exist integers m and n such that $ma + nb = 1$. Finding m and n is not easy unless you use Euclid's algorithm. Try to find, for example, integers m and n that satisfy $187m + 35n = 1$.]

Problem 2.8 Spacetime diagrams and Lorentz transformations.

Consider a spacetime diagram in which the x^0 and x^1 axes of a Lorentz frame S are represented as vertical and horizontal axes, respectively. Show that the x'^0 and x'^1 axes of the Lorentz frame S', related to S via (2.36), appear in the original spacetime diagram as oblique axes. Find the angle between the primed and unprimed axes. Show in detail how the axes appear when $\beta > 0$ and when $\beta < 0$, indicating in both cases the directions of increasing values of the coordinates.

Problem 2.9 Lightlike compactification.

The identification $x \sim x + 2\pi R$, is the statement that the coordinate x has been compactified into a circle of radius R. In this identification, the time dimension is left untouched. Consider now the strange "lightlike" compactification, in which we identify events with position and time coordinates related by

$$\begin{pmatrix} x \\ ct \end{pmatrix} \sim \begin{pmatrix} x \\ ct \end{pmatrix} + 2\pi \begin{pmatrix} R \\ -R \end{pmatrix}. \tag{1}$$

(a) Rewrite this identification using light-cone coordinates.
(b) Consider coordinates (ct', x') related to (ct, x) by a boost with velocity parameter β. Express the identifications in terms of the primed coordinates.

To interpret (1) physically, consider the family of identifications

$$\begin{pmatrix} x \\ ct \end{pmatrix} \sim \begin{pmatrix} x \\ ct \end{pmatrix} + 2\pi \begin{pmatrix} \sqrt{R^2 + R_s^2} \\ -R \end{pmatrix}, \tag{2}$$

where R_s is a length that will eventually be taken to zero, in which case (2) reduces to (1).

(c) Show that there is a boosted frame S' in which the identification (2) becomes a standard identification (i.e., the space coordinate is identified but the time coordinate is not). Find the velocity parameter of S' with respect to S and the compactification radius in this Lorentz frame S'.

(d) Represent your answer to part (c) in a spacetime diagram. Show two points related by the identification (2) and the space and time axes for the Lorentz frame S' in which the compactification is standard.

(e) Fill in the blanks in the following statement: Lightlike compactification with radius R arises by boosting a standard compactification with radius . . . with Lorentz factor $\gamma \sim R/\ldots$, in the limit as . . . $\to 0$.

Problem 2.10 A spacetime orbifold in two dimensions.

Consider a two-dimensional world with coordinates x^0 and x^1. A boost with velocity parameter β along the x^1 axis is described by the first two equations in (2.36). We want to understand the two-dimensional space that emerges if we identify

$$(x^0, x^1) \sim (x'^0, x'^1). \tag{1}$$

We are identifying spacetime points whose coordinates are related by a boost!

(a) Use the result of Problem 2.2, part (a), to recast (1) as

$$(x^+, x^-) \sim \left(e^{-\lambda} x^+, e^{\lambda} x^-\right), \quad \text{where} \quad e^{\lambda} \equiv \sqrt{\frac{1+\beta}{1-\beta}}. \tag{2}$$

What is the range of λ? What is the orbifold fixed point? Assume now that $\beta > 0$, and thus $\lambda > 0$.

(b) Draw a spacetime diagram, indicate the x^+ and x^- axes, and sketch the family of curves

$$x^+ x^- = a^2, \tag{3}$$

where $a > 0$ is a real constant that labels the various curves. Indicate which curves have small a and which have large a. For each value of a, equation (3) describes two disconnected curves. Show that the identification (2) relates points on each separate curve.

(c) Use the expression $-ds^2 = -2dx^+ dx^-$ for the interval to show that any curve in (3) is spacelike.

(d) Consider the two curves $x^+ x^- = a^2$ for some fixed a. The identification (2) makes each one of these curves into a circle. Find the invariant circumference of this circle by integrating the appropriate root of ds^2 between two neighboring identified points. Give your answer in terms of a and λ. Answer: $\sqrt{2}\, a\lambda$.

Roughly, as time goes from minus infinity to plus infinity, the parameter a goes from infinity down to zero and then back to infinity. This orbifold represents a universe where space is a circle. The circle begins large, contracts to zero size, and then expands again. This orbifold has one pathology: the curves $x^+ x^- = -a^2$ are actually closed timelike circles.

Problem 2.11 Extra dimension and statistical mechanics.

Write a double sum that represents the statistical mechanics partition function $Z(a, R)$ for the quantum mechanical system considered in Section 2.10. Note that $Z(a, R)$ factors as $Z(a, R) = Z(a)\tilde{Z}(R)$.

(a) Explicitly calculate $Z(a, R)$ in the very high temperature limit ($\beta = \frac{1}{kT} \to 0$). Prove that this partition function coincides with the partition function of a particle in a two-dimensional *box* with sides a and $2\pi R$. This shows that, at high temperatures, the effects of the extra dimension are visible.

(b) Assume that $R \ll a$ in such a way that there are temperatures that are large as far as the box dimension a is concerned, but small as far as the compact dimension is concerned. Write an inequality involving kT and other constants to express this possibility. Evaluate $Z(a, R)$ in this regime, but include the leading correction due to the small extra dimension.

Electromagnetism and gravitation in various dimensions

As a candidate theory of all interactions, string theory includes Maxwell electrodynamics and its nonlinear cousins, as well as gravitation. We review the relativistic formulation of four-dimensional electrodynamics and show how it facilitates the definition of electrodynamics in other dimensions. We give a brief description of Einstein's gravity and use the Newtonian limit to discuss the relation between Planck's length and the gravitational constant in various dimensions. We study the effect of compactification on the gravitational constant and explain how large extra dimensions could escape detection.

3.1 Classical electrodynamics

Unlike Newtonian mechanics, classical electrodynamics is a relativistic theory. In fact, Einstein was led by considerations of electrodynamics to formulate the special theory of relativity. Electrodynamics has a particularly elegant formulation in which the relativistic character of the theory is manifest. This relativistic formulation allows a natural extension of the theory to higher dimensions. Before we discuss the relativistic formulation we must review Maxwell's equations. These equations describe the dynamics of electric and magnetic fields.

Although most undergraduate and graduate courses in electrodynamics nowadays use the international system of units (SI units), the Heaviside–Lorentz system of units is far more convenient for discussions that involve relativity and extra dimensions. In this system of units, Maxwell's equations take the following form:

$$\nabla \times \vec{E} = -\frac{1}{c}\frac{\partial \vec{B}}{\partial t}, \tag{3.1}$$

$$\nabla \cdot \vec{B} = 0, \tag{3.2}$$

$$\nabla \cdot \vec{E} = \rho, \tag{3.3}$$

$$\nabla \times \vec{B} = \frac{1}{c}\vec{j} + \frac{1}{c}\frac{\partial \vec{E}}{\partial t}. \tag{3.4}$$

The above equations imply that $\vec{E}$ and $\vec{B}$ are measured with the *same* units. The first two equations are the source-free Maxwell equations. The second two involve sources: the charge density ρ, with units of charge per unit volume, and the current density $\vec{j}$, with units of current per unit area. The Lorentz force law, which gives the rate of change of the relativistic momentum of a charged particle in an electromagnetic field, takes the form

$$\frac{d\vec{p}}{dt} = q\left(\vec{E} + \frac{\vec{v}}{c} \times \vec{B}\right). \tag{3.5}$$

Since the magnetic field $\vec{B}$ is divergenceless, it can be written as the curl of a vector, the well known vector potential $\vec{A}$:

$$\vec{B} = \nabla \times \vec{A}. \tag{3.6}$$

In electrostatics the electric field $\vec{E}$ has zero curl, and it is therefore written as (minus) the gradient of a scalar, the well known scalar potential Φ. In electrodynamics, as equation (3.1) indicates, the curl of $\vec{E}$ is not always zero. Substituting (3.6) into (3.1), we find a linear combination of $\vec{E}$ and of the time derivative of $\vec{A}$ that has zero curl:

$$\nabla \times \left(\vec{E} + \frac{1}{c}\frac{\partial \vec{A}}{\partial t}\right) = 0. \tag{3.7}$$

The object inside parentheses is set equal to $-\nabla\Phi$, and the electric field $\vec{E}$ can be written in terms of the scalar potential and the vector potential:

$$\vec{E} = -\frac{1}{c}\frac{\partial \vec{A}}{\partial t} - \nabla\Phi. \tag{3.8}$$

While the potentials $(\Phi, \vec{A})$ introduced above seem to be just auxiliary quantities used to represent electric and magnetic fields, we learn in quantum mechanics that, in fact, the potentials are more fundamental than the $\vec{E}$ and $\vec{B}$ fields. The Hamiltonian that describes the motion of a charged particle uses the potentials, not the fields. It is therefore relevant to examine possible ambiguities in the definition of the potentials. As we now show, the potentials associated with a set of $\vec{E}$ and $\vec{B}$ fields are not unique.

If we change $\vec{A}$ into $\vec{A}' = \vec{A} + \nabla\epsilon$, where ϵ is an arbitrary function of space and time, the new magnetic field $\vec{B}'$ is equal to the old one:

$$\vec{B}' = \nabla \times \vec{A}' = \nabla \times A + \nabla \times \nabla\epsilon = \vec{B}, \tag{3.9}$$

noting that the curl of a gradient is zero. The change in $\vec{A}$ would not leave $\vec{E}$ unchanged, as it is clear from (3.8). We can repair this, however, by having Φ change too. In fact, letting

$$\Phi \longrightarrow \Phi' = \Phi - \frac{1}{c}\frac{\partial \epsilon}{\partial t},$$

$$\vec{A} \longrightarrow \vec{A}' = \vec{A} + \nabla\epsilon, \tag{3.10}$$

neither $\vec{B}$ nor $\vec{E}$ is changed. The changes of the potentials indicated above are called *gauge transformations* and ϵ is the gauge parameter.

Quick calculation 3.1 Verify that $\vec{E}$, as given in (3.8), is invariant under the gauge transformations (3.10).

Two sets of potentials $(\Phi, \vec{A})$ and $(\Phi', \vec{A}')$ that are related by gauge transformations are *physically* equivalent. It follows that physically equivalent sets of the potentials give identical electric and magnetic fields. It can happen, however, that potentials $(\Phi, \vec{A})$ and $(\Phi', \vec{A}')$ give the same electric and magnetic fields, but still one cannot find an ϵ such that (3.10) holds. In such case, the potentials are not gauge equivalent and must be considered physically different, even if their $\vec{E}$ and $\vec{B}$ fields are the same! This surprising situation can occur in spacetimes with compact spatial dimensions and will feature in our later studies of D-branes (Section 18.3). It does not happen in Minkowski space.

In the presence of compact spatial dimensions a related subtlety occurs. Given some $\vec{E}$ and $\vec{B}$ there may not exist potentials Φ and $\vec{A}$ that satisfy (3.6) and (3.8) and are well defined throughout the compact part of the space. The gauge transformations come to our help. It is not strictly necessary to have uniquely defined potentials $(\Phi, \vec{A})$ all over the compact space. A set of potentials defined on patches that cover fully the compact space is *admissible* if in the regions of overlap between any two patches the corresponding potentials are related by gauge transformations. Given our statement that potentials are needed in quantum mechanics, we must conclude that a configuration of $\vec{E}$ and $\vec{B}$ fields that does not arise from admissible potentials cannot be discussed.

By the introduction of potentials, the source-free Maxwell equations (3.1) and (3.2) are automatically satisfied. Equations (3.3) and (3.4) contain additional information. They are used to derive equations for Φ and $\vec{A}$.

3.2 Electromagnetism in three dimensions

What is electromagnetism in three spacetime dimensions? One way to produce a theory of electromagnetism in three dimensions is to begin with the four-dimensional theory and eliminate one spatial coordinate. This procedure is called dimensional reduction.

In four spacetime dimensions, both electric and magnetic fields have three spatial components: (E_x, E_y, E_z) and (B_x, B_y, B_z), respectively. It may seem likely that a reduction to a world without a z coordinate would require dropping the z components from the two fields. Surprisingly, this does not work! Maxwell's equations and the Lorentz force law make it impossible.

In order to construct a consistent three-dimensional theory, we must ensure that the dynamics does not depend on the z direction, the direction that we want to eliminate. If there is motion, it must remain restricted to the (x, y) plane. It is thus natural to require that *no quantity should have z-dependence*. This does *not* necessarily mean dropping quantities with a z index.

The Lorentz force law (3.5) is a useful guide to the construction of the lower-dimensional theory. Suppose that there is no magnetic field. Then, in order to keep the z component of momentum equal to zero we must have $E_z = 0$; the z component of the electric field must go. The case of the magnetic field is more surprising. Assume that the z component of the magnetic field is zero. If the velocity of the particle is a vector in the (x, y) plane, a component of the magnetic field in the plane would generate, via the cross product, a force in the z direction. On the other hand, a z component of the magnetic field would generate a force in the (x, y) plane!

We conclude that B_x and B_y must be set equal to zero, while we can keep B_z. All in all,

$$E_z = B_x = B_y = 0. \tag{3.11}$$

The left-over fields E_x, E_y, and B_z can only depend on x and y. In the three-dimensional world with coordinates t, x, and y, the z index of B_z is not a vector index. Therefore, in this reduced world, B_z behaves like a Lorentz scalar (more precisely, it is an object called a pseudo-scalar). In summary, we have a two-dimensional vector $\vec{E}$ and a scalar field B_z.

We can test the consistency of this truncation by taking a look at the x and y components of (3.1):

$$\frac{\partial E_z}{\partial y} - \frac{\partial E_y}{\partial z} = -\frac{1}{c}\frac{\partial B_x}{\partial t},$$
$$\frac{\partial E_x}{\partial z} - \frac{\partial E_z}{\partial x} = -\frac{1}{c}\frac{\partial B_y}{\partial t}. \tag{3.12}$$

Since the right-hand sides are set to zero by our truncation, the left-hand sides should vanish as well. Indeed, they do. Each term on the left-hand sides equals zero, either because it contains an E_z, or because it contains a z derivative. You may examine the consistency of the remaining equations in Problem 3.3.

While setting up three-dimensional electrodynamics was not too difficult, it is much harder to guess what five-dimensional electrodynamics should be. As we will see next, the manifestly relativistic formulation of Maxwell's equations immediately gives the appropriate generalization to other dimensions.

3.3 Manifestly relativistic electrodynamics

In the relativistic formulation of Maxwell's equations neither the electric field nor the magnetic field becomes part of a four-vector. Rather, a four-vector is obtained by combining the scalar potential Φ with the vector potential $\vec{A}$:

$$A^\mu = \left(\Phi, A^1, A^2, A^3\right). \tag{3.13}$$

The corresponding object with down indices is

$$A_\mu = \left(-\Phi, A^1, A^2, A^3\right). \tag{3.14}$$

From A_μ, we create an object known as the electromagnetic *field strength* $F_{\mu\nu}$:

$$F_{\mu\nu} \equiv \partial_\mu A_\nu - \partial_\nu A_\mu. \tag{3.15}$$

Here $\partial_\mu = \frac{\partial}{\partial x^\mu}$. Equation (3.15) implies that $F_{\mu\nu}$ is antisymmetric:

$$F_{\mu\nu} = -F_{\nu\mu}. \tag{3.16}$$

It follows from this property that all diagonal components of $F_{\mu\nu}$ vanish:

$$F_{00} = F_{11} = F_{22} = F_{33} = 0. \tag{3.17}$$

Let us calculate a few entries in $F_{\mu\nu}$. Let i denote a spatial index, that is, an index that can take the values 1, 2, and 3. Making use of (3.15) and (3.8), we find

$$F_{0i} = \frac{\partial A_i}{\partial x^0} - \frac{\partial A_0}{\partial x^i} = \frac{1}{c}\frac{\partial A^i}{\partial t} + \frac{\partial \Phi}{\partial x^i} = -E_i. \tag{3.18}$$

Similarly, we can calculate F_{12}:

$$F_{12} = \partial_1 A_2 - \partial_2 A_1 = \partial_x A_y - \partial_y A_x = B_z, \tag{3.19}$$

since $\vec{B} = \nabla \times \vec{A}$. Continuing in this manner, we can compute all the entries in the matrix $F_{\mu\nu}$:

$$F_{\mu\nu} = \begin{pmatrix} 0 & -E_x & -E_y & -E_z \\ E_x & 0 & B_z & -B_y \\ E_y & -B_z & 0 & B_x \\ E_z & B_y & -B_x & 0 \end{pmatrix}. \tag{3.20}$$

We see that the electric and magnetic fields $\vec{E}$ and $\vec{B}$ are encoded in the field strength $F_{\mu\nu}$.

The gauge transformations (3.10) discussed before can be nicely summarized with index notation as

$$A_\mu \longrightarrow A'_\mu = A_\mu + \partial_\mu \epsilon. \tag{3.21}$$

Here A_μ and A'_μ are the gauge related potentials and, as before, the gauge parameter $\epsilon(x)$ is an arbitrary function of the spacetime coordinates.

Quick calculation 3.2 Verify that the gauge transformations (3.10) are correctly summarized by (3.21).

Since gauge transformations leave the $\vec{E}$ and $\vec{B}$ fields invariant, the field strength $F_{\mu\nu}$ must be gauge *invariant*. Indeed, we readily verify that

$$\begin{aligned} F_{\mu\nu} \longrightarrow F'_{\mu\nu} &\equiv \partial_\mu A'_\nu - \partial_\nu A'_\mu \\ &= \partial_\mu (A_\nu + \partial_\nu \epsilon) - \partial_\nu (A_\mu + \partial_\mu \epsilon) \\ &= F_{\mu\nu} + \partial_\mu \partial_\nu \epsilon - \partial_\nu \partial_\mu \epsilon \\ &= F_{\mu\nu}. \end{aligned} \tag{3.22}$$

In the last step we noted that partial derivatives commute.

Recall that the use of potentials to represent $\vec{E}$ and $\vec{B}$ automatically solves the source-free Maxwell equations (3.1) and (3.2). How are these equations written in terms of the field strength $F_{\mu\nu}$? They must be written so that they hold when (3.15) holds. Consider the following combination of field strengths:

$$T_{\lambda\mu\nu} \equiv \partial_\lambda F_{\mu\nu} + \partial_\mu F_{\nu\lambda} + \partial_\nu F_{\lambda\mu}. \tag{3.23}$$

$T_{\lambda\mu\nu}$ vanishes identically on account of (3.15):

$$\partial_\lambda \left(\partial_\mu A_\nu - \partial_\nu A_\mu \right) + \partial_\mu \left(\partial_\nu A_\lambda - \partial_\lambda A_\nu \right) + \partial_\nu \left(\partial_\lambda A_\mu - \partial_\mu A_\lambda \right) = 0, \qquad (3.24)$$

using the commutativity of partial derivatives. The vanishing of $T_{\lambda\mu\nu}$,

$$\partial_\lambda F_{\mu\nu} + \partial_\mu F_{\nu\lambda} + \partial_\nu F_{\lambda\mu} = 0, \qquad (3.25)$$

is a set of differential equations for the field strength. These equations are precisely the source-free Maxwell equations. To make this clear, first note that $T_{\lambda\mu\nu}$ satisfies the antisymmetry conditions

$$T_{\lambda\mu\nu} = -T_{\mu\lambda\nu}, \quad T_{\lambda\mu\nu} = -T_{\lambda\nu\mu}. \qquad (3.26)$$

These two equations follow from (3.23) and the antisymmetry property $F_{\mu\nu} = -F_{\nu\mu}$ of the field strength. They state that T changes sign under the transposition of any two adjacent indices.

Quick calculation 3.3 Verify the equations in (3.26).

Any object, with however many indices, that changes sign under the transposition of every pair of adjacent indices, will change sign under the transposition of *any* two indices: to exchange any two indices you need an odd number of transpositions of adjacent indices (do you see why?). An object that changes sign under the transposition of any two indices is said to be *totally* antisymmetric. Therefore T is totally antisymmetric.

Since T is totally antisymmetric, it vanishes when any two of its indices take the same value. T is nonvanishing only when each of its three indices takes a different value. In such case, different orderings of these three fixed values will give T components that can differ at most by a sign. Since we are setting T to zero these various orderings do not give new conditions. Because we have four spacetime coordinates, selecting three different indices can only be done in four different ways – leaving out a different index each time. Thus the vanishing of T gives four nontrivial equations. These four equations are the three components of equation (3.1) and equation (3.2). The vanishing of T_{012}, for example, gives us

$$\partial_0 F_{12} + \partial_1 F_{20} + \partial_2 F_{01} = \frac{1}{c} \frac{\partial B_z}{\partial t} + \frac{\partial E_y}{\partial x} - \frac{\partial E_x}{\partial y} = 0. \qquad (3.27)$$

This is the z component of equation (3.1). The other three choices of indices lead to the remaining three equations (Problem 3.2).

How can we describe Maxwell equations (3.3) and (3.4) in our present framework? Since these equations have sources, we must introduce a *current* four-vector:

$$j^\mu = \left(c\rho, j^1, j^2, j^3 \right), \qquad (3.28)$$

where ρ is the charge density and $\vec{j} = (j^1, j^2, j^3)$ is the current density. In addition, we raise the indices of the field tensor to obtain the field tensor with upper indices:

$$F^{\mu\nu} = \eta^{\mu\alpha} \eta^{\nu\beta} F_{\alpha\beta}. \tag{3.29}$$

Quick calculation 3.4 Show that

$$F^{\mu\nu} = -F^{\nu\mu}, \qquad F^{0i} = -F_{0i}, \qquad F^{ij} = F_{ij}. \tag{3.30}$$

Equation (3.29), together with the definition of $F_{\mu\nu}$, gives

$$F^{\mu\nu} = \eta^{\mu\alpha} \eta^{\nu\beta} (\partial_\alpha A_\beta - \partial_\beta A_\alpha) = \eta^{\mu\alpha} \partial_\alpha (\eta^{\nu\beta} A_\beta) - \eta^{\nu\beta} \partial_\beta (\eta^{\mu\alpha} A_\alpha), \tag{3.31}$$

where the constancy of the metric components was used to move them across the derivatives. It is customary to apply the rules for raising and lowering indices to partial derivatives, so we write $\partial^\mu \equiv \eta^{\mu\alpha} \partial_\alpha$. As a result,

$$F^{\mu\nu} = \partial^\mu A^\nu - \partial^\nu A^\mu. \tag{3.32}$$

It follows from (3.30) and (3.20) that

$$F^{\mu\nu} = \begin{pmatrix} 0 & E_x & E_y & E_z \\ -E_x & 0 & B_z & -B_y \\ -E_y & -B_z & 0 & B_x \\ -E_z & B_y & -B_x & 0 \end{pmatrix}. \tag{3.33}$$

Using this equation and the current vector (3.28), we can encapsulate Maxwell's equations (3.3) and (3.4) as (Problem 3.2)

$$\frac{\partial F^{\mu\nu}}{\partial x^\nu} = \frac{1}{c} j^\mu. \tag{3.34}$$

In the absence of sources this equation becomes

$$\partial_\nu F^{\mu\nu} = 0 \longrightarrow \partial_\nu \partial^\mu A^\nu - \partial^2 A^\mu = 0, \tag{3.35}$$

where we have written $\partial^2 = \partial^\mu \partial_\mu$.

Equations (3.25) together with equations (3.34) are equivalent to Maxwell's equations in four dimensions. We will take these equations to *define* Maxwell theory in arbitrary dimensions. In d spatial dimensions the Lorentz vector A^μ has components $(\Phi, \vec{A})$ where $\vec{A}$ is a d-dimensional spatial vector.

In three-dimensional spacetime, for example, the matrix $F_{\mu\nu}$ is a 3-by-3 antisymmetric matrix, obtained from (3.20) by discarding the last row and the last column:

$$F_{\mu\nu} = \begin{pmatrix} 0 & -E_x & -E_y \\ E_x & 0 & B_z \\ E_y & -B_z & 0 \end{pmatrix}. \tag{3.36}$$

This immediately reproduces the main result of Section 3.2; B_x, B_y, and E_z are to be set to zero.

Motivated by (3.33), in arbitrary dimensions we will call F^{0i} the electric field E_i:

$$E_i \equiv F^{0i} = -F_{0i}, \quad i = 1, 2, \ldots, d, \tag{3.37}$$

where d is the number of spatial dimensions.

The electric field is a spatial vector. Equation (3.18) implies that, in any number of dimensions,

$$\vec{E} = -\frac{1}{c}\frac{\partial \vec{A}}{\partial t} - \nabla\Phi. \tag{3.38}$$

The magnetic field is identified with the F^{ij} components of the field strength. In four-dimensional spacetime F^{ij} is a 3-by-3 antisymmetric matrix. Its three independent entries are the components of the magnetic field vector (see (3.33)). In dimensions other than four, the magnetic field is no longer a spatial vector. In three spacetime dimensions the magnetic field is a single-component object. In five spacetime dimensions the magnetic field has as many entries as a 4-by-4 antisymmetric matrix, six entries. That many components do not fit into a spatial vector.

Our next goal is the determination of the electric field produced by a point charge in a spacetime with an arbitrary but fixed number of spatial dimensions. To this end, we must first learn how to calculate the volumes of higher-dimensional spheres. We turn to this subject now.

3.4 An aside on spheres in higher dimensions

Since we want to work in various numbers of dimensions we should be precise when speaking about spheres and their volumes. When we speak loosely we tend to confuse spheres and *balls*, at least in the precise sense in which they are defined in mathematics. When you say that the volume of a sphere of radius R is $\frac{4}{3}\pi R^3$, you should really be saying that this is the volume of the *three-ball* B^3 – the three-dimensional space enclosed by the two-dimensional *two-sphere* S^2. In three-dimensional space $\mathbb{R}^3$ with coordinates x_1, x_2, and x_3, we write the three-ball as the region defined by

$$B^3(R) : \quad x_1^2 + x_2^2 + x_3^2 \le R^2. \tag{3.39}$$

This region is enclosed by the two-sphere:

$$S^2(R) : \quad x_1^2 + x_2^2 + x_3^2 = R^2. \tag{3.40}$$

The superscripts in B or S denote the dimensionality of the space in question. When we drop the explicit argument R, we mean that $R = 1$. Lower-dimensional examples are also familiar. B^2 is a two-dimensional disk – the region enclosed in $\mathbb{R}^2$ by the one-dimensional unit radius circle S^1. In arbitrary dimensions we define balls and spheres as subspaces of $\mathbb{R}^d$:

$$B^d(R) : \quad x_1^2 + x_2^2 + \cdots + x_d^2 \leq R^2. \tag{3.41}$$

This is the region enclosed by the sphere $S^{d-1}(R)$:

$$S^{d-1}(R) : \quad x_1^2 + x_2^2 + \cdots + x_d^2 = R^2. \tag{3.42}$$

One last piece of terminology: to avoid confusion we will always speak of volumes. If a space is one-dimensional we take volume to mean length. If a space is two-dimensional we take volume to mean area. All higher-dimensional spaces have just volumes. The volumes of the one- and two-dimensional spheres are

$$\text{vol}\,(S^1(R)) = 2\pi R,$$
$$\text{vol}\,(S^2(R)) = 4\pi R^2. \tag{3.43}$$

Unless you have worked with other spheres before, you probably do not know what the volume of S^3 is.

Since volume has units of length to the power of the space dimension, the volume of a sphere of radius R is related to the volume of a sphere of unit radius by

$$\text{vol}\,(S^{d-1}(R)) = R^{d-1}\,\text{vol}(S^{d-1}). \tag{3.44}$$

Since the radius dependence of the volume is easily recovered, it suffices to record the volumes of unit spheres:

$$\text{vol}\,(S^1) = 2\pi,$$
$$\text{vol}\,(S^2) = 4\pi. \tag{3.45}$$

Let us now begin our calculation of the volume of the sphere S^{d-1}. For this purpose consider $\mathbb{R}^d$ with coordinates $x_1, x_2, \ldots, x_d$, and let r be the radial coordinate:

$$r^2 = x_1^2 + x_2^2 + \cdots + x_d^2. \tag{3.46}$$

We will find the desired volume by evaluating in two different ways the following integral:

$$I_d = \int_{\mathbb{R}^d} dx_1 dx_2 \ldots dx_d \, e^{-r^2}. \tag{3.47}$$

First we proceed directly. Using (3.46) in the exponential factor, the integral becomes a product of d Gaussian integrals:

$$I_d = \prod_{i=1}^{d} \int_{-\infty}^{\infty} dx_i \, e^{-x_i^2} = (\sqrt{\pi})^d = \pi^{d/2}. \tag{3.48}$$

Now we proceed indirectly. We do the integral by breaking $\mathbb{R}^d$ into thin spherical shells. Since the space of constant r is the sphere $S^{d-1}(r)$, the volume of a shell lying between r and $r + dr$ equals the volume of $S^{d-1}(r)$ times dr. Therefore,

$$I_d = \int_0^\infty dr\, \text{vol}(S^{d-1}(r))\, e^{-r^2} = \text{vol}\,(S^{d-1}) \int_0^\infty dr\, r^{d-1}\, e^{-r^2}$$
$$= \frac{1}{2}\,\text{vol}\,(S^{d-1}) \int_0^\infty dt\, e^{-t}\, t^{\frac{d}{2}-1}, \tag{3.49}$$

where use was made of (3.44), and in the final step we changed the variable of integration to $t = r^2$. The last integral on the right-hand side can be expressed in terms of the gamma function, a very useful special function. For positive x the gamma function $\Gamma(x)$ is defined by

$$\Gamma(x) = \int_0^\infty dt \, e^{-t} t^{x-1}, \quad x > 0. \tag{3.50}$$

Unless $x > 0$ the integral does not converge near $t = 0$. With this definition, equation (3.49) becomes

$$I_d = \frac{1}{2} \operatorname{vol}(S^{d-1}) \, \Gamma\left(\tfrac{d}{2}\right). \tag{3.51}$$

Comparing with the earlier evaluation (3.48), we get our final result:

$$\operatorname{vol}(S^{d-1}) = \frac{2\pi^{d/2}}{\Gamma\left(\tfrac{d}{2}\right)}. \tag{3.52}$$

It now remains to calculate the value of $\Gamma(d/2)$. Since d is an integer, we must determine the values of the gamma function for both integer and half-integer arguments. To find $\Gamma(1/2)$ we use the definition (3.50) and let $t = u^2$:

$$\Gamma\left(\frac{1}{2}\right) = \int_0^\infty dt \, e^{-t} \, t^{-1/2} = 2 \int_0^\infty du \, e^{-u^2} = \sqrt{\pi}. \tag{3.53}$$

Similarly,

$$\Gamma(1) = \int_0^\infty dt \, e^{-t} = 1. \tag{3.54}$$

For larger arguments the calculation of the gamma function is simplified using a recursion relation. To obtain this relation, begin with

$$\Gamma(x+1) = \int_0^\infty dt \, e^{-t} t^x, \quad x > 0, \tag{3.55}$$

which can be rewritten as

$$\Gamma(x+1) = -\int_0^\infty dt \left(\frac{d}{dt} e^{-t}\right) t^x = -\int_0^\infty dt \left(\frac{d}{dt}(e^{-t} t^x) - x e^{-t} t^{x-1}\right). \tag{3.56}$$

The boundary terms vanish for $x > 0$ and we find that

$$\Gamma(x+1) = x \, \Gamma(x), \quad x > 0. \tag{3.57}$$

Using this recursion relation we find, for example,

$$\Gamma\left(\frac{3}{2}\right) = \frac{1}{2} \cdot \Gamma\left(\frac{1}{2}\right) = \frac{1}{2}\sqrt{\pi}, \quad \Gamma\left(\frac{5}{2}\right) = \frac{3}{2} \cdot \Gamma\left(\frac{3}{2}\right) = \frac{3}{4}\sqrt{\pi}. \tag{3.58}$$

For integer arguments the gamma function is related to the factorial:

$$\Gamma(5) = 4 \cdot \Gamma(4) = 4 \cdot 3 \cdot \Gamma(3) = 4 \cdot 3 \cdot 2 \cdot \Gamma(2) = 4 \cdot 3 \cdot 2 \cdot 1 \cdot \Gamma(1) = 4!.$$

Therefore, for $n \in \mathbb{Z}$ and $n \geq 1$, we have

$$\Gamma(n) = (n-1)!, \tag{3.59}$$

where we recall that $0! = 1$. We can now test our formula (3.52) in the familiar cases:

$$\text{vol}(S^1) = \text{vol}(S^{2-1}) = \frac{2\pi}{\Gamma(1)} = 2\pi,$$

$$\text{vol}(S^2) = \text{vol}(S^{3-1}) = \frac{2\pi^{3/2}}{\Gamma\left(\frac{3}{2}\right)} = 4\pi, \tag{3.60}$$

in agreement with the known values. For the less familiar S^3 we find

$$\text{vol}(S^3) = \text{vol}(S^{4-1}) = \frac{2\pi^2}{\Gamma(2)} = 2\pi^2. \tag{3.61}$$

Quick calculation 3.5 Show that $\text{vol}(B^d) = \pi^{d/2}/\Gamma\left(1+\frac{d}{2}\right)$.

3.5 Electric fields in higher dimensions

In this section we calculate the electric field due to a point charge in a world with d spatial dimensions. Here d could be three, in which case the answer is familiar, or less than three, but we are particularly interested in $d > 3$. To do this calculation we will use the general version of Maxwell's equations appropriate for an arbitrary number of spatial dimensions. As you may imagine, the electric field of a point charge is radial. Our calculation will give the radial dependence and the normalization of the electric field. With minor modifications, this result will also inform us about the gravitational fields of point particles in d spatial dimensions.

Our computation is based on the zeroth component of equation (3.34):

$$\frac{\partial F^{0i}}{\partial x^i} - \rho. \tag{3.62}$$

Since $F^{0i} = E_i$ (see (3.37)), this equation is just Gauss' law:

$$\nabla \cdot \vec{E} = \rho. \tag{3.63}$$

Gauss' law is valid in all dimensions! Equation (3.63) can be used to determine the electric field of a point charge. Let us first review how this is done in the familiar setting of three spatial dimensions.

Consider a point charge q, a two-sphere $S^2(r)$ of radius r centered on the charge, and the three-ball $B^3(r)$ whose boundary is the two-sphere. We integrate both sides of equation (3.63) over the three-ball to find

$$\int_{B^3} d(\text{vol}) \, \nabla \cdot \vec{E} = \int_{B^3} d(\text{vol}) \, \rho. \tag{3.64}$$

We use the divergence theorem on the left-hand side and note that the volume integral on the right-hand side gives the total charge:

$$\int_{S^2(r)} \vec{E} \cdot d\vec{a} = q. \qquad (3.65)$$

Since the magnitude $E(r)$ of $\vec{E}$ is constant over the two-sphere, we get

$$\text{vol}(S^2(r))\, E(r) = q. \qquad (3.66)$$

The volume of the two-sphere is just its area $4\pi r^2$, so

$$E(r) = \frac{q}{4\pi r^2}. \qquad (3.67)$$

This is the familiar result for the electric field of a point charge in three spatial dimensions. The electric field magnitude falls off like $1/r^2$.

For dimensions higher than three, the starting point (3.63) is good, so we must ask if the divergence theorem also holds. It turns out that it does. We will first state the theorem in d spatial dimensions, and then we will give some justification for it.

Consider a d-dimensional subset V^d of $\mathbb{R}^d$ and let ∂V^d denote the boundary of V^d. Moreover, let $\vec{E}$ be a vector field in $\mathbb{R}^d$. The divergence theorem states that

$$\int_{V^d} d(\text{vol})\, \nabla \cdot \vec{E} = \text{Flux of } \vec{E} \text{ across } \partial V^d = \int_{\partial V^d} \vec{E} \cdot d\vec{v}. \qquad (3.68)$$

The last right-hand side requires some explanation. At any point on ∂V^d, the space ∂V^d is locally approximated by the $(d-1)$-dimensional tangent hyperplane. For a small piece of ∂V^d around this point, the associated vector $d\vec{v}$ is a vector orthogonal to the hyperplane, pointing out of the volume, and with magnitude equal to the volume of the small piece under consideration. Note that this explanation is in accord with your experience in $\mathbb{R}^3$, where $d\vec{v}$ corresponds to the area vector element $d\vec{a}$.

Let us justify the divergence theorem for the case of four space dimensions. Following a strategy used in elementary textbooks, it suffices to prove the divergence theorem for a small hypercube – the result for general subspaces follows by breaking such spaces into many small hypercubes. Because it is not easy to imagine a four-dimensional hypercube, we might as well use a three-dimensional picture with four-dimensional labels (Figure 3.1). We use Cartesian coordinates x, y, z, w, and consider a cube whose faces lie on hyperplanes selected by the condition that one of the coordinates is constant. Let one face of the cube and the face opposite to it lie on hyperplanes of constant x and constant $x + dx$, respectively. The outgoing normal vectors are $\vec{e}_x$, for the face at $x + dx$, and $(-\vec{e}_x)$, for the face at x. The volume of each of these two faces equals $dydzdw$, where dy, dz, and dw, together with dx, are the lengths of the edges of the cube. For an arbitrary electric field $\vec{E}(x, y, z, w)$, only the x component contributes to the flux through these two faces. The contribution is

$$[\, E_x(x + dx, y, z, w) - E_x(x, y, z, w)\,]\, dydzdw \simeq \frac{\partial E_x}{\partial x}\, dxdydzdw. \qquad (3.69)$$

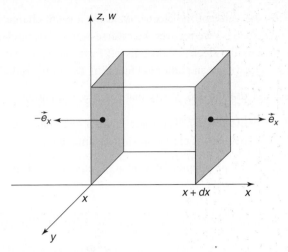

Fig. 3.1 An attempt at a representation of a four-dimensional hypercube. The two faces of constant *x* are shown (shaded) together with their outgoing normal vectors.

Analogous expressions hold for the flux across the three other pairs of faces. The total net flux from the little cube is just

$$\text{Flux of } \vec{E} = \left(\frac{\partial E_x}{\partial x} + \frac{\partial E_y}{\partial y} + \frac{\partial E_z}{\partial z} + \frac{\partial E_w}{\partial w} \right) dxdydzdw = \nabla \cdot \vec{E}\, d(\text{vol}). \tag{3.70}$$

This result is precisely the divergence theorem (3.68) applied to an infinitesimal hypercube. This is what we wanted to show.

We can now return to the computation of the electric field due to a point charge in a world with *d* spatial dimensions. Consider a point charge q, the sphere $S^{d-1}(r)$ of radius r centered on the charge (this is the sphere that surrounds the charge), and the ball $B^d(r)$ whose boundary is the sphere $S^{d-1}(r)$. Again, we integrate both sides of equation (3.63) over the ball $B^d(r)$:

$$\int_{B^d} d(\text{vol}) \nabla \cdot \vec{E} = \int_{B^d} d(\text{vol})\, \rho. \tag{3.71}$$

The volume integral on the right-hand side gives the total charge, and the divergence theorem (3.68) relates the left-hand side to a flux integral:

$$\text{Flux of } \vec{E} \text{ across } S^{d-1}(r) = q. \tag{3.72}$$

The flux equals the magnitude of the electric field times the volume of $S^{d-1}(r)$, so

$$E(r)\, \text{vol}\, (S^{d-1}(r)) = q. \tag{3.73}$$

Making use of (3.52) we find

$$E(r) = \frac{\Gamma\left(\frac{d}{2}\right)}{2\pi^{d/2}} \frac{q}{r^{d-1}}. \tag{3.74}$$

This is the value of the electric field for a point charge in a world with d spatial dimensions. For $d = 3$ we recover the inverse-squared dependence of the electric field. In higher dimensions the electric field falls off faster at large distances. For each additional spatial dimension we get an additional factor of $1/r$ in the radial dependence of the electric field.

Quick calculation 3.6 Verify that for $d = 3$ equation (3.74) coincides with (3.67).

Quick calculation 3.7 The force $\vec{F}$ on a test charge q in an electric field $\vec{E}$ is $\vec{F} = q\vec{E}$. What are the units of charge in various dimensions?

The electrostatic potential Φ is also of interest. For time independent fields (3.38) gives

$$\vec{E} = -\nabla\Phi. \tag{3.75}$$

This equation, together with Gauss' law, gives the Poisson equation:

$$\nabla^2\Phi = -\rho, \tag{3.76}$$

which can be used to calculate the potential due to a charge distribution. The two equations above hold in all dimensions using, of course, the appropriate definitions of the gradient and the Laplacian.

3.6 Gravitation and Planck's length

Einstein's theory of general relativity is a theory of gravitation. In this very elegant theory the dynamical variables encode the geometry of spacetime. When gravitational fields are sufficiently weak and velocities are small, Newtonian gravitation is accurate enough, and one need not work with the more complex machinery of general relativity. We can use Newtonian gravity to understand the definition of Planck's length in various dimensions and its relation to the gravitational constant. These are interesting issues that we will explain here and in the rest of the present chapter. Nevertheless, when gravitation emerges in string theory, it does so in the language of Einstein's theory of general relativity. To be able to recognize the appearance of gravity among the quantum vibrations of the relativistic string you need a little familiarity with the language of general relativity. Here you will take a first look at the concepts involved in this remarkable theory.

Most physicists do not expect general relativity to hold at truly small distances nor for extremely large gravitational fields. This is a realm where string theory, the first serious candidate for a quantum theory of gravitation, is necessary. General relativity is the large-distance/weak-gravity limit of string theory. String theory *modifies* general relativity; it must do so to make it consistent with quantum mechanics. The conceptual framework which underlies these modifications is not clear yet. It will no doubt emerge as we understand string theory better in the years to come.

The spacetime of special relativity, Minkowski spacetime, is the arena for physics in the *absence* of gravitational fields. The geometrical properties of Minkowski spacetime are encoded by the metric formula (2.21), which gives the invariant interval separating two nearby events:

$$-ds^2 = \eta_{\mu\nu}dx^\mu dx^\nu. \tag{3.77}$$

Here the Minkowski metric $\eta_{\mu\nu}$ is a constant metric, represented as a matrix with entries $(-1, 1, \ldots, 1)$ along the diagonal. Minkowski space is said to be a flat space. In the presence of a gravitational field, the metric becomes dynamical. We then write

$$-ds^2 = g_{\mu\nu}(x)dx^\mu dx^\nu, \tag{3.78}$$

where the constant $\eta_{\mu\nu}$ is replaced by the metric $g_{\mu\nu}(x)$. If there is a gravitational field, the metric is in general a nontrivial function of the spacetime coordinates. The metric $g_{\mu\nu}$ is defined to be symmetric

$$g_{\mu\nu}(x) = g_{\nu\mu}(x). \tag{3.79}$$

It is also customary to define $g^{\mu\nu}(x)$ as the inverse of the $g_{\mu\nu}(x)$ matrix:

$$g^{\mu\alpha}(x)\, g_{\alpha\nu}(x) = \delta^\mu_\nu. \tag{3.80}$$

For many physical phenomena gravity is very weak, and the metric $g_{\mu\nu}(x)$ can be chosen to be very close to the Minkowski metric $\eta_{\mu\nu}$. We then write,

$$g_{\mu\nu}(x) = \eta_{\mu\nu} + h_{\mu\nu}(x), \tag{3.81}$$

and we view $h_{\mu\nu}(x)$ as a small fluctuation around the Minkowski metric. This expansion is done, for example, to study gravity waves. Those waves represent small "ripples" on top of the Minkowski metric. Einstein's equations for the gravitational field are written in terms of the spacetime metric $g_{\mu\nu}(x)$. These equations imply that matter or energy sources curve the spacetime manifold. For weak gravitational fields, Einstein's equations can be expanded in powers of $h_{\mu\nu}$ using (3.81). In the absence of sources, the resulting linearized equation for $h_{\mu\nu}$ is

$$\partial^2 h^{\mu\nu} - \partial_\alpha(\partial^\mu h^{\nu\alpha} + \partial^\nu h^{\mu\alpha}) + \partial^\mu \partial^\nu h = 0. \tag{3.82}$$

Here $h^{\mu\nu} \equiv \eta^{\mu\alpha}\eta^{\nu\beta}h_{\alpha\beta}$ and $h \equiv \eta^{\mu\nu}h_{\mu\nu} = -h_{00} + h_{11} + h_{22} + h_{33}$. Equation (3.82) is the gravitational analog of equation (3.35), which describes Maxwell fields in the absence of sources. While (3.35) is exact, (3.82) is only valid for weak gravitational fields. It is the linear approximation to a nonlinear equation that includes additional terms quadratic and higher order in h.

The analogy with electromagnetism extends to the existence of gauge transformations. Einstein's gravity has gauge transformations. They arise because the use of different systems of coordinates yields equivalent descriptions of gravitational physics. In learning string theory in this book you will get to appreciate the freedom to choose coordinates on the surfaces generated by moving strings. In general relativity, an infinitesimal change of coordinates

$$x^{\mu'} = x^\mu + \epsilon^\mu(x), \tag{3.83}$$

can be viewed as an infinitesimal change of the metric $g_{\mu\nu}$ and, using (3.81), as an infinitesimal change of the fluctuating field $h^{\mu\nu}$. One can show that the change is given by

$$\delta h^{\mu\nu} = \delta_0 h^{\mu\nu} + \mathcal{O}(\epsilon, h), \quad \text{with} \quad \delta_0 h^{\mu\nu} \equiv \partial^\mu \epsilon^\nu + \partial^\nu \epsilon^\mu. \tag{3.84}$$

As indicated above, the infinitesimal change $\delta h^{\mu\nu}$ is given by $\delta_0 h^{\mu\nu}$ plus corrections, written as $\mathcal{O}(\epsilon, h)$, that are linear in ϵ and linear in the fluctuation h itself. The invariance of the full nonlinear equation of motion under the gauge transformation $\delta h^{\mu\nu}$ requires the invariance of the linearized equation of motion (3.82) under $\delta_0 h^{\mu\nu}$. Indeed, when we vary the hs in (3.82) using $\delta_0 h^{\mu\nu}$ we get terms linear in ϵ but without an h. These terms must cancel out completely because all other variations will contain at least one field h: this is clear for the variations of (3.82) using the terms represented by $\mathcal{O}(\epsilon, h)$ and for all variations of terms quadratic and higher order in h in the complete equation of motion. We will check the invariance of (3.82) under the transformation $\delta_0 h^{\mu\nu}$ in Chapter 10. In Maxwell theory the gauge parameter has no indices, but in general relativity the gauge parameter has a vector index.

As we mentioned before, Newtonian gravitation emerges from general relativity in the approximation of weak gravitational fields and motion with small velocities. For many purposes Newtonian gravity suffices. Starting now, and for the rest of this chapter, we will use Newtonian gravity to understand the definition of Planck's length in various dimensions, and to investigate how gravitational constants behave when some spatial dimensions are curled up. The results that we will obtain hold also in the full theory of general relativity.

Newton's law of gravitation in four dimensions states that the force of attraction between two masses m_1 and m_2 separated by a distance r is given by

$$|\vec{F}^{(4)}| = \frac{G m_1 m_2}{r^2}, \tag{3.85}$$

where G denotes the four-dimensional Newton constant. It follows that the units of the gravitational constant G are

$$[G] = [\text{Force}]\frac{L^2}{M^2} = \frac{ML}{T^2}\frac{L^2}{M^2} = \frac{L^3}{MT^2}. \tag{3.86}$$

The numerical value for the constant G is determined experimentally:

$$G = 6.674 \times 10^{-11} \frac{\text{m}^3}{\text{kg} \cdot \text{s}^2}. \tag{3.87}$$

Since $[c] = L/T$ and $[\hbar] = ML^2/T$, the three fundamental constants G, c, and $\hbar$ can be written as

$$G = 6.674 \times 10^{-11} \frac{\text{m}^3}{\text{kg} \cdot \text{s}^2}, \quad c = 2.998 \times 10^8 \frac{\text{m}}{\text{s}}, \quad \hbar = 1.055 \times 10^{-34} \frac{\text{kg} \cdot \text{m}^2}{\text{s}}. \tag{3.88}$$

In the study of gravitation it is sometimes convenient to use a "Planckian" system of units. Since we have three basic units, those of length, time, and mass, we can find new units of length, time, and mass such that the three fundamental constants, G, c, and $\hbar$ take the numerical value of *one* in those units. These units are called the Planck length ℓ_P, the Planck time t_P, and the Planck mass m_P, respectively. In those units

$$G = 1 \cdot \frac{\ell_P^3}{m_P \, t_P^2}, \quad c = 1 \cdot \frac{\ell_P}{t_P}, \quad \hbar = 1 \cdot \frac{m_P \, \ell_P^2}{t_P}, \tag{3.89}$$

without additional numerical constants – as opposed to equation (3.88). The above equations allow us to solve for ℓ_P, t_P, and m_P in terms of G, c, and $\hbar$. One readily finds

$$\ell_P = \sqrt{\frac{G\hbar}{c^3}} = 1.616 \times 10^{-33} \text{ cm}, \tag{3.90}$$

$$t_P = \frac{\ell_P}{c} = \sqrt{\frac{G\hbar}{c^5}} = 5.391 \times 10^{-44} \text{ s}, \tag{3.91}$$

$$m_P = \sqrt{\frac{\hbar c}{G}} = 2.176 \times 10^{-5} \text{ g}. \tag{3.92}$$

These numbers represent scales at which relativistic quantum gravity effects can be important. Indeed, the Planck length is an extremely small length, and the Planck time is an incredibly short time – the time it takes light to travel the Planck length! While Einstein's gravity can be used down to relatively small distances and back to relatively early times in the history of the universe, a quantum gravity theory (such as string theory) is needed to study gravity at distances of the order of the Planck length or to investigate the universe when it was Planck-time old.

There is an equivalent way to characterize the Planck length: ℓ_P is the unique length that can be constructed using *only* powers of G, c, and $\hbar$. One thus sets

$$\ell_P = (G)^\alpha \, (c)^\beta \, (\hbar)^\gamma, \tag{3.93}$$

and fixes the constants α, β, and γ so that the right-hand side has units of length.

Quick calculation 3.8 Show that this condition fixes uniquely $\alpha = \gamma = 1/2$, and $\beta = -3/2$, thus reproducing the result in (3.90).

It may appear that m_P is not a very large mass, but it is, in fact, a spectacularly large mass from the viewpoint of elementary particle physics. The mass m_P is roughly 10^{19} times larger than the mass of the proton. If the fundamental theory of nature is based on the basic constants G, c, and $\hbar$, it is then a great mystery why the masses of the elementary particles are so much smaller than the "obvious" mass m_P that can be built from the basic constants. This puzzle is usually called the *hierarchy* problem.

For an additional perspective on the Planck mass, consider the following question: what should be the mass M of the proton so that the gravitational force between two protons cancels the electric repulsion force between them? Equating the magnitudes of the electric and gravitational forces we get

$$\frac{GM^2}{r^2} = \frac{e^2}{4\pi r^2} \longrightarrow GM^2 = \frac{e^2}{4\pi}. \tag{3.94}$$

It is convenient to divide both sides of the equation by $\hbar c$ to find

$$\frac{GM^2}{\hbar c} = \frac{e^2}{4\pi \hbar c} \simeq \frac{1}{137} \longrightarrow \frac{M^2}{m_{\mathrm{P}}^2} \simeq \frac{1}{137}, \tag{3.95}$$

where use was made of (3.92). We thus find $M \simeq m_{\mathrm{P}}/12$, or about one-tenth of the Planck mass. The dimensionless ratio $e^2/(4\pi \hbar c)$ is called the fine structure constant. It was evaluated above using the Heaviside–Lorentz definition of electric charge where $e = \sqrt{4\pi}\, 4.8 \times 10^{-10}$ esu (see Problem 2.1(c)).

Quick calculation 3.9 The mass of the electron is $m_e = 0.9109 \times 10^{-27}$ g, and its energy equivalent is $m_e c^2 = 0.5110$ MeV. Show that the energy equivalent of the Planck mass is $m_{\mathrm{P}} c^2 = 1.221 \times 10^{19}$ GeV (1 GeV $= 10^9$ eV). This energy is called the Planck energy.

3.7 Gravitational potentials

We want to learn what happens with the gravitational constant G when we attempt to describe gravitation in spacetimes of other dimensionalities. To find out, we will examine gravitational potentials in Newtonian gravity. In this section we obtain the equation that relates the gravitational potential to the mass distribution in a spacetime with arbitrary but fixed number of spatial dimensions. In doing so we will learn how to define the relevant gravitational constant. This result will be used in the following section to define, in any dimension, the Planck length in terms of the appropriate gravitational constant.

We introduce a gravity field $\vec{g}$ with units of force per unit mass. The definition is similar to that of an electric field in terms of the force on a test particle: the force on a given test mass m at a point where the gravity field is $\vec{g}$ is given by $m\vec{g}$. We set $\vec{g}$ equal to minus the gradient of a gravitational potential V_g:

$$\vec{g} = -\nabla V_g. \tag{3.96}$$

We will take this equation to be true in all dimensions. Equation (3.96) has content: if you move a particle along a closed loop in a static gravitational field, the net work that you do against the gravitational field is zero.

Quick calculation 3.10 Prove the above statement.

What are the units for the gravitational potential? Equation (3.96) gives

$$[\vec{g}] = \frac{[\mathrm{Force}]}{M} = \frac{[V_g]}{L} \longrightarrow [V_g] = \frac{[\mathrm{Energy}]}{M}. \tag{3.97}$$

The gravitational potential has units of energy per unit mass in *any* dimension. The gravitational potential $V_g^{(4)}$ of a point mass in four dimensions is

$$V_g^{(4)} = -\frac{GM}{r}. \tag{3.98}$$

We can use the electromagnetic analogy to find the equation satisfied by the gravitational potential. In electromagnetism, we found an equation for the electrostatic potential which holds in any dimension. This is equation (3.76):

$$\nabla^2 \Phi = -\rho .$$ (3.99)

The four-dimensional scalar potential for a point charge q is

$$\Phi^{(4)} = \frac{q}{4\pi r},$$ (3.100)

and it satisfies (3.99) where ρ is the charge density for the point charge. It follows by analogy that the four-dimensional gravitational potential in (3.98) satisfies

$$\nabla^2 V_g^{(4)} = 4\pi G \rho_m,$$ (3.101)

where ρ_m is the matter density. While this equation is correct in four dimensions, a small modification is needed for other dimensions. Note that the left-hand side has the same units in any number of dimensions: the units of V_g are always the same, and the Laplacian always divides by length-squared. The right-hand side must also have the same units in any number of dimensions. Since ρ_m is mass density, it has different units in different dimensions, and, as a consequence, the units of G must change when the dimensions change. We therefore rewrite the above equation more precisely as

$$\nabla^2 V_g^{(D)} = 4\pi G^{(D)} \rho_m,$$ (3.102)

when working in D-dimensional spacetime. The superscripts shown in parentheses denote the dimensionality of *spacetime*. In particular, we identify $G^{(4)}$ as the four-dimensional Newton constant G. In general, we will use D to denote the dimensionality of spacetime, and d to denote the number of spatial dimensions. Clearly, $D = d + 1$.

Equation (3.102) defines Newtonian gravitation in an arbitrary number of dimensions. Just as the electric field of a point charge does, the gravitational field of a point mass falls off like $1/r^{d-1}$ in a world with d spatial dimensions. As a result, the force between two point masses separated by a distance r falls off like $1/r^{d-1}$. For three spatial dimensions, this is the familiar inverse-squared dependence of the gravitational force. If $D = 6$ (a world with two extra dimensions) the gravitational force falls off like $1/r^4$.

3.8 The Planck length in various dimensions

We define the Planck length in any dimension just as we did in four dimensions: the Planck length is the unique length built using *only* powers of the gravitational constant $G^{(D)}$, c, and $\hbar$. To compute the Planck length we must determine the units of $G^{(D)}$. This is easily done if we recall that the units of $G^{(D)} \rho_m$ (the right-hand side of (3.102)) are the same in all dimensions.

Comparing the cases of five and four dimensions, for example,

$$[G^{(5)}] \frac{M}{L^4} = [G] \frac{M}{L^3} \longrightarrow [G^{(5)}] = L\,[G].$$ (3.103)

The units of $G^{(5)}$ carry one more factor of length than the units of G. We use (3.90) to read the units of G in terms of units of length and units of c and $\hbar$:

$$[G] = \frac{[c]^3 \, L^2}{[\hbar]}. \tag{3.104}$$

Equation (3.103) then gives

$$[G^{(5)}] = \frac{[c]^3 \, L^3}{[\hbar]}. \tag{3.105}$$

Since the Planck length is constructed uniquely from the gravitational constant, c, and $\hbar$, we can remove the brackets in the above equation and replace L by the five-dimensional Planck length $\ell_P^{(5)}$:

$$\left(\ell_P^{(5)}\right)^3 = \frac{\hbar G^{(5)}}{c^3}. \tag{3.106}$$

Reintroducing the four-dimensional Planck length:

$$\left(\ell_P^{(5)}\right)^3 = \left(\frac{\hbar G}{c^3}\right)\frac{G^{(5)}}{G} \longrightarrow \left(\ell_P^{(5)}\right)^3 = (\ell_P)^2 \frac{G^{(5)}}{G}. \tag{3.107}$$

Since they do not have the same units, the gravitational constants in four and five dimensions cannot be compared directly. Planck lengths, however, can be compared. If the Planck length is the same in four and in five dimensions, then $G^{(5)}/G = \ell_P$; the gravitational constants differ by one factor of the common Planck length.

It is not hard to generalize the above equations to D spacetime dimensions:

Quick calculation 3.11 Show that (3.106) and (3.107) are replaced by

$$\left(\ell_P^{(D)}\right)^{D-2} = \frac{\hbar G^{(D)}}{c^3} = (\ell_P)^2 \frac{G^{(D)}}{G}. \tag{3.108}$$

3.9 Gravitational constants and compactification

If string theory is correct, our world is really higher-dimensional. The fundamental gravity theory is then defined in the higher-dimensional world, with some value for the higher-dimensional Planck length. Since we observe only four dimensions, the additional dimensions may be curled up to form a compact space with small volume. We can then ask: what is the effective value of the four-dimensional Planck length? As we shall show here, the effective four-dimensional Planck length depends on the volume of the extra dimensions, as well as on the value of the higher-dimensional Planck length.

These observations raise the possibility that the Planck length in the effectively four-dimensional world – the famous number equal to about 10^{-33} cm – may not coincide with the fundamental Planck length in the original higher-dimensional theory. Is it possible that the fundamental Planck length is much bigger than the familiar, four-dimensional one? We will answer this question in the following section. In this section we will work out the effect of compactification on gravitational constants.

How do we calculate the gravitational constant in four dimensions if we are given the gravitational constant in five? First, we recognize the need to curl up one spatial dimension, otherwise there is no effectively four-dimensional spacetime. As we will see, the size of the extra dimension enters the relationship between the gravitational constants. To explore these questions precisely, consider a five-dimensional spacetime where one dimension forms a small circle of radius R. We are given $G^{(5)}$ and we would like to calculate $G^{(4)}$.

Let (x^1, x^2, x^3) denote three spatial dimensions of infinite extent, and x^4 denote a compactified dimension of circumference $2\pi R$ (Figure 3.2). We place a uniform ring of total mass M all around the circle at $x^1 = x^2 = x^3 = 0$. This is a mass distribution which is constant along the x^4 dimension. We are interested in the gravitational potential $V_g^{(5)}$ that emerges from such a mass distribution. Alternatively, we could have placed a point mass at some fixed x^4, but this makes the calculations more involved (Problem 3.10). In the present case the gravitational potential $V_g^{(5)}$ does *not* depend on x^4. The total mass M can be written as

$$\text{total mass} = M = 2\pi R m, \tag{3.109}$$

where m is the mass per unit length.

What is the mass density in the five-dimensional world? It is only nonzero at $x^1 = x^2 = x^3 = 0$. To represent such a mass density we use delta functions. Recall that the delta function $\delta(x)$ can be viewed as a singular function whose value is zero except for $x = 0$ and such that the integral $\int_{-\infty}^{\infty} dx\,\delta(x) = 1$. This integral implies that if x has units of length, then $\delta(x)$ has units of inverse length. Since the five-dimensional mass density is concentrated at $x^1 = x^2 = x^3 = 0$, it is reasonable to include in its formula the product $\delta(x^1)\delta(x^2)\delta(x^3)$ of three delta functions. We claim that

$$\rho^{(5)} = m\,\delta(x^1)\delta(x^2)\delta(x^3). \tag{3.110}$$

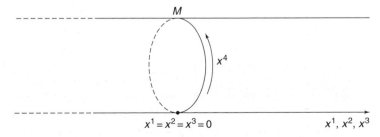

Fig. 3.2 A world with four space dimensions, one of which, x^4, is compactified into a circle of radius R. A ring of total mass M wraps around this compact dimension.

We first check the units. The mass density $\rho^{(5)}$ must have units of M/L^4. This works out since m has units of mass per unit length, and the three delta functions supply an additional factor of L^{-3}. The ansatz in (3.110) could still be off by a constant dimensionless factor, a factor of two, for example. As a final check, we integrate $\rho^{(5)}$ over all space. The result should be the total mass:

$$\int_{-\infty}^{\infty} dx^1 dx^2 dx^3 \int_0^{2\pi R} dx^4 \rho^{(5)}$$

$$= m \int_{-\infty}^{\infty} dx^1 \delta(x^1) \int_{-\infty}^{\infty} dx^2 \delta(x^2) \int_{-\infty}^{\infty} dx^3 \delta(x^3) \int_0^{2\pi R} dx^4$$

$$= m\, 2\pi R. \tag{3.111}$$

This is indeed the total mass on account of (3.109). For the effectively four-dimensional observer the mass is point-like, and it is located at $x^1 = x^2 = x^3 = 0$. So this observer writes

$$\rho^{(4)} = M\delta(x^1)\delta(x^2)\delta(x^3). \tag{3.112}$$

Note the relation

$$\rho^{(5)} = \frac{1}{2\pi R} \rho^{(4)}. \tag{3.113}$$

Let us now use this information in the equations for the gravitational potential. Using the five-dimensional version of (3.102) and (3.113), we find

$$\nabla^2 V_g^{(5)}(x^1, x^2, x^3) = 4\pi G^{(5)} \rho^{(5)} = 4\pi \frac{G^{(5)}}{2\pi R} \rho^{(4)}. \tag{3.114}$$

As we have noted before, $V_g^{(5)}$ is independent of x^4 so the Laplacian above is actually the four-dimensional one. Moreover, the four-dimensional gravitational potential, which is the average of $V_g^{(5)}$ over x^4, is just $V_g^{(5)}$. Therefore, the above equation takes the form of the gravitational equation in four dimensions, where the constant in between the 4π and the $\rho^{(4)}$ is the four-dimensional gravitational constant. We have therefore shown that

$$G = \frac{G^{(5)}}{2\pi R} \longrightarrow \boxed{\frac{G^{(5)}}{G} = 2\pi R \equiv \ell_C,} \tag{3.115}$$

where ℓ_C is the length of the extra compact dimension. This is what we were seeking: a relationship between the strength of the gravitational constants in terms of the size of the extra dimension.

The generalization of (3.115) to the case where there is more than one extra dimension is straightforward. One finds that

$$\frac{G^{(D)}}{G} = (\ell_C)^{D-4}, \tag{3.116}$$

where ℓ_C is the common length of each of the extra dimensions. When the various dimensions are curled up into circles of different lengths, the above right-hand side must be replaced by the product of the various lengths. This product is, in fact, the volume V_C of the extra dimensions, so that

$$\frac{G^{(D)}}{G} = V_C. \tag{3.117}$$

3.10 Large extra dimensions

We are now done with all the groundwork. In Section 3.8 we found the relation between the Planck length and the gravitational constant in any dimension. In Section 3.9 we determined how gravitational constants are related upon compactification. We are ready to find out how the fundamental Planck length in a higher-dimensional theory with compactification is related to the Planck length in the effectively four-dimensional theory.

To begin with, consider a five-dimensional world with Planck length $\ell_P^{(5)}$ and a single spatial coordinate curled up into a circle of circumference ℓ_C. How are these lengths related to ℓ_P? From (3.107) and (3.115) we find that

$$(\ell_P^{(5)})^3 = (\ell_P)^2 \frac{G^{(5)}}{G} = (\ell_P)^2 \, \ell_C. \tag{3.118}$$

Solving for ℓ_C, we get

$$\ell_C = \frac{(\ell_P^{(5)})^3}{(\ell_P)^2}. \tag{3.119}$$

This relation enables us to explore the possibility that the world is actually five-dimensional with a fundamental Planck length $\ell_P^{(5)}$ that is much larger than 10^{-33} cm. Of course, we must have $\ell_P \sim 10^{-33}$ cm. After all, this is the four-dimensional Planck length, whose value is given in (3.90).

Present-day accelerators explore physics down to distances of the order of 10^{-16} cm. If this distance, or a somewhat smaller one, is the fundamental length scale, we may choose $\ell_P^{(5)} \sim 10^{-18}$ cm. What would ℓ_C have to be? With $\ell_P^{(5)} \sim 10^{-18}$ cm and $\ell_P \sim 10^{-33}$ cm, equation (3.119) gives $\ell_C \sim 10^{12}$ cm $\sim 10^7$ km. This is more than twenty times the distance from the earth to the moon. Such a large extra dimension would have been detected a long time ago.

Having failed to produce a realistic scenario in five dimensions, let us try in *six* spacetime dimensions. For arbitrary D, equations (3.108) and (3.116) give

$$\left(\ell_P^{(D)}\right)^{D-2} = (\ell_P)^2 \frac{G^{(D)}}{G} = (\ell_P)^2 (\ell_C)^{D-4}. \tag{3.120}$$

Solving for ℓ_C we find

$$\ell_C = \ell_P^{(D)} \left(\frac{\ell_P^{(D)}}{\ell_P}\right)^{\frac{2}{D-4}}. \tag{3.121}$$

For $D = 6$ and $\ell_{\mathrm{P}}^{(6)} \sim 10^{-18}$ cm this formula gives

$$\ell_C = \frac{(\ell_{\mathrm{P}}^{(6)})^2}{\ell_{\mathrm{P}}} \sim 10^{-3} \text{ cm.} \tag{3.122}$$

This is a lot more interesting! A convenient unit here is the micron $\mu\mathrm{m} = 10^{-6}$ m. One-tenth of a millimeter is $100\,\mu\mathrm{m}$. We found $\ell_C \sim 10\,\mu\mathrm{m}$. Could there be extra dimensions ten microns long? You might think that this is still too big, since even microscopes probe smaller distances. Moreover, as we indicated before, accelerators probe distances of the order of 10^{-16} cm. Surprisingly, it is possible that "large extra dimensions" exist and that we have not observed them yet.

The existence of additional dimensions may be confirmed by testing the force law which gives the gravitational attraction between two masses. For distances much larger than the compactification scale ℓ_C the world is effectively four-dimensional, so the dependence of the force between two masses on their separation must follow accurately Newton's inverse-squared law. On the other hand, for distances smaller than ℓ_C, the world is effectively higher-dimensional, and the force law will change. A force between two masses that goes like $1/r^4$, where r is the separation, is consistent with the existence of two compact extra dimensions.

It turns out to be very difficult to test gravity at small distances; the force of gravity is extremely weak and spurious electrical forces must be cancelled very precisely. Motivated mainly by the possible existence of large extra dimensions, physicists set out to test the inverse-squared law at distances smaller than one millimeter. The tabletop experiments use a torsion pendulum detector or, alternatively, a micromachined cantilever with a test mass at the free end. As of 2007, experiments have found no departure from the inverse-squared law down to distances of about fifty microns. This means that extra dimensions, if they exist, must be smaller than this distance. Compact dimensions the size of ten microns, as we found in (3.122), are still consistent with experiment.

You might ask: what about forces other than gravity? Electromagnetism has been tested to much smaller distances, and we know that the electric force obeys an inverse-squared law very accurately. Rutherford scattering of alpha particles off nuclei, for example, confirms that the inverse-squared law holds down to 10^{-11} cm. Since the separation dependence of the electric force would change at distances smaller than the size of the extra dimensions, this seems to rule out large extra dimensions. The possibility of large extra dimensions, however, survives in string theory, where our spatial world could be a three-dimensional hyperplane transverse to the extra dimensions. This hyperplane is called a D3-brane. A D3-brane is a D-brane with three spatial dimensions.

Open strings have the remarkable property that their endpoints must remain attached to the D-branes. In many phenomenological models built in string theory, it is the fluctuations of open strings that give rise to the familiar leptons, quarks, and gauge fields, including the Maxwell gauge field. It follows that these fields are bound to the D3-brane and do *not* feel the extra dimensions. If the Maxwell field lives on the D-brane, the electric field lines of a charge remain on the D-brane and do not go off into the extra dimensions. The force law is not changed at any distance scale. Closed strings are not bound by D-branes, and therefore gravity, which arises from closed strings, *is* affected by the extra dimensions.

Although the Planck length ℓ_P is an important length scale in four dimensions, if there are large extra dimensions, the truly fundamental Planck length would be much bigger than the effective four-dimensional one. The possibility of large extra dimensions is slightly unnatural – why should the extra dimensions be much larger than the fundamental length scale? This is not a new problem, however, but rather the problem of a large hierarchy in another guise. We noted earlier that particle physics faces a puzzling hierarchy between the Planck mass and the masses of elementary particles. In the large-extra-dimensions scenario, the hierarchy is postulated to arise from extra dimensions that are much larger than the fundamental length scale. At any rate, the truly exciting fact is that present experimental constraints do not rule out large extra dimensions. The discovery of extra dimensions would be revolutionary.

It is possible, of course, that extra dimensions are much larger than $\ell_P \sim 10^{-33}$ cm but are still quite small. If spacetime is ten dimensional, setting $D = 10$ and $\ell_P^{(10)} \sim 10^{-18}$ cm in (3.121) gives $\ell_C \sim 10^{-13}$ cm, a distance far too small to be probed with tabletop gravitational experiments. Extra dimensions this size or smaller must be searched for using particle accelerators.

Problems

Problem 3.1 Lorentz covariance for motion in electromagnetic fields.[†]

The Lorentz force equation (3.5) can be written relativistically as

$$\frac{dp_\mu}{ds} = \frac{q}{c} F_{\mu\nu} \frac{dx^\nu}{ds}, \tag{1}$$

where p_μ is the four-momentum. Check explicitly that this equation reproduces (3.5) when μ is a spatial index. What does (1) give when $\mu = 0$? Does it make sense? Is (1) a gauge invariant equation?

Problem 3.2 Maxwell equations in four dimensions.

(a) Show explicitly that the source-free Maxwell equations emerge from $T_{\mu\lambda\nu} = 0$.
(b) Show explicitly that the Maxwell equations with sources emerge from (3.34).

Problem 3.3 Electromagnetism in three dimensions.

(a) Find the reduced Maxwell equations in three dimensions by starting with Maxwell's equations and the force law in four dimensions, using the ansatz (3.11), and assuming that no field can depend on the z direction.
(b) Repeat the analysis of three-dimensional electromagnetism starting with the Lorentz covariant formulation. Take $A^\mu = (\Phi, A^1, A^2)$, examine $F_{\mu\nu}$, the Maxwell equations (3.34), and the relativistic form of the force law derived in Problem 3.1.

Problem 3.4 Electric fields and potentials of point charges.

(a) Show that for time-independent fields, the Maxwell equation $T_{0ij} = 0$ implies that $\partial_i E_j - \partial_j E_i = 0$. Explain why this condition is satisfied by the ansatz $\vec{E} = -\nabla\Phi$.

(b) Show that with d spatial dimensions, the potential Φ due to a point charge q is given by

$$\Phi(r) = \frac{\Gamma(\frac{d}{2}-1)}{4\pi^{d/2}}\frac{q}{r^{d-2}}.$$

Problem 3.5 Calculating the divergence in higher dimensions.

Let $\vec{f} = f(r)\,\hat{\mathbf{r}}$ be a vector function in $\mathbb{R}^d$. Here $\hat{\mathbf{r}}$ is a unit radial vector, and r is the radial distance to the origin. Derive a formula for $\nabla \cdot \vec{f}$ by applying the divergence theorem to a spherical shell of radius r and width dr. Check that for $d = 3$ your answer reduces to $\nabla \cdot \vec{f} = f'(r) + \frac{2}{r}f(r)$.

Problem 3.6 Analytic continuation for gamma functions.[†]

Consider the definition of the gamma function for complex arguments z whose real part is positive:

$$\Gamma(z) = \int_0^\infty dt\, e^{-t} t^{z-1}, \quad \Re(z) > 0.$$

Use this equation to show that for $\Re(z) > 0$

$$\Gamma(z) = \int_0^1 dt\, t^{z-1}\left(e^{-t} - \sum_{n=0}^N \frac{(-t)^n}{n!}\right) + \sum_{n=0}^N \frac{(-1)^n}{n!}\frac{1}{z+n} + \int_1^\infty dt\, e^{-t} t^{z-1}.$$

Explain why the above right-hand side is well defined for $\Re(z) > -N - 1$. It follows that this right-hand side provides the analytic continuation of $\Gamma(z)$ for $\Re(z) > -N - 1$. Conclude that the gamma function has poles at $0, -1, -2, \ldots$, and give the value of the residue at $z = -n$ (with n a positive integer).

Problem 3.7 Simple quantum gravity effects are small.[†]

(a) What would be the "gravitational" Bohr radius for a hydrogen atom if the attraction binding the electron to the proton was gravitational? The standard Bohr radius is $a_0 = \frac{\hbar^2}{me^2} \simeq 5.29 \times 10^{-9}$ cm.
(b) In "units" where G, c, and $\hbar$ are set equal to one, the temperature of a black hole is given by $kT = \frac{1}{8\pi M}$. Insert back the factors of G, c, and $\hbar$ into this formula. Evaluate the temperature of a black hole of a million solar masses. What is the mass of a black hole whose temperature is room temperature?

Problem 3.8 Vacuum energy and an associated length scale.

Observations indicate that the expansion of the universe is currently accelerating possibly due to a vacuum energy density. The mass density associated with this energy is approximately $\rho_{vac} = 7.7 \times 10^{-27}$ kg/m³. Some physicists try to understand the acceleration of the universe by introducing modifications to gravity. It is then useful to know what length scales could be important. If one assumes that the only relevant parameters are ρ_{vac}, $\hbar$, and c, one can construct a length parameter ℓ_{vac} by multiplying powers:

$$\ell_{vac} = \rho_{vac}^\alpha\, \hbar^\beta\, c^\gamma.$$

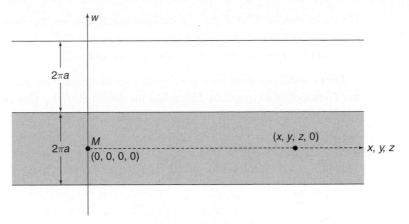

Fig. 3.3   Problem 3.10: a point mass M in a five-dimensional spacetime with one compact dimension.

What must be the values of α, β, and γ in the above equation? What is the numerical value of ℓ_{vac}? Express your answer in μm, where $1\ \mu$m $= 10^{-6}$ m.

Problem 3.9 Planetary motion in four and higher dimensions.

Consider the motion of planets in planar circular orbits around heavy stars in our four-dimensional spacetime and in spacetimes with additional spatial dimensions. We wish to study the stability of these orbits under perturbations that keep them planar. Such a perturbation would arise, for example, if a meteorite moving on the plane of the orbit hits the planet and changes its angular momentum.

Show that while planetary circular orbits in our four-dimensional world are stable under such perturbations, they are not so in five or higher dimensions. [Hint: you may find it useful to use the effective potential for motion in a central force field.]

Problem 3.10 Gravitational field of a point mass in a compactified five-dimensional world.

Consider a five-dimensional spacetime with space coordinates (x, y, z, w) *not* yet compactified. A point mass M is located at the origin $(x, y, z, w) = (0, 0, 0, 0)$.

(a) Find the gravitational potential $V_g^{(5)}(r)$. Write your answer in terms of M, $G^{(5)}$, and $r = (x^2 + y^2 + z^2 + w^2)^{1/2}$. [Hint: use $\nabla^2 V_g^{(5)} = 4\pi G^{(5)} \rho_m$ and the divergence theorem.]

Now let w become a circle with radius a while keeping the mass fixed, as shown in Figure 3.3.

(b) Write an exact expression for the gravitational potential $V_g^{(5)}(x, y, z, 0)$. This potential is a function of $R \equiv (x^2 + y^2 + z^2)^{1/2}$ and can be written as an infinite sum.

(c) Show that for $R \gg a$ the gravitational potential takes the form of a four-dimensional gravitational potential, with Newton's constant $G^{(4)}$ given in terms of $G^{(5)}$ as in (3.115). [Hint: turn the infinite sum into an integral.]

These results confirm both the relation between the four- and five-dimensional Newton constants in a compactification and the emergence of a four-dimensional potential at distances large compared to the size of the compact dimension.

Problem 3.11 Exact answer for the gravitational potential.

The infinite sum in Problem 3.10 can be evaluated exactly using the identity

$$\sum_{n=-\infty}^{\infty} \frac{1}{1+(\pi n x)^2} = \frac{1}{x} \coth\left(\frac{1}{x}\right).$$

(a) Find an exact closed-form expression for the gravitational potential $V_g^{(5)}(x, y, z, 0)$ in the compactified theory.
(b) Expand this answer to calculate the leading *correction* to the gravitational potential in the limit when $R \gg a$. For what value of R/a is the correction of order 1%?
(c) Use the exact answer in (a) to expand the potential when $R \ll a$. Give the first two terms in the expansion. Do you recognize the leading term?

4 Nonrelativistic strings

A full appreciation for the subtleties of relativistic strings requires an under-standing of the basic physics of nonrelativistic strings. These strings have mass and tension. They can vibrate both transversely and longitudinally. We study the equations of motion for nonrelativistic strings and develop the Lagrangian approach to their dynamics.

4.1 Equations of motion for transverse oscillations

We will begin our study of strings with a look at the transverse fluctuations of a stretched string. The direction along the string is called the longitudinal direction, and the directions orthogonal to the string are called the transverse directions. We consider, for notational simplicity, the case when there is only one transverse direction – the generalization to additional transverse directions is straightforward.

Working in the (x, y) plane, let the classical nonrelativistic string have its endpoints fixed at $(0, 0)$, and $(a, 0)$. In the static configuration the string is stretched along the x axis between these two points. In a transverse oscillation, the x-coordinate of any point on the string does not change in time. The transverse displacement of a point is given by its y coordinate. The x direction is longitudinal, and the y direction is transverse. To describe the classical mechanics of a homogeneous string, we need two pieces of information: the tension T_0 and the mass per unit length μ_0. The total mass of the string is then $M = \mu_0 a$.

Let us look briefly at the units. Tension has units of force, so

$$[T_0] = [\text{Force}] = \frac{[\text{Energy}]}{L}. \tag{4.1}$$

If you stretch a string an infinitesimal amount dx, its tension remains approximately con-stant through the stretching, and the change in energy equals the work done $T_0 dx$. The total mass of the string does not change. If we were considering relativistic strings, how-ever, a static string with more energy would have a larger rest mass. Using (4.1), noting that energy has units of mass times velocity squared, and that μ_0 has units of mass per unit length, we have

$$[T_0] = \frac{M}{L}[v]^2 = [\mu_0][v]^2. \tag{4.2}$$

For a nonrelativistic string, both T_0 and μ_0 are adjustable parameters, and the velocity on the right-hand side above will turn out to be the velocity of transverse waves. The above

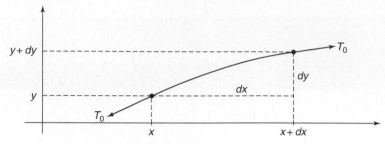

Fig. 4.1 A short piece of a classical nonrelativistic string vibrating transversely. With different slopes at the two endpoints there is a net vertical force.

equation suggests that the string tension T_0 and the linear mass density μ_0 in a relativistic string might be related by $T_0 = \mu_0 c^2$, since c is the canonical velocity in relativity. We will see in Chapter 6 that this is indeed the correct relation for a relativistic string.

Returning to our classical nonrelativistic string, let us figure out the equation of motion. Consider a small portion of the static string that extends from x to $x + dx$, with $y = 0$. This piece is shown in transverse oscillation in Figure 4.1. At time t, the transverse displacement of the string is $y(t, x)$ at x and $y(t, x + dx)$ at $x + dx$. We will assume that the oscillations are small, and by this we will mean that at all times

$$\left| \frac{\partial y}{\partial x} \right| \ll 1, \tag{4.3}$$

at any point on the string. This guarantees that the transverse displacement of the string is small compared to the length of the string. The length of the string changes little, and we can assume that the tension T_0 is unchanged.

The slope of the string is a bit different at the points x and $x + dx$. This change of slope means that the string tension changes direction and the portion of string under consideration feels a net force. For transverse oscillations we need only calculate the net vertical force; the net horizontal force is negligible (Problem 4.1). The vertical force at $(x + dx, y + dy)$ is given accurately by T_0 times $\partial y / \partial x$ evaluated at $x + dx$ and is pointing up; similarly, the vertical force at (x, y) is T_0 times $\partial y / \partial x$ evaluated at x and is pointing down. Therefore the net vertical force dF_v is

$$dF_v = T_0 \frac{\partial y}{\partial x} \bigg|_{x+dx} - T_0 \frac{\partial y}{\partial x} \bigg|_x \simeq T_0 \frac{\partial^2 y}{\partial x^2} dx. \tag{4.4}$$

The mass dm of this piece of string, originally stretched from x to $x + dx$, is given by the mass density μ_0 times dx. By Newton's law, the net vertical force equals mass times vertical acceleration. So we can simply write

$$T_0 \frac{\partial^2 y}{\partial x^2} dx = (\mu_0 dx) \frac{\partial^2 y}{\partial t^2}. \tag{4.5}$$

We cancel dx on each side and rearrange terms to get

$$\frac{\partial^2 y}{\partial x^2} - \frac{\mu_0}{T_0}\frac{\partial^2 y}{\partial t^2} = 0. \tag{4.6}$$

This is just a wave equation! Recall that for the wave equation

$$\frac{\partial^2 y}{\partial x^2} - \frac{1}{v_0^2}\frac{\partial^2 y}{\partial t^2} = 0, \tag{4.7}$$

the parameter v_0 is the velocity of the waves. Thus for the transverse waves on our stretched string, the velocity v_0 of the waves is

$$v_0 = \sqrt{T_0/\mu_0}. \tag{4.8}$$

The higher the tension or the lighter the string, the faster the waves move.

4.2 Boundary conditions and initial conditions

Since equation (4.6) is a partial differential equation involving space and time derivatives, in order to fix solutions we must in general apply both boundary conditions and initial conditions. Boundary conditions (B.C.) constrain the solution at the boundary of the system, and initial conditions constrain the solution at a given starting time. The most common types of boundary conditions are Dirichlet and Neumann boundary conditions.

For our string, Dirichlet boundary conditions specify the positions of the string endpoints. For example, if we attach each end of the string to a wall (Figure 4.2, left), we are imposing the Dirichlet boundary conditions

$$y(t, x = 0) = y(t, x = a) = 0, \quad \text{Dirichlet boundary conditions.} \tag{4.9}$$

Alternatively, if we attach a massless loop to each end of the string and the loops are allowed to slide along two frictionless poles, we are imposing Neumann boundary conditions. For our string, Neumann boundary conditions specify the values of the derivative $\partial y/\partial x$ at the endpoints. Since the loops are massless and the poles are frictionless, the derivative $\partial y/\partial x$ must vanish at the poles $x = 0, a$ (Figure 4.2, right). If this were not the case, then the slope of the string at a pole would be nonzero, and a component of the string tension would accelerate the rings in the y direction. Since each ring is massless,

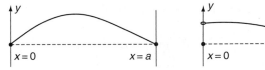

Fig. 4.2   Left: string with Dirichlet boundary conditions at the endpoints. Right: string with Neumann boundary conditions at the endpoints.

their acceleration would be infinite. This is not possible, so, in effect, we are imposing the Neumann boundary conditions

$$\frac{\partial y}{\partial x}(t, x = 0) = \frac{\partial y}{\partial x}(t, x = a) = 0, \quad \text{Neumann boundary conditions.} \quad (4.10)$$

These Neumann boundary conditions apply to strings whose endpoints are free to move along the y direction.

Let us see how we can solve the wave equation for a particular set of initial conditions. The general solution of equation (4.6) is of the form

$$y(t, x) = h_+(x - v_0 t) + h_-(x + v_0 t), \quad (4.11)$$

where h_+ and h_- are arbitrary functions of a *single* variable. This solution represents a superposition of two waves, h_+ moving to the right and h_- moving to the left. Suppose that the initial values of y and $\partial y / \partial t$ are known at time $t = 0$. Using equation (4.11) we see that this information yields the equations

$$y(0, x) = h_+(x) + h_-(x), \quad (4.12)$$

$$\frac{\partial y}{\partial t}(0, x) = -v_0 h'_+(x) + v_0 h'_-(x), \quad (4.13)$$

where the left-hand sides are known functions, and the primes denote derivatives with respect to arguments. Using (4.12) we can solve for h_- in terms of h_+. Substituting into (4.13), we get a first-order ordinary differential equation for h_+. Once we have solved for h_+ (using appropriate boundary conditions), we can use (4.12) again, this time to find the explicit form of h_-. With h_+ and h_- known, the full solution of the equations of motion is given by (4.11).

4.3 Frequencies of transverse oscillation

Suppose that we have a string where each point is oscillating in the y direction sinusoidally and in phase. This means that $y(t, x)$ is of the form

$$y(t, x) = y(x) \sin(\omega t + \phi), \quad (4.14)$$

where ω is the angular frequency of oscillation and ϕ is the constant common phase. Our aim is to find the allowed frequencies of oscillation. Substituting (4.14) into (4.6) and cancelling the common time dependence, we find

$$\frac{d^2 y(x)}{dx^2} + \omega^2 \frac{\mu_0}{T_0} y(x) = 0. \quad (4.15)$$

This is an ordinary second-order differential equation for the profile $y(x)$ of the oscillations. The allowed frequencies are selected by this equation, together with the boundary conditions. Since ω, μ_0, and T_0 are constants, the differential equation is solved in terms of trigonometric functions. With Dirichlet boundary conditions (4.9) we have the nontrivial solutions

$$y_n(x) = A_n \sin\left(\frac{n\pi x}{a}\right), \quad n = 1, 2, \ldots, \tag{4.16}$$

where A_n is an arbitrary constant. The value $n = 0$ is not included above because it represents a motionless string. Plugging $y_n(x)$ into (4.15), we find the allowed frequencies ω_n:

$$\omega_n = \sqrt{\frac{T_0}{\mu_0}}\left(\frac{n\pi}{a}\right), \quad n = 1, 2, \ldots. \tag{4.17}$$

These are the frequencies of oscillation for a Dirichlet string. The strings on a violin are Dirichlet strings. To tune a violin to the correct frequency one must adjust the string tension. The higher the tension is, the higher the pitch, as predicted by (4.17). For the case of Neumann boundary conditions (4.10), we obtain the spatial solutions

$$y_n(x) = A_n \cos\left(\frac{n\pi x}{a}\right), \quad n = 0, 1, 2, \ldots. \tag{4.18}$$

This time the $n = 0$ solution is a little less trivial: the string does not oscillate, but it is rigidly translated to $y(t, x) = A_0$. The oscillation frequencies, found by plugging (4.18) into (4.15), are the same as those in (4.17). Therefore, the oscillation frequencies are the same in the Neumann and Dirichlet problems. The Neumann case admits one extra solution not included in our oscillatory ansatz (4.14): the string can translate with constant velocity. Indeed, $y(t, x) = at + b$, with a and b arbitrary constants, satisfies both the boundary conditions and the original wave equation (4.7).

4.4 More general oscillating strings

Let us discuss briefly some problems that are closely related to the ones considered thus far. For example, we can take the mass density of the string to be a function $\mu(x)$ of position. The form (4.6) of the wave equation does not change since it is derived from local considerations: the examination of a little piece of string that can be chosen to be sufficiently small so that the mass density is approximately constant. We therefore get

$$\frac{\partial^2 y}{\partial x^2} - \frac{\mu(x)}{T_0}\frac{\partial^2 y}{\partial t^2} = 0. \tag{4.19}$$

For normal oscillations, we use the ansatz in (4.14) and find

$$\frac{d^2 y}{dx^2} + \frac{\mu(x)}{T_0}\omega^2 y(x) = 0. \tag{4.20}$$

This equation is no longer simple to solve, and it can only be studied in detail once the function $\mu(x)$ is specified. In Problems 4.3 and 4.7 you will consider some specific mass distributions, and you will explore a variational approach that gives an upper bound for the lowest oscillation frequency.

So far we have only considered strings that oscillate transversally. Strings also admit longitudinal oscillations, although the relativistic string does not. Imagine a string stretched along the x axis, and consider the infinitesimal segment which at equilibrium extends from x to $x + dx$. Suppose now that at time t the ends of this infinitesimal segment are longitudinally displaced from their equilibrium positions by distances $\eta(t, x)$ and $\eta(t, x + dx)$, respectively. If these two quantities are not the same, the piece of string is being compressed or stretched. An equation of motion can be obtained for this system, much as we did for transverse motion. It is not possible, however, to assume that the tension is constant throughout the string. For transverse oscillations the net force acting on a little piece of string arose from the different angles at which the same tension was applied on opposite ends of the piece. If the string always lies along the x axis then a net force can act on a segment only if the tension is different on its two ends. Therefore the waves on a longitudinally oscillating string are accompanied by tension waves (Problem 4.2).

4.5 A brief review of Lagrangian mechanics

The Lagrangian L of a system is defined by

$$L = T - V,\qquad(4.21)$$

where T is the kinetic energy of the system and V is the potential energy of the system. For a point particle of mass m moving along the x axis under the influence of a time-independent potential $V(x)$, the nonrelativistic Lagrangian takes the form

$$L(t) = \frac{1}{2}m\,(\dot{x}(t))^2 - V(x(t)),\qquad \dot{x}(t) \equiv \frac{dx(t)}{dt}.\qquad(4.22)$$

We must emphasize that the above Lagrangian is implicitly a function of time, but it has no explicit time dependence. All the time dependence arises from the time dependence of the position $x(t)$. The action S is defined as

$$S = \int_{\mathcal{P}} L(t)dt,\qquad(4.23)$$

where $\mathcal{P}$ is a path $x(t)$ between an initial position x_i at an initial time t_i, and a final position x_f at a final time $t_f > t_i$. One such path is shown in Figure 4.3.

The action is a *functional*. Whereas a function of a single variable takes one number – the argument – as input and gives another number as output, a functional takes a *function* as the input, and gives a number as output. Since a function is usually defined by its values at infinitely many points, we can think of a functional as a function of infinitely many variables. In our present application, the input for the action functional is the function $x(t)$ which determines the path $\mathcal{P}$. We can emphasize the argument of S by using the notation $S[x]$. Here $[x]$ represents the full function $x(t)$. It is potentially confusing to write $S[x(t)]$, since it suggests that S is ultimately a function of t, which it is not.

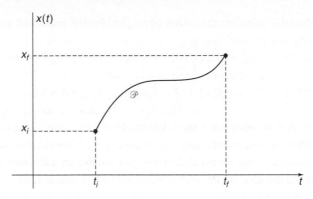

Fig. 4.3 A path $\mathcal{P}$ representing a possible one-dimensional motion $x(t)$ of a particle during the time interval $[t_i, t_f]$.

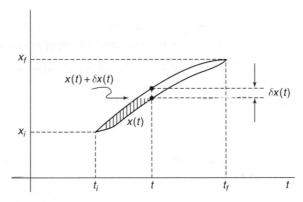

Fig. 4.4 A path $x(t)$ and its variation $x(t) + \delta x(t)$. This variation $\delta x(t)$ vanishes at $t = t_i$ and at $t = t_f$.

More explicitly, for any path $x(t)$, the action is given by

$$S[x] = \int_{t_i}^{t_f} \left\{ \frac{1}{2} m \, (\dot{x}(t))^2 - V(x(t)) \right\} dt. \tag{4.24}$$

It is very important to emphasize that the action S can be calculated for any path $x(t)$ and not only for paths that represent physically realized motion. It is because S can be calculated for all paths that it is a very powerful tool to find the paths that can be physically realized.

Hamilton's principle states that the path $\mathcal{P}$ which a system actually takes is one for which the action S is stationary. More precisely, if this path $\mathcal{P}$ is varied infinitesimally, the action does not change to first order in the variation. In terms of the function $x(t)$ which specifies the path, the perturbed path takes the form $x(t) + \delta x(t)$, as shown in Figure 4.4. For any time t, the variation $\delta x(t)$ is the vertical distance between the original path and the varied path. As in the figure, we consider variations where the initial and final positions $x_i = x(t_i)$ and $x_f = x(t_f)$ are unchanged:

$$\delta x(t_i) = \delta x(t_f) = 0. \tag{4.25}$$

We now calculate the action $S[x + \delta x]$ for the perturbed path $x(t) + \delta x(t)$:

$$S[x + \delta x] = \int_{t_i}^{t_f} \left\{ \frac{m}{2} \left(\frac{d}{dt} (x(t) + \delta x(t)) \right)^2 - V(x(t) + \delta x(t)) \right\} dt$$

$$= S[x] + \int_{t_i}^{t_f} \left\{ m\dot{x}(t) \frac{d}{dt} \delta x(t) - V'(x(t)) \delta x(t) \right\} dt + \mathcal{O}((\delta x)^2). \quad (4.26)$$

In passing to the last right-hand side we expanded V in a Taylor series about $x(t)$. The terms of order $(\delta x)^2$ and higher are unnecessary to determine whether or not the action is stationary. We have thus left them undetermined and indicated them by $\mathcal{O}((\delta x)^2)$. We can write the new action as $S + \delta S$, where δS is linear in δx. From the equation above we see that δS is given by

$$\delta S = \int_{t_i}^{t_f} \left\{ m\dot{x}(t) \frac{d}{dt} \delta x(t) - V'(x(t)) \delta x(t) \right\} dt. \quad (4.27)$$

To find the equations of motion, the variation δS must be rewritten in the form $\delta S = \int dt \, \delta x(t) \{ \ldots \}$. In particular, no derivatives must be acting on δx. This can be achieved using integration by parts:

$$\delta S = \int_{t_i}^{t_f} \left\{ \frac{d}{dt} \left(m\dot{x}(t) \delta x(t) \right) - m\ddot{x}(t) \delta x(t) - V'(x(t)) \delta x(t) \right\} dt$$

$$= m\dot{x}(t_f) \delta x(t_f) - m\dot{x}(t_i) \delta x(t_i) + \int_{t_i}^{t_f} \delta x(t) \left(-m\ddot{x}(t) - V'(x(t)) \right) dt. \quad (4.28)$$

Making use of (4.25), the variation reduces to

$$\delta S = \int_{t_i}^{t_f} \delta x(t) \left(-m\ddot{x}(t) - V'(x(t)) \right) dt. \quad (4.29)$$

The action is stationary if δS vanishes for every variation $\delta x(t)$. For this to happen, the factor multiplying $\delta x(t)$ in the integrand must vanish:

$$m\ddot{x}(t) = -V'(x(t)). \quad (4.30)$$

This is Newton's second law applied to the motion of a particle in a potential $V(x)$. We have recovered the expected equation of motion by requiring that the action be stationary under variations.

Suppose that we have determined the path that the particle takes while going from x_i to x_f. As we have seen, the action is then stationary under variations that vanish at the initial and final times. Is the action also stationary under variations that change the initial position at t_i or the final position at t_f? In general, the answer is no. This can be seen from equation (4.28). The integral term vanishes by assumption, but if $\delta x(t_f) \neq 0$, the first term on the right-hand side would not vanish unless $m\dot{x}(t_f)$, the final momentum of the particle, happens to vanish. The situation is analogous for $\delta x(t_i) \neq 0$.

Hamilton's principle states that the action is stationary about the classical solution. The classical solution does not always define a minimum of the action. It is possible to construct a simple example in which the classical solution is a saddle point of the action functional:

the action increases for some variations and decreases for others. See Problem 4.6 for the details.

4.6 The nonrelativistic string Lagrangian

Let us return now to our string with constant mass density μ_0, constant tension T_0, and ends located at $x = 0$ and $x = a$. The kinetic energy is simply the sum of the kinetic energies of all the infinitesimal segments that comprise the string. So it can be written as

$$T = \int_0^a \frac{1}{2} (\mu_0 dx) \left(\frac{\partial y}{\partial t}\right)^2 . \tag{4.31}$$

The potential energy arises from the work which must be done to stretch the segments. Consider an infinitesimal portion of string which extends from $(x, 0)$ to $(x + dx, 0)$ when the string is in equilibrium. If the string element is momentarily stretched from (x, y) to $(x + dx, y + dy)$, as in Figure 4.1, then the change in length Δl of the infinitesimal segment is given by

$$\Delta l = \sqrt{(dx)^2 + (dy)^2} - dx = dx\left(\sqrt{1 + \left(\frac{\partial y}{\partial x}\right)^2} - 1\right) \simeq dx \frac{1}{2}\left(\frac{\partial y}{\partial x}\right)^2 , \tag{4.32}$$

where we have used the small oscillation approximation (4.3) to discard higher-order terms in the expansion of the square root. Since the work done in stretching each infinitesimal segment is $T_0 \Delta l$, the total potential energy V is

$$V = \int_0^a \frac{1}{2} T_0 \left(\frac{\partial y}{\partial x}\right)^2 dx. \tag{4.33}$$

The Lagrangian for the string is given by $T - V$:

$$L(t) = \int_0^a \left[\frac{1}{2}\mu_0\left(\frac{\partial y}{\partial t}\right)^2 - \frac{1}{2}T_0\left(\frac{\partial y}{\partial x}\right)^2\right]dx \equiv \int_0^a \mathcal{L}\, dx , \tag{4.34}$$

where $\mathcal{L}$ is referred to as the *Lagrangian density*:

$$\mathcal{L}\left(\frac{\partial y}{\partial t}, \frac{\partial y}{\partial x}\right) = \frac{1}{2}\mu_0 \left(\frac{\partial y}{\partial t}\right)^2 - \frac{1}{2}T_0\left(\frac{\partial y}{\partial x}\right)^2. \tag{4.35}$$

The action for our string is therefore

$$S = \int_{t_i}^{t_f} L(t)dt = \int_{t_i}^{t_f} dt \int_0^a dx \left[\frac{1}{2}\mu_0\left(\frac{\partial y}{\partial t}\right)^2 - \frac{1}{2}T_0\left(\frac{\partial y}{\partial x}\right)^2\right]. \tag{4.36}$$

In this action the "path" is the function $y(t, x)$, defined over the region of (t, x) space shown shaded in Figure 4.5.

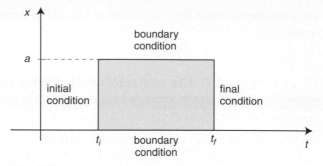

Fig. 4.5 The string motion is defined by $y(t, x)$ over the domain $t \in [t_i, t_f]$, $x \in [0, a]$. Boundary conditions apply at $x = 0$ and $x = a$ for all $t \in [t_i, t_f]$. Initial and final conditions apply for $t = t_i$ and $t = t_f$, respectively, for all $x \in [0, a]$.

To find the equations of motion, we must examine the variation of the action as we vary: $y(t, x) \rightarrow y(t, x) + \delta y(t, x)$. Performing the variation as before, we get

$$\delta S = \int_{t_i}^{t_f} dt \int_0^a dx \left[\mu_0 \frac{\partial y}{\partial t} \frac{\partial(\delta y)}{\partial t} - T_0 \frac{\partial y}{\partial x} \frac{\partial(\delta y)}{\partial x} \right]. \tag{4.37}$$

Quick calculation 4.1 Prove equation (4.37).

We must have no derivatives acting on the variations, so we rewrite each of the two terms above as a full derivative minus a term in which the derivative does not act on the variation:

$$\delta S = \int_{t_i}^{t_f} dt \int_0^a dx \left[\frac{\partial}{\partial t} \left(\mu_0 \frac{\partial y}{\partial t} \delta y \right) - \mu_0 \frac{\partial^2 y}{\partial t^2} \delta y + \frac{\partial}{\partial x} \left(-T_0 \frac{\partial y}{\partial x} \delta y \right) + T_0 \frac{\partial^2 y}{\partial x^2} \delta y \right]. \tag{4.38}$$

The time derivative reduces to evaluations at t_f and t_i, while the space derivative gives evaluations at the string endpoints:

$$\delta S = \int_0^a \left[\mu_0 \frac{\partial y}{\partial t} \delta y \right]_{t=t_i}^{t=t_f} dx + \int_{t_i}^{t_f} \left[-T_0 \frac{\partial y}{\partial x} \delta y \right]_{x=0}^{x=a} dt$$
$$- \int_{t_i}^{t_f} dt \int_0^a dx \left(\mu_0 \frac{\partial^2 y}{\partial t^2} - T_0 \frac{\partial^2 y}{\partial x^2} \right) \delta y. \tag{4.39}$$

Our final expression for δS contains three terms. Each one must vanish independently. The third term, for example, is determined by the motion of the string for $x \in (0, a)$ and $t \in (t_i, t_f)$. In this domain $\delta y(t, x)$ is not restricted by boundary conditions nor by initial or final conditions, so we set to zero the coefficient of δy and recover our original equation (4.6). The first term in (4.39) is determined by the configuration of the string at times t_i and t_f. If we specify these initial and final configurations, we are in effect setting $\delta y(t_i, x)$ and $\delta y(t_f, x)$ to zero. This causes the first term to vanish. We encountered an analogous situation in our study of the free particle.

The second term in (4.39) is new. Written out explicitly, it is

$$\int_{t_i}^{t_f} \left[-T_0 \frac{\partial y}{\partial x}(t,a)\, \delta y(t,a) + T_0 \frac{\partial y}{\partial x}(t,0)\, \delta y(t,0) \right] dt, \qquad (4.40)$$

and it concerns the motion of the string endpoints $y(t,0)$ and $y(t,a)$. We need a boundary condition for each of the two terms above. Let x_* denote the x coordinate of an endpoint; x_* can be equal to zero or equal to a. Selecting an endpoint means fixing the value of x_*. We can make each term in (4.40) vanish by specifying either Dirichlet or Neumann boundary conditions. Consider an endpoint x_* and the associated term in (4.40). If we impose a Dirichlet boundary condition the position of the chosen endpoint is fixed throughout time, and we require that the variation $\delta y(t,x_*)$ vanishes. This will cause the chosen term to vanish. If, on the other hand, we assume that the endpoint is free to move, then the variation $\delta y(t,x_*)$ is unconstrained. The term will vanish if we impose the condition

$$\frac{\partial y}{\partial x}(t,x_*) = 0, \quad \text{Neumann boundary condition.} \qquad (4.41)$$

Dirichlet boundary conditions can be written in a form where the similarity to Neumann boundary conditions is more apparent. If a string endpoint is fixed, the time derivative of the endpoint coordinate must vanish

$$\frac{\partial y}{\partial t}(t,x_*) = 0, \quad \text{Dirichlet boundary condition.} \qquad (4.42)$$

The similarity with (4.41) is quite striking. The only change is that spatial derivatives were turned into time derivatives. If we write Dirichlet boundary conditions in this form, we must still specify the values of the coordinates at the fixed endpoints. In order to appreciate further the physical import of boundary conditions, we consider the momentum p_y carried by the string. There is no other component to the momentum, because we have assumed that the motion is restricted to the y direction. This momentum is simply the sum of the momenta of each infinitesimal segment along the string:

$$p_y = \int_0^a \mu_0 \frac{\partial y}{\partial t}\, dx. \qquad (4.43)$$

Let us see if this momentum is conserved:

$$\frac{dp_y(t)}{dt} = \int_0^a \mu_0 \frac{\partial^2 y}{\partial t^2}\, dx = \int_0^a T_0 \frac{\partial^2 y}{\partial x^2}\, dx = T_0 \left[\frac{\partial y}{\partial x} \right]_{x=0}^{x=a}, \qquad (4.44)$$

where we used the wave equation (4.6). We see that momentum is conserved when Neumann boundary conditions (4.41) apply at both endpoints. For Dirichlet boundary conditions momentum is not generally conserved! Indeed, when the endpoints of a string are attached to a wall, the wall is constantly exerting a force on the string. In the lowest normal mode of a Dirichlet string, for example, the net momentum constantly oscillates between the $+y$ and $-y$ directions.

Why is this important for string theory? For a long time string theorists did not take seriously the possibility of Dirichlet boundary conditions. It seemed unphysical that the string momentum could fail to be conserved. Moreover, what could the endpoints of open strings be attached to? The answer is that they are attached to D-branes – a new kind of dynamical extended object. If a string is attached to a D-brane then momentum can be conserved – the momentum lost by the string is absorbed by the D-brane. A detailed analysis of the spatial boundary term induced by variation is crucial to recognize the possibility of D-branes in string theory.

We conclude this chapter with a more general derivation of the equation of motion for the string. For this, we use (4.35) to write the action as

$$S = \int_{t_i}^{t_f} dt \int_0^a dx \, \mathcal{L}\left(\frac{\partial y}{\partial t}, \frac{\partial y}{\partial x}\right). \tag{4.45}$$

We also define the quantities

$$\mathcal{P}^t \equiv \frac{\partial \mathcal{L}}{\partial \dot{y}}, \quad \mathcal{P}^x \equiv \frac{\partial \mathcal{L}}{\partial y'}, \tag{4.46}$$

with $y' = \partial y/\partial x$. These are simply the derivatives of $\mathcal{L}$ with respect to its first and second arguments, respectively. Explicitly, they are

$$\mathcal{P}^t = \mu_0 \frac{\partial y}{\partial t}, \quad \mathcal{P}^x = -T_0 \frac{\partial y}{\partial x}. \tag{4.47}$$

When we vary the motion by δy, the variation of the action is given by

$$\delta S = \int_{t_i}^{t_f} dt \int_0^a dx \left[\frac{\partial \mathcal{L}}{\partial \dot{y}} \delta \dot{y} + \frac{\partial \mathcal{L}}{\partial y'} \delta y'\right] = \int_{t_i}^{t_f} dt \int_0^a dx \left[\mathcal{P}^t \delta \dot{y} + \mathcal{P}^x \delta y'\right]. \tag{4.48}$$

Using the standard manipulations we find

$$\delta S = \int_0^a \left[\mathcal{P}^t \delta y\right]_{t=t_i}^{t=t_f} dx + \int_{t_i}^{t_f} \left[\mathcal{P}^x \delta y\right]_{x=0}^{x=a} dt$$
$$- \int_{t_i}^{t_f} dt \int_0^a dx \left(\frac{\partial \mathcal{P}^t}{\partial t} + \frac{\partial \mathcal{P}^x}{\partial x}\right) \delta y. \tag{4.49}$$

Quick calculation 4.2 Derive equation (4.49).

Quick calculation 4.3 Match in detail equations (4.49) and (4.39).

The variation in (4.49) gives the equation of motion

$$\frac{\partial \mathcal{P}^t}{\partial t} + \frac{\partial \mathcal{P}^x}{\partial x} = 0. \tag{4.50}$$

Using (4.47) we readily see that this is the wave equation (4.6).

Note that $\mathcal{P}^t$, as given in (4.47), coincides with the momentum density in equation (4.43). This is not an accident. In Lagrangian mechanics, the derivative of the Lagrangian with respect to a velocity is the conjugate momentum. For the string, $\dot{y}$ plays the role of a velocity, so $\mathcal{P}^t$, the derivative of the Lagrangian density with respect to $\dot{y}$, is a momentum density.

In addition, note that for string endpoints that are free to move, the vanishing of δS requires $\mathcal{P}^x = 0$. As we can see from (4.47), this is a Neumann boundary condition. Furthermore, $\mathcal{P}^t$ vanishes at the string endpoints for a Dirichlet boundary condition (4.42). A more detailed analysis of these facts will be given in Chapter 8, where $\mathcal{P}^t$ and $\mathcal{P}^x$ will be shown to have an interesting two-dimensional interpretation.

Problems

Problem 4.1 Consistency of small transverse oscillations.

Reconsider the analysis of transverse oscillations in Section 4.1. Calculate the horizontal force dF_h on the little piece of string shown in Figure 4.1. Show that for small oscillations this force is much smaller than the vertical force dF_v responsible for the transverse oscillations.

Problem 4.2 Longitudinal waves on strings.

Consider a string with uniform mass density μ_0 stretched between $x = 0$ and $x = a$. Let the equilibrium tension be T_0. Longitudinal waves are possible if the tension of the string varies as it stretches or compresses. For a piece of this string with equilibrium length L, a small change ΔL of its length is accompanied by a small change ΔT of the tension where

$$\frac{1}{\tau_0} \equiv \frac{1}{L} \frac{\Delta L}{\Delta T}.$$

Here τ_0 is a tension coefficient with units of tension. Find the equation governing the small longitudinal oscillations of this string. Give the velocity of the waves.

Problem 4.3 A configuration with two joined strings.

A string with tension T_0 is stretched from $x = 0$ to $x = 2a$. The part of the string $x \in (0, a)$ has constant mass density μ_1, and the part of the string $x \in (a, 2a)$ has constant mass density μ_2. Consider the differential equation (4.20) that determines the normal oscillations.

(a) What boundary conditions should be imposed on $y(x)$ and $\frac{dy}{dx}(x)$ at $x = a$?
(b) Write the conditions that determine the possible frequencies of oscillation.
(c) Calculate the lowest frequency of oscillation of this string when $\mu_1 = \mu_0$ and $\mu_2 = 2\mu_0$.

Problem 4.4 Evolving an initial open string configuration.

A string with tension T_0, mass density μ_0, and wave velocity $v_0 = \sqrt{T_0/\mu_0}$, is stretched from $(x, y) = (0, 0)$ to $(x, y) = (a, 0)$. The string endpoints are fixed, and the string can vibrate in the y direction.

(a) Write $y(t, x)$ as in (4.11), and prove that the above Dirichlet boundary conditions imply

$$h_+(u) = -h_-(-u) \quad \text{and} \quad h_+(u) = h_+(u + 2a). \tag{1}$$

Here $u \in (-\infty, \infty)$ is a dummy variable that stands for the argument of the functions $h_\pm$.

Now consider an initial value problem for this string. At $t = 0$ the transverse displacement is identically zero, and the velocity is

$$\frac{\partial y}{\partial t}(0, x) = v_0 \frac{x}{a}\left(1 - \frac{x}{a}\right), \quad x \in (0, a). \tag{2}$$

(b) Calculate $h_+(u)$ for $u \in (-a, a)$. Does this define $h_+(u)$ for all u?
(c) Calculate $y(t, x)$ for x and $v_0 t$ in the domain D defined by the two conditions

$$D = \{(x, v_0 t) | 0 \leq x \pm v_0 t < a\}.$$

Exhibit the domain D in a plane with axes x and $v_0 t$.
(d) At $t = 0$ the midpoint $x = a/2$ has the largest velocity of all points in the string. Show that the velocity of the midpoint reaches the value of zero at time $t_0 = a/(2v_0)$ and that $y(t_0, a/2) = a/12$. This is the maximum vertical displacement of the string.

Problem 4.5 Closed string motion.

We can describe a nonrelativistic closed string fairly accurately by having the string wrapped around a cylinder of large circumference $2\pi R$ on which it is kept taut by the string tension T_0. We assume that the string can move on the surface of the cylinder without experiencing any friction. Let x be a coordinate along the circumference of the cylinder: $x \sim x + 2\pi R$ and let y be a coordinate perpendicular to x, thus running parallel to the axis of the cylinder. As expected, the general solution for transverse motion is given by

$$y(x, t) = h_+(x - v_0 t) + h_-(x + v_0 t),$$

where $h_+(u)$ and $h_-(v)$ are arbitrary functions of single variables u and v with $-\infty < u, v < \infty$. The string has mass per unit length μ_0, and $v_0 = \sqrt{T_0/\mu_0}$.

(a) State the periodicity condition that must be satisfied by $y(x, t)$ on account of the identification that applies to the x coordinate. Show that the *derivatives* $h'_+(u)$ and $h'_-(v)$ are, respectively, periodic functions of u and v.
(b) Show that one can write

$$h_+(u) = \alpha u + f(u), \quad h_-(v) = \beta v + g(v),$$

where f and g are periodic functions and α and β are constants. Give the relation between α and β that follows from (a).

(c) Calculate the total momentum carried by the string in the y direction. Is it conserved?

Problem 4.6 Stationary action: minima and saddles.

A particle perfoming harmonic motion along the x axis can be used to show that classical solutions are not always minima of the action functional. The action for this particle is

$$S[x] = \int_0^{t_f} L\, dt = \int_0^{t_f} dt\, \frac{1}{2}m\left(\dot{x}^2 - \bar{\omega}^2 x^2\right),$$

where m is the mass of the particle, $\bar{\omega}$ is the frequency of oscillation, and the motion happens for $t \in [0, t_f]$. Consider a classical solution $\bar{x}(t)$ and a variation $\delta x(t)$ that vanishes at $t = 0$ and $t = t_f$.

(a) Show that the variation of the action is *exactly* given by

$$\Delta S[\delta x] \equiv S[\bar{x} + \delta x] - S[\bar{x}] = \frac{1}{2}m \int_0^{t_f} dt\left(\left(\frac{d\delta x}{dt}\right)^2 - \bar{\omega}^2 \delta x^2\right).$$

It is noteworthy that ΔS only depends on δx; $\bar{x}$ drops out from the answer.

(b) A complete set of variations that vanish at $t = 0$ and $t = t_f$ takes the form

$$\delta_n x = \sin \omega_n t, \quad \text{with} \quad \omega_n = \frac{\pi n}{t_f} \quad \text{and} \quad n = 1, 2, \ldots, \infty.$$

The general variation δx that vanishes at $t = 0$ and $t = t_f$ is a linear superposition of variations $\delta_n x$ with arbitrary coefficients b_n. Calculate $\Delta S[\delta_n x]$ (your answer should vanish for $\omega_n = \omega$). Prove that

$$\Delta S\left[\sum_{n=1}^{\infty} b_n \delta_n x\right] = \sum_{n=1}^{\infty} \Delta S[b_n \delta_n x].$$

(c) Show that for $t_f < \frac{\pi}{\omega}$ one gets $\Delta S[\delta_n x] > 0$ for all $n \geq 1$. Explain why this guarantees that the classical solution is a minimum of the action. Show that for $\frac{\pi}{\omega} < t_f < \frac{2\pi}{\omega}$ all variations $\delta_n x$ lead to $\Delta S > 0$, except for $\delta_1 x$, which leads to $\Delta S < 0$. In this case the classical solution is a saddle point: there are variations that increase the action and variations that decrease the action. As t_f increases, the number of variations $\delta_n x$ that decrease the action increases.

Problem 4.7 Variational problem for strings.

Consider a string stretched from $x = 0$ to $x = a$, with a tension T_0 and a position-dependent mass density $\mu(x)$. The string is fixed at the endpoints and can vibrate in the y direction. Equation (4.20) determines the oscillation frequencies ω_i and associated profiles $\psi_i(x)$ for this string.

(a) Set up a variational procedure that gives an upper bound on the lowest frequency of oscillation ω_0. (This can be done roughly as in quantum mechanics, where the ground state energy E_0 of a system with Hamiltonian H satisfies $E_0 \leq (\psi, H\psi)/(\psi, \psi)$.) As a useful first step consider the inner product

$$(\psi_i, \psi_j) \equiv \int_0^a \mu(x)\psi_i(x)\psi_j(x)dx$$

and show that it vanishes when $\omega_i \neq \omega_j$. Explain why your variational procedure works.

(b) Consider the case $\mu(x) = \mu_0 \frac{x}{a}$. Use your variational principle to find a simple bound on the lowest oscillation frequency. Compare with the answer $\omega_0^2 \simeq (18.956)\frac{T_0}{\mu_0 a^2}$ obtained by a direct numerical solution of the eigenvalue problem.

Problem 4.8 Deriving Euler–Lagrange equations.[†]

(a) Consider an action for a dynamical variable $q(t)$:

$$S = \int dt\, L(q(t), \dot{q}(t); t). \tag{1}$$

Calculate the variation δS of the action under a variation $\delta q(t)$ of the coordinate. Use the condition $\delta S = 0$ to find the equation of motion for the coordinate $q(t)$ (the Euler–Lagrange equation).

(b) Consider an action for a dynamical field variable $\phi(t, \vec{x})$. As indicated, the field is a function of space and time, and is briefly written as the spacetime function $\phi(x)$. The action is obtained by integrating the Lagrangian density $\mathcal{L}$ over spacetime. The Lagrangian density is a function of the field and the spacetime derivatives of the field:

$$S = \int d^D x\, \mathcal{L}(\phi(x), \partial_\mu \phi(x)). \tag{2}$$

Here $d^D x = dt\, dx^1 \ldots dx^d$, and $\partial_\mu \phi = \partial \phi/\partial x^\mu$. Calculate the variation δS of the action under a variation $\delta \phi(x)$ of the field. Use the condition $\delta S = 0$ to find the equation of motion for the field $\phi(x)$ (the Euler–Lagrange equation).

5 The relativistic point particle

To formulate the dynamics of a system we can write either the equations of motion or, alternatively, an action. In the case of the relativistic point particle it is rather easy to write the equations of motion. But the action is so physical and geometrical that it is worth pursuing in its own right. More importantly, while it is difficult to guess the equations of motion for the relativistic string, the action is a natural generalization of the relativistic particle action that we will study in this chapter. We conclude with a discussion of the charged relativistic particle.

5.1 Action for a relativistic point particle

In this section we learn how to formulate the relativistic theory that describes a free point particle of rest mass $m > 0$. A free particle is a particle that is not subject to any force. Our analysis begins with some preliminary remarks about units and nonrelativistic particles.

For any dynamical system, the action S is obtained by integrating the Lagrangian over time. Since the Lagrangian has units of energy, the action has units of energy times time:

$$[S] = M \frac{L^2}{T^2} T = \frac{ML^2}{T}. \tag{5.1}$$

It is worth noting that the action has the same units as $\hbar$. Indeed, a form of the quantum mechanical uncertainty principle states that the product of energy and time uncertainties is of order $\hbar$.

The action S_{nr} for a free *nonrelativistic* particle is given by the time integral of the kinetic energy:

$$S_{nr} = \int L_{nr} \, dt = \int \frac{1}{2} m v^2(t) \, dt, \quad v^2 \equiv \vec{v} \cdot \vec{v}, \quad \vec{v} = \frac{d\vec{x}}{dt}, \quad v = |\vec{v}|. \tag{5.2}$$

The equation of motion which follows by Hamilton's principle is

$$\frac{d\vec{v}}{dt} = 0. \tag{5.3}$$

The free particle moves with constant velocity. Since even a free *relativistic* particle must move with constant velocity, how do we know that the action S_{nr} is not correct in relativity? Perhaps the simplest answer is that this action allows the particle to move with *any* constant velocity, even one that exceeds the velocity of light. The velocity of light does not even appear in this action. S_{nr} cannot be the action for a relativistic point particle.

We now construct a relativistic action S for the free point particle. We will do this by making an educated guess and then showing that it works properly. Since we are interested in relativistic physics it is convenient to represent the motion of the particle in *spacetime*. The path traced out by the particle in spacetime is called the *world-line* of the particle. Even a static particle traces a line in spacetime since time always flows.

A physically consistent action must yield Lorentz invariant equations of motion. Let us elaborate on this point. Suppose that a particular Lorentz observer tells you that a particle appears to be moving in accordance to its equations of motion, plainly, that the particle is performing physical motion. Then, you should expect that any other Lorentz observer will tell you that the particle is doing physical motion. It would be inconsistent for one observer to state that a certain motion is allowed and for another observer to state that the same motion is forbidden. If the equations of motion hold in a fixed Lorentz frame, they must hold in all Lorentz frames. This is what Lorentz invariance of the equations of motion means.

We are going to write an action, and we are going to take our time to find the equations of motion. Is there any way to impose a constraint on the action that will result in the Lorentz invariance of the equations of motion? Yes, there is. We require the action to be a Lorentz scalar: for *any* particle world-line, all Lorentz observers must compute the same value for the action. Since the action has no spacetime indices, this is a reasonable requirement. If the action is a Lorentz scalar, the equations of motion will be Lorentz invariant. The reason is simple and neat. Suppose one Lorentz observer states that, for a given world-line, the action is stationary against all variations of the world-line. Since all Lorentz observers agree on the value of the action for any world-line, they will all agree that the action is stationary about the world-line in question. By Hamilton's principle, the world-line that makes the action stationary satisfies the equations of motion, and therefore all Lorentz observers will agree that the equations of motion are satisfied for the world-line in question.

Lorentz invariance imposes strong constraints on the possible forms of the action. In fact, there are valid grounds to worry that Lorentz invariance is too strong a constraint on the action. The nonrelativistic action in (5.2), for example, is *not* invariant under a Galilean boost $\vec{v} \to \vec{v} + \vec{v}_0$ with constant $\vec{v}_0$. Such a boost is a symmetry of the theory, since the equation of motion (5.3) *is* invariant. Similarly, it could happen that the equations of motion for the relativistic point particle are Lorentz invariant but that the action is not. Fortunately, this complication does not occur in this case; we will find a satisfactory fully Lorentz invariant action.

Quick calculation 5.1 Calculate explicitly the variation of the action S_{nr} under a boost.

We know that the action is a functional – it takes as input a set of functions that describe a world-line and it outputs a number S. Imagine a particle whose spacetime trajectory starts at the origin and ends at $(ct_f, \vec{x}_f)$. There are many possible world-lines between the starting and ending points, as shown in Figure 5.1 (which uses one spatial dimension

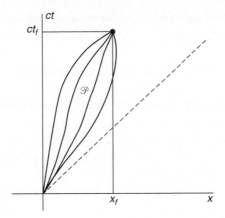

Fig. 5.1 A spacetime diagram with a series of world-lines connecting the origin to the spacetime point (ct_f, x_f).

for ease of representation). We would like that, for any world-line, all Lorentz observers compute the same value for the action. Let $\mathcal{P}$ denote one world-line. What quantity related to $\mathcal{P}$ do all Lorentz observers agree on? The elapsed proper time! All Lorentz observers agree on the amount of time that elapses on a clock carried by the moving particle. So let us take the action of the world-line $\mathcal{P}$ to be proportional to the proper time associated with it.

To formulate this idea quantitatively, we recall that

$$-ds^2 = -c^2 dt^2 + (dx^1)^2 + (dx^2)^2 + (dx^3)^2, \tag{5.4}$$

and that the infinitesimal proper time is equal to ds/c (recall that $ds^2 = (ds)^2$ since intervals are timelike). The integral of (ds/c) over the path $\mathcal{P}$ gives the proper time elapsed on $\mathcal{P}$. Since proper time has units of time, to get the units of action we need an additional multiplicative factor with units of energy or units of mass times velocity-squared. This factor should be Lorentz invariant, to preserve the Lorentz invariance of our partial guess (ds/c). For the mass we can use m, the rest mass of the particle, and for the velocity we can use c, the fundamental velocity in relativity. We cannot use the particle velocity because it is not a Lorentz invariant. The factor is then mc^2, which is, in fact, the rest energy of the particle. Therefore, our guess for the action is the integral of $mc^2 (ds/c) = mc\, ds$. Of course, there is still the possibility that a dimensionless numerical factor is missing. It turns out that there should be a minus sign, but the unit coefficient is correct. We therefore claim that the correct action is

$$S = -mc \int_{\mathcal{P}} ds. \tag{5.5}$$

The action is equal to minus the rest energy times the proper time. This action is so simple looking that it may be baffling. It probably looks nothing like the actions you have seen before. We can make its content more familiar by choosing a particular Lorentz observer and expressing the action as the integral of a Lagrangian over time. With the help of (5.4) we relate ds to dt by

$$ds = c\,dt\sqrt{1 - \frac{v^2}{c^2}}.$$ (5.6)

This allows us to write the action in (5.5) as an integral over time:

$$S = -mc^2\int_{t_i}^{t_f} dt\sqrt{1 - \frac{v^2}{c^2}},$$ (5.7)

where t_i and t_f are the values of time at the initial and final points of the world-line $\mathcal{P}$, respectively. From this version of the action, we see that the relativistic Lagrangian for the point particle is

$$L = -mc^2\sqrt{1 - \frac{v^2}{c^2}}.$$ (5.8)

The Lagrangian is equal to minus the rest energy times a relativistic factor. This Lagrangian makes no sense when $v > c$ since it ceases to be real. The constraint of maximal velocity is therefore implemented. This could have been anticipated: proper time is only defined for motion where the velocity does not exceed the velocity of light. The paths shown in Figure 5.1 all represent motion where the velocity of the particle never exceeds the velocity of light. Only for such paths is the action defined. At any point in any of those paths, the tangent vector to the path is a timelike vector.

To show that this Lagrangian gives the familiar physics in the limit of small velocities, we expand the square root assuming $v \ll c$. Keeping just the first term in the expansion gives

$$L \simeq -mc^2\left(1 - \frac{1}{2}\frac{v^2}{c^2}\right) = -mc^2 + \frac{1}{2}mv^2.$$ (5.9)

Constant terms in a Lagrangian do not affect the equations of motion, so the term $(-mc^2)$ can be ignored for this purpose. The leading significant term coincides with the nonrelativistic Lagrangian in (5.2), showing that the familiar nonrelativistic physics emerges. This also confirms that we normalized the relativistic Lagrangian correctly.

The canonical momentum is the derivative of the Lagrangian with respect to the velocity. Using (5.8) we find

$$\vec{p} = \frac{\partial L}{\partial \vec{v}} = -mc^2\left(-\frac{\vec{v}}{c^2}\right)\frac{1}{\sqrt{1 - v^2/c^2}} = \frac{m\vec{v}}{\sqrt{1 - v^2/c^2}}.$$ (5.10)

This is just the relativistic momentum of the point particle. What about the Hamiltonian? It is given by

$$H = \vec{p} \cdot \vec{v} - L = \frac{mv^2}{\sqrt{1 - v^2/c^2}} + mc^2\sqrt{1 - v^2/c^2} = \frac{mc^2}{\sqrt{1 - v^2/c^2}}, \qquad (5.11)$$

where the result was left as a function of the velocity of the particle, rather than as a function of its momentum. As expected, the answer coincides with the relativistic energy (2.68) of the point particle.

We have therefore recovered the familiar physics of a relativistic particle from the rather remarkable action (5.5). This action is very elegant: it is briefly written in terms of the geometrical quantity ds, it has a clear physical interpretation as total proper time, and it manifestly guarantees the Lorentz invariance of the physics it describes.

5.2 Reparameterization invariance

In this section we explore an important property of the point particle action (5.5). This property is called reparameterization invariance. To evaluate the integral in the action, an observer may find it useful to parameterize the particle world-line. Reparameterization invariance of the action means that the value of the action is independent of the parameterization chosen to calculate it. This should be so, since the action (5.5) is in fact defined independently of any parameterization: the integration can be done by breaking $\mathcal{P}$ into small pieces and adding the values of $mc\,ds$ for each piece. No parameterization is needed to do this. In practice, however, world-lines are described as parameterized lines, and the parameterization *is* used to compute the action.

We parameterize the world-line $\mathcal{P}$ of a point particle using a parameter τ (Figure 5.2). This parameter must be strictly increasing as the world-line goes from the initial point x_i^{μ} to the final point x_f^{μ}, but is otherwise arbitrary. As τ ranges in the interval $[\tau_i, \tau_f]$ it describes the motion of the particle. To have a parameterization of the world-line means that we have expressions for the coordinates x^{μ} as functions of τ:

$$x^{\mu} = x^{\mu}(\tau). \qquad (5.12)$$

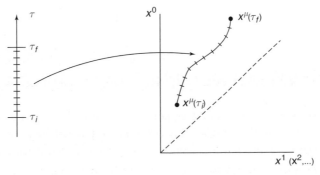

Fig. 5.2   A world-line fully parameterized by τ. All spacetime coordinates x^{μ} are functions of τ.

We also require

$$x_i^\mu = x^\mu(\tau_i), \quad x_f^\mu = x^\mu(\tau_f). \tag{5.13}$$

Note that even the time coordinate x^0 is parameterized. Normally, we use time as a parameter and describe position as a function of time. This is what we did in Section 5.1. But if we want to treat space and time coordinates on the same footing, we must parameterize both in terms of an additional parameter τ.

We now reexpress the integrand ds using the parameterized world-line. To this end, we use $ds^2 = -\eta_{\mu\nu}dx^\mu dx^\nu$ to write

$$ds^2 = -\eta_{\mu\nu}\frac{dx^\mu}{d\tau}\frac{dx^\nu}{d\tau}(d\tau)^2. \tag{5.14}$$

For any motion where the velocity does not exceed the velocity of light $ds^2 = (ds)^2$, and therefore the action (5.5) takes the form

$$S = -mc\int_{\tau_i}^{\tau_f}\sqrt{-\eta_{\mu\nu}\frac{dx^\mu}{d\tau}\frac{dx^\nu}{d\tau}}\;d\tau. \tag{5.15}$$

This is the explicit form of the action when the path has been parameterized by τ.

We have already seen that the value of the action is the same for all Lorentz observers. We have now fixed an observer, who has calculated the action using some parameter τ. Does the value of the action depend on the choice of parameter? It does not. The observer can reparameterize the world-line, and the value of the action will be the same. Thus S is *reparameterization invariant*. To see this, suppose we change the parameter from τ to τ'. Then, by the chain rule,

$$\frac{dx^\mu}{d\tau} = \frac{dx^\mu}{d\tau'}\frac{d\tau'}{d\tau}. \tag{5.16}$$

Substituting back into (5.15), we get

$$S = -mc\int_{\tau_i}^{\tau_f}\sqrt{-\eta_{\mu\nu}\frac{dx^\mu}{d\tau'}\frac{dx^\nu}{d\tau'}\frac{d\tau'}{d\tau}}d\tau = -mc\int_{\tau_i'}^{\tau_f'}\sqrt{-\eta_{\mu\nu}\frac{dx^\mu}{d\tau'}\frac{dx^\nu}{d\tau'}}d\tau', \tag{5.17}$$

which has the same form as (5.15), thus establishing the reparameterization invariance. Because the verification of this property is quite simple, we say that the action (5.15) is *manifestly* reparameterization invariant.

5.3 Equations of motion

We now move on to the equations of motion. For this we must calculate the variation δS of the action (5.5) when the world-line of the particle is varied by a small amount $\delta x^\mu(\tau)$. Here τ is an arbitrary parameter along the path. The variation is simply given by

$$\delta S = -mc\int \delta(ds). \tag{5.18}$$

The variation of ds can be found from the simpler variation of $ds^2 = (ds)^2$. Varying both sides of (5.14) we find

$$2\,ds\,\delta(ds) = -2\eta_{\mu\nu}\,\delta\!\left(\frac{dx^\mu}{d\tau}\right)\frac{dx^\nu}{d\tau}(d\tau)^2. \tag{5.19}$$

The factor of two on the right-hand side arises because, by symmetry, the variations of $\frac{dx^\mu}{d\tau}$ and $\frac{dx^\nu}{d\tau}$ give the same result. Since the variation of a velocity is equal to the time derivative of the variation of the coordinate,

$$\delta\!\left(\frac{dx^\mu}{d\tau}\right) = \frac{d(\delta x^\mu)}{d\tau}. \tag{5.20}$$

Using this result and simplifying (5.19) a bit, we get

$$\delta(ds) = -\eta_{\mu\nu}\frac{d(\delta x^\mu)}{d\tau}\frac{dx^\nu}{ds}\,d\tau = -\frac{d(\delta x^\mu)}{d\tau}\frac{dx_\mu}{ds}\,d\tau, \tag{5.21}$$

where $\eta_{\mu\nu}$ was used to lower the index of dx^ν. We can now go ahead and vary the action using (5.18):

$$\delta S = mc \int_{\tau_i}^{\tau_f} \frac{d(\delta x^\mu)}{d\tau}\frac{dx_\mu}{ds}\,d\tau. \tag{5.22}$$

Here we introduced explicit limits to the integration: τ_i and τ_f denote the values of the parameter at the initial and final points of the world-line, respectively. We recognize that

$$mc\frac{dx_\mu}{ds} = mu_\mu = p_\mu, \tag{5.23}$$

and, as a result, the variation of the action takes the form

$$\delta S = \int_{\tau_i}^{\tau_f} \frac{d(\delta x^\mu)}{d\tau}\,p_\mu\,d\tau. \tag{5.24}$$

To get an equation of motion we need to have δx^μ multiplying an object under the integral – the equation of motion is then simply the vanishing of that object. Since there are still derivatives acting on δx^μ, we rewrite the integrand as a total derivative plus additional terms where δx^μ appears multiplicatively:

$$\delta S = \int_{\tau_i}^{\tau_f} d\tau\,\frac{d}{d\tau}\Big(\delta x^\mu\,p_\mu\Big) - \int_{\tau_i}^{\tau_f} d\tau\,\delta x^\mu(\tau)\,\frac{dp_\mu}{d\tau}. \tag{5.25}$$

The first integral gives $\delta x^\mu\,p_\mu$ evaluated at the boundaries of the world-line. This term vanishes because we fix the coordinates on the boundaries. Since the second term must vanish for arbitrary $\delta x^\mu(\tau)$, we obtain the equation of motion

$$\frac{dp_\mu}{d\tau} = 0. \tag{5.26}$$

It is clear that $dp^\mu/d\tau$ also vanishes. The equation of motion states that the momentum p_μ (or p^μ) of the point particle is constant along its world-line. This is a parameterization-independent statement. It implies, of course, that the momentum is constant in time. If a function is constant over a line, its derivative with respect to any parameter used to describe the line will vanish. Indeed, the parameter τ in (5.26) is arbitrary. We obtained this equation by varying the relativistic action for the point particle using fully relativistic notation.

Quick calculation 5.2 Show that equation (5.26) implies that

$$\frac{dp_\mu}{d\tau'} = 0 \tag{5.27}$$

holds for an arbitrary parameter $\tau'(\tau)$. What should be true about $d\tau'/d\tau$ for τ' to be a good parameter when τ is one?

If we parameterize the world-line with the proper time s, equation (5.26) gives

$$\frac{dp^\mu}{ds} = 0. \tag{5.28}$$

Using (5.23) to write the momentum as a derivative of the position with respect to proper time, we find

$$\frac{d^2 x^\mu}{ds^2} = 0. \tag{5.29}$$

This is an equivalent version of the equation of motion. The constancy of dx^μ/ds means that on a path marked by equal intervals of proper time, the change in x^μ between any successive pair of marks is the same. Equation (5.29) does *not* hold when s is replaced by an arbitrary parameter τ. This is reasonable: an arbitrary parameter means arbitrarily spaced marks, so the change in x^μ between any successive pair of new marks need not be the same. It is actually possible to write a slightly more complicated version of (5.29) that uses an arbitrary parameter and is manifestly reparameterization invariant (Problem 5.2).

Our goal in this section has been achieved: we have shown how to derive the physically expected equation of motion (5.29) (or (5.26)), starting from the Lorentz invariant action (5.5). As we explained earlier, the resulting equation of motion is guaranteed to be Lorentz invariant. Let us check this explicitly.

Under a Lorentz transformation, the coordinates x^μ transform as indicated in equation (2.38): $x'^\mu = L^\mu{}_\nu x^\nu$, where the constants $L^\mu{}_\nu$ can be viewed as the entries of an invertible matrix L. Since ds is the same in all Lorentz frames, the equation of motion in primed coordinates is (5.29), with x^μ replaced by x'^μ:

$$0 = \frac{d^2 x'^\mu}{ds^2} = \frac{d^2}{ds^2}(L^\mu{}_\nu x^\nu) = L^\mu{}_\nu \frac{d^2 x^\nu}{ds^2}. \tag{5.30}$$

Since the matrix L is invertible, the above equation implies equation (5.29). Namely, if the equation of motion holds in the primed coordinates, it holds in the unprimed coordinates as well. This is the Lorentz invariance of the equations of motion.

5.4 Relativistic particle with electric charge

The point particle we have considered so far is free and it moves with constant four-velocity or four-momentum. If a point particle is electrically charged and there are nontrivial electromagnetic fields, the particle will experience forces and its four-momentum will not be constant. You know, in fact, how the momentum of such a particle varies in time. Its time derivative is governed by the Lorentz force equation (3.5), which was written in relativistic notation in Problem 3.1:

$$\frac{dp_\mu}{ds} = \frac{q}{c} F_{\mu\nu} \frac{dx^\nu}{ds}. \tag{5.31}$$

This is a relatively intricate equation which involves the field strength and the four-velocity of the particle. Since ds appears on both sides of the equation, the equation in fact holds for a general parameter τ:

$$\frac{dp_\mu}{d\tau} = \frac{q}{c} F_{\mu\nu} \frac{dx^\nu}{d\tau}. \tag{5.32}$$

In the spirit of our previous analysis we try to write an action that gives this equation upon variation. The action turns out to be remarkably simple.

Since the Maxwell field couples to the point particle along its world-line $\mathcal{P}$, we should add to the action (5.5) an integral over $\mathcal{P}$ representing the interaction of the particle with the electromagnetic field. The integral must be Lorentz invariant, and the form of (5.32) suggests that it involves the four-velocity of the particle. Since the four-velocity has one spacetime index, to obtain a Lorentz scalar we must multiply it against another object with one index. The natural candidate is the gauge potential A_μ. We claim that the interaction term in the action is

$$\frac{q}{c} \int_{\mathcal{P}} d\tau \, A_\mu(x(\tau)) \frac{dx^\mu}{d\tau}(\tau). \tag{5.33}$$

Here q is the electric charge, and the integral is over the world-line $\mathcal{P}$, parameterized with the arbitrary parameter τ. At each τ, the vector $(dx^\mu/d\tau)$ is dot multiplied against the gauge potential A_μ, evaluated at the position $x(\tau)$ of the particle. The integrand can be written more briefly as $A_\mu dx^\mu$, by cancelling the factors of $d\tau$. In this form, the interaction term is manifestly independent of parameterization. The world-line of the particle is a one-dimensional space, and the natural field that can couple to a particle in a Lorentz invariant way is a field with one index. This will have an interesting generalization when we consider the motion of strings. Since strings are one-dimensional, they trace out two-dimensional world-sheets in spacetime. We will see that they couple naturally to fields with two Lorentz indices!

The full action for the electrically charged point particle is obtained by adding the term in (5.33) to (5.5):

$$S = -mc \int_{\mathcal{P}} ds + \frac{q}{c} \int_{\mathcal{P}} A_\mu(x) dx^\mu. \qquad (5.34)$$

This Lorentz invariant action is simple and elegant. The equation of motion (5.32) arises by setting to zero the variation of S under a change δx^μ of the particle world-line. I do not want to take away from you the satisfaction of deriving this important result. I have therefore left to Problem 5.5 the task of varying the action (5.34) and deriving the equation of motion.

Problems

Problem 5.1 Point particle equation of motion and reparameterizations.

If the path of a point particle is parameterized by proper time, the equation of motion is (5.29). Consider now a new parameter $\tau = f(s)$. Find the most general function f for which (5.29) implies

$$\frac{d^2 x^\mu}{d\tau^2} = 0.$$

Problem 5.2 Particle equation of motion with arbitrary parameterization.

Vary the point particle action (5.15) to find a *manifestly* reparameterization invariant form of the free particle equation of motion.

Problem 5.3 Current of a charged point particle.

Consider a point particle with charge q whose motion in a $D = d + 1$-dimensional space-time is described by functions $x^\mu(\tau) = \{x^0(\tau), \vec{x}(\tau)\}$, where τ is a parameter. The moving particle generates an electromagnetic current $j^\mu = (c\rho, \vec{j})$.

(a) Use delta functions to write expressions for the current components $j^0(\vec{x}, t)$ and $j^i(\vec{x}, t)$.

(b) Show that your answers in (a) arise from the integral representation

$$j^\mu(t, \vec{x}) = qc \int d\tau \, \delta^D(x - x(\tau)) \frac{dx^\mu(\tau)}{d\tau}.$$

Here $\delta^D(x) \equiv \delta(x^0)\delta(x^1)\ldots\delta(x^d)$.

Problem 5.4 Hamiltonian for a nonrelativistic charged particle.[†]

The action for a nonrelativistic particle of mass m and charge q coupled to an electromagnetic field is obtained by replacing the first term in (5.34) by the nonrelativistic action for a free point particle:

$$S = \int \frac{1}{2} m v^2 \, dt + \frac{q}{c} \int A_\mu(x) \frac{dx^\mu}{dt} dt.$$

We have also chosen to use time to parameterize the second integral.

(a) Rewrite the action S in terms of the potentials $(\Phi, \vec{A})$ and the ordinary velocity $\vec{v}$. What is the Lagrangian?

(b) Calculate the canonical momentum $\vec{p}$ conjugate to the position of the particle and show that it is given by $\vec{p} = m\vec{v} + \frac{q}{c}\vec{A}$.

(c) Show that the Hamiltonian for the charged particle is

$$H = \frac{1}{2m}\left(\vec{p} - \frac{q}{c}\vec{A}\right)^2 + q\,\Phi.$$

Problem 5.5 Equations of motion for a charged point particle.

Consider the variation of the action (5.34) under a variation $\delta x^\mu(x)$ of the particle trajectory. The variation of the first term in the action was obtained in Section 5.3. Vary the second term (written more explicitly in (5.33)) and show that the equation of motion is (5.32). Begin your calculation by explaining why

$$\delta A_\mu(x(\tau)) = \frac{\partial A_\mu}{\partial x^\nu}(x(\tau))\,\delta x^\nu(\tau).$$

Problem 5.6 Electromagnetic field dynamics with a charged particle.[†]

The action for the dynamics of *both* a charged point particle and the electromagnetic field is given by

$$S' = -mc\int_\mathcal{P} ds + \frac{q}{c}\int_\mathcal{P} A_\mu(x)dx^\mu - \frac{1}{4c}\int d^D x\ F_{\mu\nu}F^{\mu\nu}.$$

Here $d^D x = dx^0 dx^1 \ldots dx^d$. Note that the action S' is a hybrid; the last term is an integral over spacetime, and the first two terms are integrals over the particle world-line. While included for completeness, the first term will play no role here. Obtain the equation of motion for the electromagnetic field in the presence of the charged particle by calculating the variation of S' under a variation δA_μ of the gauge potential. The answer should be equation (3.34), where the current is the one calculated in Problem 5.3. [Hint: to vary $A_\mu(x)$ in the world-line action it is useful to rewrite this term as a full spacetime integral with the help of delta functions.]

Problem 5.7 Point particle action in curved space.

In Section 3.6 we considered the invariant interval $ds^2 = -g_{\mu\nu}(x)dx^\mu dx^\nu$ in a curved space with metric $g_{\mu\nu}(x)$. The motion of a point particle of mass m on curved space is studied using the action

$$S = -mc\int ds.$$

Show that the equation of motion obtained by variation of the world-line is

$$\frac{d}{ds}\left[g_{\mu\rho}\frac{dx^\mu}{ds}\right] = \frac{1}{2}\frac{\partial g_{\mu\nu}}{\partial x^\rho}\frac{dx^\mu}{ds}\frac{dx^\nu}{ds}.$$

This is called the geodesic equation. When the metric is constant we recover the equation of motion of a free point particle.

6

Relativistic strings

We now begin our study of the classical relativistic string – a string that is, in many ways, much more elegant than the nonrelativistic one considered before. Inspired by the point particle case, we focus our attention on the surface traced out by the string in spacetime. We use the proper area of this surface as the action; this is the Nambu–Goto action. We study the reparameterization property of this action, identify the string tension, and find the equations of motion. For open strings, we focus on the motion of the endpoints and introduce the concept of D-branes. Finally, we see that the only physical motion is transverse to the string.

6.1 Area functional for spatial surfaces

The action for a relativistic string must be a functional of the string trajectory. Just as a particle traces out a line in spacetime, a string traces out a surface. The line traced out by the particle in spacetime is called the world-line. The two-dimensional surface traced out by a string in spacetime will be called the *world-sheet*. A closed string, for example, will trace out a tube, while an open string will trace out a strip. These two-dimensional world-sheets are shown in the spacetime diagram of Figure 6.1. The lines of constant x^0 in these surfaces are the strings. These are the objects an observer sees at the fixed time x^0. They are open curves for the surface describing the open string evolution (left), and they are closed curves for the surface describing the closed string evolution (right).

In Chapter 5 we learned that the point particle action is proportional to the proper time elapsed on the point particle world-line. The proper time, multiplied by c, is the Lorentz invariant "proper length" of the world-line. For strings we will define the Lorentz invariant "proper area" of a world-sheet. The relativistic string action will be proportional to this proper area, and is called the Nambu–Goto action.

Area functionals are useful in other applications: a soap film held between two rings, for example, automatically constructs the surface of minimal area which joins one ring to the other (Figure 6.2). The string world-sheet and the soap bubble between two rings are very different types of surfaces. At any given instant of time a Lorentz observer will see the full two-dimensional surface of the soap film, but he or she can only see one string from the two-dimensional world-sheet. Imagine that the soap film is static in some Lorentz frame. In this case, time is not relevant to the description of the film, and we think of the film as a *spatial surface*, namely, a surface that extends along two spatial dimensions. The surface

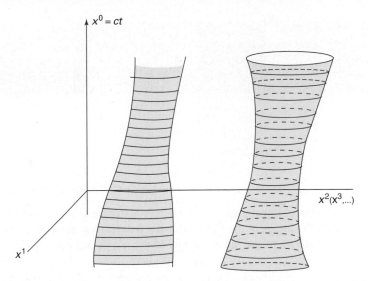

Fig. 6.1   The world-sheets traced out by an open string (left) and by a closed string (right).

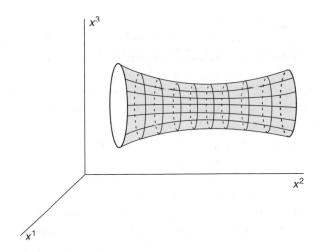

Fig. 6.2   A spatial surface stretching between two rings. If the surface were a soap film, it would be a minimal area surface.

exists in its entirety at any instant of time. We will first study these familiar surfaces, and then we will apply our experience to the case of surfaces in spacetime.

A line in space can be parameterized using only one parameter. A surface in space is two-dimensional, so it requires two parameters ξ^1 and ξ^2. Given a parameterized surface, we can draw on that surface the lines of constant ξ^1 and the lines of constant ξ^2. These lines cover the surface with a grid. We call *target space* the world where the two-dimensional surface lives. In the case of a soap bubble in three dimensions, the target space is the three-dimensional space x^1, x^2, and x^3. The parameterized surface is described by the collection of functions

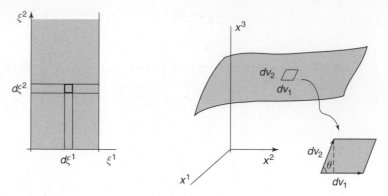

Fig. 6.3 Left: the parameter space, with a little rectangle selected. Right: the target space surface with the image of the little rectangle, a parallelogram whose sides are the vectors $d\vec{v}_1$ and $d\vec{v}_2$ (shown magnified at the end of the wiggly arrow).

$$\vec{x}(\xi^1, \xi^2) = \left(x^1(\xi^1, \xi^2),\ x^2(\xi^1, \xi^2),\ x^3(\xi^1, \xi^2) \right). \tag{6.1}$$

The parameter space is defined by the ranges of the parameters ξ^1 and ξ^2. It may be a square, for example, if we use parameters $\xi^1, \xi^2 \in [0, \pi]$. The physical surface is the image of the parameter space under the map $\vec{x}(\xi^1, \xi^2)$; it is a surface in target space. Alternatively, we can view the parameters ξ^1 and ξ^2 as *coordinates* on the physical surface, at least locally. The map inverse to $\vec{x}$ takes the surface to the parameter space. Locally this map is one-to-one and it assigns to each point on the surface two coordinates: the values of the parameters ξ^1 and ξ^2.

We want to calculate the area of a small element of the target space surface. Let us start by looking at an infinitesimal rectangle on the parameter space. Denote the sides of the rectangle by $d\xi^1$ and $d\xi^2$. We want to find dA, the area of the image of this little rectangle in the target space. As shown in Figure 6.3, this is the area of the actual piece of surface that corresponds to the infinitesimal rectangle on parameter space.

Of course, there is no reason why that infinitesimal area element in target space should be a rectangle. In general, it is a parallelogram. Let us call the sides of this parallelogram $d\vec{v}_1$ and $d\vec{v}_2$. They are the images under the map $\vec{x}$ of the vectors $(d\xi^1, 0)$ and $(0, d\xi^2)$, respectively. We can write them as

$$d\vec{v}_1 = \frac{\partial \vec{x}}{\partial \xi^1}\, d\xi^1, \quad d\vec{v}_2 = \frac{\partial \vec{x}}{\partial \xi^2}\, d\xi^2. \tag{6.2}$$

This makes sense: $\partial \vec{x}/\partial \xi^1$, for example, represents the rate of variation of the space coordinates with respect to ξ^1. Multiplying this rate by the length $d\xi^1$ of the horizontal side of the tiny parameter-space rectangle, gives us the vector $d\vec{v}_1$ that represents this side in the target space. Now let us calculate the area dA. Using the formula for the area of a parallelogram,

$$dA = |d\vec{v}_1||d\vec{v}_2||\sin\theta| = |d\vec{v}_1||d\vec{v}_2|\sqrt{1 - \cos^2\theta}$$

$$= \sqrt{|d\vec{v}_1|^2|d\vec{v}_2|^2 - |d\vec{v}_1|^2|d\vec{v}_2|^2\cos^2\theta}, \tag{6.3}$$

where θ is the angle between the vectors $d\vec{v}_1$ and $d\vec{v}_2$. In terms of spatial dot products, we have

$$dA = \sqrt{(d\vec{v}_1 \cdot d\vec{v}_1)(d\vec{v}_2 \cdot d\vec{v}_2) - (d\vec{v}_1 \cdot d\vec{v}_2)^2}. \tag{6.4}$$

Finally, using (6.2),

$$dA = d\xi^1 d\xi^2 \sqrt{\left(\frac{\partial\vec{x}}{\partial\xi^1} \cdot \frac{\partial\vec{x}}{\partial\xi^1}\right)\left(\frac{\partial\vec{x}}{\partial\xi^2} \cdot \frac{\partial\vec{x}}{\partial\xi^2}\right) - \left(\frac{\partial\vec{x}}{\partial\xi^1} \cdot \frac{\partial\vec{x}}{\partial\xi^2}\right)^2}. \tag{6.5}$$

This is the general expression for the area element of a parameterized spatial surface. The full area functional A is given by

$$A = \int d\xi^1 d\xi^2 \sqrt{\left(\frac{\partial\vec{x}}{\partial\xi^1} \cdot \frac{\partial\vec{x}}{\partial\xi^1}\right)\left(\frac{\partial\vec{x}}{\partial\xi^2} \cdot \frac{\partial\vec{x}}{\partial\xi^2}\right) - \left(\frac{\partial\vec{x}}{\partial\xi^1} \cdot \frac{\partial\vec{x}}{\partial\xi^2}\right)^2}. \tag{6.6}$$

The integral extends over the relevant ranges of the parameters ξ^1 and ξ^2. The solution of a minimal area problem for a spatial surface is the function $\vec{x}(\xi^1, \xi^2)$ that minimizes the functional A.

6.2 Reparameterization invariance of the area

As we have seen, the parameterization of a surface allows us to write the area element in an explicit form. The area of the surface, or even more, the area of any piece of the surface, should be independent of the parameterization chosen to calculate it. This is what we mean when we say that the area must be reparameterization invariant.

Because we will soon equate the relativistic string action to some notion of proper area, it, too, will be reparameterization invariant. This means that we will be free to choose the most useful parameterization without changing the underlying physics. A good choice of parameterization will enable us to solve the equations of motion of the relativistic string in an elegant way.

Reparameterization invariance is thus an important concept so it should be understood thoroughly. To this end we will try to make it manifest in our formulae. The aim of the following analysis is to show how this can be done.

Let us begin by asking: is the area functional A in (6.6) reparameterization invariant? We would certainly hope it is. In fact, at first glance it appears to be manifestly reparameterization invariant. After all, if one reparameterizes the surface with $\tilde{\xi}^1(\xi^1)$ and $\tilde{\xi}^2(\xi^2)$, then all of the derivatives introduced by the chain rule cancel appropriately.

Quick calculation 6.1 Verify the above statement. That is, show that (6.6), written fully with tilde parameters $(\tilde{\xi}^1, \tilde{\xi}^2)$, equals (6.6) when $\tilde{\xi}^1 = \tilde{\xi}^1(\xi^1)$ and $\tilde{\xi}^2 = \tilde{\xi}^2(\xi^2)$.

The above reparameterization, however, is not completely general for it fails to mix the ξ^1 and ξ^2 coordinates. Suppose, instead, that we make a reparameterization $\tilde{\xi}^1(\xi^1, \xi^2)$ and $\tilde{\xi}^2(\xi^1, \xi^2)$. This time we can verify, using a somewhat laborious computation, that (6.6) is invariant under such a reparameterization. But the invariance is no longer intuitively clear. To make the reparameterization invariance of (6.6) manifest we will have to rewrite the area functional in a different way.

We begin by observing how the measure of integration transforms. The change-of-variable theorem from calculus tells us that

$$d\xi^1 d\xi^2 = \left| \det\left(\frac{\partial \xi^i}{\partial \tilde{\xi}^j}\right) \right| d\tilde{\xi}^1 d\tilde{\xi}^2 = |\det M| \, d\tilde{\xi}^1 d\tilde{\xi}^2 \,, \tag{6.7}$$

where $M = [M_{ij}]$ is the matrix defined by $M_{ij} = \partial \xi^i / \partial \tilde{\xi}^j$. Similarly,

$$d\tilde{\xi}^1 d\tilde{\xi}^2 = \left| \det\left(\frac{\partial \tilde{\xi}^i}{\partial \xi^j}\right) \right| d\xi^1 d\xi^2 = |\det \tilde{M}| \, d\xi^1 d\xi^2 \,, \tag{6.8}$$

where $\tilde{M} = [\tilde{M}_{ij}]$ is the matrix defined by $\tilde{M}_{ij} = \partial \tilde{\xi}^i / \partial \xi^j$. Combining equations (6.7) and (6.8), we see that

$$|\det M| |\det \tilde{M}| = 1 \,. \tag{6.9}$$

Let us now consider a target space surface S described by the mapping functions $\vec{x}(\xi^1, \xi^2)$. Given a vector $d\vec{x}$ tangent to the surface, let ds denote its length. Then we can write

$$ds^2 \equiv (ds)^2 = d\vec{x} \cdot d\vec{x}. \tag{6.10}$$

For surfaces in space, as we are considering now, it is *not* customary to include a minus sign in front of ds^2 (compare with (2.21)). The vector $d\vec{x}$ can be expressed in terms of partial derivatives and the differentials $d\xi^1$, $d\xi^2$:

$$d\vec{x} = \frac{\partial \vec{x}}{\partial \xi^1} d\xi^1 + \frac{\partial \vec{x}}{\partial \xi^2} d\xi^2 = \frac{\partial \vec{x}}{\partial \xi^i} \, d\xi^i \,. \tag{6.11}$$

The repeated index i is summed over its possible values 1 and 2. Back in (6.10),

$$ds^2 = \left(\frac{\partial \vec{x}}{\partial \xi^i} d\xi^i\right) \cdot \left(\frac{\partial \vec{x}}{\partial \xi^j} d\xi^j\right) = \frac{\partial \vec{x}}{\partial \xi^i} \cdot \frac{\partial \vec{x}}{\partial \xi^j} \, d\xi^i d\xi^j \,. \tag{6.12}$$

This can be neatly summarized as

$$ds^2 = g_{ij}(\xi) \, d\xi^i d\xi^j \,, \tag{6.13}$$

where $g_{ij}(\xi)$ is defined as

$$g_{ij}(\xi) \equiv \frac{\partial \vec{x}}{\partial \xi^i} \cdot \frac{\partial \vec{x}}{\partial \xi^j} \,. \tag{6.14}$$

The quantity $g_{ij}(\xi)$ is known as the *induced metric on* S. It is called a metric because (6.13) takes, up to a sign, the form of equation (3.78), where we introduced the general concept of a metric. It is a metric on S because, with ξ^i playing the role of coordinates on S, equation (6.13) determines distances on S. It is said to be induced because it uses the metric on the ambient space in which S *lives* to determine distances on S. Indeed, the dot product which appears in (6.14) is to be performed in the space where $\mathcal{S}$ lives and therefore presupposes that a metric exists on that space. We only have two parameters ξ^1 and ξ^2, so the full matrix g_{ij} takes the form:

$$g_{ij} = \begin{pmatrix} \dfrac{\partial \vec{x}}{\partial \xi^1} \cdot \dfrac{\partial \vec{x}}{\partial \xi^1} & \dfrac{\partial \vec{x}}{\partial \xi^1} \cdot \dfrac{\partial \vec{x}}{\partial \xi^2} \\[2mm] \dfrac{\partial \vec{x}}{\partial \xi^2} \cdot \dfrac{\partial \vec{x}}{\partial \xi^1} & \dfrac{\partial \vec{x}}{\partial \xi^2} \cdot \dfrac{\partial \vec{x}}{\partial \xi^2} \end{pmatrix} . \tag{6.15}$$

Now we see something truly nice! The determinant of g_{ij} is precisely the quantity which appears under the square root in (6.6). Letting

$$g \equiv \det(g_{ij}), \tag{6.16}$$

we can write

$$A = \int d\xi^1 d\xi^2 \sqrt{g} . \tag{6.17}$$

This is an elegant formula for the area in terms of the determinant of the induced metric. Instead of trying to understand the reparameterization invariance of (6.6), we now focus on the equivalent but simpler expression (6.17).

We are now in position to understand the invariance of the area in terms of the transformation properties of the metric g_{ij}. The key to this lies in equation (6.13). The length-squared ds^2 is a geometrical property of the vector $d\vec{x}$ that must not depend upon the particular parameterization used to calculate it. For another set of parameters $\tilde{\xi}$ and metric $\tilde{g}(\tilde{\xi})$, the following equality must therefore hold:

$$g_{ij}(\xi) \, d\xi^i d\xi^j = \tilde{g}_{pq}(\tilde{\xi}) \, d\tilde{\xi}^p d\tilde{\xi}^q . \tag{6.18}$$

Making use of the chain rule to express the differentials $d\tilde{\xi}$ in terms of differentials $d\xi$,

$$g_{ij}(\xi) \, d\xi^i d\xi^j = \tilde{g}_{pq}(\tilde{\xi}) \, \frac{\partial \tilde{\xi}^p}{\partial \xi^i} \frac{\partial \tilde{\xi}^q}{\partial \xi^j} \, d\xi^i d\xi^j . \tag{6.19}$$

Since this result holds for any choice of differentials $d\xi$, we find a relation between the metric in ξ and $\tilde{\xi}$ coordinates:

$$g_{ij}(\xi) = \tilde{g}_{pq}(\tilde{\xi}) \, \frac{\partial \tilde{\xi}^p}{\partial \xi^i} \frac{\partial \tilde{\xi}^q}{\partial \xi^j} . \tag{6.20}$$

Making use of the definition of $\tilde{M}$ below (6.8), we rewrite the above equation as

$$g_{ij}(\xi) = \tilde{g}_{pq} \, \tilde{M}_{pi} \, \tilde{M}_{qj} = (\tilde{M}^{\mathrm{T}})_{ip} \, \tilde{g}_{pq} \, \tilde{M}_{qj} . \tag{6.21}$$

In matrix notation, the right-hand side is the product of three matrices. Taking the determinant and using the notation in (6.16) gives

$$g = (\det \tilde{M}^{\mathrm{T}}) \, \tilde{g} \, (\det \tilde{M}) = \tilde{g} (\det \tilde{M})^2 . \tag{6.22}$$

Taking a square root

$$\sqrt{g} = \sqrt{\tilde{g}}\,|\det \tilde{M}|\,,\tag{6.23}$$

we obtain the transformation property for the square root of the determinant of the metric.

We are finally ready to appreciate the reparameterization invariance of (6.17). Making use of (6.7), (6.23), and (6.9) we have

$$\int d\xi^1 d\xi^2 \sqrt{g} = \int d\tilde{\xi}^1 d\tilde{\xi}^2 |\det M| \sqrt{\tilde{g}}\,|\det \tilde{M}| = \int d\tilde{\xi}^1 d\tilde{\xi}^2 \sqrt{\tilde{g}}\,,\tag{6.24}$$

which proves the reparameterization invariance of the area functional. To the trained eye the area formula in (6.17) is *manifestly* reparameterization invariant. That is, once you know how metrics transform, the invariance is reasonably simple to establish. No cumbersome calculation is necessary.

Quick calculation 6.2 Consider the equation $\partial \xi^i / \partial \xi^j = \delta^i_j$ and use the chain rule to show the matrix property

$$M\tilde{M} = 1\,.\tag{6.25}$$

Show that $\tilde{M}M = 1$ holds as well. Finally, note that $\det M \det \tilde{M} = 1$, a result stronger than the one we proved in (6.9).

6.3 Area functional for spacetime surfaces

Let us now move to our case of interest, the case of surfaces in *spacetime*. These surfaces are obtained by representing in spacetime the history of strings, in the same way as a spacetime world-line is obtained by representing the history of a particle. For the case of strings, we obtain a two-dimensional surface called the world-sheet of the string. Spacetime surfaces, such as string world-sheets, are not all that different from the spatial surfaces we considered in the previous section. They are two-dimensional and require two parameters. Instead of calling the parameters ξ^1 and ξ^2, we give them special names: τ and σ.

Given our usual spacetime coordinates $x^\mu = (x^0, x^1, \dots, x^d)$, the surface is described by the mapping functions

$$x^\mu(\tau, \sigma)\,,\tag{6.26}$$

which take some region of the (τ, σ) parameter space into spacetime. Following a standard convention in string theory, we change the notation slightly. We will denote the above mapping functions with the capitalized symbols

$$X^\mu(\tau, \sigma)\,.\tag{6.27}$$

We are not changing the meaning of the functions. Given a fixed point (τ, σ) in the parameter space, this point is mapped to a point with spacetime coordinates

$$(X^0(\tau, \sigma),\ X^1(\tau, \sigma), \dots,\ X^d(\tau, \sigma))\,.\tag{6.28}$$

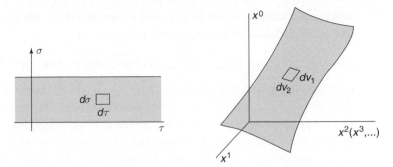

Fig. 6.4 Left: the parameter space (τ, σ), with a little square selected. Right: the surface in target spacetime with the image of the little square, a parallelogram whose sides are the vectors dv_1^μ and dv_2^μ.

Why do we capitalize the X? Suppose we used the same symbol to denote spacetime coordinates and mapping functions. Then we could still distinguish between them by writing x^μ or $x^\mu(\tau, \sigma)$, but we would not have the luxury of dropping the (τ, σ) arguments. On the other hand, with X^μ we can drop the (τ, σ) arguments and still know that we are talking about the mapping functions of the string. We will call X^μ the *string coordinates*.

As before, the parameters τ and σ can be viewed as coordinates on the world-sheet, at least locally. The map inverse to X^μ takes the world-sheet to the parameter space, and locally it assigns to each point on the surface two coordinates: the values of the parameters τ and σ. Introducing some potential for confusion, physicists also use the term world-sheet to denote the two-dimensional parameter space whose image under X^μ gives us the... world-sheet! Unless explicitly stated, we will reserve the use of the term world-sheet for the spacetime surface. In Figure 6.4 we consider an open string: to the left, you see the parameter space surface and, to the right, you see the spacetime surface. In this parameter space, σ ranges over a finite interval, while τ may extend from minus infinity to plus infinity. The parameter τ is roughly related to time on the strings – much more on this later – and the parameter σ is roughly related to positions along the strings. The world-lines of the string endpoints have constant σ, so they are parameterized by τ. As τ flows, time must flow. Thus, at least at the endpoints

$$\frac{\partial X^0}{\partial \tau}\bigg|_{\text{endpoint}} \neq 0. \tag{6.29}$$

We will assume that this also holds for other values of σ.

To find the area element we proceed as in the case of the spatial surface, this time using relativistic notation. The situation is illustrated in Figure 6.4. A little rectangle of sides $d\tau$ and $d\sigma$ in parameter-space becomes a quadrilateral area element in spacetime. This quadrilateral is spanned by the vectors dv_1^μ and dv_2^μ. Furthermore,

$$dv_1^\mu = \frac{\partial X^\mu}{\partial \tau}d\tau, \quad dv_2^\mu = \frac{\partial X^\mu}{\partial \sigma}d\sigma, \tag{6.30}$$

which are analogous to our earlier spatial formulae (6.2). We can now use the analog of
(6.4) as a candidate for the area element dA:

$$dA \overset{?}{=} \sqrt{(dv_1 \cdot dv_1)(dv_2 \cdot dv_2) - (dv_1 \cdot dv_2)^2}\,, \tag{6.31}$$

where the dot is the relativistic dot product. Using this dot product guarantees that the area
element is Lorentz invariant: it is a proper area element. We wrote a question mark on top
of the equal sign because there is one problem. Even though this is not obvious to us yet,
the sign of the object under the square root is negative. To be able to take the square root we
must exchange the two terms under the square root. This change of sign has no effect on
the Lorentz invariance. Doing this, and using (6.30), we find that the proper area is given as

$$A = \int d\tau d\sigma \sqrt{\left(\frac{\partial X^\mu}{\partial \tau}\frac{\partial X_\mu}{\partial \sigma}\right)^2 - \left(\frac{\partial X^\mu}{\partial \tau}\frac{\partial X_\mu}{\partial \tau}\right)\left(\frac{\partial X^\nu}{\partial \sigma}\frac{\partial X_\nu}{\partial \sigma}\right)}\,. \tag{6.32}$$

Using the relativistic dot product notation,

$$A = \int d\tau d\sigma \sqrt{\left(\frac{\partial X}{\partial \tau}\cdot\frac{\partial X}{\partial \sigma}\right)^2 - \left(\frac{\partial X}{\partial \tau}\right)^2\left(\frac{\partial X}{\partial \sigma}\right)^2}\,. \tag{6.33}$$

To understand why the above sign is correct we must convince ourselves that the expression
under the square root is greater than or equal to zero at any point on the world-sheet of a
string.

What characterizes locally the spacetime surface traced by a string? The answer is quite
interesting. Consider a point on the world-sheet and the set of all vectors tangent to the
surface at that point. These vectors form a two-dimensional vector space. We claim that
in this vector space there is a basis made by two vectors, one of which is spacelike and
one of which is timelike. This implies that at each point on the world-sheet there are both
timelike and spacelike tangent directions. There is a small caveat: on each fixed-time string
there can be a finite set of exceptional points where the tangents to the world-sheet do not
include a timelike vector. At those points, as we will see, the string moves with the speed
of light.

   The string we are trying to define cannot have *finite* size pieces moving with the speed
of light. In such case, a string would contain a continuous set of points, not just a finite
set, where the tangents to the world-sheet do not include timelike vectors. Imagine, for
example, a straight piece of string along the x axis moving with the speed of light in the
y direction. At any point on this string all vectors tangent to the world-sheet are spacelike,
except for null vectors in one particular direction. To see this consider Figure 6.5 where we
show this string at various closely separated times. Tangent vectors at P are well approximated by vectors that joint the event P to a nearby event P' on the *world-sheet*. Since P
and P' are points on the same string at different times, the spatial part of the tangent vector
appears in the figure as the arrow joining P to P'. Consider all arrows joining P to points
on a semicircle around P. Any world-sheet tangent direction at P is represented by one
of these arrows. The world-sheet tangent vector associated with PQ is clearly spacelike,
since P and Q occur at the same time. The typical tangent, that associated with the arrow
PR is still spacelike: in the elapsed time P can get to $\bar{R}$ moving at the speed of light, but to

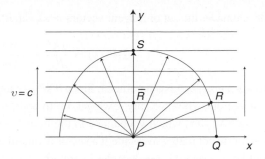

Fig. 6.5 A string along the *x* direction moving with the velocity of light along the *y* direction. This is not allowed motion. All tangent vectors to the world-sheet at any point *P* on the string are either spacelike or null.

get to R it must move faster than light. All tangent directions are spacelike, except for that associated with PS, which is null. This shows that there is no timelike tangent at P. Since P is a generic point along the original piece of string, nowhere on this piece of string is there a timelike vector tangent to the world-sheet.

At any point on a world-sheet a spacelike direction exists and is easy to visualize: if you take a photograph of the string at some time, every tangent vector along the length of the string points in a spacelike direction. Indeed, in your frame, the events defining the string are simultaneous but spatially separated.

To appreciate the need for a timelike vector at regular points on the world-sheet, consider first the world-line of a point particle. The tangent vector to the world-line is timelike. At each point on the world-line this tangent vector can be used to describe an instantaneous Lorentz observer who sees the particle at rest. A spacelike tangent vector to the world-line is unphysical: it describes a particle moving faster than the speed of light. Strings are a little more subtle since there is no way to tell how individual points on them move. As we shall make abundantly clear, the string is not made of constituents whose position we can keep track of (exception: one can keep track of the endpoints of an open string). Still, a timelike tangent to the world-sheet at a given point of a string allows us to describe an instantaneous Lorentz observer who sees the point at rest. If there is a timelike tangent at a point, there are many, by continuity. Each of these timelike tangents defines a different instantaneous Lorentz observer who sees the point at rest. This is consistent with our inability to track unambiguously the motion of points on a string. If we watch a string at two closely separated times we cannot tell which point went where, but for each point p on the final string we must be able to find some point p' on the initial string that could reach p moving with speed less than or at most equal to c.

The existence of both timelike directions and spacelike directions at any regular point on the world-sheet is our criterion for physical motion. It guarantees that equation (6.33) makes sense.

Claim: At any point P on the world-sheet where there is both a timelike direction and a spacelike direction, the quantity under the square root in (6.33) is positive:

$$\left(\frac{\partial X}{\partial \tau} \cdot \frac{\partial X}{\partial \sigma}\right)^2 - \left(\frac{\partial X}{\partial \sigma}\right)^2 \left(\frac{\partial X}{\partial \tau}\right)^2 > 0. \qquad (6.34)$$

Proof: Consider the set of tangent vectors $v^\mu(\lambda)$ at P obtained as:

$$v^\mu(\lambda) = \frac{\partial X^\mu}{\partial \tau} + \lambda \frac{\partial X^\mu}{\partial \sigma}, \tag{6.35}$$

where $\lambda \in (-\infty, \infty)$ is a parameter. Since $\partial X^\mu/\partial \tau$ and $\partial X^\mu/\partial \sigma$ are linearly independent tangent vectors, when we vary λ we get, up to constant scalings, all tangent vectors at P, including $\partial X^\mu/\partial \sigma$, which is obtained in the limit $\lambda \to \infty$ (Figure 6.6). Constant scalings of a vector do not matter to decide if a vector is timelike or spacelike. To determine if $v^\mu(\lambda)$ is timelike or spacelike, we consider its square:

$$v^2(\lambda) = v^\mu(\lambda)v_\mu(\lambda) = \lambda^2\left(\frac{\partial X}{\partial \sigma}\right)^2 + 2\lambda\left(\frac{\partial X}{\partial \tau} \cdot \frac{\partial X}{\partial \sigma}\right) + \left(\frac{\partial X}{\partial \tau}\right)^2. \tag{6.36}$$

The dot products appearing on the final right-hand side are just numbers, so we have a quadratic polynomial in λ. To have both timelike and spacelike tangent vectors at P, $v^2(\lambda)$ must take both negative and positive values as we vary λ. In other words, the equation $v^2(\lambda) = 0$ must have two real roots. For this to happen, the discriminant of the quadratic equation $v^2(\lambda) = 0$ must be positive. From (6.36) we see that this requires

$$\left(\frac{\partial X}{\partial \tau} \cdot \frac{\partial X}{\partial \sigma}\right)^2 - \left(\frac{\partial X}{\partial \sigma}\right)^2\left(\frac{\partial X}{\partial \tau}\right)^2 > 0, \tag{6.37}$$

which is precisely the condition (6.34) we set out to prove!

Since we always have spacelike tangents, the plot of $v^2(\lambda)$ in Figure 6.6 must include a region where $v^2(\lambda) > 0$. Consider a point P on the world-sheet where all tangent directions are spacelike with the exception of one that is null. Then $v^2(\lambda) > 0$, except for one value of λ where v^2 vanishes. The equation $v^2(\lambda) = 0$ must have a single root and the associated discriminant is zero. It follows that the quantity under the square root in the action (6.33) is zero at P. Any possible motion of the string at P must be associated with a world-sheet tangent at P. Since motion along spacelike directions is unphysical, only the null vector provides an acceptable answer: the string is moving with the speed of light at P.

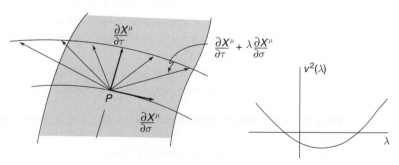

Fig. 6.6 Left: a set of tangent vectors $v(\lambda)$ at a point P on the world-sheet. Right: a plot of $v^2(\lambda)$ as a function of λ. The vector $v(\lambda)$ may be spacelike or timelike depending on the value of λ.

6.4 The Nambu–Goto string action

Now that we are sure that the proper area functional in (6.33) is correctly defined, we can introduce the action for the relativistic string. This action is proportional to the proper area of the world-sheet. To have the units of action we must multiply the area functional by some suitable constants.

The area functional in (6.33) has units of length-squared, as it must be. This is because X^μ has units of length, and each term under the square root has four X. The units of τ and σ cancel out. Each term in the square root has two σ derivatives and two τ derivatives. Their units cancel against the units of the differentials. Nevertheless, we will take σ to have units of length and τ to have units of time. We do this anticipating a relation between τ and time and between σ and positions on strings. To summarize:

$$[\tau] = T, \quad [\sigma] = L, \quad [X^\mu] = L, \quad [A] = L^2. \tag{6.38}$$

Since S must have units of ML^2/T and A has units of L^2, we must multiply the proper area by a quantity with units of M/T. The string tension T_0 has units of force, and force divided by velocity has the desired units of M/T. We can therefore multiply the proper area by T_0/c to get a quantity with the units of action. Making use of (6.33) we set the string action equal to

$$S = -\frac{T_0}{c} \int_{\tau_i}^{\tau_f} d\tau \int_0^{\sigma_1} d\sigma \sqrt{(\dot{X} \cdot X')^2 - (\dot{X})^2 (X')^2}. \tag{6.39}$$

Here $\sigma_1 > 0$ is some constant, and we have introduced some notation for derivatives:

$$\dot{X}^\mu \equiv \frac{\partial X^\mu}{\partial \tau}, \quad X^{\mu\prime} \equiv \frac{\partial X^\mu}{\partial \sigma}. \tag{6.40}$$

Of course, we have not yet confirmed that the symbol T_0 in the string action has the precise interpretation of tension, but we will do so in Section 6.7. We will also confirm there that the overall negative sign which multiplies the action is correct. The action S is the Nambu–Goto action for the relativistic string.

It is crucial that this action be reparameterization invariant. We can proceed just as we did with spatial surfaces to write the Nambu–Goto action in a manifestly reparameterization invariant way. In this case we have

$$-ds^2 = dX^\mu dX_\mu = \eta_{\mu\nu} dX^\mu dX^\nu = \eta_{\mu\nu} \frac{\partial X^\mu}{\partial \xi^\alpha} \frac{\partial X^\nu}{\partial \xi^\beta} d\xi^\alpha d\xi^\beta. \tag{6.41}$$

Here $\eta_{\mu\nu}$ is the target-space Minkowski metric. The indices α and β run over two values, 1 and 2, and we have taken $\xi^1 = \tau, \xi^2 = \sigma$. Just as we did for spatial surfaces, we define an induced metric $\gamma_{\alpha\beta}$ on the world-sheet:

$$\gamma_{\alpha\beta} \equiv \eta_{\mu\nu} \frac{\partial X^\mu}{\partial \xi^\alpha} \frac{\partial X^\nu}{\partial \xi^\beta} = \frac{\partial X}{\partial \xi^\alpha} \cdot \frac{\partial X}{\partial \xi^\beta}. \tag{6.42}$$

More explicitly, the 2-by-2 matrix $\gamma_{\alpha\beta}$ is

$$\gamma_{\alpha\beta} = \begin{bmatrix} (\dot{X})^2 & \dot{X} \cdot X' \\ \dot{X} \cdot X' & (X')^2 \end{bmatrix}. \tag{6.43}$$

With the help of this metric we can write the Nambu–Goto action in the manifestly reparameterization invariant form

$$S = -\frac{T_0}{c} \int d\tau d\sigma \sqrt{-\gamma}, \quad \gamma = \det(\gamma_{\alpha\beta}). \tag{6.44}$$

The analysis in Section 6.2 of reparameterization invariance for spatial surfaces holds, without change, in the present case. Not only is the action (6.44) manifestly reparameterization invariant, it is also more compact than (6.39). In this form, one can readily generalize the Nambu–Goto action to describe the dynamics of objects that have more dimensions than strings. An action of this kind is useful as a first approximation to the dynamics of D-branes.

6.5 Equations of motion, boundary conditions, and D-branes

In this section we will obtain the equations of motion that follow by variation of the string action. In doing so we will also have an opportunity to discuss the various boundary conditions that can be imposed on the ends of open strings. Dirichlet boundary conditions will be interpreted to arise owing to the existence of D-branes.

Let us begin by writing the Nambu–Goto action (6.39) as the double integral of a Lagrangian density $\mathcal{L}$:

$$S = \int_{\tau_i}^{\tau_f} d\tau L = \int_{\tau_i}^{\tau_f} d\tau \int_0^{\sigma_1} d\sigma \, \mathcal{L}(\dot{X}^\mu, X^{\mu\prime}), \tag{6.45}$$

where $\mathcal{L}$ is given by

$$\mathcal{L}(\dot{X}^\mu, X^{\mu\prime}) = -\frac{T_0}{c}\sqrt{(\dot{X} \cdot X')^2 - (\dot{X})^2(X')^2}. \tag{6.46}$$

We can obtain the equations of motion for the relativistic string by setting the variation of the action (6.45) equal to zero. The variation is simply

$$\delta S = \int_{\tau_i}^{\tau_f} d\tau \int_0^{\sigma_1} d\sigma \left[\frac{\partial \mathcal{L}}{\partial \dot{X}^\mu} \frac{\partial(\delta X^\mu)}{\partial\tau} + \frac{\partial \mathcal{L}}{\partial X^{\mu\prime}} \frac{\partial(\delta X^\mu)}{\partial\sigma} \right], \tag{6.47}$$

where we have used

$$\delta \dot{X}^\mu = \delta\left(\frac{\partial X^\mu}{\partial \tau}\right) = \frac{\partial(\delta X^\mu)}{\partial \tau}\,, \tag{6.48}$$

and an analogous equation for $\delta X^{\mu\prime}$.

The quantities $\partial \mathcal{L}/\partial \dot{X}^\mu$ and $\partial \mathcal{L}/\partial X^{\prime\mu}$ will appear frequently throughout the remainder of our discussion, so it is useful to introduce new symbols for them. This is just what we did when we studied the nonrelativistic string in Section 4.6. This time we find

$$\mathcal{P}^\tau_\mu \equiv \frac{\partial \mathcal{L}}{\partial \dot{X}^\mu} = -\frac{T_0}{c}\frac{(\dot{X}\cdot X')X'_\mu - (X')^2\dot{X}_\mu}{\sqrt{(\dot{X}\cdot X')^2 - (\dot{X})^2(X')^2}}\,, \tag{6.49}$$

$$\mathcal{P}^\sigma_\mu \equiv \frac{\partial \mathcal{L}}{\partial X^{\mu\prime}} = -\frac{T_0}{c}\frac{(\dot{X}\cdot X')\dot{X}_\mu - (\dot{X})^2 X'_\mu}{\sqrt{(\dot{X}\cdot X')^2 - (\dot{X})^2(X')^2}}\,. \tag{6.50}$$

Quick calculation 6.3 Verify equations (6.49) and (6.50).

Using this notation, the variation δS in (6.47) becomes

$$\delta S = \int_{\tau_i}^{\tau_f} d\tau \int_0^{\sigma_1} d\sigma \left[\frac{\partial}{\partial \tau}(\delta X^\mu \mathcal{P}^\tau_\mu) + \frac{\partial}{\partial \sigma}\left(\delta X^\mu \mathcal{P}^\sigma_\mu\right) - \delta X^\mu \left(\frac{\partial \mathcal{P}^\tau_\mu}{\partial \tau} + \frac{\partial \mathcal{P}^\sigma_\mu}{\partial \sigma}\right)\right]. \tag{6.51}$$

The first term on the right-hand side, being a full derivative in τ, will contribute terms proportional to $\delta X^\mu(\tau_f,\sigma)$ and $\delta X^\mu(\tau_i,\sigma)$. Since the flow of τ implies the flow of time, we can imagine specifying the initial and final states of the string, and we restrict ourselves to variations for which $\delta X^\mu(\tau_f,\sigma) = \delta X^\mu(\tau_i,\sigma) = 0$. We will always assume such variations, so we can forget about these terms. The variation then becomes

$$\delta S = \int_{\tau_i}^{\tau_f} d\tau \left[\delta X^\mu \mathcal{P}^\sigma_\mu\right]_0^{\sigma_1} - \int_{\tau_i}^{\tau_f} d\tau \int_0^{\sigma_1} d\sigma\, \delta X^\mu \left(\frac{\partial \mathcal{P}^\tau_\mu}{\partial \tau} + \frac{\partial \mathcal{P}^\sigma_\mu}{\partial \sigma}\right). \tag{6.52}$$

Since the second term on the right-hand side must vanish for all variations δX^μ of the motion, we set

$$\frac{\partial \mathcal{P}^\tau_\mu}{\partial \tau} + \frac{\partial \mathcal{P}^\sigma_\mu}{\partial \sigma} = 0\,. \tag{6.53}$$

This is the equation of motion for the relativistic string, open or closed. A quick glance at definitions (6.49) and (6.50) shows that this equation is incredibly complicated. The key to its solution will lie in the reparameterization invariance of the Nambu–Goto action. Choosing a clever parameterization will simplify our work enormously.

The first term on the right-hand side of (6.52) has to do with the string endpoints. It is, in fact, a collection of terms that includes two terms for each value of the index μ. More explicitly, the list is

$$\int_{\tau_i}^{\tau_f} d\tau \Big(\delta X^0(\tau, \sigma_1) \, \mathcal{P}_0^\sigma(\tau, \sigma_1) - \delta X^0(\tau, 0) \mathcal{P}_0^\sigma(\tau, 0)$$

$$+ \, \delta X^1(\tau, \sigma_1) \mathcal{P}_1^\sigma(\tau, \sigma_1) - \delta X^1(\tau, 0) \mathcal{P}_1^\sigma(\tau, 0)$$

$$\vdots \quad \vdots \qquad\qquad \vdots \quad \vdots$$

$$+ \, \delta X^d(\tau, \sigma_1) \mathcal{P}_d^\sigma(\tau, \sigma_1) - \delta X^d(\tau, 0) \mathcal{P}_d^\sigma(\tau, 0) \Big). \tag{6.54}$$

We need a boundary condition for *each* term in the above list. This is a total of $2D = 2(d + 1)$ boundary conditions.

Let us focus on a single term, that is, we fix μ and select one endpoint. Let σ_* denote the σ coordinate of an endpoint; σ_* can be equal to zero or equal to σ_1. Selecting an endpoint means fixing the value of σ_*. As before, there are two natural boundary conditions that one can impose at an endpoint. The first is a Dirichlet boundary condition, in which the endpoint of the string remains fixed throughout the motion:

$$\text{Dirichlet boundary condition:} \quad \frac{\partial X^\mu}{\partial \tau}(\tau, \sigma_*) = 0, \quad \mu \neq 0. \tag{6.55}$$

Since time varies as τ varies (see (6.29)), the value $\mu = 0$ must be excluded. Dirichlet boundary conditions are only possible for space directions. Given that constancy in τ means constancy in time, equation (6.55) implies that the μ coordinate of the selected string endpoint is fixed in time. Alternatively, rather than requiring that the τ derivative vanishes, we could simply specify a constant value for $X^\mu(\tau, \sigma_*)$. If the string endpoint is fixed, the variations are set to vanish there: $\delta X^\mu(\tau, \sigma_*) = 0$. This guarantees that the relevant term in (6.54) vanishes.

The second possible boundary condition is a free endpoint condition:

$$\text{free endpoint condition:} \quad \mathcal{P}_\mu^\sigma(\tau, \sigma_*) = 0. \tag{6.56}$$

This condition, as needed, also results in the vanishing of the relevant term in (6.54). This is called a free endpoint condition because it does not impose any constraint on the variation $\delta X^\mu(\tau, \sigma_*)$ of the string coordinate at the endpoint. The endpoint is free to do whatever is needed to get the variation of the action to vanish. The free endpoint boundary condition must apply for $\mu = 0$:

$$\mathcal{P}_0^\sigma(\tau, \sigma_1) = \mathcal{P}_0^\sigma(\tau, 0) = 0. \tag{6.57}$$

For the nonrelativistic string, the free endpoint boundary condition implies the vanishing of $\mathcal{P}^x$, which imposes a Neumann boundary condition on the string coordinate (see (4.47)).

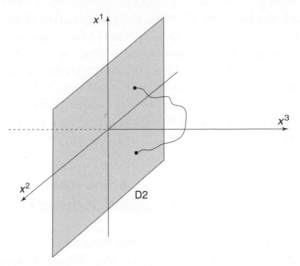

Fig. 6.7 **A D2-brane stretched over the (x^1, x^2) plane. The endpoints of the open string can move freely on the plane, but must remain attached to it. The coordinate x^3 of the endpoints must vanish at all times. This is a Dirichlet boundary condition for the string coordinate X^3.**

We will eventually understand (6.56) in terms of a Neumann boundary condition. Similarly, the Dirichlet boundary (6.55) will be shown to imply the vanishing of $\mathcal{P}_\mu^\tau$ at the string endpoints.

The boundary conditions (6.55) and (6.56) can be imposed in many possible ways. For each spatial direction, and at each endpoint, we can choose either a Dirichlet or a free-endpoint boundary condition. Since closed strings have no endpoints, they do not require boundary conditions.

Let us elaborate on the case of Dirichlet boundary conditions. It is clear from the study of nonrelativistic strings that Dirichlet boundary conditions arise if string endpoints are attached to some physical objects. Consider, for example, Figure 4.2. On the left, the string is attached to two points. On the right, the string is free to slide up and down at the endpoints; the string endpoints are forced to stay on one-dimensional lines and horizontal motion of the endpoints is forbidden. The objects on which open string endpoints must lie are characterized by their dimensionality, more precisely, by the number of spatial dimensions that they have. They are called D-branes, where the letter D stands for Dirichlet. The objects which fix the string endpoints on the left side of Figure 4.2 are zero-dimensional. They are called D0-branes. The lines which constrain the string endpoints on the right side of the figure are one dimensional. They are called D1-branes.

A Dp-brane is an object with p spatial dimensions. Since the string endpoints must lie on the Dp-brane, a set of Dirichlet boundary conditions is specified. A flat D2-brane in a three-dimensional space, for example, is specified by one condition, say $x^3 = 0$ (Figure 6.7). This means that the D2-brane extends over the (x^1, x^2) plane. The Dirichlet boundary condition applies to the string coordinate X^3, which must vanish at the string endpoints. Since the motion of open string endpoints is free along the directions of the brane, the string coordinates X^1 and X^2 satisfy free boundary conditions. When the open string endpoints have free boundary conditions along all spatial directions, we still have a D-brane, but this time it is a *space-filling* D-brane. The D-brane extends all over space, and since open string endpoints can be anywhere on the D-brane, open string endpoints are completely free.

For (quantum) relativistic strings the consistency of Dirichlet boundary conditions allows one to discover the properties of D-branes. D-branes are physical objects that exist in a theory of strings and are not introduced by hand. D-branes need not have infinite extent nor are they necessarily hyperplanes. They have calculable energy densities and a host of remarkable properties. We will study D-branes in detail beginning in Chapter 15.

6.6 The static gauge

To make progress in understanding the action for the relativistic string, we must parameterize the world-sheet in a useful way. We are allowed to choose the parameterization freely because of the reparameterization invariance of the string action. Reparameterization invariance in string theory is analogous to gauge invariance in electrodynamics. Maxwell's equations possess a symmetry under gauge transformations that allows us to use different potentials A_μ to represent the same electromagnetic fields $\vec{E}$ and $\vec{B}$. A suitable choice of gauge helps to uncover the physics. Similarly, we may use many different grids on the world-sheet to describe the same physical motion of the string. A suitable choice of grid can make this task much easier. A good choice of parameterization was useful even for the relativistic point particle – its equation of motion is simplest when the trajectory is parameterized by proper time.

In this section, we will discuss only a partial parameterization on the world-sheet. We will fix the lines of constant τ by relating τ to the time coordinate $X^0 = ct$ in some chosen Lorentz frame.

Consider the constant time hyperplane $t = t_0$ in the target space (Figure 6.8). This plane will intersect the world-sheet along a curve – *the string at time t_0* according to observers in our chosen Lorentz frame. We declare this curve to be a curve of constant τ; in fact, we declare it to be the curve $\tau = t_0$. Extending this definition to all times t, we declare that for any point Q on the world-sheet

$$\tau(Q) = t(Q). \tag{6.58}$$

This choice of τ parameterization is called the *static gauge* because lines of constant τ are "static strings" in the chosen Lorentz frame.

We will not try to make a sophisticated choice of σ at this time. For an open string, we will choose one edge of the world-sheet to be the curve $\sigma = 0$ and the other edge to be the curve $\sigma = \sigma_1$:

$$\sigma \in [0, \sigma_1], \quad \text{for an open string.} \tag{6.59}$$

We draw lines of constant σ on the surface quite arbitrarily, provided, of course, that constant σ lines vary smoothly, do not intersect, and are consistent with the two curves which

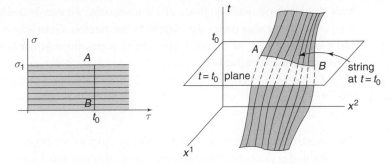

Left: the parameter space strip for an open string. The vertical segment AB is the line $\tau = t_0$. Right: the open string world-sheet in target space. The string at time $t = t_0$ is the intersection of the world-sheet with the hyperplane $t = t_0$. In the static gauge, the string at time $t = t_0$ is the image of the $\tau = t_0$ segment AB.

are the boundary of the world-sheet (Figure 6.8). Drawing lines of constant σ is equivalent to giving an explicit σ parameterization to all the strings. For closed strings the same ideas apply, but there is a significant proviso: there must be an identification in the (τ, σ) parameter space. The σ direction must be made into a circle, making the (τ, σ) parameter space into a cylinder. This is needed because the closed string world-sheet is topologically a cylinder. Letting σ_c denote the circumference of the σ circle, the identification is

$$(\tau, \sigma) \sim (\tau, \sigma + \sigma_c). \tag{6.60}$$

Points that are identified by this relation on the parameter space map to the same point on the closed string world-sheet. The closed strings can be parameterized using any σ interval of length σ_c, for example

$$\sigma \in [0, \sigma_c], \quad \text{for a closed string.} \tag{6.61}$$

Let us now explore some implications of our choice of τ. We can write (6.58) as

$$X^0(\tau, \sigma) \equiv c\,t(\tau, \sigma) = c\,\tau, \tag{6.62}$$

or simply

$$\tau = t. \tag{6.63}$$

We can thus describe the collection of string coordinates X^μ as

$$X^\mu(\tau, \sigma) = X^\mu(t, \sigma) = \left(c\,t, \vec{X}(t, \sigma) \right), \tag{6.64}$$

letting the vector $\vec{X}$ represent the spatial string coordinates. We then find

$$\frac{\partial X^\mu}{\partial \sigma} = \left(\frac{\partial X^0}{\partial \sigma}, \frac{\partial \vec{X}}{\partial \sigma} \right) = \left(0, \frac{\partial \vec{X}}{\partial \sigma} \right),$$
$$\frac{\partial X^\mu}{\partial \tau} = \left(\frac{\partial X^0}{\partial t}, \frac{\partial \vec{X}}{\partial t} \right) = \left(c, \frac{\partial \vec{X}}{\partial t} \right). \tag{6.65}$$

As you can see, this parameterization separates the time and space components quite neatly.

Now that we have made a choice of τ coordinates, we can do a simple test to confirm that we got the right sign under the radical in the Nambu–Goto action (6.39). Imagine a little piece of string with no velocity. Because it is not moving, $\partial \vec{X}/\partial t = 0$, and using (6.65), the square root in (6.39) becomes

$$\sqrt{0 - \left(\frac{\partial \vec{X}}{\partial \sigma}\right)^2 (-c^2)}.$$ (6.66)

The quantity under the square root is positive, just as we expected. If some day you forget the sign under the radical in the string action, this is a good way to check it quickly.

6.7 Tension and energy of a stretched string

Let us now do our first calculation with the Nambu–Goto action – our first calculation in string theory! We are going to analyze a stretched relativistic string. The endpoints of the string are fixed at $x^1 = 0$, and at $x^1 = a > 0$, with vanishing values for the coordinates of the additional spatial dimensions. We therefore denote the spatial coordinates of the endpoints as $(0, \vec{0})$ and $(a, \vec{0})$. The inclusion of the common $(d-1)$-dimensional vector $\vec{0}$ tells us that the string is only stretched along the first spatial coordinate.

We evaluate the string action for this stretched string using the static gauge $X^0 = c\tau$. Because this is a static string stretched from $x^1 = 0$ to $x^1 = a$, we can write

$$X^1(t, \sigma) = f(\sigma), \quad X^2 = X^3 = \cdots = X^d = 0,$$ (6.67)

where

$$f(0) = 0, \quad f(\sigma_1) = a,$$ (6.68)

and the function $f(\sigma)$ is strictly increasing and continuous on the interval $\sigma \in [0, \sigma_1]$. The setup is illustrated in Figure 6.9. The function f must be strictly increasing to ensure that each point along the string is assigned a unique σ coordinate.

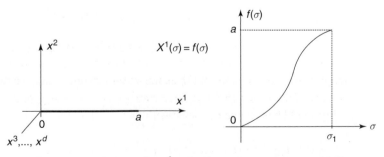

Fig. 6.9   A string of length a stretched along the x^1 axis. The string is parameterized as $X^1(t, \sigma) = f(\sigma)$.

It now follows that

$$\dot{X}^{\mu} = (c, 0, \vec{0}), \quad X^{\mu\prime} = (0, f', \vec{0}),$$ (6.69)

with $f' = df/d\sigma > 0$. Therefore

$$(\dot{X})^2 = -c^2, \quad (X')^2 = (f')^2, \quad \dot{X} \cdot X' = 0.$$ (6.70)

We can now evaluate the action (6.39):

$$S = -\frac{T_0}{c} \int_{t_i}^{t_f} dt \int_0^{\sigma_1} d\sigma \sqrt{0 - (-c^2)(f')^2} = -T_0 \int_{t_i}^{t_f} dt \int_0^{\sigma_1} d\sigma \frac{df}{d\sigma}.$$ (6.71)

The σ integrand is a total derivative, so

$$S = -T_0 \int_{t_i}^{t_f} dt \, (f(\sigma_1) - f(0)) = \int_{t_i}^{t_f} dt \, (-T_0 a),$$ (6.72)

where we used (6.68). Note that the value of the action does not depend on the function f used to parameterize the string. This is an explicit confirmation of the reparameterization invariance of the string action.

   We would like to interpret our result. For this, recall that the action is the time integral of the Lagrangian L. When the kinetic energy vanishes, $L = -V$, where V is the potential energy. Since our string is static, there is no kinetic energy, so

$$S = \int_{t_i}^{t_f} dt \, (-V).$$ (6.73)

Comparing this with (6.72) we conclude that

$$V = T_0 \, a.$$ (6.74)

The potential energy of our stretched string is just $T_0 a$. What does this mean? If the tension of a static string is T_0, regardless of its length, then $T_0 a$ is the amount of energy that you must spend to create a string of length a. Imagine that you start with an infinitesimal string and you start pulling it. As you do work you are giving energy to the string, in fact, you are creating rest energy, or rest mass. The rest mass μ_0 per unit length is

$$\mu_0 c^2 = \frac{V}{a} = T_0 \longrightarrow \mu_0 = \frac{T_0}{c^2}.$$ (6.75)

The mass (or rest energy) arises only because the string has a tension. Because of this, the relativistic string is sometimes referred to as a massless string. The above calculation supports the identification of T_0 as the string tension. It also confirms that the minus sign in front of the action (6.39) is necessary – otherwise the potential energy of the stretched string would have come out negative.

There is one point that we have glossed over. We assumed in our analysis that the configuration (6.67) satisfies the string equations of motion. If it does not, then the configuration cannot be physically realized. Let us check that the equations of motion are satisfied.

First note that on account of (6.69) neither $\dot{X}^\mu$ nor $X^{\mu\prime}$ has τ dependence. Therefore neither $\mathcal{P}^\tau$ nor $\mathcal{P}^\sigma$ has τ dependence (see (6.49) and (6.50)). This being the case, the equation of motion (6.53) reduces to

$$\frac{\partial \mathcal{P}_\mu^\sigma}{\partial \sigma} = 0 . \tag{6.76}$$

This requires that $\mathcal{P}_\mu^\sigma$ be σ-independent. We look again at (6.50) and use (6.70) to find

$$\mathcal{P}_\mu^\sigma = -\frac{T_0}{c} \frac{c^2 X_\mu'}{\sqrt{c^2(f')^2}} = -T_0 \frac{X_\mu'}{f'} . \tag{6.77}$$

This is nonvanishing only for $\mu = 1$, in which case $X_1' = f'$, so $\mathcal{P}^\sigma$ is indeed σ-independent. Thus the equation of motion is satisfied. Even the boundary conditions are satisfied. As we discussed in Section 6.5, there is no condition to check for string coordinates that satisfy Dirichlet boundary conditions at the endpoints. In our problem this means that there are no extra conditions to be checked for any of the spatial coordinates. For the zeroth coordinate, equation (6.57) requires the free boundary condition $\mathcal{P}_0^\sigma = 0$. This holds on account of (6.77).

6.8 Action in terms of transverse velocity

We have chosen a partial parameterization of the world-sheet by imposing the condition $X^0 = ct = c\tau$. With this choice, a line of constant τ on the world-sheet corresponds to the string, as seen by our chosen Lorentz observer, at the particular time $t = \tau$.

Can we define some sort of string velocity? Since the components of $\vec{X}(t, \sigma)$ are the string spatial coordinates, the derivative $\partial \vec{X}/\partial t$ seems to be the closest thing we have to a velocity. This velocity, however, depends upon the choice of σ. Its direction, for example, goes along the lines of constant σ. Since σ can be chosen quite arbitrarily, keeping σ constant in taking the derivative is clearly not very physically significant!

Fixing physically the σ parameterization of a string is subtle because the string is an object with no substructure. When comparing a string at two nearby times, it is not possible to say that a point moved from one location to the next. To speak of points on the string we need a σ parameterization, and reparameterization invariance makes it clear to us that this parameterization is not unique. This suggests that longitudinal motion on the string is not physically meaningful.

There is a reparameterization invariant velocity that can be defined on the string. This is, however, a *transverse* velocity. We consider the string motion in *space*, and imagine that each point on the string moves transversely to the string (Figure 6.10). Consider a string at some fixed time t and pick a point p on it. Draw the hyperplane orthogonal to the string at p. At time $t + dt$, with dt infinitesimal, the string has moved, but it will still intersect the plane, this time at a point p'. The transverse velocity is what we get if we presume that the point p moved to p'. No string parameterization is needed to define this velocity.

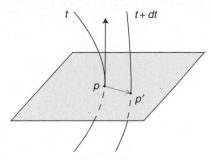

Fig. 6.10 A string at time t and the plane orthogonal to the string at p. At time $t + dt$ the string intersects the plane at p'. To define the transverse velocity we assume that p moved to p'.

When speaking of evolving strings there are two surfaces we can discuss. One is the world-sheet, the surface in spacetime which represents the history of the string. The other is a surface in *space*. This *spatial surface* is put together by combining the strings that we observe at all times. This is the surface that would be generated if the string were to leave a wake as it moved. The transverse velocity $\vec{v}_\perp$ at any point on the string is a vector orthogonal to the string and tangent to the string spatial surface. Since $\vec{v}_\perp$ is a string reparameterization invariant notion of velocity, we expect it to enter naturally into the evaluation of the string action.

In order to define the transverse velocity $\vec{v}_\perp$, it is useful to have a unit vector tangent to the string. To this end, we now introduce a parameter s which is more physical than our nearly arbitrary σ: s measures length along the string. Let us work with the string at a *fixed* time, and define $s(\sigma)$ to be the length of the string in the interval $[0, \sigma]$. Thus, for example, $s(0) = 0$, and $s(\sigma_1)$ is the length of an entire open string. Since ds is the length of the infinitesimal vector $d\vec{X}$ which arises from an interval $d\sigma$ along the string, we have:

$$ds = |d\vec{X}| = \left| \frac{\partial \vec{X}}{\partial \sigma} \right| |d\sigma|. \tag{6.78}$$

Now consider the quantity $\partial \vec{X}/\partial s$, which is the rate of change of $\vec{X}$ with respect to the length of the string. First note that it is a unit vector:

$$\frac{\partial \vec{X}}{\partial s} \cdot \frac{\partial \vec{X}}{\partial s} = \frac{\partial \vec{X}}{\partial \sigma} \cdot \frac{\partial \vec{X}}{\partial \sigma} \left(\frac{d\sigma}{ds} \right)^2 = \left| \frac{\partial \vec{X}}{\partial \sigma} \right|^2 \left(\frac{d\sigma}{ds} \right)^2 = 1. \tag{6.79}$$

The derivative $\partial \vec{X}/\partial \sigma$ is taken with t held fixed, so it lies along a line of constant t. Since the lines of constant t are precisely the strings, it is tangent to the string. In addition

$$\frac{\partial \vec{X}}{\partial s} = \frac{\partial \vec{X}}{\partial \sigma} \frac{d\sigma}{ds}, \tag{6.80}$$

and thus $\partial \vec{X}/\partial s$ is also tangent to the string. Because it has unit length,

$$\frac{\partial \vec{X}}{\partial s} \quad \text{is a unit vector tangent to the string.} \tag{6.81}$$

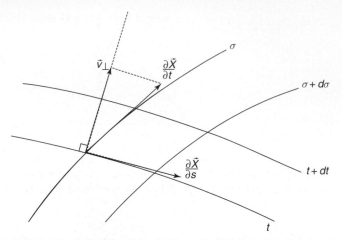

A small piece of the world-sheet showing the vector $\partial\vec{X}/\partial t$, the transverse velocity $\vec{v}_\perp$, and the unit vector $\partial\vec{X}/\partial s$.

We define $\vec{v}_\perp$ to be the component of the velocity $\partial\vec{X}/\partial t$, in the direction perpendicular to the string (see Figure 6.11). For any vector $\vec{u}$, its component perpendicular to a unit vector $\vec{n}$ is $\vec{u} - (\vec{u} \cdot \vec{n})\vec{n}$. Therefore, using our unit vector $\partial\vec{X}/\partial s$ along the string, we have

$$\vec{v}_\perp = \frac{\partial\vec{X}}{\partial t} - \left(\frac{\partial\vec{X}}{\partial t} \cdot \frac{\partial\vec{X}}{\partial s}\right)\frac{\partial\vec{X}}{\partial s} \,. \tag{6.82}$$

For future use, we calculate $v_\perp^2$. A small computation gives

$$v_\perp^2 = \left(\frac{\partial\vec{X}}{\partial t}\right)^2 - \left(\frac{\partial\vec{X}}{\partial t} \cdot \frac{\partial\vec{X}}{\partial s}\right)^2 \,. \tag{6.83}$$

Our goal now is to write the string action in terms of $\vec{v}_\perp$ and other quantities, if necessary. Using the static gauge $\tau = t$, and equations (6.65), we find

$$(\dot{X})^2 = -c^2 + \left(\frac{\partial\vec{X}}{\partial t}\right)^2, \quad (X')^2 = \left(\frac{\partial\vec{X}}{\partial\sigma}\right)^2, \quad \dot{X} \cdot X' = \frac{\partial\vec{X}}{\partial t} \cdot \frac{\partial\vec{X}}{\partial\sigma} \,. \tag{6.84}$$

With these relations we simplify the argument of the square root in the string action:

$$(\dot{X} \cdot X')^2 - (\dot{X})^2(X')^2 = \left(\frac{\partial\vec{X}}{\partial t} \cdot \frac{\partial\vec{X}}{\partial\sigma}\right)^2 + \left[c^2 - \left(\frac{\partial\vec{X}}{\partial t}\right)^2\right]\left(\frac{\partial\vec{X}}{\partial\sigma}\right)^2$$

$$= \left(\frac{ds}{d\sigma}\right)^2\left[\left(\frac{\partial\vec{X}}{\partial t} \cdot \frac{\partial\vec{X}}{\partial s}\right)^2 + c^2 - \left(\frac{\partial\vec{X}}{\partial t}\right)^2\right]. \tag{6.85}$$

The terms on the right-hand side above can be neatly expressed in terms of $v_\perp^2$. Making use of (6.83),

$$(\dot{X} \cdot X')^2 - (\dot{X})^2(X')^2 = \left(\frac{ds}{d\sigma}\right)^2\left(c^2 - v_\perp^2\right) \,, \tag{6.86}$$

or, alternatively,

$$\sqrt{(\dot{X} \cdot X')^2 - (\dot{X})^2(X')^2} = c\,\frac{ds}{d\sigma}\sqrt{1 - \frac{v_\perp^2}{c^2}}\,. \tag{6.87}$$

This simple expression for the string Lagrangian density shows that $\vec{v}_\perp$ is a natural dynamical variable. Moreover, the longitudinal component of the velocity is completely irrelevant. Now we can write the string action as

$$S = -T_0 \int dt \int_0^{\sigma_1} d\sigma \left(\frac{ds}{d\sigma}\right)\sqrt{1 - \frac{v_\perp^2}{c^2}}\,. \tag{6.88}$$

Here $ds/d\sigma = |\partial \vec{X}/\partial \sigma|$. We did not cancel the $d\sigma$ because it is typically useful to have an integral over a fixed parameter range. While the range of σ is constant, the total length of a string is time dependent. We introduced s as the function of σ that gives length along the string at a fixed time. This definition, used at different times, endows s with a time dependence that can be relevant if we compare strings at different times.

The associated Lagrangian is given by

$$L = -T_0 \int ds \sqrt{1 - \frac{v_\perp^2}{c^2}}\,. \tag{6.89}$$

This formula was written as an integral over the length parameter in order to give an interpretation. For each infinitesimal piece of string, $T_0 ds$ is its rest energy. As a result, the Lagrangian is an integral over the string of (minus) the rest energy times a local relativistic factor. In this form, we recognize (6.89) as the natural generalization of the relativistic particle Lagrangian (5.8).

The action (6.88) is valid both for open strings and for closed strings. Although relatively simple, it still leads to rather complicated equations of motion in all but the most symmetrical situations. In order to obtain simple equations of motion, we will have to be clever in our choice of σ. For open strings, in addition, we must understand how the endpoints move.

We conclude this section by simplifying our expressions (6.49) and (6.50) for $\mathcal{P}^{\tau\mu}$ and $\mathcal{P}^{\sigma\mu}$ in the static gauge. Let us begin with $\mathcal{P}^{\sigma\mu}$. Its denominator is given in (6.87) and its numerator is simplified using relations (6.84). We find

$$\mathcal{P}^{\sigma\mu} = -\frac{T_0}{c}\,\frac{\left(\dfrac{\partial\vec{X}}{\partial\sigma}\cdot\dfrac{\partial\vec{X}}{\partial t}\right)\dot{X}^\mu - \left(-c^2 + \left(\dfrac{\partial\vec{X}}{\partial t}\right)^2\right)X^{\mu\prime}}{c\,\dfrac{ds}{d\sigma}\sqrt{1 - \dfrac{v_\perp^2}{c^2}}}\,. \tag{6.90}$$

Bringing the $ds/d\sigma$ from the denominator up to the numerator, we can turn derivatives with respect to σ into derivatives with respect to s:

$$\mathcal{P}^{\sigma\mu} = -\frac{T_0}{c^2} \frac{\left(\frac{\partial \vec{X}}{\partial s} \cdot \frac{\partial \vec{X}}{\partial t}\right)\dot{X}^\mu + \left(c^2 - \left(\frac{\partial \vec{X}}{\partial t}\right)^2\right)\frac{\partial X^\mu}{\partial s}}{\sqrt{1 - \frac{v_\perp^2}{c^2}}}. \tag{6.91}$$

The $\mu = 0$ component of this quantity simplifies considerably. Since $\dot{X}^0 = c$ and $\partial X^0/\partial s = c\,\partial t/\partial s = 0$, we find

$$\mathcal{P}^{\sigma 0} = -\frac{T_0}{c} \frac{\left(\frac{\partial \vec{X}}{\partial s} \cdot \frac{\partial \vec{X}}{\partial t}\right)}{\sqrt{1 - \frac{v_\perp^2}{c^2}}}. \tag{6.92}$$

A rather similar calculation for $\mathcal{P}^{\tau\mu}$ gives

$$\mathcal{P}^{\tau\mu} = \frac{T_0}{c^2}\frac{ds}{d\sigma} \frac{\dot{X}^\mu - \left(\frac{\partial \vec{X}}{\partial s} \cdot \frac{\partial \vec{X}}{\partial t}\right)\frac{\partial X^\mu}{\partial s}}{\sqrt{1 - \frac{v_\perp^2}{c^2}}}. \tag{6.93}$$

Quick calculation 6.4 Prove (6.93).

It follows from (6.93) that $\mathcal{P}^{\tau 0}$ and $\vec{\mathcal{P}}^\tau$ are given by

$$\mathcal{P}^{\tau 0} = \frac{T_0}{c}\frac{ds}{d\sigma}\frac{1}{\sqrt{1 - \frac{v_\perp^2}{c^2}}}, \qquad \vec{\mathcal{P}}^\tau = \frac{T_0}{c^2}\frac{ds}{d\sigma}\frac{\vec{v}_\perp}{\sqrt{1 - \frac{v_\perp^2}{c^2}}}. \tag{6.94}$$

6.9 Motion of open string endpoints

We will now analyze the motion of the endpoints of an open relativistic string. We consider endpoints that are free to move in all directions. Given our discussion in Section 6.5, this means that we have a space-filling D-brane. Free endpoints are specified by the boundary conditions (6.56), which require the vanishing of $\mathcal{P}^\sigma_\mu$ at the endpoints. We will discover two important properties of the free motion of open string endpoints.

- The endpoints move with the speed of light.
- The endpoints move transversely to the string.

On the interior of the string the notion of a velocity was ambiguous. For the string endpoints, however, the velocity is well defined – there is no ambiguity defining the velocity of points! Therefore, our statements about endpoint motion have content. In the second state-ment, motion transverse to the string means that the velocity of an endpoint is orthogonal to the tangent to the string at the endpoint.

To prove the above properties we first recall that $\mathcal{P}^{\sigma 0}$ must vanish at the endpoints. Using (6.92) and noting that the square root in the denominator cannot be infinite, we deduce that the numerator must vanish:

$$\frac{\partial \vec{X}}{\partial s} \cdot \frac{\partial \vec{X}}{\partial t} = 0 \quad \text{at the endpoints} . \tag{6.95}$$

Since $\partial \vec{X}/\partial s$ is a unit vector tangent to the string, and $\partial \vec{X}/\partial t$ is the endpoint veloc-ity, this equation proves that the endpoints move transversely to the string – one of our two claims. In agreement with this interpretation, using (6.95) in (6.82) we see that at the endpoints $\vec{v}_\perp = \partial \vec{X}/\partial t \equiv \vec{v}$. Equation (6.95) actually allows for vanishing endpoint velocity, in which case the transversality property would be trivially satisfied. But this cannot happen; the endpoints move with the speed of light, as we now show.

Using (6.95), we simplify the expression (6.91) for $\mathcal{P}^{\sigma \mu}$ at the endpoints:

$$\mathcal{P}^{\sigma \mu} = -T_0 \sqrt{1 - \frac{v^2}{c^2}} \frac{\partial X^\mu}{\partial s} \quad \text{at the endpoints.} \tag{6.96}$$

For the space coordinates, $\mu = 1, \ldots, d$, equation (6.96) gives

$$\vec{\mathcal{P}}^\sigma = -T_0 \sqrt{1 - \frac{v^2}{c^2}} \frac{\partial \vec{X}}{\partial s} = 0 \quad \text{at the endpoints.} \tag{6.97}$$

Since $\partial \vec{X}/\partial s$ is a unit vector, we conclude that

$$v^2 = c^2. \tag{6.98}$$

Free open string endpoints move with the speed of light.

This conclusion implies that the denominator in (6.92) actually vanishes at the endpoints. Equation (6.95) must still hold, otherwise $\mathcal{P}^{\sigma 0}$ would diverge, rather than vanish at the endpoints. In fact, as we approach the string endpoints, the numerator in (6.92) must vanish faster than the denominator to ensure that the ratio goes to zero.

Problems

Problem 6.1 The induced metric on a two-dimensional surface.

A two-dimensional surface in flat three-dimensional space is described by the height func-tion $z = h(x, y)$, so that points on the surface take the form $(x, y, h(x, y))$. This surface is naturally parameterized by x and y. Recall (6.14), which gives the metric $g_{ij}(\xi)$ induced on a surface parameterized by (ξ^1, ξ^2).

(a) Calculate the components of the metric g_{ij} in terms of h and its derivatives. Write a formula for g_{ij} of the form $g_{ij} = \delta_{ij} + \cdots$.

(b) Write the fully simplified appropriate form of the area integral $(A = \int d\xi^1 d\xi^2 \sqrt{g})$.

Problem 6.2 Metric on S^2 from stereographic parameterization.

Consider a unit sphere S^2 in $\mathbb{R}^3$ centered at the origin: $x^2 + y^2 + z^2 = 1$. Denote a point on the sphere by $\vec{x} = (x, y, z)$. In the stereographic parameterization of the sphere we use parameters ξ^1 and ξ^2 and points on the sphere are (see (6.1)):

$$\vec{x}(\xi^1, \xi^2) = \left(x(\xi^1, \xi^2), \; y(\xi^1, \xi^2), \; z(\xi^1, \xi^2) \right).$$

Given parameters (ξ^1, ξ^2), the corresponding point on the sphere is that which lies on the line that goes through the north pole $N = (0, 0, 1)$ and the point $(\xi^1, \xi^2, 0)$. Note that the north pole itself is not attained for any finite values of the parameters.

(a) Draw a sketch for the above construction. What are the required ranges for ξ^1 and ξ^2 if we wish to parameterize the full sphere (except for the north pole)?
(b) Calculate the functions $x(\xi^1, \xi^2)$, $y(\xi^1, \xi^2)$, and $z(\xi^1, \xi^2)$.
(c) Calculate the four components of the induced metric $g_{ij}(\xi)$. This is the metric on the sphere, described using the ξ parameters. The algebra is a bit messy, but the result is quite simple (use a symbolic manipulator!).
(d) Check your result by computing the area of the sphere using (6.17).

Problem 6.3 Schwarz inequality in $\mathbb{R}^{1,1}$.

Consider a two-dimensional vector space V with a constant metric such that there is a timelike vector t' $(t'^2 = t' \cdot t' < 0)$ and a spacelike vector s' $(s'^2 = s' \cdot s' > 0)$.

(a) Show that you can construct vectors t and s such that $t \cdot t = -1$, $s \cdot s = 1$, and $t \cdot s = 0$. [Hint: choose t to be in the direction of t'.] The vectors t and s provide a canonical basis for V, which is now identified as the space $\mathbb{R}^{1,1}$.
(b) Consider now two arbitrary vectors v_1 and v_2 in V. Use the basis in (a) to prove that

$$(v_1 \cdot v_2)^2 \geq v_1^2 v_2^2,$$

where the equality only holds if and only if the vectors v_1 and v_2 are parallel. This result gives some additional perspective on our proof of (6.34).

Problem 6.4 Stretched string and a nonrelativistic limit.

Examine the action (6.88) for a relativistic string with endpoints attached at $(0, \vec{0})$ and $(a, \vec{0})$, as in Section 6.7. Consider the nonrelativistic approximation where $|\vec{v}_\perp| \ll c$ and the oscillations are small (see (4.3)). You may denote by $\vec{y}$ the collection of transverse coordinates $X^2, \ldots, X^d$ and write $\vec{y}(t, x)$, where x is the coordinate corresponding to X^1. Explain why the following relations hold:

$$ds^2 = dx^2 + d\vec{y} \cdot d\vec{y}, \quad \vec{v}_\perp \simeq \frac{\partial \vec{y}}{\partial t}.$$

Show that the action reduces, up to an additive constant, to the *action* for a nonrelativistic string performing small transverse oscillations. What are the tension and the linear mass density of the resulting string? What is the additive constant?

Problem 6.5 Alternative derivation of the nonrelativistic limit.

Consider the same setup and nonrelativistic approximation discussed in Problem 6.4. This time, however, start your analysis with the Nambu–Goto action (6.39) and work in the static gauge. Moreover, parameterize the strings using $X^1 = x = a\sigma/\sigma_1$. This parameterization is allowed for small oscillations. In fact, it is allowed for any motion in which X^1 is an increasing function along the string.

Problem 6.6 Planar motion for open string with attached endpoints.

Consider the motion of a relativistic open string on the (x, y) plane. The string endpoints are attached to $(x, y) = (0, 0)$ and $(x, y) = (a, 0)$, where $a > 0$. We want to study motions that can be described using the function $y(t, x)$ which gives the vertical displacement of the string for $x \in [0, a]$. Show that the Lagrangian (6.89) can be written in the form:

$$ L = -T_0 \int_0^a dx \sqrt{1 + y'^2 - \frac{\dot{y}^2}{c^2}} . $$

Here $y' = \partial y/\partial x$ and $\dot{y} = \partial y/\partial t$.

Problem 6.7 Time evolution of a closed circular string.

At $t = 0$, a closed string forms a circle of radius R on the (x, y) plane and has zero velocity. The time development of this string can be studied using the action (6.88). The string will remain circular, but its radius will be a time-dependent function $R(t)$. Give the Lagrangian L as a function of $R(t)$ and its time derivative. Calculate the radius and velocity as functions of time. Sketch the spacetime surface traced by the string in a three-dimensional plot with x, y, and ct axes. [Hint: calculate the Hamiltonian associated with L and use energy conservation.]

Problem 6.8 Covariant analysis of open string endpoint motion.

Use the explicit form of $\mathcal{P}_\mu^\sigma$ to calculate $\mathcal{P}_\mu^\sigma \mathcal{P}^{\sigma\mu}$, and use the result of this calculation to prove that free open string endpoints move with the speed of light.

Problem 6.9 Hamiltonian density for relativistic strings.[†]

Consider the string Lagrangian density $\mathcal{L}$ in the static gauge and written in terms of $\partial_\sigma \vec{X}$ and $\partial_t \vec{X}$. Show that the canonical momentum density $\vec{\mathcal{P}}(t, \sigma)$ is given by

$$ \vec{\mathcal{P}}(t, \sigma) \equiv \frac{\partial \mathcal{L}}{\partial(\partial_t \vec{X})} = \frac{T_0}{c^2} \frac{\vec{v}_\perp}{\sqrt{1 - \frac{v_\perp^2}{c^2}}} \frac{ds}{d\sigma} . $$

Calculate the Hamiltonian density $\mathcal{H}$, again in terms of $\vec{v}_\perp$ and $\frac{ds}{d\sigma}$. Write the total Hamiltonian as $H = \int d\sigma \mathcal{H} = \int ds(\ldots)$ and show that your answer is consistent with the interpretation that the energy of the string arises as the energy of transverse motion of a string whose rest mass arises solely from the tension.

Problem 6.10 Circular string in de Sitter space.

The Nambu–Goto action on a curved spacetime with metric $g_{\mu\nu}(x)$ can be written as in (6.39) with the convention that all dot products use the metric $g_{\mu\nu}$:

$$\dot{X} \cdot X' = g_{\mu\nu}(X)\dot{X}^\mu X^{\nu\prime}, \quad (\dot{X})^2 = g_{\mu\nu}(X)\dot{X}^\mu \dot{X}^\nu, \quad (X')^2 = g_{\mu\nu}(X)X'^\mu X'^\nu.$$

Consider strings in an expanding four-dimensional de Sitter spacetime, for which the metric $g_{\mu\nu}$ can be taken to be diagonal, with values

$$g_{00} = -1, \quad g_{11} = g_{22} = g_{33} = e^{2Ht}.$$

The Hubble constant H has units of one over time so that Ht is dimensionless.

(a) Assume $X^0 \equiv ct = c\tau$ and write $X^\mu = \{X^0, \vec{X}\}$. Write the Nambu–Goto action in terms of t and σ derivatives of $\vec{X}$.

(b) Consider now a circular string on the (x^1, x^2) plane, namely

$$X^1(t, \sigma) = r(t) \cos \sigma, \quad X^2(t, \sigma) = r(t) \sin \sigma, \quad \sigma \in [0, 2\pi],$$

where $r(t)$ is a radius function to be determined. Use this ansatz to simplify the string action and perform the integration over σ. Write your result as

$$S = \int dt \, L(\dot{r}(t), r(t); t),$$

and determine $L(\dot{r}(t), r(t); t)$, which is explicitly time dependent. Because of the e^{2Ht} factors in the metric, the physically measured radius of the string is actually $R(t) = e^{Ht} r(t)$. Write the Lagrangian in terms of R and $\dot{R}$.

(c) Consider strings with constant R and use the Lagrangian to give the potential $V(R)$ for such strings. Plot this potential and verify that it is well-defined only if $R \leq c/H$. Find a critical point of the potential and the corresponding value of R. Is this static string in stable equilibrium?

(d) Use the Lagrangian for R and $\dot{R}$ to calculate the corresponding Hamiltonian, expressed as a function of R and $\dot{R}$. Simplify your answer. Is this Hamiltonian function conserved for physical motion?

Problem 6.11 Open strings ending on D-branes of various dimensions.

Consider a world with d spatial dimensions. A Dp-brane is an extended object with p spatial dimensions: a p-dimensional hyperplane inside the d-dimensional space. We will examine properties of strings ending on a Dp-brane, where $0 \leq p < d$. The case $p = d$, where the D-brane is space filling was discussed in Section 6.9.

For a Dp-brane, let x^i, with $i = 1, \ldots, p$, correspond to directions on the Dp-brane, and x^a with $a = p + 1, \ldots, d$, correspond to directions orthogonal to the Dp-brane. The Dp-brane position is specified by $x^a = 0$, for $a = p + 1, \ldots, d$. Open string endpoints must lie on the Dp-brane. Focusing on the $\sigma = 0$ endpoint, we thus have

$$X^a(t, \sigma = 0) = 0, \quad a = p + 1, \ldots, d.$$

The motion of the endpoint along the D-brane directions x^i is free. We work in the static gauge.

(a) State the conditions satisfied by $\mathcal{P}_0^\sigma$, $\mathcal{P}_i^\sigma$, and $\mathcal{P}_a^\sigma$ at the endpoint (no condition is a possibility!).

Prove the following.

(b) All boundary conditions are automatically satisfied if the string ends on a D0-brane.
(c) For a string ending on a D1-brane, the tangent to the string at the endpoint is orthogonal to the D1-brane, and the endpoint velocity is unconstrained.
(d) For a string ending on a Dp-brane, with $p \geq 2$, there are two possibilities:
  (i) the string is orthogonal to the Dp-brane at the endpoint, and the endpoint velocity is unconstrained, or,
  (ii) the string is not orthogonal to the Dp-brane at the endpoint, and the endpoint moves with the speed of light transversely to the string.

7

String parameterization and classical motion

We construct lines of constant σ that are perpendicular to the lines of constant τ and use the energy density carried by the string to fix the σ parameterization completely. The resulting system of string equations of motion includes wave equations and two nonlinear constraints. We find the general solution which describes the motion of open strings with free endpoints. We examine the free evolution of closed strings and find that cusps that appear and later disappear are generic. We discuss cosmic strings and the gravitational lensing they can produce.

7.1 Choosing a σ parameterization

We have already learned a few facts about the motion of relativistic strings. In particular, we learned that free open string endpoints move with the speed of light transversely to the string. This result was obtained using the static gauge $X^0 \equiv ct = c\tau$, which partially fixed the parameterization of the world-sheet. Once we have chosen this gauge, the string motion is defined by the functions $\vec{X}(t, \sigma)$. As we vary t and σ, $\vec{X}(t, \sigma)$ describes the string spatial surface – the surface in space consisting of the strings at all times. In this chapter, the string spatial surface will simply be called the string surface. We will also use the static gauge throughout. The statement that the open string endpoints move transversely to the string implies that the vectors tangent to the boundary of the string surface are orthogonal to the strings. Our goal is to find a useful σ parameterization of the string surface. If we know this parameterization, we also know the σ parameterization of the world-sheet and, in fact, the complete parameterization of the world-sheet.

We will now show how to use a particular σ parameterization of a single string to construct a useful σ parameterization of all strings, and thus of the entire string surface. Suppose that the $t = 0$ string is given some σ parameterization with $\sigma \in [0, \sigma_1]$ (see Figure 7.1). Now, consider the string at $t = \epsilon$, with ϵ infinitesimal. On the string surface we can draw short segments perpendicular to the $t = 0$ string. Let these segments intersect the $t = \epsilon$ string. Consider a point σ_0 on the $t = 0$ string, and the short perpendicular above it. We declare the intersection of this perpendicular with the $t = \epsilon$ string to also have $\sigma = \sigma_0$. We do this all over the $t = 0$ string, obtaining a parameterization of the $t = \epsilon$ string. We then repeat this procedure, using the $t = \epsilon$ string to parameterize the $t = 2\epsilon$ string. We continue in this way, working in the limit of very small ϵ. The result is a set of lines of constant σ that are

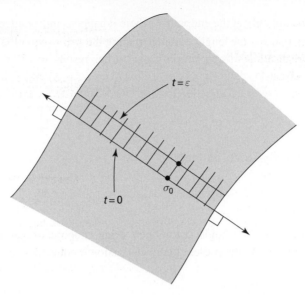

Fig. 7.1 Using the parameterization of the $t = 0$ string to construct the parameterization of the $t = \epsilon$ string. On the string surface, the lines of constant σ are chosen to be orthogonal to the lines of constant t.

everywhere orthogonal to the strings (that is, orthogonal to the lines of constant t). This construction can be done both for open and for closed strings. For closed strings the σ range $[0, \sigma_c]$ of the $t = 0$ string becomes the range of all other strings. For open strings the σ range $[0, \sigma_1]$ of the $t = 0$ string also becomes the range of all other strings. This happens because the boundaries of the string surface are orthogonal to the strings, and, as a result, the boundaries are lines of constant σ.

In summary, the σ parameterization of a given string can be used to construct lines of constant σ that are always perpendicular to the lines of constant t. In this parameterization of the string surface, the tangent $\partial \vec{X}/\partial \sigma$ to the strings and the tangent $\partial \vec{X}/\partial t$ to the lines of constant σ are perpendicular to each other at any point:

$$\frac{\partial \vec{X}}{\partial \sigma} \cdot \frac{\partial \vec{X}}{\partial t} = 0. \tag{7.1}$$

Since the velocity $\partial \vec{X}/\partial t$ is perpendicular to the string, it coincides with $\vec{v}_\perp$ (see (6.82)):

$$\vec{v}_\perp = \frac{\partial \vec{X}}{\partial t} \quad \text{at all points,} \tag{7.2}$$

and not only at the endpoints, where it happens independently of parameterization. Recalling that s is the length parameter along the string, equation (7.1) implies that $\frac{\partial \vec{X}}{\partial s} \cdot \frac{\partial \vec{X}}{\partial t} = 0$, which allows us to simplify our earlier results for $\mathcal{P}^{\tau\mu}$ and $\mathcal{P}^{\sigma\mu}$. We find that (6.93) becomes

$$\mathcal{P}^{\tau\mu} = \frac{T_0}{c^2} \frac{\frac{ds}{d\sigma}}{\sqrt{1 - \frac{v_\perp^2}{c^2}}} \frac{\partial X^\mu}{\partial t}. \tag{7.3}$$

Similarly, (6.91) becomes

$$\mathcal{P}^{\sigma\mu} = -T_0 \sqrt{1 - \frac{v_\perp^2}{c^2}} \frac{\partial X^\mu}{\partial s}. \tag{7.4}$$

Equation (7.4) holds at the open string endpoints regardless of σ-parameterization (see (6.96)); with the present parameterization it holds all over the string.

7.2 Physical interpretation of the string equation of motion

Having used the σ parameterization to simplify some of our previous expressions, we now look into the string equations of motion (6.53). With $t = \tau$ we have

$$\frac{\partial \mathcal{P}^{\tau\mu}}{\partial t} = -\frac{\partial \mathcal{P}^{\sigma\mu}}{\partial \sigma}. \tag{7.5}$$

Let us first consider the $\mu = 0$ component of this equation. It follows from (7.4) that $\mathcal{P}^{\sigma 0} = 0$. Furthermore, equation (7.3) gives

$$\mathcal{P}^{\tau 0} = \frac{T_0}{c} \frac{\frac{ds}{d\sigma}}{\sqrt{1 - \frac{v_\perp^2}{c^2}}}. \tag{7.6}$$

Back into the equation of motion (7.5), we obtain

$$\frac{\partial \mathcal{P}^{\tau 0}}{\partial t} = \frac{\partial}{\partial t} \left(\frac{T_0 \frac{ds}{d\sigma}}{c\sqrt{1 - \frac{v_\perp^2}{c^2}}} \right) = 0. \tag{7.7}$$

To understand this result physically, consider a little piece of string associated with a $d\sigma$ that is a small fixed number. The motion of this particular piece of string is well defined now that we have fixed the lines of constant σ. Since ds denotes the length of the $d\sigma$ piece of string, ds can depend on time. If we multiply equation (7.7) by the constant $d\sigma$, we conclude that

$$\frac{T_0 \, ds}{\sqrt{1 - \frac{v_\perp^2}{c^2}}}, \tag{7.8}$$

is constant in time. Expression (7.8) has units of energy, which suggests that it may be the relativistic energy associated with this piece of the string. Indeed, this is in agreement with our findings in Section 6.7 where we saw that the rest energy of a static stretched string is given by its length times the tension T_0. The rest energy in the above expression is $T_0 ds$, and the relativistic factor in the denominator makes (7.8) the total energy. Equation (7.7) therefore states that the energy in *each* piece $d\sigma$ of the string is conserved. This is a very interesting fact. It implies, for example, that the energy in the range $[0, \sigma_0]$ of an open string is constant in time for any fixed σ_0.

The above interpretation is also confirmed by a calculation of the energy E of a relativistic string (Problem 6.9). You found that the Hamiltonian is

$$H = \int \frac{T_0 \, ds}{\sqrt{1 - \frac{v_\perp^2}{c^2}}},$$

(7.9)

which shows that (7.8) is the energy in a piece of string.

Now we turn to the space components of the string equation of motion. The space components $\vec{\mathcal{P}}^\tau$ of $\mathcal{P}^{\tau\mu}$ can be read from (7.3):

$$\vec{\mathcal{P}}^\tau = \frac{T_0}{c^2} \frac{\frac{ds}{d\sigma}}{\sqrt{1 - \frac{v_\perp^2}{c^2}}} \vec{v}_\perp,$$

(7.10)

and similarly, from (7.4),

$$\vec{\mathcal{P}}^\sigma = -T_0 \sqrt{1 - \frac{v_\perp^2}{c^2}} \frac{\partial \vec{X}}{\partial s}.$$

(7.11)

Now we can substitute these expressions back into (7.5) to find that

$$\frac{\partial}{\partial \sigma}\left[T_0 \sqrt{1 - \frac{v_\perp^2}{c^2}} \frac{\partial \vec{X}}{\partial s} \right] = \frac{\partial}{\partial t}\left[\frac{T_0}{c^2} \frac{\frac{ds}{d\sigma}}{\sqrt{1 - \frac{v_\perp^2}{c^2}}} \vec{v}_\perp \right]$$

$$= \frac{T_0}{c^2} \frac{\frac{ds}{d\sigma}}{\sqrt{1 - \frac{v_\perp^2}{c^2}}} \frac{\partial \vec{v}_\perp}{\partial t},$$

(7.12)

where the final step used equation (7.7). It is possible to interpret this equation loosely in terms of an "effective" nonrelativistic string. Recall that the equations of motion for a classical nonrelativistic string are

$$\mu_0 \frac{\partial^2 \vec{y}}{\partial t^2} = T_0 \frac{\partial^2 \vec{y}}{\partial x^2} = \frac{\partial}{\partial x}\left[T_0 \frac{\partial \vec{y}}{\partial x} \right],$$

(7.13)

where x is a length parameter along the direction defined by the static stretched string, and $\vec{y}$ is the transverse displacement. How do we recast (7.12) to resemble (7.13)? We use the

$ds/d\sigma$ factor on the right-hand side to transform the σ-derivative on the left-hand side into an s-derivative:

$$\frac{T_0}{c^2} \frac{1}{\sqrt{1 - \frac{v_\perp^2}{c^2}}} \frac{\partial \vec{v}_\perp}{\partial t} = \frac{\partial}{\partial s}\left[T_0 \sqrt{1 - \frac{v_\perp^2}{c^2}} \left(\frac{\partial \vec{X}}{\partial s}\right)\right]. \tag{7.14}$$

For small oscillations the length parameter s along the vibrating string is roughly equal to the parameter x along the direction of the static string. We can then compare this equation with (7.13) and conclude that the relativistic string has a velocity-dependent effective tension T_{eff} and a velocity-dependent effective mass density μ_{eff} given by

$$T_{\text{eff}} = T_0 \sqrt{1 - \frac{v_\perp^2}{c^2}}, \quad \mu_{\text{eff}} = \frac{T_0}{c^2} \frac{1}{\sqrt{1 - \frac{v_\perp^2}{c^2}}}. \tag{7.15}$$

Since free open string endpoints move with $v_\perp = c$, the effective tension of the string goes to zero at the endpoints. One could say that the endpoints have to move at the speed of light in order to make the tension vanish at the endpoints. This is the only way that the relativistic string can have a tension and still make sense with free open ends. The effective mass density diverges at the endpoints. This is not a problem; the same divergence is present for the energy density which appears as the integrand in (7.9). Despite the singular behavior at the endpoints, the integral turns out to be finite, as required by consistency since, after all, we are describing strings with finite energy.

7.3 Wave equation and constraints

Equation (7.14) is still fairly complicated. It may seem that, having fixed the lines of constant σ, we have run out of reparameterizations that can help us simplify the equations of motion. This is not the case, however. We showed how to construct the lines of constant σ *if* we have one parameterized string. We must now try to parameterize this first string in the best possible way!

Here is the physical way to do so: we will parameterize the string so that each string segment of equal parameter length σ carries the same amount of energy. We will parameterize the string using the energy! This parameterization will yield simple equations of motion. To see this, we first rewrite (7.14) suggestively by changing s-derivatives into σ-derivatives:

$$\frac{1}{c^2} \frac{\partial^2 \vec{X}}{\partial t^2} = \frac{\sqrt{1 - \frac{v_\perp^2}{c^2}}}{\frac{ds}{d\sigma}} \frac{\partial}{\partial \sigma}\left[\frac{\sqrt{1 - \frac{v_\perp^2}{c^2}}}{\frac{ds}{d\sigma}} \frac{\partial \vec{X}}{\partial \sigma}\right]. \tag{7.16}$$

Let $A(\sigma)$ denote the ratio that appears prominently in the above equation:

$$A(\sigma) = \frac{\frac{ds}{d\sigma}}{\sqrt{1 - \frac{v_\perp^2}{c^2}}}. \tag{7.17}$$

We already showed in (7.7) that $A(\sigma)$ is independent of time. We will now choose σ in such a way that $A = 1$. If we do so, the equation of motion (7.16) becomes the familiar wave equation:

$$\frac{1}{c^2}\frac{\partial^2 \vec{X}}{\partial t^2} = \frac{\partial^2 \vec{X}}{\partial \sigma^2}. \tag{7.18}$$

This is a great simplification. To find a σ that leads to $A = 1$ we assign $\sigma = 0$ to one endpoint of the open string and work our way along the string assigning to each piece ds of string the interval $d\sigma$ given by

$$d\sigma = \frac{ds}{\sqrt{1 - \frac{v_\perp^2}{c^2}}} = \frac{1}{T_0} dE. \tag{7.19}$$

The first equality implies $A = 1$, and the second equality follows from the identification of (7.8) with the energy dE carried by the little piece of string. In this parameterization the energy density $dE/d\sigma$ is a constant equal to the tension. Equation (7.19) can be integrated from the $\sigma = 0$ endpoint up to a point Q, giving

$$\sigma(Q) = \frac{E(Q)}{T_0}. \tag{7.20}$$

The coordinate $\sigma(Q)$ assigned to Q equals the energy $E(Q)$ carried by the portion of the string stretching from the selected endpoint up to Q, divided by the tension. It also follows from the above equation that

$$\sigma \in [0, \sigma_1], \quad \sigma_1 = \frac{E}{T_0}, \tag{7.21}$$

where E is the total energy of the string.

Our choice of σ, done for all strings, is consistent with the orthogonality condition (7.1). In fact, the lines of constant σ have not been changed, only the values of σ assigned to them have. Our constant σ lines still ensure that the energy in the $[0, \sigma]$ portion of the strings is constant and such lines are orthogonal to the strings.

The parameterization condition (7.19) is actually equivalent to a differential constraint on the coordinates $\vec{X}$. We first rewrite the first equality in (7.19) as

$$\left(\frac{ds}{d\sigma}\right)^2 + \frac{1}{c^2} v_\perp^2 = 1. \tag{7.22}$$

Recalling that $\partial\vec{X}/\partial s$ is a unit vector, and making use of (7.2), we find

$$\left(\frac{\partial\vec{X}}{\partial\sigma}\right)^2 + \frac{1}{c^2}\left(\frac{\partial\vec{X}}{\partial t}\right)^2 = 1. \tag{7.23}$$

Finally, let us examine the boundary conditions. From (7.11) we have

$$\vec{\mathcal{P}}^{\sigma} = -T_0 \sqrt{1 - \frac{v_{\perp}^2}{c^2}} \frac{d\sigma}{ds} \frac{\partial \vec{X}}{\partial \sigma} = -T_0 \frac{\partial \vec{X}}{\partial \sigma}. \tag{7.24}$$

Therefore, the free endpoint boundary condition is very simple:

$$\frac{\partial \vec{X}}{\partial \sigma} = 0 \quad \text{at the endpoints.} \tag{7.25}$$

With the chosen parameterization, the free endpoint boundary condition, first introduced in (6.56), is simply a Neumann boundary condition.

All in all we have four equations to solve in order to find the motion of a relativistic string: (7.18), (7.1), (7.23), and (7.25). We collect them here:

$$\text{wave equation:} \quad \frac{\partial^2 \vec{X}}{\partial \sigma^2} - \frac{1}{c^2} \frac{\partial^2 \vec{X}}{\partial t^2} = 0, \tag{7.26}$$

$$\text{parameterization condition:} \quad \frac{\partial \vec{X}}{\partial t} \cdot \frac{\partial \vec{X}}{\partial \sigma} = 0, \tag{7.27}$$

$$\text{parameterization condition:} \quad \left(\frac{\partial \vec{X}}{\partial \sigma}\right)^2 + \frac{1}{c^2} \left(\frac{\partial \vec{X}}{\partial t}\right)^2 = 1, \tag{7.28}$$

$$\text{boundary condition:} \quad \frac{\partial \vec{X}}{\partial \sigma}\bigg|_{\sigma=0} = \frac{\partial \vec{X}}{\partial \sigma}\bigg|_{\sigma=\sigma_1} = 0. \tag{7.29}$$

For a string with energy E, the above equations require $\sigma_1 = E/T_0$. For the record, we also include (7.19), written as

$$\frac{1}{T_0} \frac{dE}{d\sigma} = \frac{\frac{ds}{d\sigma}}{\sqrt{1 - \frac{v_{\perp}^2}{c^2}}} = 1. \tag{7.30}$$

Finally, from equations (7.3) and (7.4) we get

$$\mathcal{P}^{\tau\mu} = \frac{T_0}{c^2} \frac{\partial X^{\mu}}{\partial t}, \tag{7.31}$$

$$\mathcal{P}^{\sigma\mu} = -T_0 \frac{\partial X^{\mu}}{\partial \sigma}. \tag{7.32}$$

7.4 General motion of an open string

In this section our goal is to describe the general motion of open strings with free boundary conditions. We will therefore examine in detail how to solve equations (7.26)–(7.29).

Let us first consider the wave equation for $\vec{X}$. This equation is readily solved in terms of arbitrary vector functions of $(ct \pm \sigma)$. We thus write

$$\vec{X}(t, \sigma) = \frac{1}{2}\left(\vec{F}(ct + \sigma) + \vec{G}(ct - \sigma)\right). \tag{7.33}$$

The boundary condition at the $\sigma = 0$ endpoint demands that

$$\frac{\partial \vec{X}}{\partial \sigma}\Big|_{\sigma=0} = 0 \longrightarrow \vec{F}'(ct) - \vec{G}'(ct) = 0, \tag{7.34}$$

where prime denotes derivative with respect to the argument. Since ct takes all possible values, the above equation holds for all values of the argument. Calling u the argument, we write

$$\frac{d\vec{F}(u)}{du} = \frac{d\vec{G}(u)}{du} \longrightarrow \vec{G}(u) = \vec{F}(u) + \vec{a}_0, \tag{7.35}$$

where $\vec{a}_0$ is a constant vector. Back in (7.33) we now have

$$\vec{X}(t, \sigma) = \frac{1}{2}\left(\vec{F}(ct + \sigma) + \vec{F}(ct - \sigma) + \vec{a}_0\right). \tag{7.36}$$

We can absorb the constant vector $\vec{a}_0$ into the definition of $\vec{F}$ (call $\vec{F}(u) + \vec{a}_0/2$ the new $\vec{F}$), and therefore our solution so far takes the form

$$\vec{X}(t, \sigma) = \frac{1}{2}\left(\vec{F}(ct + \sigma) + \vec{F}(ct - \sigma)\right). \tag{7.37}$$

Consider now the boundary condition at $\sigma = \sigma_1$:

$$\frac{\partial \vec{X}}{\partial \sigma}\Big|_{\sigma=\sigma_1} = 0 \longrightarrow \vec{F}'(ct + \sigma_1) - \vec{F}'(ct - \sigma_1) = 0. \tag{7.38}$$

Letting $u = ct - \sigma_1$, the above condition becomes

$$\frac{d\vec{F}}{du}(u + 2\sigma_1) = \frac{d\vec{F}}{du}(u). \tag{7.39}$$

This equation tells us that the derivative of $\vec{F}$ is periodic with period $2\sigma_1$. Integrating, we find that the function $\vec{F}$ is quasi-periodic: after a period $2\sigma_1$ it changes by a fixed constant. We write this as

$$\vec{F}(u + 2\sigma_1) = \vec{F}(u) + 2\sigma_1 \frac{\vec{v}_0}{c}, \tag{7.40}$$

where $\vec{v}_0$ is a vector constant of integration with units of velocity, and the constants have been added for convenience. This concludes our analysis of the boundary conditions.

We now examine what restrictions the parameterization conditions (7.27) and (7.28) impose on the function $\vec{F}$. A standard trick is to add and subtract the first equation to the second, as in

$$\left(\frac{\partial \vec{X}}{\partial \sigma}\right)^2 \pm 2 \frac{\partial \vec{X}}{\partial \sigma} \cdot \frac{1}{c}\frac{\partial \vec{X}}{\partial t} + \frac{1}{c^2}\left(\frac{\partial \vec{X}}{\partial t}\right)^2 = 1, \tag{7.41}$$

which can then be written more briefly as

$$\left(\frac{\partial \vec{X}}{\partial \sigma} \pm \frac{1}{c} \frac{\partial \vec{X}}{\partial t} \right)^2 = 1. \tag{7.42}$$

Note that this is equivalent to the two constraints (7.27) and (7.28). Using (7.37) we can evaluate the derivatives that enter into the constraints:

$$\frac{\partial \vec{X}}{\partial \sigma} = \frac{1}{2}\left(\vec{F}'(ct + \sigma) - \vec{F}'(ct - \sigma) \right),$$

$$\frac{1}{c} \frac{\partial \vec{X}}{\partial t} = \frac{1}{2}\left(\vec{F}'(ct + \sigma) + \vec{F}'(ct - \sigma) \right). \tag{7.43}$$

As a result,

$$\frac{\partial \vec{X}}{\partial \sigma} \pm \frac{1}{c} \frac{\partial \vec{X}}{\partial t} = \pm \vec{F}'(ct \pm \sigma). \tag{7.44}$$

It follows that the constraints (7.42) require $\vec{F}' \cdot \vec{F}' = 1$ for all arguments of $\vec{F}$. In other words, the vector $\vec{F}'(u)$ is a unit vector:

$$\left| \frac{d\vec{F}(u)}{du} \right|^2 = 1. \tag{7.45}$$

This is good progress: the constraints give a simple condition for $\vec{F}(u)$. Since $\vec{F}$ is a vector function of the parameter u, we can visualize $\vec{F}(u)$ as a parameterized curve in space. Equation (7.45) has a simple interpretation:

$$u \text{ is a length parameter along the curve } \vec{F}(u). \tag{7.46}$$

This is explained as follows. Consider two nearby points $\vec{F}(u + du)$ and $\vec{F}(u)$ on the curve. Their vector separation $d\vec{F} = \vec{F}(u + du) - \vec{F}(u)$ has length $|d\vec{F}|$. Equation (7.45) implies that $|d\vec{F}| = |du|$, showing that the parameter change $|du|$ is the distance between the two nearby points.

We can now summarize our analysis of the equations of motion. The general solution which describes the motion of an open string with free endpoints is given by

$$\vec{X}(t, \sigma) = \frac{1}{2}\left(\vec{F}(ct + \sigma) + \vec{F}(ct - \sigma) \right), \quad \sigma \in [0, \sigma_1], \tag{7.47}$$

where $\sigma_1 = E/T_0$, E is the energy of the string, and $\vec{F}$ satisfies the conditions

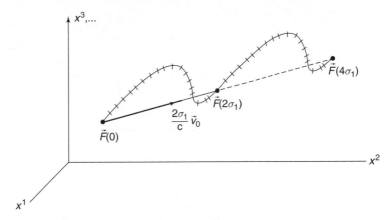

Fig. 7.2 The vector function $\vec{F}(u)$ changes by a constant vector $(2\sigma_1\vec{v}_0/c)$ when $u \to u + 2\sigma_1$. The parameterized curve $\vec{F}(u)$ encodes the full motion of an open string with free endpoints.

$$\left|\frac{d\vec{F}(u)}{du}\right|^2 = 1 \quad \text{and} \quad \vec{F}(u + 2\sigma_1) = \vec{F}(u) + 2\sigma_1 \frac{\vec{v}_0}{c}. \qquad (7.48)$$

The problem has been reduced to that of finding a vector function $\vec{F}$ which satisfies equations (7.48). The second of these equations tells us that it suffices to find $\vec{F}(u)$ for $u \in [0, 2\sigma_1]$. This determines $\vec{F}(u)$ for all u, and thus it determines $\vec{X}(t, \sigma)$ completely. The interpretation of $\vec{v}_0$ will be given below. An illustration of $\vec{F}$ is shown in Figure 7.2.

We can give a physical interpretation to $\vec{F}(u)$. It follows from (7.47) that the motion of the $\sigma = 0$ endpoint of the open string is described by

$$\vec{X}(t, 0) = \vec{F}(ct). \qquad (7.49)$$

Therefore, we see that:

$$\vec{F}(u) \text{ is the position of the } \sigma = 0 \text{ endpoint at time } u/c. \qquad (7.50)$$

Additionally, we can give a physical interpretation to the constant velocity $\vec{v}_0$. From the second equation in (7.48) we have

$$\vec{F}(2\sigma_1) = \vec{F}(0) + 2\sigma_1 \frac{\vec{v}_0}{c}, \qquad (7.51)$$

and expressing the $\vec{F}$ in terms of the position of the $\sigma = 0$ endpoint, we find

$$\vec{X}\left(t = \frac{2\sigma_1}{c}, 0\right) = \vec{X}(t = 0, 0) + \left(\frac{2\sigma_1}{c}\right)\vec{v}_0. \qquad (7.52)$$

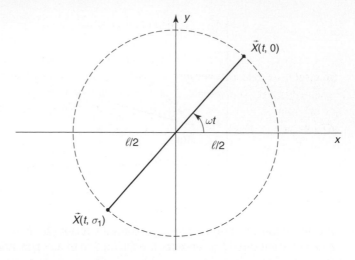

Fig. 7.3   An open string of length ℓ rotating in the (x, y) plane with angular velocity ω.

This shows that $\vec{v}_0$ is the average velocity of the $\sigma = 0$ endpoint during the time interval $[0, 2\sigma_1/c]$.

Quick calculation 7.1 Show that $\vec{v}_0$ is, in fact, the average velocity of any point σ on the string calculated over any time interval of duration $2\sigma_1/c$.

Quick calculation 7.2 Show that the velocity of any point on the string is a periodic function of time with period $2\sigma_1/c$, namely, $\dot{\vec{X}}(t, \sigma) = \dot{\vec{X}}(t + \frac{2\sigma_1}{c}, \sigma)$.

Since $\vec{F}$ can be reconstructed by looking only at the $\sigma = 0$ endpoint, we may ask: how long do we need to observe this endpoint in order to determine the full motion, past and future, of an open string with energy E? Since the motion is determined if we know $\vec{F}(u)$ from $u = 0$ to $u = 2\sigma_1$, we must observe $\vec{X}(t, 0)$ from $t = 0$ to $t = 2\sigma_1/c$. Since $\sigma_1 = E/T_0$, we need to observe the endpoint for a time interval $\Delta t = 2E/cT_0$. This is twice the time that light takes to travel a length E/T_0, the length of a static string of energy E.

We now use the above construction to describe the motion of a straight open string with energy E which rotates rigidly about its fixed midpoint in the (x, y) plane (Figure 7.3). Our first goal is to produce the function $\vec{F}(u)$. This function is easily constructed because we know all about the motion of the endpoints. Assuming that the string is of length ℓ and rotates with angular frequency ω, we describe the motion of the $\sigma = 0$ endpoint by

$$\vec{X}(t, 0) = \frac{\ell}{2} \left(\cos \omega t, \sin \omega t \right), \tag{7.53}$$

where we use vector notation with two components, since the motion is restricted to the (x, y) plane. Given that $\vec{F}(ct) = \vec{X}(t, 0)$ we have

$$\vec{F}(u) = \frac{\ell}{2}\left(\cos\frac{\omega u}{c}, \sin\frac{\omega u}{c}\right). \tag{7.54}$$

The function $\vec{F}$ is periodic, suggesting that the vector $\vec{v}_0$ in (7.48) must vanish. More precisely, $\vec{v}_0$ vanishes because it is the average velocity during the period in which the velocity repeats, the velocity repeats after precisely one full turn, and the average velocity of any point during one full turn is zero. With $\vec{v}_0 = 0$ equation (7.48) imposes the condition $\vec{F}(u + 2\sigma_1) = \vec{F}(u)$. Using (7.54), this gives

$$\frac{\omega}{c}(2\sigma_1) = 2\pi m \longrightarrow \frac{\omega}{c} = \frac{\pi}{\sigma_1}m, \tag{7.55}$$

where m is an integer. It is simple to see that we must choose $m = 1$. For this we just calculate $\vec{X}(0, \sigma)$, which gives the string at time equal to zero:

$$\vec{X}(0, \sigma) = \frac{1}{2}\left(\vec{F}(\sigma) + \vec{F}(-\sigma)\right) = \frac{\ell}{2}\left(\cos\frac{\pi m \sigma}{\sigma_1}, 0\right). \tag{7.56}$$

For $m = 1$, the string is recovered when $\sigma \in [0, \sigma_1]$. For arbitrary m, the function $\vec{X}(0, \sigma)$ traces out the string m times when $\sigma \in [0, \sigma_1]$. Choosing $m = 1$, we find

$$\frac{\omega}{c} = \frac{\pi}{\sigma_1} = \frac{\pi T_0}{E}. \tag{7.57}$$

This gives the angular frequency of the motion in terms of the energy. The first condition in (7.48) determines the length ℓ. Indeed,

$$\frac{d\vec{F}}{du} = \frac{\omega\ell}{2c}\left(-\sin\frac{\omega u}{c}, \cos\frac{\omega u}{c}\right), \tag{7.58}$$

and, as a result,

$$\left|\frac{d\vec{F}}{du}\right|^2 = \left(\frac{\omega\ell}{2c}\right)^2 = 1 \longrightarrow \ell = \frac{2c}{\omega} = \frac{2\sigma_1}{\pi} = \frac{2}{\pi}\frac{E}{T_0}. \tag{7.59}$$

This length is smaller, by a factor of $2/\pi$, than the length of a static string with energy E. This is sensible since this string has kinetic energy. Alternatively,

$$E = \frac{\pi}{2}T_0\ell, \tag{7.60}$$

which states that the energy of a rotating string is a factor of $\pi/2$ bigger than the energy of a static string with the same length. Note also that $\omega(\ell/2) = c$, telling us that the endpoints move with the speed of light. In this solution the energy is proportional to the length of the string. Perhaps surprisingly, as the energy of the string increases, the angular velocity ω decreases. This happens because the string endpoints have to move at the speed of light, so as the length of the string grows the angular velocity has to decrease.

Having determined ω and ℓ, we really know the motion of the string. It is of interest, however, to write the complete expression for the motion of the *parameterized* string $\vec{X}(t, \sigma)$. In terms of σ_1, the vector $\vec{F}$ in (7.54) is now given by

$$\vec{F}(u) = \frac{\sigma_1}{\pi}\left(\cos\frac{\pi u}{\sigma_1}, \sin\frac{\pi u}{\sigma_1}\right). \tag{7.61}$$

Finally, using (7.47) we obtain

$$\vec{X}(t,\sigma) = \frac{\sigma_1}{2\pi}\left(\cos\frac{\pi(ct+\sigma)}{\sigma_1} + \cos\frac{\pi(ct-\sigma)}{\sigma_1}, \sin\frac{\pi(ct+\sigma)}{\sigma_1} + \sin\frac{\pi(ct-\sigma)}{\sigma_1}\right), \tag{7.62}$$

and after simplification,

$$\vec{X}(t,\sigma) = \frac{\sigma_1}{\pi}\cos\frac{\pi\sigma}{\sigma_1}\left(\cos\frac{\pi ct}{\sigma_1}, \sin\frac{\pi ct}{\sigma_1}\right). \tag{7.63}$$

The parameterized string is of interest because the energy density is a constant function of the parameter σ. In Problem 7.2 you will calculate the energy $\mathcal{E}(s)$ per unit length on the string as a function of the distance s to the center:

$$\mathcal{E}(s) = \frac{T_0}{\sqrt{1 - \frac{4s^2}{\ell^2}}}. \tag{7.64}$$

At the center of the string the energy density $\mathcal{E}(0)$ coincides with T_0. This had to be so, because the string does not move at the center. The energy density diverges at the string endpoints $s = \pm\ell/2$. The total energy, however, is finite.

7.5 Motion of closed strings and cusps

Let us now consider the general motion of a free closed string. Just like for open strings, both the wave equation (7.26) and the parameterization constraints (7.27) and (7.28) apply. The wave equation is solved by the expression

$$\vec{X}(t,\sigma) = \frac{1}{2}\left(\vec{F}(u) + \vec{G}(v)\right), \tag{7.65}$$

where we have introduced variables u and v defined by

$$\begin{aligned} u &\equiv ct + \sigma, \\ v &\equiv ct - \sigma. \end{aligned} \tag{7.66}$$

Taking derivatives of $\vec{X}$ we immediately find

$$\frac{1}{c}\frac{\partial\vec{X}}{\partial t} = \frac{1}{2}\left(\vec{F}'(u) + \vec{G}'(v)\right), \tag{7.67}$$

$$\frac{\partial\vec{X}}{\partial\sigma} = \frac{1}{2}\left(\vec{F}'(u) - \vec{G}'(v)\right), \tag{7.68}$$

where primes denote derivatives with respect to argument. We can now form the linear combinations

$$\frac{\partial\vec{X}}{\partial\sigma} + \frac{1}{c}\frac{\partial\vec{X}}{\partial t} = \vec{F}'(u) \quad \text{and} \quad \frac{\partial\vec{X}}{\partial\sigma} - \frac{1}{c}\frac{\partial\vec{X}}{\partial t} = -\vec{G}'(v). \tag{7.69}$$

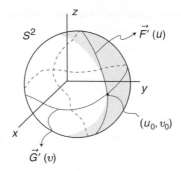

Fig. 7.4 The tips of the unit vectors $\vec{F}'(u)$ and $\vec{G}'(v)$ define two closed parameterized curves that, in this case, intersect for $u = u_0$ and $v = v_0$.

The parameterization constraints, summarized in (7.42), then give

$$|\vec{F}'(u)|^2 = |\vec{G}'(v)|^2 = 1, \quad \text{for all } u,\, v. \tag{7.70}$$

For closed strings we do not have boundary conditions but rather a periodicity condition. Since $d\sigma = dE/T_0$ on account of (7.30), the closed string parameter σ has the identification

$$\sigma \sim \sigma + \sigma_1, \quad \text{where} \quad \sigma_1 = E/T_0, \tag{7.71}$$

and E is the energy of the string. As σ increases by σ_1 we are back to the same point on the closed string and therefore

$$\vec{X}(t, \sigma + \sigma_1) = \vec{X}(t, \sigma) \tag{7.72}$$

is the periodicity condition. Making use of (7.65), this condition gives

$$\vec{F}(u + \sigma_1) + \vec{G}(v - \sigma_1) = \vec{F}(u) + \vec{G}(v), \tag{7.73}$$

or, equivalently,

$$\vec{F}(u + \sigma_1) - \vec{F}(u) = \vec{G}(v) - \vec{G}(v - \sigma_1). \tag{7.74}$$

The functions $\vec{F}(u)$ and $\vec{G}(v)$ need not be periodic with period σ_1, but they must change by the same vector when their arguments are increased by σ_1. Since u and v are independent variables, partial derivatives of (7.74) with respect to u and v give

$$\vec{F}'(u + \sigma_1) = \vec{F}'(u) \quad \text{and} \quad \vec{G}'(v + \sigma_1) = \vec{G}'(v). \tag{7.75}$$

Equations (7.70) and (7.75) imply that $\vec{F}'(u)$ and $\vec{G}'(v)$ are periodic unit vectors; they can be described as two independent parameterized closed curves on the surface of a unit two-sphere (Figure 7.4). A closed string motion is fully specified, up to a constant translation, by these two parameterized curves. Indeed, the curves fix $\vec{F}'(u)$ and $\vec{G}'(v)$ and, by integration, $\vec{F}(u)$ and $\vec{G}(v)$ up to integration constants, resulting in $\vec{X}(t, \sigma)$ fixed up to the addition of a constant vector.

An interesting situation arises quite generically. As shown in Figure 7.4, the two parameterized curves $\vec{F}'(u)$ and $\vec{G}'(v)$ may intersect for some values u_0 and v_0 of the parameters u and v:

$$\vec{F}'(u_0) = \vec{G}'(v_0). \tag{7.76}$$

Let t_0 and σ_0 be the values of t and σ defined by u_0 and v_0 through (7.66). It now follows from (7.67) that

$$\frac{1}{c}\frac{\partial \vec{X}}{\partial t}(t_0, \sigma_0) = \frac{1}{2}\left(\vec{F}'(u_0) + \vec{G}'(v_0)\right) = \vec{F}'(u_0). \tag{7.77}$$

Since $\vec{F}'$ is a unit vector we learn that at $t = t_0$ the point $\sigma = \sigma_0$ on the string reaches the speed of light! Moreover, the motion of the point is in the direction of $\vec{F}'(u_0)$. Additional information follows from (7.68):

$$\frac{\partial \vec{X}}{\partial \sigma}(t_0, \sigma_0) = \frac{1}{2}\left(\vec{F}'(u_0) - \vec{G}'(v_0)\right) = \vec{0}. \tag{7.78}$$

This means that the parameterization of the $t = t_0$ string becomes singular at $\sigma = \sigma_0$. To examine the shape of the string near $\sigma = \sigma_0$ we fix $t = t_0$ and use a Taylor expansion around $\sigma = \sigma_0$:

$$\vec{X}(t_0, \sigma) = \vec{X}(t_0, \sigma_0) + (\sigma - \sigma_0)\frac{\partial \vec{X}}{\partial \sigma}(t_0, \sigma_0) + \frac{1}{2}(\sigma - \sigma_0)^2\frac{\partial^2 \vec{X}}{\partial \sigma^2}(t_0, \sigma_0)$$
$$+ \frac{1}{3!}(\sigma - \sigma_0)^3\frac{\partial^3 \vec{X}}{\partial \sigma^3}(t_0, \sigma_0) + \cdots. \tag{7.79}$$

Using (7.78) and the definitions

$$\vec{X}_0 = \vec{X}(t_0, \sigma_0), \quad \vec{T} \equiv \frac{\partial^2 \vec{X}}{\partial \sigma^2}(t_0, \sigma_0), \quad \vec{R} \equiv \frac{\partial^3 \vec{X}}{\partial \sigma^3}(t_0, \sigma_0), \tag{7.80}$$

we find that the expansion (7.79) loses the term linear in $\sigma - \sigma_0$ and becomes

$$\vec{X}(t_0, \sigma) = \vec{X}_0 + \frac{1}{2}(\sigma - \sigma_0)^2\vec{T} + \frac{1}{3!}(\sigma - \sigma_0)^3\vec{R} + \cdots. \tag{7.81}$$

For generic situations $\vec{T}$ and $\vec{R}$ are nonvanishing and nonparallel. It then follows that for $\sigma = \sigma_0$ we have a *cusp*. A cusp on a string is a point where the two outgoing string directions form zero angle. Equivalently, at a cusp the oriented tangent to the string reverses direction. Equation (7.81) describes a cusp at $\vec{X}_0$: as σ grows from values right below σ_0 to values right above σ_0 the string approaches $\vec{X}_0$ along $\vec{T}$ and recedes from $\vec{X}_0$ along $\vec{T}$. Indeed, near $\vec{X}_0$, $\sigma - \sigma_0$ is very small and the term that contains $\vec{T}$ dominates the expansion (7.81). As we move away from $\vec{X}_0$ the cusp opens up due to the term that contains $\vec{R}$. If we choose a coordinate system in which the tip of the cusp is at the origin, $\vec{T}$ points along the positive y axis, and $\vec{R}$ lies on the (x, y) plane, the cusp traces the curve $y \sim x^{2/3}$, as you will show in Problem 7.7.

If a cusp occurs for (u_0, v_0) it will also occur for $(u_0 + m\sigma_1, v_0 + n\sigma_1)$, with m and n arbitrary integers. As a result, as time goes by, cusps will appear and disappear periodically at various points on the string. Given any two parameterized paths $\vec{F}'(u)$ and $\vec{G}'(v)$ on the two-sphere, there may be several different intersection points. Each will give rise to a set of cusps.

7.6 Cosmic strings

Our analysis of the classical motion of relativistic strings may be applied to the study of cosmic strings. It is conceivable that physics in the early universe results in strings that are not microscopic but rather expand with the universe to become very large. The motion of cosmic size strings can be studied in the classical approximation. As of 2007 no cosmic string has been detected. The discovery of a cosmic string would be a remarkable event. As it turns out, cosmic strings may arise from phenomena unrelated to string theory so, even if discovered, much work will be needed to tell if a cosmic string is a string theory string.

The most direct way to detect a cosmic string is through gravitational lensing. To understand this we begin by discussing the gravitational effects of a straight, infinitely long relativistic string. Suppose you place a massive particle some distance away from such a string. Since the string has rest energy one may imagine that the particle would experience gravitational attraction. This is not the case; the particle would experience exactly zero force. This is a result in general relativity and would not hold in Newtonian gravitation where only the effective mass density μ_0 of the string contributes to the gravitational attraction. In general relativity the tension of the string gives an additional contribution, in fact, a gravitational repulsion. The total attractive force is proportional to $(\mu_0 - \frac{T_0}{c^2})$, a combination that precisely vanishes for relativistic strings.

Although a string does not exert gravitational attraction it affects the geometry of the planes orthogonal to the string. Suppose you go around the string keeping your distance r to the string constant. The circumference $\mathcal{C}$ of the traced circle would be less than the expected value of $2\pi r$. More precisely, for any radial distance r,

$$\frac{\mathcal{C}}{r} = 2\pi - \Delta, \tag{7.82}$$

where the constant Δ is called the *deficit angle*. The two-dimensional spaces orthogonal to the string are in fact cones with deficit angle Δ. The string runs along the apexes of the cones. Indeed cones are spaces in which circular loops at constant distance from the apex satisfy (7.82). This can be made manifest in the construction of a cone starting from the plane. If we represent the plane using the complex variable $z = x + iy$ (as we did when studying orbifolds in Section 2.8) the cone arises by cutting out the region $0 \leq \arg(z) \leq \Delta$ and identifying the resulting boundaries via $z \sim e^{i\Delta}z$. This is shown in Figure 7.5, where we also show a circle with constant distance to the apex $z = 0$. In our discussion of lensing we will assume, for simplicity, that the observer O and the light source S lie on the same cone, namely, they share the same value of the coordinate along the string.

The deficit angle Δ produced by the relativistic string depends on the tension of the string and the value of Newton's constant G. A calculation in general relativity shows that

$$\Delta = \frac{8\pi G T_0}{c^4} = \frac{8\pi G \mu_0}{c^2}. \tag{7.83}$$

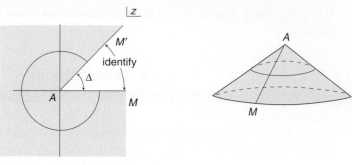

Fig. 7.5 **Left:** the complex z plane with the set of points whose argument is greater than zero and less than Δ removed. **Right:** the familiar picture of a cone obtained by identifying the lines AM and AM' on the left.

The angle Δ, as we will see soon, is closely related to lensing angles, whose typical values are measured in arc seconds. Noting that $1'' = 4.85 \times 10^{-6}$ rad, we can write

$$\Delta = 5.18'' \times \left(\frac{G\mu_0/c^2}{10^{-6}}\right). \tag{7.84}$$

Quick calculation 7.3 Use the values of G and c to show that

$$\Delta = 3.85'' \times \left(\frac{\mu_0}{10^{21}\,\text{kg/m}}\right). \tag{7.85}$$

Six kilometers of a string with $\Delta = 3.85''$ would pack a mass equal to that of Earth.

Further insight into the value of Δ can be obtained by writing (7.83) as a manifestly unit-less ratio of the Planck mass m_P and a string mass m_s, defined as the unique mass that can be written using only powers of μ_0, c, and $\hbar$.

Quick calculation 7.4 Show that

$$m_s = \sqrt{\frac{\hbar\mu_0}{c}}. \tag{7.86}$$

Recalling that $m_P = \sqrt{\hbar c/G}$ we find

$$\frac{m_s}{m_P} = \sqrt{\frac{G\mu_0}{c^2}}, \tag{7.87}$$

and, as a result,

$$\Delta = 8\pi \left(\frac{m_s}{m_P}\right)^2. \tag{7.88}$$

A small Δ arises for m_s small compared to m_P.

To discuss lensing we first review some properties of geodesics. A curve that joins two fixed points is a geodesic if its length is stationary under arbitrary infinitesimal deformations that vanish at the fixed points; the length does not change to first order in such deformations of the curve. In the plane there is just one geodesic between any two points: the straight line that joins them. This geodesic is also the shortest path between the two points. On more complicated spaces geodesics need not be unique nor have minimum length. There are infinitely many geodesics joining the north pole and the south pole of a two-sphere, each one a half-circle of constant longitude. For two generic points on the two-sphere you get two geodesics, a short one and a long one. Indeed, the two points determine a great circle, and the geodesics are the two complementary arcs bounded by the points. The short geodesic is the shortest path between the two points. The long geodesic is only a path of stationary length, a saddle point for the length function: some deformations make it longer and others make it shorter. In fact, the long geodesic can be continuously deformed into the short geodesic. On a cylinder any two points can be joined by an infinite number of geodesics. The geodesics are characterized by the number of times they wrap around the cylinder while getting from one point to the other.

Quick calculation 7.5 Sketch and find the length of the five shortest geodesics that join the points $(0, 0)$ and $(1, 0)$ on the cylinder $(x, y) \sim (x, y + 1)$.

On a cone there is an intricate pattern of geodesics. Consider two points P and Q and the geodesics from P to Q. The number of geodesics depends on the deficit angle Δ of the cone at the apex A and the angle ϕ between P and Q, defined as the positive angle that by clockwise rotation takes AP to AQ. The angle ϕ must necessarily be smaller than the total angle α at the apex:

$$\phi \leq \alpha \equiv 2\pi - \Delta. \tag{7.89}$$

For a complete calculation of the geodesics see Problem 7.8. Here we restrict ourselves to the case $\Delta \ll 1$, relevant to possible astrophysical situations. Assume that the observer O is a distance d_O from the string and that the light source S is a distance d_S from the string. Lensing occurs if the source S is roughly opposite to O across the string. In this case there are two geodesics that join S and O, one to each side of the string. Light uses both geodesics to reach O, who detects two identical images of the source S. Two identical images is a signature of lensing by strings and occurs because cones have no curvature away from the apex. Lensing by compact objects can produce more than two images and, since the relevant geometry is curved, the images are distorted and can be quite different.

To visualize the geodesics we represent the cone by placing the source on the two radial lines that are to be identified across an angle Δ (Figure 7.6). The source appears as the points S and S', and the two geodesics that join the source and the observer are the straight lines SO and $S'O$. The angle $\delta\phi$ between SO and $S'O$ at O is the lensing angle – the angle between the images seen by the observer. From the figure we see that

$$\delta\phi = \alpha + \beta. \tag{7.90}$$

Moreover, since Δ is the sum of exterior angles for the triangles SAO and $S'AO$, it equals

$$\Delta = \alpha + \beta + \alpha' + \beta'. \tag{7.91}$$

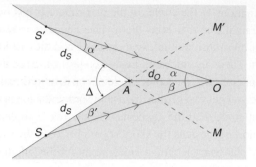

Fig. 7.6   A source S and an observer O at distances d_S and d_O, respectively, from the apex A of a cone with deficit angle Δ. For simplicity, the cut is along the source, which appears as points S and S' to be identified. Rays SO and $S'O$ from the source reach the observer in directions separated by the lensing angle $\delta\phi = \alpha + \beta$.

Since each angle that appears on the above right-hand side is positive and $\Delta \ll 1$, each one of them is small: $\alpha, \alpha', \beta, \beta' \ll 1$. The law of sines then gives

$$\frac{\sin\alpha}{d_S} = \frac{\sin\alpha'}{d_O}, \quad \frac{\sin\beta}{d_S} = \frac{\sin\beta'}{d_O} \quad \rightarrow \quad \frac{\alpha}{d_S} = \frac{\alpha'}{d_O}, \quad \frac{\beta}{d_S} = \frac{\beta'}{d_O}, \tag{7.92}$$

using the small angle approximation. Solving for α' and β' and substituting back in (7.91), we have

$$\Delta = \alpha + \beta + \frac{d_O}{d_S}(\alpha + \beta) = \delta\phi\left(1 + \frac{d_O}{d_S}\right), \tag{7.93}$$

where we also used (7.90). We finally get

$$\delta\phi = \frac{\Delta}{\left(1 + \frac{d_O}{d_S}\right)}. \tag{7.94}$$

We see that the lensing angle $\delta\phi$ is bounded above by Δ. It approaches this upper bound as $d_S \to \infty$. Figure 7.6 also makes clear that lensing occurs only if O lies within the sector MAM'. If O lies outside this sector, there is just one geodesic and O sees a single image.

In the past few years images of galaxies that were candidates for string lensing were later shown, in higher resolution, to be the images of two similar but different galaxies. Since cosmic strings possibly move with relativistic speeds, it may be that lenses exist only for limited periods of time, making their detection challenging.

It may be possible to infer the existence of cosmic strings indirectly. Cosmic strings can give contributions to temperature anisotropies in the cosmic microwave background. The lack of observed evidence to this effect suggests the bound

$$\frac{G\mu_0}{c^2} < 3 \times 10^{-7}. \tag{7.95}$$

Equation (7.84) shows that for strings obeying this bound, lensing angles will be smaller than a couple of arc-seconds. Additionally, the motion of cosmic strings produces gravitational waves. In fact, the cusps we studied in the previous section are efficient generators of gravity waves. It seems possible that such waves could be observed by gravitational wave detectors even for strings of $G\mu_0/c^2 \sim 10^{-13}$. While it remains a long shot, cosmic strings may furnish the first experimental evidence for string theory.

Problems

Problem 7.1 Short proofs concerning open strings with completely free endpoints.

Prove the following four statements that refer to open strings with free endpoints.

(a) If one endpoint of an open string happens to lie at all times on a hyperplane, the full open string lies at all times on the same hyperplane.

(b) If one endpoint of an open string happens to lie at all times within a distance R from a point P_0, the full open string lies at all times within a distance R from P_0.

(c) If one endpoint of an open string happens to lie at all times within a convex subspace, the full open string lies at all times within that convex subspace. [This is the general version of the results proven in (a) and (b).]

(d) Show that the length ℓ of an open string parameterized with energy (Section 7.3) is given by

$$\ell = \int_0^{\sigma_1} \sqrt{1 - \frac{v_\perp^2}{c^2}}\, d\sigma.$$

Problem 7.2 Exploring further the rigidly rotating string.

Let $s \in (-\ell/2, \ell/2)$ be a length parameter on the rigidly rotating string studied in Section 7.4, with $s = 0$ chosen to be the fixed center of the string. Let $\mathcal{E}(s)$ denote the energy per unit length as a function of s.

(a) Show that $\mathcal{E}(s) = T_0/\sqrt{1 - (4s^2/\ell^2)}$. Plot $\mathcal{E}(s)$ as a function of s. Note that $\mathcal{E}(s)$ has integrable singularities at the string endpoints, and confirm that the total energy is $\frac{\pi}{2}\ell T_0$.

(b) For what points on the string is the local energy density equal to the average energy density?

(c) Calculate the energy $\hat{E}(s)$ carried by the string on the interval $[-s, s]$. What is the value of $s/(\ell/2)$ for this energy to be half of the total energy of the string? 90% of the energy?

Problem 7.3 Time evolution of an initially static closed relativistic string.

The time development of closed strings in the static gauge is governed by equations (7.26), (7.27), and (7.28).

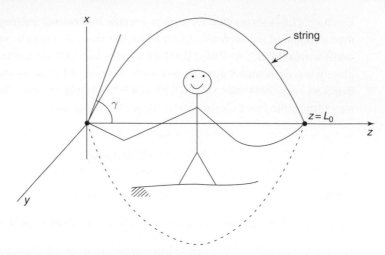

Fig. 7.7 Problem 7.4: Kasey's relativistic jumping rope.

(a) Assume that $\frac{\partial \vec{X}}{\partial t}(0, \sigma) = 0$, and write the general solution for $\vec{X}(t, \sigma)$ in terms of a vector function $\vec{F}(u)$ of a single variable. What do the parameterization conditions require on $\vec{F}$?

(b) On a closed string the parameter σ lives on the circle $\sigma \sim \sigma + \sigma_1$. What condition would you impose on $\vec{X}(t, \sigma)$ to implement this feature? What are the implications for $\vec{F}$?

(c) Consider a closed string which at $t = 0$ is static and traces a closed curve γ of length ℓ. How are σ_1 and ℓ related? Find a time $t_P > 0$ and smaller than ℓ/c for which the closed string traces the curve γ again. Relate $\vec{X}(t_P, \sigma)$ to the time equal to zero string $\vec{X}(0, \sigma)$.

(d) List the steps you would take with a computer (that can do integrals and invert functions) to produce the time evolution of an initially static closed string of arbitrary shape lying on the (x, y) plane. Assume that the initial string is given to you as the parameterized closed curve $(x(\lambda), y(\lambda))$, with some parameter $\lambda \in [0, \lambda_0]$.

Problem 7.4 Kasey's relativistic jumping rope.

Consider a relativistic open string with fixed endpoints:

$$\vec{X}(t, 0) = \vec{x}_1, \quad \vec{X}(t, \sigma_1) = \vec{x}_2. \tag{1}$$

The boundary condition at $\sigma = 0$ is satisfied by the following solution to the wave equation:

$$\vec{X}(t, \sigma) = \vec{x}_1 + \frac{1}{2}\Big(\vec{F}(ct + \sigma) - \vec{F}(ct - \sigma) \Big). \tag{2}$$

Here $\vec{F}$ is a vector function of a single variable.

(a) Use (2) and the boundary condition at $\sigma = \sigma_1$ to find a condition on $\vec{F}(u)$.

(b) Write down the constraint on $\vec{F}(u)$ that arises from the parameterization conditions (7.42).

As an application, consider Kasey's attempts to use a relativistic open string as a jumping rope. For this purpose, she holds the open string (in three spatial dimensions) with her right hand at the origin $\vec{x}_1 = (0, 0, 0)$ and with her left hand at the point $z = L_0$ on the z axis, or $\vec{x}_2 = (0, 0, L_0)$ (Figure 7.7). As she starts jumping we observe that the tangent vector $\vec{X}'$ to the string at the origin rotates with constant angular velocity around the z axis forming an angle γ with it.

(c) Use the above information to write an expression for $\vec{F}'(u)$.
(d) Find σ_1 in terms of the length L_0 and the angle γ.
(e) Calculate $\vec{X}(t, \sigma)$ for the motion of Kasey's relativistic jumping rope.
(f) How is the energy distributed in the string as a function of z?

Problem 7.5 Planar motion for an open string with fixed endpoints.

Consider the motion of a relativistic open string on the (x, y) plane. The string endpoints are attached to $(x, y) = (0, 0)$ and $(x, y) = (a, 0)$, where $a > 0$. As opposed to the relativistic jumping rope, the string now remains in the (x, y) plane. The motion is described by

$$\vec{X}(t, \sigma) = \frac{1}{2}\Big(\vec{F}(ct + \sigma) - \vec{F}(ct - \sigma)\Big), \tag{1}$$

where $\vec{F}(u)$ is a vector function of a single variable which satisfies

$$\left|\frac{d\vec{F}}{du}\right|^2 = 1 \quad \text{and} \quad \vec{F}(u + 2\sigma_1) = \vec{F}(u) + (2a, 0). \tag{2}$$

Consider an ansatz of the form

$$\vec{F}'(u) \equiv \frac{d\vec{F}}{du} = \left(\cos\Big[\gamma \cos\frac{\pi u}{\sigma_1}\Big], \ \sin\Big[\gamma \cos\frac{\pi u}{\sigma_1}\Big]\right). \tag{3}$$

(a) Is this ansatz consistent with the conditions in (2)?
(b) Calculate $\vec{X}'(0, \sigma)$. Letting $\vec{X}(0, \sigma) \equiv (x(\sigma), y(\sigma))$, give $dy/d\sigma$ and plot it as a function of $\sigma \in [0, \sigma_1]$, assuming, for convenience, that $0 < \gamma < \pi/2$. Use this to make a rough sketch of the string position $y(\sigma)$ as a function of σ at $t = 0$.
(c) Calculate $\vec{X}'(t, 0)$ and use it to describe the motion of the string near the origin. What is the interpretation of γ?
(d) Use the second condition in (2) to find an integral relation between a, σ_1 and γ. Assume that γ is small, and find an approximate explicit relation between these three variables keeping terms of order γ^2.
(e) Show that $a/\sigma_1 = J_0(\gamma)$, where J_0 is the Bessel function of order zero. [Hint: look up integral representations of Bessel functions.]

Problem 7.6 Planar motion (*continued*) and the formation of cusps.

We investigate further the solution obtained in Problem 7.5, where it was shown that $a/\sigma_1 = J_0(\gamma)$. Here a is the distance between the fixed endpoints of the open string, and σ_1 is the length parameter of the string: it equals the length of the string at any time when the string has zero velocity (why?). The relation between a and the angular variable γ is unusual since J_0 is not a periodic function. This means that the open string motions corresponding to γ and $2\pi + \gamma$ are not the same.

(a) Show that the instantaneous slope of the string is described by

$$\vec{X}'(t, \sigma) = \cos\left(\gamma \, \sin\frac{\pi ct}{\sigma_1} \, \sin\frac{\pi\sigma}{\sigma_1}\right) (\cos\beta, \sin\beta),$$

where

$$\beta = \gamma \cos\frac{\pi ct}{\sigma_1} \cos\frac{\pi\sigma}{\sigma_1}.$$

Show that at $ct = \sigma_1/2$ the string is horizontal.

(b) Prove that the instantaneous (transverse) velocity of the string satisfies

$$\left|\frac{1}{c}\frac{\partial\vec{X}}{\partial t}\right| = \left|\sin\left(\gamma \, \sin\frac{\pi ct}{\sigma_1} \, \sin\frac{\pi\sigma}{\sigma_1}\right)\right|.$$

Note that at $t = 0$ the string has zero velocity. Conclude that whenever $\gamma < \pi/2$ no point on the string ever reaches the speed of light. Moreover, show that for $\gamma = \pi/2$ the string midpoint $\sigma = \sigma_1/2$ acquires the speed of light when the string is horizontal.

(c) A tractable case is obtained for $\gamma = \sqrt{2}(\pi/2)$ ($\simeq 127.3°$). Show that at $ct = \sigma_1/4$ one point on the string reaches the speed of light. Examine the string at the slightly later time $ct = \sigma_1/3$, show that there are *two* points that have the speed of light, and find the corresponding values of σ. Analyze $\vec{X}'$ as a function of σ to show that the string has a cusp at each of these points.

(d) Use your favorite mathematical software package to generate the picture of the string considered in (c) at various times (you will use numerical integration). Assume that $a = 1$ and verify that $\sigma_1 \simeq 10.155$. Show the string for $ct = 0$, $\sigma_1/4$, and $\sigma_1/3$.

Problem 7.7 Cusps in the evolution of closed strings.

In this problem we derive a few properties of the cusps that appear generically in the evolution of free closed strings. For this we examine in more detail equation (7.81).

(a) Use the Taylor expansions of $\vec{F}(u)$ and $\vec{G}(v)$ around u_0 and v_0 to prove that

$$\vec{T} = \frac{1}{2}\left(\vec{F}''(u_0) + \vec{G}''(v_0)\right), \quad \vec{R} = \frac{1}{2}\left(\vec{F}'''(u_0) - \vec{G}'''(v_0)\right), \qquad (1)$$

where primes denote derivatives with respect to the argument. Assume that the intersection of the paths on the two-sphere indicated in equation (7.76) is regular: the paths are not parallel at the intersection and neither $\vec{F}''(u_0)$ nor $\vec{G}''(v_0)$ vanishes. Explain why $\vec{T}$ is non-zero and orthogonal to $\vec{F}'(u_0)$. In general $\vec{R}$ does not vanish, but it may under special conditions.

(b) Equation (7.81) shows that the cusp opens up along the direction of the vector $\vec{T}$ and, locally, is contained in the plane spanned by $\vec{T}$ and $\vec{R}$. Fix the origin of the coordinate system at $\vec{X}_0$, align the positive y axis along $\vec{T}$, and position the x axis so that $\vec{R}$ lies on the (x, y) plane. Demonstrate that near the cusp $y \sim x^{2/3}$. In what plane does the velocity of the cusp lie?

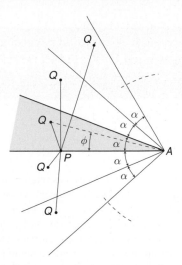

Fig. 7.8 Problem 7.8: a cone, shown shaded, with total internal angle α at the apex and two points P and Q separated by an angle ϕ.

(c) Consider the functions $\vec{F}(u)$ and $\vec{G}(v)$ given by

$$\vec{F}(u) = \frac{\sigma_1}{2\pi}\left(\sin\frac{2\pi u}{\sigma_1}, -\cos\frac{2\pi u}{\sigma_1}, 0\right), \quad \vec{G}(v) = \frac{\sigma_1}{4\pi}\left(\sin\frac{4\pi v}{\sigma_1}, 0, -\cos\frac{4\pi v}{\sigma_1}\right).$$

(2)

Verify that conditions (7.70) and (7.74) are satisfied. For the cusp at $t = \sigma = 0$ give its direction, the plane it lies on, and its velocity. Draw a sketch.

(d) Show that the motion of the closed string has period $\sigma_1/(4c)$. (Remember, as you found in Problem 7.3, that the string returns to its original position in less time than the function $\vec{F}(ct + \sigma)$ takes to repeat itself.) How many cusps are formed during a period?

Problem 7.8 Counting geodesics on a cone.

Let $\Delta < 2\pi$ be the deficit angle of a cone and $\alpha = 2\pi - \Delta$ the angle at the apex A. Consider two points P and Q on the cone separated by an angle $\phi > 0$: the ray from the apex through Q is obtained from the ray from the apex through P by a clockwise rotation through an angle ϕ. The number N of geodesics joining the points P and Q is given by

$$N = \left[\frac{\pi - \phi}{\alpha}\right] + \left[\frac{\pi + \phi}{\alpha}\right] + 1,$$

(1)

where $[x]$ denotes the largest integer less than or equal to x. A convenient depiction of the conical region and copies of it is shown in Figure 7.8.

(a) Convince yourself of the validity of (1) by counting the geodesics from P to Q and from P to the images of Q developing in the clockwise direction, as well as those from P to the images of Q developing in the counterclockwise direction.

(b) Verify that N is invariant under $\phi \to \alpha - \phi$ and explain why it should be so.

(c) $\Delta < \pi$ is the case relevant for gravitational lensing. What are the possible values of N as a function of ϕ?

8

World-sheet currents

For physical insight, physicists often turn to ideas of symmetry and invariance. Symmetry properties of dynamical systems and conserved quantities are closely related. We will learn that in string theory there are currents that flow on the two-dimensional world-sheet traced out by the string in spacetime. The conserved charges associated with these currents are key quantities that characterize the free motion of strings. We give a simple physical interpretation of the objects $\mathcal{P}^\tau$ and $\mathcal{P}^\sigma$ which we encountered earlier.

8.1 Electric charge conservation

We begin our study by reviewing the physics and the mathematics of charge conservation in the context of Maxwell theory. This classic example will help us to develop a more general understanding of the concept of conserved currents.

In electromagnetism, the conserved current is the four-vector $j^\alpha = (c\rho, \vec{j})$, where ρ is the electric charge density and $\vec{j}$ is the current density. Why do we say that j^α is a conserved current? By definition, j^α is a conserved current because it satisfies the equation

$$\partial_\alpha j^\alpha = 0. \tag{8.1}$$

Any four-vector which satisfies this equation is called a conserved current. The term "conserved current" is a little misleading, but it is a convention. More precisely, we should say that we have a conserved *charge*, because it is really the charge associated with the current that is conserved. Let us see how this conservation arises.

When we separate the space and time indices in (8.1) we get

$$\partial_0 j^0 + \partial_i j^i = \frac{\partial j^0}{\partial x^0} + \nabla \cdot \vec{j} = 0. \tag{8.2}$$

Why is this equation a statement of charge conservation? In electromagnetism, the total electric charge $Q(t)$ in a fixed volume V is just the integral of the charge density ρ over the volume:

$$Q(t) = \int_V \rho(t, \vec{x})\, d^3x = \int_V \frac{j^0(t, \vec{x})}{c}\, d^3x. \tag{8.3}$$

Up to a constant, the charge is the integral over space of the first component of the current. Its time derivative is given by

$$\frac{dQ}{dt} = \int_V \frac{\partial j^0}{\partial x^0} \, d^3x.$$

(8.4)

Using equation (8.2), we can write

$$\frac{dQ}{dt} = -\int_V \nabla \cdot \vec{j} \, d^3x.$$

(8.5)

Letting S denote the boundary of V, the divergence theorem gives

$$\frac{dQ}{dt} = -\int_S \vec{j} \cdot d\vec{a}.$$

(8.6)

This equation encodes the physical statement of charge conservation: the charge inside a volume V can only change if there is a flux of current across the surface S which bounds the volume. In many cases, we take V to be so large that the current $\vec{j}$ vanishes on the surface S. In these cases,

$$\frac{dQ}{dt} = 0.$$

(8.7)

The charge Q is then time independent and is said to be "conserved." It is also well established that electric charge is a Lorentz invariant: all inertial observers measuring charge obtain the same number. Not all conserved quantities are Lorentz invariant. Energy, for example, is conserved, but energy is *not* Lorentz invariant. These facts are examined in more detail in Problems 8.1 and 8.2.

8.2 Conserved charges from Lagrangian symmetries

One of the most useful properties of Lagrangians is that they can be used to deduce the existence of conserved quantities. Conserved quantities can help us to understand the dynamics of a system. In this section we begin our work in the context of Lagrangian mechanics, learning how to construct the conserved quantity associated with a symmetry. We then turn to Lagrangian densities, and show how to construct the conserved current associated with a symmetry.

Let $L(q(t), \dot{q}(t); t)$ be a Lagrangian that depends on a coordinate $q(t)$, the velocity $\dot{q}(t)$, and may even have explicit time dependence. Consider, moreover, a variation of the coordinate $q(t)$:

$$q(t) \rightarrow q(t) + \delta q(t),$$

(8.8)

where $\delta q(t)$ is some specific infinitesimal variation. If $q(t)$ represents the path of a particle, for example, the variation above is an instruction on how to change the path: the position at time t is changed by $\delta q(t)$. Suppose that we have a rule that tells us how to vary *any* path $q(t)$. This means that given any $q(t)$, we know how to construct the corresponding variation $\delta q(t)$. Such a rule can be written in the form

$$\delta q(t) = \epsilon \, h\big(q(t); t\big), \tag{8.9}$$

where ϵ is an infinitesimal constant and h is some function.

As a result of the change (8.8) of the path, there is an induced change in the velocity $\dot{q}(t)$:

$$\dot{q}(t) \to \dot{q}(t) + \frac{d(\delta q(t))}{dt}. \tag{8.10}$$

In order to determine how $L(q(t), \dot{q}(t); t)$ changes as a result of the variation (8.8), we must also vary the velocity $\dot{q}(t)$ according to (8.10). We will generally speak of the variation of $q(t)$ leaving it implicit that the velocity $\dot{q}(t)$ is also varied accordingly. Because δq is infinitesimal, the variation of the Lagrangian consists only of terms which are linear in δq. If those terms vanish, the Lagrangian is said to be invariant. Moreover, the transformation in (8.8) is then said to be a *symmetry transformation*. A rule that tells us how to vary any path in such a way that the Lagrangian is not changed is a symmetry transformation. The rule is specified by the function h in (8.9).

We now state our claim: if the Lagrangian L is invariant under the variation (8.8), then the quantity Q, defined by

$$\epsilon \, Q \equiv \frac{\partial L}{\partial \dot{q}} \, \delta q \,, \tag{8.11}$$

is conserved in time for *physical motion*. That is, for any motion $q(t)$ which satisfies the equations of motion, the "charge" Q is a constant:

$$\frac{dQ}{dt} = 0. \tag{8.12}$$

Note that the ϵ on the left-hand side of (8.11) cancels with the ϵ that appears in δq (see (8.9)).

To prove the conservation of Q, consider the Euler–Lagrange equations that follow from the variation of the action $S = \int dt \, L$ (you may have derived them in Problem 4.8). These equations are:

$$\frac{d}{dt}\left(\frac{\partial L}{\partial \dot{q}}\right) - \frac{\partial L}{\partial q} = 0. \tag{8.13}$$

Since the Lagrangian does not change under the coordinate and velocity variations (8.8) and (8.10), we must have

$$\frac{\partial L}{\partial q}\delta q + \frac{\partial L}{\partial \dot{q}}\frac{d}{dt}(\delta q) = 0. \tag{8.14}$$

Using (8.13) to eliminate $\frac{\partial L}{\partial q}$ we obtain

$$\frac{d}{dt}\left(\frac{\partial L}{\partial \dot{q}}\right)\delta q + \frac{\partial L}{\partial \dot{q}}\frac{d}{dt}(\delta q) = \frac{d}{dt}\left(\frac{\partial L}{\partial \dot{q}}\,\delta q\right) = 0, \tag{8.15}$$

proving that (8.12) holds for Q defined as in (8.11).

Let us apply this new perspective on conservation to a Lagrangian $L(\dot{q}(t))$ that depends only on the velocity. How can we vary $q(t)$ leaving the Lagrangian unchanged? One way is to apply $q(t) \rightarrow q(t) + \epsilon$, where ϵ is any constant. In this variation, $\delta q(t) = \epsilon$, and the function h of (8.9) is just equal to one. This is a uniform space translation: at any time t the position of the particle is changed by the same amount ϵ. Correspondingly, the velocity does not change: $\dot{q}(t) \rightarrow \dot{q}(t) + d\epsilon/dt = \dot{q}(t)$. Since the Lagrangian only depends on the velocity, it does not change, and we have a symmetry. Making use of (8.11), we find

$$\epsilon Q = \frac{\partial L}{\partial \dot{q}}\delta q = \frac{\partial L}{\partial \dot{q}}\epsilon \longrightarrow Q = \frac{\partial L}{\partial \dot{q}} = p. \tag{8.16}$$

We recognize Q as the momentum associated with the coordinate q. This quantity is conserved because the position q does not appear in the Lagrangian. Note that the conservation equation $dQ/dt = 0$ coincides with the Euler–Lagrange equation (8.13). This example illustrates a familiar result in Lagrangian mechanics: if the Lagrangian of a system does not contain a given coordinate, the momentum conjugate to that coordinate is conserved. For a free nonrelativistic particle, for example, $L = \frac{1}{2}m(\dot{q})^2$. In this case $Q = m\dot{q}$.

Now let us consider Lagrangian *densities* with symmetries. While symmetries of a Lagrangian guarantee the existence of conserved *charges*, symmetries of a Lagrangian density guarantee the existence of conserved *currents*. We write the action as the integral of the Lagrangian density over the full set of coordinates ξ^α of some relevant "world":

$$S = \int d\xi^0 d\xi^1 \ldots d\xi^k \; \mathcal{L}(\phi^a, \partial_\alpha \phi^a). \tag{8.17}$$

Here k denotes the number of space dimensions of the world. The world could be the full Minkowski spacetime, some subspace thereof, or, for example, the two-dimensional parameter space of the string world-sheet. The fields $\phi^a(\xi)$ are functions of the coordinates, and

$$\partial_\alpha \phi^a = \frac{\partial \phi^a}{\partial \xi^\alpha} \tag{8.18}$$

are derivatives of the fields with respect to the coordinates. Each value of the index a corresponds to a field. Consider now the infinitesimal variation

$$\phi^a(\xi) \rightarrow \phi^a(\xi) + \delta\phi^a(\xi), \tag{8.19}$$

and the associated variation for the field derivatives $\partial_\alpha \phi^a$. The infinitesimal variations are conveniently written as the rule

$$\delta\phi^a = \epsilon^i\, h_i^a(\phi), \tag{8.20}$$

that enables us to vary any arbitrary field configuration. Here, ϵ^i are a set of infinitesimal constants, and, for brevity, we omitted all indices in the arguments of h_i^a. We have included an index in ϵ^i because the variation may involve several parameters. A spacetime translation, for example, involves as many parameters as there are spacetime dimensions. Since the index i is repeated in (8.20), it is summed over. You must distinguish clearly the various types of indices we are working with:

α index to label world coordinates ξ^α, or vector components,

i index to label parameters in the symmetry transformation,

a index to label fields in the Lagrangian. $\qquad(8.21)$

If $\mathcal{L}$ is invariant under (8.19) and the associated variations of the field derivatives, then the quantities j_i^α defined by

$$\epsilon^i \, j_i^\alpha \equiv \frac{\partial \mathcal{L}}{\partial(\partial_\alpha \phi^a)} \delta\phi^a, \qquad (8.22)$$

are conserved currents:

$$\partial_\alpha \, j_i^\alpha = 0. \qquad (8.23)$$

We will prove this shortly. Equation (8.23) holds for any field configuration that satisfies the Lagrangian equations of motion. In (8.22) the repeated field index a is summed over. If the index i is present in (8.20), then we have *several* currents indexed by i, one for each parameter of the variation. The components of the currents, as many as the number of dimensions in the world, are indexed by α. It is important not to confuse the very different roles that these two kinds of indices play:

j_i^α: i labels the various currents,

α labels the components of the currents. $\qquad(8.24)$

We showed in Section 8.1 that a conserved current gives rise to a conserved charge. The charge is the integral over space of the zeroth component of the current. It follows that the currents j_i^α give rise to the conserved charges

$$Q_i = \int d\xi^1 d\xi^2 \ldots d\xi^k \, j_i^0. \qquad (8.25)$$

We have as many conserved charges as there are parameters in the symmetry transformation.

Quick calculation 8.1 Verify that (8.23) implies that

$$\frac{dQ_i}{d\xi^0} = 0, \qquad (8.26)$$

when the currents j_i^α vanish sufficiently rapidly at spatial infinity.

To prove (8.23), we write both the Euler–Lagrange equations associated with the action (8.17) (see Problem 4.8) and the statement of invariance:

$$\partial_\alpha \left(\frac{\partial \mathcal{L}}{\partial(\partial_\alpha \phi^a)} \right) - \frac{\partial \mathcal{L}}{\partial \phi^a} = 0, \tag{8.27}$$

$$\frac{\partial \mathcal{L}}{\partial \phi^a} \delta \phi^a + \frac{\partial \mathcal{L}}{\partial(\partial_\alpha \phi^a)} \partial_\alpha(\delta \phi^a) = 0. \tag{8.28}$$

Using the first equation to eliminate $\frac{\partial \mathcal{L}}{\partial \phi^a}$ from the second, we get

$$\partial_\alpha \left(\frac{\partial \mathcal{L}}{\partial(\partial_\alpha \phi^a)} \right) \delta \phi^a + \frac{\partial \mathcal{L}}{\partial(\partial_\alpha \phi^a)} \partial_\alpha(\delta \phi^a) = \partial_\alpha \left(\frac{\partial \mathcal{L}}{\partial(\partial_\alpha \phi^a)} \delta \phi^a \right) = 0, \tag{8.29}$$

which demonstrates that (8.23) holds for currents defined as in (8.22).

We will use (8.22) to construct conserved currents which live on the string world-sheet. Actually, conserved charges and conserved currents exist under conditions less stringent than those stated in this section. A transformation can be considered a symmetry if the Lagrangian, or the Lagrangian density, if appropriate, is changed by a total derivative. These ideas are explored in Problem 8.9 and Problem 8.10.

8.3 Conserved currents on the world-sheet

To each string we would like to assign a relativistic momentum p_μ which is conserved if the string is moving freely. Even though the momentum p_μ carries an index, it is not a current, but rather a charge. In fact, since each component of p_μ should be separately conserved, this is a case where we have a set of conserved charges.

In the notation of the previous section, Q_i denote the various charges, so we see that the μ index in p_μ is playing the role of the i index in Q_i; it labels the various charges. What then is the α index in j_i^α? We will see that this index labels the coordinates on the world-sheet. The currents live on the world-sheet!

In the string action (6.39), the Lagrangian density is integrated over the world-sheet coordinates τ and σ, and not over the spacetime coordinates x^μ. In this example, the world of (8.17) is two-dimensional, and the index α in (8.21) takes two values. As a result, the conserved currents live on the world-sheet: they have two components, and they are functions of the world-sheet coordinates. More explicitly,

$$S = \int d\xi^0 d\xi^1 \mathcal{L}(\partial_0 X^\mu, \partial_1 X^\mu), \quad \text{with} \quad (\xi^0, \xi^1) = (\tau, \sigma), \tag{8.30}$$

and $\partial_\alpha = \partial/\partial \xi^\alpha$. Comparing with (8.17), we see that the field variables ϕ^a are simply the string coordinates X^μ. Note that the string action only depends on derivatives of the string coordinates.

To find conserved currents, we need a field variation δX^μ that does not change the Lagrangian density. One such variation is given by

$$\delta X^\mu(\tau, \sigma) = \epsilon^\mu, \tag{8.31}$$

where ϵ^μ is a constant – that is, it does not depend on τ or σ. This transformation is a constant *spacetime* translation: each point on the world-sheet is displaced by the same vector ϵ^μ. The Lagrangian density is invariant because it only depends on derivatives $\partial_\alpha X^\mu$, and their variations vanish: $\delta(\partial_\alpha X^\mu) = \partial_\alpha(\delta X^\mu) = \partial_\alpha \epsilon^\mu = 0$. The role of the various indices in (8.21) is now clear: α is a world-sheet index, i is a spacetime index, and so is a.

Now let us construct the conserved current. Using (8.22), and letting i and a indices run over the values of μ, we have

$$\epsilon^\mu j_\mu^\alpha = \frac{\partial \mathcal{L}}{\partial(\partial_\alpha X^\mu)} \delta X^\mu = \frac{\partial \mathcal{L}}{\partial(\partial_\alpha X^\mu)} \epsilon^\mu. \tag{8.32}$$

Cancelling the common ϵ^μ factor on the two sides of this equation, we find an expression for the currents:

$$j_\mu^\alpha = \frac{\partial \mathcal{L}}{\partial(\partial_\alpha X^\mu)} \longrightarrow (j_\mu^0, j_\mu^1) = \left(\frac{\partial \mathcal{L}}{\partial \dot{X}^\mu}, \frac{\partial \mathcal{L}}{\partial X^{\mu\prime}} \right). \tag{8.33}$$

We have seen such derivatives of $\mathcal{L}$ before; they appeared in equations (6.49) and (6.50). We can in fact identify

$$j_\mu^\alpha = \mathcal{P}_\mu^\alpha \longrightarrow (j_\mu^0, j_\mu^1) = (\mathcal{P}_\mu^\tau, \mathcal{P}_\mu^\sigma). \tag{8.34}$$

This is really interesting: the τ and σ superscripts in $\mathcal{P}_\mu$ label the *components* of a world-sheet current. The equation for current conservation is

$$\partial_\alpha \mathcal{P}_\mu^\alpha = \frac{\partial \mathcal{P}_\mu^\tau}{\partial \tau} + \frac{\partial \mathcal{P}_\mu^\sigma}{\partial \sigma} = 0. \tag{8.35}$$

This is simply the equation of motion (6.53) for the relativistic string.

Since the currents $\mathcal{P}_\mu^\alpha$ are indexed by μ, the conserved charges are also indexed by μ. Following (8.25), to get the charges we integrate the zeroth components $\mathcal{P}_\mu^\tau$ of the currents over space. In the present case, this means integrating over σ:

$$p_\mu(\tau) = \int_0^{\sigma_1} \mathcal{P}_\mu^\tau(\tau, \sigma) \, d\sigma. \tag{8.36}$$

This integral is done with τ held constant. We have called the conserved charges p_μ because they give the spacetime momentum carried by the string. Indeed, we have seen that these charges arise from spacetime translational invariance. Since (8.36) gives the total spacetime momentum as an integral over a string parameterized by σ, we learn that

$\mathcal{P}_\mu^\tau$ is the σ density of spacetime momentum carried by the string.   (8.37)

$\mathcal{P}_\mu^\tau$ is the derivative of the Lagrangian density with respect to the velocity $\dot{X}^\mu$, so it has the interpretation of a canonical momentum density, in agreement with (8.37). A similar identification was obtained for nonrelativistic strings: the quantity $\mathcal{P}^t$, which we defined in (4.46), has the interpretation of momentum density on account of (4.43).

To check conservation, we differentiate (8.36) with respect to τ and use the equation of motion (8.35):

$$\frac{dp_\mu}{d\tau} = \int_0^{\sigma_1} \frac{\partial \mathcal{P}_\mu^\tau}{\partial \tau}\, d\sigma = -\int_0^{\sigma_1} \frac{\partial \mathcal{P}_\mu^\sigma}{\partial \sigma}\, d\sigma = -\mathcal{P}_\mu^\sigma \Big|_0^{\sigma_1}. \qquad (8.38)$$

For a closed string, the coordinates $\sigma = 0$ and $\sigma = \sigma_1$ represent the same point on the world-sheet, so the right-hand side vanishes. For an open string with free endpoints, the boundary condition (6.56) states that $\mathcal{P}_\mu^\sigma$ vanishes at the endpoints, so, again, the right-hand side vanishes. In both of these cases, p_μ is conserved:

$$\frac{dp_\mu}{d\tau} = 0. \qquad (8.39)$$

There is more to this equation than our previous statement (8.12) of charge conservation. The derivative in (8.39) is with respect to τ and *not* with respect to t. So we may ask: is p_μ conserved in world-sheet time or in Minkowski time? We will discuss this question at length in the following section. The short answer is: in both.

For open strings with Dirichlet boundary conditions along some space directions, the momentum carried by the string along those directions may fail to be conserved. Indeed, the boundary condition (6.55) does not guarantee that the right-hand side of (8.38) vanishes. We already noted this possible nonconservation for nonrelativistic strings (Section 4.6). In open string theory, Dirichlet boundary conditions appear when we have D-branes that are not space filling. The momentum of the string can then fail to be conserved, but the total momentum of the string *and* the D-brane is conserved.

8.4 The complete momentum current

Equation (8.39) is very intriguing. It is a conservation law on the world-sheet rather than in spacetime. We could not have expected otherwise – the currents, after all, live on the world-sheet: their indices are world-sheet indices, and their arguments are world-sheet coordinates. As seen in spacetime, the currents vanish everywhere except on the surface traced out by the string.

If we trust the reparameterization invariance of the physics, we can easily obtain a standard *spacetime* conservation law by choosing the static gauge $t = \tau$. The integral in (8.36) is then an integral over the strings, the lines of constant time as seen by the chosen Lorentz observer. The conservation law (8.39) becomes

$$\frac{dp_\mu}{dt} = 0 \,. \tag{8.40}$$

The Lorentz observer confirms that momentum is conserved in *time*.

Equation (8.39), together with (8.36), tells us that for any fixed world-sheet parameterization, we can compute a unique quantity p_μ using any line of constant τ. This quantity must coincide with the time-independent momentum p_μ obtained using the static gauge, as we now explain.

Consider a moving string, a fixed Lorentz observer, and a particular choice of parameterization on the world-sheet. In this parameterization, over some region of the world-sheet the lines of constant τ are lines of constant t. As a result, over this region we have the static gauge. Over the rest of the world-sheet the parameterization changes smoothly: the lines of constant τ are no longer lines of constant time. On the static gauge region the integral (8.36) gives us the time-independent momentum carried by the strings. Because of (8.39), on the rest of the world-sheet the integral (8.36) must still give the same value for p_μ, even though the lines of constant τ are not strings.

This argument suggests that *any* curve on the world-sheet can be used to calculate the conserved momentum p_μ. Equation (8.36), however, only tells us how to calculate the momentum if the curve is a curve of constant τ. We now show how to generalize (8.36) to be able to compute the momentum p_μ using an (almost) arbitrary curve on the world-sheet *together* with an arbitrary parameterization of the world-sheet. When we deal with open strings, the curve must stretch from one boundary to the other. When we deal with closed strings, the curve must be a closed noncontractible curve.

Reconsider equation (8.36), which integrates the τ-component of the two-dimensional current $(\mathcal{P}^\tau_\mu, \mathcal{P}^\sigma_\mu)$ over a curve of constant τ. The quantity that this integral computes is actually the *flux* of the current across the curve. Since the σ-component $\mathcal{P}^\sigma_\mu$ of the current is tangent to the curve of constant τ, it does not contribute to the flux. More generally, consider an infinitesimal segment $(d\tau, d\sigma)$ along an oriented closed curve Γ that encloses a simply connected region $\mathcal{R}$ of the world-sheet (Figure 8.1). Since $(d\tau, d\sigma)$ is parallel to the oriented tangent, the outgoing normal to the segment is $(d\sigma, -d\tau)$. It is reasonable to define the outgoing flux of the current across the segment as the scalar product of the current vector and the outgoing normal vector:

$$\text{Infinitesimal flux} = (\mathcal{P}^\tau_\mu, \mathcal{P}^\sigma_\mu) \cdot (d\sigma, -d\tau) = \mathcal{P}^\tau_\mu \, d\sigma - \mathcal{P}^\sigma_\mu \, d\tau. \tag{8.41}$$

We now show that the flux, so defined, vanishes when computed across a contractible closed curve Γ on the world-sheet. This is a reasonable result because a contractible curve

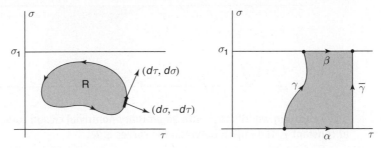

Fig. 8.1 Left: the total momentum flux out of a simply connected region $\mathcal{R}$ of the world-sheet is zero. Right: a simply connected region whose boundary includes an arbitrary curve γ and a string $\bar{\gamma}$.

surrounds a domain $\mathcal{R}$, and we do not expect a domain to be a momentum source or sink. The outgoing flux across Γ is written as

$$p_\mu(\Gamma) = \oint_\Gamma \left(\mathcal{P}_\mu^\tau \, d\sigma - \mathcal{P}_\mu^\sigma \, d\tau\right). \tag{8.42}$$

By the two-dimensional version of the divergence theorem, the flux of the current $\mathcal{P}_\mu^\alpha$ out of $\mathcal{R}$ is equal to the integral of the divergence of $\mathcal{P}_\mu^\alpha$ over $\mathcal{R}$:

$$p_\mu(\Gamma) = \int_\mathcal{R} \left(\frac{\partial \mathcal{P}_\mu^\tau}{\partial \tau} + \frac{\partial \mathcal{P}_\mu^\sigma}{\partial \sigma}\right) d\tau d\sigma = 0, \tag{8.43}$$

since $\mathcal{P}_\mu^\alpha$ is a conserved current. This is what we wanted to show.

Quick calculation 8.2 Consider $\mathbb{R}^2$ with coordinates (x, y) and a simply connected region M surrounded by a boundary Γ with counterclockwise orientation. The divergence theorem for a vector (A^x, A^y) reads

$$\oint_\Gamma (A^x dy - A^y dx) = \iint_M \left(\frac{\partial A^x}{\partial x} + \frac{\partial A^y}{\partial y}\right) dx dy. \tag{8.44}$$

Verify that this equation holds for a tiny rectangle with corners (x_0, y_0), $(x_0 + dx, y_0)$, $(x_0 + dx, y_0 + dy)$, and $(x_0, y_0 + dy)$. This suffices to prove (8.44) by breaking M into a collection of tiny rectangles. We used (8.44), for an $\mathbb{R}^2$ with coordinates (τ, σ) and a vector $(\mathcal{P}_\mu^\tau, \mathcal{P}_\mu^\sigma)$, to pass from (8.42) to (8.43).

We now generalize (8.36) as follows. For an arbitrary curve γ that starts on the $\sigma = 0$ boundary of the world-sheet and ends on the $\sigma = \sigma_1$ boundary, we define

$$p_\mu(\gamma) = \int_\gamma \left(\mathcal{P}_\mu^\tau \, d\sigma - \mathcal{P}_\mu^\sigma \, d\tau\right). \tag{8.45}$$

If γ is a curve of constant τ, then $d\tau = 0$ all along γ, and $p_\mu(\gamma)$ reduces to (8.36). We now prove that $p_\mu(\gamma)$, as defined in all generality by (8.45), actually coincides with p_μ, as defined in (8.36). Consider a curve γ that stretches from one side of the world-sheet to the other, and a curve $\bar{\gamma}$ of constant τ, as shown to the right of Figure 8.1. Let α and β denote

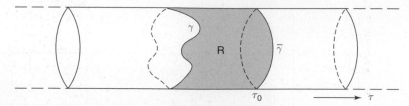

Fig. 8.2 A closed string world-sheet with an arbitrary nontrivial closed curve γ and a closed curve $\bar{\gamma}$ at constant τ. The region between the curves is $\mathcal{R}$.

oriented paths along the world-sheet boundary such that the closed curve Γ surrounding the shaded region is given by

$$\Gamma = \bar{\gamma} - \beta - \gamma + \alpha. \tag{8.46}$$

The curve Γ is oriented counterclockwise and it is contractible. As a result, the flux $p_\mu(\Gamma)$ vanishes:

$$p_\mu(\Gamma) = \int_\Gamma \left(\mathcal{P}_\mu^\tau \, d\sigma - \mathcal{P}_\mu^\sigma \, d\tau \right) = \left(\int_{\bar{\gamma}} - \int_\gamma + \int_\alpha - \int_\beta \right) \left(\mathcal{P}_\mu^\tau \, d\sigma - \mathcal{P}_\mu^\sigma \, d\tau \right) = 0. \tag{8.47}$$

Since α and β are curves where $d\sigma$ vanishes, only $\mathcal{P}_\mu^\sigma \, d\tau$ contributes to the integrals. But $\mathcal{P}_\mu^\sigma$ vanishes at the string endpoints (for free endpoints), so these integrals vanish identically. Only the integrals over γ and $\bar{\gamma}$ survive, so we have

$$\int_\gamma \left(\mathcal{P}_\mu^\tau \, d\sigma - \mathcal{P}_\mu^\sigma \, d\tau \right) = \int_{\bar{\gamma}} \left(\mathcal{P}_\mu^\tau \, d\sigma - \mathcal{P}_\mu^\sigma \, d\tau \right) = \int_{\bar{\gamma}} \mathcal{P}_\mu^\tau \, d\sigma = p_\mu, \tag{8.48}$$

where we noted that $d\tau = 0$ on $\bar{\gamma}$ and used (8.36). This proves that $p_\mu(\gamma) = p_\mu$ for any contour γ which connects the $\sigma = 0$ and $\sigma = \sigma_1$ boundaries of the world-sheet. We can thus rewrite (8.45) as

$$p_\mu = \int_\gamma \left(\mathcal{P}_\mu^\tau \, d\sigma - \mathcal{P}_\mu^\sigma \, d\tau \right). \tag{8.49}$$

Conservation is now the statement that the integral above is independent of the chosen curve γ, as long as the endpoints of γ lie on the boundary components of the world-sheet.

The arguments are similar in the case of closed strings. We consider an arbitrary nontrivial closed curve γ winding once around the world-sheet, and another similarly nontrivial closed curve $\bar{\gamma}$ of constant τ, for some arbitrary, but fixed, parameterization. The two curves form the boundary of an annular region $\mathcal{R}$ (Figure 8.2). A completely analogous argument shows that both contours give the same result for p_μ. Therefore, we can calculate the momentum of the closed string using any closed curve that winds once around the world-sheet.

How does an arbitrary Lorentz observer use equation (8.49)? The observer looks at a string at some time t, and asks for its momentum. This requires the use of (8.49), with a curve γ

which corresponds to the string in question. With an arbitrary parameterization γ need not be a constant τ curve. At some later time t', the observer asks again for the momentum. At this time, the string corresponds to a curve γ', generically different from γ. By the curve independence of (8.49), the observer concludes that the momentum did not change. The momentum was conserved in time.

We conclude this section by making contact with familiar results. Using a constant τ string γ to evaluate (8.49) we have

$$p^0 = \int_\gamma \mathcal{P}^{\tau 0}\, d\sigma, \quad \vec{p} = \int_\gamma \vec{\mathcal{P}}^\tau\, d\sigma. \tag{8.50}$$

The values of the momentum densities in static gauge can be read from (6.94). We thus find that the energy and the spatial momentum of the string are given by:

$$p^0 \equiv \frac{E}{c} = \frac{1}{c}\int_\gamma \frac{T_0 ds}{\sqrt{1 - \frac{v_\perp^2}{c^2}}}, \quad \vec{p} = \int_\gamma \frac{T_0 ds}{c^2}\, \frac{\vec{v}_\perp}{\sqrt{1 - \frac{v_\perp^2}{c^2}}}. \tag{8.51}$$

The above equation for the energy of the string coincides with (7.9). The expression for the momentum is reasonable: the momentum carried by a little piece of string is given by its rest mass $T_0 ds/c^2$ times its velocity, times its relativistic factor γ.

8.5 Lorentz symmetry and associated currents

By construction, the action of the relativistic string is Lorentz invariant. It is written in terms of Lorentz vectors that are contracted to build Lorentz scalars. Concretely, this means that Lorentz transformations of the coordinates X^μ leave the action invariant. In this section we will construct the conserved charges associated with Lorentz symmetry.

These charges will be particularly useful when we study quantum string theory in Chapter 12. Whenever we quantize a classical system, there is the possibility that crucial symmetries of the classical theory will be lost. If Lorentz invariance were lost upon quantization, quantum string theory would be very problematic, to say the least. We will have to make sure that the quantum theory is Lorentz invariant. To calculate the conserved charges, we first need the infinitesimal form of the general Lorentz transformations. We recall (Section 2.2) that Lorentz transformations are linear transformations of the coordinates X^μ that leave the quadratic form $\eta_{\mu\nu} X^\mu X^\nu$ invariant. Every infinitesimal linear transformation is of the form $X^\mu \to X^\mu + \delta X^\mu$, where

$$\delta X^\mu = \epsilon^{\mu\nu} X_\nu. \tag{8.52}$$

Here $\epsilon^{\mu\nu}$ is a matrix of infinitesimal constants. Lorentz invariance imposes conditions on the constants $\epsilon^{\mu\nu}$. We require $\delta(\eta_{\mu\nu} X^\mu X^\nu) = 0$, and therefore

$$2\eta_{\mu\nu}(\delta X^\mu)X^\nu = 2\eta_{\mu\nu}(\epsilon^{\mu\rho} X_\rho)\, X^\nu = 2\,\epsilon^{\mu\rho} X_\rho X_\mu = 0. \tag{8.53}$$

Imagine decomposing the matrix ϵ into its antisymmetric part and its symmetric part. The antisymmetric part would not contribute to $\epsilon^{\mu\rho} X_\rho X_\mu$. The vanishing of $\epsilon^{\mu\rho} X_\rho X_\mu$, for all values of X_μ, implies that the symmetric part of ϵ is zero. It follows that the general solution is represented by an antisymmetric $\epsilon^{\mu\nu}$:

$$\epsilon^{\mu\nu} = -\epsilon^{\nu\mu}. \tag{8.54}$$

Infinitesimal Lorentz transformations are thus very simple: they are transformations of the form $\delta X^\mu = \epsilon^{\mu\nu} X_\nu$, with $\epsilon^{\mu\nu}$ antisymmetric. This result applies for any number of spacetime dimensions since Lorentz transformations always leave $\eta_{\mu\nu} X^\mu X^\nu$ invariant.

Quick calculation 8.3 Consider a fixed 2-by-2 matrix A^{ab} $(a, b = 1, 2)$ that satisfies $A^{ab} v_a v_b = 0$ for *all* values of v_1 and v_2. Write out the four terms on the left-hand side of the vanishing condition, and confirm explicitly that the matrix A^{ab} must be antisymmetric.

Quick calculation 8.4 Repeat the above exercise for a 4-by-4 matrix $\epsilon^{\mu\nu}$ $(\mu, \nu = 0, 1, 2, 3)$ that satisfies $\epsilon^{\mu\nu} v_\mu v_\nu = 0$ for *all* values of v_0, v_1, v_2, and v_3.

Quick calculation 8.5 Show that (8.54) implies $\epsilon_{\mu\nu} = -\epsilon_{\nu\mu}$.

Quick calculation 8.6 Examine the boost in (2.36) for very small β. Write $x'^\mu = x^\mu + \epsilon^{\mu\nu} x_\nu$ and calculate the entries of the matrix $\epsilon^{\mu\nu}$. Show that $\epsilon^{10} = -\epsilon^{01} = \beta$, and that all other entries are zero. This confirms that $\epsilon^{\mu\nu}$ is antisymmetric for an infinitesimal boost.

Let us now show explicitly that the string Lagrangian density is invariant under Lorentz transformations. All terms that appear in this density are of the form

$$\eta_{\mu\nu} \frac{\partial X^\mu}{\partial \xi^\alpha} \frac{\partial X^\nu}{\partial \xi^\beta}, \tag{8.55}$$

where ξ^α and ξ^β are either τ or σ. We claim that any such term is Lorentz invariant. Indeed,

$$\delta\left(\eta_{\mu\nu} \frac{\partial X^\mu}{\partial \xi^\alpha} \frac{\partial X^\nu}{\partial \xi^\beta}\right) = \eta_{\mu\nu}\left(\frac{\partial \delta X^\mu}{\partial \xi^\alpha} \frac{\partial X^\nu}{\partial \xi^\beta} + \frac{\partial X^\mu}{\partial \xi^\alpha} \frac{\partial \delta X^\nu}{\partial \xi^\beta}\right)$$
$$= \eta_{\mu\nu}\left(\epsilon^{\mu\rho} \frac{\partial X_\rho}{\partial \xi^\alpha} \frac{\partial X^\nu}{\partial \xi^\beta} + \epsilon^{\nu\rho} \frac{\partial X^\mu}{\partial \xi^\alpha} \frac{\partial X_\rho}{\partial \xi^\beta}\right)$$
$$= \epsilon_{\nu\rho} \frac{\partial X^\rho}{\partial \xi^\alpha} \frac{\partial X^\nu}{\partial \xi^\beta} + \epsilon_{\mu\rho} \frac{\partial X^\mu}{\partial \xi^\alpha} \frac{\partial X^\rho}{\partial \xi^\beta}, \tag{8.56}$$

where we used η to lower the first index on the ϵ constants. Letting $\mu \to \rho$ and $\rho \to \nu$ in the second term we get

$$\delta\left(\eta_{\mu\nu} \frac{\partial X^\mu}{\partial \xi^\alpha} \frac{\partial X^\nu}{\partial \xi^\beta}\right) = (\epsilon_{\nu\rho} + \epsilon_{\rho\nu}) \frac{\partial X^\rho}{\partial \xi^\alpha} \frac{\partial X^\nu}{\partial \xi^\beta} = 0, \tag{8.57}$$

because of the antisymmetry of ϵ. This explicitly proves the Lorentz invariance of the string action.

We can now use equation (8.22) to write the currents. It follows from (8.52) that the role of the small parameter ϵ^i is played by $\epsilon^{\mu\nu}$. We therefore have

$$\epsilon^{\mu\nu} j^\alpha_{\mu\nu} = \frac{\partial \mathcal{L}}{\partial(\partial_\alpha X^\mu)} \delta X^\mu = \mathcal{P}^\alpha_\mu \epsilon^{\mu\nu} X_\nu. \tag{8.58}$$

The current $j^{\alpha}_{\mu\nu}$, since it is multiplied by the antisymmetric matrix $\epsilon^{\mu\nu}$, can be defined to be antisymmetric – any symmetric part would drop out of the left-hand side. Using the antisymmetry of $\epsilon^{\mu\nu}$, the right-hand side is written as

$$\epsilon^{\mu\nu} j^{\alpha}_{\mu\nu} = (-\tfrac{1}{2}\epsilon^{\mu\nu}) \left(X_{\mu} \mathcal{P}^{\alpha}_{\nu} - X_{\nu} \mathcal{P}^{\alpha}_{\mu} \right). \tag{8.59}$$

The currents can be read directly from this equation because the factor multiplying $\epsilon^{\mu\nu}$ on the right-hand side is explicitly antisymmetric. Since the overall normalization of the currents is for us to choose, we define the currents $\mathcal{M}^{\alpha}_{\mu\nu}$ by

$$\mathcal{M}^{\alpha}_{\mu\nu} = X_{\mu}\, \mathcal{P}^{\alpha}_{\nu} - X_{\nu}\, \mathcal{P}^{\alpha}_{\mu}. \tag{8.60}$$

By construction,

$$\mathcal{M}^{\alpha}_{\mu\nu} = -\mathcal{M}^{\alpha}_{\nu\mu}. \tag{8.61}$$

The equation of current conservation is

$$\frac{\partial \mathcal{M}^{\tau}_{\mu\nu}}{\partial \tau} + \frac{\partial \mathcal{M}^{\sigma}_{\mu\nu}}{\partial \sigma} = 0, \tag{8.62}$$

and the charges, in analogy with (8.49), are given by

$$M_{\mu\nu} = \int_{\gamma} \left(\mathcal{M}^{\tau}_{\mu\nu} d\sigma - \mathcal{M}^{\sigma}_{\mu\nu} d\tau \right). \tag{8.63}$$

The charges, just like the currents, are antisymmetric:

$$M_{\mu\nu} = -M_{\nu\mu}. \tag{8.64}$$

The conservation of $M_{\mu\nu}$ is a result of the contour independence of the definition (8.63). For closed strings, contour independence is guaranteed by the argument given earlier in the case of momentum charges. For free open strings, one point must be addressed to ensure contour independence. The $M_{\mu\nu}$ integrals must receive no contributions from lines on the boundary of the world-sheet. This, in turn, requires the vanishing of $\mathcal{M}^{\sigma}_{\mu\nu}$ on the boundary. This condition is satisfied because $\mathcal{M}^{\sigma}_{\mu\nu}$ involves $\mathcal{P}^{\sigma}$ multiplicatively, and $\mathcal{P}^{\sigma}$ vanishes on the world-sheet boundary. As explained for the case of the momentum charges, a Lorentz observer measuring $M_{\mu\nu}$ using strings at different times will conclude that $dM_{\mu\nu}/dt = 0$.

We can also compute the Lorentz charges $M_{\mu\nu}$ using constant τ lines. In that case,

$$M_{\mu\nu} = \int \mathcal{M}^{\tau}_{\mu\nu}(\tau, \sigma)\, d\sigma = \int (X_{\mu}\mathcal{P}^{\tau}_{\nu} - X_{\nu}\mathcal{P}^{\tau}_{\mu})\, d\sigma. \tag{8.65}$$

Since the $M_{\mu\nu}$ are antisymmetric, we have six conserved charges in four dimensions. Letting i and j denote space indices, M_{0i} are the three charges associated with the three basic boosts, and M_{ij} are the three charges associated with the three basic rotations. Given that $\vec{\mathcal{P}}^\tau$ is the momentum density, the normalization chosen in (8.60) ensures that $\mathcal{M}_{ij}^\tau$ is precisely the angular momentum density. As a consequence, the components M_{ij} measure the string angular momentum $\vec{L}$ via the usual relations $L_i = \frac{1}{2}\epsilon_{ijk}M_{jk}$. Here ϵ_{ijk} is the totally antisymmetric symbol which satisfies $\epsilon_{123} = 1$. More explicitly, we have $L_1 = M_{23}$, $L_2 = M_{31}$, and $L_3 = M_{12}$.

The charges associated with the boosts are:

$$M^{0i} = \int d\sigma \left(ct\,\mathcal{P}^{\tau i} - X^i \mathcal{P}^{\tau 0} \right) = ct\, p^i - \int d\sigma X^i \mathcal{P}^{\tau 0}. \tag{8.66}$$

Multiplying by c/E, where E is the conserved energy of the string,

$$\frac{cM^{0i}}{E} = t\,\frac{c^2 p^i}{E} - \frac{1}{E} \int d\sigma\, X^i\, c\mathcal{P}^{\tau 0}. \tag{8.67}$$

Since $c\mathcal{P}^{\tau 0}$ is the energy density along the string, the last term on the above right-hand side can be identified with the time-dependent position $X_{\text{cm}}^i(t)$ of the center of mass (energy) of the string. We thus obtain,

$$X_{\text{cm}}^i(t) = -\frac{cM^{0i}}{E} + t\,\frac{c^2 p^i}{E}. \tag{8.68}$$

The quantity $c^2 p^i/E$ has the interpretation of center of mass velocity because it coincides with the velocity of a point particle with momentum p^i and energy E. Equation (8.68) describes the motion of the center of mass. The conserved charges M^{0i}, together with E, determine the $t = 0$ position of the center of mass.

8.6 The slope parameter α'

The string tension T_0 is the only dimensionful parameter in the string action. In this section we will motivate the definition of an alternative parameter: the slope parameter α'. These two parameters are related, we can use one or the other. The parameter α' has an interesting physical interpretation, used since the early days of string theory. If we consider a rigidly rotating open string, α' is the proportionality constant that relates the angular momentum J of the string, measured in units of $\hbar$, to the square of its energy E. More explicitly,

$$\frac{J}{\hbar} = \alpha' E^2. \tag{8.69}$$

Since the left-hand side has no units, the units of α' are those of inverse energy-squared:

$$[\alpha'] = \frac{1}{[E]^2}. \tag{8.70}$$

The appearance of $\hbar$ in (8.69) is just a convention. The constant α' was introduced in the quantum theory of strings, and its relation to the string tension involves $\hbar$. For our present purposes, however, the important fact is just the proportionality between J and E^2. This relation does *not* involve $\hbar$ when we use the string tension T_0.

To verify the proportionality implied in equation (8.69), we consider a straight open string of energy E which rotates rigidly on the (x, y) plane. This is precisely the problem examined in Section 7.4. The only nonvanishing component of angular momentum is M_{12}, and its magnitude is denoted by $J = |M_{12}|$. Equation (8.65) tells us that

$$M_{12} = \int_0^{\sigma_1} (X_1 \mathcal{P}_2^\tau - X_2 \mathcal{P}_1^\tau) \, d\sigma. \tag{8.71}$$

To evaluate this integral we need formulae for the position and momenta of the rotating string. We recall equation (7.63),

$$\vec{X}(t, \sigma) = \frac{\sigma_1}{\pi} \cos \frac{\pi \sigma}{\sigma_1} \left(\cos \frac{\pi ct}{\sigma_1}, \, \sin \frac{\pi ct}{\sigma_1} \right), \tag{8.72}$$

which records the components (X_1, X_2) of the rotating string. Using equation (7.31), we find

$$\vec{\mathcal{P}}^\tau = \frac{T_0}{c^2} \frac{\partial \vec{X}}{\partial t} = \frac{T_0}{c} \cos \frac{\pi \sigma}{\sigma_1} \left(-\sin \frac{\pi ct}{\sigma_1}, \, \cos \frac{\pi ct}{\sigma_1} \right), \tag{8.73}$$

where the right-hand side gives the components $(\mathcal{P}_1^\tau, \mathcal{P}_2^\tau)$. The integral in (8.71) is thus given by

$$M_{12} = \frac{\sigma_1}{\pi} \frac{T_0}{c} \int_0^{\sigma_1} \cos^2 \frac{\pi \sigma}{\sigma_1} \, d\sigma = \frac{\sigma_1^2 T_0}{2\pi c}. \tag{8.74}$$

The time dependence disappeared, as expected for a conserved charge. Since $J = |M_{12}|$ and $\sigma_1 = E/T_0$, we have found that

$$J = \frac{1}{2\pi T_0 c} E^2. \tag{8.75}$$

As anticipated, the angular momentum is proportional to the square of the energy of the string. Comparing with equation (8.69) we deduce that

$$\alpha' = \frac{1}{2\pi T_0 \hbar c} \quad \text{and} \quad T_0 = \frac{1}{2\pi \alpha' \hbar c}. \tag{8.76}$$

These equations relate the slope parameter α' to the string tension T_0.

Quick calculation 8.7 To appreciate how unusual the relation $J \sim E^2$ is, consider a one-dimensional straight bar of fixed length and fixed (uniform) mass that is rotating around its midpoint. Show that the nonrelativistic energy and the angular momentum are related by $J \sim \sqrt{E}$.

One may have anticipated the relationship $J \sim E^2$ by the following admittedly rough argument. We know that for rigid rotations $J = I\omega$, where I is the moment of inertia. For an object of mass M and length scale L, we have $I \sim ML^2$. So, $J \sim ML^2\omega$. For a rotating relativistic string $M \sim E$, $L \sim E$, and $\omega \sim 1/E$ (see (7.57)). As a result $J \sim E^2$.

The name slope parameter arises because α' is the slope of the lines of $J/\hbar$, when plotted as a function of energy-squared. In fact, Regge trajectories are approximate lines that arise when plotting angular momentum as a function of energy-squared for hadronic excitations. In the early 1970s, when string theory was investigated as a theory of strong interactions, the slope parameter α' was the experimentally measured quantity that entered into the string action. The action (6.39), which we wrote in terms of T_0, takes the form

$$S = -\frac{1}{2\pi\alpha'\hbar c^2} \int_{\tau_i}^{\tau_f} d\tau \int_0^{\sigma_1} d\sigma \sqrt{(\dot{X} \cdot X')^2 - (\dot{X})^2 (X')^2} . \tag{8.77}$$

It turns out that most of the modern work on string theory uses the slope parameter α' as opposed to the string tension T_0. In Section 3.6 we used $\hbar$, c, and Newton's constant G to calculate a characteristic length ℓ_P called the Planck length. In string theory, we can use $\hbar$, c, and the dimensionful parameter α' to construct a characteristic length ℓ_s called the string length.

Quick calculation 8.8 Show that

$$\ell_s = \hbar c \sqrt{\alpha'} . \tag{8.78}$$

Up to factors of $\hbar$ and c, the string length ℓ_s is the square root of α'. This connection to a fundamental length scale provides an alternative physical interpretation for the slope parameter α'.

Problems

Problem 8.1 Lorentz invariants and their densities.

We want to understand how Lorentz invariance of electric charge implies that charge *density* is the zeroth component of a Lorentz vector.

Consider two Lorentz frames S and S' with S' boosted along the $+x$ direction of S with velocity v. In frame S we have a cubic box of volume L^3 with edges, all of length L, aligned with the axes of the coordinate system. The box is at rest and is filled by a material, also at rest, that has uniform charge density ρ_0. In this frame the current density vanishes $\vec{j} = \vec{0}$. Note that S' sees the box Lorentz contracted along the x direction.

(a) Use the Lorentz invariance of charge to calculate the charge density and current density $(c\rho', \vec{j}')$ that describe the material in the box according to frame S'. Verify that the values $(c\rho_0, \vec{0})$ and $(c\rho', \vec{j}')$ behave as four-vectors $j^\mu \equiv (c\rho_0, \vec{0})$ and $j'^\mu \equiv (c\rho', \vec{j}')$, in the sense that the quantities $(c\rho', \vec{j}')$ are obtained from the rest frame values $(c\rho_0, \vec{0})$ by a Lorentz boost.

(b) Write a formula for the current j^μ in terms of ρ_0 and the four-velocity of the box, and check that your result works both for frames S and S'.

Problem 8.2 Lorentz invariance of electric charge.

The explicit test of Lorentz invariance of electric charge is quite subtle, so this is a *challenging* problem! We consider, as usual, a frame S and a frame S' moving along the $+x$ direction of S with velocity v. In frame S the total charge is given by integration over all of space at a fixed time $t = 0$:

$$Q \equiv \int d^3x\, \rho(t=0, \vec{x}),$$

and in the primed frame S' it is given by an integral over all of space at $t' = 0$:

$$Q' \equiv \int d^3x'\, \rho'(t'=0, \vec{x}').$$

We want to prove that $Q' = Q$. The complication is that S and S' do not agree on what is equal time. We will assume that $j^\alpha = (c\rho, \vec{j})$ is a four-vector and it is conserved: $\partial_\alpha j^\alpha = 0$. Moreover, charge densities and currents are taken to vanish at infinity for all times.

(a) Write explicitly the transformations that relate $(t, \vec{x})$ to $(t', \vec{x}')$ and $(c\rho, \vec{j})$ to $(c\rho', \vec{j}')$. Think of ρ and $\vec{j}$ as functions of arguments (t, x, y, z). Prove that

$$Q' = \int d^3x'\, \gamma \left[\rho \left(\gamma \frac{vx'}{c^2}, \gamma x', y', z' \right) - \frac{v}{c^2} j^x \left(\gamma \frac{vx'}{c^2}, \gamma x', y', z' \right) \right].$$

To make Q' look more like the expression for Q, change integration variables using $x = \gamma x'$, $y = y'$, and $z = z'$. Write the resulting expression for Q'. Confirm that for $v = 0$ you get $Q' = Q$.

(b) The strategy is to show that Q' is independent of the velocity v. In that case, the equality of Q' and Q for $v = 0$ implies the equality of Q' and Q for all velocities. Show that the derivative of Q' with respect to the velocity can be written as

$$\frac{dQ'}{dv} = \frac{1}{c^2} \int d^3x \left[x \left(\frac{\partial\rho}{\partial t} - \frac{v}{c^2} \frac{\partial j^x}{\partial t} \right) - j^x \right] \Bigg|_{t=\frac{vx}{c^2}, x, y, z}. \qquad (1)$$

(c) Use the conservation equation $\partial_\alpha j^\alpha = 0$ to show that the integrand in (1) is a total derivative. It then follows from our conditions at infinity that $\frac{dQ'}{dv} = 0$, completing the proof that $Q' = Q$. You may find the following equation useful:

$$\left(\frac{\partial}{\partial x} j^x \left(\frac{vx}{c^2}, x, y, z \right) \right)_{y,z} = \left[\frac{v}{c^2} \frac{\partial j^x}{\partial t} + \frac{\partial j^x}{\partial x} \right] \Bigg|_{t=\frac{vx}{c^2}, x, y, z}, \qquad (2)$$

where the subscripts y, z on the left-hand side indicate that the partial derivative is taken with these variables (and not t) held fixed. Equation (2) is just a statement about derivatives and should be clear if you understand the notation.

Problem 8.3 Angular momentum as a conserved charge.

Consider a Lagrangian L that depends only on the *magnitude* of the velocity $\dot{\vec{q}}(t)$ of a particle which moves in ordinary three-dimensional space.

(a) Write an infinitesimal variation $\delta\vec{q}(t)$ that represents a small rotation of the vector $\vec{q}(t)$. Explain why it leaves the Lagrangian invariant.
(b) Construct the conserved charge associated with this symmetry transformation. Verify that this conserved quantity is the (vector) angular momentum.

Problem 8.4 A generalization for charges and a special case for currents.

(a) Generalize equations (8.9) and (8.11) to the case in which there are various symmetry parameters ϵ^i and a set of coordinates q^a. Verify that your charges are conserved.
(b) Consider the setup of (8.17) for a "world" with no spatial dimensions. What are the possible values of the index α? What do equations (8.22), (8.23), and (8.25) give? Compare with your results in part (a).

Problem 8.5 Lorentz charges for the relativistic point particle.[†]

Consider the point particle action (5.15): $S = -mc \int d\tau \sqrt{-\eta_{\mu\nu}\dot{x}^\mu \dot{x}^\nu}$, where $\dot{x}^\mu(\tau) = dx^\mu(\tau)/d\tau$. This action can be treated as a mechanics action where x^μ is a coordinate and τ plays the role of time, or as a field action where x^μ is a field in a world with no spatial dimensions and with $\xi^0 = \tau$.

(a) Show that $x^\mu(\tau) \to x^\mu(\tau) + \epsilon^\mu$, with ϵ^μ constant, is a symmetry. Find the conserved charges associated with this symmetry and verify explicitly their conservation. Compare with the momenta $p_\mu(\tau)$ defined canonically from the Lagrangian.
(b) Write the infinitesimal Lorentz transformations of $x^\mu(\tau)$, and explain why the action is invariant under such transformations.
(c) Find an expression for the Lorentz charges in terms of $x^\mu(\tau)$ and $p^\mu(\tau)$. Make sure that your Lorentz charges coincide with angular momentum charges for the appropriate values of the indices. Verify explicitly their conservation.

Problem 8.6 Simple estimates regarding α', T_0, and ℓ_s.

(a) In hadronic physics $\alpha' \simeq 0.95$ GeV^{-2}. Calculate the hadronic string tension in tons and the string length in centimeters.
(b) Assume that the string length is $\ell_s \simeq 10^{-30}$ cm. Calculate α' in GeV^{-2} and the string tension in tons.

Problem 8.7 Angular momentum of a rotating string.

Equation (8.51) tells us that the momentum $d\vec{p}$ carried by a small piece of string is given by

$$d\vec{p} = \frac{T_0 ds}{c^2} \frac{\vec{v}_\perp}{\sqrt{1 - \frac{v_\perp^2}{c^2}}}$$

Use this result to calculate by direct integration the angular momentum J carried by a rotating open string. Make sure that your result coincides with (8.75).

Problem 8.8 Angular momentum of Kasey's jumping rope.

Consider the relativistic jumping rope solution of Problem 7.4. Calculate the magnitude J_z of the z-component of angular momentum (measured with respect to the origin) as a function of the separation L_0 and the angle γ. Show that $J_z/\hbar = (\sin^2 \gamma)\alpha' E^2$, where E is the energy of the string.

Problem 8.9 Generalizing the construction of conserved charges.

We reconsider the setup that led to the conserved charge Q defined in (8.11). Assume now that the transformation (8.8) does not leave the Lagrangian L invariant, but rather the change is a total time derivative:

$$\delta L = \frac{d}{dt}(\epsilon \Lambda), \tag{1}$$

where Λ is some calculable function of coordinates, velocities, and possibly of time. Show that there is a modified conserved charge which takes the form

$$\epsilon Q = \frac{\partial L}{\partial \dot{q}} \delta q - \epsilon \Lambda. \tag{2}$$

Since it leads to a conserved charge, a transformation that changes L by a total derivative is said to be a symmetry transformation.

   As an application consider a Lagrangian $L(q(t), \dot{q}(t))$ that has no explicit time dependence. A transformation of the form

$$q(t) \longrightarrow q(t + \epsilon) \simeq q(t) + \epsilon \dot{q}(t), \tag{3}$$

represents the effect of a constant infinitesimal time translation. Show that the transformation (3) is a symmetry in the sense of (1). Calculate Λ and construct the conserved charge Q. Is the result familiar?

Problem 8.10 Generalizing the construction of conserved currents.

Assume that the transformation (8.19) does not leave the Lagrangian density $\mathcal{L}$ invariant, but rather the change is a total derivative:

$$\delta \mathcal{L} = \frac{\partial}{\partial \xi^\alpha}(\epsilon^i \Lambda_i^\alpha), \tag{1}$$

where the Λ_i^α are a set of calculable functions of fields, field derivatives, and possibly coordinates. Show that there is a conserved current which takes the form

$$\epsilon^i j_i^\alpha = \frac{\partial \mathcal{L}}{\partial(\partial_\alpha \phi^a)} \delta \phi^a - \epsilon^i \Lambda_i^\alpha. \tag{2}$$

Since it leads to a conserved current, a transformation that changes $\mathcal{L}$ by a total derivative is said to be a symmetry transformation.

As an application consider a Lagrangian density $\mathcal{L}(\phi^a, \partial_\alpha \phi^a)$ that has no explicit dependence on the world coordinates ξ^α. The transformation

$$\phi^a(\xi^\beta) \longrightarrow \phi^a(\xi^\beta + \epsilon^\beta) \simeq \phi^a + \epsilon^\beta \partial_\beta \phi^a, \tag{3}$$

where ϵ^β is an infinitesimal constant vector, represents the effect of a constant translation. Show that (3) is a symmetry in the sense of (1), with

$$\Lambda^\alpha_\beta = \delta^\alpha_\beta \mathcal{L}. \tag{4}$$

Show that the conserved currents j^α_β take the form

$$j^\alpha_\beta = \frac{\partial \mathcal{L}}{\partial(\partial_\alpha \phi^a)} \partial_\beta \phi^a - \delta^\alpha_\beta \mathcal{L}. \tag{5}$$

Is j^0_0 familiar? The quantities j^α_β actually define the energy-momentum tensor T^α_β.

9

Light-cone relativistic strings

A class of gauges is introduced that fix the parameterization of the world-sheet, lead to a pair of constraints, and give equations of motion that are wave equations. One of these gauges, the light-cone gauge, sets X^+ proportional to τ and allows for a complete and explicit solution to the equations of motion. In this gauge, the dynamics of the string is determined by the motion along the transverse directions and the values of two zero modes. We encounter the classical Virasoro modes as the oscillation modes of the X^- coordinate, and we learn how to calculate the mass of an arbitrary string configuration.

9.1 A class of choices for τ

Our first encounter with classical string dynamics was simplified by our use of the static gauge. In this gauge the world-sheet time τ is identified with the spacetime time coordinate X^0 by

$$X^0(\tau, \sigma) = c\tau. \tag{9.1}$$

We are now going to consider more general gauges. Among the class of gauges that we will examine, one of them, the light-cone gauge, will turn out to be particularly useful. Using this gauge we will solve the equations of motion of the string in a complete and explicit fashion. Our solution in the static gauge was not fully explicit; the motion was characterized by a constrained vector function (see (7.48)).

The gauges we will focus on are those for which τ is set equal to a linear combination of the string coordinates. This condition can be written as

$$n_\mu X^\mu(\tau, \sigma) = \lambda\tau. \tag{9.2}$$

If we choose $n_\mu = (1, 0, \ldots, 0)$ and $\lambda = c$, this equation becomes (9.1). To understand the meaning of (9.2), we introduce the related equation

$$n_\mu x^\mu = \lambda\tau, \tag{9.3}$$

where we write x^μ, as opposed to X^μ, to emphasize that we are dealing with the general spacetime coordinates. Consider now the two equations above with the same *fixed* value of τ. If x_1^μ and x_2^μ are two points which satisfy (9.3), then $n_\mu(x_1^\mu - x_2^\mu) = 0$. This shows

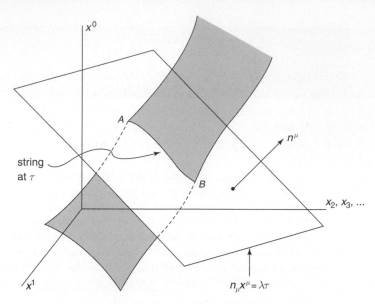

Fig. 9.1 The gauge condition $n \cdot X = \lambda\tau$ fixes the strings to be the curves at the intersection of the world-sheet with the hyperplanes orthogonal to the vector n^μ.

that any vector that joins points in the space (9.3) is orthogonal to the vector n^μ. The set of all points which satisfy (9.3) forms a hyperplane normal to n^μ.

We can now make clear the meaning of equation (9.2). The points X^μ that satisfy $n_\mu X^\mu = \lambda\tau$ are points that lie *both* on the world-sheet and on the hyperplane (9.3). Equation (9.2) states that all these points must be assigned the same value τ of the world-sheet time. If we define a string to be the set of points $X^\mu(\tau, \sigma)$ with constant τ, in the gauge (9.2), strings lie on hyperplanes of the form (9.3). The string with world-sheet time τ is the intersection of the world-sheet with the hyperplane $n \cdot x = \lambda\tau$, as illustrated in Figure 9.1.

We want strings to be spacelike objects. More precisely, the interval ΔX^μ between any two points on a string should be spacelike, perhaps null in some limit, but certainly never timelike. We will now see that in the gauge (9.2), a timelike n^μ guarantees that the string is spacelike. In this gauge any interval ΔX^μ along the string satisfies $n \cdot \Delta X = 0$. Since this condition is Lorentz invariant, it can be analyzed in a Lorentz frame where the only nonzero component of n^μ is the time component. In this frame ΔX^μ cannot have a time component. It is therefore a spacelike vector.

If n^μ is a null vector $((1, 1, 0, \ldots, 0)$, for example), one can show that $n \cdot \Delta X = 0$ implies that ΔX^μ is generally spacelike and occassionally null (Problem 9.1). We will allow n^μ to be null in (9.2). This choice can be viewed as the limit of a sequence of choices all of which involve a timelike n^μ.

The gauges defined by (9.2) are not Lorentz covariant gauges for any choice of n^μ. Choosing n^μ selects a particular linear combination of spacetime coordinates to be set equal to τ. There is no linear combination of coordinates that is left invariant by arbitrary

Lorentz transformations. Therefore, the gauge condition takes different forms in different Lorentz frames – the gauge is not Lorentz covariant.

At this point, it is convenient to streamline the way that we deal with units. While we have been using the convention that τ and σ have units of time and length, respectively, starting now τ and σ will be *dimensionless*. We proved in Chapter 8 that for open strings with free endpoints, there is a well defined conserved momentum p^μ. We will use this Lorentz vector to rewrite our gauge condition (9.2) as follows:

$$n \cdot X(\tau, \sigma) = \tilde{\lambda} \, (n \cdot p) \, \tau. \tag{9.4}$$

Since $n \cdot p$ is a constant, the net effect is that we have traded λ for another constant $\tilde{\lambda}$. When open strings are attached to D-branes not all components of the string momentum are conserved. Since we want our analysis to hold even in this case, we will assume that the vector n^μ is chosen in such a way that $n \cdot p$ *is* conserved. This condition is weaker than the condition of momentum conservation. We will assume that $n \cdot \mathcal{P}^\sigma = 0$ at the open string endpoints since this condition naturally guarantees the conservation of $n \cdot p$ (consider equation (8.38) dotted with n^μ).

By involving the vector n^μ explicitly on both sides of equation (9.4), the scale of n^μ has been made irrelevant. Only the direction of n^μ matters. Bearing in mind that $n \cdot X$ has units of length, $n \cdot p$ has units of momentum, and τ is dimensionless, imagine dividing both sides of this equation by the unit of time. We then see that $\tilde{\lambda}$ has units of velocity divided by force. The canonical choice for velocity is c, and the canonical choice for force is the string tension T_0. It is therefore natural to set

$$\tilde{\lambda} \sim \frac{c}{T_0} = 2\pi\alpha' hc^2, \tag{9.5}$$

where we used (8.76) to relate the string tension to α'. Before fixing $\tilde{\lambda}$ precisely, let us further simplify our treatment of units.

At stake is our ability to track the units of different physical quantities. We can simplify this matter by deciding to track just *one* unit, instead of the three units of length, time, and mass. By convention, we do this by setting two of the fundamental constants equal to one:

$$\hbar = c = 1, \tag{9.6}$$

as if these constants had no units! This has two consequences. First, any $\hbar$ or c in our formulae disappears without leaving a trace. This is not a serious problem, if we know the full units of an expression where $\hbar$ and c have been set equal to one, we can reconstruct the dependence on $\hbar$ and c unambiguously. Second, the units become dependent, and we are left with just one independent unit. Since $[c] = L/T$, the condition $c = 1$ implies that

$$L = T. \tag{9.7}$$

At this stage $[\hbar] = ML^2/T$ becomes $[\hbar] = ML$. With $\hbar = 1$ we get

$$M = 1/L. \tag{9.8}$$

Thus we can write all units in terms of mass or length (nobody uses time!). We will say that we are working with *natural units* when we set $\hbar = c = 1$ and track just one unit.

Back to (9.5), the units of α' have now become

$$[\alpha'] = \frac{1}{[T_0]} = \frac{L}{M} = L^2. \tag{9.9}$$

While the complete units of α' are those of inverse energy-squared, we see that in natural units, α' has units of length-squared. This is in agreement with our result in (8.78). In natural units the string length is

$$\ell_s = \sqrt{\alpha'}. \tag{9.10}$$

For reference, in natural units the Nambu–Goto action (8.77) takes the form

$$S = -\frac{1}{2\pi\alpha'} \int_{\tau_i}^{\tau_f} d\tau \int_0^{\sigma_1} d\sigma \sqrt{(\dot{X} \cdot X')^2 - (\dot{X})^2 (X')^2}. \tag{9.11}$$

In natural units equation (9.5) sets $\tilde{\lambda}$ proportional to α'. For open strings we choose $\tilde{\lambda} = 2\alpha'$, and equation (9.4) becomes

$$n \cdot X(\tau, \sigma) = 2\alpha'(n \cdot p)\,\tau \quad \text{(open strings)}. \tag{9.12}$$

This is the final form of the gauge condition which fixes the τ parameterization of the world-sheet.

When we use natural units, length scales can be expressed in terms of mass or energy scales. Given a length ℓ we can construct a unique mass m using only ℓ, $\hbar$, and c. This unique mass is $m = \hbar/(\ell c)$, or equivalently, $mc^2 = \hbar c/\ell$. For quick estimates we use $\hbar c \simeq 200\ \text{MeV} \times 10^{-15}\ \text{m}$.

Quick calculation 9.1 Show that the energy equivalent of the length 10^{-18} cm of a large extra dimension is roughly 20 TeV (1 TeV$= 10^{12}$ eV).

9.2 The associated σ parameterization

Having fixed the τ parameterization, let us determine the appropriate σ parameterization. In the static gauge the σ parameterization was fixed by the condition of constant energy density $\mathcal{P}^{\tau 0}$ over the strings (the curves of constant τ). Indeed, in setting $A(\sigma)$ in (7.17) equal to one by a suitable choice of σ, we made $\mathcal{P}^{\tau 0}$ in (7.6) constant. Since the static gauge uses $n_\mu = (1, 0, \ldots, 0)$, we were actually demanding the constancy of $n_\mu \mathcal{P}^{\tau\mu}$.

The proper generalization to situations where n^μ is arbitrary is to demand the constancy of $n_\mu \mathcal{P}^{\tau\mu} = n \cdot \mathcal{P}^\tau$ over the strings. Additionally, we require a parameterization range $\sigma \in [0, \pi]$ for all open strings. To show that we can satisfy these conditions we examine how $\mathcal{P}^{\tau\mu}(\tau, \sigma)$ transforms under σ reparameterizations. Looking at (6.49) we note that the numerator has two σ derivatives while the denominator has effectively one, making $\mathcal{P}^{\tau\mu}$ transform like $\frac{d}{d\sigma}$ under reparameterizations:

$$\mathcal{P}^{\tau\mu}(\tau, \sigma) = \frac{d\tilde{\sigma}}{d\sigma} \mathcal{P}^{\tau\mu}(\tau, \tilde{\sigma}) \quad \rightarrow \quad n \cdot \mathcal{P}^\tau(\tau, \sigma) = \frac{d\tilde{\sigma}}{d\sigma} n \cdot \mathcal{P}^\tau(\tau, \tilde{\sigma}). \tag{9.13}$$

If we are handed a string with $\tilde{\sigma}$ parameterization in which $n \cdot \mathcal{P}^\tau(\tau, \tilde{\sigma})$ depends on $\tilde{\sigma}$, we can choose a σ parameter so that $n \cdot \mathcal{P}^\tau(\tau, \sigma)$ does not depend on σ. This is simply done by adjusting the value of $\frac{d\tilde{\sigma}}{d\sigma}$ so that the final right-hand side above is set equal to a number that can only depend on τ. We then do a further reparameterization that scales σ by a constant factor b: $\sigma \rightarrow b\sigma$. This preserves the σ-independence of $n \cdot \mathcal{P}^\tau$ but allows us to have $\sigma \in [0, \pi]$, by adjusting b suitably. In this final parameterization we have

$$n \cdot \mathcal{P}^\tau(\tau, \sigma) = a(\tau), \tag{9.14}$$

where $a(\tau)$ is some function of τ. In fact $a(\tau)$ cannot depend on τ, and its value is already fixed by the conditions we have imposed. Integration of (9.14) over the string $\sigma \in [0, \pi]$ gives

$$\int_0^\pi d\sigma \, n \cdot \mathcal{P}^\tau(\tau, \sigma) = n \cdot p = \pi a(\tau) \quad \rightarrow \quad a(\tau) = \frac{n \cdot p}{\pi}. \tag{9.15}$$

Since $n \cdot p$ is conserved, $a(\tau)$ does not depend on τ. Back in (9.14) we have learned that

$$n \cdot \mathcal{P}^\tau = \frac{n \cdot p}{\pi} \quad \text{is an open string world-sheet constant.} \tag{9.16}$$

In this parameterization the momentum density $n \cdot \mathcal{P}^\tau$ is constant so the σ value assigned to a point is proportional to the amount of $n \cdot p$ momentum carried by the portion of string from the endpoint $\sigma = 0$ to the point.

Let us now examine the equation of motion $\partial_\tau \mathcal{P}^\tau_\mu + \partial_\sigma \mathcal{P}^\sigma_\mu = 0$. Dotting this equation with n^μ we arrive at the condition

$$\frac{\partial}{\partial\tau}(n \cdot \mathcal{P}^\tau) + \frac{\partial}{\partial\sigma}(n \cdot \mathcal{P}^\sigma) = 0. \tag{9.17}$$

The first term vanishes on account of (9.16), and we are left with

$$\frac{\partial}{\partial\sigma}(n \cdot \mathcal{P}^\sigma) = 0. \tag{9.18}$$

This means that $n \cdot \mathcal{P}^\sigma$ is independent of σ.

Although we were able to define the $\sigma = 0$ line on the closed string world-sheet consistently, our construction has an obvious ambiguity. We had to choose one arbitrary point on one string. Any other point on that one string could have been used as $\sigma = 0$. This means that the parameterization of the closed string world-sheet can be shifted rigidly along the σ direction. There is no way to avoid this ambiguity. The gauge conditions do not fix uniquely the parameterization of the closed string world-sheet. This will have implications for the theory of closed strings.

9.3 Constraints and wave equations

Let us now explore the constraints on X' and $\dot{X}$ that are implied by our chosen parameterization. The vanishing of $n \cdot \mathcal{P}^\sigma$, together with (9.23) and the recognition that $\partial_\tau (n \cdot X)$ is a nonvanishing constant, leads to

$$\dot{X} \cdot X' = 0. \tag{9.29}$$

In the static gauge, $X^{0\prime} = 0$, and (9.29) reduces to $\dot{\vec{X}} \cdot \vec{X}' = 0$, which we obtained before in (7.1). Equation (9.29) is a constraint that follows from our parameterization.

We now use (9.29) to simplify the expression (6.49) for $\mathcal{P}^\tau$:

$$\mathcal{P}^{\tau\mu} = \frac{1}{2\pi\alpha'} \frac{X'^2 \dot{X}^\mu}{\sqrt{-\dot{X}^2 X'^2}}. \tag{9.30}$$

With the help of this result, the second equation in (9.27) gives

$$n \cdot p = \frac{1}{\beta\alpha'} \frac{X'^2 (n \cdot \dot{X})}{\sqrt{-\dot{X}^2 X'^2}}. \tag{9.31}$$

Since $n \cdot \dot{X} = \beta\alpha' (n \cdot p)$ (see (9.27)), the factors of β cancel, and we find

$$1 = \frac{X'^2}{\sqrt{-\dot{X}^2 X'^2}} \quad \longrightarrow \quad \dot{X}^2 + X'^2 = 0, \tag{9.32}$$

where we used $X'^2 \neq 0$. Aside from the units that are now different, this is consistent with the earlier result (7.23), which we obtained using the static gauge and involves only spatial components. Equations (9.29) and (9.32) are the constraint equations that follow from our choice of parameterization. Together they read

$$\dot{X} \cdot X' = 0, \quad \dot{X}^2 + X'^2 = 0. \tag{9.33}$$

These two conditions are conveniently packaged together as

$$(\dot{X} \pm X')^2 = 0. \tag{9.34}$$

Note that the constraints take this form for *any* value of β in (9.27). Our choices of β for open and closed strings give convenient σ ranges, but other choices are possible.

Given the above constraints, the momentum densities $\mathcal{P}^{\tau\mu}$ and $\mathcal{P}^{\sigma\mu}$ simplify considerably. To use the constraints to simplify (9.30), we must take the *positive* square root in the denominator. Since $X'^2 > 0$, using (9.32) gives

$$\sqrt{-\dot{X}^2 X'^2} = \sqrt{X'^2 X'^2} = X'^2. \tag{9.35}$$

Back in (9.30) we therefore have

$$\mathcal{P}^{\tau\mu} = \frac{1}{2\pi\alpha'}\,\dot{X}^\mu. \tag{9.36}$$

The momentum density $\mathcal{P}^{\sigma\mu}$, recorded in (6.50), simplifies down to

$$\mathcal{P}^{\sigma\mu} = \frac{1}{2\pi\alpha'}\frac{\dot{X}^2 X^{\mu\prime}}{\sqrt{-\dot{X}^2 X'^2}} = \frac{1}{2\pi\alpha'}\frac{\dot{X}^2 X^{\mu\prime}}{X'^2}, \tag{9.37}$$

and, using (9.32),

$$\mathcal{P}^{\sigma\mu} = -\frac{1}{2\pi\alpha'}\,X^{\mu\prime}. \tag{9.38}$$

The momentum densities are simple derivatives of the coordinates. Using these expressions in the field equation $\partial_\tau \mathcal{P}^{\tau\mu} + \partial_\sigma \mathcal{P}^{\sigma\mu} = 0$, we find

$$\ddot{X}^\mu - X^{\mu\prime\prime} = 0. \tag{9.39}$$

With our parameterization, the equations of motion are just wave equations! Notice that the minus sign on the right-hand side of (9.38) was necessary in order to get a wave equation. For open strings with free endpoints, the wave equations are supplemented by the requirement that the $\mathcal{P}^{\sigma\mu}$, and therefore the $X^{\mu\prime}$, vanish at the endpoints.

9.4 Wave equation and mode expansions

We will now explicitly solve the wave equation (9.39) in full generality for the case of open strings. In doing so we will introduce some of the basic notation that is used in string theory. We will assume that we have a space-filling D-brane. As a result, all string coordinates X^μ satisfy free boundary conditions at the endpoints. We know that the most general $X^\mu(\tau, \sigma)$ that solves the wave equation (9.39) is

$$X^\mu(\tau, \sigma) = \frac{1}{2}\Big(f^\mu(\tau + \sigma) + g^\mu(\tau - \sigma)\Big), \tag{9.40}$$

where f^μ and g^μ are arbitrary functions of a single argument. Bearing in mind (9.38), the free endpoint boundary conditions $\mathcal{P}^{\sigma\mu} = 0$ imply the Neumann boundary conditions

$$\frac{\partial X^\mu}{\partial \sigma} = 0, \quad \text{at} \quad \sigma = 0, \pi. \tag{9.41}$$

The boundary condition at $\sigma = 0$ gives us

$$\frac{\partial X^\mu}{\partial \sigma}(\tau, 0) = \frac{1}{2}\Big(f^{\mu\prime}(\tau) - g^{\mu\prime}(\tau)\Big) = 0. \tag{9.42}$$

Since the derivatives of f^μ and g^μ coincide, f^μ and g^μ can differ only by a constant c^μ. After replacing $g^\mu = f^\mu + c^\mu$ in (9.40), the constant c^μ can be reabsorbed into the definition of f^μ. The result is

$$X^\mu(\tau, \sigma) = \frac{1}{2}\Big(f^\mu(\tau + \sigma) + f^\mu(\tau - \sigma)\Big). \tag{9.43}$$

Now let us consider the boundary condition at $\sigma = \pi$:

$$\frac{\partial X^\mu}{\partial \sigma}(\tau, \pi) = \frac{1}{2}\Big(f^{\mu\prime}(\tau + \pi) - f^{\mu\prime}(\tau - \pi)\Big) = 0. \tag{9.44}$$

Since this equation must hold for all τ, we learn that $f^{\mu\prime}$ *is periodic with period* 2π. Since 2π is a natural period, our decision to parameterize the open string with $\sigma \in [0, \pi]$ has paid off.

We now write the general Fourier series for the periodic function $f^{\mu\prime}(u)$:

$$f^{\mu\prime}(u) = f_1^\mu + \sum_{n=1}^{\infty}\Big(a_n^\mu \cos nu + b_n^\mu \sin nu\Big). \tag{9.45}$$

Integrating this equation we get the expansion of $f^\mu(u)$:

$$f^\mu(u) = f_0^\mu + f_1^\mu u + \sum_{n=1}^{\infty}\Big(A_n^\mu \cos nu + B_n^\mu \sin nu\Big), \tag{9.46}$$

where we have absorbed the constants arising from integration into new coefficients. We substitute this expression for $f(u)$ back into (9.43) and simplify to get

$$X^\mu(\tau, \sigma) = f_0^\mu + f_1^\mu \tau + \sum_{n=1}^{\infty}\Big(A_n^\mu \cos n\tau + B_n^\mu \sin n\tau\Big) \cos n\sigma. \tag{9.47}$$

We want to replace the coefficients in equation (9.47) by new coefficients that have a simple physical interpretation. Our first step is introducing constants a_n^μ through the relations

$$A_n^\mu \cos n\tau + B_n^\mu \sin n\tau = -\frac{i}{2}\Big((B_n^\mu + iA_n^\mu)e^{in\tau} - (B_n^\mu - iA_n^\mu)e^{-in\tau}\Big)$$

$$\equiv -i\frac{\sqrt{2\alpha'}}{\sqrt{n}}\Big(a_n^{\mu*}e^{in\tau} - a_n^\mu e^{-in\tau}\Big). \tag{9.48}$$

Here $*$ denotes complex conjugation. The purpose of the $\sqrt{2\alpha'}$ factor is to make the constants a_n^μ dimensionless. These constants, and their complex conjugates, will turn into annihilation and creation operators when we consider the quantum theory. Equation (9.48) introduces the notation that string theorists conventionally use.

In equation (9.47) the constant f_1^μ has a simple physical interpretation. Using (9.36), the momentum density is given by

$$\mathcal{P}^{\tau\mu} = \frac{1}{2\pi\alpha'}\,\dot{X}^\mu = \frac{1}{2\pi\alpha'}\,f_1^\mu + \cdots, \tag{9.49}$$

where the dots denote terms with $\cos n\sigma$ dependence ($n \neq 0$). To find the total momentum p^μ, we integrate $\mathcal{P}^{\tau\mu}$ over $\sigma \in [0, \pi]$. Happily, the terms represented by the dots do not contribute as the integral of $\cos n\sigma$ vanishes. We get

$$p^\mu = \int_0^\pi \mathcal{P}^{\tau\mu} d\sigma = \frac{1}{2\pi\alpha'}\,\pi f_1^\mu \longrightarrow f_1^\mu = 2\alpha' p^\mu. \tag{9.50}$$

This identifies f_1^μ as a quantity proportional to the spacetime momentum carried by the string. Declaring $f_0^\mu = x_0^\mu$, and collecting all the above information, equation (9.47) now takes the conventional form

$$X^\mu(\tau, \sigma) = x_0^\mu + 2\alpha' p^\mu \tau - i\sqrt{2\alpha'}\sum_{n=1}^\infty \left(a_n^{\mu*} e^{in\tau} - a_n^\mu e^{-in\tau}\right)\frac{\cos n\sigma}{\sqrt{n}}. \tag{9.51}$$

The terms on the right-hand side clearly correspond to the zero mode, to the momentum, and to the oscillations of the string. If all the coefficients a_n^μ of the oscillations vanish, the equation represents the motion of a point particle.

Quick calculation 9.3 Verify explicitly that $X^\mu(\tau, \sigma)$ is real.

Let us now introduce some notation that will allow us to write simple expressions for the τ and σ derivatives of $X^\mu(\tau, \sigma)$. We start by defining

$$\alpha_0^\mu = \sqrt{2\alpha'}p^\mu. \tag{9.52}$$

Furthermore, we define

$$\alpha_n^\mu = a_n^\mu\sqrt{n}, \quad \alpha_{-n}^\mu = a_n^{\mu*}\sqrt{n}, \quad n \geq 1. \tag{9.53}$$

It is important to note that

$$\alpha_{-n}^\mu = (\alpha_n^\mu)^*. \tag{9.54}$$

Moreover, while the a_n^μ are only defined when n is a positive integer, the α_n^μ are defined for any integer n, including zero. Using these new names, we can rewrite X^μ as

$$X^\mu(\tau, \sigma) = x_0^\mu + \sqrt{2\alpha'}\, \alpha_0^\mu \tau - i\sqrt{2\alpha'} \sum_{n=1}^{\infty} \frac{1}{n}\left(\alpha_{-n}^\mu e^{in\tau} - \alpha_n^\mu e^{-in\tau}\right)\cos n\sigma. \qquad (9.55)$$

It is convenient to sum over all integers except zero:

$$X^\mu(\tau, \sigma) = x_0^\mu + \sqrt{2\alpha'}\, \alpha_0^\mu \tau + i\sqrt{2\alpha'} \sum_{n\neq 0} \frac{1}{n}\, \alpha_n^\mu \, e^{-in\tau}\cos n\sigma. \qquad (9.56)$$

This completes the solution of the wave equations with Neumann boundary conditions. In the above equation, a solution is defined once we specify the constants x_0^μ and α_n^μ for $n \geq 0$.

It is convenient to record here the τ and σ derivatives of X^μ. From (9.56) we see that

$$\dot{X}^\mu = \sqrt{2\alpha'} \sum_{n\in\mathbb{Z}} \alpha_n^\mu \cos n\sigma\, e^{-in\tau}, \qquad (9.57)$$

$$X^{\mu\prime} = -i\sqrt{2\alpha'} \sum_{n\in\mathbb{Z}} \alpha_n^\mu \sin n\sigma\, e^{-in\tau}, \qquad (9.58)$$

where $\mathbb{Z}$ denotes the set of all integers (positive, negative, and zero). Finally, two linear combinations of the above derivatives are particularly nice:

$$\dot{X}^\mu \pm X^{\mu\prime} = \sqrt{2\alpha'} \sum_{n\in\mathbb{Z}} \alpha_n^\mu \, e^{-in(\tau\pm\sigma)}. \qquad (9.59)$$

We have found solutions of the wave equations that satisfy the relevant boundary conditions, but we must also ensure that the constraints (9.33) are satisfied. If we specify arbitrarily all constants α_n^μ, the constraints will not be satisfied. We will use the light-cone gauge to find a solution that satisfies the wave equations as well as the constraints.

9.5 Light-cone solution of equations of motion

The light-cone solution of the equations of motion involves using light-cone coordinates to represent the motion of strings, and imposing a set of conditions that defines the light-cone gauge. We have seen in Chapter 2 that using light-cone coordinates means using x^+ and x^- instead of x^0 and x^1 – this is just a change of coordinates. Imposing a light-cone gauge condition is a more substantial step. The gauges we have examined in this chapter represent very specific choices of world-sheet coordinates. One of these choices is the light-cone gauge.

Selecting the light-cone gauge means imposing the conditions (9.27) with a vector n^μ that gives $n \cdot X = X^+$. Taking

$$n_\mu = \left(\frac{1}{\sqrt{2}}, \frac{1}{\sqrt{2}}, 0, \ldots, 0 \right), \tag{9.60}$$

we indeed find

$$n \cdot X = \frac{X^0 + X^1}{\sqrt{2}} = X^+, \quad n \cdot p = \frac{p^0 + p^1}{\sqrt{2}} = p^+. \tag{9.61}$$

Using these relations in (9.27) we have

$$X^+(\tau, \sigma) = \beta \alpha' p^+ \tau, \quad p^+ = \frac{2\pi}{\beta} \mathcal{P}^{\tau +}, \tag{9.62}$$

where $\beta = 2$ for open strings and $\beta = 1$ for closed strings. The second equation tells us that the density of p^+ is constant along the string.

The strategy behind the light-cone gauge is to use the especially simple form of X^+ to show that there is no dynamics in X^- (up to a zero mode) and that all the dynamics is in the *transverse* coordinates $X^2, X^3, \ldots, X^d$. These transverse coordinates will be denoted by X^I, where the transverse index I runs from 2 up to d:

$$X^I = (X^2, X^3, \ldots, X^d). \tag{9.63}$$

In order to proceed we look at the constraint equations (9.34). Using the definition (2.59) of the relativistic dot product in light-cone coordinates, we can write these constraints as

$$-2(\dot{X}^+ \pm X^{+\prime})(\dot{X}^- \pm X^{-\prime}) + (\dot{X}^I \pm X^{I\prime})^2 = 0, \tag{9.64}$$

where $(a^I)^2 = a^I a^I$ and, as usual, repeated indices imply summation. Since $X^{+\prime} = 0$ and $\dot{X}^+ = \beta \alpha' p^+$, we in fact have

$$\dot{X}^- \pm X^{-\prime} = \frac{1}{\beta \alpha'} \frac{1}{2 p^+} (\dot{X}^I \pm X^{I\prime})^2. \tag{9.65}$$

In writing the above we have assumed that $p^+ \neq 0$. While p^+ certainly satisfies $p^+ \geq 0$, it can happen that p^+ is equal to zero. For this, the momentum p^1 must cancel the energy, and this can only occur if we have a massless particle traveling exactly in the negative x^1 direction. Since the vanishing of p^+ is thus not a common occurrence, we will take p^+ to be always positive. If we come across a situation where p^+ is zero, the light-cone formalism will not apply.

   Note the crucial role played by both the choice of light-cone coordinates and the choice of light-cone gauge in allowing us to solve for the derivatives of X^-. Light-cone coordinates were useful because the off-diagonal metric in the $(+, -)$ sector allowed us to solve for the derivatives of X^- without having to take a square root! We just had to divide by $\dot{X}^+$. Here the light-cone *gauge* was useful, since it made $\dot{X}^+$ equal to a constant.

Equations (9.65) determine both $\dot{X}^-$ and $X^{-\prime}$ in terms of the X^I, and therefore they determine X^- up to a single integration constant. All that is needed is the value of X^- at some point P on the world-sheet and then we can integrate the relation

$$dX^- = \frac{\partial X^-}{\partial \tau} d\tau + \frac{\partial X^-}{\partial \sigma} d\sigma, \qquad (9.66)$$

to find the value of X^- at any other point Q. On an open string world-sheet we can choose any path from P to Q to perform the integration and the result for $X^-(Q)$ will not depend on the path, as discussed in Problem 9.2. On a closed string world-sheet there is a further consistency condition. Imagine a contour of integration that starts at P and ends at P after going around the world-sheet. It is not guaranteed that the integration of dX^- on this contour gives zero, a result necessary for X^- to be well-defined. If we choose the contour to be at constant τ, we must require

$$\int_0^{2\pi} d\sigma \, \frac{\partial X^-}{\partial \sigma} = 0. \qquad (9.67)$$

This is a nontrivial constraint; see Problem 9.5.

Our analysis indicates that the full evolution of the string is determined by the following set of objects:

$$X^I(\tau, \sigma), \quad p^+, \; x_0^-, \qquad (9.68)$$

where x_0^- is the constant of integration needed for X^-.

Let us focus on the case of open strings ($\beta = 2$). We consider the explicit solution for the transverse coordinates X^I and calculate the associated X^-. Making use of the general solution in (9.56) we have

$$X^I(\tau, \sigma) = x_0^I + \sqrt{2\alpha'} \, \alpha_0^I \tau + i\sqrt{2\alpha'} \sum_{n \neq 0} \frac{1}{n} \alpha_n^I e^{-in\tau} \cos n\sigma. \qquad (9.69)$$

Moreover, for the X^+ coordinate the gauge condition gives

$$X^+(\tau, \sigma) = 2\alpha' p^+ \tau = \sqrt{2\alpha'} \, \alpha_0^+ \tau. \qquad (9.70)$$

As we can see, the position zero mode and the oscillations of the X^+ coordinate have been set to equal to zero:

$$x_0^+ = 0, \quad \alpha_n^+ = \alpha_{-n}^+ = 0, \quad n = 1, 2, \ldots, \infty. \qquad (9.71)$$

What about X^-? Being a linear combination of X^0 and X^1, the coordinate X^- satisfies the same wave equation and the same boundary conditions as all the other coordinates. We can therefore use the same expansion as in (9.56) to write

$$X^-(\tau, \sigma) = x_0^- + \sqrt{2\alpha'} \, \alpha_0^- \tau + i\sqrt{2\alpha'} \sum_{n \neq 0} \frac{1}{n} \alpha_n^- e^{-in\tau} \cos n\sigma. \qquad (9.72)$$

Using equation (9.59) with $\mu = -$ and $\mu = I$, we find

$$\dot{X}^- \pm X^{-\prime} = \sqrt{2\alpha'} \sum_{n\in\mathbb{Z}} \alpha_n^- \, e^{-in(\tau\pm\sigma)}, \tag{9.73}$$

$$\dot{X}^I \pm X^{I\prime} = \sqrt{2\alpha'} \sum_{n\in\mathbb{Z}} \alpha_n^I \, e^{-in(\tau\pm\sigma)}. \tag{9.74}$$

We use these equations and (9.65) to solve for the minus oscillators:

$$\sqrt{2\alpha'} \sum_{n\in\mathbb{Z}} \alpha_n^- \, e^{-in(\tau\pm\sigma)} = \frac{1}{2p^+} \sum_{p,q\in\mathbb{Z}} \alpha_p^I \alpha_q^I e^{-i(p+q)(\tau\pm\sigma)}$$

$$= \frac{1}{2p^+} \sum_{n,p\in\mathbb{Z}} \alpha_p^I \alpha_{n-p}^I e^{-in(\tau\pm\sigma)}$$

$$= \frac{1}{2p^+} \sum_{n\in\mathbb{Z}} \left(\sum_{p\in\mathbb{Z}} \alpha_p^I \alpha_{n-p}^I \right) e^{-in(\tau\pm\sigma)}. \tag{9.75}$$

It follows that we can identify α_n^- consistently as

$$\sqrt{2\alpha'}\,\alpha_n^- = \frac{1}{2p^+} \sum_{p\in\mathbb{Z}} \alpha_{n-p}^I \alpha_p^I. \tag{9.76}$$

This represents a complete solution! We now have explicit expressions for the minus oscillators α_n^- in terms of the transverse oscillators. On the right-hand side the spacetime indices are only to be summed over the labels of the transverse coordinates.

The general solution which represents an allowed motion is fixed by specifying arbitrary values for p^+, x_0^-, x_0^I, and for all the constants α_n^I. This clearly determines $X^I(\tau, \sigma)$ in (9.69), and $X^+(\tau, \sigma)$ in (9.70). Using (9.76) we can calculate the constants α_n^-, which together with x_0^- determine $X^-(\tau, \sigma)$ in (9.72). The full solution is thus constructed.

The quadratic combination of oscillators on the right-hand side of (9.76) is remarkably useful, so it has been given a name. It is the *transverse Virasoro mode* $L_n^\perp$:

$$\sqrt{2\alpha'}\,\alpha_n^- = \frac{1}{p^+} L_n^\perp, \quad L_n^\perp \equiv \frac{1}{2} \sum_{p\in\mathbb{Z}} \alpha_{n-p}^I \alpha_p^I. \tag{9.77}$$

In particular, for $n = 0$ we use (9.52) and find

$$\sqrt{2\alpha'}\,\alpha_0^- = 2\alpha'p^- = \frac{1}{p^+} L_0^\perp \longrightarrow 2p^+ p^- = \frac{1}{\alpha'} L_0^\perp. \tag{9.78}$$

Using the value of α_n^- given in (9.77), equations (9.73) and (9.65) are written as

$$\dot{X}^- \pm X^{-\prime} = \frac{1}{p^+} \sum_{n \in \mathbb{Z}} L_n^\perp e^{-in(\tau \pm \sigma)} = \frac{1}{4\alpha' p^+} (\dot{X}^I \pm X^{I\prime})^2. \tag{9.79}$$

Quick calculation 9.4 Show that

$$X^-(\tau, \sigma) = x_0^- + \frac{1}{p^+} L_0^\perp \tau + \frac{i}{p^+} \sum_{n \neq 0} \frac{1}{n} L_n^\perp e^{-in\tau} \cos n\sigma. \tag{9.80}$$

This equation explicitly demonstrates the claim that the Virasoro modes are the expansion modes of the coordinate $X^-(\tau, \sigma)$.

It is instructive to compute the mass of a string which is performing an arbitrary motion. The mass can be calculated using the relativistic equation

$$M^2 = -p^2 = 2p^+ p^- - p^I p^I. \tag{9.81}$$

Since the mass is a constant of the motion, we anticipate that it depends on the constant coefficients a_n^I introduced to define a classical solution. To evaluate the mass we start with (9.78), substitute the value of $L_0^\perp$ from (9.77), and use the definitions in (9.52) and (9.53):

$$2p^+ p^- = \frac{1}{\alpha'} L_0^\perp = \frac{1}{\alpha'} \left(\frac{1}{2} \alpha_0^I \alpha_0^I + \sum_{n=1}^\infty \alpha_n^{I*} \alpha_n^I \right) = p^I p^I + \frac{1}{\alpha'} \sum_{n=1}^\infty n \, a_n^{I*} a_n^I. \tag{9.82}$$

Replacing this result on the right-hand side of (9.81) we finally find

$$M^2 = \frac{1}{\alpha'} \sum_{n=1}^\infty n \, a_n^{I*} a_n^I. \tag{9.83}$$

This is a very interesting result. The mass-squared is written as a sum of terms each of which is of the form $a^* a = |a|^2 \geq 0$. So, we find that $M^2 \geq 0$. This shows that the classical string mass $M = \sqrt{M^2}$ is a real number (conventionally taken to be positive). Such a result is actually hard to obtain without using the light-cone gauge. We also see that we can adjust the coefficients a_n^I to obtain classical string solutions with arbitrary values of the mass. If all the coefficients a_n^I vanish, the result is a massless object $M^2 = 0$. Indeed, when all a_n^I vanish the string collapses to a moving point: equation (9.69) gives $X^I(\tau, \sigma) = x_0^I + \sqrt{2\alpha'} \, \alpha_0^I \tau$, and the σ dependence disappears.

Quick calculation 9.5 Calculate $X^-(\tau, \sigma)$ when all a_n^I vanish. Note that the σ dependence of X^- disappears.

The classical result (9.83) for M^2 does not survive quantization. First, M^2 will become quantized, and string states will not exhibit a continuous spectrum of masses. This is good because we do not observe in nature particle states that take continuous values of the mass. Even more, equation (9.83) does not give enough massless states. The few massless states that are obtained do not behave at all like the massless states of Maxwell theory. For closed strings, the few massless states that are obtained by a similar analysis do not behave at all like the massless states of gravitation. Quantum mechanics will add an extra constant to the formula for M^2, both for open and for closed strings. These additive constants will enable us to find states that correspond to those of physical theories. String theory has a chance to describe gauge fields and gravity because quantization changes (9.83) and the corresponding formula for closed strings.

Problems

Problem 9.1 Vectors orthogonal to null vectors are null or spacelike.

Let n^μ be a nonzero null vector $(n_\mu n^\mu = 0)$ in D-dimensional Minkowski space. In addition, let b^μ be a vector that satisfies $n_\mu b^\mu = 0$. Prove the following.

(a) The vector b^μ is either spacelike or null.
(b) If b^μ is null, then $b^\mu = \lambda n^\mu$ for some constant λ.
(c) The set of vectors b^μ that satisfies $n_\mu b^\mu = 0$ is a vector space V of dimension $(D-1)$. The subset of null vectors b^μ is a vector subspace of V of dimension one.

This result shows that for gauges (9.2) with n^μ null and for $D > 2$, strings are almost always spacelike objects. Moreover, the hyperplane orthogonal to n^μ contains n^μ. This is readily confirmed in two dimensions:

(d) Let $D = 2$ and consider a spacetime diagram such as the one in Figure 2.2. What is the null vector n^μ such that $n \cdot X = X^+$? Confirm that n^μ points along the lines of constant X^+.

Problem 9.2 Consistency checks on the solution for X^-.

(a) Use (9.65) to find $\partial_\tau X^-$ and $\partial_\sigma X^-$. Show that the consistency condition $\partial_\sigma(\partial_\tau X^-) = \partial_\tau(\partial_\sigma X^-)$ holds if the transverse coordinates X^I satisfy the wave equation. Prove that this condition guarantees that X^-, determined by integration of (9.66), is independent of the chosen path.
(b) Show that X^-, as calculated in (9.65), satisfies the wave equation if the transverse coordinates X^I satisfy the wave equation.
(c) Assume that at the open string endpoints some of the transverse light-cone coordinates X^I satisfy Neumann boundary conditions and some satisfy Dirichlet boundary conditions. Prove that X^-, as calculated in (9.65), will always satisfy Neumann boundary conditions.

Problem 9.3 Rotating open string in the light-cone gauge.

Consider string motion defined by $x_0^- = x_0^I = 0$, and the vanishing of all coefficients α_n^I with the exception of

$$\alpha_1^{(2)} = \alpha_{-1}^{(2)*} = a, \quad \alpha_1^{(3)} = \alpha_{-1}^{(3)*} = ia. \tag{1}$$

Here a is a dimensionless real constant. We want to construct a solution that represents an open string that is rotating on the (x^2, x^3) plane.

(a) What is the mass (or energy) of this string?
(b) Construct the explicit functions $X^2(\tau, \sigma)$ and $X^3(\tau, \sigma)$. What is the length of the string in terms of a and α'?
(c) Calculate the $L_n^\perp$ modes for all n. Use your result to construct $X^-(\tau, \sigma)$. Your answer should be σ-independent!
(d) Determine the value of p^+ using the condition that for this string $X^1(\tau, \sigma) = 0$. Find the relation between t and τ.
(e) Confirm that in your solution the energy of the string and its angular frequency of rotation are related to its length as in (7.59).

Problem 9.4 Generating consistent open string motion: How does an open string collapse?

Consider string motion defined by $x_0^- = x_0^I = 0$, and the vanishing of all coefficients α_n^I with the exception of

$$\alpha_1^{(2)} = \alpha_{-1}^{(2)*} = a.$$

Here a is a dimensionless real constant. We want to construct a solution that represents an open string oscillating on the (x^1, x^2) plane and having *zero* momentum in this plane.

(a) Show that the string motion is described by

$$\frac{1}{\sqrt{2\alpha'}} \frac{1}{a} X^0(\tau, \sigma) = \sqrt{2} \left(\tau + \frac{1}{4} \sin 2\tau \cos 2\sigma \right),$$

$$\frac{1}{\sqrt{2\alpha'}} \frac{1}{a} X^1(\tau, \sigma) = -\frac{1}{2\sqrt{2}} \sin 2\tau \cos 2\sigma,$$

$$\frac{1}{\sqrt{2\alpha'}} \frac{1}{a} X^2(\tau, \sigma) = 2 \sin \tau \cos \sigma.$$

(b) Confirm that τ flows as t flows. In the chosen Lorentz frame, strings are lines on the world-sheet that lie at constant time X^0. Find the values of τ for which constant τ lines are strings. Describe those strings.
(c) At $\tau = 0$ the string has zero length. Study in detail the motion for $\tau \ll 1$. Calculate $\tau = \tau(t, \sigma)$ and use this result to find $X^1(t, \sigma)$ and $X^2(t, \sigma)$. Prove that as the string expands from zero size, it lies on the portion $\cos \theta \geq -1/3$ of a circle centered at the origin, whose radius grows at the speed of light (θ is measured with respect to the positive x^1 axis). Note that the endpoints move transversely to the string.
(d) Use your favorite software package to do a parametric plot of the string world-sheet as a surface in three dimensions. Use X^1, X^2, and X^0 as x, y, and z axes, respectively, and parameters τ and σ. For further help visualizing the motion of the string, plot the string on the (x^1, x^2) plane at various values of the time X^0. This requires solving (numerically) for τ as a function of X^0 and σ.

Problem 9.5 A closed string in the light-cone gauge.

Consider a closed string for which

$$X^{(2)}(\tau, \sigma) = \sqrt{2\alpha'} \left(a \sin(\tau - \sigma) + b \cos(\tau - \sigma) + \bar{a} \sin(\tau + \sigma) + \bar{b} \cos(\tau + \sigma) \right),$$

and all other transverse coordinates $X^I(\tau, \sigma)$ vanish. In the above, $a, b, \bar{a}$, and $\bar{b}$ are real constants. Note that a and b are the coefficients of waves that propagate towards increasing σ while $\bar{a}$ and $\bar{b}$ are the coefficients of waves that propagate towards decreasing σ. As usual for closed strings, we set $X^+(\tau, \sigma) = \alpha' p^+ \tau$ and use $\sigma \in [0, 2\pi]$.

(a) As it turns out, it is not possible to generate a solution of the equations of motion for completely arbitrary values of the constants $a, b, \bar{a}$, and $\bar{b}$. Examine the calculation of $X^-(\tau, \sigma)$ and derive the constraint that the constants must satisfy.

(b) Your constraint must allow $a = b = \bar{a} = \bar{b} = r$. Use these values to calculate $X^-(\tau, \sigma)$ and to determine the mass of this string.

We study the classical equations of motion for scalar fields, Maxwell fields, and gravitational fields. We use the light-cone gauge to find plane-wave solutions to their equations of motion and the number of degrees of freedom that characterize them. We explain how the quantization of such classical field configurations gives rise to particle states – scalar particles, photons, and gravitons. In doing so we prepare the ground for the later identification of such states among the quantum states of relativistic strings.

10.1 Introduction

In our investigation of classical string motion we had a great deal of freedom in choosing the coordinates on the world-sheet. This freedom was a direct consequence of the reparameterization invariance of the action, and we exploited it to simplify the equations of motion tremendously. Reparameterization invariance is an example of a *gauge invariance*, and a choice of parameterization is an example of a choice of *gauge*. We saw that the light-cone gauge – a particular parameterization in which τ is related to the light-cone time X^+ and σ is chosen so that the p^+-density is constant – was useful to obtain a complete and explicit solution of the equations of motion.

Classical field theories sometimes have gauge invariances. Classical electrodynamics, for example, is described in terms of gauge potentials A_μ. The gauge invariance of this description is often used to great advantage. The classical theory of a *scalar* field is simpler than classical electromagnetism. This theory is not studied at the undergraduate level, however, because elementary scalar particles – the kind of particles associated with the quantum theory of scalar fields – have not been detected yet. On the other hand, photons – the particles associated with the quantum theory of the electromagnetic field – are found everywhere! Scalar particles may play an important role in the Standard Model of particle physics, where they can help trigger symmetry breaking. Thus physicists may detect scalar particles some time in the future. The field theory of a single scalar field has no gauge invariance. We will study it because it is the simplest field theory and because scalar particles arise in string theory. The most famous scalar particle in string theory is the tachyon. Also important is the dilaton, a massless scalar particle.

Einstein's classical field theory of gravitation is more complicated than classical electromagnetism. In gravity, as we explained in Section 3.6, the dynamical variable is the two-index metric field $g_{\mu\nu}(x)$. Gravitation has a very large gauge invariance. The gauge transformations involve reparameterizations of spacetime.

We will consider scalar fields, electromagnetic fields, and gravitational fields. The light-cone gauge will allow us to simplify dramatically the (linearized) equations of motion, find their plane-wave solutions, and count the number of degrees of freedom that characterize the solutions. We will also briefly consider how the quantization of plane-wave solutions gives rise to particle states. These are the quantum states associated with the field theories. We quantize the relativistic string in Chapter 12. There we relate the quantum states of the string to the particle states of the field theories that we study in the present chapter. We use the light-cone gauge here because we will quantize the relativistic string in the light-cone gauge.

10.2 An action for scalar fields

A scalar field is simply a single real function of spacetime. It is written as $\phi(t, \vec{x})$ or, more briefly, as $\phi(x)$. The term scalar means scalar under Lorentz transformations: all Lorentz observers will agree on the value of the scalar field at any fixed point in spacetime. Scalar fields have no Lorentz indices.

Let us now motivate the simplest kind of action principle that can be used to define the dynamics of a scalar field. Consider first the kinetic energy. In mechanics, the kinetic energy of a particle is proportional to its velocity squared. For a scalar field, the kinetic energy density T is declared to be proportional to the square of the rate of change of the field with time:

$$T = \frac{1}{2} (\partial_0 \phi)^2 . \tag{10.1}$$

We speak of densities because, at any fixed time, T is a function of position. The total kinetic energy will be the integral of the density T over space.

Now consider the potential energy density. There is one class of term that is natural. Suppose the equilibrium value of the field is $\phi = 0$. For a simple harmonic oscillator with equilibrium position $x = 0$, the potential energy goes like $V \sim x^2$. If we want the field to prefer its equilibrium state, then this must be encoded in the potential. The simplest potential which does this is quadratic:

$$V = \frac{1}{2} m^2 \phi^2 . \tag{10.2}$$

It is interesting to note that the constant m introduced here has the units of mass. Indeed, since the expressions on the right-hand side of the two equations above must have the same units ($[T] = [V]$), and both have two factors of ϕ, we require $[m] = [\partial_0] = L^{-1} = M$.

We could now attempt to form a Lagrangian density by combining the two energies above and setting

$$\mathcal{L} \overset{?}{=} T - V = \frac{1}{2} (\partial_0 \phi)^2 - \frac{1}{2} m^2 \phi^2 . \tag{10.3}$$

This Lagrangian density, however, is not Lorentz invariant. The second term on the right-hand side is a scalar, but the first term is not, for it treats time as special. We are missing a contribution representing the energy cost when the scalar field varies in space. This is eminently reasonable in special relativity: if it costs energy to have the field vary in time, it must also cost energy to have the field vary in space. The extra contribution is therefore associated with spatial derivatives of the scalar field, and can be written as

$$V' = \frac{1}{2} \sum_i (\partial_i \phi)^2 = \frac{1}{2} (\nabla \phi)^2 \,, \tag{10.4}$$

where ∂_i are derivatives with respect to spatial coordinates. We have written this term as a new contribution V' to the potential energy, rather than as a contribution to the kinetic energy. There are several reasons for this. First, it is needed for Lorentz invariance. The two options lead to contributions of opposite signs in the Lagrangian, and only one sign is consistent with Lorentz invariance. Second, kinetic energy is always associated with time derivatives. Third, the calculation of the total energy vindicates the correctness of the choice. Indeed, with this additional term the Lagrangian density becomes

$$\mathcal{L} = T - V' - V = \frac{1}{2} \partial_0 \phi \, \partial_0 \phi - \frac{1}{2} \partial_i \phi \, \partial_i \phi - \frac{1}{2} m^2 \phi^2 \,, \tag{10.5}$$

where the repeated spatial index i denotes summation. The relative sign between the first two terms on the right-hand side allows us to rewrite them as a single term which uses the Minkowski metric $\eta^{\mu\nu}$:

$$\mathcal{L} = -\frac{1}{2} \eta^{\mu\nu} \partial_\mu \phi \, \partial_\nu \phi - \frac{1}{2} m^2 \phi^2 \,. \tag{10.6}$$

Since all the indices are matched, the Lagrangian density is Lorentz invariant. The associated action is

$$S = \int d^D x \left(-\frac{1}{2} \eta^{\mu\nu} \partial_\mu \phi \partial_\nu \phi - \frac{1}{2} m^2 \phi^2 \right), \tag{10.7}$$

where $d^D x = dx^0 dx^1 \ldots dx^d$, and $D = d + 1$, is the number of spacetime dimensions. This is the action for a *free* scalar field with mass m. A field is said to be free when its equations of motion are linear. If each term in the action is quadratic in the field, as is the case in (10.7), the equations of motion will be linear in the field. A field that is not free is said to be interacting, in which case the action contains terms of order three or higher in the field.

To find the energy density in this field we calculate the Hamiltonian density $\mathcal{H}$. The momentum Π conjugate to the field is given by

$$\Pi \equiv \frac{\partial \mathcal{L}}{\partial(\partial_0 \phi)} = \partial_0 \phi \,, \tag{10.8}$$

where we used (10.5) to evaluate the derivative. The Hamiltonian density is then constructed as

$$\mathcal{H} = \Pi \partial_0 \phi - \mathcal{L}. \tag{10.9}$$

Quick calculation 10.1 Show that the Hamiltonian density takes the form

$$\mathcal{H} = \frac{1}{2}\,\Pi^2 + \frac{1}{2}\,(\nabla\phi)^2 + \frac{1}{2}\,m^2\phi^2. \tag{10.10}$$

The three terms in $\mathcal{H}$ are identified as T, V', and V, respectively. This is what we expected physically for the energy density. The total energy E is given by the Hamiltonian H, which in turn, is the spatial integral of the Hamiltonian density $\mathcal{H}$:

$$E = H = \int d^d x \left(\frac{1}{2}\,\partial_0\phi\,\partial_0\phi + \frac{1}{2}\,(\nabla\phi)^2 + \frac{1}{2}\,m^2\phi^2\right). \tag{10.11}$$

To find the equations of motion from the action (10.7), we consider a variation $\delta\phi$ of the field and set the variation of the action equal to zero. After discarding a total derivative we find

$$\delta S = \int d^D x \left(-\eta^{\mu\nu}\partial_\mu(\delta\phi)\partial_\nu\phi - m^2\phi\delta\phi\right) = \int d^D x\,\delta\phi\left(\eta^{\mu\nu}\partial_\mu\partial_\nu\phi - m^2\phi\right) = 0. \tag{10.12}$$

The equation of motion for ϕ is therefore

$$\eta^{\mu\nu}\partial_\mu\partial_\nu\phi - m^2\phi = 0. \tag{10.13}$$

If we define $\partial^2 \equiv \eta^{\mu\nu}\partial_\mu\partial_\nu$, then we have

$$\left(\partial^2 - m^2\right)\phi = 0. \tag{10.14}$$

Separating out time and space derivatives, this equation is recognized as the Klein–Gordon equation:

$$-\frac{\partial^2\phi}{\partial t^2} + \nabla^2\phi - m^2\phi = 0. \tag{10.15}$$

We will now study some classical solutions of this equation.

10.3 Classical plane-wave solutions

We can find plane-wave solutions to the classical scalar field equation (10.15). Consider, for example, the expression

$$\phi(t,\vec{x}) = a\,e^{-iEt + i\vec{p}\cdot\vec{x}}, \tag{10.16}$$

where a and E are constants and $\vec{p}$ is an arbitrary vector. The field equation (10.15) fixes the possible values of E in terms of $\vec{p}$ and m:

$$E^2 - \vec{p}^2 - m^2 = 0 \longrightarrow E = \pm E_p, \quad E_p \equiv \sqrt{\vec{p}^2 + m^2}. \tag{10.17}$$

The square root is chosen to be positive, so $E_p > 0$. There is a small problem with the solution in (10.16). While ϕ is a real field, the solution (10.16) is not real. To make it real, we just add to it its complex conjugate:

$$\phi(t, \vec{x}) = a\, e^{-iE_p t + i\vec{p}\cdot\vec{x}} + a^*\, e^{iE_p t - i\vec{p}\cdot\vec{x}}\,. \tag{10.18}$$

This solution depends on the complex number a. A general solution to the equation of motion (10.15) is obtained by superimposing solutions, such as those above, for all values of $\vec{p}$. Since $\vec{p}$ can be varied continuously, the general superposition is actually an integral. The classical field does not have a simple quantum mechanical interpretation. If the two terms above were to be thought of as wavefunctions, the first would represent the wavefunction of a particle with momentum $\vec{p}$ and positive energy E_p. The second would represent the wavefunction of a particle with momentum $(-\vec{p})$ and *negative* energy $(-E_p)$. This is not acceptable. To do quantum mechanics with a classical field one must quantize the field. The result is particle states with positive energy, as we will discuss briefly in the following section.

An analysis of the classical field equation (in a practical way that applies elsewhere) uses the Fourier transformation of the scalar field $\phi(x)$:

$$\phi(x) = \int \frac{d^D p}{(2\pi)^D}\, e^{ip\cdot x}\, \phi(p)\,. \tag{10.19}$$

Here $\phi(p)$ is the Fourier transform of $\phi(x)$. We will always show the argument of ϕ so no confusion should arise between the spacetime field and the momentum-space field. Note that we are performing the Fourier transform over all spacetime coordinates, time included: $p \cdot x = -p^0 x^0 + \vec{p}\cdot\vec{x}$. The reality of $\phi(x)$ means that $\phi(x) = (\phi(x))^*$. Using equation (10.19), this condition yields

$$\int \frac{d^D p}{(2\pi)^D}\, e^{ip\cdot x}\, \phi(p) = \int \frac{d^D p}{(2\pi)^D}\, e^{-ip\cdot x}\, (\phi(p))^*\,. \tag{10.20}$$

We let $p \to -p$ on the left-hand side of this equation. This change of integration variable does not affect the integration $\int d^D p$, and results in

$$\int \frac{d^D p}{(2\pi)^D}\, e^{-ip\cdot x} \left(\phi(-p) - (\phi(p))^* \right) = 0\,, \tag{10.21}$$

where we collected all terms on the left-hand side. This left-hand side is a function of x that must vanish identically. It is also the Fourier transform of the momentum-space function in parentheses. This function must therefore vanish:

$$(\phi(p))^* = \phi(-p)\,. \tag{10.22}$$

This is the reality condition in momentum space.

Substituting (10.19) into (10.14) and letting ∂^2 act on $e^{ip\cdot x}$ we find

$$\int \frac{d^D p}{(2\pi)^D} \left(-p^2 - m^2 \right) \phi(p)\, e^{ipx} = 0\,. \tag{10.23}$$

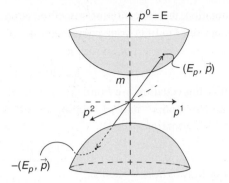

Fig. 10.1 The mass-shell hyperboloid $E^2 - p^2 = m^2$ (only two spatial directions shown). The equation of motion sets the mass m scalar field $\phi(p)$ equal to zero away from the hyperboloid. On the hyperboloid the field is arbitrary up to a reality condition that relates the field values at antipodal points.

Since this must hold for all values of x, this equation requires

$$(p^2 + m^2)\,\phi(p) = 0 \quad \text{for all } p . \tag{10.24}$$

This is a simple equation: $(p^2 + m^2)$ is just a number multiplying $\phi(p)$ and the product must vanish. Solving this equation means specifying the values of $\phi(p)$ for all values of p. Since either factor may vanish we must consider two cases.

(i) $p^2 + m^2 \neq 0$. In this case the scalar field vanishes: $\phi(p) = 0$.

(ii) $p^2 + m^2 = 0$. In this case the scalar field $\phi(p)$ is arbitrary.

In momentum space the hypersurface $p^2 + m^2 = 0$ is called the *mass-shell*. With $p^\mu = (E, \vec{p}\,)$, the mass-shell is the locus of points in momentum space where $E^2 = \vec{p}^2 + m^2$, the hyperboloid sketched in Figure 10.1. The mass-shell is therefore the set of points $(\pm E_p, \vec{p}\,)$, for all values of $\vec{p}$. We have learned that $\phi(p)$ vanishes off the mass-shell and is arbitrary (up to the reality condition) on the mass-shell.

We now introduce the idea of *classical degrees of freedom*. For a point p^μ on the mass-shell, the solution is determined by specifying the complex number $\phi(p)$. This number determines as well the solution at the point $(-p^\mu)$, also belonging to the mass-shell: $\phi(-p) = (\phi(p))^*$. So, a complex number fixes the values of the field for *two* points on the mass-shell. We need, on average, one real number for each point on the mass-shell. We will say that a field satisfying equation (10.24) represents *one degree of freedom per point on the mass-shell*.

We conclude this section by writing the scalar field equation of motion in light-cone coordinates. Let $\vec{x}_T$ denote a vector whose components are the transverse coordinates x^I:

$$\vec{x}_T = (x^2, x^3, \ldots, x^d) . \tag{10.25}$$

In this notation, the collection of spacetime coordinates becomes $(x^+, x^-, \vec{x}_T)$. Equation (10.14) expanded in light-cone coordinates is

$$\left(-2\frac{\partial}{\partial x^+}\frac{\partial}{\partial x^-} + \frac{\partial}{\partial x^I}\frac{\partial}{\partial x^I} - m^2\right)\phi(x^+, x^-, \vec{x}_T) = 0. \tag{10.26}$$

To simplify this equation, we Fourier transform the *spatial* dependence of the field, changing x^- into p^+ and x^I into p^I. Letting $\vec{p}_T$ denote the vector whose components are the transverse momenta p^I

$$\vec{p}_T = (p^2, p^3, \ldots, p^d), \tag{10.27}$$

the Fourier transform is written as

$$\phi(x^+, x^-, \vec{x}_T) = \int \frac{dp^+}{2\pi} \int \frac{d^{D-2}\vec{p}_T}{(2\pi)^{D-2}} \, e^{-ix^- p^+ + i\vec{x}_T \cdot \vec{p}_T} \, \phi(x^+, p^+, \vec{p}_T). \tag{10.28}$$

We now substitute this form of the scalar field into (10.26) to get

$$\left(-2\frac{\partial}{\partial x^+}(-ip^+) - p^I p^I - m^2\right)\phi(x^+, p^+, \vec{p}_T) = 0, \tag{10.29}$$

and dividing by $2p^+$ we find

$$\left(i\frac{\partial}{\partial x^+} - \frac{1}{2p^+}(p^I p^I + m^2)\right)\phi(x^+, p^+, \vec{p}_T) = 0. \tag{10.30}$$

This is the equation we were after. As opposed to the original Lorentz covariant equation of motion, which has second-order derivatives with respect to time, the light-cone equation is a first-order differential equation in light-cone time. Equation (10.30) has the formal structure of a Schrödinger equation, which is also first order in time. This fact will prove useful when we study how the quantum point particle is related to the scalar field.

Another version of equation (10.30) will be needed in our later work. Using a new time parameter τ related to x^+ as $x^+ = p^+\tau/m^2$ we obtain

$$\left(i\frac{\partial}{\partial \tau} - \frac{1}{2m^2}(p^I p^I + m^2)\right)\phi(\tau, p^+, \vec{p}_T) = 0. \tag{10.31}$$

Quick calculation 10.2 Consider the mass-shell condition $p^2 + m^2 = 0$ in light-cone coordinates. Show that

$$p^- = \frac{1}{2p^+}(p^I p^I + m^2). \tag{10.32}$$

10.4 Quantum scalar fields and particle states

Quantum field theory is a natural language to describe the quantum behavior of elementary particles and their interactions. Quantum field theory is quantum mechanics applied to classical fields. In quantum mechanics, classical dynamical variables turn into operators. The position and momentum of a classical particle, for example, turn into position and momentum operators. If our dynamical variables are classical fields, the quantum operators will be *field operators*. Thus, in quantum field theory, the fields are operators. The state space in a quantum field theory is typically described using a set of *particle states*. Quantum field theory also has energy and momentum operators. When they act on a particle state, these operators give the energy and momentum of the particles described by the state.

In this section, we examine briefly how the features just described arise concretely. Inspired by the plane-wave solution (10.18), which describes a superposition of complex waves with momenta $\vec{p}$ and $-\vec{p}$ (with $\vec{p} \neq 0$), we consider a classical field configuration $\phi_p(t, \vec{x})$ of the form

$$\phi_p(t, \vec{x}) = \frac{1}{\sqrt{V}} \frac{1}{\sqrt{2E_p}} \left(a(t)\, e^{i\vec{p}\cdot\vec{x}} + a^*(t)\, e^{-i\vec{p}\cdot\vec{x}} \right). \tag{10.33}$$

There have been two changes. First, the time dependence has been made more general by introducing the function $a(t)$ and its complex conjugate $a^*(t)$. The function $a(t)$ is the dynamical variable that determines the field configuration. Second, we have placed a normalization factor $\sqrt{V}$, where V is assumed to be the volume of space. The normalization factor also includes a square root of the energy E_p defined in equation (10.17).

We can imagine space as a box with sides $L_1, L_2, \ldots, L_d$, in which case $V = L_1 L_2 \ldots L_d$. When we put a field on a box we usually require it to be periodic. The field ϕ_p is periodic if each component p_i of $\vec{p}$ satisfies

$$p_i L_i = 2\pi n_i, \quad i = 1, 2, \ldots, d. \tag{10.34}$$

Here the n_i are integers. Each component of the momentum is quantized.

We now try to do quantum mechanics with the field configuration (10.33). To this end, we evaluate the scalar field action (10.7) for $\phi = \phi_p(t, \vec{x})$:

$$S = \int dt \int d^d x \left(\frac{1}{2}(\partial_0 \phi_p)^2 - \frac{1}{2}(\nabla \phi_p)^2 - \frac{1}{2}m^2 \phi_p^2 \right). \tag{10.35}$$

The evaluation involves squaring the field, squaring its time derivative, and squaring its gradient. In squaring any of these, we obtain from the cross multiplication two types of terms: those with spatial dependence $\exp(\pm 2i\,\vec{p}\cdot\vec{x})$ and those without spatial dependence. We claim that the spatial integral $\int d^d x$ of the terms with spatial dependence is zero, so these terms cannot contribute. Indeed, the quantization conditions in (10.34) imply that

$$\int_0^{L_1} dx^1 \ldots \int_0^{L_d} dx^d \, \exp(\pm 2i\,\vec{p}\cdot\vec{x}) = 0. \tag{10.36}$$

For the terms without spatial dependence, the spatial integral gives a factor of the volume V, which cancels with the product of $\sqrt{V}$ factors we introduced in (10.33). The result is

$$S = \int dt \left(\frac{1}{2E_p} \dot{a}^*(t)\dot{a}(t) - \frac{1}{2} E_p\, a^*(t)a(t) \right). \tag{10.37}$$

Quick calculation 10.3 Verify that equation (10.37) is correct.

Similarly, we use equation (10.11) to evaluate the field energy H:

$$H = \frac{1}{2E_p} \dot{a}^*(t)\dot{a}(t) + \frac{1}{2} E_p\, a^*(t)a(t). \tag{10.38}$$

Quick calculation 10.4 Verify that equation (10.38) is correct.

The action (10.37) describes the dynamics of two simple harmonic oscillators. Since $a(t)$ is a complex dynamical variable we can write

$$a(t) = q_1(t) + i\, q_2(t), \tag{10.39}$$

where $q_1(t)$ and $q_2(t)$ are real coordinates. In terms of $q_1(t)$ and $q_2(t)$, the action becomes

$$S = \int dt\, L = \sum_{i=1}^{2} \int dt \left(\frac{1}{2E_p}\, \dot{q}_i^2(t) - \frac{1}{2} E_p\, q_i^2(t) \right). \tag{10.40}$$

We see here that $q_1(t)$ and $q_2(t)$ are indeed the coordinates of identical simple harmonic oscillators. Their canonical momenta are

$$p_i(t) = \frac{\partial L}{\partial \dot{q}_i} = \frac{\dot{q}_i(t)}{E_p} \longrightarrow p_1(t) + ip_2(t) = \frac{1}{E_p}\, \dot{a}(t). \tag{10.41}$$

The equations of motion that follow from the variation of the action (10.40) are

$$\ddot{q}_i(t) = -E_p^2\, q_i(t), \quad i = 1, 2. \tag{10.42}$$

It is actually convenient to work with the complex combination $a(t)$. Using (10.39) the equations of motion become the single complex equation

$$\ddot{a}(t) = -E_p^2\, a(t). \tag{10.43}$$

Equation (10.43) is readily solved in terms of exponentials. Since it is a second-order equation, there are two solutions:

$$a(t) = a_p\, e^{-iE_p t} + a_{-p}^*\, e^{iE_p t}. \tag{10.44}$$

There is no reality condition since $a(t)$ is complex. In writing the above solution we introduced two independent complex constants a_p and a_{-p}^*. Substituting this solution into (10.38), we find

$$H = E_p \left(a_p^* a_p + a_{-p}^* a_{-p} \right). \tag{10.45}$$

Note that the time dependence disappeared; the energy is conserved.

In classical scalar field theory there is an integral expression that gives the spacetime momentum $\vec{P}$ carried by the field. In the present case, this field momentum is conserved, and a formula can be obtained from the analysis of Problem 8.10. We will not discuss this here, but the answer is

$$\vec{P} = -\int d^d x\, (\partial_0 \phi) \nabla \phi. \tag{10.46}$$

In Problem 10.1 you are asked to evaluate $\vec{P}$ for the field configuration (10.33), when $a(t)$ is given by (10.44). The result that you will find is

$$\vec{P} = \vec{p} \left(a_p^* a_p - a_{-p}^* a_{-p} \right). \tag{10.47}$$

The system at hand consists of two harmonic oscillators. The expression for H suggests that in the quantum theory a_p and a_{-p} become annihilation operators, while a_p^* and a_{-p}^* become the creation operators $a_p^\dagger$ and $a_{-p}^\dagger$, respectively. We will verify the correctness of this assumption shortly. These oscillators are required to satisfy the standard commutation relations:

$$[a_p, a_p^\dagger] = 1, \quad [a_{-p}, a_{-p}^\dagger] = 1. \tag{10.48}$$

Any commutator which involves an operator with subscript p and an operator with subscript $(-p)$ is declared to vanish. The variable $a(t)$ in (10.44) now becomes an operator. We record its form, together with that for its Hermitian conjugate $a^\dagger(t)$, as well as those for their time derivatives:

$$a(t) = a_p e^{-iE_p t} + a_{-p}^\dagger e^{iE_p t},$$
$$a^\dagger(t) = a_p^\dagger e^{iE_p t} + a_{-p} e^{-iE_p t},$$
$$\dot{a}(t) = -iE_p \left(a_p e^{-iE_p t} - a_{-p}^\dagger e^{iE_p t} \right),$$
$$\dot{a}^\dagger(t) = iE_p \left(a_p^\dagger e^{iE_p t} - a_{-p} e^{-iE_p t} \right). \tag{10.49}$$

It takes a short computation to verify that the only nonvanishing commutators among $a(t), a^\dagger(t), \dot{a}(t)$, and $\dot{a}^\dagger(t)$ are

$$\left[a(t), \dot{a}^\dagger(t) \right] = \left[a^\dagger(t), \dot{a}(t) \right] = 2iE_p. \tag{10.50}$$

We can now show that these commutation relations imply the commutation relations that we would naturally impose between the coordinates and their corresponding conjugate momenta:

$$[q_1(t), p_1(t)] = [q_2(t), p_2(t)] = i. \tag{10.51}$$

To do this, we can use equations (10.39) and (10.41) to solve for the coordinates and the momenta:

$$q_1(t) = \frac{1}{2}(a(t) + a^\dagger(t)), \qquad p_1(t) = \frac{1}{2E_p}(\dot{a}(t) + \dot{a}^\dagger(t)), \tag{10.52}$$

$$q_2(t) = \frac{1}{2i}(a(t) - a^\dagger(t)), \qquad p_2(t) = \frac{1}{2iE_p}(\dot{a}(t) - \dot{a}^\dagger(t)). \tag{10.53}$$

We can now check that, for example,

$$[q_1(t), p_1(t)] = \frac{1}{4E_p} \left[a(t) + a^\dagger(t), \dot{a}(t) + \dot{a}^\dagger(t) \right] = \frac{1}{4E_p}(2iE_p + 2iE_p) = i, \tag{10.54}$$

as expected. The other commutators can be checked similarly, thus confirming that the postulated commutation relations (10.48) are correct.

Quick calculation 10.5 Check that $[q_2(t), p_2(t)] = i$ and that $[q_1(t), p_2(t)] = 0$.

At the quantum level, the Hamiltonian (10.45) becomes the operator

$$H = E_p \left(a_p^\dagger a_p + a_{-p}^\dagger a_{-p} \right), \tag{10.55}$$

describing a pair of simple harmonic oscillators, each of which has frequency E_p. The momentum (10.46) becomes the operator

$$\vec{P} = \vec{p}\left(a_p^\dagger a_p - a_{-p}^\dagger a_{-p}\right). \tag{10.56}$$

Note that the oscillators with subscripts $(-p)$ contribute with a negative sign to the momentum. This equation will help our interpretation below.

The first two equations in (10.49) can be used back in (10.33) to obtain the operator version of the field configuration:

$$\phi_p(t, \vec{x}) = \frac{1}{\sqrt{V}}\frac{1}{\sqrt{2E_p}}\left(a_p\, e^{-iE_p t + i\vec{p}\cdot\vec{x}} + a_p^\dagger\, e^{iE_p t - i\vec{p}\cdot\vec{x}}\right)$$
$$+ \frac{1}{\sqrt{V}}\frac{1}{\sqrt{2E_p}}\left(a_{-p}\, e^{-iE_p t - i\vec{p}\cdot\vec{x}} + a_{-p}^\dagger\, e^{iE_p t + i\vec{p}\cdot\vec{x}}\right). \tag{10.57}$$

We see that ϕ_p is in fact a spacetime dependent operator, or a field operator. The second line in (10.57) is obtained from the first line by the replacement $\vec{p} \to -\vec{p}$, which does not affect E_p. In full generality, the quantum field $\phi(x)$ includes contributions from all values of the spatial momentum $\vec{p}$:

$$\phi(t, \vec{x}) = \frac{1}{\sqrt{V}} \sum_{\vec{p}} \frac{1}{\sqrt{2E_p}}\left(a_p\, e^{-iE_p t + i\vec{p}\cdot\vec{x}} + a_p^\dagger\, e^{iE_p t - i\vec{p}\cdot\vec{x}}\right). \tag{10.58}$$

The commutation relations for the oscillators are the natural generalization of those in (10.48):

$$[a_p, a_k^\dagger] = \delta_{p,k}, \quad [a_p, a_k] = [a_p^\dagger, a_k^\dagger] = 0. \tag{10.59}$$

All subscripts here are spatial vectors, written without the arrows to avoid cluttering the equations. The Kronecker delta $\delta_{p,k}$ is zero unless $\vec{p} = \vec{k}$, in which case it equals one. Once we consider contributions from all values of the momenta, the previous expression for the Hamiltonian in (10.55) and that for the momentum operator in (10.56) must be changed. One can show that

$$H = \sum_{\vec{p}} E_p\, a_p^\dagger a_p, \tag{10.60}$$

$$\vec{P} = \sum_{\vec{p}} \vec{p}\, a_p^\dagger a_p. \tag{10.61}$$

We will not derive these expressions, but they should seem quite plausible.

The state space of this quantum system is built in the same way as the state space of the simple harmonic oscillator. We assume the existence of a vacuum state $|\Omega\rangle$, which acts just as the simple harmonic oscillator ground state $|0\rangle$ in that it is annihilated by the annihilation operators a_p: $a_p|\Omega\rangle = 0$ for all $\vec{p}$. It follows that $H|\Omega\rangle = 0$, which makes the vacuum a zero-energy state. This vacuum state is interpreted as a state in which there are no particles. On the other hand, the state

$$a_p^\dagger |\Omega\rangle \tag{10.62}$$

is interpreted as a state with precisely one particle. We claim that it is a particle with momentum $\vec{p}$. To verify this, we act on the state with the momentum operator (10.61) and use (10.59) to find

$$\vec{P} a_p^\dagger |\Omega\rangle = \sum_{\vec{k}} \vec{k} a_k^\dagger [a_k, a_p^\dagger] |\Omega\rangle = \vec{p} \, a_p^\dagger |\Omega\rangle. \tag{10.63}$$

The energy of the state is similarly computed by acting on the state with the Hamiltonian H:

$$H a_p^\dagger |\Omega\rangle = \sum_{\vec{k}} E_k \, a_k^\dagger [a_k, a_p^\dagger] |\Omega\rangle = E_p \, a_p^\dagger |\Omega\rangle. \tag{10.64}$$

The state $a_p^\dagger|\Omega\rangle$ has positive energy. While the quantum field operator has both positive and negative energy components, the states that represent the particles have positive energy. The states $a_p^\dagger |\Omega\rangle$ are the *one-particle* states.

The state space contains multiparticle states as well. These are states built by acting on the vacuum with a collection of creation operators:

$$a_{p_1}^\dagger \, a_{p_2}^\dagger \ldots a_{p_k}^\dagger |\Omega\rangle. \tag{10.65}$$

This state, with k creation operators acting on the vacuum, represents a state with k particles. The particles have momenta $\vec{p}_1, \vec{p}_2, \ldots, \vec{p}_k$ and energies $E_{p_1}, E_{p_2}, \ldots, E_{p_k}$. The various momenta $\vec{p}_i$ need not be all different.

Quick calculation 10.6 Show that the eigenvalues of $\vec{P}$ and H acting on (10.65) are $\sum_{n=1}^{k} \vec{p}_n$ and $\sum_{n=1}^{k} E_{p_n}$, respectively.

Quick calculation 10.7 Convince yourself that $N = \sum_{\vec{p}} a_p^\dagger a_p$ is a number operator: acting on a state it gives us the number of particles contained in the state.

Our analysis of classical solutions in the previous section led to the conclusion that there is one degree of freedom per point on the mass-shell. At the quantum level, we focus on the one-particle states. Consequently, we restrict ourselves to the physical part of the mass-shell, the part where the energy is positive ($p^0 = E > 0$). We have a single one-particle state for each point on the physical mass-shell. This state is labeled by its spatial momentum $\vec{p}$.

To describe the particle states in light-cone coordinates, the changes are minimal. The physical part of the mass-shell is parameterized by the transverse momenta $\vec{p}_T$ and the light-cone momenta p^+ for which $p^+ > 0$. The value of the light-cone energy p^- is then fixed and positive. Thus, instead of labeling the oscillators with $\vec{p}$, we simply label them with $p^+ > 0$ and $\vec{p}_T$. The one-particle states are written as

$$\text{One-particle states of a scalar field:} \quad a^\dagger_{p^+, \, p_T} \, |\Omega\rangle. \qquad (10.66)$$

The momentum operator given in (10.61) has a natural light-cone version. The various components of the operator take the form

$$\hat{p}^+ = \sum_{p^+, p_T} p^+ \, a^\dagger_{p^+, \, p_T} a_{p^+, \, p_T} \,,$$

$$\hat{p}^I = \sum_{p^+, p_T} p^I \, a^\dagger_{p^+, \, p_T} a_{p^+, \, p_T} \,, \qquad (10.67)$$

$$\hat{p}^- = \sum_{p^+, p_T} \frac{1}{2p^+} \left(p^I p^I + m^2 \right) a^\dagger_{p^+, \, p_T} a_{p^+, \, p_T} \,.$$

In the last equation, the factor which multiplies the oscillators is the value of p^- determined from the mass-shell condition (10.32). This last equation is analogous to (10.60), where E_p is the energy determined from the mass-shell condition.

10.5 Maxwell fields and photon states

We now turn to an analysis of Maxwell fields and their corresponding quantum states. As opposed to the case of the scalar field, where there is no gauge invariance, electromagnetic fields have a gauge invariance that will make our analysis more subtle and interesting. In order to study the field equations in a convenient way, we will impose the gauge condition that defines the light-cone gauge. We will then be able to describe the quantum states of the Maxwell field.

The field equations for electromagnetism are written in terms of the electromagnetic vector potential $A_\mu(x)$. As we reviewed in Section 3.3, the field strength $F_{\mu\nu} = \partial_\mu A_\nu - \partial_\nu A_\mu$ is invariant under the gauge transformation

$$\delta A_\mu = \partial_\mu \epsilon \,, \qquad (10.68)$$

where ϵ is the gauge parameter. The field equations take the form

$$\partial_\nu F^{\mu\nu} = 0 \; \longrightarrow \; \partial_\nu (\partial^\mu A^\nu - \partial^\nu A^\mu) = 0 \,, \qquad (10.69)$$

and can be written as

$$\partial^2 A^\mu - \partial^\mu (\partial \cdot A) = 0 \,. \qquad (10.70)$$

Compare this equation with equation (10.14) for the scalar field. There is no indication of a mass term for the Maxwell field – such a term would be recognized as one without spacetime derivatives. We will confirm below that the Maxwell field is, indeed, massless.

We Fourier transform all the components of the vector potential in order to determine the equation of motion in momentum space:

$$A^\mu(x) = \int \frac{d^D p}{(2\pi)^D}\, e^{ip\cdot x}\, A^\mu(p)\,, \qquad (10.71)$$

where reality of $A^\mu(x)$ implies $A^\mu(-p) = (A^\mu(p))^*$. Substituting (10.71) into (10.70) we obtain the equation:

$$p^2 A^\mu - p^\mu(p\cdot A) = 0\,. \qquad (10.72)$$

The gauge transformation (10.68) can also be Fourier transformed. In momentum space, the gauge transformation relates $\delta A_\mu(p)$ to the Fourier transform $\epsilon(p)$ of the gauge parameter:

$$\delta A_\mu(p) = ip_\mu\, \epsilon(p)\,. \qquad (10.73)$$

Since the gauge parameter $\epsilon(x)$ is real, we have $\epsilon(-p) = \epsilon^*(p)$. The gauge transformation (10.73) is consistent with the reality of $\delta A_\mu(x)$. Indeed,

$$(\delta A_\mu(p))^* = -ip_\mu(\epsilon(p))^* = i(-p_\mu)\epsilon(-p) = \delta A_\mu(-p)\,. \qquad (10.74)$$

Note the role of the factor of i in getting the signs to work out.

Being done with preliminaries, we can analyze (10.72) subject to the gauge transformations (10.73). At this point, it is more convenient to work with the light-cone components of the gauge field:

$$A^+(p)\,, \quad A^-(p)\,, \quad A^I(p)\,. \qquad (10.75)$$

The gauge transformations (10.73) then read

$$\delta A^+ = ip^+\epsilon\,, \quad \delta A^- = ip^-\epsilon\,, \quad \delta A^I = ip^I\epsilon\,. \qquad (10.76)$$

We now impose a gauge condition. As we have emphasized before, when working with the light-cone formalism we always assume $p^+ \neq 0$. The above gauge transformations now make it clear that we can set A^+ to zero by choosing ϵ correctly. Indeed, if we apply a gauge transformation

$$A^+ \to A'^+ = A^+ + ip^+\epsilon\,, \qquad (10.77)$$

then the $+$ component of the new gauge field A' vanishes if we choose $\epsilon = iA^+/p^+$. In other words, we can always make the $+$ component of the Maxwell field zero by applying a gauge transformation. This will be our defining condition for the light-cone gauge in Maxwell theory:

$$\text{light-cone gauge condition}: \quad A^+(p) = 0\,. \qquad (10.78)$$

Setting A^+ to zero determines the gauge parameter ϵ, and no additional gauge transformations are possible: if $A^+ = 0$, any further gauge transformation will make it nonzero.

There is one minor exception: any parameter of the form $\epsilon(p) = \epsilon(p^-, p^I)\delta(p^+)$ will keep $A^+ = 0$ because $p^+\epsilon(p) = 0$. The similarities with light-cone gauge string theory are noteworthy. In light-cone gauge open string theory all world-sheet reparameterization invariances are fixed. Moreover, while not equal to zero, X^+ is very simple: the corresponding zero mode and oscillators vanish.

The gauge condition (10.78) simplifies the equation of motion (10.72) considerably. Taking $\mu = +$ we find

$$p^+(p \cdot A) = 0 \;\longrightarrow\; p \cdot A = 0. \qquad (10.79)$$

This equation can be expanded out using light-cone indices:

$$-p^+ A^- - p^- A^+ + p^I A^I = 0. \qquad (10.80)$$

Since $A^+ = 0$, this equation determines A^- in terms of the transverse A^I:

$$A^- = \frac{1}{p^+}(p^I A^I). \qquad (10.81)$$

This is reminiscent of our light-cone analysis of the string, where X^- was solved for in terms of the transverse coordinates (and a zero mode). Using (10.79) back in (10.72), all that remains from the field equation is

$$p^2 A^\mu(p) = 0. \qquad (10.82)$$

For $\mu = +$ this equation is trivially satisfied, since $A^+ = 0$. For $\mu = I$ we get a set of nontrivial conditions:

$$p^2 A^I(p) = 0. \qquad (10.83)$$

For $\mu = -$, we get $p^2 A^-(p) = 0$. This is automatically satisfied on account of (10.81) and (10.83).

For each value of I, equation (10.83) takes the form of the equation of motion for a massless scalar. Thus, $A^I(p) = 0$ when $p^2 \neq 0$. This makes $A^- = 0$, and since A^+ is zero, the full gauge field vanishes. For $p^2 = 0$, the $A^I(p)$ are unconstrained, and the $A^-(p)$ are determined as a function of the A^I (see (10.81)). The degrees of freedom of the Maxwell field are thus carried by the $(D - 2)$ transverse fields $A^I(p)$, for $p^2 = 0$. We say that we have $(D - 2)$ degrees of freedom per point on the mass-shell.

It is actually possible to show that there are no degrees of freedom for $p^2 \neq 0$, without having to make a choice of gauge. Although not every field is zero, every field is *gauge equivalent* to the zero field when $p^2 \neq 0$. If a field differs from the zero field by only a gauge transformation, we say that the field is *pure gauge*. Recall that fields A_μ and A'_μ are gauge equivalent if $A_\mu = A'_\mu + \partial_\mu \chi$, for some scalar function χ. Taking $A'_\mu = 0$, we learn that $A_\mu = \partial_\mu \chi$ is gauge equivalent to the zero field, and is therefore pure gauge. The term pure gauge is suitable: A_μ takes the form of a gauge transformation. In momentum space, a pure gauge is a field that can be written as

$$\text{pure gauge:} \quad A_\mu(p) = i p_\mu \chi(p), \qquad (10.84)$$

for some choice of χ. Rewrite now the equation of motion (10.72) as

$$p^2 A_\mu = p_\mu (p \cdot A) \,. \tag{10.85}$$

Since $p^2 \neq 0$, we can write

$$A_\mu = i p_\mu \left(\frac{-i p \cdot A}{p^2} \right) \,. \tag{10.86}$$

Comparing with (10.84), we see that A_μ is pure gauge. This means that there are no degrees of freedom in the Maxwell field when $p^2 \neq 0$. For all intents and purposes, there is no field.

Let us now discuss briefly photon states. Each of the independent classical fields A^I can be expanded as we did for the scalar field in (10.58). To do so, we introduce – as you can infer by analogy – oscillators a_p^I and $a_p^{I\dagger}$, where the subscripts p represent the values of p^+ and $\vec{p}_T$. We thus get $(D-2)$ species of oscillators. Introducing a vacuum $|\Omega\rangle$, the one-photon states would be written as

$$a_{p^+, \, p_T}^{I\dagger} |\Omega\rangle \,. \tag{10.87}$$

Here the label I is a polarization label. The photon state (10.87) is said to be polarized in the Ith direction. Since we have $(D-2)$ possible polarizations, *we have $(D-2)$ linearly independent one-photon states for each point on the physical sector of the mass-shell.* A general one-photon state with momentum $(p^+, \vec{p}_T)$ is a linear superposition of the above states:

$$\text{one-photon states:} \quad \sum_{I=2}^{D-1} \xi_I \, a_{p^+, \, p_T}^{I\dagger} |\Omega\rangle \,. \tag{10.88}$$

Here the transverse vector ξ_I is called the polarization vector.

In four-dimensional spacetime, Maxwell theory gives rise to $D - 2 = 2$ single-photon states for any fixed spatial momentum. This is indeed familiar to you, at least classically. An electromagnetic plane wave which propagates in a fixed direction and has some fixed wavelength (i.e., fixed momentum), can be written as a superposition of two plane waves that represent independent polarization states.

10.6 Gravitational fields and graviton states

Gravitation emerges in string theory in the language of Einstein's theory of general relativity. We discussed this language briefly in Section 3.6. The dynamical field variable is the spacetime metric $g_{\mu\nu}(x)$, which in the approximation of weak gravitational fields can be taken to be of the form $g_{\mu\nu}(x) = \eta_{\mu\nu} + h_{\mu\nu}(x)$. Both $g_{\mu\nu}$ and $h_{\mu\nu}$ are symmetric under the exchange of their indices. The field equations for $g_{\mu\nu}$ – Einstein's equations – can be used to derive a linearized equation of motion for the fluctuations $h_{\mu\nu}$. This equation was given in (3.82). Defining $h_{\mu\nu}(p)$ to be the Fourier transform of $h_{\mu\nu}(x)$, the momentum-space version of this equation is

$$S^{\mu\nu}(p) \equiv p^2 h^{\mu\nu} - p_\alpha(p^\mu h^{\nu\alpha} + p^\nu h^{\mu\alpha}) + p^\mu p^\nu h = 0. \tag{10.89}$$

If we were considering Einstein's equations in the presence of sources, the right-hand side of this equation would include terms which represent the energy-momentum tensor of the sources. In the above equation $h = \eta^{\mu\nu} h_{\mu\nu} = h^\mu_\mu$, and indices on $h_{\mu\nu}$ can be raised or lowered using the Minkowski metric $\eta_{\mu\nu}$ and its inverse $\eta^{\mu\nu}$. Since every term in (10.89) contains two derivatives, this suggests that the fluctuations $h_{\mu\nu}$ are associated with massless excitations.

As we will see shortly, the equation of motion (10.89) is invariant under the gauge transformations discussed in Section 3.6:

$$\delta_0 h^{\mu\nu}(p) = ip^\mu \epsilon^\nu(p) + ip^\nu \epsilon^\mu(p). \tag{10.90}$$

The infinitesimal gauge parameter $\epsilon^\mu(p)$ is a vector. In gravitation, the gauge invariance is reparameterization invariance: the choice of coordinate system used to parameterize spacetime does not affect the physics.

Let us verify that (10.89) is invariant under the gauge transformation (10.90). First we compute $\delta_0 h$ and find that

$$\delta_0 h = \eta_{\mu\nu}\delta_0 h^{\mu\nu} = i\eta_{\mu\nu}(p^\mu \epsilon^\nu + p^\nu \epsilon^\mu) = 2ip \cdot \epsilon. \tag{10.91}$$

The resulting variation in $S^{\mu\nu}$ is therefore given by

$$\delta_0 S^{\mu\nu} = ip^2(p^\mu \epsilon^\nu + p^\nu \epsilon^\mu) - ip_\alpha p^\mu(p^\nu \epsilon^\alpha + p^\alpha \epsilon^\nu)$$
$$- ip_\alpha p^\nu(p^\mu \epsilon^\alpha + p^\alpha \epsilon^\mu) + 2ip^\mu p^\nu p \cdot \epsilon. \tag{10.92}$$

But we can rewrite

$$\delta_0 S^{\mu\nu} = ip^2(p^\mu \epsilon^\nu + p^\nu \epsilon^\mu) - ip^\mu p^\nu(p \cdot \epsilon) - ip^2 p^\mu \epsilon^\nu$$
$$- ip^\mu p^\nu(p \cdot \epsilon) - ip^2 p^\nu \epsilon^\mu + 2ip^\mu p^\nu p \cdot \epsilon. \tag{10.93}$$

It is readily seen that all the terms in (10.93) cancel, so $\delta_0 S^{\mu\nu} = 0$. The equation of motion exhibits the claimed gauge invariance.

Since the metric $h^{\mu\nu}$ is symmetric and has two indices, each running over $(+, -, I)$, we have the following objects to consider:

$$(h^{IJ}, h^{+I}, h^{-I}, h^{+-}, h^{++}, h^{--}). \tag{10.94}$$

We shall try to set to zero all the fields in (10.94) that contain a $+$ index. For this, we use (10.90) to examine their gauge transformations:

$$\delta_0 h^{++} = 2ip^+ \epsilon^+, \tag{10.95}$$
$$\delta_0 h^{+-} = ip^+ \epsilon^- + ip^- \epsilon^+, \tag{10.96}$$
$$\delta_0 h^{+I} = ip^+ \epsilon^I + ip^I \epsilon^+. \tag{10.97}$$

As before, we assume $p^+ \neq 0$. From (10.95), we see that a judicious choice of ϵ^+ will permit us to gauge away h^{++}. This fixes our choice of ϵ^+. Looking at equation (10.96), we see that although we have fixed ϵ^+, we can still find an ϵ^- that will gauge away h^{+-}. This fixes ϵ^-. Similarly, we can use (10.97) and a suitable choice of ϵ^I to set h^{+I} to zero. We have used the full gauge freedom to set to zero all of the entries in $h^{\mu\nu}$ with $+$ indices. This defines the light-cone gauge for the gravity field:

$$\text{light-cone gauge conditions:} \quad h^{++} = h^{+-} = h^{+I} = 0. \tag{10.98}$$

The remaining degrees of freedom are carried by

$$(h^{IJ}, h^{-I}, h^{--}). \tag{10.99}$$

We must now see what is implied by the equations of motion (10.89). Bearing in mind the gauge conditions (10.98), when $\mu = \nu = +$ we find

$$(p^+)^2 h = 0 \longrightarrow h = 0. \tag{10.100}$$

More explicitly,

$$h = \eta_{\mu\nu} h^{\mu\nu} = -2h^{+-} + h^{II} = 0 \longrightarrow h^{II} = 0, \tag{10.101}$$

since $h^{+-} = 0$ in our gauge. The equation $h^{II} = 0$ means that the matrix h^{IJ} is traceless. With $h = 0$, the equation of motion (10.89) reduces to

$$p^2 h^{\mu\nu} - p^\mu (p_\alpha h^{\nu\alpha}) - p^\nu (p_\alpha h^{\mu\alpha}) = 0. \tag{10.102}$$

Now set $\mu = +$. We obtain $p^+(p_\alpha h^{\nu\alpha}) = 0$, and as a result

$$p_\alpha h^{\nu\alpha} = 0. \tag{10.103}$$

If (10.103) holds, equation (10.102) reduces to

$$p^2 h^{\mu\nu} = 0. \tag{10.104}$$

This is all that remains of the equation of motion! Before delving into this familiar equation, let us investigate the implications of (10.103). The only free index here is ν. For $\nu = +$ the equation is trivial, since the $h^{+\alpha}$ are zero in our gauge. Consider now $\nu = I$. This gives $p_\alpha h^{I\alpha} = 0$, which we can expand as

$$-p^+ h^{I-} - p^- h^{I+} + p_J h^{IJ} = 0 \longrightarrow h^{I-} = \frac{1}{p^+} p_J h^{IJ}. \tag{10.105}$$

Similarly, with $\nu = -$ we get $p_\alpha h^{-\alpha} = 0$, which expands to

$$-p^+ h^{--} - p^- h^{-+} + p_I h^{-I} = 0 \longrightarrow h^{--} = \frac{1}{p^+} p_I h^{-I}. \tag{10.106}$$

Equations (10.105) and (10.106) give us the h with $-$ indices in terms of the transverse h^{IJ}. There is no additional content to (10.103).

We now return to equation (10.104). This equation holds trivially for any field with a $+$ index. The equation is nontrivial for transverse indices:

$$p^2 h^{IJ}(p) = 0. \tag{10.107}$$

Equations $p^2 h^{I-} = 0$ and $p^2 h^{--} = 0$ are automatically satisfied on account of our solutions (10.105) and (10.106), together with (10.107). Equation (10.107) implies that $h^{IJ}(p) = 0$ for $p^2 \neq 0$, in which case all other components of $h^{\mu\nu}$ also vanish. For $p^2 = 0$, the $h^{IJ}(p)$ are unconstrained, except for the tracelessness condition $h_{II}(p) = 0$. All other components are determined in terms of the transverse components.

We conclude that the degrees of freedom of the classical D-dimensional gravitational field are carried by a *symmetric, traceless, transverse* tensor field h^{IJ}, the components of which satisfy the equation of motion of a massless scalar. This tensor has as many components as a symmetric traceless square matrix of size $(D-2)$. The number of components $n(D)$ in this matrix is

$$n(D) = \frac{1}{2}(D-2)(D-1) - 1 = \frac{1}{2}D(D-3). \tag{10.108}$$

Moreover, as before, we count a massless scalar as one degree of freedom per point on the mass-shell. Therefore we say that a classical gravity wave has $n(D)$ degrees of freedom per point on the mass-shell. In four-dimensional spacetime there are two transverse directions, and a symmetric traceless 2×2 matrix has two independent components. In four dimensions we thus have $n(4) = 2$ degrees of freedom. In five dimensions we have $n(5) = 5$ degrees of freedom, in ten dimensions $n(10) = 35$ degrees of freedom, and in twenty-six dimensions $n(26) = 299$ degrees of freedom. To obtain graviton states, each of the independent classical fields h^{IJ} fields is expanded in terms of creation and annihilation operators, just as we did for the scalar field in (10.58). To do so we need oscillators $a^{IJ}_{p^+, p_T}$ and $a^{IJ\dagger}_{p^+, p_T}$. We introduce a vacuum $|\Omega\rangle$, and a basis of states

$$a^{IJ\dagger}_{p^+, p_T} |\Omega\rangle. \tag{10.109}$$

A one-graviton state with momentum $(p^+, \vec{p}_T)$ is a linear superposition of the above states:

$$\text{one-graviton states:} \quad \sum_{I,J=2}^{D-1} \xi_{IJ}\, a^{IJ\dagger}_{p^+, p_T} |\Omega\rangle, \quad \xi_{II} = 0. \tag{10.110}$$

Here ξ_{IJ} is the graviton polarization tensor. The classical tracelessness condition that we found earlier becomes in the quantum theory the tracelessness $\xi_{II} = 0$ of the polarization tensor. Since ξ_{IJ} is a traceless symmetric matrix of size $(D-2)$, we have $n(D)$ linearly independent graviton states for each point on the physical mass-shell.

Problems

Problem 10.1 Momentum for the classical scalar field.

Show that the integral (10.46), when evaluated for the field configuration (10.33) gives

$$\vec{P} = -\frac{i\,\vec{p}}{2E_p}\left(\dot{a}^*a - a^*\dot{a}\right).$$

Use (10.44) to show that $\vec{P} = \vec{p}\left(a_p^* a_p - a_{-p}^* a_{-p}\right)$, as quoted in (10.47).

Problem 10.2 Commutator for the quantum scalar field.

(a) Consider a periodic function $f(\vec{x})$ on the box described above equation (10.34). Such a function can be expanded as the Fourier series

$$f(\vec{x}) = \sum_{\vec{p}} f(\vec{p})\,e^{i\vec{p}\cdot\vec{x}}. \tag{1}$$

Show that

$$f(\vec{p}) = \frac{1}{V}\int d\vec{x}'\,f(\vec{x}')e^{-i\vec{p}\cdot\vec{x}'}. \tag{2}$$

Plug (2) back into (1) to obtain an infinite sum representation for the d-dimensional delta function $\delta^d(\vec{x} - \vec{x}')$.

(b) Consider the complete scalar field expansion in (10.58). Calculate the corresponding expansion of $\Pi(t, \vec{x}) = \partial_0\phi(t, \vec{x})$. Show that

$$[\phi(t, \vec{x}),\ \Pi(t, \vec{x}')] = i\,\delta^d(\vec{x} - \vec{x}'). \tag{3}$$

This is the equal-time commutator between the field operator and its corresponding conjugate momentum. Most discussions of quantum field theory begin by postulating this commutator.

Problem 10.3 Light-cone components of Lorentz tensors.[†]

(a) Verify that the Lorentz covariant equation $A^\mu = B^\mu$, for $\mu = 0, 1, \ldots, d$, implies that $A^+ = B^+$, $A^- = B^-$, and $A^I = B^I$.

Given a Lorentz tensor $R^{\mu\nu}$, how do we define the light-cone components $R^{+-}, R^{++}, \ldots$? To find out, note that the definition must work for *any* tensor, so it must work when $R^{\mu\nu} = A^\mu B^\nu$. Thus, for example, $R^{+-} = A^+ B^-$, and writing A^+ and B^- in terms of Lorentz components, you can determine R^{+-} in terms of R^{00}, R^{01}, R^{10}, and R^{11}.

(b) Calculate R^{++}, R^{+-}, R^{-+}, and R^{--} in terms of the Lorentz components of $R^{\mu\nu}$. Explain why an equality $R^{\mu\nu} = S^{\mu\nu}$ between Lorentz tensors implies the equality of the light-cone components.

(c) Check that your result in (b) gives the expected answers for the light-cone components of the Minkowski metric: η^{++}, η^{+-}, η^{-+}, and η^{--}.

(d) Consider the antisymmetric electromagnetic field strength $F^{\mu\nu}$ in four dimensions. Calculate the light-cone components F^{+-}, F^{+I}, F^{-I}, and F^{IJ} in terms of the Lorentz components of $F^{\mu\nu}$. Rewrite your answers in terms of $\vec{E}$ and $\vec{B}$ fields.

Problem 10.4 Constant electric field in light-cone gauge.

Find potentials that describe a uniform constant electric field $\vec{E} = E_0\vec{e}_x$ in the light-cone gauge $(A^+ = (A^0 + A^1)/\sqrt{2} = 0)$. Write A^- and A^I in terms of the light-cone coordinates x^+, x^-, and x^I.

Problem 10.5 Gravitational fields that are pure gauge.

Following the discussion of Maxwell fields that are pure gauge, define gravitational fields that are pure gauge. Prove that any gravitational field $h_{\mu\nu}(p)$ which satisfies the equations of motion is pure gauge when $p^2 \neq 0$.

Problem 10.6 Field equations and particle states for the Kalb–Ramond field $B_{\mu\nu}$.[†]

Here we examine the field theory of a massless antisymmetric tensor gauge field $B_{\mu\nu} = -B_{\nu\mu}$. This gauge field is a tensor analog of the Maxwell gauge field A_μ. In Maxwell theory we defined the field strength $F_{\mu\nu} = \partial_\mu A_\nu - \partial_\nu A_\mu$. For $B_{\mu\nu}$ we define a field strength $H_{\mu\nu\rho}$:

$$H_{\mu\nu\rho} \equiv \partial_\mu B_{\nu\rho} + \partial_\nu B_{\rho\mu} + \partial_\rho B_{\mu\nu}.$$

(a) Show that $H_{\mu\nu\rho}$ is totally antisymmetric. Prove that $H_{\mu\nu\rho}$ is invariant under the *gauge transformations*

$$\delta B_{\mu\nu} = \partial_\mu \epsilon_\nu - \partial_\nu \epsilon_\mu.$$

(b) The above gauge transformations are peculiar: the gauge parameters themselves have a gauge invariance! Show that ϵ'_μ given as

$$\epsilon'_\mu = \epsilon_\mu + \partial_\mu \lambda,$$

generates the *same* gauge transformations as ϵ_μ.

(c) Use light-cone coordinates and momentum space to argue that $\epsilon^+(p)$ can be set to zero for a suitable choice of $\lambda(p)$. Thus, the effective gauge symmetry of the Kalb–Ramond field is generated by the gauge parameters $\epsilon^I(p)$ and $\epsilon^-(p)$.

(d) Consider the spacetime action principle

$$S \sim \int d^D x \left(-\frac{1}{6} H_{\mu\nu\rho} H^{\mu\nu\rho}\right).$$

Find the field equation for $B_{\mu\nu}$, and write it in momentum space.

(e) What are the suitable light-cone gauge conditions for $B^{\mu\nu}$? Bearing in mind the results of part (c), show that these gauge conditions can be implemented using the gauge invariance. Analyze the equations of motion, show that $p^2 B^{\mu\nu} = 0$, and find the components of $B^{\mu\nu}$ which represent truly independent degrees of freedom.

(f) Argue that the one-particle states of the Kalb–Ramond field are

$$\sum_{I,J=2}^{D-1} \zeta_{IJ}\, a^{IJ\dagger}_{p+,\, PT}\, |\Omega\rangle .\tag{1}$$

What kind of matrix is ζ_{IJ}?

Problem 10.7 Massive vector field.[†]

The purpose of this problem is to understand the massive version of Maxwell fields. We will see that in D-dimensional spacetime a massive vector field has $(D-1)$ degrees of freedom.

Consider the action $S = \int d^D x\, \mathcal{L}$ with

$$\mathcal{L} = -\frac{1}{4} F_{\mu\nu} F^{\mu\nu} - \frac{1}{2} m^2 A_\mu A^\mu - \frac{1}{2} \partial_\mu \phi\, \partial^\mu \phi + m A^\mu\, \partial_\mu \phi .$$

The first term in $\mathcal{L}$ is the familiar one for the Maxwell field. The second looks like a mass term for the Maxwell field, but alone would not suffice. The extra terms show the real scalar field ϕ that, as we shall see, is *eaten* to give the gauge field a mass.

(a) Show that the Lagrangian $\mathcal{L}$ is invariant under the infinitesimal gauge transformation

$$\delta A_\mu = \partial_\mu \epsilon , \qquad \delta \phi = \dots,$$

where the dots denote an expression that you must determine. While the gauge field has the familiar Maxwell gauge transformation, it is unusual to have a real scalar field with a gauge transformation.

(b) Vary the action and write down the field equations for A^μ and for ϕ.
(c) Argue that the gauge transformations allow us to set $\phi = 0$. Since the field ϕ disappears from sight, we say it is *eaten*. What do the field equations in part (b) simplify into?
(d) Write the simplified equations in momentum space and show that for $p^2 \neq -m^2$ there are no nontrivial solutions, while for $p^2 = -m^2$ the solution implies that there are $D - 1$ degrees of freedom. (It may be useful to use a Lorentz transformation to represent the vector p^μ which satisfies $p^2 = -m^2$ as a vector that has a component only in one direction.)

The relativistic quantum point particle

To prepare ourselves for quantizing the string, we study the light-cone gauge quantization of the relativistic point particle. We set up the quantum theory by requiring that the Heisenberg operators satisfy the classical equations of motion. We show that the quantum states of the relativistic point particle coincide with the one-particle states of the quantum scalar field. Moreover, the Schrödinger equation for the particle wavefunctions coincides with the classical scalar field equations. Finally, we set up light-cone gauge Lorentz generators.

11.1 Light-cone point particle

In this section we study the classical relativistic point particle using the light-cone gauge. This is, in fact, a much easier task than the one we faced in Chapter 9, where we examined the classical relativistic string in the light-cone gauge. Our present discussion will allow us to face the complications of quantization in the simpler context of the particle. Many of the ideas needed to quantize the string are also needed to quantize the point particle.

The action for the relativistic point particle was studied in Chapter 5. Let us begin our analysis with the expression given in equation (5.15), where an arbitrary parameter τ is used to parameterize the motion of the particle:

$$S = -m \int_{\tau_i}^{\tau_f} \sqrt{-\eta_{\mu\nu} \frac{dx^\mu}{d\tau} \frac{dx^\nu}{d\tau}} \, d\tau . \tag{11.1}$$

In writing the above action, we have set $c = 1$. We will also set $\hbar = 1$ when appropriate. Finally, the parameter τ will be dimensionless, just as it was chosen to be for the relativistic string starting in Chapter 9. We can simplify our notation by writing

$$\eta_{\mu\nu} \frac{dx^\mu}{d\tau} \frac{dx^\nu}{d\tau} = \eta_{\mu\nu} \dot{x}^\mu \dot{x}^\nu = \dot{x}^2 . \tag{11.2}$$

Thinking of τ as a time variable and of the $x^\mu(\tau)$ as coordinates, the action S defines a Lagrangian L as

$$S = \int_{\tau_i}^{\tau_f} L \, d\tau , \quad L = -m\sqrt{-\dot{x}^2} . \tag{11.3}$$

As usual, the momentum is obtained by differentiating the Lagrangian with respect to the velocity:

$$p_\mu = \frac{\partial L}{\partial \dot{x}^\mu} = \frac{m \dot{x}_\mu}{\sqrt{-\dot{x}^2}} \,. \tag{11.4}$$

The Euler–Lagrange equations which arise from the Lagrangian are

$$\frac{dp_\mu}{d\tau} = 0 \,. \tag{11.5}$$

All components of the momentum are constants of the motion. Given (11.4) one readily checks that the momentum components satisfy the constraint

$$p^2 + m^2 = 0 \,. \tag{11.6}$$

To define the light-cone gauge for the particle, we set the coordinate x^+ of the particle proportional to τ:

$$\text{light-cone gauge condition:} \quad x^+ = \frac{1}{m^2} p^+ \tau \,. \tag{11.7}$$

The factor of m^2 on the right-hand side is needed to get the units to work. Now consider the $+$ component of equation (11.4):

$$p^+ = \frac{m}{\sqrt{-\dot{x}^2}} \, \dot{x}^+ = \frac{1}{\sqrt{-\dot{x}^2}} \frac{p^+}{m} \,. \tag{11.8}$$

Cancelling the common factor of p^+, and squaring, we find

$$\dot{x}^2 = -\frac{1}{m^2} \,. \tag{11.9}$$

This result helps us simplify the expression (11.4) for the momentum:

$$p_\mu = m^2 \dot{x}_\mu \,. \tag{11.10}$$

The appearance of m^2, as opposed to m, is due to our choice of unitless τ. The equation of motion (11.5) then gives

$$\ddot{x}_\mu = 0 \,. \tag{11.11}$$

Expanding the constraint (11.6) in light-cone components,

$$-2p^+ p^- + p^I p^I + m^2 = 0 \quad \longrightarrow \quad p^- = \frac{1}{2p^+} (p^I p^I + m^2) \,. \tag{11.12}$$

The value of p^- is determined by p^+ and the components p^I of the transverse momentum $\vec{p}_T$. Having solved for p^-, equation (11.10) gives

$$\frac{dx^-}{d\tau} = \frac{1}{m^2} p^- \,, \tag{11.13}$$

which is integrated to find

$$x^-(\tau) = x_0^- + \frac{p^-}{m^2}\,\tau\,, \tag{11.14}$$

where x_0^- is a constant of integration. Equation (11.10) also gives $dx^I/d\tau = p^I/m^2$, which is integrated to give

$$x^I(\tau) = x_0^I + \frac{p^I}{m^2}\,\tau\,, \tag{11.15}$$

where x_0^I is a constant of integration. Note that the light-cone gauge condition (11.7) implies that $x^+(\tau)$ has no constant piece x_0^+.

The specification of the motion of the point particle is now complete. Equation (11.12) tells us that the momentum is completely determined once we fix p^+ and the components p^I of the transverse momentum $\vec{p}_T$. The motion in the x^- direction is determined by (11.14), once we fix the value of x_0^-. The transverse motion is determined by the $x^I(\tau)$, or the x_0^I, since we presume to know the p^I. For a symmetric treatment of coordinates versus momenta in the quantum theory, we choose the x^I as dynamical variables. Our independent dynamical variables for the point particle are therefore

$$\text{Dynamical variables:} \quad \left(x^I,\quad x_0^-,\quad p^I,\quad p^+\right). \tag{11.16}$$

11.2 Heisenberg and Schrödinger pictures

Traditionally, there are two main approaches to the understanding of time evolution in quantum mechanics. In the Schrödinger picture, the state of a system evolves in time, while operators remain unchanged. In the Heisenberg picture, it is the operators which evolve in time, while the state remains unchanged. Of the two, the Heisenberg picture is more closely related to classical mechanics, where the dynamical variables (which become operators in quantum mechanics) evolve in time. Both the Schrödinger and the Heisenberg pictures will be useful in developing the quantum theories of the relativistic point particle and the relativistic string. Because we would like to exploit our understanding of classical dynamics in developing the quantum theories, we will begin by focusing on the Heisenberg picture.

Both the Heisenberg and the Schrödinger picture make use of the same state space. Whereas in the Heisenberg picture the state which represents a particular physical system is fixed in time, in the Schrödinger picture the state of a system is constantly changing direction in the state space in a manner which is determined by the Schrödinger equation. Although we generally think of the operators in the Schrödinger picture as being time independent, there are those which depend *explicitly* on time and therefore have time dependence. These operators are formed from time-independent operators and the variable t. For example, the position and momentum operators q and p are time independent. But the operator $\mathcal{O} = q + p\,t$ has explicit time dependence. If it has explicit time dependence, even the Hamiltonian $H(p, q; t)$ can be a time-dependent operator.

Now, as we move from the Schrödinger to the Heisenberg picture, we will encounter operators with two types of time dependence. As we noted earlier, Heisenberg operators have time dependence, but this time dependence can be both *implicit* and *explicit*. The Heisenberg equivalent of a time-independent Schrödinger operator is said to have implicit time dependence. This implicit time dependence is due to our folding into the operators the time dependence which, in the Schrödinger picture, is present in the state. If a Heisenberg operator is explicitly time dependent it is because the explicit time dependence of the corresponding Schrödinger operator has been carried over.

When we pass from the Schrödinger to the Heisenberg picture, the time-independent Schrödinger operators q and p, for example, become $q(t)$ and $p(t)$, respectively. The Schrödinger commutator $[q, p] = i$ turns into the commutator

$$[\,q(t),\ p(t)\,] = i\,. \tag{11.17}$$

Although $q(t)$ and $p(t)$ depend on time, their time dependence is implicit. If $\xi(t)$ is a Heisenberg operator arising from a time-*independent* Schrödinger operator, the time evolution of $\xi(t)$ is governed by

$$i\frac{d\xi(t)}{dt} = \Big[\,\xi(t),\ H(p(t), q(t); t)\,\Big]\,. \tag{11.18}$$

Here $H(p(t), q(t); t)$ is the Heisenberg Hamiltonian corresponding to the possibly time-dependent Schrödinger Hamiltonian $H(p, q; t)$.

If $\mathcal{O}(t)$ is the Heisenberg operator which corresponds to an explicitly time-*dependent* Schrödinger operator, then the time evolution of $\mathcal{O}(t)$ is given by

$$i\frac{d\mathcal{O}(t)}{dt} = i\,\frac{\partial\mathcal{O}}{\partial t} + \Big[\,\mathcal{O}(t),\ H(p(t), q(t); t)\,\Big]\,. \tag{11.19}$$

This equation reduces to (11.18) when the operator has no explicit time dependence. If the Hamiltonian $H(p(t), q(t))$ has no explicit time dependence, then we can use (11.18) with $\xi = H$, to find

$$\frac{d}{dt}H(p(t), q(t)) = 0\,. \tag{11.20}$$

In this case the Hamiltonian is a constant of the motion.

The discussion above must be supplemented by rules to pass from Schrödinger to Heisenberg operators. Assume that the Schrödinger Hamiltonian $H(p, q)$ is time independent. In this case a state $|\Psi\rangle$ at time $t = 0$ evolves in time, and is given at time t by

$$|\Psi, t\rangle = e^{-iHt}|\Psi\rangle\,. \tag{11.21}$$

Quick calculation 11.1 Confirm that $|\Psi, t\rangle$ satisfies the Schrödinger equation

$$i\frac{d}{dt}|\Psi, t\rangle = H|\Psi, t\rangle. \tag{11.22}$$

It is clear from (11.21) that the operator e^{iHt} brings time-dependent states to rest:

$$e^{iHt}|\Psi, t\rangle = |\Psi\rangle. \tag{11.23}$$

If we act with this operator on the product $\alpha|\Psi, t\rangle$, where α is a Schrödinger operator, we find

$$e^{iHt}\alpha|\Psi, t\rangle = e^{iHt}\alpha\, e^{-iHt}|\Psi\rangle \equiv \alpha(t)|\Psi\rangle, \tag{11.24}$$

where $\alpha(t) = e^{iHt}\alpha\, e^{-iHt}$ is the Heisenberg operator corresponding to the Schrödinger operator α. This definition also applies for a Schrödinger operator $\alpha_S(t)$ that has explicit time dependence, in which case the corresponding Heisenberg operator is $\alpha_H(t) = e^{iHt}\alpha_S(t)\, e^{-iHt}$. The construction ensures that given a set of Schrödinger operators that satisfy certain commutation relations, the corresponding Heisenberg operators satisfy the same commutation relations.

Quick calculation 11.2 If $[\alpha_1, \alpha_2] = \alpha_3$ holds for Schrödinger operators α_1, α_2, and α_3, show that $[\alpha_1(t), \alpha_2(t)] = \alpha_3(t)$ holds for the corresponding Heisenberg operators.

This result holds even if the Hamiltonian is time dependent (Problem 11.2). It justifies the commutator in (11.17), noting that the constant right-hand side is not affected by the rule turning a Schrödinger operator into a Heisenberg operator.

11.3 Quantization of the point particle

We now develop a quantum theory from the classical theory of the relativistic point particle. We will define the relevant Schrödinger and Heisenberg operators, including the Hamiltonian, and describe the state space. All of this will be done in the light-cone gauge.

Our first step is to choose a set of time-independent Schrödinger operators. A reasonable choice is provided by the dynamical variables in (11.16):

$$\text{time-independent Schrödinger operators}: \quad \left(x^I, \quad x_0^-, \quad p^I, \quad p^+\right). \tag{11.25}$$

We could include hats to distinguish the operators from their eigenvalues, but this will not be necessary in most cases. We parameterize the trajectory of a point particle using τ, so the associated Heisenberg operators are:

$$\text{Heisenberg operators:} \quad \left(x^I(\tau), \quad x_0^-(\tau), \quad p^I(\tau), \quad p^+(\tau) \right). \qquad (11.26)$$

We postulate the following commutation relations for the Schrödinger operators:

$$[x^I, p^J] = i\,\eta^{IJ}, \quad [x_0^-, p^+] = i\,\eta^{-+} = -i\,, \qquad (11.27)$$

with all other commutators set equal to zero. The first commutator is the familiar commutator of spatial coordinates with the corresponding spatial momenta (recall that $\eta^{IJ} = \delta^{IJ}$). The second commutator is well motivated, after all, x_0^- is treated as a spatial coordinate in the light-cone, and p^+ is the corresponding conjugate momentum. The second commutator, just like the first one, has an η carrying the indices of the coordinate and the momentum.

The Heisenberg operators, as explained earlier, satisfy the same commutation relations as the Schrödinger operators:

$$[x^I(\tau), p^J(\tau)] = i\,\eta^{IJ}, \quad [x_0^-(\tau), p^+(\tau)] = -i\,, \qquad (11.28)$$

with all other commutators set equal to zero.

We have discussed the operators that correspond to the independent observables of the classical theory. But just as there are classical observables which depend on those independent ones, there are also quantum operators which are constructed from the set of independent Schrödinger operators, and time. These additional operators are $x^+(\tau), x^-(\tau)$ and p^-. These operators are *defined* using the quantum analogs of equations (11.7), (11.14), and (11.12):

$$x^+(\tau) \equiv \frac{p^+}{m^2}\,\tau\,, \qquad (11.29)$$

$$x^-(\tau) \equiv x_0^- + \frac{p^-}{m^2}\,\tau\,, \qquad (11.30)$$

$$p^- \equiv \frac{1}{2p^+}\left(p^I p^I + m^2 \right). \qquad (11.31)$$

Note that p^- is time independent. Both $x^+(\tau)$ and $x^-(\tau)$ are time-dependent Schrödinger operators.

The commutation relations involving the operators $x^+(\tau), x^-(\tau)$, and p^- are determined by the postulated commutation relations in (11.27), along with the defining equations (11.29)–(11.31). The decision to choose the operators in (11.25) as the independent operators of our quantum theory was very significant. For example, if we had chosen x^+ and p^- to be independent operators, we might have been led to write a commutation relation $[x^+, p^-] = -i$. In our present framework, however, this quantity vanishes, since $[p^+, p^I] = 0$.

We have not yet determined the Hamiltonian H. Since p^- is the light-cone energy (see (2.94)), we expect it to generate x^+ evolution:

$$\frac{\partial}{\partial x^+} \longleftrightarrow p^- .$$

(11.32)

Although x^+ is light-cone time, we are parameterizing our operators with τ, so we expect H to generate τ evolution, which is related, but is not the same as x^+ evolution. Since $x^+ = p^+ \tau / m^2$, we can anticipate that τ evolution will be generated by

$$\frac{\partial}{\partial \tau} = \frac{p^+}{m^2} \frac{\partial}{\partial x^+} \longleftrightarrow \frac{p^+}{m^2} p^- .$$

(11.33)

We therefore postulate the Heisenberg Hamiltonian

$$H(\tau) = \frac{p^+(\tau)}{m^2} \, p^-(\tau) = \frac{1}{2m^2}\Big(p^I(\tau) p^I(\tau) + m^2\Big).$$

(11.34)

Note that $H(\tau)$ has no explicit time dependence. Equation (11.20) applies, and as a result, the Hamiltonian is actually time independent.

Let us now make sure that this Hamiltonian generates the expected equations of motion. First we check that H gives the correct time evolution of the Heisenberg operators (11.26) which arise from the time-independent Schrödinger operators. The equation governing the time evolution of those operators is (11.18). Let us begin with p^+ and p^I:

$$i \frac{dp^+(\tau)}{d\tau} = \big[p^+(\tau), H(\tau) \big] = 0,$$

$$i \frac{dp^I(\tau)}{d\tau} = \big[p^I(\tau), H(\tau) \big] = 0.$$

(11.35)

Both of these commutators vanish because H is a function of $p^I(\tau)$ alone, and all the momenta commute. Equations (11.35) are good news, because the classical momenta p^+ and p^I are constants of the motion. This allows us to write $p^I(\tau) = p^I$ and $p^+(\tau) = p^+$. We now test the τ development of the Heisenberg operator $x^I(\tau)$:

$$i \frac{dx^I(\tau)}{d\tau} = \Big[x^I(\tau), \frac{1}{2m^2}\big(p^J p^J + m^2 \big) \Big] = i \frac{p^I}{m^2} .$$

(11.36)

Here, we have used $[x^I, p^J p^J] = [x^I, p^J]p^J + p^J[x^I, p^J] = 2i\, p^I$. Cancelling the common factor of i in (11.36), we find

$$\frac{dx^I(\tau)}{d\tau} = \frac{p^I}{m^2} .$$

(11.37)

This result is in accord with our classical expectations and allows us to write

$$x^I(\tau) = x_0^I + \frac{p^I}{m^2} \tau ,$$

(11.38)

where x_0^I is an operator without any time dependence. Finally, we must examine $x_0^-(\tau)$. Since $x_0^-(\tau)$ commutes with $p^I(\tau)$,

$$i\frac{dx_0^-(\tau)}{d\tau} = \left[x_0^-(\tau), \frac{1}{2m^2}\left(p^I p^I + m^2\right)\right] = 0. \qquad (11.39)$$

As expected, this operator is a constant of the motion, and we can write $x_0^-(\tau) = x_0^-$. So as far as the operators in (11.26) are concerned, our ansatz for H functions properly as a Hamiltonian.

We now turn to the remaining operators $x^+(\tau)$, $x^-(\tau)$, and $p^-(\tau)$. Of these, $p^-(\tau)$ is a function of the p^I and p^+ only and is therefore time independent. It is easy to to see that the commutator with H vanishes, so we have nothing left to check for this operator. The Heisenberg operators $x^+(\tau)$ and $x^-(\tau)$ both arise from Schrödinger operators with explicit time dependence, so we use (11.19) to calculate their time evolution. For example:

$$i\frac{dx^-(\tau)}{d\tau} = i\frac{\partial x^-}{\partial \tau} + \left[x^-(\tau), H(\tau)\right]. \qquad (11.40)$$

Since $x^-(\tau) \equiv x_0^- + p^- \tau/m^2$ and both x_0^- and p^- commute with the p^I, we see that $[x^-(\tau), H(\tau)] = 0$. Consequently,

$$\frac{dx^-(\tau)}{d\tau} = \frac{p^-}{m^2}, \qquad (11.41)$$

which is the expected result. Similarly, since $x^+(\tau) = p^+\tau/m^2$, we find that $[x^+(\tau), H(\tau)] = 0$, and therefore

$$\frac{dx^+(\tau)}{d\tau} = \frac{\partial x^+}{\partial \tau} = \frac{p^+}{m^2}. \qquad (11.42)$$

These computations show that our ansatz (11.34) for the Hamiltonian generates the expected equations of operator evolution.

Quick calculation 11.3 We introduced x_0^I in (11.38) as a constant operator. Show that $dx_0^I/d\tau$ must be calculated by viewing x_0^I as the explicitly time-dependent Heisenberg operator defined by (11.38).

To complete our construction of the point particle quantum theory we must develop the state space, set up the Schrödinger equation, and define physical states. The time-independent states of the quantum theory are labeled by the eigenvalues of a maximal set of commuting operators. For the set of operators introduced in (11.25), a maximal commuting subset can include only one element from the pair (x^-, p^+), and one element from each of the pairs (x^I, p^I). Because it is usually convenient to work in momentum space, we will work with the operators p^+ and p^I. So we write the states as

$$\text{states of the quantum point particle:} \quad |p^+, \vec{p}_T\rangle, \qquad (11.43)$$

where p^+ is the eigenvalue of the p^+ operator, and $\vec{p}_T$ is the transverse momentum, the components of which are the eigenvalues of the p^I operators:

$$\hat{p}^+|p^+, \vec{p}_T\rangle = p^+|p^+, \vec{p}_T\rangle, \quad \hat{p}^I|p^+, \vec{p}_T\rangle = p^I|p^+, \vec{p}_T\rangle. \tag{11.44}$$

In light of (11.31), these equations imply

$$\hat{p}^-|p^+, \vec{p}_T\rangle = \frac{1}{2p^+}\left(p^I p^I + m^2\right)|p^+, \vec{p}_T\rangle. \tag{11.45}$$

In addition, the Hamiltonian (11.34) acting on the states gives

$$H|p^+, \vec{p}_T\rangle = \frac{1}{2m^2}\left(p^I p^I + m^2\right)|p^+, \vec{p}_T\rangle. \tag{11.46}$$

It then follows that the time-dependent states

$$\exp\left(-i\frac{1}{2m^2}\left(p^I p^I + m^2\right)\tau\right)|p^+, \vec{p}_T\rangle \tag{11.47}$$

satisfy the Schrödinger equation. They are the *physical* time-dependent states associated with the states in (11.43).

More generally, we consider time-dependent superpositions of the basis states in (11.43):

$$|\Psi, \tau\rangle = \int dp^+ d\vec{p}_T \, \psi(\tau, p^+, \vec{p}_T)|p^+, \vec{p}_T\rangle. \tag{11.48}$$

Since p^+ and $\vec{p}_T$ are continuous variables, an integral is necessary. To produce a general τ-dependent superposition, we introduced the arbitrary function $\psi(\tau, p^+, \vec{p}_T)$. In fact, this function is the momentum-space wavefunction associated with the state $|\Psi, \tau\rangle$. Indeed, with dual bras $\langle p^+, \vec{p}_T|$ defined to satisfy

$$\langle p'^+, \vec{p}_T'|p^+, \vec{p}_T\rangle = \delta(p'^+ - p^+)\,\delta\,(\vec{p}_T' - \vec{p}_T), \tag{11.49}$$

we see that

$$\langle p^+, \vec{p}_T|\Psi, \tau\rangle = \psi(\tau, p^+, \vec{p}_T). \tag{11.50}$$

The Schrödinger equation for the state $|\Psi, \tau\rangle$ is

$$i\frac{\partial}{\partial\tau}|\Psi, \tau\rangle = H|\Psi, \tau\rangle. \tag{11.51}$$

Using the state in (11.48) and the Hamiltonian in (11.34), we find

$$\int dp^+ d\vec{p}_T \left[i\frac{\partial}{\partial\tau}\psi(\tau, p^+, \vec{p}_T) - \frac{1}{2m^2}\left(p^I p^I + m^2\right)\psi(\tau, p^+, \vec{p}_T)\right]|p^+, \vec{p}_T\rangle = 0. \tag{11.52}$$

Since the basis vectors $|p^+, \vec{p}_T\rangle$ are all linearly independent, the expression within brackets must vanish for all values of the momenta:

$$i\frac{\partial}{\partial\tau}\psi\,(\tau, p^+, \vec{p}_T) = \frac{1}{2m^2}\left(p^I p^I + m^2\right)\psi(\tau, p^+, \vec{p}_T). \tag{11.53}$$

We recognize this equation as a Schrödinger equation for the momentum-space wavefunction $\psi(\tau, p^+, \vec{p}_T)$. If the wavefunction satisfies the Schrödinger equation, the state $|\Psi, \tau\rangle$ is a physical time-dependent state. This completes our development of the point particle quantum theory.

11.4 Quantum particle and scalar particles

The states of the quantum point particle given in (11.43) may remind you of the one-particle states (10.66) in the quantum theory of the scalar field. This is actually a fundamental correspondence.

> There is a natural identification of the quantum states of a relativistic point particle of mass m with the one-particle states of the quantum theory of a scalar field of mass m:
>
> $$|p^+, \vec{p}_T\rangle \longleftrightarrow a^\dagger_{p^+, p_T} |\Omega\rangle. \tag{11.54}$$

The identification is possible because the labels of the point particle states match with the labels of the creation operators which generate the one-particle states of the scalar quantum field theory. The correspondence between the quantum point particle and the quantum scalar field theory can be extended from the state space to the operators that act on the state space. The quantum point particle theory has operators p^+, p^I, and p^-, and so does the quantum field theory, as shown in (10.67). If we identify the state spaces using (11.54), then the two sets of operators give the same eigenvalues. This makes the identification natural.

The above observations lead us to conclude that the states of the quantum point particle and the one-particle states of the scalar field theory are indistinguishable. Because it contains creation operators that can act multiple times on the vacuum state, the scalar field theory has multiparticle states that did not arise in our quantization of the point particle. Indeed, there are no creation operators in the theory of the quantum point particle. Because it provides a natural description of multiparticle states, the scalar field theory can be said to be a more complete theory.

How could we have anticipated that the one-particle states of a quantum *scalar* field theory would match those of the quantum point particle? The answer is quite interesting: the Schrödinger equation for the quantum point particle wavefunctions has the form of the classical field equation for the scalar field. More precisely:

> There is a canonical correspondence between the quantum point particle wave-functions and the classical scalar field, such that the Schrödinger equation for the quantum point particle wavefunctions becomes the classical field equation for the scalar field.

One element of this correspondence is the classical field equation for the scalar field. In light-cone gauge, this equation takes the form (10.31):

$$\left(i \frac{\partial}{\partial \tau} - \frac{1}{2m^2} (p^I p^I + m^2) \right) \phi(\tau, p^+, \vec{p}_T) = 0. \tag{11.55}$$

This differential equation is first order in τ. The other element in the correspondence is this Schrödinger equation (11.53). The two equations are identical once we identify the wavefunction $\psi(\tau, p^+, \vec{p}_T)$ and the scalar field $\phi(\tau, p^+, \vec{p}_T)$:

$$\psi(\tau, p^+, \vec{p}_T) \longleftrightarrow \phi(\tau, p^+, \vec{p}_T). \qquad (11.56)$$

This is the claimed correspondence.

The quantization of the point particle is an example of *first quantization*. In first quantization, the coordinates and momenta of classical mechanics are turned into quantum operators and a state space is constructed. Generically, the result is a set of one-particle states. *Second quantization* refers to the quantization of a classical field theory, the result of which is a quantum field theory with field operators and multiparticle states. Our analysis allows us to see how second quantization follows after first quantization. A first quantization of the classical point particle mechanics gives one-particle states. We then reinterpret the Schrödinger equation for the associated wavefunctions as the classical field equation for a scalar field. A second quantization, this time of the classical field theory, gives us the set of multiparticle states.

So far we have only quantized the *free* relativistic point particle. All quantum states, including the multiparticle ones obtained by second quantization, represent free particles. How do we get interactions between the particles? Such processes are included in the scalar field theory by adding interaction terms to the action. All the terms that we have included so far are quadratic in the fields. The interaction terms include three or more fields. Since the quantum point particle state space does not include multiparticle states, the description of interactions in the language of first quantization is not straightforward. On the other hand, in the framework of quantum field theory interactions are dealt with very naturally.

11.5 Light-cone momentum operators

Since the point particle Lagrangian L in (11.3) depends only on τ derivatives of the coordinates, it is invariant under the translations

$$\delta x^\mu(\tau) = \epsilon^\mu, \qquad (11.57)$$

with ϵ^μ constant. The conserved charge associated with this symmetry transformation is the momentum p_μ of the particle. This follows from (8.16) and (11.4).

What happens to conserved charges in the quantum theory? They become quantum operators with a remarkable property: they generate, via commutation, a quantum version of the symmetry transformation that gave rise to them classically!

This property is most apparent if we use a framework where the manifest Lorentz invariance of the classical theory is preserved in the quantization. This is *not* the framework we have used to quantize the point particle. In light-cone gauge quantization, the x^0 and x^1 coordinates of the particle are afforded special treatment. This hides the Lorentz invariance of the theory from plain view. We will not discuss fully the Lorentz covariant quantization of the point particle. A few remarks will suffice for our present purposes. The covariant quantization of the string is discussed in some detail in Chapter 24.

In the Lorentz covariant quantization of the point particle, we have Heisenberg operators $x^\mu(\tau)$ and $p^\mu(\tau)$. Note that even the time coordinate $x^0(\tau)$ becomes an operator! The commutation relations are

$$[x^\mu(\tau), p^\nu(\tau)] = i\,\eta^{\mu\nu}, \tag{11.58}$$

as well as

$$[x^\mu(\tau), x^\nu(\tau)] = 0 \quad \text{and} \quad [p^\mu(\tau), p^\nu(\tau)] = 0. \tag{11.59}$$

Equation (11.58) is reasonable. The indices match, which ensures consistency with Lorentz covariance. Moreover, when μ and ν take spatial values, the commutation relations are the familiar ones. We already know that (11.58) is not consistent with the light-cone gauge commutators of Section 11.3. We saw there that $[x^+(\tau), p^-(\tau)] = 0$, while (11.58) would predict a nonzero result. An equality of two objects which carry Lorentz indices applies when the indices run over the light-cone values $+$, $-$, and I. The equation $R^{\mu\nu} = S^{\mu\nu}$, gives, for example, $R^{+-} = S^{+-}$ (Problem 10.3). As a result, equation (11.58) indeed gives $[x^+(\tau), p^-(\tau)] = i\eta^{+-} = -i$. Let us now check that the operator $p^\mu(\tau)$ generates translations. More precisely, we check that $i\epsilon_\rho\, p^\rho(\tau)$ generates the translation (11.57):

$$\delta x^\mu(\tau) = \left[i\epsilon_\rho\, p^\rho(\tau), x^\mu(\tau)\right] = i\epsilon_\rho\,(-i\eta^{\rho\mu}) = \epsilon^\mu. \tag{11.60}$$

This is an elegant result, but it is by no means clear that it carries over to our light-cone gauge quantization. We must find out whether the light-cone gauge momentum operators generate translations.

For this purpose, we expand the generator $i\epsilon_\rho\, p^\rho(\tau)$ in light-cone components:

$$i\epsilon_\rho\, p^\rho(\tau) = -i\epsilon^-\, p^+ - i\epsilon^+\, p^- + i\epsilon^I\, p^I. \tag{11.61}$$

We have dropped the τ arguments from the momenta because they are τ-independent. Note that here, p^- is given by (11.31). Let us test (11.60) with $\epsilon^I \neq 0$, and $\epsilon^+ = \epsilon^- = 0$:

$$\delta x^\mu(\tau) = i\epsilon^I [\, p^I, x^\mu(\tau)\,]. \tag{11.62}$$

We would expect that $\delta x^J(\tau) = \epsilon^J$ and that $\delta x^+(\tau) = \delta x^-(\tau) = 0$. All these expectations are realized. Choosing $\mu = J$, and using the commutator (11.28) we find $\delta x^J(\tau) = \epsilon^J$. To compute the action on $x^+(\tau)$ and $x^-(\tau)$, we must use their definitions:

$$x^+(\tau) = \frac{p^+}{m^2}\tau, \quad x^-(\tau) = x_0^- + \frac{p^-}{m^2}\tau. \tag{11.63}$$

Recalling that p^I commutes with all momenta and with x_0^-, we confirm that $\delta x^+(\tau) = \delta x^-(\tau) = 0$.

Quick calculation 11.4 Test (11.60) with $\epsilon^- \neq 0$ and $\epsilon^+ = \epsilon^I = 0$. To do this compute $\delta x^\mu(\tau) = -i\epsilon^-[\, p^+, x^\mu(\tau)\,]$. Confirm that $\delta x^-(\tau) = \epsilon^-$ and that all other coordinates are not changed.

It remains to see whether p^- generates the expected translations. Since p^- is a nontrivial function of other momenta, there is some scope for complications! This time we consider the transformations that are generated using (11.60) with $\epsilon^+ \neq 0$ and $\epsilon^- = \epsilon^I = 0$:

$$\delta x^\mu(\tau) = -i\epsilon^+ [\, p^- , x^\mu(\tau)\,]. \tag{11.64}$$

The naive expectation $\delta x^+(\tau) = \epsilon^+$ is not realized: choosing $\mu = +$ and using (11.63) we see that

$$\delta x^+(\tau) = -i\epsilon^+ \left[\, p^- , p^+ \frac{\tau}{m^2} \right] = 0. \tag{11.65}$$

Not only is $x^+(\tau)$ left unchanged, but the other components, which naively should be left unchanged, are not:

$$\delta x^I(\tau) = -i\epsilon^+ \left[\, p^- , x^I(\tau) \right] = -i\epsilon^+ \frac{1}{2p^+}(-2ip^I) = -\epsilon^+ \frac{p^I}{p^+}, \tag{11.66}$$

$$\delta x^-(\tau) = -i\epsilon^+ \left[\, p^- , x_0^- + \frac{p^-}{m^2}\tau \right] = -i\epsilon^+ [\, p^- , x_0^-] = -\epsilon^+ \frac{p^-}{p^+}. \tag{11.67}$$

In these calculations only one step requires some explanation. How do we find $[\, p^- , x_0^-]$? The only reason p^- does not commute with x_0^- is that p^- depends on p^+. In fact, what we need to know is the commutator $[x_0^-, 1/p^+]$. This can be found as follows:

$$\left[x_0^- , \frac{1}{p^+} \right] = x_0^- \frac{1}{p^+} - \frac{1}{p^+}x_0^- = \frac{1}{p^+}p^+x_0^- \frac{1}{p^+} - \frac{1}{p^+}x_0^- p^+ \frac{1}{p^+}$$
$$= \frac{1}{p^+}[\, p^+, x_0^-]\frac{1}{p^+} = \frac{i}{p^{+2}}. \tag{11.68}$$

Quick calculation 11.5 Use (11.68) to show that

$$[x_0^- , p^-] = i\frac{p^-}{p^+}. \tag{11.69}$$

Equations (11.65), (11.66), and (11.67) show that p^- does not generate the expected transformations. What happened? It turns out that p^- actually generates both a translation *and* a reparameterization of the world-line of the particle. We know that the particle action is invariant under changes of parameterization $\tau \to \tau'(\tau)$. When we described symmetries in Chapter 8, however, we exhibited them as changes in the dynamical variables of the system. A change in parameterization can also be described in that way. Writing $\tau \to \tau' = \tau + \lambda(\tau)$, with λ infinitesimal, we note that the plausible change

$$x^\mu(\tau) \longrightarrow x^\mu(\tau + \lambda(\tau)) = x^\mu(\tau) + \lambda(\tau)\partial_\tau x^\mu(\tau), \tag{11.70}$$

leads us to write

$$\delta x^\mu(\tau) = \lambda(\tau)\partial_\tau x^\mu(\tau). \tag{11.71}$$

We claim that these are *symmetries* of the point particle theory. Actually, the variation (11.71) does not leave the point particle Lagrangian invariant. The Lagrangian changes into a total τ derivative (Problem 11.4), but this, in fact, suffices to have a symmetry (see Problem 8.9).

Let us now show that p^- generates a translation plus a reparameterization. The expected translation was $\delta x^+ = \epsilon^+$. On the other hand, from (11.71), a reparameterization of x^+ gives $\delta x^+ = \lambda \partial_\tau x^+$. Bearing in mind (11.65), the expected translation plus the reparameterization give zero variation, so,

$$0 = \epsilon^+ + \lambda \partial_\tau x^+(\tau) = \epsilon^+ + \lambda \frac{p^+}{m^2} \quad \longrightarrow \quad \lambda = -\frac{m^2}{p^+}\epsilon^+. \tag{11.72}$$

The reparameterization parameter λ turns out to be a constant. We can now use this result to "explain" the transformations (11.66) and (11.67) that p^- generates on x^I and on x^-. For these coordinates there is no translation, but the reparameterization still applies. Therefore,

$$\delta x^I(\tau) = \lambda \, \partial_\tau x^I(\tau) = -\frac{m^2}{p^+}\epsilon^+ \frac{p^I}{m^2} = -\epsilon^+ \frac{p^I}{p^+}, \tag{11.73}$$

$$\delta x^-(\tau) = \lambda \, \partial_\tau x^-(\tau) = -\frac{m^2}{p^+}\epsilon^+ \frac{p^-}{m^2} = -\epsilon^+ \frac{p^-}{p^+}, \tag{11.74}$$

in perfect agreement with the transformations generated by p^-. We can also understand why p^- does not change x^+. If x^+ had been changed by a constant ϵ^+, the new x^+ coordinate would not satisfy the light-cone gauge condition whereby x^+ is just proportional to τ. In fact, p^- generates a translation plus the compensating transformation needed to preserve the light-cone gauge condition! That transformation turned out to be a reparameterization of the world-line.

One final remark about momentum operators. The Lorentz covariant momentum operators that we used to motivate our analysis generate simple translations and commute among each other. It follows directly that, using light-cone *coordinates*, the operators $p^\pm = (p^0 \pm p^1)/\sqrt{2}$ and the transverse p^I all commute. The light-cone *gauge* momentum operators we discussed above are completely different objects. They had an intricate action on coordinates, and p^- was defined in terms of the transverse momenta and p^+. Nevertheless, all the light-cone gauge momentum operators still commute. They obey the same commutation relations that the covariant operators do when expressed using light-cone coordinates.

11.6 Light-cone Lorentz generators

In Section 8.5 we determined the conserved charges that are associated with the Lorentz invariance of the relativistic string Lagrangian. Similar charges exist for the relativistic point particle. As we found in (8.52), the infinitesimal Lorentz transformations of the point particle coordinates $x^\mu(\tau)$ take the form

$$\delta x^\mu(\tau) = \epsilon^{\mu\nu} x_\nu(\tau), \tag{11.75}$$

where $\epsilon^{\mu\nu} = -\epsilon^{\nu\mu}$ are a set of infinitesimal constants. The associated Lorentz charges are given by

$$M^{\mu\nu} = x^\mu(\tau)p^\nu(\tau) - x^\nu(\tau)p^\mu(\tau), \qquad (11.76)$$

as you may have derived in Problem 8.5. These charges are conserved classically. The quantum charges are expected to generate Lorentz transformations of the coordinates. Again, it is straightforward to see this using the operators of Lorentz covariant quantization. In this case, the quantum charges are given by (11.76) with $x^\mu(\tau)$ and $p^\mu(\tau)$ taken to be the Heisenberg operators introduced earlier and satisfying the commutation relations (11.58) and (11.59). Both $x^\mu(\tau)$ and $p^\mu(\tau)$ are Hermitian operators. The Lorentz charges $M^{\mu\nu}$ are Hermitian as well:

$$(M^{\mu\nu})^\dagger = p^\nu(\tau)x^\mu(\tau) - p^\mu(\tau)x^\nu(\tau) = M^{\mu\nu}, \qquad (11.77)$$

since the two constants induced by rearranging the coordinates and momenta back to the original form cancel out.

Quick calculation 11.6 Show that

$$\left[M^{\mu\nu}, x^\rho(\tau)\right] = i\,\eta^{\mu\rho}\,x^\nu(\tau) - i\,\eta^{\nu\rho}\,x^\mu(\tau). \qquad (11.78)$$

This commutator helps us check that the quantum Lorentz charges generate Lorentz transformations:

$$\delta x^\rho(\tau) = \left[-\frac{i}{2}\,\epsilon_{\mu\nu}M^{\mu\nu}, x^\rho(\tau)\right]$$

$$= \frac{1}{2}\,\epsilon_{\mu\nu}\left(\eta^{\mu\rho}\,x^\nu(\tau) - \eta^{\nu\rho}\,x^\mu(\tau)\right)$$

$$= \frac{1}{2}\,\epsilon^{\rho\nu}x_\nu(\tau) + \frac{1}{2}\,\epsilon^{\rho\mu}x_\mu(\tau) = \epsilon^{\rho\nu}x_\nu(\tau). \qquad (11.79)$$

Equation (11.78) can be used in light-cone *coordinates* by simply using light-cone indices. For example,

$$[M^{-I}, x^+(\tau)] = i\eta^{-+}x^I(\tau) - i\eta^{I+}x^-(\tau) = -i\,x^I(\tau), \qquad (11.80)$$

since $\eta^{I+} = 0$. The operator M^{-I} here is a Lorentz covariant generator expressed in light-cone coordinates. It is *not* a light-cone gauge Lorentz generator. Those we have not yet constructed.

Given a set of quantum operators, it is interesting to calculate their commutators. In quantum mechanics, for example, you learned that the components L_x, L_y, and L_z of the angular momentum satisfy a set of commutation relations ($[L_x, L_y] = iL_z$, and others) that define the Lie algebra of angular momentum. The momentum operators p^μ considered earlier define a very simple Lie algebra: they all commute. We would like to know what is the commutator of two Lorentz generators. The computation takes a few steps (Problem 11.5). Using equation (11.78), and a similar equation for $[M^{\mu\nu}, p^\rho]$, one finds that the commutator can be written as a linear combination of four Lorentz generators:

$$[M^{\mu\nu}, M^{\rho\sigma}] = i\eta^{\mu\rho}M^{\nu\sigma} - i\eta^{\nu\rho}M^{\mu\sigma} + i\eta^{\mu\sigma}M^{\rho\nu} - i\eta^{\nu\sigma}M^{\rho\mu}. \qquad (11.81)$$

This result defines the Lorentz Lie algebra. Equation (11.81) must be satisfied by the analogous operators $M^{\mu\nu}$ of any Lorentz invariant quantum theory. If it is not possible to construct such operators, the theory is not Lorentz invariant. This will be crucial to our quantization of the string, for requiring that (11.81) holds imposes additional restrictions, which have significant physical consequences.

Quick calculation 11.7 Since $M^{\mu\nu} = -M^{\nu\mu}$, the left-hand side of (11.81) changes sign under the exchange of μ and ν. Verify that the right-hand side also changes sign under this exchange.

We can now use (11.81) to determine the commutators of Lorentz charges in light-cone *coordinates*. The Lorentz generators are given by

$$M^{IJ}, \quad M^{+I}, \quad M^{-I}, \quad \text{and} \quad M^{+-} . \tag{11.82}$$

Consider, for example, the commutator $[M^{+-}, M^{+I}]$. To use (11.81) notice the structure of its right-hand side: each η contains one index from each of the generators in the left-hand side. For $[M^{+-}, M^{+I}]$, the only way to get a nonvanishing η is to use the $-$ from the first generator and the $+$ from the second generator. The nonvanishing term is the second one on the right-hand side of (11.81), and we find

$$[M^{+-}, M^{+I}] = -i\,\eta^{-+}M^{+I} = iM^{+I} . \tag{11.83}$$

Similarly,

$$[M^{-I}, M^{-J}] = 0 . \tag{11.84}$$

Here η must use the I and J indices, but then the other two indices must go into M giving us M^{--}, which vanishes by antisymmetry.

Quick calculation 11.8 Show that $M^{+-} = M^{10}$. This shows that M^{+-} generates a boost along the x^1 direction.

So far, we have considered the covariant Lorentz charges in light-cone *coordinates*. We must now find Lorentz charges for our light-cone *gauge* quantization of the particle. Our earlier discussion of the momenta suggests that we really face three questions.

(1) How are these charges going to be defined?
(2) What kind of transformations will they generate?
(3) Which commutation relations will they satisfy?

   In the remaining part of this section we will explore question (1) in detail. Before doing so, let us give brief answers to questions (2) and (3), leaving further analysis of these questions to Problems 11.6 and 11.7. The light-cone gauge Lorentz generators are expected to generate Lorentz transformations of coordinates and momentum, but in some cases, these transformations will be accompanied by reparameterizations of the world-line. Regarding (3), the light-cone gauge Lorentz generators will satisfy the same commutation relations that the covariant operators in light-cone coordinates do. This establishes that Lorentz symmetry holds in the light-cone theory of the quantum point particle. The success of the construction is not obvious *a priori*. It is not clear that the reduced set of light-cone gauge operators suffices to construct quantum Lorentz charges that generate Lorentz transformations (plus other transformations) and satisfy the Lorentz algebra.

The simplest guess for the light-cone gauge generators is to use light-cone coordinates in the covariant formula (11.76) and then replace $x^+(\tau)$, $x^-(\tau)$, and p^- using their light-cone gauge definitions in (11.29), (11.30), and (11.31). Let us try this prescription with M^{+-}:

$$M^{+-} \stackrel{?}{=} x^+(\tau)\, p^-(\tau) - x^-(\tau)\, p^+(\tau)$$

$$\stackrel{?}{=} \frac{p^+\tau}{m^2}\, p^- - \left(x_0^- + \frac{p^-}{m^2}\tau\right)p^+$$

$$\stackrel{?}{=} -x_0^-\, p^+ . \tag{11.85}$$

Since x_0^- and p^+ are τ-independent, so too is M^{+-}. We have a minor complication, however. The operator M^{+-} is not Hermitian: $(M^{+-})^\dagger - M^{+-} = [x_0^-, p^+] \neq 0$. This failure of Hermiticity illustrates how the use of the light-cone gauge can affect basic properties of operators. The covariant Lorentz generators were automatically Hermitian, the light-cone gauge generators are not. We are therefore motivated to define a Hermitian M^{+-} as

$$M^{+-} = -\frac{1}{2}(x_0^-\, p^+ + p^+ x_0^-) . \tag{11.86}$$

We take this to be the light-cone gauge Lorentz generator M^{+-}.

The most complicated of all generators is M^{-I}. It is the most interesting one as well. The prescription used for M^{+-} this time gives

$$M^{-I} \stackrel{?}{=} x^-(\tau)\, p^I - x^I(\tau)\, p^-$$

$$\stackrel{?}{=} \left(x_0^- + \frac{p^-}{m^2}\tau\right)p^I - \left(x_0^I + \frac{p^I\tau}{m^2}\right)p^-$$

$$\stackrel{?}{=} x_0^-\, p^I - x_0^I\, p^- . \tag{11.87}$$

As before, the τ-dependence vanishes, but we are left with a complicated result since p^- is a nontrivial function of the other momenta. We define M^{-I} as the Hermitian version of the operator obtained above:

$$M^{-I} \equiv x_0^-\, p^I - \frac{1}{2}\left(x_0^I\, p^- + p^-\, x_0^I\right) . \tag{11.88}$$

If the light-cone gauge Lorentz charges are to satisfy the Lorentz algebra we must have

$$[\, M^{-I},\, M^{-J}\,] = 0 , \tag{11.89}$$

as we noted in (11.84). Does M^{-I}, as defined by (11.88), satisfy this equation? The answer is yes, as you will see for yourself in Problem 11.6. This result is necessary to ensure Lorentz invariance of the quantum theory. All other commutators of Lorentz generators also work out correctly.

The calculation of $[\, M^{-I},\, M^{-J}\,]$ in quantum string theory is fairly complicated, but the answer is very interesting. It turns out that this commutator is zero if and only if the string propagates in a spacetime of some particular dimension and, furthermore, only if the definition of mass is changed in such a way that we can find massless gauge fields in the spectrum of the open string! String theory is such a constrained theory that it is only Lorentz invariant for a fixed spacetime dimensionality.

Problems

Problem 11.1 Equation of motion for Heisenberg operators.

Assume that the Schrödinger Hamiltonian $H = H(p, q)$ is time independent. In this case the time-independent Schrödinger operator ξ yields a Heisenberg operator $\xi(t) = e^{iHt}\xi e^{-iHt}$. Show that this operator satisfies the equation

$$i\frac{d\xi(t)}{dt} = \left[\xi(t),\, H(p(t), q(t))\right].$$

This computation proves that equation (11.18) holds for time-independent Hamiltonians.

Problem 11.2 Heisenberg operators and time-dependent Hamiltonians.

When the Schrödinger Hamiltonian $H = H(p, q; t)$ is time dependent, time evolution of states is generated by a unitary operator $U(t)$:

$$|\Psi, t\rangle = U(t)|\Psi\rangle, \tag{1}$$

where $U(t)$ bears some nontrivial relation to H. Here $|\Psi\rangle$ denotes the state at zero time and $U(0) = 1$.

(a) Use the Schrödinger equation to show that

$$i\frac{dU(t)}{dt} = HU(t). \tag{2}$$

Let $U \equiv U(t)$, for brevity. Since U^{-1} acting on $|\Psi, t\rangle$ gives a time-independent state, considerations similar to those given for (11.24) lead us to define the Heisenberg operator corresponding to the Schrödinger operator α as

$$\alpha(t) = U^{-1}\alpha\, U. \tag{3}$$

(b) Let ξ be a time-independent Schrödinger operator, and let $\xi(t)$ be the corresponding Heisenberg operator, defined using (3). Show that

$$i\frac{d\xi(t)}{dt} = \left[\xi(t),\, H(p(t), q(t); t)\right].$$

This computation proves that equation (11.18) holds for time-dependent Hamiltonians.

(c) If $[\alpha_1, \alpha_2] = \alpha_3$ holds for Schrödinger operators α_1, α_2, and α_3, show that $[\alpha_1(t), \alpha_2(t)] = \alpha_3(t)$ holds for the corresponding Heisenberg operators.

Problem 11.3 Classical dynamics in Hamiltonian language.

Consider a classical phase space (q, p), a trajectory $(q(t), p(t))$, and an observable $v(q(t), p(t); t)$. From the standard rules of differentiation,

$$\frac{dv}{dt} = \frac{\partial v}{\partial t} + \frac{\partial v}{\partial p}\frac{dp}{dt} + \frac{\partial v}{\partial q}\frac{dq}{dt}. \tag{1}$$

With the Poisson bracket defined as

$$\{A, B\} = \frac{\partial A}{\partial q}\frac{\partial B}{\partial p} - \frac{\partial A}{\partial p}\frac{\partial B}{\partial q}, \tag{2}$$

show that

$$\frac{dv}{dt} = \frac{\partial v}{\partial t} + \{v, H\}. \tag{3}$$

Comparing this result to (11.19) we see the parallel between the time evolution of a general operator $\mathcal{O}$ and the classical Hamiltonian evolution of an observable v in phase space.

To derive (3) you need the classical equations of motion in Hamiltonian language. These can be obtained by demanding that

$$\int dt \Big(p(t)\dot{q}(t) - H(p(t), q(t); t)\Big)$$

be stationary for independent variations $\delta q(t)$ and $\delta p(t)$.

Problem 11.4 Reparameterization symmetries of the point particle.

Show that the variation $\delta x^\mu(\tau) = \lambda(\tau)\partial_\tau x^\mu(\tau)$ induces a variation δL of the point particle Lagrangian that can be written as

$$\delta L(\tau) = \partial_\tau\Big(\lambda(\tau)L(\tau)\Big).$$

This proves that the reparameterizations δx^μ are symmetries in the sense defined in Problem 8.9. Show, however, that the charges associated with these reparameterization symmetries vanish. When λ is τ-independent, the reparameterization is an infinitesimal constant τ translation. The conserved charge is then the Hamiltonian. Show directly that the Hamiltonian defined canonically from the point particle Lagrangian vanishes.

Problem 11.5 Lorentz generators and Lorentz algebra.

In this problem we consider the Lorentz covariant charges (11.76).

(a) Calculate the commutator $[M^{\mu\nu}, p^\rho]$.
(b) Calculate the commutator $[M^{\mu\nu}, M^{\rho\sigma}]$ and verify that (11.81) holds.
(c) Consider the Lorentz algebra in light cone *coordinates*. Give

$$[M^{\pm I}, M^{JK}], \quad [M^{\pm I}, M^{\mp J}], \quad [M^{+-}, M^{\pm I}], \quad \text{and} \quad [M^{\pm I}, M^{\pm J}].$$

Problem 11.6 Light-cone gauge commutator $[M^{-I}, M^{-J}]$ for the particle.

The purpose of the present calculation is to show that

$$[M^{-I}, M^{-J}] = 0. \tag{1}$$

(a) Verify that the light-cone gauge operator M^{-I} takes the form

$$M^{-I} = (x_0^- p^I - x_0^I p^-) + \frac{i}{2} \frac{p^I}{p^+}. \tag{2}$$

Set up now the computation of (1) distinguishing the two kinds of terms in (2). Calculate the contributions to the commutator from mixed terms and from the last term.

(b) Complete the computation of (1) by finding the contribution from the first term on the right-hand side of (2).

Problem 11.7 Transformations generated by the light-cone gauge Lorentz generators M^{+-} and M^{-I}.

(a) Calculate the commutator of M^{+-} (defined in (11.86)) with the light-cone coordinates $x^+(\tau)$, $x^-(\tau)$, and $x^I(\tau)$. Show that M^{+-} generates the expected Lorentz transformations of these coordinates.

(b) Calculate the commutator of M^{-I} with the light-cone coordinates $x^+(\tau)$, $x^-(\tau)$, and $x^J(\tau)$. Show that M^{-I} generates the expected Lorentz transformations together with a compensating reparameterization of the world-line. Calculate the parameter λ for this reparameterization. [*Hint:* the reparameterization takes the "hermiticized" form $\delta x^\mu(\tau) = \frac{1}{2}(\lambda \, \partial_\tau x^\mu + \partial_\tau x^\mu \, \lambda).$]

We finally quantize the relativistic open string. We use the light-cone gauge to set up commutation relations and to define a Hamiltonian in the Heisenberg picture. We discover an infinite set of creation and annihilation operators, labeled by an integer and a transverse vector index. The oscillators corresponding to the X^- direction are transverse Virasoro operators. The ambiguities we encounter in defining the quantum theory are fixed by requiring that the theory be Lorentz invariant. Among these ambiguities, the dimensionality of spacetime is fixed to the value 26, and the mass formula is shifted slightly from its classical counterpart such that the spectrum admits massless photon states. The spectrum also contains a tachyon state, which indicates the instability of the D25-brane.

12.1 Light-cone Hamiltonian and commutators

We are at long last in a position to quantize the relativistic string. We have acquired considerable intuition for the dynamics of classical relativistic strings, and we have examined in detail how to quantize the simpler, but still nontrivial, relativistic point particle. Moreover, having taken a brief look into the basics of scalar, electromagnetic, and gravitational quantum fields in the light-cone gauge, we will be able to appreciate the implications of quantum open string theory. In this chapter we will deal with open strings. We will assume throughout the presence of a space-filling D-brane. In the next chapter we will quantize the closed string.

Just as before, we will interpret the classical equations of motion in the light-cone gauge as equations for the appropriate Heisenberg operators. It is therefore necessary for us to review the results of our light-cone analysis of the classical relativistic string.

We found a class of world-sheet parameterizations (9.27) for which the equations of motion are wave equations $\ddot{X}^\mu - X^{\mu''} = 0$. This remarkable simplification came at the expense of two constraints: $(\dot{X} \pm X')^2 = 0$. With these constraints, the momentum densities are simple derivatives of the coordinates:

$$\mathcal{P}^{\sigma\mu} = -\frac{1}{2\pi\alpha'}X^{\mu'}, \quad \mathcal{P}^{\tau\mu} = \frac{1}{2\pi\alpha'}\dot{X}^\mu. \tag{12.1}$$

These equations hold in all gauges within the class we considered. In particular, they are true in the light-cone gauge. For open strings in the light-cone gauge, we set $X^+ = 2\alpha' p^+ \tau$

and solved for X^- in terms of the transverse coordinates X^I. Indeed, using (9.65) with $\beta = 2$, we have

$$\dot{X}^- = \frac{1}{2\alpha'}\frac{1}{2p^+}(\dot{X}^I\dot{X}^I + X^{I'}X^{I'}).$$ (12.2)

This gives us an explicit expression for $\mathcal{P}^{\tau-}$:

$$\mathcal{P}^{\tau-} = \frac{1}{2\pi\alpha'}\dot{X}^- = \frac{1}{2\pi\alpha'}\frac{1}{2\alpha'}\frac{1}{2p^+}(2\pi\alpha')^2\left(\mathcal{P}^{\tau I}\mathcal{P}^{\tau I} + \frac{X^{I'}X^{I'}}{(2\pi\alpha')^2}\right)$$

$$= \frac{\pi}{2p^+}\left(\mathcal{P}^{\tau I}\mathcal{P}^{\tau I} + \frac{X^{I'}X^{I'}}{(2\pi\alpha')^2}\right).$$ (12.3)

These equations will soon become useful.

As a first step in defining a quantum theory of the light-cone relativistic string, we must give the list of Schrödinger operators. Motivated by the list (11.25) of Schrödinger operators for the quantum point particle, we choose our τ-independent Schrödinger operators to be

$$\text{Schrödinger operators:} \left(X^I(\sigma), \quad x_0^-, \quad \mathcal{P}^{\tau I}(\sigma), \quad p^+ \right).$$ (12.4)

The associated Heisenberg operators are then

$$\text{Heisenberg operators:} \left(X^I(\tau, \sigma), \quad x_0^-(\tau), \quad \mathcal{P}^{\tau I}(\tau, \sigma), \quad p^+(\tau) \right).$$ (12.5)

Because the operators (12.4) have no explicit τ dependence, neither do the Heisenberg operators (12.5). As in the case of the point particle, we expect x_0^- and p^+ to be fully τ-independent Heisenberg operators.

Now we set up the commutation relations. For the Schrödinger operators $X^I(\sigma)$ and $\mathcal{P}^{\tau I}(\sigma)$ we must face the fact that these operators have σ dependence. It is reasonable to demand that such operators fail to commute only if they are at the same point along the string. We do not expect (simultaneous) measurements at different points on the string to interfere with each other. Therefore we set

$$\left[X^I(\sigma), \mathcal{P}^{\tau J}(\sigma') \right] = i\eta^{IJ}\delta(\sigma - \sigma').$$ (12.6)

Here the delta function is being used to implement the constraint that the commutator must vanish when $\sigma \neq \sigma'$. We had to use a Dirac delta function, as opposed to a Kronecker delta, since σ is a continuous variable. Equation (12.6) is naturally supplemented with the commutation relations

$$\left[X^I(\sigma), X^J(\sigma') \right] = \left[\mathcal{P}^{\tau I}(\sigma), \mathcal{P}^{\tau J}(\sigma') \right] = 0,$$ (12.7)

and

$$[x_0^-, p^+] = -i. \tag{12.8}$$

The operators x_0^- and p^+ commute with all of the other Schrödinger operators:

$$[x_0^-, X^I(\sigma)] = [x_0^-, \mathcal{P}^{\tau I}(\sigma)] = [p^+, X^I(\sigma)] = [p^+, \mathcal{P}^{\tau I}(\sigma)] = 0. \tag{12.9}$$

For the associated Heisenberg operators, the only nonvanishing equal-time commutation relations are therefore

$$[X^I(\tau, \sigma), \mathcal{P}^{\tau J}(\tau, \sigma')] = i\,\eta^{IJ}\,\delta(\sigma - \sigma'), \tag{12.10}$$

as well as

$$[x_0^-(\tau), p^+(\tau)] = -i. \tag{12.11}$$

All other commutators vanish:

$$[X^I(\tau, \sigma), X^J(\tau, \sigma')] = [\mathcal{P}^{\tau I}(\tau, \sigma), \mathcal{P}^{\tau J}(\tau, \sigma')] = 0,$$
$$[x_0^-(\tau), X^I(\tau, \sigma)] = [x_0^-(\tau), \mathcal{P}^{\tau I}(\tau, \sigma)] = 0,$$
$$[p^+(\tau), X^I(\tau, \sigma)] = [p^+(\tau), \mathcal{P}^{\tau I}(\tau, \sigma)] = 0. \tag{12.12}$$

We must now invent the Hamiltonian. Our Hamiltonian should generate τ translation. From our experience with the point particle, we know that p^- generates X^+ translation. But in the light-cone gauge $X^+ = 2\alpha' p^+ \tau$, so

$$\frac{\partial}{\partial \tau} = \frac{\partial X^+}{\partial \tau} \frac{\partial}{\partial X^+} = 2\alpha' p^+ \frac{\partial}{\partial X^+}. \tag{12.13}$$

It follows that the Hamiltonian that generates change in τ is expected to be

$$H = 2\alpha' p^+ p^- = 2\alpha' p^+ \int_0^\pi d\sigma\, \mathcal{P}^{\tau -}. \tag{12.14}$$

This will indeed turn out to be the correct string Hamiltonian. Using (12.3), the Hamiltonian can be written more explicitly as the Heisenberg operator

$$H(\tau) = \pi\alpha' \int_0^\pi d\sigma \left(\mathcal{P}^{\tau I}(\tau, \sigma) \mathcal{P}^{\tau I}(\tau, \sigma) + \frac{X^{I\prime}(\tau, \sigma) X^{I\prime}(\tau, \sigma)}{(2\pi\alpha')^2} \right). \tag{12.15}$$

H must generate quantum equations of motion that are operator versions of the classical equations of motion. H is very simple when expressed in terms of the transverse

Virasoro modes of Chapter 9. There we saw that $L_0^\perp = 2\alpha' p^+ p^-$ (equation (9.78)), so (12.14) immediately gives

$$H = L_0^\perp . \tag{12.16}$$

This expression for the Hamiltonian is perhaps the most memorable one, though, as we will see later on, the true Hamiltonian is slightly different. The operator products $\mathcal{P}\mathcal{P}$ and $X'X'$ in (12.15) are actually ambiguous operators and need careful definition. Additionally, Lorentz invariance will require the subtraction of a calculable constant from H.

Now that we have a plausible candidate for the Hamiltonian, we have to derive the equations of motion. Any Heisenberg operator $\xi(\tau, \sigma)$ which arises from a time-independent Schrödinger operator $\xi(\sigma)$ must satisfy

$$i\,\dot{\xi}(\tau, \sigma) = [\,\xi(\tau, \sigma)\,,\,H(\tau)\,]\,, \tag{12.17}$$

where $H(\tau)$ is given in (12.15). Since $H(\tau)$ is built from Heisenberg operators that have no explicit time dependence, we can substitute $H(\tau)$ for $\xi(\tau, \sigma)$ in (12.17). We conclude that it is completely time independent: $H(\tau) = H$. Furthermore, we can see that $x_0^-(\tau)$ and $p^+(\tau)$ commute with H. They are therefore time independent operators, so we will henceforth denote them by x_0^- and p^+. The commutator (12.11) then becomes

$$\left[\,x_0^-\,,\,p^+\right] = -i\,. \tag{12.18}$$

The Heisenberg equation of motion for $X^I(\tau, \sigma)$ is

$$i\dot{X}^I(\tau, \sigma) = [X^I(\tau, \sigma)\,,\,H(\tau)] = \left[\,X^I(\tau, \sigma)\,,\,\pi\alpha' \int_0^\pi d\sigma'\,\mathcal{P}^{\tau J}(\tau, \sigma')\mathcal{P}^{\tau J}(\tau, \sigma')\right],$$

where we dropped the second term in H since it commutes with $X^I(\tau, \sigma)$:

$$\left[X^I(\tau, \sigma)\,,\,X^{J\prime}(\tau, \sigma')\right] = \frac{\partial}{\partial\sigma'}\left[X^I(\tau, \sigma)\,,\,X^J(\tau, \sigma')\right] = 0\,. \tag{12.19}$$

We also reinserted the time parameter in $H(\tau)$, choosing a time that gives easily evaluated equal-time commutators. Making use of (12.10) we find

$$i\dot{X}^I(\tau, \sigma) = \pi\alpha' \cdot 2 \cdot \int_0^\pi d\sigma'\,\mathcal{P}^{\tau J}(\tau, \sigma')\,i\,\eta^{IJ}\delta(\sigma - \sigma')\,. \tag{12.20}$$

Performing the integral and cancelling the common factor of i, we find

$$\dot{X}^I(\tau, \sigma) = 2\pi\alpha'\,\mathcal{P}^{\tau I}(\tau, \sigma)\,. \tag{12.21}$$

Happily, this coincides with the classical equation of motion (12.1). The other equations of motion can be checked in a similar fashion. For example, you can calculate $\dot{\mathcal{P}}^{\tau I}$ and use the result to verify that

$$\ddot{X}^I - X^{I\prime\prime} = 0 \tag{12.22}$$

is the quantum equation of motion (Problem 12.1). As we turn classical string theory into a quantum theory, the classical boundary conditions become operator equations. For example, the Neumann boundary conditions

$$\partial_\sigma X^I(\tau, \sigma) = 0, \quad \sigma = 0, \pi, \tag{12.23}$$

are taken literally as the condition that the operator $\partial_\sigma X^I(\tau, \sigma)$ vanishes at the open string endpoints.

We learned in Chapter 9 that the linear combinations of derivatives $(\dot{X}^I \pm X^{I'})$ are particularly simple and useful. We conclude this section by calculating commutators of these derivatives. We begin by using (12.21) to rewrite the commutator (12.10) as

$$\left[X^I(\tau, \sigma), \, \dot{X}^J(\tau, \sigma') \right] = 2\pi\alpha' i \, \eta^{IJ} \, \delta(\sigma - \sigma'). \tag{12.24}$$

Taking the σ derivative of this equation yields

$$\left[X^{I'}(\tau, \sigma), \, \dot{X}^J(\tau, \sigma') \right] = 2\pi\alpha' i \, \eta^{IJ} \, \frac{d}{d\sigma}\delta(\sigma - \sigma'). \tag{12.25}$$

Differentiating $\left[X^I(\tau, \sigma), X^J(\tau, \sigma') \right] = 0$ with respect to σ and σ' and recalling that $\left[\mathcal{P}^{\tau I}(\tau, \sigma), \mathcal{P}^{\tau J}(\tau, \sigma') \right] = 0$, we find that τ and σ derivatives of the coordinates separately commute among themselves:

$$\left[X^{I'}(\tau, \sigma), X^{J'}(\tau, \sigma') \right] = \left[\dot{X}^I(\tau, \sigma), \, \dot{X}^J(\tau, \sigma') \right] = 0. \tag{12.26}$$

Now we examine the commutator

$$\left[(\dot{X}^I + X^{I'})(\tau, \sigma), \, (\dot{X}^J + X^{J'})(\tau, \sigma') \right], \tag{12.27}$$

which as a consequence of (12.26) equals

$$\left[\dot{X}^I(\tau, \sigma), \, X^{J'}(\tau, \sigma') \right] + \left[X^{I'}(\tau, \sigma), \, \dot{X}^J(\tau, \sigma') \right]. \tag{12.28}$$

The second term is given by (12.25). The first term equals

$$-\left[X^{J'}(\tau, \sigma'), \, \dot{X}^I(\tau, \sigma) \right] = -(2\pi\alpha')i\eta^{JI} \frac{d}{d\sigma'}\delta(\sigma' - \sigma) = 2\pi\alpha' i\eta^{IJ} \frac{d}{d\sigma}\delta(\sigma - \sigma').$$

To obtain this result we noted that a σ' derivative can be traded for minus a σ derivative when it acts on a function of $(\sigma - \sigma')$. Moreover, we used $\delta(x) = \delta(-x)$. We now see that both terms in (12.28) are equal, so

$$\left[(\dot{X}^I + X^{I'})(\tau, \sigma), \, (\dot{X}^J + X^{J'})(\tau, \sigma') \right] = 4\pi\alpha' i\eta^{IJ} \frac{d}{d\sigma}\delta(\sigma - \sigma'). \tag{12.29}$$

In fact, more generally, we have found that

$$\left[(\dot{X}^I \pm X^{I'})(\tau, \sigma), \, (\dot{X}^J \pm X^{J'})(\tau, \sigma') \right] = \pm 4\pi\alpha' i\eta^{IJ} \frac{d}{d\sigma}\delta(\sigma - \sigma'), \tag{12.30}$$

since only cross terms contribute. Finally,

$$\left[(\dot{X}^I \pm X^{I'})(\tau, \sigma) , \ (\dot{X}^J \mp X^{J'})(\tau, \sigma') \right] = 0 \,. \tag{12.31}$$

Equations (12.30) and (12.31) hold for $\sigma, \sigma' \in [0, \pi]$.

12.2 Commutation relations for oscillators

The commutation relations written so far are delicate to handle because they involve field operators and use delta functions. They are an infinite set of relations which hold for continuous values of σ and σ'. It is therefore useful to recast them in discrete form, namely, as a denumerable set of commutation relations. For this purpose we will examine the mode expansions of Section 9.4. These followed from the classical wave equations and the boundary conditions for a space-filling D-brane. Since the wave equations and the boundary conditions continue to hold in the quantum theory, we can use the mode expansions in the quantum theory. The classical modes α_n^I, however, become quantum operators with nontrivial commutation relations.

Recall our solution (9.69) to the wave equation with Neumann boundary conditions:

$$X^I(\tau, \sigma) = x_0^I + \sqrt{2\alpha'}\, \alpha_0^I \tau + i\sqrt{2\alpha'} \sum_{n \neq 0} \frac{1}{n} \alpha_n^I \cos n\sigma \, e^{-in\tau} \,. \tag{12.32}$$

In addition, from (9.74) we have

$$(\dot{X}^I + X^{I'})(\tau, \sigma) = \sqrt{2\alpha'} \sum_{n \in \mathbb{Z}} \alpha_n^I e^{-in(\tau+\sigma)} \,, \quad \sigma \in [0, \pi] \,,$$
$$(\dot{X}^I - X^{I'})(\tau, \sigma) = \sqrt{2\alpha'} \sum_{n \in \mathbb{Z}} \alpha_n^I e^{-in(\tau-\sigma)} \,, \quad \sigma \in [0, \pi] \,. \tag{12.33}$$

The above equalities hold for $\sigma \in [0, \pi]$ because the open string coordinates are only defined for $\sigma \in [0, \pi]$. We will now construct a function of σ with period 2π which is naturally expressed in terms of open string coordinates. To do this, we *evaluate* the second equation above at $-\sigma$:

$$(\dot{X}^I - X^{I'})(\tau, -\sigma) = \sqrt{2\alpha'} \sum_{n \in \mathbb{Z}} \alpha_n^I e^{-in(\tau+\sigma)}, \quad \sigma \in [-\pi, 0] \,. \tag{12.34}$$

The range $\sigma \in [-\pi, 0]$ is required because otherwise the left-hand side is not defined. We now define the operator $A^I(\tau, \sigma)$ by

$$A^I(\tau, \sigma) \equiv \sqrt{2\alpha'} \sum_{n \in \mathbb{Z}} \alpha_n^I e^{-in(\tau+\sigma)} \,, \quad A^I(\tau, \sigma + 2\pi) = A^I(\tau, \sigma) \,. \tag{12.35}$$

The indicated periodicity is a direct consequence of the definition. It is now simple to relate A^I to open string coordinates over the length-2π interval $\sigma \in [-\pi, \pi]$. For $\sigma \in [0, \pi]$ we use the top equation in (12.33) and for $\sigma \in [-\pi, 0]$ we use (12.34):

$$A^I(\tau, \sigma) = \begin{cases} (\dot{X}^I + X^{I'})(\tau, \sigma) & \sigma \in [0, \pi] \\ (\dot{X}^I - X^{I'})(\tau, -\sigma) & \sigma \in [-\pi, 0]. \end{cases} \tag{12.36}$$

The operator A^I will be useful to determine the commutation relations for the α_n^I oscillators. For this, we must compute the commutator $[A^I(\tau, \sigma), A^J(\tau, \sigma')]$. Given (12.36) the evaluation for the full range $\sigma, \sigma' \in [-\pi, \pi]$ requires four computations:

$$\left[(\dot{X}^I + X^{I'})(\tau, \sigma), \ (\dot{X}^J + X^{J'})(\tau, \sigma') \right], \quad \sigma, \sigma' \in [0, \pi],$$

$$\left[(\dot{X}^I + X^{I'})(\tau, \sigma), \ (\dot{X}^J - X^{J'})(\tau, -\sigma') \right], \quad \sigma \in [0, \pi], \ \sigma' \in [-\pi, 0],$$

$$\left[(\dot{X}^I - X^{I'})(\tau, -\sigma), \ (\dot{X}^J + X^{J'})(\tau, \sigma') \right], \quad \sigma \in [-\pi, 0], \ \sigma' \in [0, \pi],$$

$$\left[(\dot{X}^I - X^{I'})(\tau, -\sigma), \ (\dot{X}^J - X^{J'})(\tau, -\sigma') \right], \quad \sigma, \sigma' \in [-\pi, 0]. \tag{12.37}$$

The first commutator, for $\sigma, \sigma' \in [0, \pi]$, is simply read from (12.30):

$$\left[(\dot{X}^I + X^{I'})(\tau, \sigma), \ (\dot{X}^J + X^{J'})(\tau, \sigma') \right] = 4\pi\alpha' i\eta^{IJ} \frac{d}{d\sigma} \delta(\sigma - \sigma'). \tag{12.38}$$

The last commutator, for $\sigma, \sigma' \in [-\pi, 0]$, is also obtained from (12.30):

$$\left[(\dot{X}^I - X^{I'})(\tau, -\sigma), \ (\dot{X}^J - X^{J'})(\tau, -\sigma') \right] = -4\pi\alpha' i\eta^{IJ} \frac{d}{d(-\sigma)} \delta(-\sigma + \sigma')$$

$$= 4\pi\alpha' i\eta^{IJ} \frac{d}{d\sigma} \delta(\sigma - \sigma'),$$

and coincides with the result (12.38) of the first commutator. For the second and third commutators in (12.2) we can use (12.31) to conclude that both of them vanish. In fact, the right-hand side of (12.38) also vanishes since in those cases σ and σ' cannot be equal. It follows that the result of all four commutators can be summarized as

$$\left[A^I(\tau, \sigma), A^J(\tau, \sigma') \right] = 4\pi\alpha' i\eta^{IJ} \frac{d}{d\sigma} \delta(\sigma - \sigma'), \quad \sigma, \sigma' \in [-\pi, \pi]. \tag{12.39}$$

Using (12.35) and cancelling a common factor of $2\alpha'$, the above result gives

$$\sum_{m', n' \in \mathbb{Z}} e^{-im'(\tau+\sigma)} e^{-in'(\tau+\sigma')} \left[\alpha_{m'}^I, \alpha_{n'}^J \right] = 2\pi i\eta^{IJ} \frac{d}{d\sigma} \delta(\sigma - \sigma'). \tag{12.40}$$

This equation holds for $\sigma, \sigma' \in [-\pi, \pi]$. In order to extract information we apply on both sides the integral operations

$$\frac{1}{2\pi} \int_0^{2\pi} d\sigma \, e^{im\sigma} \cdot \frac{1}{2\pi} \int_0^{2\pi} d\sigma' \, e^{in\sigma'}. \tag{12.41}$$

On the left-hand side of (12.40) the integrals pick the term with $m' = m$ and $n' = n$:

$$e^{-i(m+n)\tau} \left[\alpha_m^I, \alpha_n^J \right]. \tag{12.42}$$

On the right-hand side of (12.40) the integrals give

$$i \eta^{IJ} \frac{1}{2\pi} \int_0^{2\pi} d\sigma \, e^{im\sigma} \frac{d}{d\sigma} \int_0^{2\pi} d\sigma' \, e^{in\sigma'} \delta(\sigma - \sigma')$$

$$= i \eta^{IJ} \frac{1}{2\pi} \int_0^{2\pi} d\sigma \, e^{im\sigma} \frac{d}{d\sigma} e^{in\sigma} = -n \, \eta^{IJ} \frac{1}{2\pi} \int_0^{2\pi} d\sigma \, e^{i(m+n)\sigma} \quad (12.43)$$

$$= -n \, \eta^{IJ} \delta_{m+n,0} = m \, \eta^{IJ} \delta_{m+n,0}.$$

Equating our results (12.42) and (12.43), we find

$$[\alpha_m^I, \alpha_n^J] = m \, \eta^{IJ} \delta_{m+n,0} \, e^{+i(m+n)\tau} = m \, \eta^{IJ} \delta_{m+n,0}, \qquad (12.44)$$

since the Kronecker delta can be used to set $m = -n$. Therefore, the commutation relation is

$$\left[\alpha_m^I, \alpha_n^J \right] = m \, \eta^{IJ} \delta_{m+n,0}. \qquad (12.45)$$

This is the fundamental commutation relation between α modes. Note that α_0^I commutes with all other oscillators. This is quite reasonable: as shown in (9.52), α_0^I is proportional to the momentum of the string

$$\alpha_0^I = \sqrt{2\alpha'} \, p^I, \qquad (12.46)$$

and it is expected to have a nontrivial commutator with x_0^J only.

To complete the list of all possible commutators we must find the commutators between x_0^I and the oscillators α_n^J. For this, we consider equation (12.24), and integrate both sides of the equation over $\sigma \in [0, \pi]$. On the left-hand side, the terms with oscillators in $X^I(\tau, \sigma)$ give no contribution, and on the right-hand side the delta function disappears giving a factor of one:

$$\left[x_0^I + \sqrt{2\alpha'} \, \alpha_0^I \, \tau, \, \dot{X}^J(\tau, \sigma') \right] = 2\alpha' i \, \eta^{IJ}. \qquad (12.47)$$

Since $\dot{X}^J$ is a sum of terms that contain α_n^J, we have $[\alpha_0^I, \dot{X}^I] = 0$. Additionally, using the mode expansion of $\dot{X}^J$, equation (12.47) becomes

$$\sum_{n' \in \mathbb{Z}} \left[x_0^I, \alpha_{n'}^J \right] \cos n'\sigma' \, e^{-in'\tau} = \sqrt{2\alpha'} i \, \eta^{IJ}. \qquad (12.48)$$

Reorganizing the left-hand side of this equation we find

$$[x_0^I, \alpha_0^J] + \sum_{n'=1}^{\infty} \left[x_0^I, \, \alpha_{n'}^J e^{-in'\tau} + \alpha_{-n'}^J e^{in'\tau} \right] \cos n'\sigma = \sqrt{2\alpha'} i \, \eta^{IJ}. \qquad (12.49)$$

We apply to both sides of this equation the integral operation $\frac{1}{\pi} \int_0^\pi d\sigma \cos n\sigma$, with $n \geq 1$. We then find that

$$\left[x_0^I , \, \alpha_n^J e^{-in\tau} + \alpha_{-n}^J e^{in\tau} \right] = 0, \tag{12.50}$$

or equivalently

$$\left[x_0^I, \alpha_n^J \right] e^{-in\tau} + \left[x_0^I, \alpha_{-n}^J \right] e^{in\tau} = 0. \tag{12.51}$$

Since the left-hand side must vanish for all values of τ, each term must vanish separately (prove this!). It follows that

$$\left[x_0^I, \alpha_n^J \right] = 0 \quad \text{for} \quad n \neq 0. \tag{12.52}$$

Additionally, equation (12.49) gives

$$\left[x_0^I, \alpha_0^J \right] = \sqrt{2\alpha'}\, i \eta^{IJ}. \tag{12.53}$$

This, together with (12.46), gives the expected commutator

$$\left[x_0^I, \, p^J \right] = i \eta^{IJ}. \tag{12.54}$$

As in familiar quantum mechanics, the operators x_0^I and p^I are Hermitian:

$$(x_0^I)^\dagger = x_0^I, \quad (p^I)^\dagger = p^I. \tag{12.55}$$

The calculations we performed to obtain the commutation relations took quite a few steps, which we explained in detail. When we discuss closed strings, or open strings on general D-brane configurations, similar computations will be required. We will be able to carry them out using, with minimal modifications, the calculations we just did.

It is useful at this point to examine in detail the commutation relations (12.45) for the α_n^I modes. As we will show below, they are equivalent to those of an infinite set of creation and annihilation operators. To see this, we begin by defining *oscillators*, taking our inspiration from the classical variables introduced in (9.53):

$$\alpha_n^\mu = a_n^\mu \sqrt{n}, \quad \alpha_{-n}^\mu = a_n^{\mu*} \sqrt{n}, \quad n \geq 1. \tag{12.56}$$

In these equations, both the α and the a are classical variables. Now they become operators. Classical variables that are complex conjugates of each other become operators that are Hermitian conjugates of each other in the quantum theory. We can therefore preserve the first of the above definitions, but the second must be changed. For our light-cone modes $\mu = I$ we take

$$\alpha_n^I = a_n^I \sqrt{n} \quad \text{and} \quad \alpha_{-n}^I = a_n^{I\dagger} \sqrt{n}, \quad n \geq 1. \tag{12.57}$$

Note that, with this definition,

$$(\alpha_n^I)^\dagger = \alpha_{-n}^I, \quad n \in \mathbb{Z}. \tag{12.58}$$

This equation holds for $n = 0$ because α_0^I, being proportional to p^I, is also Hermitian. It is useful to emphasize that, while the α_n^I modes are defined for all integers n, the a_n^I and $a_n^{I\dagger}$ operators are only defined for positive n.

An important consequence of the above Hermiticity properties is that $X^I(\tau, \sigma)$, which used to be real in the classical theory, is now a Hermitian operator.

Quick calculation 12.1 Use the expansion (12.32) and the Hermiticity conditions (12.55) and (12.58) to show that

$$(X^I(\tau, \sigma))^\dagger = X^I(\tau, \sigma). \tag{12.59}$$

The i factor in front of the sum in (12.32) is needed for this calculation to work out.

We can now rephrase the commutation relation for the α modes in terms of the oscillators $(a_n^I, a_n^{I\dagger})$. For this purpose, rewrite (12.45) as

$$\left[\alpha_m^I, \alpha_{-n}^J \right] = m\, \delta_{m,n} \eta^{IJ}. \tag{12.60}$$

When m and n are integers of opposite signs the right-hand side vanishes, and the two operators in the commutator have mode numbers of the same sign. Therefore, we learn that

$$\left[a_m^I, a_n^J \right] = \left[a_m^{I\dagger}, a_n^{J\dagger} \right] = 0. \tag{12.61}$$

If both m and n are positive in (12.60) we find

$$\left[\sqrt{m}\, a_m^I, \sqrt{n}\, a_n^{J\dagger} \right] = m\delta_{m,n} \eta^{IJ}. \tag{12.62}$$

Moving the square roots to the right-hand side

$$\left[a_m^I, a_n^{J\dagger} \right] = \frac{m}{\sqrt{mn}} \delta_{m,n} \eta^{IJ}. \tag{12.63}$$

Since the right-hand side vanishes unless $m = n$, it simplifies to

$$\left[a_m^I, a_n^{J\dagger} \right] = \delta_{m,n}\, \eta^{IJ}. \tag{12.64}$$

This, together with (12.61), shows that $(a_m^I, a_m^{I\dagger})$ satisfy the commutation relations of the canonical annihilation and creation operators of a quantum simple harmonic oscillator. There is a pair of creation and annihilation operators for each value $m \geq 1$ of the mode number and for each transverse light-cone direction I. The commutation relations are diagonal: oscillators corresponding to different mode numbers, or to different light-cone coordinates, commute. If the mode numbers and the coordinate labels agree, the commutator is equal to one. In terms of the α operators, with $n \geq 1$:

α_n^I are annihilation operators,

α_{-n}^I are creation operators $(n \geq 1)$.   (12.65)

For future reference let us rewrite the expansion of $X^I(\tau, \sigma)$ in (12.32) in terms of creation and annihilation operators. Separating out the sum over all integers into sums over positive and negative integers, and using (12.46), we find

$$X^I(\tau, \sigma) = x_0^I + 2\alpha' p^I \tau + i\sqrt{2\alpha'} \sum_{n=1}^{\infty} \left(\alpha_n^I e^{-in\tau} - \alpha_{-n}^I e^{in\tau} \right) \frac{\cos n\sigma}{n}. \qquad (12.66)$$

Replacing α modes by the corresponding oscillators we obtain

$$X^I(\tau, \sigma) = x_0^I + 2\alpha' p^I \tau + i\sqrt{2\alpha'} \sum_{n=1}^{\infty} \left(a_n^I e^{-in\tau} - a_n^{I\dagger} e^{in\tau} \right) \frac{\cos n\sigma}{\sqrt{n}}. \qquad (12.67)$$

This is the expansion of the coordinate operator in terms of creation and annihilation operators.

Let us take stock of what we have learned. The list of operators we started with was given in (12.5). We have seen that the operators $X^I(\tau, \sigma)$ and $\mathcal{P}^{\tau I}(\tau, \sigma)$ can be traded for an infinite collection of oscillators, plus pairs of zero modes (x_0^I, p^I). Since the other two operators in the list, x_0^- and p^+, are also zero modes, the full set of basic operators of string theory is a collection of zero modes plus an infinite set of creation and annihilation operators. This result is so important that we will now derive it in a different way, showing explicitly how the quantum simple harmonic oscillators arise.

12.3 Strings as harmonic oscillators

Our aim here is to give a more physical derivation of the results obtained in the previous section. In particular, we will rederive the mode expansion (12.67) and the commutation relations between the operators in that expansion. These results followed from the fundamental commutation relation (12.10) together with the operator equations of motion (12.22) and the operator boundary conditions (12.23). Of these, the commutation relations (12.10) are perhaps the least intuitive, as they involve a delta function. In the derivation below there will be no delta function.

Here is our strategy. We will invent a simple Lagrangian that describes the dynamics of the light-cone coordinates X^I. This is not such a difficult task since we know the equations of motion of the X^I, their boundary conditions, and the definition of the canonical momenta $\mathcal{P}^{\tau I}$. Then we will expand the coordinate $X^I(\tau, \sigma)$ as a function of σ but with τ-dependent expansion coefficients. Using the Lagrangian, we will show that those expansion coefficients are actually the coordinates of harmonic oscillators that have ever-increasing energy! We will conclude by relating these oscillators to the creation and annihilation operators obtained in the previous analysis.

To set up the notation, we begin by reviewing the basic properties of the quantum harmonic oscillator. Let $q_n(t)$ be the coordinate of a classical simple harmonic oscillator, and let the action be given by

$$S_n = \int L_n(t)\, dt = \int dt \left(\frac{1}{2n}\dot{q}_n^2(t) - \frac{n}{2}q_n^2(t) \right). \tag{12.68}$$

We recognize this as a harmonic oscillator because the kinetic energy is proportional to the velocity-squared, and the potential energy is proportional to the coordinate-squared. For this Lagrangian, the momentum p_n conjugate to the coordinate q_n is

$$p_n = \frac{\partial L}{\partial \dot{q}_n} = \frac{1}{n}\dot{q}_n. \tag{12.69}$$

A little calculation now gives the Hamiltonian as

$$H_n(p_n, q_n) = p_n\dot{q}_n - L_n = \frac{n}{2}(p_n^2 + q_n^2). \tag{12.70}$$

In this equation, n plays the role of the frequency ω of the harmonic oscillator. To define the quantum oscillator, we introduce Schrödinger operators q_n and p_n, with the canonical commutation relation

$$[q_n, p_n] = i. \tag{12.71}$$

Creation and annihilation operators can be introduced as

$$a_n = \frac{1}{\sqrt{2}}(p_n - iq_n), \quad a_n^\dagger = \frac{1}{\sqrt{2}}(p_n + iq_n). \tag{12.72}$$

You should check that as a consequence of (12.71) the creation and annihilation operators satisfy the commutation relation

$$[a_n, a_n^\dagger] = 1. \tag{12.73}$$

Inverting the relations in (12.72), we find

$$q_n = \frac{i}{\sqrt{2}}(a_n - a_n^\dagger), \quad p_n = \frac{1}{\sqrt{2}}(a_n + a_n^\dagger). \tag{12.74}$$

These can be used to rewrite the Hamiltonian H_n in terms of the creation and annihilation operators. We find the familiar result

$$H_n = n\left(a_n^\dagger a_n + \frac{1}{2} \right). \tag{12.75}$$

We can now consider the Heisenberg operators $(a_n(t), a_n^\dagger(t))$ that are associated with the Schrödinger operators $(a_n, a_n^\dagger)$. As emphasized in Section 11.2, the Heisenberg operators satisfy the same commutation relations as the Schrödinger operators:

$$\left[a_n(t), a_n^\dagger(t) \right] = 1. \tag{12.76}$$

The Heisenberg equation of motion for $a_n(t)$ is

$$\dot{a}_n(t) = i\,[\,H_n(t)\,,a_n(t)\,] = in\,\Big[a_n^\dagger(t)a_n(t)\,,a_n(t)\Big] = -in\,a_n(t)\,. \qquad (12.77)$$

This differential equation is solved by

$$a_n(t) = e^{-int}a_n(0) = e^{-int}a_n\,, \qquad (12.78)$$

where a_n is the constant Heisenberg operator that equals $a_n(t)$ at $t=0$. A similar calculation gives

$$a_n^\dagger(t) = e^{int}a_n^\dagger(0) = e^{int}a_n^\dagger\,. \qquad (12.79)$$

As you can see, the angular frequency of oscillation is indeed equal to n. Finally, with these results and (12.74), we can find the explicit time dependence of the operator $q_n(t)$:

$$q_n(t) = \frac{i}{\sqrt{2}}(a_n(t) - a_n^\dagger(t)) = \frac{i}{\sqrt{2}}\Big(a_n\,e^{-int} - a_n^\dagger\,e^{int}\Big)\,. \qquad (12.80)$$

This concludes our review of the quantum simple harmonic oscillator.

We now turn to the discussion of an action that encodes the dynamics of the transverse light-cone coordinates $X^I(\tau,\sigma)$. We claim that the action is simply given by

$$S = \int d\tau d\sigma\,\mathcal{L} = \frac{1}{4\pi\alpha'}\int d\tau \int_0^\pi d\sigma\Big(\dot{X}^I\dot{X}^I - X^{I'}X^{I'}\Big)\,. \qquad (12.81)$$

This action is much simpler than the Nambu–Goto action; it has no square root, for example. The first term, which contains time derivatives, represents kinetic energy. The second term, which contains spatial derivatives, represents potential energy. The canonical momentum associated with X^I coincides with the momentum density $\mathcal{P}^{\tau I}$:

$$\frac{\partial\mathcal{L}}{\partial\dot{X}^I} = \frac{1}{2\pi\alpha'}\dot{X}^I = \mathcal{P}^{\tau I}\,, \qquad (12.82)$$

as we see comparing with (12.1). This confirms that $\mathcal{L}$ is correctly normalized. The equations of motion for X^I follow by variation:

$$\delta S = \frac{1}{2\pi\alpha'}\int d\tau \int_0^\pi d\sigma\Big(\partial_\tau(\delta X^I)\dot{X}^I - \partial_\sigma(\delta X^I)X^{I'}\Big)\,. \qquad (12.83)$$

Restricting ourselves to variations where the initial and final positions are fixed, we can drop the total τ derivatives and find

$$\delta S = -\frac{1}{2\pi\alpha'}\int d\tau\Big[\,(X^{I'}\delta X^I)\Big|_0^\pi + \int_0^\pi d\sigma\,\delta X^I\Big(\ddot{X}^I - X^{I''}\Big)\Big]\,. \qquad (12.84)$$

It is clear that the requirement that the action be stationary gives us both the wave equation (12.22) for the coordinates and the boundary conditions at the string endpoints. As a final check of the consistency of the action we calculate the Hamiltonian:

$$H = \int_0^\pi d\sigma\,\mathcal{H} = \int_0^\pi d\sigma\Big(\mathcal{P}^{\tau I}\dot{X}^I - \mathcal{L}\Big)\,. \qquad (12.85)$$

Writing the τ derivative of X^I in terms of $\mathcal{P}^{\tau I}$ we find

$$H = \int_0^\pi d\sigma \left(\pi\alpha' \mathcal{P}^{\tau I} \mathcal{P}^{\tau I} + \frac{1}{4\pi\alpha'} X^{I'} X^{I'} \right). \tag{12.86}$$

This Hamiltonian coincides with the one we postulated and tested in Section 12.1.

Let us now use the action (12.81) to quantize the theory. For this purpose we replace the dynamical variable $X^I(\tau, \sigma)$ by a collection of dynamical variables that have no σ dependence. This is done by writing the expansion

$$X^I(\tau, \sigma) = q^I(\tau) + 2\sqrt{\alpha'} \sum_{n=1}^\infty q_n^I(\tau) \frac{\cos n\sigma}{\sqrt{n}}. \tag{12.87}$$

This is the most general expression that satisfies Neumann boundary conditions at the endpoints. The particular normalization used to introduce the expansion coefficients was chosen for convenience.

Our next step is to evaluate the action (12.81) using the above expansion for $X^I(\tau, \sigma)$. For this we use

$$\dot{X}^I = \dot{q}^I(\tau) + 2\sqrt{\alpha'} \sum_{n=1}^\infty \dot{q}_n^I(\tau) \frac{\cos n\sigma}{\sqrt{n}},$$

$$X^{I'} = -2\sqrt{\alpha'} \sum_{n=1}^\infty q_n^I(\tau) \sqrt{n} \sin n\sigma. \tag{12.88}$$

The evaluation of the action S using the above expansions is quite straightforward because the σ integrals of $(\cos n\sigma \cos m\sigma)$ and $(\sin n\sigma \sin m\sigma)$ vanish unless $n = m$. We find,

$$S = \int d\tau \left[\frac{1}{4\alpha'} \dot{q}^I(\tau) \dot{q}^I(\tau) + \sum_{n=1}^\infty \left(\frac{1}{2n} \dot{q}_n^I(\tau) \dot{q}_n^I(\tau) - \frac{n}{2} q_n^I(\tau) q_n^I(\tau) \right) \right]. \tag{12.89}$$

Quick calculation 12.2 Prove equation (12.89).

Comparing the action (12.89) with the one recorded in (12.68) we see that the $q_n^I(\tau)$, with $n \geq 1$, are the coordinates of simple harmonic oscillators. The frequency of oscillation of $q_n^I(\tau)$ is n. This is the physical interpretation of the expansion coefficients in (12.87). Since the action for $q_n^I(\tau)$ coincides exactly with the action S_n, no new work is necessary to compute the Hamiltonian, except for the zero mode q^I:

$$p^I = \frac{\partial L}{\partial \dot{q}^I} = \frac{1}{2\alpha'} \dot{q}^I \quad \text{and} \quad [q^I, p^J] = i\eta^{IJ}. \tag{12.90}$$

The Hamiltonian is then given by

$$H = \alpha' p^I p^I + \sum_{n=1}^\infty \frac{n}{2} \left(p_n^I p_n^I + q_n^I q_n^I \right), \tag{12.91}$$

where we used (12.70) to write the part of the Hamiltonian that arises from the oscillators. The earlier analysis of the Heisenberg operator $q_n(t)$ led to the solution (12.80). This means that for the $q_n^I(\tau)$ oscillators we have

$$q_n^I(\tau) = \frac{i}{\sqrt{2}}\left(a_n^I e^{-in\tau} - a_n^{I\dagger} e^{in\tau}\right),$$ (12.92)

where $(a_n^I, a_n^{I\dagger})$ are canonically normalized annihilation and creation operators. For the Heisenberg operator $q^I(\tau)$ we find

$$\dot{q}^I(\tau) = i\,[\,H, q^I(\tau)] = i\alpha'\,[\,p^J\,p^J(\tau), q^I(\tau)] = 2\alpha'\,p^I(\tau)\,.$$ (12.93)

Note that p^I is a τ-independent Heisenberg operator. We solve this differential equation for $q^I(\tau)$ by writing

$$q^I(\tau) = x_0^I + 2\alpha'\,p^I\,\tau\,.$$ (12.94)

Here x_0^I is a constant operator that on account of (12.90) satisfies $[x_0^I, p^J] = i\eta^{IJ}$. Finally, we can substitute our solutions (12.92) and (12.94) into the expansion (12.87) for X^I to find

$$X^I(\tau, \sigma) = x_0^I + 2\alpha'\,p^I\tau + i\sqrt{2\alpha'}\sum_{n=1}^{\infty}\left(a_n^I e^{-in\tau} - a_n^{I\dagger} e^{in\tau}\right)\frac{\cos n\sigma}{\sqrt{n}}\,,$$ (12.95)

in exact agreement with the previously derived (12.67). We have therefore given a physical derivation of the mode expansion and commutation relations. We identified the classical variables that become oscillators, and we did not have to use delta functions. Having done so in this case, when we quantize other string configurations we will simply use the abstract approach of the previous section. It gives a direct and quick route to the desired answers.

12.4 Transverse Virasoro operators

We have written mode expansions for the transverse coordinates $X^I(\tau, \sigma)$ and we have seen quite explicitly the connection to harmonic oscillators. How about the other light-cone coordinates, $X^+(\tau, \sigma)$ and $X^-(\tau, \sigma)$? The expansion of X^+ is truly simple:

$$X^+(\tau, \sigma) = 2\alpha'p^+\,\tau = \sqrt{2\alpha'}\,\alpha_0^+\,\tau\,.$$ (12.96)

As discussed for the classical case in Section 9.5, this means that we are setting

$$x_0^+ = 0,\quad \alpha_n^+ = 0,\quad n \neq 0\,.$$ (12.97)

For the X^- coordinate a mode expansion was provided by (9.72):

$$X^-(\tau, \sigma) = x_0^- + \sqrt{2\alpha'}\,\alpha_0^-\tau + i\sqrt{2\alpha'}\sum_{n\neq 0}\frac{1}{n}\alpha_n^- e^{-in\tau}\cos n\sigma\,.$$ (12.98)

Moreover, we used the constraints to solve for X^- in terms of X^I, p^+, and a constant of integration x_0^-. This meant that the α_n^- modes could be written in terms of the α_n^I modes, as shown in equation (9.77):

$$\sqrt{2\alpha'}\,\alpha_n^- = \frac{1}{p^+}L_n^\perp\,, \tag{12.99}$$

where

$$L_n^\perp \equiv \frac{1}{2}\sum_{p\in\mathbb{Z}}\alpha_{n-p}^I\alpha_p^I\,. \tag{12.100}$$

The repeated index I is summed over the transverse light-cone directions. In Chapter 9 the $L_n^\perp$ were called transverse Virasoro *modes*. Having seen that the α modes become operators, the $L_n^\perp$ will now be called transverse Virasoro *operators*. The steps that led to (12.100) remain valid in the quantum theory, except for the fact that the α modes were treated as commuting classical variables. We now know that the α operators do not commute. We must therefore question whether the ordering of the two α operators appearing in (12.100) is the correct one. A better question is whether the ordering matters. Since two α operators fail to commute only when their mode numbers add up to zero, the two operators in $L_n^\perp$ fail to commute only when $n=0$. So $L_0^\perp$ is the only ambiguous operator.

There is plenty at stake in ordering $L_0^\perp$ correctly. The operator $L_0^\perp$ is in fact the light-cone Hamiltonian, as we showed in equation (12.16). Moreover, we saw at the end of Chapter 9 that $L_0^\perp$ enters directly into the calculation of the mass of string states. We also mentioned at that time that the quantum theory would bring a subtlety into the calculation of the mass. Well, the subtlety has arrived: we must define the quantum operator $L_0^\perp$! Let us therefore look at $L_0^\perp$ in more detail:

$$L_0^\perp = \frac{1}{2}\sum_{p\in\mathbb{Z}}\alpha_{-p}^I\alpha_p^I = \frac{1}{2}\alpha_0^I\alpha_0^I + \frac{1}{2}\sum_{p=1}^{\infty}\alpha_{-p}^I\alpha_p^I + \frac{1}{2}\sum_{p=1}^{\infty}\alpha_p^I\alpha_{-p}^I\,. \tag{12.101}$$

The first sum on the right-hand side is normal-ordered: annihilation operators appear to the right of the creation operators. It is useful to work with normal-ordered operators since they act in a simple manner on the vacuum state. We cannot use operators that do not have a well defined action on the vacuum state. Since the last sum on the right-hand side of (12.101) is not a normal-ordered operator, we rewrite it as

$$\frac{1}{2}\sum_{p=1}^{\infty}\alpha_p^I\alpha_{-p}^I = \frac{1}{2}\sum_{p=1}^{\infty}\left(\alpha_{-p}^I\alpha_p^I + [\alpha_p^I\,,\alpha_{-p}^I]\right)$$

$$= \frac{1}{2}\sum_{p=1}^{\infty}\alpha_{-p}^I\alpha_p^I + \frac{1}{2}\sum_{p=1}^{\infty}p\,\eta^{II}$$

$$= \frac{1}{2}\sum_{p=1}^{\infty}\alpha_{-p}^I\alpha_p^I + \frac{1}{2}(D-2)\sum_{p=1}^{\infty}p\,. \tag{12.102}$$

If you look at the last term of the above equation you will note that it is divergent; it involves the sum of all positive integers! This is clearly problematic. How do we deal with this? One option is simply to ignore this difficulty, claiming that it is really up to us how we define $L_0^\perp$. There is a kernel of truth to this option, but it is not completely correct. Adding a constant to $L_0^\perp$ changes the values of the masses of the string states, and if anything, the above computation has alerted us to the fact that this additive constant could be nonzero, or even infinite. Taken at face value, the above computation gives

$$L_0^\perp = \frac{1}{2}\alpha_0^I \alpha_0^I + \sum_{p=1}^{\infty} \alpha_{-p}^I \alpha_p^I + \frac{1}{2}(D-2)\sum_{p=1}^{\infty} p \,. \tag{12.103}$$

The operator $L_0^\perp$ enters into our computation of the mass via the definition of p^-. From (12.99), with $n=0$,

$$\sqrt{2\alpha'}\,\alpha_0^- = 2\alpha' p^- = \frac{1}{p^+}L_0^\perp \,. \tag{12.104}$$

This suggests a strategy. First, we *define*, once and for all, $L_0^\perp$ to be the normal-ordered operator in (12.103) *without* including the ordering constant:

$$L_0^\perp \equiv \frac{1}{2}\alpha_0^I \alpha_0^I + \sum_{p=1}^{\infty} \alpha_{-p}^I \alpha_p^I = \alpha' p^I p^I + \sum_{p=1}^{\infty} p\, a_p^{I\dagger} a_p^I \,. \tag{12.105}$$

Note that $L_0^\perp$ is Hermitian: $(L_0^\perp)^\dagger = L_0^\perp$. Second, we introduce an ordering constant a into (12.104):

$$2\alpha' p^- \equiv \frac{1}{p^+}\left(L_0^\perp + a\right). \tag{12.106}$$

If we took seriously our attempt to order $L_0^\perp$, we would have to conclude that

$$a \overset{?}{=} \frac{1}{2}(D-2)\sum_{p=1}^{\infty} p \,. \tag{12.107}$$

We will discuss below one remarkable interpretation of this equation which does, in fact, give the correct result. More pragmatically, we will take a to be an undetermined constant. As we will show in Section 12.5, the quantum consistency of string theory fixes the constant a to an interesting finite value. Before proceeding further, let us investigate how the inclusion of a modifies the computation of the mass-squared operator. Working from the definition $M^2 = -p^2$, and using (12.105) and (12.106), we find

$$M^2 = -p^2 = 2p^+ p^- - p^I p^I = \frac{1}{\alpha'}(L_0^\perp + a) - p^I p^I$$

$$= \frac{1}{\alpha'}\left(a + \sum_{n=1}^{\infty} n\, a_n^{I\dagger} a_n^I\right). \tag{12.108}$$

As expected, a introduces a constant shift into the mass-squared operator.

It is impossible to resist the temptation to interpret (12.107). An important result in mathematics suggests a finite value for the right-hand side. For this we consider the zeta function $\zeta(s)$, which is defined as the infinite sum

$$\zeta(s) = \sum_{n=1}^{\infty} \frac{1}{n^s}, \quad \Re(s) > 1. \tag{12.109}$$

The argument s of the zeta function is assumed to be a complex number, but as indicated, the above sum only converges if the real part of the argument is greater than one. We can use analytic continuation to define the zeta function for all possible values of the argument. $\zeta(s)$ turns out to be finite for all values of s except $s = 1$. In particular, as you will see in Problem 12.4, $\zeta(-1) = -1/12$. On account of (12.109) this suggests that

$$\zeta(-1) = -\frac{1}{12} \stackrel{?}{=} 1 + 2 + 3 + 4 + \cdots. \tag{12.110}$$

This is a surprising interpretation for the infinite sum $\sum_{p=1}^{\infty} p$. Not only is the result finite, but it is also negative! Substituting back in (12.107), it gives us

$$a = -\frac{1}{24}(D - 2). \tag{12.111}$$

This is actually the correct value of a, as we will explain in Section 12.5. The inspired guess gave the right answer. We will also see that consistency requires $D = 26$, so that, in fact, $a = -1$. This value for the shift in the mass-squared operator is precisely what is needed for the spectrum of open strings to include massless photon states!

Having discussed $L_0^{\perp}$ in detail, let us consider the other transverse Virasoro operators. Since $(\alpha_n^J)^{\dagger} = \alpha_{-n}^J$, we may expect that $(\alpha_n^-)^{\dagger} = \alpha_{-n}^-$, or equivalently, on account of (12.99), that

$$(L_n^{\perp})^{\dagger} = L_{-n}^{\perp}. \tag{12.112}$$

We have already verified this equation for $n = 0$. For $n \neq 0$, we can easily prove this Hermiticity property using (12.100):

$$(L_n^{\perp})^{\dagger} = \frac{1}{2} \sum_{p \in \mathbb{Z}} (\alpha_{n-p}^I \alpha_p^I)^{\dagger} = \frac{1}{2} \sum_{p \in \mathbb{Z}} (\alpha_p^I)^{\dagger} (\alpha_{n-p}^I)^{\dagger} = \frac{1}{2} \sum_{p \in \mathbb{Z}} \alpha_{-p}^I \alpha_{-n+p}^I. \tag{12.113}$$

Since the oscillators in each term of the sum commute, we can exchange them. By also letting $p \to -p$, we get the expected result:

$$(L_n^{\perp})^{\dagger} = \frac{1}{2} \sum_{p \in \mathbb{Z}} \alpha_{-n-p}^I \alpha_p^I = L_{-n}^{\perp}. \tag{12.114}$$

Perhaps the most interesting property of the Virasoro operators is that they do not commute. We have seen that α_m^I and α_n^I commute except when $m + n$ equals zero. This is not the case with the α_n^- modes. Two Virasoro operators $L_m^{\perp}$ and $L_n^{\perp}$ never commute when $m \neq n$. The commutation properties of Virasoro operators are a bit intricate, so we will consider them in steps of increasing generality.

As a warmup, let us consider the commutator between a Virasoro operator and an oscillator α_n^J. We have

$$\left[L_m^\perp, \alpha_n^J \right] = \frac{1}{2} \sum_{p\in\mathbb{Z}} \left[\alpha_{m-p}^I \alpha_p^I, \alpha_n^J \right] = \frac{1}{2} \sum_{p\in\mathbb{Z}} \left(\alpha_{m-p}^I \left[\alpha_p^I, \alpha_n^J \right] + \left[\alpha_{m-p}^I, \alpha_n^J \right] \alpha_p^I \right).$$

(12.115)

Evaluating the commutators and recalling that $\eta^{IJ} = \delta^{IJ}$, we find

$$\left[L_m^\perp, \alpha_n^J \right] = \frac{1}{2} \sum_{p\in\mathbb{Z}} \left(p\, \delta_{p+n,0}\, \alpha_{m-p}^J + (m-p)\delta_{m-p+n,0}\, \alpha_p^J \right).$$

(12.116)

Because of the Kronecker deltas only one term contributes in each sum: $p = -n$ in the first and $p = m + n$ in the second. We thus find

$$\left[L_m^\perp, \alpha_n^J \right] = \frac{1}{2} \left(-n\, \alpha_{m+n}^J - n\, \alpha_{m+n}^J \right).$$

(12.117)

Our final result is therefore

$$\left[L_m^\perp, \alpha_n^J \right] = -n\, \alpha_{m+n}^J.$$

(12.118)

The mode number on the right-hand side is the sum of the mode numbers on the left-hand side. This could not be otherwise, since the basic commutator of α operators trades two operators with opposite mode number for a constant, and in doing so, the total mode number is conserved. Moreover, the spatial index on the oscillator is preserved. Equation (12.118) holds for all values of m, including $m = 0$. Indeed, (12.100), which we used, gives $L_0^\perp$ up to a constant that, although infinite, cannot affect the commutator. It is worth anyway to check the result directly.

Quick calculation 12.3 Calculate $[L_0^\perp, \alpha_n^J]$ using (12.105) and confirm that (12.118) holds for $m = 0$.

Quick calculation 12.4 Show that

$$\left[L_m^\perp, x_0^I \right] = -i\sqrt{2\alpha'}\, \alpha_m^I.$$

(12.119)

Let us now consider the commutator of two Virasoro operators $L_m^\perp$ and $L_n^\perp$. The computation is a bit subtle, and it is easy to get incorrect answers. We will avoid subtleties by checking, at every stage of the calculation, that our expressions are normal ordered. For this, we begin by rewriting the Visaroso operator (12.100) in such a way that it gives the correct result even for L_0. For this, the sum is split as:

$$L_m^\perp = \frac{1}{2} \sum_{k\geq 0} \alpha_{m-k}^I \alpha_k^I + \frac{1}{2} \sum_{k<0} \alpha_k^I \alpha_{m-k}^I.$$

(12.120)

For any value of m the above right-hand side is normal-ordered: in the first sum the α to the right is an annihilation operator (or a zero mode) and in the second sum the α to the left is a creation operator. We now evaluate

$$
\begin{aligned}
\left[L_m^\perp, L_n^\perp \right] &= \frac{1}{2} \sum_{k \ge 0} \left[\alpha_{m-k}^I \alpha_k^I , L_n^\perp \right] + \frac{1}{2} \sum_{k < 0} \left[\alpha_k^I \alpha_{m-k}^I , L_n^\perp \right] , \\
&= \frac{1}{2} \sum_{k \ge 0} \left[\alpha_{m-k}^I, L_n^\perp \right] \alpha_k^I + \frac{1}{2} \sum_{k < 0} \alpha_k^I \left[\alpha_{m-k}^I , L_n^\perp \right] \\
&\quad + \frac{1}{2} \sum_{k \ge 0} \alpha_{m-k}^I \left[\alpha_k^I , L_n^\perp \right] + \frac{1}{2} \sum_{k < 0} \left[\alpha_k^I , L_n^\perp \right] \alpha_{m-k}^I .
\end{aligned}
\tag{12.121}
$$

Evaluating the commutators we get

$$
\begin{aligned}
\left[L_m^\perp, L_n^\perp \right] &= \frac{1}{2} \sum_{k \ge 0} (m-k) \alpha_{m+n-k}^I \alpha_k^I + \frac{1}{2} \sum_{k < 0} (m-k) \alpha_k^I \alpha_{m+n-k}^I \\
&\quad + \frac{1}{2} \sum_{k \ge 0} k \, \alpha_{m-k}^I \alpha_{k+n}^I + \frac{1}{2} \sum_{k < 0} k \, \alpha_{k+n}^I \alpha_{m-k}^I .
\end{aligned}
\tag{12.122}
$$

The terms on the first line of the right-hand side are always normal ordered. The terms on the second line can require ordering, depending on the values of m and n. Let us now consider two cases: $m + n \ne 0$ and $m + n = 0$.

Case m+n ≠ 0. All pairs of oscillators on the right-hand side of (12.122) commute so we find

$$
\begin{aligned}
\left[L_m^\perp, L_n^\perp \right] &= \frac{1}{2} \sum_{k \in \mathbb{Z}} (m-k) \alpha_{m+n-k}^I \alpha_k^I + \frac{1}{2} \sum_{k \in \mathbb{Z}} k \, \alpha_{m-k}^I \alpha_{k+n}^I , \\
&= \frac{1}{2} \sum_{k \in \mathbb{Z}} (m-k) \alpha_{m+n-k}^I \alpha_k^I + \frac{1}{2} \sum_{k \in \mathbb{Z}} (k-n) \, \alpha_{m+n-k}^I \alpha_k^I , \\
&= (m-n) \frac{1}{2} \sum_{k \in \mathbb{Z}} \alpha_{m+n-k}^I \alpha_k^I .
\end{aligned}
\tag{12.123}
$$

In passing from the first to the second line we let $k \to k - n$ in the second sum. Since $m + n \ne 0$ the final operator requires no ordering and is recognized to be $L_{m+n}^\perp$. We have therefore shown that

$$
\left[L_m^\perp, L_n^\perp \right] = (m-n) L_{m+n}^\perp , \quad m + n \ne 0 .
\tag{12.124}
$$

The commutator of two Virasoro operators is a Virasoro operator with mode number equal to the sum of the mode numbers of the operators which enter the commutator. The above result does not hold when $m + n = 0$, in which case the answer is different. Nevertheless, as a mathematical construct, a set of operators $L_n^\perp$ with $n \in \mathbb{Z}$, satisfying (12.124) for *all m* and *n*, defines an interesting Lie algebra (Problem 12.5). This algebra is called the *Virasoro algebra without central extension* or the Witt algebra.

Case m+n =0. In this case we write $n = -m$ and equation (12.122) becomes

$$\left[L_m^\perp, L_{-m}^\perp\right] = \frac{1}{2}\sum_{k=0}^{\infty}(m-k)\alpha_{-k}^I\alpha_k^I + \frac{1}{2}\sum_{k<0}(m-k)\alpha_k^I\alpha_{-k}^I$$

$$+ \frac{1}{2}\sum_{k=0}^{\infty}k\,\alpha_{m-k}^I\alpha_{k-m}^I + \frac{1}{2}\sum_{k<0}k\,\alpha_{k-m}^I\alpha_{m-k}^I . \quad (12.125)$$

In order to compare the various terms it is useful to relabel the summation variables so that the rightmost oscillator is always α_k^I. For this we let $k \to -k$ in the second term of the first line, $k \to m+k$ in the first term of the second line, and $k \to m-k$ in the second term of the second line:

$$\left[L_m^\perp, L_{-m}^\perp\right] = \frac{1}{2}\sum_{k=0}^{\infty}(m-k)\alpha_{-k}^I\alpha_k^I + \frac{1}{2}\sum_{k=1}^{\infty}(m+k)\alpha_{-k}^I\alpha_k^I$$

$$+ \frac{1}{2}\underline{\sum_{k=-m}^{\infty}(m+k)\,\alpha_{-k}^I\alpha_k^I} + \frac{1}{2}\sum_{k=m+1}^{\infty}(m-k)\,\alpha_{-k}^I\alpha_k^I . \quad (12.126)$$

We now assume, without loss of generality, that $m > 0$. It then follows that all terms are normal-ordered except those in the underlined summand for which $-m \leq k \leq 0$. Splitting the sum, the underlined term becomes equal to

$$\frac{1}{2}\sum_{k=0}^{m}(m-k)\,\alpha_k^I\alpha_{-k}^I + \frac{1}{2}\sum_{k=1}^{\infty}(m+k)\,\alpha_{-k}^I\alpha_k^I$$

$$= \frac{1}{2}\sum_{k=0}^{m}(m-k)\,[\alpha_k^I,\alpha_{-k}^I] + \frac{1}{2}\sum_{k=0}^{m}(m-k)\,\alpha_{-k}^I\alpha_k^I + \frac{1}{2}\sum_{k=1}^{\infty}(m+k)\,\alpha_{-k}^I\alpha_k^I .$$

Evaluating the commutator and substituting the result back into (12.126) we get

$$\left[L_m^\perp, L_{-m}^\perp\right] = \sum_{k=0}^{\infty}(m-k)\alpha_{-k}^I\alpha_k^I + \sum_{k=1}^{\infty}(m+k)\alpha_{-k}^I\alpha_k^I + (D-2)A(m) , \quad (12.127)$$

where $A(m)$ is the constant

$$A(m) = \frac{1}{2}\sum_{k=0}^{m}k(m-k) = \frac{1}{2}m\sum_{k=1}^{m}k - \frac{1}{2}\sum_{k=1}^{m}k^2 . \quad (12.128)$$

To evaluate $A(m)$ we need the following result:

Quick calculation 12.5 Use mathematical induction to prove that

$$\sum_{k=1}^{m}k^2 = \frac{1}{6}(2m^3 + 3m^2 + m) . \quad (12.129)$$

Making use of (12.129) we find

$$A(m) = \frac{1}{4}m^2(m+1) - \frac{1}{12}(2m^3 + 3m^2 + m) = \frac{1}{12}(m^3 - m) . \quad (12.130)$$

Expanding out the sums in (12.127) and substituting the value of $A(m)$, we find

$$\left[L_m^\perp, L_{-m}^\perp\right] = 2m\left(\frac{1}{2}\alpha_0^I\alpha_0^I + \sum_{k=1}^\infty \alpha_{-k}^I\alpha_k^I\right) + \frac{1}{12}(D-2)(m^3 - m)\,. \tag{12.131}$$

We recognize that the operator in parenthesis is $L_0^\perp$. Therefore, our final result is

$$\left[L_m^\perp, L_{-m}^\perp\right] = 2m\,L_0^\perp + \frac{1}{12}(D-2)(m^3 - m)\,. \tag{12.132}$$

This completes the computation for the case $m + n = 0$.

The general result for the commutator of two Virasoro operators is obtained by writing a formula that gives (12.124) for $m + n \neq 0$ and gives (12.132) for $m + n = 0$. This is easily done and the result is

$$\left[L_m^\perp, L_n^\perp\right] = (m-n)\,L_{m+n}^\perp + \frac{D-2}{12}\left(m^3 - m\right)\delta_{m+n,0}\,. \tag{12.133}$$

A set of operators $L_n^\perp$, with $n \in \mathbb{Z}$, satisfying (12.133) defines the *centrally extended Virasoro algebra*. The second term on the above right-hand side is called the central extension. It is a constant or, more properly, a constant times the identity operator (the operator that acting on any state gives back the state). This term is said to be central because it commutes with all other operators in the algebra. The central term vanishes for $m = 0$ and for $m = \pm 1$. There is therefore no central term in the commutator $[L_1^\perp, L_{-1}^\perp]$. The Virasoro algebra is perhaps the most important algebra in string theory. In the light-cone quantization of string theory – our subject in this chapter – the Virasoro operators enter into the definition of the Lorentz generators, as we will see in Section 12.5.

We conclude this section by studying how the Virasoro operators act on the string coordinates. Since quantum operators act via commutators, we must find the commutator of a Virasoro operator with the coordinate operator $X^I(\tau, \sigma)$. We will see that the Virasoro operators generate reparameterizations of the world-sheet.

Making use of the coordinate expansion (12.32) we find

$$[L_m^\perp, X^I(\tau, \sigma)] = [L_m^\perp, x_0^I] + i\sqrt{2\alpha'}\sum_{n\neq 0}\frac{1}{n}\cos n\sigma\, e^{-in\tau}[L_m^\perp, \alpha_n^I]$$
$$= -i\sqrt{2\alpha'}\alpha_m^I - i\sqrt{2\alpha'}\sum_{n\neq 0}\cos n\sigma\, e^{-in\tau}\alpha_{m+n}^I\,, \tag{12.134}$$

where we used (12.118) and (12.119) to evaluate the commutators. The right-hand side above can be written as a single sum:

$$[L_m^\perp, X^I(\tau, \sigma)] = -i\sqrt{2\alpha'} \sum_{n\in\mathbb{Z}} \cos n\sigma \, e^{-in\tau} \, \alpha_{m+n}^I$$

$$= -i\sqrt{2\alpha'} \frac{1}{2} \sum_{n\in\mathbb{Z}} (e^{-in(\tau-\sigma)} + e^{-in(\tau+\sigma)}) \alpha_{m+n}^I$$

$$= -i\sqrt{2\alpha'} \frac{1}{2} \sum_{n\in\mathbb{Z}} (e^{-i(n-m)(\tau-\sigma)} + e^{-i(n-m)(\tau+\sigma)}) \alpha_n^I,$$

where in the last step we let $n \to n - m$. Finally,

$$[L_m^\perp, X^I(\tau, \sigma)] = -\frac{i}{2} e^{im(\tau-\sigma)} \sqrt{2\alpha'} \sum_{n\in\mathbb{Z}} e^{-in(\tau-\sigma)} \alpha_n^I$$

$$- \frac{i}{2} e^{im(\tau+\sigma)} \sqrt{2\alpha'} \sum_{n\in\mathbb{Z}} e^{-in(\tau+\sigma)} \alpha_n^I.$$

To interpret this result it is necessary to express the right-hand side in terms of derivatives of the string coordinates. This is done with the help of (12.33):

$$[L_m^\perp, X^I(\tau, \sigma)] = -\frac{i}{2} e^{im(\tau-\sigma)} (\dot{X}^I - X^{I'}) - \frac{i}{2} e^{im(\tau+\sigma)} (\dot{X}^I + X^{I'})$$

$$= -i e^{im\tau} \cos m\sigma \, \dot{X}^I + e^{im\tau} \sin m\sigma \, X^{I'}. \tag{12.135}$$

This equation has taken the form

$$[L_m^\perp, X^I(\tau, \sigma)] = \xi_m^\tau \dot{X}^I + \xi_m^\sigma X^{I'}, \tag{12.136}$$

where

$$\xi_m^\tau(\tau, \sigma) = -i e^{im\tau} \cos m\sigma,$$

$$\xi_m^\sigma(\tau, \sigma) = e^{im\tau} \sin m\sigma. \tag{12.137}$$

The interpretation of (12.136), we claim, is that the Virasoro operators generate reparameterizations of the world-sheet. In particular, they change the τ and σ coordinates as

$$\tau \longrightarrow \tau + \epsilon \, \xi_m^\tau(\tau, \sigma),$$

$$\sigma \longrightarrow \sigma + \epsilon \, \xi_m^\sigma(\tau, \sigma), \tag{12.138}$$

where ϵ is an infinitesimal parameter. In order to see this, note that Taylor expansion gives

$$X^I(\tau + \epsilon \, \xi_m^\tau, \sigma + \epsilon \, \xi_m^\sigma) = X^I(\tau, \sigma) + \epsilon \, (\xi_m^\tau \dot{X}^I + \xi_m^\sigma X^{I'})$$

$$= X^I(\tau, \sigma) + \epsilon \left[L_m^\perp, X^I(\tau, \sigma) \right]. \tag{12.139}$$

This equation states that the action of the Virasoro operators on the string coordinates generates the same change that would occur as a result of a reparameterization of the world-sheet. This is what we wanted to show. Problem 12.10 shows that the above reparameterizations preserve the constraints (9.33).

What is the reparameterization generated by $L_0^\perp$? Setting $m = 0$ in (12.137), we find $\xi_0^\tau = -i$ and $\xi_0^\sigma = 0$. As a result, (12.136) gives

$$[L_0^\perp, X^I] = -i\partial_\tau X^I, \tag{12.140}$$

which we recognize as the Heisenberg equation of motion for X^I. Indeed, $L_0^\perp$ is, up to an additive constant, the string Hamiltonian, and as such, it must generate time translations. It is also interesting to note that for all m, ξ_m^σ vanishes at $\sigma = 0$ and at $\sigma = \pi$. This means that the reparameterizations generated by the Virasoro operators do not change the σ coordinates of the string endpoints. The range of σ remains $[0, \pi]$.

The functions ξ_m^τ and ξ_m^σ in (12.137) are not real, and used in (12.138) they spoil the reality of the coordinates τ and σ. This complication is familiar to you from quantum mechanics. Real transformations are generated by anti-Hermitian operators. The momentum operator $\vec{p} = -i\nabla$ is Hermitian, and therefore it is the anti-Hermitian combination $i\vec{p} = \nabla$ that generates real translations. Out of the operators $L_m^\perp$ and $L_{-m}^\perp$, we can generate two anti-Hermitian combinations:

$$L_m^\perp - L_{-m}^\perp \quad \text{and} \quad i\,(L_m^\perp + L_{-m}^\perp). \tag{12.141}$$

Consider the first combination. It follows from (12.136) that the parameters for the transformation generated by $(L_m^\perp - L_{-m}^\perp)$ are

$$\xi^\tau = \xi_m^\tau - \xi_{-m}^\tau = 2\sin m\tau \, \cos m\sigma,$$
$$\xi^\sigma = \xi_m^\sigma - \xi_{-m}^\sigma = 2\cos m\tau \, \sin m\sigma. \tag{12.142}$$

Quick calculation 12.6 Show that the parameters for the transformation generated by $i\,(L_m^\perp + L_{-m}^\perp)$ are

$$\xi^\tau = 2\cos m\tau \, \cos m\sigma,$$
$$\xi^\sigma = -2\sin m\tau \, \sin m\sigma. \tag{12.143}$$

Our discussion of the Virasoro operators has been quite detailed. We have examined their precise definition, and we have seen how they affect the computation of the mass. We have determined their commutator algebra, and we have exhibited how they act on string coordinates. In the following section we will see that the Virasoro operators also enter into the definition of the operators that generate Lorentz transformations.

12.5 Lorentz generators

In Chapter 8, the Lorentz invariance of the string action allowed us to find a set of conserved world-sheet currents $\mathcal{M}_{\mu\nu}^\alpha$ labeled by the indices μ and ν, with $\mu \neq \nu$. The resulting conserved charges $M_{\mu\nu}$ were given in (8.65), and for open strings with $\sigma \in [0, \pi]$ they read

$$M_{\mu\nu} = \int_0^\pi \mathcal{M}_{\mu\nu}^\tau(\tau, \sigma)\, d\sigma = \int_0^\pi (X_\mu \mathcal{P}_\nu^\tau - X_\nu \mathcal{P}_\mu^\tau)\, d\sigma. \tag{12.144}$$

Making use of (12.1) and raising the spacetime indices,

$$M^{\mu\nu} = \frac{1}{2\pi\alpha'} \int_0^\pi \left(X^\mu \dot{X}^\nu - X^\nu \dot{X}^\mu \right) d\sigma. \tag{12.145}$$

Constructing suitable quantum operators can be delicate, so let us gain some intuition by thinking classically. Explicit mode expansions for X^μ and $\dot{X}^\nu$ are given in equations (9.56) and (9.57). Since $M^{\mu\nu}$ is guaranteed to be τ-independent, to evaluate (12.145) it suffices to pick up the τ-independent terms that arise in the products. For example,

$$X^\mu \dot{X}^\nu = x_0^\mu (\sqrt{2\alpha'}\, \alpha_0^\nu) + i\, 2\alpha' \sum_{n\neq 0} \frac{1}{n} \alpha_n^\mu \alpha_{-n}^\nu \cos^2 n\sigma + \cdots, \tag{12.146}$$

where the dots represent τ-dependent terms that must fail to contribute in the calculation of $M^{\mu\nu}$. With this equation, and a similar one with μ and ν exchanged, we find that upon integration (12.145) gives

$$M^{\mu\nu} = x_0^\mu p^\nu - x_0^\nu p^\mu - i \sum_{n=1}^\infty \frac{1}{n} \left(\alpha_{-n}^\mu \alpha_n^\nu - \alpha_{-n}^\nu \alpha_n^\mu \right). \tag{12.147}$$

Quick calculation 12.7 Prove equation (12.147).

Equation (12.147) gives the classical Lorentz generators in terms of oscillation modes. We should ask ourselves whether we can use it, with the α recognized as operators, to define the quantum Lorentz generators. We will use (12.147) to *suggest* the form of the quantum Lorentz generators in light-cone gauge string theory. Since the canonical structure of the theory in the light-cone gauge is unusual, there is no guarantee that we can build consistent quantum Lorentz generators. An inability to construct quantum Lorentz generators would mean that quantum string theory fails to be physically Lorentz invariant.

In light-cone gauge, the most delicate quantum Lorentz generator is M^{-I} because the X^- coordinate is a rather nontrivial function of the transverse coordinates. A consistent M^{-I} must generate Lorentz transformations on the string coordinates, possibly accompanied by world-sheet reparameterizations. Indeed, in the simpler context of the point particle, the action of M^{-I} includes world-line reparameterizations. The generator M^{-I} must also satisfy the commutation relation

$$\left[M^{-I}, M^{-J} \right] = 0. \tag{12.148}$$

To find a candidate for M^{-I}, we consider equation (12.147) and write

$$M^{-I} \stackrel{?}{=} x_0^- p^I - x_0^I p^- - i \sum_{n=1}^\infty \frac{1}{n} \left(\alpha_{-n}^- \alpha_n^I - \alpha_{-n}^I \alpha_n^- \right). \tag{12.149}$$

This is just a first guess, though it is a pretty good one. A satisfactory M^{-I} should be both Hermitian and normal-ordered. Let us consider Hermiticity first. The first term in the right-hand side of (12.149) is Hermitian since x_0^- and p^I commute. The second term, however, is not, since x_0^I and p^- do not commute. A simple solution is to symmetrize the term by writing

$$M^{-I} \stackrel{?}{=} x_0^- \, p^I - \frac{1}{2} \left(x_0^I \, p^- + p^- x_0^I \right) - i \sum_{n=1}^{\infty} \frac{1}{n} \left(\alpha_{-n}^- \alpha_n^I - \alpha_{-n}^I \alpha_n^- \right). \tag{12.150}$$

The last term above is fully Hermitian since $(\alpha_n^I)^\dagger = \alpha_{-n}^I$ and $(\alpha_n^-)^\dagger = \alpha_{-n}^-$. Consider now normal ordering. Do all of the annihilation operators appear to the right of the creation operators? They do, because the α^- oscillators are normal-ordered Virasoro operators. Finally, to be complete, we must give the definition of p^-. As stated in (12.106), p^- includes an undetermined constant a that reflects our difficulties in ordering the Virasoro operator $L_0^\perp$. With this definition, and writing the other minus oscillators in terms of Virasoro operators, we find

$$M^{-I} = x_0^- \, p^I - \frac{1}{4\alpha' p^+} \left(x_0^I \left(L_0^\perp + a \right) + \left(L_0^\perp + a \right) x_0^I \right)$$

$$- \frac{i}{\sqrt{2\alpha'} \, p^+} \sum_{n=1}^{\infty} \frac{1}{n} \left(L_{-n}^\perp \alpha_n^I - \alpha_{-n}^I L_n^\perp \right). \tag{12.151}$$

Now that we have a candidate for the quantum Lorentz charge M^{-I}, we can discuss the computation of $[M^{-I}, M^{-J}]$.

There is much at stake in this calculation. It is in fact, one of the most important calculations in string theory. Our Lorentz charge has two undetermined parameters: the dimension D of spacetime, implicit in the sums over transverse directions, and the constant a affecting the mass of the particles. The calculation is long and uses many of our previously derived results, including the Virasoro commutation relations. We will not attempt to do it here, but the result is

$$\left[M^{-I}, M^{-J} \right] = -\frac{1}{\alpha' p^{+2}} \sum_{m=1}^{\infty} \left(\alpha_{-m}^I \alpha_m^J - \alpha_{-m}^J \alpha_m^I \right)$$

$$\times \left\{ m \left[1 - \frac{1}{24} (D-2) \right] + \frac{1}{m} \left[\frac{1}{24} (D-2) + a \right] \right\}. \tag{12.152}$$

The right-hand side is a sum of terms, each of which contains the operator $(\alpha_{-m}^I \alpha_m^J - \alpha_{-m}^J \alpha_m^I)$ for a different value of m. Such terms cannot cancel each other, so the commutator above vanishes if and only if the coefficient in large braces vanishes for all positive integers m:

$$m \left[1 - \frac{1}{24} (D-2) \right] + \frac{1}{m} \left[\frac{1}{24} (D-2) + a \right] = 0, \quad \forall m \in \mathbb{Z}^+. \tag{12.153}$$

It suffices to examine this condition for $m = 1$ and $m = 2$ to conclude that each of the terms in brackets must simply vanish. We therefore have

$$1 - \frac{1}{24} (D-2) = 0 \quad \text{and} \quad \frac{1}{24} (D-2) + a = 0. \tag{12.154}$$

The first equation fixes the dimension of spacetime:

$$D = 26. \tag{12.155}$$

The second equation then fixes the constant a:

$$a = -\frac{1}{24}(D - 2) = -\frac{24}{24} = -1. \tag{12.156}$$

This value of a coincides with the one obtained in (12.111) by ordering $L_0^{\perp}$ and using the zeta function to interpret the resulting infinity. For future reference, with $a = -1$ the expression for p^- in (12.106) becomes

$$2\alpha' p^- \equiv \frac{1}{p^+} \left(L_0^{\perp} - 1\right). \tag{12.157}$$

In addition, because of (12.14) the string Hamiltonian is now just

$$H = L_0^{\perp} - 1. \tag{12.158}$$

Here, of course, $L_0^{\perp}$ is the normal-ordered operator without additional constants. The above equation is the precise version of equation (12.16).

In summary, we have seen that the condition of Lorentz invariance of quantum string theory simultaneously fixes the dimension of spacetime and the constant shift in the masses of the particles. In superstring theory a similar calculation fixes the dimensionality of spacetime to the value $D = 10$. The fact that string theory cannot be a good Lorentz invariant quantum theory in any arbitrary dimension shows that string theory is very constrained. Even more, since the dimension of spacetime is uniquely selected by the requirement of consistency, we can say that string theory predicts the dimension of spacetime!

12.6 Constructing the state space

The classical open string does not provide a reasonable theory of physics because the mass of string states assumes a continuous range of values. Only the ground state is massless in the classical theory, and this ground state does not include any polarization labels. As a result, classical open strings have no states that can be identified as photons. The miracle of quantum string theory is that both of these problems are solved. The continuous spectrum disappears after quantization. Candidate photon states emerge because the downward shift of the squared masses gives us massless states with polarization labels.

Let us begin by introducing the ground states of the quantum string. The quantum string shares with the quantum point particle the same set of zero modes. We have the canonical pairs (x_0^I, p^I) and (x_0^-, p^+). Therefore, just as for the point particle (see (11.43)), we introduce states

$$|p^+, \vec{p}_T\rangle. \tag{12.159}$$

The above states are called *ground states* for all values of the momenta indicated by the labels. They are also declared to be vacuum states for all the oscillators in string theory. Thus, by definition, they are annihilated by all the a_n^I:

$$a_n^I |p^+, \vec{p}_T\rangle = 0, \quad n \geq 1, \quad I = 2, \ldots, 25. \tag{12.160}$$

How do we create states from the $|p^+, \vec{p}_T\rangle$? We simply act on them with the creation operators. There are infinitely many of them, and we can operate on each state arbitrarily many times with any particular creation operator. The list of creation operators at our disposal is infinite, but it can be organized as follows:

$$
\begin{array}{cccc}
a_1^{(2)\,\dagger} & a_1^{(3)\,\dagger} & \cdots & a_1^{(25)\,\dagger} \\
a_2^{(2)\,\dagger} & a_2^{(3)\,\dagger} & \cdots & a_2^{(25)\,\dagger} \\
\vdots & \vdots & \vdots & \vdots \\
a_n^{(2)\,\dagger} & a_n^{(3)\,\dagger} & \cdots & a_n^{(25)\,\dagger} \\
\vdots & \vdots & \vdots & \vdots
\end{array}
\tag{12.161}
$$

Above, the polarization index I has been enclosed by parentheses. The general basis state $|\lambda\rangle$ of the state space can be written as

$$|\lambda\rangle = \prod_{n=1}^{\infty} \prod_{I=2}^{25} \left(a_n^{I\,\dagger} \right)^{\lambda_{n,I}} |p^+, \vec{p}_T\rangle. \tag{12.162}$$

Here the non-negative integer $\lambda_{n,I}$ denotes the number of times that the creation operator $a_n^{I\,\dagger}$ appears. As you can see, the state $|\lambda\rangle$ is specified by stating how many times each of the oscillators in the list (12.161) acts on the ground state. This information is given by the list of non-negative integers $\lambda_{n,I}$ for all $n \geq 1$ and all $I = 2, \ldots, 25$. Since all the creation operators commute among each other, the order in which they appear is irrelevant. We restrict ourselves to the case where states only have a finite number of creation operators acting on the ground states. This means that for each state $|\lambda\rangle$ only a finite number of $\lambda_{n,I}$ are different from zero. The string Hilbert space is an infinite-dimensional vector space: it is spanned by an infinite set of linearly independent basis states $|\lambda\rangle$. This is why string theory describes an infinite number of different particles! A general state in the Hilbert space is a linear superposition of the basis states $|\lambda\rangle$.

To understand the physical significance of the above states, consider the mass-squared operator (12.108), with our new found knowledge that $a = -1$:

$$M^2 = \frac{1}{\alpha'}\left(-1 + \sum_{n=1}^{\infty} n a_n^{I\dagger} a_n^{I}\right). \tag{12.163}$$

The sum appearing in (12.163) is important enough to have its own name; it is called the number operator $N^{\perp}$:

$$N^{\perp} \equiv \sum_{n=1}^{\infty} n a_n^{I\dagger} a_n^{I}, \quad M^2 = \frac{1}{\alpha'}(-1 + N^{\perp}). \tag{12.164}$$

$N^{\perp}$ is the sum of standard number operators, one for each harmonic oscillator in the string. The main property of $N^{\perp}$ is that its commutator with a creation operator gives the mode number of that operator:

$$\left[N^{\perp}, a_n^{I\dagger}\right] = n\, a_n^{I\dagger}, \tag{12.165}$$

as you can readily verify. In addition,

$$\left[N^{\perp}, a_n^{I}\right] = -n\, a_n^{I}. \tag{12.166}$$

Since the number operator is normal-ordered it annihilates the ground states:

$$N^{\perp}|p^+, \vec{p}_T\rangle = 0. \tag{12.167}$$

Note, incidentally, that the number operator $N^{\perp}$ enters into the definition of $L_0^{\perp}$ in (12.105). We can write

$$L_0^{\perp} = \alpha' p^I p^I + N^{\perp}. \tag{12.168}$$

Let us get some practice using $N^{\perp}$ by computing its action on some basis states. Consider, for example, its action on $a_2^{I\dagger}|p^+, \vec{p}_T\rangle$:

$$N^{\perp} a_2^{I\dagger}|p^+, \vec{p}_T\rangle = \left[N^{\perp}, a_2^{I\dagger}\right]|p^+, \vec{p}_T\rangle + a_2^{I\dagger} N^{\perp}|p^+, \vec{p}_T\rangle = 2 a_2^{I\dagger}|p^+, \vec{p}_T\rangle.$$

The state is an eigenstate of $N^{\perp}$ with eigenvalue 2. Now let us try a more complicated state:

$$N^{\perp} a_3^{J\dagger} a_2^{I\dagger}|p^+, \vec{p}_T\rangle = \left[N^{\perp}, a_3^{J\dagger}\right] a_2^{I\dagger}|p^+, \vec{p}_T\rangle + a_3^{J\dagger} N^{\perp} a_2^{I\dagger}|p^+, \vec{p}_T\rangle$$
$$= 5 a_3^{J\dagger} a_2^{I\dagger}|p^+, \vec{p}_T\rangle. \tag{12.169}$$

It is clear that when the number operator acts on a basis state, the eigenvalue is the sum of the mode numbers of the creation operators appearing in the state. In general, for the basis state $|\lambda\rangle$ in (12.162) we have

$$N^{\perp}|\lambda\rangle = N_{\lambda}^{\perp}|\lambda\rangle, \quad \text{with} \quad N_{\lambda}^{\perp} = \sum_{n=1}^{\infty}\sum_{I=2}^{25} n\lambda_{n,I} \,. \tag{12.170}$$

Since $N^{\perp}$ enters additively into the mass-squared operator (12.164), we see that the oscillator with mode number n contributes n units of $1/\alpha'$ to M^2. The eigenvalues of $N^{\perp}$ are non-negative integers, so for all string states $M^2 \geq -1/\alpha'$.

The open string state space is naturally equipped with an inner product. To define this inner product we introduce bras $\langle p^{+}, \vec{p}_T|$ which are the Hermitian conjugates of the kets $|p^{+}, \vec{p}_T\rangle$, and declare that

$$\langle p'^{+}, \vec{p}_T'| p^{+}, \vec{p}_T\rangle = \delta(p'^{+} - p^{+})\,\delta(\vec{p}_T' - \vec{p}_T)\,. \tag{12.171}$$

The delta functions here are necessary because the Hermiticity of the operators p^{+} and $\vec{p}_T$ guarantees that the overlap above vanishes unless the p^{+} and $\vec{p}_T$ eigenvalues of the states are identical. For the basis state $|\lambda\rangle$ in (12.162), we introduce the Hermitian conjugate bra $\langle\lambda|$ defined by

$$\langle\lambda| = \langle p^{+}, \vec{p}_T| \prod_{n=1}^{\infty}\prod_{I=2}^{25} \left(a_n^{I}\right)^{\lambda_{n,I}}. \tag{12.172}$$

The Hermitian conjugate of $|\lambda\rangle b$, where b is a number, is simply $b^*\langle\lambda|$. The inner product (λ', λ) between two basis states $|\lambda'\rangle$ and $|\lambda\rangle$ is defined as

$$(\lambda', \lambda) = \langle\lambda'|\lambda\rangle\,. \tag{12.173}$$

For arbitrary states, the inner product is defined by declaring it to be linear on the second argument and antilinear on the first argument. To evaluate the overlap $\langle\lambda'|\lambda\rangle$ one moves the annihilation operators in $\langle\lambda'|$ towards the ground state in $|\lambda\rangle$ and the creation operators in $|\lambda\rangle$ towards the ground state in $\langle\lambda'|$. In moving these operators across each other we use the commutation relations (12.64), which appear in the standard form because in the overlap the annihilation operators are to the left of the creation operators. The commutators give $+1$ whenever they are nonvanishing, so the result of simplifying the oscillator content of the overlap is a positive number times the basic overlap in (12.171). For example, take $|\lambda'\rangle = a_1^{I\dagger}|p'^{+}, \vec{p}_T'\rangle$ and $|\lambda\rangle = a_1^{J\dagger}|p^{+}, \vec{p}_T\rangle$. We then find

$$\langle\lambda'|\lambda\rangle = \langle p'^{+}, \vec{p}_T'| a_1^{I} a_1^{J\dagger}|p^{+}, \vec{p}_T\rangle = \delta^{IJ}\delta(p'^{+} - p^{+})\,\vec{\delta}\,(\vec{p}_T' - \vec{p}_T)\,, \tag{12.174}$$

where we evaluated the overlap by replacing the product of oscillators by their commutator and using (12.171). Since the delta functions are positive functions, we conclude that all basis states $|\lambda\rangle$ have positive norm: $(\lambda, \lambda) > 0$. This inner product does not vanish because the annihilation operators on the bra match with the creation operators on the ket. Technically, (λ, λ) is infinite because of the delta functions, but this infinity becomes harmless when we consider continuous superpositions of states. Using the linearity and antilinearity of the inner product, we conclude that any state has positive norm. This is consistent with our statement that the open string state space is a Hilbert space.

Quick calculation 12.8 Explain why $(\lambda', \lambda) = 0$ whenever the basis states $|\lambda'\rangle$ and $|\lambda\rangle$ are different.

For each state $|\lambda\rangle$ in (12.162), we can construct the corresponding time-dependent physical state

$$\exp\left(-i(L_0^{\perp} - 1)\,\tau\right)|\lambda\rangle, \tag{12.175}$$

which satisfies the Schrödinger equation with Hamiltonian (12.158). In the light-cone quantization of the string a time-dependent state is physical if it satisfies the Schrödinger equation. We will consider more general superpositions of time-dependent states in the following section.

We are now ready to discuss particular states in some detail. We will begin with the simplest ones, the ground states. These are the unique states with $N^{\perp} = 0$. As in the case of the point particle, the states $|p^+, \vec{p}_T\rangle$ are the one-particle states of a scalar field. They are states of a scalar particle. What is the mass of this particle? To find out, we act on the states with M^2:

$$M^2|\,p^+,\ \vec{p}_T\rangle = \frac{1}{\alpha'}(-1 + N^{\perp})|\,p^+,\ \vec{p}_T\rangle = -\frac{1}{\alpha'}|\,p^+,\ \vec{p}_T\rangle. \tag{12.176}$$

The value of M^2 is all due to the ordering constant. If this constant had vanished the mass would have been zero. In fact, massless scalars are problematic – they have not been observed in nature. The result, however, is strange: $M^2 = -1/\alpha' < 0$. The wavefunction of the state tells the same story: $\psi(\tau, p^+, \vec{p}_T)$ can be set in correspondence with a classical scalar field. This scalar field, which has a negative mass-squared, is called a *tachyon*. A negative mass-squared is a sign of instability: the potential for a scalar field goes like $V = \frac{1}{2}M^2\phi^2$ (see (10.2)), so a negative M^2 simply means that the stationary point $\phi = 0$ is unstable. The energy can be reduced by having $\phi \neq 0$. We will study this further in Section 12.8.

Let us consider now the excited states with lowest M^2. Those arise when $N^{\perp}$ takes the smallest possible nonzero value, $N^{\perp} = 1$. Remarkably, due to the ordering constant, the $N^{\perp} = 1$ states have $M^2 = 0$. They are massless states. Had the ordering constant taken a noninteger value, quantum string theory would have no massless states. We get states with $N^{\perp} = 1$ when we act with any of the transverse oscillators $a_1^{I\dagger}$ on the ground states $|p^+, \vec{p}_T\rangle$. That means that we have $D - 2 = 24$ massless states:

$$a_1^{I\dagger}|p^+,\ \vec{p}_T\rangle, \quad M^2 a_1^{I\dagger}|p^+,\ \vec{p}_T\rangle = 0. \tag{12.177}$$

The general massless state is a linear combination of the above basis states:

$$\sum_{I=2}^{25} \xi_I\, a_1^{I\dagger}|p^+,\ \vec{p}_T\rangle. \tag{12.178}$$

The above expression may remind you of the photon states (10.88) that we found in our light-cone analysis of Maxwell theory:

$$\sum_{I=2}^{D-1} \xi_I \, a^{I\dagger}_{p^+, \, p_T} |\Omega\rangle \, . \tag{12.179}$$

We have a matching of states: in both cases ξ_I is an arbitrary transverse vector, and the states correspond to one another:

$$a^{I\dagger}_1 |p^+, \vec{p}_T\rangle \longleftrightarrow a^{I\dagger}_{p^+, \, p_T} |\Omega\rangle \, . \tag{12.180}$$

Both states have exactly the same Lorentz labels, they carry the same momenta, and have the same mass. This proves a remarkable result. The open string theory quantum states include photon states! Our study of open string theory started from the Nambu–Goto action. This action has no hint whatsoever of electromagnetic gauge invariance. Nevertheless, we have shown that open string theory contains Maxwell field excitations. This astonishing result is largely due to the mass shift encountered in passing from the classical to the quantum theory of the open string.

It is worth belaboring the point. In Chapter 10 we showed that the quantum states of free Maxwell theory – the photon states – are $(D-2)$ massless states, labeled by a transverse Lorentz index. The index is important: it indicates that these states transform into each other under Lorentz transformations. Exactly these kind of states have appeared in our quantization of the string. Additionally, the collection of wavefunctions $\psi_I(\tau, p^+, \vec{p}_T)$ associated with the states in (12.177) matches with the components A_I of the Maxwell gauge field. Finally, the Schrödinger equation for these wavefunctions matches the light-cone gauge field equation for the Maxwell field. We will show this in the following section.

Let us conclude with an examination of the states with $N^\perp = 2$. They are built by acting on the ground states with $a^{I\dagger}_1 a^{J\dagger}_1$ or with $a^{I\dagger}_2$. The number of states built with $a^{I\dagger}_1 a^{J\dagger}_1$ is the same as the number of independent entries in a symmetric matrix of size $(D-2)$, namely $\frac{1}{2}(D-2)(D-1)$. The number of states built with $a^{I\dagger}_2$ is $(D-2)$. The total number of states is therefore

$$\frac{1}{2}(D-2)(D-1) + (D-2) = \frac{1}{2}(D-2)(D+1) \, , \tag{12.181}$$

and their mass-squared is given by $M^2 = 1/\alpha'$. These particles are known as massive tensors, and in $D = 26$ there are 324 such states. Our results for all states with $N^\perp \leq 2$ are summarized in Table 12.1. A formula useful to count states at higher values of $N^\perp$ is derived in Problem 12.11. The formula tells you that at $N^\perp = 3$, for example, you get 2600 states by acting with $a^{I\dagger}_1 a^{J\dagger}_1 a^{K\dagger}_1$ on the ground states.

Each state $|\lambda\rangle$ of the quantum string represents a one-particle state of fixed momentum. Thus, the $a^{I\dagger}_1 |p^+, \vec{p}_T\rangle$ are one-photon states, and the $a^{I\dagger}_1 a^{J\dagger}_1 |p^+, \vec{p}_T\rangle$ are one-particle tensor states (*not* two-photon states). Each state $|\lambda\rangle$ has discrete labels $\lambda_{n,I}$ and continuous labels p^+ and $\vec{p}_T$. There is one wavefunction for each set of discrete labels, as you can see in the table. Accordingly, there is one quantum field for each set of discrete labels. The multiparticle states are described using these quantum fields. The total quantum field

Table 12.1		List of open string states with $N^\perp \leq 2$		
$N^\perp$	$\|\lambda\rangle$	$\alpha' M^2$	Number of states	Wavefunction
0	$\|p^+, \vec{p}_T\rangle$	-1	1	$\psi(\tau, p^+, \vec{p}_T)$
1	$a_1^{I\dagger}\|p^+, \vec{p}_T\rangle$	0	$D-2$	$\psi_I(\tau, p^+, \vec{p}_T)$
2	$a_1^{I\dagger}a_1^{J\dagger}\|p^+, \vec{p}_T\rangle$	1	$\frac{1}{2}(D-2)(D+1)$	$\psi_{IJ}(\tau, p^+, \vec{p}_T)$
	$a_2^{I\dagger}\|p^+, \vec{p}_T\rangle$			$\psi_I(\tau, p^+, \vec{p}_T)$

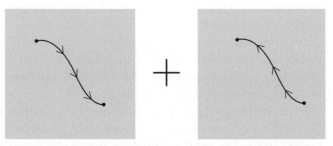

Fig. 12.1   An unoriented open string is a quantum state that can be obtained by superposition of states that differ only by orientation. The unoriented state is invariant under orientation reversal.

theory which describes the whole set of quantum fields associated with the one-particle states of the string is called *string field theory*.

The quantum theory we have discussed so far is that of *oriented* open strings. The quantum operator $X^I(\tau, \sigma)$ involves a parameter $\sigma \in [0, \pi]$ and thus an orientation, defined as the direction of increasing σ. It is possible to define a theory of *unoriented* strings (Problem 12.12). The key idea is that one can define an operator Ω that is a symmetry of the theory (commutes with the Hamiltonian) and reverses the orientation of strings. The theory of unoriented strings is obtained by restricting the oriented string spectrum to the set of states that are invariant under the action of Ω. Unoriented strings are not strings without orientation: they should be viewed as quantum superposition of states that, as a whole, are invariant under orientation reversal. We can imagine an unoriented state as the superposition of a string state and the same state with opposite orientation, as shown in Figure 12.1.

12.7 Equations of motion

To elaborate on the correspondence between string states and quantum fields we now consider the Schrödinger equations satisfied by the string wavefunctions. We saw in Section 11.4 that the Schrödinger equation for the point particle wavefunction is isomorphic to the classical field equation of a scalar field. We want to repeat such an analysis for the string.

To construct general time-dependent states from the string basis states we need wavefunctions. Consider, for example, a basis state

$$a_{n_1}^{I_1\,\dagger}\ldots a_{n_k}^{I_k\,\dagger}|p^+,\vec{p}_T\rangle\,.\qquad(12.182)$$

The general time-dependent state built by superposition is

$$|\Psi,\tau\rangle=\int dp^+\,d\vec{p}_T\,\psi_{I_1\ldots I_k}(\tau,p^+,\vec{p}_T)\,a_{n_1}^{I_1\,\dagger}\ldots a_{n_k}^{I_k\,\dagger}|p^+,\vec{p}_T\rangle\,.\qquad(12.183)$$

The polarization indices carried by the oscillators match with the index labels of the wavefunction $\psi_{I_1\ldots I_k}(\tau,p^+,\vec{p}_T)$. This equation is the string analog of (11.48), which gives the general time-dependent state of the point particle. For general tachyon states (12.183) becomes

$$|\text{tachyon},\tau\rangle=\int dp^+\,d\vec{p}_T\,\psi(\tau,p^+,\vec{p}_T)|p^+,\vec{p}_T\rangle\,.\qquad(12.184)$$

For photon states we write

$$|\text{photon},\tau\rangle=\int dp^+\,d\vec{p}_T\,\psi_I(\tau,p^+,\vec{p}_T)\,a_1^{I\,\dagger}|p^+,\vec{p}_T\rangle\,.\qquad(12.185)$$

The Schrödinger equation satisfied by the general states (12.183) is

$$i\frac{\partial}{\partial\tau}|\Psi,\tau\rangle=H|\Psi,\tau\rangle\,.\qquad(12.186)$$

Here, the Hamiltonian is given by

$$H=(L_0^\perp-1)=\alpha'p^Ip^I+N^\perp-1=\alpha'(p^Ip^I+M^2)\,,\qquad(12.187)$$

where we used (12.158) and (12.168). Using the explicit expression (12.183) for the states, equation (12.186) gives:

$$i\frac{\partial}{\partial\tau}\psi_{I_1\ldots I_k}=\left(\alpha'p^Ip^I+N^\perp-1\right)\psi_{I_1\ldots I_k}\,,\qquad(12.188)$$

where $N^\perp$ denotes the eigenvalue of the operator $N^\perp$ for the state (12.183).

Quick calculation 12.9 Show that equation (12.188) emerges from the Schrödinger equation (12.186). The calculation parallels that which gave (11.53).

For the tachyon states (12.184), $N^\perp=0$, and we get

$$i\frac{\partial\psi}{\partial\tau}=\left(\alpha'p^Ip^I-1\right)\psi\,.\qquad(12.189)$$

For the photon states (12.185), $N^\perp=1$, and we get

$$i\frac{\partial\psi_I}{\partial\tau}=\alpha'p^Jp^J\psi_I\,.\qquad(12.190)$$

Let us now compare these Schrödinger equations with the relevant classical field equations. We showed in Chapter 10 that the scalar field equation

$$\left(\partial^2-m^2\right)\phi=0\,,\qquad(12.191)$$

could be written as (10.30):

$$\left(i\,\frac{\partial}{\partial x^+} - \frac{1}{2p^+}(p^I p^I + m^2) \right) \phi(x^+, p^+, \vec{p}_T) = 0 . \tag{12.192}$$

Letting $x^+ = 2\alpha' p^+ \tau$ we now have

$$\left(i\,\frac{\partial}{\partial \tau} - \alpha'(p^I p^I + m^2) \right) \phi(\tau, p^+, \vec{p}_T) = 0 . \tag{12.193}$$

This equation is precisely the same as (12.189) when $m^2 = -1/\alpha'$, confirming the identification of the tachyon with a scalar field. Perhaps more surprisingly, equation (12.193) is structurally equivalent to the Schrödinger equation (12.188) satisfied by *any* string wavefunction. The only difference is that the wavefunctions carry indices. As a result, if the correspondence is to hold, the classical field equation for the field associated with any string state must take the form (12.191), with the field carrying some indices.

This may seem strange: is not the Maxwell classical field equation, for example, more complicated than the field equation for a scalar? Not in the light-cone gauge. We noticed this before: equation (10.83) showed that the transverse components of the gauge field satisfy $p^2 A^I(p) = 0$. This is of the form (12.191) with $m^2 = 0$. The steps leading from (12.191) to (12.193), when applied to $\partial^2 A^I = 0$ give

$$\left(i\,\frac{\partial}{\partial \tau} - \alpha' p^J p^J \right) A^I(\tau, p^+, \vec{p}_T) = 0 . \tag{12.194}$$

This classical field equation for the Maxwell field is in complete correspondence with the Schrödinger equation (12.190) for the $N^\perp = 1$ wavefunctions.

12.8 Tachyons and D-brane decay

We conclude this chapter by discussing the physics of the tachyon. We explained earlier that the tachyon state has the lowest value of M^2:

$$M^2 |p^+, \vec{p}_T\rangle = -\frac{1}{\alpha'}|p^+, \vec{p}_T\rangle . \tag{12.195}$$

The field associated with this state is a scalar field. What does it mean for this scalar to have a negative M^2? The physics of the open string tachyon was a mystery ever since the discovery of string theory. A series of developments starting in 1999 have essentially elucidated the role of the open string tachyon. Let us discuss what has been learned.

Our first goal is to understand the instability of a theory with a tachyon. For this purpose we consider the Lagrangian density for a classical scalar field, along the lines of Section 10.2. In some generality,

$$\mathcal{L} = -\frac{1}{2}\eta^{\mu\nu}\,\partial_\mu\phi\,\partial_\nu\phi - V(\phi) = \frac{1}{2}(\partial_0\phi)^2 - \frac{1}{2}|\nabla\phi|^2 - V(\phi) , \tag{12.196}$$

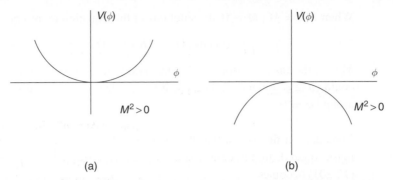

Fig. 12.2 (a) A potential $V(\phi) = \frac{1}{2}M^2\phi^2$ with positive mass-squared M^2. The value $\phi = 0$ is a stable critical point. (b) A potential $V(\phi) = \frac{1}{2}M^2\phi^2$ with negative mass-squared M^2. The value $\phi = 0$ is an unstable critical point.

where $V(\phi)$ is the *potential* for the scalar field. For spatially homogeneous field configurations, $\nabla\phi = 0$, and the potential energy density is given by the potential $V(\phi)$. The equation of motion following from variation is

$$\partial^2\phi - V'(\phi) = 0, \tag{12.197}$$

where prime denotes the derivative with respect to the argument. More explicitly,

$$-\frac{\partial^2\phi}{\partial t^2} + \nabla^2\phi - V'(\phi) = 0. \tag{12.198}$$

Quick calculation 12.10 Prove that equation (12.197) arises from variation of the action $S = \int d^D x \mathcal{L}$.

To understand the instability of the tachyon scalar field theory it suffices to consider the free part of the tachyon Lagrangian; interactions will feature later. For a free scalar field theory, the potential $V(\phi)$ takes the form

$$V(\phi) = \frac{1}{2}M^2\phi^2. \tag{12.199}$$

Here M^2 is the mass-squared of the scalar field (this potential will change with the inclusion of interactions). When $M^2 > 0$, the potential $V(\phi)$ has a stable minimum at $\phi = 0$. When $M^2 < 0$, $V(\phi)$ has an unstable maximum at $\phi = 0$ (Figure 12.2). We can understand the implications of such potentials by studying the equation of motion for the field. Using the specified form of V, equation (12.198) gives

$$-\frac{\partial^2\phi}{\partial t^2} + \nabla^2\phi - M^2\phi = 0. \tag{12.200}$$

To make our analysis simpler, let us assume that the field ϕ depends only on time. The equation of motion then becomes

$$\frac{d^2\phi(t)}{dt^2} + M^2\phi(t) = 0. \tag{12.201}$$

When $M^2 = M \cdot M > 0$, the solutions of this equation represent oscillations:

$$\phi = A \cos(Mt) + B \sin(Mt) = C \sin(Mt + \alpha_0). \qquad (12.202)$$

This is the interpretation of a scalar field with a "good" mass-squared. The scalar field could sit at $\phi = 0$ forever because it is a stable point; if it is displaced, it simply oscillates around $\phi = 0$.

Consider, on the other hand, the tachyon, which is an example of a scalar with negative mass-squared. In this case it is convenient to write $M^2 = -\beta^2 = -\beta \cdot \beta$, and equation (12.201) becomes

$$\frac{d^2\phi(t)}{dt^2} - \beta^2 \phi(t) = 0, \qquad (12.203)$$

with $\beta^2 > 0$. This time the solutions are

$$\phi(t) = A \cosh(\beta t) + B \sinh(\beta t). \qquad (12.204)$$

Consider the solution $\phi(t) = \sinh(\beta t)$. At time zero ϕ is zero, but as time goes to infinity ϕ also goes to infinity. We can imagine this as the field rolling to the right of the potential in Figure 12.2(b). In fact, for any nontrivial solution, ϕ necessarily reaches infinite absolute value either in the far past or in the far future. The tachyon could stay at $\phi = 0$ forever, using the trivial solution $\phi(t) = 0$, but any infinitesimal perturbation would set it on a course to a dramatic roll-off. The value $\phi = 0$ is an allowed critical point, but it is unstable. We cannot realistically expect the tachyon to stay near $\phi = 0$ for an indefinite length of time. This is the instability of a theory that contains a tachyon. Since the mass-squared of the open string tachyon is equal to $(-1/\alpha')$, the free part of the tachyon potential is

$$V_{\text{tach}}^{\text{free}}(\phi) = -\frac{1}{2\alpha'}\phi^2. \qquad (12.205)$$

A mechanical analogy works for arbitrary potentials $V(\phi)$. You can visualize the spatially homogeneous rolling of a scalar field on a potential $V(\phi)$ by considering the motion of a particle on the potential $V(x)$, where x, replacing ϕ, is the coordinate along the motion. Indeed, the relevant equations match. For an arbitrary potential $V(\phi)$, homogeneous rolling is governed by

$$\frac{d^2\phi}{dt^2} = -V'(\phi), \qquad (12.206)$$

while the rolling of a unit-mass particle on a potential $V(x)$ is governed by Newton's second law:

$$\frac{d^2x}{dt^2} = -V'(x). \qquad (12.207)$$

The presence of a tachyon signals an instability of open string theory. More precisely, there is some instability in the theory of open strings on the background of a space-filling D25-brane. It is clear that we should try to understand the fate of this instability: once the tachyon begins to roll, where does it end? For a while, not everyone agreed that this was an urgent question. Some argued that along with the lack of fermions, the tachyon was another good reason to consider this open string theory unrealistic and not worth much study. Some even saw the tachyon as a sign that open bosonic string theory is simply inconsistent. For quite a few years, superstring theories, the kind of string theories that also include fermions, seemed blessedly devoid of tachyons. Later studies, however, showed that tachyons can appear when we construct realistic models based on superstrings. It then became clear that we must try to understand tachyons.

The open string theory we have in our hands is the theory of strings on a D25-brane, a D-brane that fills all of the space dimensions. The D25-brane is a physical object, not just a mathematical construct, so it has a constant energy density T_{25} which, in fact, can be calculated exactly. The key insight can now be stated: the theory of open strings is, in some sense, the theory of the D25-brane itself! We have viewed tachyons as states of strings attached to a D-brane. A D-brane with open strings attached, it turns out, is an excited state of the D-brane. If this is so, a tachyon state represents an excitation that can lower the energy of the D-brane. The existence of the tachyon is telling us that the D25-brane is unstable!

Since the tachyon describes the physics of the D25-brane, the energy density of this brane is a contribution to the potential energy of the system, and it must be incorporated into the tachyon potential. As a result, the potential in (12.205) is changed into

$$V_{\text{tach}}(\phi) = T_{25} - \frac{1}{2\alpha'}\phi^2 + \beta\phi^3 + \cdots. \tag{12.208}$$

We have also included in the tachyon potential a cubic term and represented other possible terms by dots. All the terms that are cubic or higher order in the field represent the effect of interactions. The above potential describes correctly our statements about the D25-brane. The unstable point $\phi = 0$ represents the world with a D25-brane, and therefore has an energy density T_{25}. To find out what happens when the tachyon starts rolling down, we need to calculate the full tachyon potential $V_{\text{tach}}(\phi)$.

The physics can be anticipated before computing this potential. If the D25-brane is unstable, it will decay. The stable endpoint of this process would be a world without the D25-brane. If this is so, the tachyon potential must have a stable critical point at some $\phi = \phi^*$ with $V_{\text{tach}}(\phi^*) = 0$. That stable critical point would represent a background with zero energy, a background where the D25-brane, rendered unstable by the tachyon, has disappeared completely! The expected form of the tachyon potential is shown in Figure 12.3.

This proposal has now been verified convincingly. The complete tachyon potential was calculated using the field theory of open strings and, remarkably, it was possible to display a critical point with zero energy. With this result, and additional evidence obtained by other means, physicists have demonstrated that the tachyon instability is the instability of the D25-brane.

What happens when the tachyon rolls down to the stable minimum and the D25-brane disappears? All the open strings must also disappear, because open string endpoints are confined to D-branes. Closed strings, however, can exist in the absence of D-branes. All

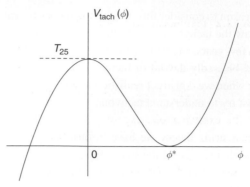

Fig. 12.3 The tachyon potential for the open string theory based on a D25-brane. The configuration $\phi = 0$ represents the unstable D-brane. The stable critical point ϕ^* has zero energy.

the energy initially stored in the D25-brane goes into closed strings. At the stable critical point ϕ^* all particles that arise as open string excitations, including the tachyon, must disappear. This shows that the theory near ϕ^* is quite subtle. *Vacuum string field theory* is an attempt to formulate string theory at the vacuum ϕ^* where both D-branes and open strings disappear.

Further interesting facts about tachyons and D-branes have emerged. It has been shown that Dp-branes with $p < 25$ are themselves large and coherent excitations of the tachyon field of the D25-brane. In some sense, D-branes are made of tachyons! This is also true, with minor modifications, in superstring theory. In superstring theory certain D-branes carry charge and therefore charge conservation ensures that they are stable against decay. In fact, the open string theory in the background of any such D-brane has no tachyons. However, a configuration consisting of a D-brane and a coincident, oppositely charged anti-D-brane, is unstable: the two objects can annihilate without violating charge conservation. The open strings which stretch from the D-brane to the anti-D-brane contain a tachyon – a superstring tachyon! This tachyon describes the instability of the D-brane/anti-D-brane pair. The study of D-brane/anti-D-brane annihilation plays an important role in attempts to use string theory to describe the early universe. It is thus possible that the tachyon will end up playing a prominent role in string cosmology.

Problems

Problem 12.1 Heisenberg equation for the momentum density.

We verified in (12.21) that $\dot{X}^I = 2\pi\alpha'\mathcal{P}^{\tau I}$ follows from the Heisenberg equation $i\dot{\xi} = [\xi, H]$, when $\xi = X^I$. Calculate $\dot{\mathcal{P}}^{\tau I}$, and use the result to verify that the classical equation of motion $\ddot{X}^I - X^{I''} = 0$ holds as an operator equation.

Problem 12.2 Testing explicitly some vanishing commutators.

Use the mode expansion (12.32) and the commutation relations of the α to check explicitly that equation (12.26) holds: $[X^{I'}(\tau,\sigma), X^{J'}(\tau,\sigma')] = [\dot{X}^I(\tau,\sigma), \dot{X}^J(\tau,\sigma')] = 0$.

Problem 12.3 Testing explicitly the main commutator.

(a) Use the explicit mode expansions of X^I and $\mathcal{P}^{\tau J}$, together with the commutation relations (12.45) and (12.54), to show that

$$\left[X^I(\tau,\sigma), \mathcal{P}^{\tau J}(\tau,\sigma')\right] = i\,\eta^{IJ}\,\frac{1}{\pi}\sum_{n\in\mathbb{Z}}\cos n\sigma\,\cos n\sigma'.$$

(b) If the above result agrees with (12.10), we must have

$$\delta(\sigma-\sigma') = \frac{1}{\pi}\sum_{n\in\mathbb{Z}}\cos n\sigma\,\cos n\sigma'. \tag{1}$$

This equation follows from the completeness of the functions $\cos n\sigma$ with $n \geq 0$ on the interval $\sigma \in [0,\pi]$. The completeness is readily explained: any function $f(\sigma)$ defined over $\sigma \in [0,\pi]$ can be extended to a function over $\sigma \in [-\pi,\pi]$ by letting $f(-\sigma) \equiv f(\sigma)$ for $\sigma \in [0,\pi]$. The resulting function is an even function of σ and by the basic result of Fourier series it can be expanded in terms of cosines. We can therefore expand any function $f(\sigma)$ with $\sigma \in [0,\pi]$ as

$$f(\sigma) = \sum_{n=0}^{\infty} A_n \cos n\sigma. \tag{2}$$

Prove (1) by calculating A_n and substituting the result back into the right-hand side of (2).

Problem 12.4 Analytic continuation of the zeta function.

Consider the definition of the gamma function $\Gamma(s) = \int_0^\infty dt\, e^{-t} t^{s-1}$. Let $t \to nt$ in this integral, and use the resulting equation to prove that

$$\Gamma(s)\,\zeta(s) = \int_0^\infty dt\,\frac{t^{s-1}}{e^t-1}, \qquad \Re(s) > 1. \tag{1}$$

Verify also the small t expansion

$$\frac{1}{e^t-1} = \frac{1}{t} - \frac{1}{2} + \frac{t}{12} + \mathcal{O}(t^2). \tag{2}$$

Use the above equations to show that for $\Re(s) > 1$

$$\Gamma(s)\,\zeta(s) = \int_0^1 dt\,t^{s-1}\left(\frac{1}{e^t-1} - \frac{1}{t} + \frac{1}{2} - \frac{t}{12}\right) + \frac{1}{s-1} - \frac{1}{2s} + \frac{1}{12(s+1)}$$
$$+ \int_1^\infty dt\,\frac{t^{s-1}}{e^t-1}.$$

Explain why the right-hand side above is well defined for $\Re(s) > -2$. It follows that this right-hand side defines the analytic continuation of the left-hand side to $\Re(s) > -2$. Recall the pole structure of $\Gamma(s)$ (Problem 3.6) and use it to show that $\zeta(0) = -1/2$ and that $\zeta(-1) = -1/12$.

Problem 12.5 The Virasoro algebra is a Lie algebra.

A vector space L with elements $x, y, z, \ldots$ and a bilinear bracket $[\cdot, \cdot]$, that takes two elements of L and yields another element of L, is a Lie algebra if the two following conditions hold.

(i) Antisymmetry: $[x, y] = -[y, x]$ for all elements x and y of L.
(ii) Jacobi identity: $[x, [y, z]] + [y, [z, x]] + [z, [x, y]] = 0$, for all elements x, y, and z of L.

Consider the vector space L spanned by the Virasoro operators with modes $n \in \mathbb{Z}$. Show that the commutators in (12.124), assumed to hold for all values of m and n, define a Lie algebra. Then, consider the commutators in (12.133), and show that they also define a Lie algebra.

Problem 12.6 Consistency conditions on the Virasoro algebra central terms.

The Virasoro commutation relations take the form

$$[L_m^{\perp}, L_n^{\perp}] = (m - n)L_{m+n}^{\perp} + C(m)\delta_{m+n,0}, \tag{1}$$

where $C(m)$ is a function of m that was calculated directly in this chapter. The purpose of this problem is to find the constraints on $C(m)$ that follow from the condition that (1) defines a Lie algebra.

(a) What does the antisymmetry requirement on a Lie algebra tell you about $C(m)$? What is $C(0)$?
(b) Consider now the Jacobi identity for generators $L_m^{\perp}, L_n^{\perp}$, and $L_k^{\perp}$ with $m + n + k = 0$. Show that

$$(m - n)C(k) + (n - k)C(m) + (k - m)C(n) = 0. \tag{2}$$

(c) Use equation (2) to show that $C(m) = \alpha m$ and $C(m) = \beta m^3$, for constants α and β, yield consistent central extensions.
(d) Consider equation (2) with $k = 1$. Show that $C(1)$ and $C(2)$ determine all $C(n)$.

Problem 12.7 Exercises with Virasoro operators.

(a) Use the Virasoro algebra (12.133) to show that if a state is annihilated by L_1 and L_2 it is annihilated by all L_n with $n \geq 1$.
(b) Consider the Virasoro operators L_0, L_1, and L_{-1}. Write out the three relevant commutators. Do these operators form a subalgebra of the Virasoro algebra? Is there a central term in here? Calculate the result of acting with each of these three operators on the zero-momentum vacuum state $|0\rangle$.

Problem 12.8 Virasoro operators acting on states.

(a) Use (12.100) to write the state $L_{-6}^{\perp}|0\rangle$ as a finite sum of terms with oscillators acting on the vacuum. Do the same for $L_{-2}^{\perp}L_{-2}^{\perp}|0\rangle$.

(b) Write $L_2^{\perp}L_{-2}^{\perp}|0\rangle$ and $L_{-2}^{\perp}L_2^{\perp}|0\rangle$ as finite sums of terms with oscillators acting on the vacuum. Use the results to evaluate $[L_2^{\perp}, L_{-2}^{\perp}]|0\rangle$. Confirm that your evaluation is consistent with the Virasoro algebra (12.133).

Problem 12.9 Reparameterizations generated by Virasoro operators.

(a) Consider the string at $\tau = 0$. Which of the combinations in (12.141) reparameterizes the σ coordinate of the string while keeping $\tau = 0$? When $\tau = 0$ is preserved, the world-sheet reparameterization is actually a *string* reparameterization. Show that the generators of these reparameterizations form a subalgebra of the Virasoro algebra.

(b) Describe the general *world-sheet* reparameterization that leaves the midpoint $\sigma = \pi/2$ of the $\tau = 0$ open string fixed. Express this reparameterization using an infinite set of constrained parameters.

Problem 12.10 Reparameterizations and constraints.

(a) Verify that the reparameterization parameters in (12.137) satisfy the relations (omitting the subscript m for convenience)

$$\dot{\xi}^{\tau} = \xi^{\sigma\prime}, \quad \dot{\xi}^{\sigma} = \xi^{\tau\prime}.$$

(b) Think of the reparameterizations (12.138) generated by the Virasoro operators as a change of coordinates

$$\tau' = \tau + \epsilon\,\xi^{\tau}(\tau,\sigma), \quad \sigma' = \sigma + \epsilon\,\xi^{\sigma}(\tau,\sigma).$$

Note that for infinitesimal ϵ the above equations also imply that

$$\tau = \tau' - \epsilon\,\xi^{\tau}(\tau',\sigma'), \quad \sigma = \sigma' - \epsilon\,\xi^{\sigma}(\tau',\sigma').$$

Show that the classical constraints

$$\partial_{\tau}X \cdot \partial_{\sigma}X = 0, \quad (\partial_{\tau}X)^2 + (\partial_{\sigma}X)^2 = 0,$$

assumed to hold in (τ,σ) coordinates, also hold in (τ',σ') coordinates (to order ϵ).

Problem 12.11 Counting symmetric products.[†]

Prove that the number N of different products of the form

$$a^{I_1}a^{I_2}\ldots a^{I_n},$$

where the superscripts $I_1, I_2, \ldots, I_n$ run over k values $1, 2, \ldots, k$, is given by

$$N = \frac{(n+k-1)!}{n!\,(k-1)!} = \frac{k}{1}\cdot\frac{k+1}{2}\cdot\frac{k+2}{3}\cdots\frac{k+n-1}{n}.$$

Hint: one can represent each product using n identical balls and $k-1$ identical dividers. For $k=6$ and $n=9$, for example, we have:

$$\bullet | \bullet\bullet | \quad | \bullet\bullet\bullet | \bullet | \bullet\bullet \quad \longleftrightarrow \quad a^1\, a^2 a^2\, a^4 a^4 a^4\, a^5\, a^6 a^6\,.$$

Problem 12.12 Unoriented open strings.

The open string $X^\mu(\tau,\sigma)$, with $\sigma \in [0,\pi]$ and fixed τ, is a parameterized curve in spacetime. The orientation of a string is the direction of increasing σ on this curve.

(a) Consider now the open string $X^\mu(\tau, \pi-\sigma)$ at the same τ. How is this second string related to the first string above? How are their endpoints and orientations related? Make a rough sketch showing the original string as a continuous curve in spacetime, and the second string as a dashed curve in spacetime.

Introduce an orientation reversing *twist* operator Ω such that

$$\Omega\, X^I(\tau,\sigma)\, \Omega^{-1} = X^I(\tau, \pi-\sigma)\,. \tag{1}$$

Moreover, declare that

$$\Omega\, x_0^-\, \Omega^{-1} = x_0^-\,, \qquad \Omega\, p^+\, \Omega^{-1} = p^+\,. \tag{2}$$

(b) Use the open string oscillator expansion (12.32) to calculate

$$\Omega\, x_0^I\, \Omega^{-1}\,, \qquad \Omega\, \alpha_0^I\, \Omega^{-1}\,, \quad \text{and} \quad \Omega\, \alpha_n^I\, \Omega^{-1}\ (n \neq 0)\,.$$

(c) Show that $\Omega\, X^-(\tau,\sigma)\, \Omega^{-1} = X^-(\tau, \pi-\sigma)$. Since $\Omega\, X^+(\tau,\sigma)\, \Omega^{-1} = X^+(\tau, \pi-\sigma)$, equation (1) actually holds for all string coordinates. We say that orientation reversal is a symmetry of open string theory because it leaves the open string Hamiltonian H invariant: $\Omega H \Omega^{-1} = H$. Explain why this is true.

(d) Assume that the ground states are twist invariant:

$$\Omega\, |p^+, \vec{p}_T\rangle = \Omega^{-1}\, |p^+, \vec{p}_T\rangle = |p^+, \vec{p}_T\rangle\,.$$

List the open string states for $N^\perp \leq 3$, and give their twist eigenvalues. Prove that, in general,

$$\Omega = (-1)^{N^\perp}\,.$$

(e) A state is said to be *unoriented* if it is invariant under twist. If you are commissioned to build a theory of unoriented open strings, which of the states in part (d) would you have to discard? In general, which levels of the original string state space must be discarded?

Problem 12.13 Tachyon potentials.

Consider scalar field theories of the form

$$S = \int d^D x\left(-\frac{1}{2}\partial_\mu\phi\partial^\mu\phi - V(\phi)\right)\,. \tag{1}$$

We will examine three different scalar potentials:

$$V_1(\phi) = \frac{1}{\alpha'} \frac{1}{3\phi_0} (\phi - \phi_0)^2 \left(\phi + \frac{1}{2}\phi_0\right), \qquad (2a)$$

$$V_2(\phi) = -\frac{1}{4\alpha'} \phi^2 \ln\left(\frac{\phi^2}{\phi_0^2}\right), \qquad (2b)$$

$$V_3(\phi) = \frac{1}{8\alpha'} \phi_0^2 \left(\frac{\phi^2}{\phi_0^2} - 1\right)^2, \qquad (2c)$$

where ϕ_0 is a (positive) constant. For each of the three potentials V_i, do the following.

(a) Plot $V_i(\phi)$ as a function of ϕ.
(b) Find the critical points of the potential and the values of the potentials at those points. Each critical point represents a possible background for the scalar field theory.
(c) At each critical point $\bar{\phi}$ expand the action for fluctuations of ϕ around this point, that is, let $\phi = \bar{\phi} + \psi$ where the fluctuation ψ is small. The quadratic term in ψ (with no derivatives) can be used to read the mass of the scalar particle. Give the mass of the scalar particle for each critical point.

The potential V_1 is a rough model for the tachyon potential on an unstable D-brane. V_2 is an exact (effective) tachyon potential on an unstable D-brane. The potential V_3 is a rough model for the superstring tachyon potential on the world-volume of a D-brane and a coincident anti-D-brane.

Except for shared position and momentum zero modes, the operator content of quantum closed strings can be viewed as two commuting copies of the open string operators. Even in the light-cone gauge, the reparameterization invariance cannot be fully fixed: there is no natural way to choose a starting point for a closed string. As a result, the closed string spectrum is subject to the constraint $L_0^\perp - \tilde{L}_0^\perp = 0$, which selects the states that are invariant under rigid rotations of the string. We find that the massless closed string quantum states include one-particle graviton states, making string theory a quantum gravity theory. Additionally, we find massless Kalb–Ramond and dilaton states. The dilaton state controls the strength of string interactions. We study closed strings on the orbifold $\mathbb{R}^1/\mathbb{Z}_2$.

13.1 Mode expansions and commutation relations

When it was first discovered, string theory was thought to be a theory of strongly interacting particles – a theory of hadrons. The consistency of open string theory required the inclusion of closed strings. But there was a problem with closed strings: among the excitations of closed strings there were massless states with spin two. No known hadron had these properties. Despite much effort, all attempts to eliminate these closed string states from the spectrum failed.

It turns out that these massless states can be identified as graviton states, and physicists eventually realized that closed string theory could be a theory of quantum gravity. In this chapter we quantize the relativistic closed string and see how graviton states emerge. Much of the quantization procedure will resemble our quantization of the open string in Chapter 12, but there are a number of new features. Let us begin by recalling some of the important facts about closed strings that we learned in Chapter 9. We considered at that time a family of gauges (see (9.27)) defined by the conditions

$$n \cdot X = \alpha'(n \cdot p)\, \tau , \quad n \cdot p = 2\pi\, n \cdot \mathcal{P}^\tau . \tag{13.1}$$

The second condition implies that the parameter σ spans an interval of length 2π:

$$\sigma \in [0, 2\pi] . \tag{13.2}$$

Here, $\sigma = 0$ and $\sigma = 2\pi$ represent the *same* point on the closed string. The range $\sigma \in [0, 2\pi]$ for closed strings is twice the range $\sigma \in [0, \pi]$ used for open strings. We found that the conditions (13.1) do not fully fix the parameterization of the closed strings. Unlike open strings, closed strings do not have a special point that can be selected as $\sigma = 0$. We used this arbitrariness to our advantage: once we selected some $\sigma = 0$ on one closed string, we could impose the constraint $X' \cdot \dot{X} = 0$ by suitably choosing the $\sigma = 0$ point on all of the other closed strings on the world-sheet. After this, we still had the ability to let $\sigma \to \sigma + \sigma_0$, with some constant σ_0 that is the same for all strings. This rigid rotation of the lines of constant σ is a reparameterization invariance of the closed string action that cannot be fixed. When we build the quantum states of the closed string, this will result in a constraint on states.

The condition $X' \cdot \dot{X} = 0$, together with the parameterization conditions (13.1), implied $X'^2 + \dot{X}^2 = 0$. We thus obtained the familiar conditions

$$(\dot{X} \pm X')^2 = 0\,, \tag{13.3}$$

and the momentum densities became simple derivatives of the coordinates:

$$\mathcal{P}^{\sigma\mu} = -\frac{1}{2\pi\alpha'} X'^{\mu}\,, \quad \mathcal{P}^{\tau\mu} = \frac{1}{2\pi\alpha'} \dot{X}^{\mu}\,. \tag{13.4}$$

Finally, all the string coordinates were seen to satisfy the wave equation:

$$\left(\frac{\partial^2}{\partial\tau^2} - \frac{\partial^2}{\partial\sigma^2}\right) X^{\mu} = 0\,. \tag{13.5}$$

Let us now consider the classical solution to the equation of motion for the closed string. The general solution to the wave equation is

$$X^{\mu}(\tau, \sigma) = X^{\mu}_{L}(\tau + \sigma) + X^{\mu}_{R}(\tau - \sigma)\,, \tag{13.6}$$

where X^{μ}_{L} (the L stands for left-moving) is a wave moving towards more negative σ and X^{μ}_{R} (the R stands for right-moving) is a wave moving towards more positive σ. For open strings, the left-moving and right-moving waves were related to each other by the boundary conditions at the endpoints. The closed string has no endpoints, but we have a periodicity condition to work with. The parameter space (τ, σ) for closed strings is a cylinder, so, to describe closed strings properly we compactify the world-sheet coordinate σ:

$$\sigma \sim \sigma + 2\pi\,. \tag{13.7}$$

Two points on the world-sheet whose difference of σ coordinates is a multiple of 2π are the same point. We can in fact use any interval of the form $[\sigma_0, \sigma_0 + 2\pi]$ to describe the closed strings; the choice in (13.2) is one of the possible choices. When we include the τ coordinate, the identification of points on the parameter space is given by

$$(\tau, \sigma) \sim (\tau, \sigma + 2\pi)\,. \tag{13.8}$$

We demand that X^{μ} assumes the same value at any two coordinates that represent the same point on the parameter space:

$$X^\mu(\tau, \sigma) = X^\mu(\tau, \sigma + 2\pi), \quad \text{for all } \tau \text{ and } \sigma. \tag{13.9}$$

This condition is both easier to deal with and easier to interpret than the naive condition $X^\mu(\tau, 0) = X^\mu(\tau, 2\pi)$. The periodicity condition (13.9) is appropriate for strings that propagate in a simply connected space, a space in which every closed string can be continuously shrunk to a point. Minkowski space, for example, is simply connected. If a spatial direction is curled up into a circle, closed strings wrapped around the circle cannot be shrunk away. Space is then not simply connected, and the coordinate along the circle is not single valued. For such a coordinate, the periodicity condition (13.9) must be modified (see Chapter 17).

We will now show that the periodicity condition (13.9) induces a small but significant constraint that relates the left-moving and the right-moving waves. Let us define two new variables,

$$\begin{aligned} u &= \tau + \sigma, \\ v &= \tau - \sigma. \end{aligned} \tag{13.10}$$

In terms of these variables, equation (13.6) becomes

$$X^\mu = X_L^\mu(u) + X_R^\mu(v). \tag{13.11}$$

When $\sigma \to \sigma + 2\pi$, the variables u and v increase and decrease by 2π, respectively. As a result, the periodicity condition (13.9) gives

$$X_L^\mu(u) + X_R^\mu(v) = X_L^\mu(u + 2\pi) + X_R^\mu(v - 2\pi), \tag{13.12}$$

or, equivalently,

$$X_L^\mu(u + 2\pi) - X_L^\mu(u) = X_R^\mu(v) - X_R^\mu(v - 2\pi). \tag{13.13}$$

This equation establishes that the left-moving and right-moving waves are in fact dependent on each other: if one fails to be periodic, the other has to fail by the same amount. Since u and v are independent variables, both the u derivative of the right-hand side and the v derivative of the left-hand side must vanish. As a consequence, we find that both $X_L^{\mu\prime}(u)$ and $X_R^{\mu\prime}(v)$ are strictly periodic functions with period 2π (for functions of a single variable primes denote derivatives with respect to the argument). We can therefore write the mode expansions

$$X_L^{\mu\prime}(u) = \sqrt{\frac{\alpha'}{2}} \sum_{n \in \mathbb{Z}} \bar{\alpha}_n^\mu e^{-inu},$$

$$X_R^{\mu\prime}(v) = \sqrt{\frac{\alpha'}{2}} \sum_{n \in \mathbb{Z}} \alpha_n^\mu e^{-inv}. \tag{13.14}$$

A set of barred α modes was introduced for the expansion of $X_L^{\mu\prime}(u)$. Even though they are written identically, the unbarred α modes used in the expansion of $X_R^{\mu\prime}(v)$ have no relation to the open string modes of Chapter 12. In closed string theory we need two sets of α modes, barred and unbarred. We integrate equations (13.14) to find

$$X_L^\mu(u) = \frac{1}{2} x_0^{L\mu} + \sqrt{\frac{\alpha'}{2}}\, \bar\alpha_0^\mu \, u + i\sqrt{\frac{\alpha'}{2}} \sum_{n\neq 0} \frac{\bar\alpha_n^\mu}{n} e^{-inu},$$

$$X_R^\mu(v) = \frac{1}{2} x_0^{R\mu} + \sqrt{\frac{\alpha'}{2}}\, \alpha_0^\mu \, v + i\sqrt{\frac{\alpha'}{2}} \sum_{n\neq 0} \frac{\alpha_n^\mu}{n} e^{-inv}, \tag{13.15}$$

where the coordinate zero modes $x_0^{L\mu}$ and $x_0^{R\mu}$ have appeared as constants of integration. These are somewhat puzzling, after all, in open string theory there was a single coordinate zero mode whose canonical conjugate was the momentum of the string. We will see that only the sum of the two zero modes plays a role here. If the space is not simply connected, however, each of the coordinate zero modes plays a role, as we will see in Chapter 17.

The aperiodicity of X_R^μ and of X_L^μ is a consequence of the linear terms appearing in (13.15). Condition (13.13) constrains these terms giving

$$2\pi \sqrt{\frac{\alpha'}{2}}\, \bar\alpha_0^\mu = 2\pi \sqrt{\frac{\alpha'}{2}}\, \alpha_0^\mu, \tag{13.16}$$

and therefore

$$\bar\alpha_0^\mu = \alpha_0^\mu. \tag{13.17}$$

Owing to this equality, quantum closed string theory has only *one* momentum operator. As we will soon see, this means that canonical quantization works consistently with only *one* coordinate zero mode operator.

We can now assemble the mode expansion for $X^\mu(\tau, \sigma)$ by substituting (13.15) into (13.6):

$$X^\mu(\tau, \sigma) = \frac{1}{2} x_0^{L\mu} + \sqrt{\frac{\alpha'}{2}}\, \bar\alpha_0^\mu (\tau + \sigma) + i\sqrt{\frac{\alpha'}{2}} \sum_{n\neq 0} \frac{\bar\alpha_n^\mu}{n} e^{-in(\tau+\sigma)}$$

$$+ \frac{1}{2} x_0^{R\mu} + \sqrt{\frac{\alpha'}{2}}\, \alpha_0^\mu (\tau - \sigma) + i\sqrt{\frac{\alpha'}{2}} \sum_{n\neq 0} \frac{\alpha_n^\mu}{n} e^{-in(\tau-\sigma)}. \tag{13.18}$$

With $\bar\alpha_0^\mu = \alpha_0^\mu$, we find

$$X^\mu(\tau, \sigma) = \frac{1}{2}(x_0^{L\mu} + x_0^{R\mu}) + \sqrt{2\alpha'}\, \alpha_0^\mu \, \tau + i\sqrt{\frac{\alpha'}{2}} \sum_{n\neq 0} \frac{e^{-in\tau}}{n}(\alpha_n^\mu e^{in\sigma} + \bar\alpha_n^\mu e^{-in\sigma}). \tag{13.19}$$

The reality of $X^\mu(\tau, \sigma)$ requires $\alpha_0^\mu, x_0^{L\mu}, x_0^{R\mu}$ to be real, as well as $\alpha_{-n}^\mu = (\alpha_n^\mu)^*$ and $\bar\alpha_{-n}^\mu = (\bar\alpha_n^\mu)^*$, for $n \geq 1$.

As expected, X^μ is a periodic function of σ with period 2π. The canonically conjugate momentum density is

$$\mathcal{P}^{\tau\mu}(\tau, \sigma) = \frac{1}{2\pi\alpha'}\dot{X}^\mu(\tau, \sigma) = \frac{1}{2\pi\alpha'}(\sqrt{2\alpha'}\, \alpha_0^\mu + \cdots), \tag{13.20}$$

where the dots represent the terms in $\dot{X}^\mu$ that integrate to zero over the interval $\sigma \in [0, 2\pi]$ and therefore do not contribute to the evaluation of the total momentum:

$$p^\mu = \int_0^{2\pi} \mathcal{P}^{\tau\mu}(\tau, \sigma) d\sigma = \frac{1}{2\pi\alpha'} \int_0^{2\pi} d\sigma \sqrt{2\alpha'}\, \alpha_0^\mu = \sqrt{\frac{2}{\alpha'}}\alpha_0^\mu. \tag{13.21}$$

Thus we have the relation

$$\alpha_0^\mu = \sqrt{\frac{\alpha'}{2}}\, p^\mu. \tag{13.22}$$

This differs from the corresponding open string result (12.46) by a factor of two, but the idea is the same: α_0^μ is proportional to the spacetime momentum carried by the string.

There is only one momentum variable, and thus in the quantum theory there is only one momentum operator. We should also have only one conjugate coordinate zero mode. Thus, despite our left–right decomposition of the solution to the wave equation, x_0^L and x_0^R cannot both be independent variables. Only their sum appears in (13.19), so it must be the sum that is the relevant coordinate zero mode. Without loss of generality, we set

$$x_0^{L\mu} = x_0^{R\mu} \equiv x_0^\mu, \tag{13.23}$$

and equation (13.19) takes its final form:

$$X^\mu(\tau, \sigma) = x_0^\mu + \sqrt{2\alpha'}\,\alpha_0^\mu \tau + i\sqrt{\frac{\alpha'}{2}} \sum_{n \neq 0} \frac{e^{-in\tau}}{n}(\alpha_n^\mu e^{in\sigma} + \bar{\alpha}_n^\mu e^{-in\sigma}). \tag{13.24}$$

It is convenient at this stage to record the τ and σ derivatives of the coordinates. With the help of (13.6) we note that

$$\dot{X}^\mu = X_L^{\mu\prime}(\tau + \sigma) + X_R^{\mu\prime}(\tau - \sigma),$$
$$X^{\mu\prime} = X_L^{\mu\prime}(\tau + \sigma) - X_R^{\mu\prime}(\tau - \sigma). \tag{13.25}$$

Adding and subtracting these equations, and using (13.14), we find

$$\dot{X}^\mu + X^{\mu\prime} = 2X_L^{\mu\prime}(\tau + \sigma) = \sqrt{2\alpha'} \sum_{n \in \mathbb{Z}} \bar{\alpha}_n^\mu\, e^{-in(\tau+\sigma)},$$
$$\dot{X}^\mu - X^{\mu\prime} = 2X_R^{\mu\prime}(\tau - \sigma) = \sqrt{2\alpha'} \sum_{n \in \mathbb{Z}} \alpha_n^\mu e^{-in(\tau-\sigma)}. \tag{13.26}$$

Note that the barred oscillators do not mix with the unbarred oscillators in these combinations of derivatives. We have tailored the normalization constants to arrive at the above relations. They are completely analogous to the open string expansions (12.33). This will allow us to obtain some closed string commutators without doing any new computations.

Let us now turn to the quantization of the closed string theory. The canonical commutation relations take the same form as in open string theory. For the transverse light-cone coordinates and momenta we set

$$\left[X^I(\tau, \sigma), \mathcal{P}^{\tau J}(\tau, \sigma')\right] = i\delta(\sigma - \sigma')\eta^{IJ}, \tag{13.27}$$

and, as usual, we set to zero the commutator of coordinates with coordinates and the commutator of momenta with momenta. For zero modes, we also have $[x_0^-, p^+] = -i$. Since the commutation relations did not change, equations (12.30) and (12.31) are valid here. The first of these is

$$[(\dot{X}^I \pm X^{I'})(\tau, \sigma), (\dot{X}^J \pm X^{J'})(\tau, \sigma')] = \pm 4\pi\alpha' i\eta^{IJ}\frac{d}{d\sigma}\delta(\sigma - \sigma'). \tag{13.28}$$

In fact, this time the situation is simpler. Equations (13.28) hold for $\sigma, \sigma' \in [0, 2\pi]$ since the string coordinates are defined for this full interval. Moreover, the mode expansions (13.26) also hold for the full interval. Since the combinations of derivatives take the same exact form as they did for open strings, the above equation leads to identical looking commutation relations. The oscillators, however, are barred when we use the top sign and unbarred when we use the lower sign. The result is therefore

$$\left[\bar{\alpha}_m^I, \bar{\alpha}_n^J\right] = m\,\delta_{m+n,0}\,\eta^{IJ},$$

$$\left[\alpha_m^I, \alpha_n^J\right] = m\,\delta_{m+n,0}\,\eta^{IJ}. \tag{13.29}$$

On account of the expansions (13.26), we call the $\bar{\alpha}$ operators left-moving operators, and we call the α operators right-moving operators. Each of these sets matches the operator content of an open string theory. The commutation relations also take the form we would expect from open string theory. Closed string theory thus has the operator content of two copies of open string theory, except for zero modes. The momentum zero modes are equal ($\alpha_0^I = \bar{\alpha}_0^I$), and there is only one set of coordinate zero modes x_0^I, x_0^-.

Equation (12.31) states that combinations of derivatives with opposite signs commute. In the present case this leads to the result that left-moving and right-moving oscillators commute:

$$\left[\alpha_m^I, \bar{\alpha}_n^J\right] = 0. \tag{13.30}$$

We can define canonical creation and annihilation operators just as we did for open strings:

$$\alpha_n^I = a_n^I\sqrt{n} \quad \text{and} \quad \alpha_{-n}^I = a_n^{I\dagger}\sqrt{n}, \quad n \geq 1,$$

$$\bar{\alpha}_n^I = \bar{a}_n^I\sqrt{n} \quad \text{and} \quad \bar{\alpha}_{-n}^I = \bar{a}_n^{I\dagger}\sqrt{n}, \quad n \geq 1. \tag{13.31}$$

Together with the hermiticity of x_0^μ and α_0^μ, these make $X^\mu(\tau, \sigma)$ in (13.24) Hermitian.

The nonvanishing commutation relations are then the expected ones:

$$\left[\bar{a}_m^I, \bar{a}_n^{J\dagger}\right] = \delta_{m,n}\,\eta^{IJ}, \qquad \left[a_m^I, a_n^{J\dagger}\right] = \delta_{m,n}\,\eta^{IJ}. \tag{13.32}$$

The commutators that involve x_0^I can be found following steps analogous to those used for open strings. This time (Problem 13.1) we find that $[x_0^I, \alpha_n^J]$ and $[x_0^I, \bar{\alpha}_n^J]$ vanish when $n \neq 0$, and

$$\left[x_0^I, \alpha_0^J\right] = \left[x_0^I, \bar{\alpha}_0^J\right] = i\sqrt{\frac{\alpha'}{2}}\,\eta^{IJ} \longrightarrow \left[x_0^I, p^J\right] = i\eta^{IJ}, \tag{13.33}$$

where the expression to the right arises because of (13.22).

What is the light-cone closed string Hamiltonian? We know that p^- generates X^+ translations, and that, in addition, $X^+ = \alpha' p^+ \tau$. As a result, $\partial_\tau = \alpha' p^+ \partial_{X^+}$, and the Hamiltonian must be given by

$$H = \alpha' p^+ p^-. \tag{13.34}$$

In order to find the normal-ordered version of this Hamiltonian, we now turn to the transverse Virasoro operators of closed string theory.

13.2 Closed string Virasoro operators

We learned in Chapter 12 that the open string transverse Virasoro operators are essentially the modes α_n^- of the light-cone coordinate X^-. For closed string coordinates we have two types of modes, barred and unbarred. This is also true for the closed string X^- coordinates: we have α_n^- and $\bar{\alpha}_n^-$ modes, and therefore we expect to have two sets of Virasoro operators. On account of (13.17), however, we have $\alpha_0^- = \bar{\alpha}_0^-$, so a surprise awaits us with regards to the Virasoro operators with mode number zero.

To begin our analysis we need an expression that relates X^- to the transverse coordinates. The requisite formula is (9.65), with $\beta = 1$, as appropriate for closed strings:

$$\dot{X}^- \pm X^{-\prime} = \frac{1}{\alpha'}\frac{1}{2p^+}(\dot{X}^I \pm X^{I\prime})^2. \tag{13.35}$$

We define Virasoro operators following the pattern in equation (9.79):

$$(\dot{X}^I + X^{I\prime})^2 = 4\alpha'\sum_{n\in\mathbb{Z}}\left(\frac{1}{2}\sum_{p\in\mathbb{Z}}\bar{\alpha}_p^I\bar{\alpha}_{n-p}^I\right)e^{-in(\tau+\sigma)} \equiv 4\alpha'\sum_{n\in\mathbb{Z}}\bar{L}_n^\perp e^{-in(\tau+\sigma)},$$

$$(\dot{X}^I - X^{I\prime})^2 = 4\alpha'\sum_{n\in\mathbb{Z}}\left(\frac{1}{2}\sum_{p\in\mathbb{Z}}\alpha_p^I\alpha_{n-p}^I\right)e^{-in(\tau-\sigma)} \equiv 4\alpha'\sum_{n\in\mathbb{Z}}L_n^\perp e^{-in(\tau-\sigma)}. \tag{13.36}$$

In each of the above lines, the equality requires a small calculation using (13.26), and the second relation is a definition. More explicitly,

$$\bar{L}_n^{\perp} = \frac{1}{2} \sum_{p \in \mathbb{Z}} \bar{\alpha}_p^I \bar{\alpha}_{n-p}^I \,, \qquad L_n^{\perp} = \frac{1}{2} \sum_{p \in \mathbb{Z}} \alpha_p^I \alpha_{n-p}^I \,. \tag{13.37}$$

These are the two sets of Virasoro operators of closed string theory. Plugging the definitions in (13.36) back into (13.35) we obtain

$$\dot{X}^- + X^{-\prime} = \frac{2}{p^+} \sum_{n \in \mathbb{Z}} \bar{L}_n^{\perp} e^{-in(\tau+\sigma)} \,, \qquad \dot{X}^- - X^{-\prime} = \frac{2}{p^+} \sum_{n \in \mathbb{Z}} L_n^{\perp} e^{-in(\tau-\sigma)} \,. \tag{13.38}$$

On the other hand, the derivatives of X^-, as those of any other closed string coordinate, can be expanded along the lines of (13.26) to give

$$\dot{X}^- + X^{-\prime} = \sqrt{2\alpha'} \sum_{n \in \mathbb{Z}} \bar{\alpha}_n^- e^{-in(\tau+\sigma)} \,, \qquad \dot{X}^- - X^{-\prime} = \sqrt{2\alpha'} \sum_{n \in \mathbb{Z}} \alpha_n^- e^{-in(\tau-\sigma)} \,. \tag{13.39}$$

We compare equations (13.38) and (13.39) to read the expressions for the minus oscillators:

$$\sqrt{2\alpha'}\, \bar{\alpha}_n^- = \frac{2}{p^+}\, \bar{L}_n^{\perp} \,, \qquad \sqrt{2\alpha'}\, \alpha_n^- = \frac{2}{p^+}\, L_n^{\perp} \,. \tag{13.40}$$

For $n = 0$, however, there is a constraint. Since $\alpha_0^- = \bar{\alpha}_0^-$, we have the *level-matching* condition

$$L_0^{\perp} = \bar{L}_0^{\perp} \,. \tag{13.41}$$

If you look at the definitions of $\bar{L}_0^{\perp}$ and $L_0^{\perp}$ in (13.37), you will realize that these two operators are clearly very different from each other. What does it mean that they must be equal? Since operators are ultimately defined by their action on states, the meaning of the equality (13.41) is that any state $|\lambda, \bar{\lambda}\rangle$ of the closed string must satisfy $L_0^{\perp}|\lambda, \bar{\lambda}\rangle = \bar{L}_0^{\perp}|\lambda, \bar{\lambda}\rangle$. This is therefore a constraint on the state space of the theory: "states" that do not satisfy this constraint do not in fact belong to the state space.

To fix the ordering ambiguities in the operators $\bar{L}_0^{\perp}$ and $L_0^{\perp}$ we define them to be ordered operators without any additional constants:

$$\bar{L}_0^{\perp} = \frac{\alpha'}{4} p^I p^I + \bar{N}^{\perp} \,, \qquad L_0^{\perp} = \frac{\alpha'}{4} p^I p^I + N^{\perp} \,. \tag{13.42}$$

Here $\bar{N}^{\perp}$ and $N^{\perp}$ are the number operators that are associated with the barred and un-barred operators, respectively:

$$\bar{N}^{\perp} \equiv \sum_{n=1}^{\infty} n \bar{a}_n^{I\dagger} \bar{a}_n^I \,, \qquad N^{\perp} \equiv \sum_{n=1}^{\infty} n a_n^{I\dagger} a_n^I \,. \tag{13.43}$$

While we will not go through the trouble of proving it, the critical dimension for closed strings turns out to be $D = 26$. This follows from the requirement that the quantum theory be Lorentz invariant. It is no coincidence that the critical dimension for closed strings coincides with the critical dimension for open strings. It means that both types of strings can coexist. In fact, since open strings can, in general, close to form closed strings, it would have been quite strange if the critical dimensions did not agree.

The constant ambiguities due to the ordering of $\bar{L}_0^\perp$ and $L_0^\perp$ are also fixed by the condition of Lorentz invariance, just as it happened for open strings. The answer could be anticipated, since the left and right sectors of closed string theory behave like open strings. In addition, the naive argument based on zeta functions suggests that the ordering constants for $L_0^\perp$ and $\bar{L}_0^\perp$ are the same and equal to that for the $L_0^\perp$ operator of the open string. These constants are included in the relation between α_0^- and $L_0^\perp$ and in the corresponding barred relation. Therefore equations (13.40), for $n = 0$, become

$$\sqrt{2\alpha'}\,\bar{\alpha}_0^- = \frac{2}{p^+}(\bar{L}_0^\perp - 1)\,, \qquad \sqrt{2\alpha'}\,\alpha_0^- = \frac{2}{p^+}(L_0^\perp - 1)\,. \tag{13.44}$$

The level-matching constraint $L_0^\perp = \bar{L}_0^\perp$, which emerged from $\alpha_0^- = \bar{\alpha}_0^-$, remains unchanged by the constant shifts. Using (13.42), this constraint can be written more simply as

$$N^\perp = \bar{N}^\perp\,. \tag{13.45}$$

Averaging the two expressions for α_0^- in (13.44), we can find a symmetric expression:

$$\sqrt{2\alpha'}\alpha_0^- \equiv \frac{1}{p^+}(L_0^\perp + \bar{L}_0^\perp - 2) = \alpha' p^-\,, \tag{13.46}$$

where the relation to p^- follows from (13.22). With p^- known, we can calculate the mass-squared:

$$M^2 = -p^2 = 2p^+ p^- - p^I p^I = \frac{2}{\alpha'}(L_0^\perp + \bar{L}_0^\perp - 2) - p^I p^I\,. \tag{13.47}$$

Substituting the values of $\bar{L}_0^\perp$ and $L_0^\perp$ given in (13.42) yields

$$M^2 = \frac{2}{\alpha'}\left(N^\perp + \bar{N}^\perp - 2\right)\,. \tag{13.48}$$

This is the mass formula for closed string states. The closed string Hamiltonian (13.34) can be written in terms of Virasoro operators using (13.46). The result is very simple:

$$H = \alpha' p^+ p^- = L_0^\perp + \bar{L}_0^\perp - 2\,. \tag{13.49}$$

This Hamiltonian is the sum of an "open string" Hamiltonian $L_0^\perp - 1$ for the right-moving operators, and an "open string" Hamiltonian $\bar{L}_0^\perp - 1$ for the left-moving operators. Using (13.42) the Hamiltonian can be written as

$$H = \frac{\alpha'}{2} p^I p^I + N^\perp + \bar{N}^\perp - 2 \,. \tag{13.50}$$

Both the $L_m^\perp$ operators and the $\bar{L}_m^\perp$ operators satisfy the Virasoro algebra (12.133). In addition, the commutators between barred and unbarred Virasoro operators vanish. So, the full set of closed string Virasoro operators define two commuting Virasoro algebras.

We conclude this section with a study of the Virasoro action on closed string coordinates. The commutation of the closed string Virasoro operators with the oscillators follows the pattern of equation (12.118). We have

$$\left[\bar{L}_m^\perp, \bar{\alpha}_n^J \right] = -n\,\bar{\alpha}_{m+n}^J \,, \qquad \left[L_m^\perp, \alpha_n^J \right] = -n\,\alpha_{m+n}^J \,, \tag{13.51}$$

and, in addition,

$$\left[L_m^\perp, \bar{\alpha}_n^J \right] = \left[\bar{L}_m^\perp, \alpha_n^J \right] = 0 \,. \tag{13.52}$$

On the other hand, both $\bar{L}_m^\perp$ and $L_m^\perp$ have a nontrivial commutator with x_0^I:

Quick calculation 13.1 Verify that

$$\left[\bar{L}_m^\perp, x_0^I \right] = -i\sqrt{\frac{\alpha'}{2}}\,\bar{\alpha}_m^I \,, \qquad \left[L_m^\perp, x_0^I \right] = -i\sqrt{\frac{\alpha'}{2}}\,\alpha_m^I \,. \tag{13.53}$$

Let us focus here only on the action of $L_0^\perp$ and $\bar{L}_0^\perp$ on the string coordinates. The required formulae are obtained in the following exercise:

Quick calculation 13.2 Verify that

$$\left[\bar{L}_0^\perp, X^I(\tau, \sigma) \right] = -\frac{i}{2}(\dot{X}^I + X^{I\prime}) \,, \qquad \left[L_0^\perp, X^I(\tau, \sigma) \right] = -\frac{i}{2}(\dot{X}^I - X^{I\prime}) \,. \tag{13.54}$$

Adding the two equations in (13.54), we find

$$\left[L_0^\perp + \bar{L}_0^\perp, X^I(\tau, \sigma) \right] = -i\frac{\partial X^I}{\partial \tau} \,. \tag{13.55}$$

This equation is consistent with the Heisenberg equation of motion for X^I, since the closed string Hamiltonian is $(L_0^\perp + \bar{L}_0^\perp - 2)$. Subtracting the two equations in (13.54), we find a more surprising result:

$$\left[L_0^\perp - \bar{L}_0^\perp, X^I(\tau, \sigma) \right] = i\frac{\partial X^I}{\partial \sigma} \,. \tag{13.56}$$

This equation shows that $L_0^\perp - \bar{L}_0^\perp$ generates constant translations along the string. Indeed, for infinitesimal ϵ,

$$X^I(\tau, \sigma) + \left[-i\epsilon(L_0^\perp - \bar{L}_0^\perp), X^I(\tau, \sigma) \right] = X^I(\tau, \sigma + \epsilon) \,. \tag{13.57}$$

More generally, a finite translation along the string can be obtained by acting on the string coordinate with exponentials of $L_0^\perp - \bar{L}_0^\perp$. Writing

$$P \equiv L_0^\perp - \bar{L}_0^\perp \,, \tag{13.58}$$

we find that (Problem 13.3)

$$e^{-iP\sigma_0}\, X^I(\tau,\sigma)\, e^{iP\,\sigma_0} = X^I(\tau, \sigma+\sigma_0)\,, \tag{13.59}$$

for any finite σ_0. For $\sigma_0 = \epsilon$, infinitesimal, this general result reduces to (13.57). The operator P generates the one reparameterization symmetry that cannot be fixed even in the light-cone gauge. Since P annihilates all closed string states (see (13.41)), we conclude that closed string states are invariant under rigid σ translations. By this we mean that $\exp(-i\,P\sigma_0)|\Psi\rangle = |\Psi\rangle$, for any closed string state $|\Psi\rangle$.

In the two-dimensional (τ, σ) parameter space of the closed string world-sheet, the operator $L_0^\perp + \bar{L}_0^\perp$ is a generator of τ translations. It is therefore a world-sheet energy. Since the gauge condition relates τ to the light-cone time, this world-sheet energy turns out to give us the spacetime Hamiltonian, the generator of light-cone time evolution. The other combination $L_0^\perp - \bar{L}_0^\perp = P$ generates translations along the world-sheet coordinate σ. It can therefore be viewed as a *world-sheet* momentum. This momentum should not be confused with the spacetime momentum of the string. For closed string states the world-sheet momentum must in fact vanish, and this is a nontrivial constraint. States with nonzero momentum along σ can be built, but they do not belong to the closed string state space.

13.3 Closed string state space

We are now ready to build the state space of the quantum closed string. The ground states are $|p^+, \vec{p}_T\rangle$ and they are annihilated by both the left-moving and the right-moving annihilation operators. To generate all of the basis states we must act on the ground states with the creation operators $a_n^{I\dagger}$ and $\bar{a}_n^{I\dagger}$. The general *candidate* basis vector is

$$|\lambda, \bar{\lambda}\rangle = \left[\prod_{n=1}^{\infty}\prod_{I=2}^{25}(a_n^{I\dagger})^{\lambda_{n,I}}\right] \times \left[\prod_{m=1}^{\infty}\prod_{J=2}^{25}(\bar{a}_m^{J\dagger})^{\bar{\lambda}_{m,J}}\right]|p^+, \vec{p}_T\rangle. \tag{13.60}$$

Just as with open strings, the occupation numbers $\lambda_{n,I}$ and $\bar{\lambda}_{n,I}$ are non-negative integers. The number operators act on $|\lambda, \bar{\lambda}\rangle$ with eigenvalues

$$N^\perp = \sum_{n=1}^{\infty}\sum_{I=2}^{25} n\lambda_{n,I}\,, \qquad \bar{N}^\perp = \sum_{m=1}^{\infty}\sum_{J=2}^{25} m\bar{\lambda}_{m,J}\,. \tag{13.61}$$

Except for the momentum labels, the above states are those that one would obtain by combining *multiplicatively* arbitrary states built from the left-moving and from the right-moving operators (compare with (12.162)). Not all of the states in (13.60) belong to the closed string state space. The constraint (13.45) must be satisfied by the true states of the theory. A basis vector $|\lambda, \bar{\lambda}\rangle$ belongs to the state space *if and only if* it satisfies the level-matching constraint

$$N^\perp = \bar{N}^\perp\,. \tag{13.62}$$

Table 13.1		The states with $N^\perp + \bar{N}^\perp \leq 2$ in the closed string spectrum			
$N^\perp, \bar{N}^\perp$	$\lvert \lambda, \bar{\lambda} \rangle$	$\frac{1}{2}\alpha' M^2$	Number of states	Wavefunction
0, 0	$\lvert p^+, \vec{p}_T \rangle$	-2	1	$\psi(\tau, p^+, \vec{p}_T)$
1, 1	$a_1^{I\dagger} \bar{a}_1^{J\dagger} \lvert p^+, \vec{p}_T \rangle$	0	$(D-2)^2$	$\psi_{IJ}(\tau, p^+, \vec{p}_T)$

This constraint cannot be "solved". It is clear that eliminating some oscillator from the list of operators that can act on the ground states is no help whatsoever. A better strategy would be to select a particular oscillator and attempt to use it, as many times as needed, to fix any state that does not satisfy (13.62). Not even this is possible, because any oscillator can only contribute a positive amount to the number operator. The constraint (13.62) must be implemented in a case-by-case fashion. The masses of the states are obtained from (13.48):

$$\frac{1}{2}\alpha' M^2 = N^\perp + \bar{N}^\perp - 2 \,. \tag{13.63}$$

Let us identify the first few basis states, give their masses, and explain what fields they represent. The results are tabulated in Table 13.1.

The ground states in the first row of Table 13.1 are the one-particle states of a quantum scalar field. For such states $N^\perp = \bar{N}^\perp = 0$ and $M^2 = -4/\alpha' < 0$, so these are closed string tachyons; they are in fact completely analogous to the tachyons of open string theory. The mass-squared of the closed string tachyon is four times larger than that of the open string tachyon. The closed string tachyon is far less understood than the open string tachyon. In particular, the closed string tachyon potential has not yet been calculated reliably. The instabilities associated with closed string tachyons are expected to be instabilities of spacetime itself. They remain largely mysterious.

The next excited states must be built with *two* oscillators acting on the ground states. This is because we must satisfy the constraint $N^\perp = \bar{N}^\perp$. One oscillator must be from the left-sector and one from the right-sector, both with the lowest possible mode number – mode number one. This gives the states described in the second line of the table. All these states have $M^2 = 0$, so they are of great interest. Since I and J are completely arbitrary labels attached to *different* oscillators, the number of states is $(D-2)^2$.

Let us consider the general state of fixed momentum at the massless level. We write it as

$$\sum_{I,J} R_{IJ}\, a_1^{I\dagger} \bar{a}_1^{J\dagger} \lvert p^+, \vec{p}_T \rangle \,. \tag{13.64}$$

Here R_{IJ} are the elements of an arbitrary square matrix of size $(D-2)$. Any square matrix can be decomposed into its symmetric part and its antisymmetric part:

$$R_{IJ} = \frac{1}{2}(R_{IJ} + R_{JI}) + \frac{1}{2}(R_{IJ} - R_{JI}) \equiv S_{IJ} + A_{IJ} \,, \tag{13.65}$$

where S_{IJ} and A_{IJ} are the symmetric and antisymmetric parts of R_{IJ}, respectively. The symmetric part S_{IJ} can in fact be decomposed further:

$$S_{IJ} = \left(S_{IJ} - \frac{1}{D-2} \delta_{IJ} S \right) + \frac{1}{D-2} \, \delta_{IJ} \, S, \quad S \equiv S^{II} = \delta^{IJ} S_{IJ} \,. \tag{13.66}$$

The first term on the right-hand side is traceless:

$$\delta^{IJ} \left(S_{IJ} - \frac{1}{D-2} \delta_{IJ} \, S \right) = S - \frac{1}{D-2} \, \delta_{IJ} \, \delta^{IJ} S = 0 \,, \tag{13.67}$$

since $\delta_{IJ} \, \delta^{IJ} = D - 2$. Therefore (13.66) decomposes S_{IJ} into a traceless matrix plus a multiple of the unit matrix. Let $\hat{S}_{IJ}$ denote the traceless part of S_{IJ} and let $S' = S/(D-2)$. All in all, we have decomposed R_{IJ} as

$$R_{IJ} = \hat{S}_{IJ} + A_{IJ} + S' \delta_{IJ} \,. \tag{13.68}$$

This is the standard decomposition of a matrix into a symmetric-traceless part, an antisymmetric part, and a trace part. Each of the three pieces can be specified independently when writing a general matrix R_{IJ}. Therefore, the states in (13.64) can be split into three groups of linearly independent states:

$$\sum_{I,J} \hat{S}_{IJ} \, a_1^{I \dagger} \bar{a}_1^{J \dagger} |p^+, \vec{p}_T \rangle \,, \tag{13.69}$$

$$\sum_{I,J} A_{IJ} \, a_1^{I \dagger} \bar{a}_1^{J \dagger} |p^+, \vec{p}_T \rangle \,, \tag{13.70}$$

$$S' \, a_1^{I \dagger} \bar{a}_1^{I \dagger} \, |p^+, \vec{p}_T \rangle \,. \tag{13.71}$$

We now make a remarkable claim: the states (13.69) represent one-particle graviton states! We examined one-particle graviton states in Section 10.6. In the quantum theory of the free gravitational field these states are given by (10.110):

$$\sum_{I,J=2}^{D-1} \xi_{IJ} \, a_{p^+, \, p_T}^{IJ \dagger} |\Omega \rangle \,, \tag{13.72}$$

where ξ_{IJ} is an arbitrary symmetric traceless matrix. Since $\hat{S}_{IJ}$ is also a symmetric traceless matrix, the identification of states is possible if we identify the basis states:

$$a_1^{I \dagger} \bar{a}_1^{J \dagger} |p^+, \vec{p}_T \rangle \longleftrightarrow a_{p^+, \, p_T}^{IJ \dagger} |\Omega \rangle \,. \tag{13.73}$$

This identification is possible because the two sets of states have the same Lorentz labels, they carry the same momentum, and they have the same mass (both zero). This shows that the closed string has graviton states. Gravity has appeared in string theory! We never put in a dynamical metric and we never spoke about general covariance, yet somehow, the quantum states of the gravitational field have emerged!

The set of states in (13.70) corresponds to the one-particle states of the Kalb–Ramond field, an antisymmetric tensor field $B_{\mu\nu}$ with two indices. The light-cone analysis of this field was discussed in Problem 10.6 (see, in particular, parts (e) and (f)). The $B_{\mu\nu}$ field is in many ways the tensor generalization of the Maxwell gauge field A_μ. The Kalb–Ramond field couples to strings in a way that is analogous to the way that the Maxwell field couples to particles. Thus, as we will see in Chapter 16, strings carry Kalb–Ramond charge.

There is one state that remains to be examined. The oscillator part of (13.71) has no free indices (I is summed over), so it represents one state. It corresponds to a one-particle state of a massless scalar field. This field is called the *dilaton*.

The above discussion of particle states is supplemented by an analysis of wavefunctions and field equations. Such analysis follows closely the treatment in Section 12.7. Wavefunctions $\psi_{IJ}(\tau, p^+, p_T)$ describe the general time-dependent states at the massless level of the closed string:

$$|\Psi, \tau\rangle = \int dp^+ d\vec{p}_T\ \psi_{IJ}(\tau, p^+, \vec{p}_T)\, a_1^{I\dagger}\bar{a}_1^{J\dagger}|p^+, \vec{p}_T\rangle. \tag{13.74}$$

The Schrödinger equation satisfied by the states is $i\partial_\tau|\Psi, \tau\rangle = H|\Psi, \tau\rangle$. Using (13.50) and noting that for the states in question $N^\perp = \bar{N}^\perp = 1$, we find

$$i\frac{\partial \psi_{IJ}}{\partial \tau} = \frac{\alpha'}{2}\, p^K p^K\, \psi_{IJ}. \tag{13.75}$$

The wavefunctions $\psi_{IJ}(\tau, p^+, p_T)$ become the fields of the classical field theories, with the Schrödinger equations interpreted as classical field equations. The symmetric-traceless part of ψ_{IJ} becomes the graviton field, the antisymmetric part becomes the Kalb–Ramond field, and the trace part becomes the dilaton field. The Schrödinger equations for the wavefunctions of the graviton states, the Kalb–Ramond states, and the dilaton state, are all included in (13.75), and can be separated by selecting the symmetric-traceless components, the antisymmetric components, and the trace component of ψ_{IJ}. On the other hand, in light-cone coordinates the massless scalar field equation $\partial^2\phi = 0$ takes the form (10.30):

$$\left(i\frac{\partial}{\partial x^+} - \frac{1}{2p^+}\, p^K p^K\right)\phi(x^+, p^+, \vec{p}_T) = 0. \tag{13.76}$$

Setting $x^+ = \alpha' p^+ \tau$, we find

$$\left(i\frac{\partial}{\partial \tau} - \frac{\alpha'}{2}\, p^K p^K\right)\phi(\tau, p^+, \vec{p}_T) = 0. \tag{13.77}$$

This equation takes the same form as (13.75). In fact, in the light-cone gauge, graviton fields, Kalb–Ramond fields, and the dilaton field, all satisfy the simple equation $\partial^2\phi^{\cdots} = 0$, where the dots refer to the relevant indices. This is manifestly true for the massless dilaton, which is a scalar. For graviton fields, it was demonstrated in equation (10.107). For Kalb–Ramond fields, the equation $p^2 B^{\mu\nu} = 0$ was shown to hold in the light-cone gauge as part of the analysis in Problem 10.6.

In summary, at the massless level of the closed string we found gravity fields, Kalb–Ramond fields, and dilaton fields. Each of these fields deserves intense study. The gravity field is studied in general relativity. We will focus on the Kalb–Ramond field in Chapter 16.

The dilaton is a massless scalar field with surprising properties. A proper study of the dilaton belongs to an advanced course, but in the following section we discuss the role that it plays in string theory.

13.4 String coupling and the dilaton

The massless scalar field called the dilaton has a fascinating property: its expectation value controls the string coupling! This coupling is a dimensionless number that sets the strength of string interactions.

The classic example of a coupling constant is the electromagnetic fine-structure constant $\alpha \equiv e^2/(4\pi\hbar c) \simeq 1/137$. This dimensionless coupling constant controls the strength of electromagnetic interactions. In the hydrogen atom Hamiltonian, for example, α appears in the term which gives the electrostatic interaction energy between the proton and the electron. The mass-scale in the hydrogen atom is set by the mass m of the electron. Physical, dimensionful quantities, such as the binding energy E of the ground state, depend on the dimensionful parameter m of the theory, the fundamental constants $\hbar$ and c, *and* the dimensionless coupling α:

$$E = \frac{e^2}{4\pi}\frac{1}{2a_0} = \frac{1}{2}\left(\frac{e^2}{4\pi\hbar c}\right)^2 mc^2 = \frac{1}{2}\alpha^2\,(mc^2)\,, \tag{13.78}$$

where $a_0 = 4\pi\hbar^2/me^2$ is the Bohr radius. In the hypothetical limit $\alpha \to 0$, the binding energy vanishes, and the Bohr radius is infinite. This is what becomes of the hydrogen atom when we turn off electromagnetic interactions.

In string theory the story is not so different, at first. The dimensionful parameter can be taken to be α', which defines the string length $\ell_s = \sqrt{\alpha'}$ (working with $\hbar = c = 1$). Let g denote the dimensionless coupling for closed string interactions. If g was set to zero, then strings would not interact. Interactions in gravity are determined by the value of Newton's gravitational constant. If closed strings do not interact, gravitation would emerge without interactions, and the value of Newton's constant in string theory would be zero. Since it vanishes when $g \to 0$, Newton's constant is naturally expected to be proportional to some positive power of g; it turns out that it is proportional to g^2. Dimensional analysis fixes the α' dependence of Newton's constant. Equation (3.108) shows that the D-dimensional Newton constant $G^{(D)}$ has (natural) units of L^{D-2}. In fact, in natural units, $G^{(D)}$ is equal to the D-dimensional Planck length $\ell_P^{(D)}$ to the power $(D-2)$. So $[G^{(26)}] = L^{24}$, and since $[\alpha'] = L^2$, we find

$$G^{(26)} \sim g^2\,(\alpha')^{12}\,. \tag{13.79}$$

Most phenomenological studies of string theory begin with ten-dimensional superstring theories. These theories contain both bosonic and fermionic excitations, so they include the two types of particles that we observe in nature. The ten-dimensional Newton constant $G^{(10)}$ in superstring theory is given by

$$G^{(10)} = \left(\ell_P^{(10)}\right)^8 \sim g^2 \left(\alpha'\right)^4. \tag{13.80}$$

It then follows that

$$\ell_P^{(10)} \sim g^{1/4}\sqrt{\alpha'} = g^{1/4}\,\ell_s. \tag{13.81}$$

If the string coupling g is a small number, the string length is larger than the Planck length. If g is of order unity, the string length and the Planck length are comparable. To find four-dimensional implications, we use the relation (3.116) between higher- and lower-dimensional Newton constants in a compactification. Assume that six of the dimensions of the ten-dimensional world are curled up into a space with volume $V^{(6)}$. Then, the four-dimensional Newton constant G is related to the ten-dimensional one by

$$G = \frac{G^{(10)}}{V^{(6)}} \sim g^2\,\alpha'\,\frac{1}{V^{(6)}/(\alpha')^3}. \tag{13.82}$$

The ratio $V^{(6)}/(\alpha')^3$ is a dimensionless number that is typically assumed to be large. For a compactification of fixed volume, as measured in units of the string length, the four-dimensional Newton constant behaves like

$$G \sim g^2\,\alpha'. \tag{13.83}$$

In a theory with both open and closed strings, the open string coupling g_o is actually determined in terms of the closed string coupling g. One can prove that

$$g_o^2 \sim g. \tag{13.84}$$

This relation arises due to certain topological properties of two-dimensional world-sheets.

The coupling which controls the strength of an interaction may sometimes fail to be constant. Consider adding to a free Hamiltonian H_0 an interaction term $g H_{\text{int}}$ that is proportional to a dimensionless coupling g. If g is declared to be a constant, you must specify its value by hand in order to define the complete Hamiltonian $H_0 + g H_{\text{int}}$. But suppose that g is not a constant but rather a dynamical variable $g(t)$, and that the full Hamiltonian includes an additional term H_g that gives dynamics to g. In this case you cannot specify the coupling g arbitrarily by hand. The coupling would be determined, perhaps uniquely, or perhaps not, by the Hamiltonian equations derived from $H_0 + g H_{\text{int}} + H_g$. If $g(t)$ is uniquely determined, no choice is needed. If $g(t)$ is not uniquely determined, some other criterion may be needed to select the physically realized solution.

The situation in string theory is similar; the string coupling can fail to be a constant. The closed string coupling g is *determined* by the value of the dilaton field $\phi(x)$:

$$g \sim e^\phi. \tag{13.85}$$

It follows that, in principle, the string coupling g is not an adjustable parameter of string theory. Rather, it is a *dynamical* parameter – a field, in fact. This is an ideal property in a unified theory of all interactions, for it holds the promise that the string coupling may be calculable. On the other hand, it seems clear that the string coupling is not uniquely determined in string theory. Depending on the values taken by the other fields in the theory, the dilaton field may evolve in different ways. Under certain circumstances, the dilaton expectation value may even be an adjustable constant. One attractive possibility is that the other fields in the theory generate a potential for the dilaton. If this potential has a stable critical point the dilaton may be set equal to the critical value. Even more, the dilaton field would then acquire a mass. This is necessary for a realistic model of physics because there are no known massless scalars in nature.

If the string coupling is small, the quantum mechanical amplitudes for string interactions can be calculated accurately using known results about Riemann surfaces. The fascinating properties of Riemann surfaces allow one to understand why the infinities that plague quantum amplitudes in general relativity do not appear in string theory. This subject will be discussed in Chapter 26.

13.5 Closed strings on the $\mathbb{R}^1/\mathbb{Z}_2$ orbifold

We introduced orbifolds in §2.8 and stated that they are nontrivial spaces on which string propagation is tractable. In this and the following section we demonstrate this claim for the case of closed strings on the simplest orbifold, the half-line $\mathbb{R}^1/\mathbb{Z}_2$. Since we must deal with the full quantum string theory, the orbifold direction is just one direction in the 26-dimensional spacetime. This direction $x^{25} \equiv x$, is effectively restricted to $x \geq 0$ by the $\mathbb{Z}_2$ identification

$$x \sim -x. \tag{13.86}$$

No other coordinate is affected. The orbifold theory does not require boundary conditions at $x = 0$. The orbifold theory is defined by imposing a natural restriction on states of closed strings that live on the spacetime before orbifolding – a restriction on the spectrum of the original *parent* theory.

Writing $X^{25}(\tau, \sigma) \equiv X(\tau, \sigma)$, the collection of string coordinates is X^+, X^-, X^i, with $i = 2, \ldots, 24$, and X. We introduce an operator U that implements on the string coordinates the $\mathbb{Z}_2$ transfomation that defines the orbifold. Thus, acting on X we must have

$$U X(\tau, \sigma) U^{-1} = -X(\tau, \sigma). \tag{13.87}$$

Since U should not transform any other string coordinate we demand

$$U X^i(\tau, \sigma) U^{-1} = X^i(\tau, \sigma). \tag{13.88}$$

Moreover, we demand the invariance of p^+ and x_0^-:

$$U p^+ U^{-1} = p^+, \quad \text{and} \quad U x_0^- U^{-1} = x_0^-, \tag{13.89}$$

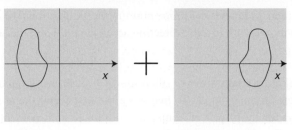

Fig. 13.1 A closed string state invariant under $x \to -x$ is obtained by superposition of two states that go into each other as $x \to -x$.

since these, together with (13.87) and (13.88), imply the invariance of X^+ and X^-:

$$U X^{\pm}(\tau, \sigma) U^{-1} = X^{\pm}(\tau, \sigma). \qquad (13.90)$$

Indeed, the invariance of $X^+ = \alpha' p^+ \tau$ follows directly from the invariance of p^+. The invariance of X^- holds because it is the sum of the invariant mode x_0^-, terms quadratic in the invariant coordinates, and terms quadratic in $\dot{X}$, X' (see (13.35)). It follows that p^- is invariant under the U action. Since p^+ is also invariant, the Hamiltonian $H = \alpha' p^+ p^-$ (see (13.49)) is invariant: $U H U^{-1} = H$. This means that U is a symmetry of the closed string theory.

The *orbifold* closed string theory keeps only U-invariant states of the parent theory. Since U is a symmetry, the truncation to U-invariant states is a consistent reduction: if the Hamiltonian were not U invariant, states that are U invariant at one time need not remain U invariant for all times. Intuitively, for U-invariant string states, the physics at $-x$ is determined by the physics at x, thus effectively making half of the space irrelevant. U-invariant states are naturally constructed by quantum superposition (Figure 13.1).

To implement the restriction to U-invariant states it is convenient to determine the action of U on the oscillators. The coordinate $X(\tau, \sigma)$ has the usual mode expansion (13.24) of a closed string coordinate

$$X(\tau, \sigma) = x_0 + \alpha' p \tau + i \sqrt{\frac{\alpha'}{2}} \sum_{n \neq 0} \frac{e^{-in\tau}}{n} \left(\alpha_n e^{in\sigma} + \bar{\alpha}_n e^{-in\sigma} \right). \qquad (13.91)$$

If we are to have equation (13.87) for all values of τ and σ we must have

$$\begin{aligned} U x_0 U^{-1} &= -x_0, & U p U^{-1} &= -p, \\ U \alpha_n U^{-1} &= -\alpha_n, & U \bar{\alpha}_n U^{-1} &= -\bar{\alpha}_n. \end{aligned} \qquad (13.92)$$

All operators in the expansion of X change sign under the action of U. All modes in the expansion of the coordinates X^i are left invariant by the action of U.

Quick calculation 13.3 Justify $U H U^{-1} = H$ directly from the oscillator expansion of $H = L_0^{\perp} + \bar{L}_0^{\perp} - 2$.

Let us now discuss the states of the theory, beginning with the ground states. We denote the ground states of the parent theory by $|p^+, \vec{p}, p\rangle$, where $\vec{p}$ is a vector with components

p^i and p denotes the momentum in the x direction. We also assume that states that have no momentum in the x direction are invariant under the action of U:

$$U|p^+, \vec{p}, 0\rangle = |p^+, \vec{p}, 0\rangle. \qquad (13.93)$$

Being invariant under U, the state $|p^+, \vec{p}, 0\rangle$ is a ground state of the orbifold theory, but there are many more. To find these, we first derive the U action on $|p^+, \vec{p}, p\rangle$. We assume that for every pair of conjugate coordinate and momentum operators (q, p) momentum states are defined to satisfy $|p + \delta p\rangle = \exp(i\delta p\, q)|p\rangle$, with δp a constant. Applied to our pair (x_0, p), we find $|p^+, \vec{p}, p\rangle = e^{ix_0 p}|p^+, \vec{p}, 0\rangle$ and, therefore,

$$U|p^+, \vec{p}, p\rangle = Ue^{ix_0 p}U^{-1}U|p^+, \vec{p}, 0\rangle = e^{i(-x_0)p}|p^+, \vec{p}, 0\rangle = |p^+, \vec{p}, -p\rangle, \quad (13.94)$$

namely,

$$U|p^+, \vec{p}, p\rangle = |p^+, \vec{p}, -p\rangle. \qquad (13.95)$$

It is now easy to form U-invariant states by linear combinations:

$$\text{Orbifold ground states:} \quad |p^+, \vec{p}, p\rangle + |p^+, \vec{p}, -p\rangle. \qquad (13.96)$$

More precisely, orbifold ground states are general time-dependent superpositions of the above states:

$$\int \psi(\tau, p^+, \vec{p}, p)|p^+, \vec{p}, p\rangle\, dp^+ d\vec{p}\, dp, \quad \psi(\tau, p^+, \vec{p}, -p) = \psi(\tau, p^+, \vec{p}, p). \qquad (13.97)$$

The Fourier transformed ground state wavefunctions $\psi(\tau, p^+, \vec{p}, x)$ are even functions of x. The massless states of the orbifold theory require $N^\perp = \bar{N}^\perp = 1$. Indeed, orbifolding does not change the mode expansion of X and the formula for M^2 is not changed. To build the states we need two oscillators, one barred one unbarred, acting on suitable ground states. Basis massless states are given by

$$\alpha^i_{-1}\bar{\alpha}^j_{-1}\left(|p^+, \vec{p}, p\rangle + |p^+, \vec{p}, -p\rangle\right),$$
$$\alpha^i_{-1}\bar{\alpha}_{-1}\left(|p^+, \vec{p}, p\rangle - |p^+, \vec{p}, -p\rangle\right),$$
$$\alpha_{-1}\bar{\alpha}^i_{-1}\left(|p^+, \vec{p}, p\rangle - |p^+, \vec{p}, -p\rangle\right),$$
$$\alpha_{-1}\bar{\alpha}_{-1}\left(|p^+, \vec{p}, p\rangle + |p^+, \vec{p}, -p\rangle\right).$$

All these states are U-invariant. For those that have an odd number of X oscillators we use a combination of vacuum states with $U = -1$ so that the full state has $U = +1$. Massive states can be constructed similarly. It would seem that this is the end of the story, but there is a surprise. The orbifold theory includes more states than the ones discussed above. It contains a *twisted* sector with a new kind of closed strings. Let us examine these strings now.

13.6 The twisted sector of the orbifold

The additional closed strings that appear in the so-called twisted sector can be imagined as *open* strings in the parent theory, but with endpoints at locations identified by the orbifold condition (13.86). More precisely, we write

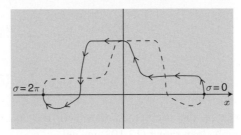

Fig. 13.2 A "twisted" closed string satisfies $X(\tau, \sigma + 2\pi) = -X(\tau, \sigma)$. The string for $\sigma \in [0, 2\pi]$ is shown as a continuous line and the part $\sigma \in [2\pi, 4\pi]$ as a dashed line. The vertical coordinate represents any x^i.

$$X(\tau, \sigma + 2\pi) = -X(\tau, \sigma). \tag{13.98}$$

This equation tells us that $X(\tau, 2\pi) = -X(\tau, 0)$, which means that the orbifold identification makes them the same point and the string becomes effectively closed. For arbitrary σ the equation tells us that the string effectively repeats its position when σ is increased by 2π. In the parent space the string only closes after $\sigma \to \sigma + 4\pi$, as is clear from two applications of (13.98). A two-dimensional representation of a string obeying (13.98) is shown in Figure 13.2. We note that a twisted closed string must go through $x = 0$. This is clear because the continuous function $X(\tau, \sigma)$ in (13.98) takes both positive and negative values.

To develop the quantum theory of this sector we first find the appropriate oscillator expansion for the coordinate X that satisfies (13.98). As usual, to solve the wave equation, we write

$$X(\tau, \sigma) = X_L(u) + X_R(v), \qquad u = \tau + \sigma, \; v = \tau - \sigma. \tag{13.99}$$

The constraint (13.98) implies that

$$X_L(u + 2\pi) + X_R(v - 2\pi) = -X_L(u) - X_R(v), \tag{13.100}$$

or, equivalently,

$$X_L(u + 2\pi) + X_L(u) = -\big(X_R(v) + X_R(v - 2\pi)\big). \tag{13.101}$$

Taking derivatives with respect to u and v we find

$$X'_L(u + 2\pi) = -X'_L(u), \qquad X'_R(v + 2\pi) = -X'_R(v). \tag{13.102}$$

For an ordinary string coordinate $X_L^{i\,\prime}$ and $X_R^{i\,\prime}$ are periodic functions with period 2π. For X the derivatives reverse sign when the argument changes by 2π. To write a mode expansion that is conveniently normalized we mimic (13.14). In order to get a sign change when $u \to u + 2\pi$ we need exponentials of the form $\exp(iku)$ with k half-integer. We therefore write

$$X'_L(u) = \sqrt{\frac{\alpha'}{2}} \sum_{n \in \mathbb{Z}_{\text{odd}}} \bar{\alpha}_{\frac{n}{2}} \, e^{-i\frac{n}{2}u} \, ,$$

$$X'_R(v) = \sqrt{\frac{\alpha'}{2}} \sum_{n \in \mathbb{Z}_{\text{odd}}} \alpha_{\frac{n}{2}} \, e^{-i\frac{n}{2}v} \, . \qquad (13.103)$$

The labels on the oscillators, now half-integers, match the factors of u or v in the exponentials. Integrating (13.103) we get

$$X_L(u) = x_L + i\sqrt{\frac{\alpha'}{2}} \sum_{n \in \mathbb{Z}_{\text{odd}}} \frac{2}{n} \bar{\alpha}_{\frac{n}{2}} \, e^{-i\frac{n}{2}u}$$

$$X_R(v) = x_R + i\sqrt{\frac{\alpha'}{2}} \sum_{n \in \mathbb{Z}_{\text{odd}}} \frac{2}{n} \alpha_{\frac{n}{2}} \, e^{-i\frac{n}{2}v} \, . \qquad (13.104)$$

Equation (13.101) now gives $x_L = -x_R$, or $x_L + x_R = 0$. Putting the left- and right-moving parts together we arrive at the mode expansion for the twisted sector closed strings:

$$X(\tau, \sigma) = i\sqrt{\frac{\alpha'}{2}} \sum_{n \in \mathbb{Z}_{\text{odd}}} \frac{2}{n} e^{-i\frac{n}{2}\tau} \left(\bar{\alpha}_{\frac{n}{2}} e^{-i\frac{n}{2}\sigma} + \alpha_{\frac{n}{2}} e^{i\frac{n}{2}\sigma} \right) . \qquad (13.105)$$

The expansion contains neither coordinate nor momentum zero modes. It costs energy to get away from $x = 0$ since at least one point on the string must remain at $x = 0$. As we will see later, the fields associated with twisted states are localized at the fixed point $x = 0$ and depend only on x^+, x^-, and x^i.

The commutation relations for the modes can be derived in complete analogy to the usual case. The coordinate obeys $[X(\tau, \sigma), \mathcal{P}^\tau(\tau, \sigma')] = i\delta(\sigma - \sigma')$. Moreover, given our mode expansions the coordinate X satisfies (13.26) with $n \to n/2$ and sums over odd numbers:

$$\dot{X} + X' = 2X'_L(\tau + \sigma) = \sqrt{2\alpha'} \sum_{n \in \mathbb{Z}_{\text{odd}}} \bar{\alpha}_{\frac{n}{2}} \, e^{-i\frac{n}{2}(\tau+\sigma)} \, ,$$

$$\dot{X} - X' = 2X'_R(\tau - \sigma) = \sqrt{2\alpha'} \sum_{n \in \mathbb{Z}_{\text{odd}}} \alpha_{\frac{n}{2}} \, e^{-i\frac{n}{2}(\tau-\sigma)} \, . \qquad (13.106)$$

Applying the top-sign version of (13.28) to X, we get

$$\sum_{m',n' \in \mathbb{Z}_{\text{odd}}} e^{-i\frac{m'}{2}(\tau+\sigma)} e^{-i\frac{n'}{2}(\tau+\sigma')} \left[\bar{\alpha}_{\frac{m'}{2}}, \bar{\alpha}_{\frac{n'}{2}} \right] = 2\pi \, i \frac{d}{d\sigma} \delta(\sigma - \sigma') . \qquad (13.107)$$

To extract the commutators we apply the following integral operators to the left and right sides of equation (13.107):

$$\frac{1}{2\pi} \int_0^{2\pi} d\sigma \, e^{i\frac{m}{2}\sigma} \cdot \frac{1}{2\pi} \int_0^{2\pi} d\sigma' \, e^{i\frac{n}{2}\sigma'} \, , \quad m, n \in \mathbb{Z}_{\text{odd}} \, . \qquad (13.108)$$

Two functions $e^{i\frac{k}{2}\sigma}$ and $e^{i\frac{k'}{2}\sigma}$ with $k, k' \in \mathbb{Z}_{\text{odd}}$ and $k + k' \neq 0$ are orthogonal over the interval $[0, 2\pi]$ because $k + k'$ is an even integer. The integrations in (13.108) therefore select a single commutator on the left-hand side, and one can show that

$$\left[\bar{\alpha}_{\frac{m}{2}}, \bar{\alpha}_{\frac{n}{2}}\right] = \frac{m}{2}\delta_{m+n,0}. \tag{13.109}$$

This is the expected form of the commutation relations.

Quick calculation 13.4 Prove (13.109).

Similar commutation relations hold for the right-moving oscillators and, as usual, left-moving and right-moving oscillators commute:

$$\left[\alpha_{\frac{m}{2}}, \alpha_{\frac{n}{2}}\right] = \frac{m}{2}\delta_{m+n,0}, \quad \left[\alpha_{\frac{m}{2}}, \bar{\alpha}_{\frac{n}{2}}\right] = 0. \tag{13.110}$$

We must still implement the orbifold identification, so equation (13.87) must hold for the twisted X in (13.105). We can readily read off the action of U on the new oscillators:

$$U\alpha_{\frac{n}{2}}U^{-1} = -\alpha_{\frac{n}{2}}, \quad U\bar{\alpha}_{\frac{n}{2}}U^{-1} = -\bar{\alpha}_{\frac{n}{2}}. \tag{13.111}$$

The absence of a momentum zero mode in the expansion of X means that twisted states do not have a conserved momentum along the X direction. U-invariant ground states are labeled by p^+ and the transverse momentum $\vec{p}$ along the X^i:

$$\text{twisted sector ground states:} \quad |p^+, \vec{p}\rangle. \tag{13.112}$$

These are not to be confused with the ground states (13.96) in the "untwisted" sector that have zero momentum along X.

To discuss excited states in the twisted sector we need the relevant formula for $\alpha' M^2$. This time we can expect changes because the half-integer moding of the X oscillators can alter the ordering constant. To find out, we recall that in open string theory we anticipated the value of the ordering constant by performing the naive ordering of $L_0^\perp$ and using the ζ-function motivated rule $1 + 2 + 3 + \cdots \to -\frac{1}{12}$. Let us do the same now for the closed string operator $\bar{L}_0^\perp$. We see from the top equation in (13.36) that $\bar{L}_0^\perp$ is determined by contributions from $(\dot{X}^I + X^{I'})^2$, that is, $(\dot{X}^i + X^{i'})^2 + (\dot{X} + X')^2$. Since the mode expansions of $\dot{X}^i + X^{i'}$ and those of $\dot{X} + X'$ are completely analogous, the formula for $\bar{L}_0^\perp$ in (13.37) will be modified to read

$$\bar{L}_0^\perp = \frac{1}{2}\sum_{p\in\mathbb{Z}}\bar{\alpha}_p^i\bar{\alpha}_{-p}^i + \frac{1}{2}\sum_{k\in\mathbb{Z}_{\text{odd}}}\bar{\alpha}_{\frac{k}{2}}\bar{\alpha}_{-\frac{k}{2}}. \tag{13.113}$$

Each of the twenty-three directions i in the first sum contributes an ordering constant of $\frac{1}{2} \cdot (-\frac{1}{12}) = -\frac{1}{24}$. The new contribution arises from ordering the second summand:

$$
\frac{1}{2} \sum_{k \in \mathbb{Z}_{\text{odd}}} \bar{\alpha}_{\frac{k}{2}} \bar{\alpha}_{-\frac{k}{2}} = \frac{1}{2} \sum_{k \in \mathbb{Z}_{\text{odd}}^+} \bar{\alpha}_{\frac{k}{2}} \bar{\alpha}_{-\frac{k}{2}} + \frac{1}{2} \sum_{k \in \mathbb{Z}_{\text{odd}}^+} \bar{\alpha}_{-\frac{k}{2}} \bar{\alpha}_{\frac{k}{2}}
$$

$$
= \sum_{k \in \mathbb{Z}_{\text{odd}}^+} \bar{\alpha}_{-\frac{k}{2}} \bar{\alpha}_{\frac{k}{2}} + \frac{1}{2} \sum_{k \in \mathbb{Z}_{\text{odd}}^+} \left[\bar{\alpha}_{\frac{k}{2}}, \bar{\alpha}_{-\frac{k}{2}} \right]
$$

$$
= \sum_{k \in \mathbb{Z}_{\text{odd}}^+} \bar{\alpha}_{-\frac{k}{2}} \bar{\alpha}_{\frac{k}{2}} + \frac{1}{4} \sum_{k \in \mathbb{Z}_{\text{odd}}^+} k . \tag{13.114}
$$

To evaluate this we need to calculate the "sum" of all the odd positive integers. This is done as follows:

$$
\sum_{k=1}^{\infty} k = \sum_{k \in \mathbb{Z}_{\text{odd}}^+} k + \sum_{k \in \mathbb{Z}_{\text{even}}^+} k = \sum_{k \in \mathbb{Z}_{\text{odd}}^+} k + 2 \sum_{k=1}^{\infty} k . \tag{13.115}
$$

It then follows that

$$
\sum_{k \in \mathbb{Z}_{\text{odd}}^+} k = -\sum_{k=1}^{\infty} k = \frac{1}{12} . \tag{13.116}
$$

We can now write the precise quantum form of (13.113):

$$
\bar{L}_0^{\perp} = \frac{1}{4} \alpha' p^i p^i + \sum_{p=1}^{\infty} \bar{\alpha}_{-p}^i \bar{\alpha}_p^i - 23 \cdot \frac{1}{24} + \sum_{k \in \mathbb{Z}_{\text{odd}}^+} \bar{\alpha}_{-\frac{k}{2}} \bar{\alpha}_{\frac{k}{2}} + \frac{1}{4} \cdot \frac{1}{12} , \tag{13.117}
$$

which we write as

$$
\bar{L}_0^{\perp} = \frac{1}{4} \alpha' p^i p^i + \bar{N}^{\perp} - \frac{15}{16} , \quad \bar{N}^{\perp} = \sum_{p=1}^{\infty} \bar{\alpha}_{-p}^i \bar{\alpha}_p^i + \sum_{k \in \mathbb{Z}_{\text{odd}}^+} \bar{\alpha}_{-\frac{k}{2}} \bar{\alpha}_{\frac{k}{2}} . \tag{13.118}
$$

A completely analogous formula holds for $L_0^{\perp}$.

Quick calculation 13.5 Show that $[\bar{N}^{\perp}, \bar{\alpha}_{-\frac{q}{2}}] = \frac{q}{2} \bar{\alpha}_{-\frac{q}{2}}$ and explain why $\bar{N}^{\perp}$ is properly called a number operator.

To write the mass-squared formula recall that $\frac{1}{2} \alpha' M^2$ is equal to the sum of $N^{\perp}$ plus $\bar{N}^{\perp}$ plus the ordering constants, as in (13.48). We thus have

$$
\frac{1}{2} \alpha' M^2 = N^{\perp} + \bar{N}^{\perp} - \frac{15}{8} . \tag{13.119}
$$

The twisted sector ground states $|p^+, \vec{p}\rangle$ have $N^{\perp} = \bar{N}^{\perp} = 0$. They are tachyonic states with $\frac{1}{2} \alpha' M^2 = -\frac{15}{8}$. The first excited states are built using the lowest moded twisted oscillators:

$$
\alpha_{-\frac{1}{2}} \bar{\alpha}_{-\frac{1}{2}} |p^+, \vec{p}\rangle , \qquad \frac{1}{2} \alpha' M^2 = \frac{1}{2} + \frac{1}{2} - \frac{15}{8} = -\frac{7}{8} . \tag{13.120}
$$

The momentum labels on twisted states indicate that the wavefunctions are of the form $\psi(\tau, p^+, \bar{p})$, or $\psi(\tau, x^-, \bar{x})$ in coordinate space. Given the correspondence between wavefunctions and fields, we conclude that the fields associated with twisted states have no x argument. These fields must live at some specific value of x. Given that energetics forces twisted states to localize around $x = 0$, we reasonably conclude that the fields live on the reduced 25-dimensional spacetime defined by $x = 0$. The above ground states and first excited states are associated with scalar fields because the states have no indices along directions in the reduced spacetime.

Orbifolds always have a twisted sector thus orbifolding has a double effect – the so-called *double strike*. One effect is that states of the parent theory that are not invariant under the orbifold action are projected out. The other effect is that we gain a whole new sector of states that satisfy "twisted" boundary conditions.

Problems

Problem 13.1 Commutation relations for oscillators.

(a) Use the lower-sign version of equation (13.28) and the appropriate mode expansion to verify explicitly that the unbarred commutation relations of (13.29) emerge.

(b) The set of functions $e^{in\sigma}$, with $n \in \mathbb{Z}$, is complete on the interval $\sigma \in [0, 2\pi]$. Use this fact to prove that

$$\delta(\sigma - \sigma') = \frac{1}{2\pi} \sum_{n \in \mathbb{Z}} e^{in(\sigma - \sigma')} . \tag{1}$$

(c) Compute explicitly the commutator $[X^I(\tau, \sigma), \mathcal{P}^{\tau J}(\tau, \sigma')]$ using the mode expansions of X and $\mathcal{P}$ and the commutation relations (13.29), (13.30), and (13.33). Use equation (1) to confirm that the expected answer (13.27) emerges.

(d) Prove the zero mode commutation relations (13.33), starting with a derivation of

$$\left[x_0^I + \sqrt{2\alpha'}\alpha_0^I \tau , \ \dot{X}^J(\tau, \sigma') \right] = i\alpha' \eta^{IJ} ,$$

which is the closed string analog of equation (12.47).

Problem 13.2 A projector into physical states.

Consider the vector space $\mathcal{H}$ spanned by the set of states $|\lambda, \bar{\lambda}\rangle$ in equation (13.60). Explain why, for any state $|\lambda, \bar{\lambda}\rangle \in \mathcal{H}$, the eigenvalue of $P = L_0^\perp - \bar{L}_0^\perp$ is an integer. Show that

$$\mathcal{P}_0 = \int_0^{2\pi} \frac{d\theta}{2\pi} e^{-i(L_0^\perp - \bar{L}_0^\perp)\theta} ,$$

is a projector from $\mathcal{H}$ into the vector subspace where $P = 0$. Thus $\mathcal{P}_0$ projects into the state space of closed strings.

Problem 13.3 Action of $L_0^\perp - \bar{L}_0^\perp$.

(a) Prove that equation (13.59) holds for finite σ_0. You may find it useful to define $f(\sigma_0) = e^{-iP\sigma_0} X^I(\tau, \sigma) e^{iP\sigma_0}$ and to calculate multiple derivatives of f, evaluated at $\sigma_0 = 0$.

(b) Explain why

$$e^{-iP\sigma_0} (\dot{X}^I \pm X^{I\prime})(\tau, \sigma) e^{iP\sigma_0} = (\dot{X}^I \pm X^{I\prime})(\tau, \sigma + \sigma_0). \tag{1}$$

(c) Use equation (1) to calculate $e^{-iP\sigma_0} \alpha_n^I e^{iP\sigma_0}$ and $e^{-iP\sigma_0} \bar{\alpha}_n^I e^{iP\sigma_0}$. In doing so, you are finding the action of a σ translation on the oscillators.

(d) Consider the state

$$|U\rangle = \alpha_{-m}^I \bar{\alpha}_{-n}^J |p^+, \vec{p}_T\rangle, \quad m, n > 0.$$

Use the results of (c) to calculate $e^{-iP\sigma_0} |U\rangle$. What is the condition that makes the state $|U\rangle$ invariant under σ translations?

Problem 13.4 $L_0^\perp - \bar{L}_0^\perp$ as world-sheet momentum.

(a) Use equations (13.36) to show that

$$L_0^\perp - \bar{L}_0^\perp = -\frac{1}{2\pi\alpha'} \int_0^{2\pi} d\sigma \, \dot{X}^I X^{I\prime}. \tag{1}$$

Also prove that

$$L_0^\perp - \bar{L}_0^\perp = -\frac{p^+}{2\pi} \int_0^{2\pi} d\sigma \, \frac{\partial X^-}{\partial \sigma},$$

which explains that $L_0^\perp - \bar{L}_0^\perp$ vanishes classically because X^-, just like any other string coordinate, must satisfy the closed string periodicity condition.

(b) The dynamics of the transverse light-cone string coordinates is governed by the Lagrangian density (12.81):

$$\mathcal{L} = \frac{1}{4\pi\alpha'} \left(\dot{X}^I \dot{X}^I - X^{I\prime} X^{I\prime} \right).$$

Show that the infinitesimal constant σ translation $\delta X^I = \epsilon \, \partial_\sigma X^I$ is a symmetry of $\mathcal{L}$ in the sense of Problem 8.10. Calculate the associated charge and show that it is proportional to $L_0^\perp - \bar{L}_0^\perp$, as given in (1).

Problem 13.5 Unoriented closed strings.

This problem is the closed string version of Problem 12.12. The closed string $X^\mu(\tau, \sigma)$ with $\sigma \in [0, 2\pi]$ and fixed τ is a parameterized closed curve in spacetime. The orientation of a string is the direction of increasing σ on this curve.

(a) Consider now the closed string $X^\mu(\tau, 2\pi - \sigma)$ with the same τ as above. How is this second string related to the first string above? How are their orientations related? Make a rough sketch, showing the original string as a continuous line and the second string as a dashed line.

Introduce an orientation reversing *twist* operator Ω such that

$$\Omega\, X^I(\tau, \sigma)\, \Omega^{-1} = X^I(\tau, 2\pi - \sigma)\,. \tag{1}$$

Moreover, declare that

$$\Omega\, x_0^-\, \Omega^{-1} = x_0^-\,, \quad \Omega\, p^+\, \Omega^{-1} = p^+\,. \tag{2}$$

(b) Use the closed string oscillator expansion (13.24) to calculate

$$\Omega\, x_0^I\, \Omega^{-1}\,, \quad \Omega\, \alpha_0^I\, \Omega^{-1}\,, \quad \Omega\, \alpha_n^I\, \Omega^{-1}\,, \quad \text{and} \quad \Omega\, \bar{\alpha}_n^I\, \Omega^{-1}\,.$$

(c) Show that $\Omega\, X^-(\tau, \sigma)\, \Omega^{-1} = X^-(\tau, 2\pi - \sigma)$. Since $\Omega\, X^+(\tau, \sigma)\, \Omega^{-1} = X^+(\tau, 2\pi - \sigma)$, equation (1) actually holds for all string coordinates. We say that orientation reversal is a symmetry of closed string theory because it leaves the closed string Hamiltonian H invariant: $\Omega H \Omega^{-1} = H$. Explain why this is true.

(d) Assume that the ground states are twist invariant. List the closed string states for $N^\perp \le 2$, and give their twist eigenvalues. If you were commissioned to build a theory of *unoriented* closed strings, which of the states would you have to discard? What are the massless fields of unoriented closed string theory?

Problem 13.6 Orientifold Op-planes.

An orientifold Op-plane is a a hyperplane with p spatial dimensions, just as a Dp-brane has p spatial dimensions. The Op-plane arises when we perform a truncation which keeps only the closed string states that are invariant under the symmetry transformation which simultaneously reverses the string orientation and reflects the coordinates normal to the Op-plane.

For an Op-plane, let $x^1, \ldots, x^p$ be the directions along the Op-plane, and let $x^{p+1}, \ldots, x^d$ with $d = 25$ be the directions orthogonal to the Op-plane. The Op-plane position is defined by $x^a = 0$ for $a = p + 1, \ldots, d$. We will organize the string coordinates as $X^+, X^-, \{X^i\}, \{X^a\}$ with $i = 2, \ldots, p$, and $a = p + 1, \ldots, d$. Let Ω_p denote the operator generating the transformation

$$\Omega_p\, X^a(\tau, \sigma)\, \Omega_p^{-1} = -X^a(\tau, 2\pi - \sigma)\,, \tag{1}$$

$$\Omega_p\, X^i(\tau, \sigma)\, \Omega_p^{-1} = X^i(\tau, 2\pi - \sigma)\,. \tag{2}$$

Moreover, assume that

$$\Omega_p\, x_0^-\, \Omega_p^{-1} = x_0^-\,, \quad \Omega_p\, p^+\, \Omega_p^{-1} = p^+\,. \tag{3}$$

(a) For an O23-plane the two normal directions x^{24}, x^{25} can be represented by a plane. A closed string at a fixed τ appears as a parameterized closed curve $X^a(\tau, \sigma)$ in this plane. Draw such an oriented closed string that lies fully in the first quadrant of the (x^{24}, x^{25}) plane. Draw also the string $\tilde{X}^a(\tau, \sigma) = -X^a(\tau, 2\pi - \sigma)$.

(b) For any operator $\mathcal{O}$ the action of Ω_p is defined by $\mathcal{O} \rightarrow \Omega_p \mathcal{O} \Omega_p^{-1}$. Use the expansion (13.24) to calculate the action of Ω_p on the operators

$$x_0^a, \ p^a, \ \alpha_n^a, \ \bar{\alpha}_n^a, \ x_0^i, \ p^i, \ \alpha_n^i, \quad \text{and} \quad \bar{\alpha}_n^i.$$

Show that $\Omega_p \, X^{\pm}(\tau, \sigma) \, \Omega_p^{-1} = X^{\pm}(\tau, 2\pi - \sigma)$. Explain why the orientifold transformations are symmetries of closed string theory.

(c) Denote the ground states by $|p^+, p^i, p^a\rangle$. Assume that the states $|p^+, p^i, \vec{0}\rangle$ are invariant under Ω_p: $\Omega_p |p^+, p^i, \vec{0}\rangle = |p^+, p^i, \vec{0}\rangle$. Prove that

$$\Omega_p |p^+, p^i, p^a\rangle = |p^+, p^i, -p^a\rangle.$$

Assume that, for every pair of conjugate coordinate and momentum operators (q, p), the momentum states are defined to satisfy $|p + \delta p\rangle = \exp(i \delta p \, q)|p\rangle$, with δp a constant.

The massless states of the oriented closed string theory are given by

$$|\Phi\rangle = \int dp^+ d\vec{p}^{\,i} \, d\vec{p}^{\,a} \ \Phi_{IJ}^{\pm}(\tau, p^+, p^i, p^a) \left(\alpha_{-1}^I \bar{\alpha}_{-1}^J \pm \alpha_{-1}^J \bar{\alpha}_{-1}^I \right) |p^+, p^i, p^a\rangle,$$

where $\Phi_{IJ}^{\pm}$ are wavefunctions, and the transverse light-cone indices I, J run over all the values taken by the indices i and a.

Now consider the truncation to Ω_p invariant states: $\Omega_p |\Phi\rangle = |\Phi\rangle$. The resulting string theory is the theory in the presence of the orientifold Op-plane. Intuitively, for an invariant state the amplitudes for the string to lie along each of the two related curves of part (a) are equal.

(d) Find the conditions that must be satisfied by $\Phi_{ab}^{\pm}$, $\Phi_{ia}^{\pm}$, $\Phi_{ij}^{\pm}$ to guarantee Ω_p invariance. All of the conditions are of the form

$$\Phi_{IJ}^{\pm}(\tau, p^+, p^i, p^a) = \cdots \Phi_{IJ}^{\pm}(\tau, p^+, p^i, -p^a),$$

where the dots represent a sign factor that you must determine for each case.

Remarks: in coordinate space, letting $x^m = \{x^0, \ldots, x^p\}$, the above invariance conditions require that

$$\Phi_{IJ}^{\pm}(x^m, x^a) = \cdots \Phi_{IJ}^{\pm}(x^m, -x^a).$$

In the presence of an orientifold plane, the values of the fields at (x^m, x^a) determine the values of the fields at the reflected point $(x^m, -x^a)$. The fields are either even or odd under $x^a \rightarrow -x^a$. The orientifold plane is some kind of mirror that relates the physics at reflected points, effectively cutting the space in half. In one half of the space, and away from the orientifold, there are no constraints, so one has the full set of fields of oriented closed strings. An O25-plane is space filling. Since it has no normal directions, the orientifold symmetry includes only string orientation reversal. This case was studied in Problem 13.5.

14

A look at relativistic superstrings

Realistic string theories must contain fermionic states, like the states of electrons or quarks. Superstrings include anticommuting dynamical variables in addition to the commuting coordinates X^μ that describe the position of strings. For open superstrings, quantization gives a state space with a Neveu–Schwarz (NS) sector that contains bosonic states and a Ramond (R) sector that contains fermionic states. The theory has supersymmetry, a symmetry that ensures that the number of bosonic and fermionic degrees of freedom are the same at any mass level. We examine type II closed string theories, which arise by tensoring the state spaces of open superstrings.

14.1 Introduction

We have so far studied bosonic string theories, both open and closed. These string theories live in 26-dimensional spacetime, and all of their quantum states represent bosonic particle states. Among them we found important bosonic particles, such as the photon and the graviton. Non-Abelian gauge bosons, needed to transmit the strong and weak forces, also arise in bosonic string theory, as we will see in Chapter 15.

Realistic string theories, however, must also contain the states of fermionic particles. You may recall that a quantum state of identical bosonic particles is symmetric under the exchange of any two of the particles. A quantum state of identical fermionic particles, on the other hand, is antisymmetric under the exchange of any two of the particles. Quarks and leptons are fermionic particles. To obtain them we need *superstring* theories. We will not study superstrings in detail in this book. A proper explanation would require a detailed discussion of spinors and the Dirac equation in various numbers of dimensions, as well as other technicalities. This would take us too long. Here we give you a brief discussion of the basics of superstrings, enough to appreciate that they are natural generalizations of the bosonic string with some interesting new ingredients. Certain applications that are discussed in this book involve superstrings; this chapter provides the required background material.

The superstring spectrum contains no tachyon. As a result, the theory of open superstrings can describe D-branes that are stable. Since bosonic string theory D-branes are always unstable, this affords new and interesting possibilities. In addition, superstrings have supersymmetry. Supersymmetry is a symmetry that relates the bosonic and fermionic quantum states of the theory: in a superstring theory we find an equal number of fermionic and bosonic states at every mass level. If supersymmetry exists in nature it must be spontaneously broken: we do not observe degeneracies between fermions and bosons. Many physicists believe that supersymmetry is an attractive candidate for the physics that lies beyond the presently established Standard Model of particle physics.

14.2 Anticommuting variables and operators

We describe the position of classical bosonic strings using the string coordinates $X^\mu(\tau, \sigma)$. The $X^\mu(\tau, \sigma)$ are classical *commuting* variables: products of them are independent of the order of the factors. In the quantum theory the X^μs become operators that do not generally commute. The failure of two operators A and B to commute is encoded in the commutator $[A, B] \equiv AB - BA$.

To get fermions in string theory we introduce new dynamical world-sheet variables $\psi_1^\mu(\tau, \sigma)$ and $\psi_2^\mu(\tau, \sigma)$. The classical variables ψ_α^μ ($\alpha = 1, 2$) are not ordinary commuting variables, but rather *anticommuting* variables. Since this is an important concept let us digress.

Two variables b_1 and b_2 are said to anticommute with one another if

$$b_1 b_2 = -b_2 b_1. \tag{14.1}$$

If b_1 and b_2 are anticommuting variables, more is true. The variables must anticommute with themselves. Thus $b_1 b_1 = -b_1 b_1$, which means that

$$b_1 b_1 = 0, \qquad b_2 b_2 = 0. \tag{14.2}$$

For classical anticommuting variables the order of factors is important. If we have a set of anticommuting variables b_i indexed by i, then we have

$$b_i b_j = -b_j b_i, \quad \forall i, j. \tag{14.3}$$

Quick calculation 14.1 Verify that the matrices γ^1 and γ^2 defined by

$$\gamma^1 = \begin{pmatrix} 0 & -1 \\ 1 & 0 \end{pmatrix}, \qquad \gamma^2 = \begin{pmatrix} 0 & 1 \\ 1 & 0 \end{pmatrix} \tag{14.4}$$

anticommute, but are not anticommuting variables.

In a quantum theory classical anticommuting variables become quantum operators that can sometimes fail to anticommute. Given two such quantum operators f_1 and f_2, the failure to anticommute is measured by the *anticommutator* $\{f_1, f_2\}$, defined by

$$\{f_1, f_2\} \equiv f_1 f_2 + f_2 f_1. \tag{14.5}$$

If two quantum operators anticommute, their anticommutator is zero.

We can explain what anticommuting operators have to do with fermions. Recall that the quantization of a scalar field, for example, gave us creation and annihilation operators that create particles. The creation operators $a_p^\dagger$ all commute among themselves and help us construct the multiparticle states

$$(a_{p_1}^\dagger)^{n_1} (a_{p_2}^\dagger)^{n_2} \ldots (a_{p_k}^\dagger)^{n_k} |\Omega\rangle, \qquad (14.6)$$

that contain n_1 particles with momentum $\vec{p}_1$, n_2 particles with momentum $\vec{p}_2$, and so on. The n_is are arbitrary positive integers.

To describe the relativistic electron and its antiparticle, the positron, one uses the classical Dirac field, a classical anticommuting field variable. The quantization of the Dirac field gives rise to creation and annihilation operators for electrons and different creation and annihilation operators for positrons. For convenience, let us focus on electrons. The creation operators $f_{p,s}^\dagger$ are labeled by momentum $\vec{p}$ and spin s. Electrons are spin one-half particles, so the spin label can only take two values. All the electron creation operators are anticommuting variables and they all anticommute. In particular, this means that $f_{p,s}^\dagger f_{p,s}^\dagger = 0$, for any $\vec{p}$ and s. A multiparticle state of electrons takes the form

$$f_{p_1,s_1}^\dagger f_{p_2,s_2}^\dagger \ldots f_{p_k,s_k}^\dagger |\Omega\rangle \qquad (14.7)$$

and describes a state with an electron with momentum $\vec{p}_1$ and spin s_1, an electron with momentum $\vec{p}_2$ and spin s_2, and so on. Note that we *cannot* get a state with two electrons that have the same momentum and the same spin because $f_{p,s}^\dagger f_{p,s}^\dagger |\Omega\rangle = 0$. The anticommuting creation operators automatically implement Fermi's exclusion principle. In fact, if we use a wavefunction to create a superposition of states

$$\sum_{s_1,s_2} \int d\vec{p} \; \psi(p_1, s_1; p_2, s_2) \, f_{p_1,s_1}^\dagger f_{p_2,s_2}^\dagger |\Omega\rangle, \qquad (14.8)$$

we can quickly conclude that only the part of the wavefunction ψ that is antisymmetric under the simultaneous exchange $p_1 \leftrightarrow p_2$ and $s_1 \leftrightarrow s_2$ contributes to the above state.

Quick calculation 14.2 Prove the above claim.

14.3 World-sheet fermions

As we mentioned before, classical superstrings require anticommuting dynamical variables $\psi_\alpha^\mu(\tau, \sigma)$, with $\alpha = 1, 2$. Recall that for each value of μ the dynamical variable $X^\mu(\tau, \sigma)$ is a world-sheet boson. As it turns out, for each μ, the two components ψ_1^μ and ψ_2^μ comprise the variables needed to describe a fermion on the (τ, σ) world, that is, a *world-sheet* fermion. Remarkably, the quantization of such objects results in particle states that behave as *spacetime* fermions, which is what we need.

In light-cone quantization X^+ was set proportional to τ and X^- was solved for in terms of other quantities. With superstrings this remains true but, in addition, the light-cone gauge condition sets $\psi_\alpha^+ = 0$ and allows one to solve for ψ_α^-. Both X^- and ψ_α^- receive contributions from the transverse X^I and ψ_α^I. Since both X^μ and ψ_α^μ are spacetime Lorentz vectors, both enter into the definition of the light-cone Lorentz generator M^{-I}. It follows that the commutator $[M^{-I}, M^{-J}]$ receives contributions from both X^{μ} and ψ_{α}^{μ}. The requirement that this commutator vanishes gives constraints different from those obtained for bosonic strings. The number of spacetime dimensions is no longer twenty-six, but rather ten. The downward shift of the mass-squared equals minus one-half, rather than minus one.

In the light-cone gauge we need only concern ourselves with the transverse fields $\psi_{\alpha}^{I}(\tau, \sigma)$. To study their quantization it is convenient to have an action, denoted as S_{ψ}, that describes their dynamics. This action is the fermion analog of the action (12.81) that describes the dynamics of the transverse coordinates X^{I}. So, all in all, the action S that describes the full set of degrees of freedom takes the form

$$S = \frac{1}{4\pi\alpha'} \int d\tau \int_0^{\pi} d\sigma \left(\dot{X}^I \dot{X}^I - X^{I'} X^{I'} \right) + S_{\psi}, \tag{14.9}$$

with

$$S_{\psi} = \frac{1}{2\pi} \int d\tau \int_0^{\pi} d\sigma \left[\psi_1^I (\partial_{\tau} + \partial_{\sigma}) \psi_1^I + \psi_2^I (\partial_{\tau} - \partial_{\sigma}) \psi_2^I \right]. \tag{14.10}$$

The action S_{ψ} is the Dirac action for a fermion that lives in the two-dimensional (τ, σ) world. Note that each term in S_{ψ} contains just one derivative. For bosons, like the X^I, terms in the action contain two derivatives. A term in the Lagrangian that couples a field to itself and has just one derivative risks being an irrelevant total derivative. Indeed, consider an arbitrary field h and a coupling $h(\partial_{\tau} h)$. We have

$$h(\partial_{\tau} h) = \partial_{\tau}(hh) - (\partial_{\tau} h)h, \tag{14.11}$$

or, equivalently,

$$h(\partial_{\tau} h) + (\partial_{\tau} h)h = \partial_{\tau}(hh). \tag{14.12}$$

If the field h is commuting, the two terms on the left-hand side are identical and, indeed, $h(\partial_{\tau} h)$ is a total derivative. If the field h is anticommuting, $(\partial_{\tau} h)h = -h(\partial_{\tau} h)$ and both the left-hand side and the right-hand side of the equation vanish. We learn that $h(\partial_{\tau} h)$, for h anticommuting, is not a total derivative. This means that the action S_{ψ} is nontrivial only because the ψ_{α}^I fields are anticommuting.

Let us vary the fields ψ_{α}^I in S_{ψ} in order to find the equations of motion and the boundary conditions. We have

$$\delta S_{\psi} = \frac{1}{2\pi} \int d\tau \int_0^{\pi} d\sigma \left[\delta\psi_1^I (\partial_{\tau} + \partial_{\sigma})\psi_1^I + \psi_1^I (\partial_{\tau} + \partial_{\sigma})\delta\psi_1^I \right. $$
$$\left. + \delta\psi_2^I (\partial_{\tau} - \partial_{\sigma})\psi_2^I + \psi_2^I (\partial_{\tau} - \partial_{\sigma})\delta\psi_2^I \right]. \tag{14.13}$$

Consider the second term on the first line of the above right-hand side:

$$\psi_1^I (\partial_{\tau} + \partial_{\sigma})\delta\psi_1^I = \partial_{\tau}(\psi_1^I \delta\psi_1^I) + \partial_{\sigma}(\psi_1^I \delta\psi_1^I) - [(\partial_{\tau} + \partial_{\sigma})\psi_1^I]\delta\psi_1^I$$
$$= \partial_{\tau}(\psi_1^I \delta\psi_1^I) + \partial_{\sigma}(\psi_1^I \delta\psi_1^I) + \delta\psi_1^I (\partial_{\tau} + \partial_{\sigma})\psi_1^I. \tag{14.14}$$

A similar calculation can be done for the second term on the second line of (14.13). Back into this equation and, as usual, ignoring the total derivatives along time,

$$\delta S_\psi = \frac{1}{\pi} \int d\tau \int_0^\pi d\sigma \Big[\delta\psi_1^I(\partial_\tau + \partial_\sigma)\psi_1^I + \delta\psi_2^I(\partial_\tau - \partial_\sigma)\psi_2^I \Big]$$
$$+ \frac{1}{2\pi} \int d\tau \Big[\psi_1^I \delta\psi_1^I - \psi_2^I \delta\psi_2^I \Big]_{\sigma=0}^{\sigma=\pi} . \tag{14.15}$$

We can now read immediately the equations of motion

$$(\partial_\tau + \partial_\sigma)\psi_1^I = 0, \qquad (\partial_\tau - \partial_\sigma)\psi_2^I = 0, \tag{14.16}$$

as well as the boundary conditions

$$\psi_1^I(\tau, \sigma_*)\, \delta\psi_1^I(\tau, \sigma_*) - \psi_2^I(\tau, \sigma_*)\, \delta\psi_2^I(\tau, \sigma_*) = 0, \tag{14.17}$$

that must hold both at $\sigma_* = 0$ and at $\sigma_* = \pi$ for all times.

We will use the above equations of motion and boundary conditions to discuss the quantization and the state space of the theory. Our analysis will not be complete. In particular, we will not discuss the anticommutation relations satisfied by the fields ψ_α^I and, as a result, we will not justify the anticommutation relations of the corresponding oscillators. Nevertheless, the discussion will illuminate the main features of the quantization, and all results should seem plausible.

We begin by noting that the equations of motion (14.16) imply that ψ_1^I is right-moving and ψ_2^I is left-moving:

$$\psi_1^I(\tau, \sigma) = \Psi_1^I(\tau - \sigma),$$
$$\psi_2^I(\tau, \sigma) = \Psi_2^I(\tau + \sigma). \tag{14.18}$$

Let us now consider the boundary conditions (14.17). What can we do to satisfy these conditions? A little thought shows that attempts to make each term in (14.17) vanish do not work. Try, for example, setting $\psi_1^I(\tau, 0) = 0$. Our solution in (14.18) then implies $\Psi_1^I(\tau) = 0$, for all τ, which results in $\psi_1^I(\tau, \sigma) \equiv 0$.

The situation improves if we impose boundary conditions that relate ψ_1^I and ψ_2^I at the endpoints. For each σ_* we take

$$\psi_1^I(\tau, \sigma_*) = \pm\psi_2^I(\tau, \sigma_*), \tag{14.19}$$

where the choice of sign is still to be determined. If the fields are so constrained at an endpoint, their variations must also respect this condition:

$$\delta\psi_1^I(\tau, \sigma_*) = \pm\delta\psi_2^I(\tau, \sigma_*). \tag{14.20}$$

It then follows by multiplication of the last two equations that (14.17) holds for either choice of sign.

Let us now discuss the signs. Since both ψ_1^I and ψ_2^I appear quadratically in the action, their signs can be changed without physical consequence. This arbitrariness is used conventionally to demand that at the endpoint $\sigma_* = 0$,

$$\psi_1^I(\tau, 0) = \psi_2^I(\tau, 0). \tag{14.21}$$

Since we cannot change the sign of ψ_1^I or ψ_2^I without changing this condition, the choice of sign at the other endpoint is physically relevant:

$$\psi_1^I(\tau, \pi) = \pm\psi_2^I(\tau, \pi). \tag{14.22}$$

The full superstring theory state space breaks into two subspaces or, as they are typically called, two *sectors*: a Ramond (R) sector which contains the states that arise using the top choice of sign, and a Neveu–Schwarz (NS) sector which contains the states that arise using the lower choice of sign. The boundary conditions are better understood by assembling a fermion field Ψ^I defined over the full interval $\sigma \in [-\pi, \pi]$:

$$\Psi^I(\tau, \sigma) \equiv \begin{cases} \psi_1^I(\tau, \sigma) & \sigma \in [0, \pi] \\ \psi_2^I(\tau, -\sigma) & \sigma \in [-\pi, 0]. \end{cases} \tag{14.23}$$

This construction is reminiscent of (12.36), which defined a field over $\sigma \in [-\pi, \pi]$ for bosonic open strings. The boundary condition (14.21) guarantees that Ψ^I is continuous at $\sigma = 0$. Moreover, on account of (14.18), $\Psi^I(\tau, \sigma)$ is a function of $\tau - \sigma$:

$$\Psi^I(\tau, \sigma) = \chi^I(\tau - \sigma). \tag{14.24}$$

Finally, the boundary condition (14.22) gives

$$\Psi^I(\tau, \pi) = \psi_1^I(\tau, \pi) = \pm\psi_2^I(\tau, \pi) = \pm\Psi^I(\tau, -\pi). \tag{14.25}$$

We thus learn that a periodic fermion Ψ^I corresponds to Ramond boundary conditions and an antiperiodic fermion Ψ^I corresponds to Neveu–Schwarz boundary conditions:

$$\begin{aligned} \Psi^I(\tau, \pi) &= +\Psi^I(\tau, -\pi) \quad \text{Ramond boundary condition,} \\ \Psi^I(\tau, \pi) &= -\Psi^I(\tau, -\pi) \quad \text{Neveu–Schwarz boundary condition.} \end{aligned} \tag{14.26}$$

Let us consider both cases in detail now.

14.4 Neveu–Schwarz sector

Since the Neveu–Schwarz fermion Ψ^I is a function of $\tau - \sigma$ and changes sign when $\sigma \to \sigma + 2\pi$, it must be expanded with fractionally moded exponentials:

$$\Psi^I(\tau, \sigma) \sim \sum_{r \in \mathbb{Z}+1/2} b_r^I \, e^{-ir(\tau-\sigma)}, \tag{14.27}$$

up to a normalization factor that will not enter our discussion. Indeed, for any $r = n + \frac{1}{2}$, with n integer, we have

$$e^{ir(\sigma+2\pi)} = e^{ir\sigma} e^{i\left(n+\frac{1}{2}\right)2\pi} = e^{ir\sigma} e^{i\pi} = -e^{ir\sigma}, \tag{14.28}$$

which guarantees that Ψ^I is antiperiodic. Since Ψ^I is anticommuting, the expansion coefficients b_r^I are anticommuting operators. Following our usual notation, the negatively moded coefficients $b_{-1/2}^I, b_{-3/2}^I, b_{-5/2}^I, \ldots$, are creation operators, while the positively moded ones $b_{1/2}^I$, $b_{3/2}^I$, $b_{5/2}^I$, ... are annihilation operators. These operators act on a vacuum, henceforth called the Neveu–Schwarz vacuum $|NS\rangle$. These operators satisfy the anticommutation relation

$$\{b_r^I, b_s^J\} = \delta_{r+s,0}\eta^{IJ}. \tag{14.29}$$

Given that all creation operators anticommute with one another and square to zero, each b_{-r}^I can appear at most once in any state. Since the $X^I(\tau, \sigma)$ are quantized as usual, we still have the α_{-n}^I creation operators. Thus, the states in the NS sector are of the form:

$$\text{NS sector: } |\lambda\rangle = \prod_{I=2}^{9}\prod_{n=1}^{\infty}(\alpha_{-n}^I)^{\lambda_{n,I}}\prod_{J=2}^{9}\prod_{r=\frac{1}{2},\frac{3}{2},\dots}(b_{-r}^J)^{\rho_{r,J}}|NS\rangle \otimes |p^+, \vec{p}_T\rangle. \tag{14.30}$$

Here the $\rho_{r,J}$ are either zero or one. We have written the full ground state as a "product" $\otimes$ of the ground state $|NS\rangle$ for the b_{-r}^J operators and the ground state $|p^+, \vec{p}_T\rangle$ for the α_{-n}^I operators. The order in which the b operators appear in the state does not matter when we consider a single state. Since all the bs anticommute, different orderings can only differ by an overall sign, and no new states are obtained.

The mass-squared operator in the NS sector, before normal ordering, is given by

$$M^2 = \frac{1}{\alpha'}\left(\frac{1}{2}\sum_{p\neq 0}\alpha_{-p}^I\alpha_p^I + \frac{1}{2}\sum_{r\in\mathbb{Z}+\frac{1}{2}}rb_{-r}^I b_r^I\right). \tag{14.31}$$

We can use the heuristic method based on ζ-functions to find the ordering constant "a" in M^2 – the constant that must be added inside the above parentheses when the sums are replaced by their normal-ordered versions. For the bosonic α^I oscillators we know that each coordinate contributes $-\frac{1}{24}$ to a. We record this piece of information as

$$a_B = -\frac{1}{24}. \tag{14.32}$$

For the NS fermions the terms in M^2 that require reordering are

$$\frac{1}{2}\sum_{r=-\frac{1}{2},-\frac{3}{2},\dots}rb_{-r}^I b_r^I = \frac{1}{2}\sum_{r=\frac{1}{2},\frac{3}{2},\dots}(-r)b_r^I b_{-r}^I$$

$$= \frac{1}{2}\sum_{r=\frac{1}{2},\frac{3}{2},\dots}rb_{-r}^I b_r^I - \frac{1}{2}(D-2)\left(\frac{1}{2}+\frac{3}{2}+\frac{5}{2}+\cdots\right). \tag{14.33}$$

In the first step we let $r \to -r$ and in the second step we used the anticommutator (14.29). The sum over odd positive integers was evaluated in (13.116) and gives $+\frac{1}{12}$. It then follows that

$$\frac{1}{2}\sum_{r=-\frac{1}{2},-\frac{3}{2},\dots}rb_{-r}^I b_r^I = \frac{1}{2}\sum_{r=\frac{1}{2},\frac{3}{2},\dots}rb_{-r}^I b_r^I - \frac{1}{48}(D-2). \tag{14.34}$$

We have thus learned that each NS fermion (antiperiodic fermion) contributes to a the constant a_{NS} given by

$$a_{NS} = -\frac{1}{48}. \tag{14.35}$$

The full ordering constant for M^2 is therefore

$$a = (D-2)(a_B + a_{\text{NS}}) = (D-2)\left(-\frac{1}{24} - \frac{1}{48}\right) = -(D-2)\frac{1}{16}. \quad (14.36)$$

For $D = 10$ we find $a = -\frac{1}{2}$ and therefore (14.31) gives

$$M^2 = \frac{1}{\alpha'}\left(-\frac{1}{2} + N^\perp\right), \quad \text{with} \quad N^\perp = \sum_{p=1}^{\infty} \alpha_{-p}^I \alpha_p^I + \sum_{r=\frac{1}{2},\frac{3}{2}\cdots} r b_{-r}^I b_r^I. \quad (14.37)$$

$N^\perp$ is a number operator that counts the contributions from the α^Is and the b^Is.

Quick calculation 14.3 To test that $N^\perp$ includes the fermionic contribution to the number, show that the eigenvalue of $N^\perp$ on $b_{-r_1}^I b_{-r_2}^J |\text{NS}\rangle$, with $r_1, r_2 > 0$, is $r_1 + r_2$.

We can now list the first few levels of states in the NS sector. Organized by the number eigenvalue or, equivalently, mass-squared, we have

$$\alpha' M^2 = -\tfrac{1}{2}, \ N^\perp = 0 : |\text{NS}\rangle \otimes |p^+, \vec{p}_T\rangle,$$

$$\alpha' M^2 = 0, \ N^\perp = \tfrac{1}{2} : b_{-1/2}^I |\text{NS}\rangle \otimes |p^+, \vec{p}_T\rangle,$$

$$\alpha' M^2 = \tfrac{1}{2}, \ N^\perp = 1 : \big\{\alpha_{-1}^I, \ b_{-1/2}^I b_{-1/2}^J\big\} |\text{NS}\rangle \otimes |p^+, \vec{p}_T\rangle,$$

$$\alpha' M^2 = 1, \ N^\perp = \tfrac{3}{2} : \big\{\alpha_{-1}^I b_{-1/2}^J, \ b_{-3/2}^I, \ b_{-1/2}^I b_{-1/2}^J b_{-1/2}^K\big\} |\text{NS}\rangle \otimes |p^+, \vec{p}_T\rangle. \quad (14.38)$$

The states with $N^\perp = 0$ have $\alpha' M^2 = -\tfrac{1}{2}$. The states with $N^\perp = \tfrac{1}{2}$ are massless. There are eight of them, labeled by the vector index I.

Quick calculation 14.4 How many states are there at $N^\perp = \tfrac{3}{2}$?

It is useful to have an operator whose value on states is $+1$ if the state is bosonic and -1 if the state is fermionic. This operator is usually called $(-1)^F$, where F stands for fermion number. This is reasonable, states with even fermion number are bosonic and states with odd fermion number are fermionic. To calculate $(-1)^F$ on any state we must first give the eigenvalue of $(-1)^F$ on the Neveu–Schwarz ground states $|\text{NS}\rangle \otimes |p^+, \vec{p}_T\rangle$. Let us declare that number to be minus one, thus making the ground states fermionic:

$$(-1)^F |\text{NS}\rangle \otimes |p^+, \vec{p}_T\rangle = -|\text{NS}\rangle \otimes |p^+, \vec{p}_T\rangle. \quad (14.39)$$

The eigenvalue of $(-1)^F$ on a state is equal to minus one times a sequence of factors of minus one, one for each fermionic oscillator that appears in the state. Thus, acting on the generic state (14.30) we get

$$(-1)^F |\lambda\rangle = -(-1)^{\sum_{r,J} \rho_{r,J}} |\lambda\rangle. \quad (14.40)$$

Operationally, this result follows if we take $(-1)^F$ to anticommute with all of the fermionic operators

$$\big\{(-1)^F, b_r^I\big\} = 0. \quad (14.41)$$

Consider again our list of states (14.38). Since fermionic oscillators contribute half-integers to $N^\perp$, states with integer $N^\perp$ must have an even number of fermionic oscillators. So all states with integer $N^\perp$ have $(-1)^F = -1$; they are fermionic states. All states with half-integer $N^\perp$ have an odd number of fermionic oscillators and therefore $(-1)^F = +1$. These are bosonic states, and they include the eight massless states on the second line of the table.

The fermion or boson character of the states we have discussed so far is restricted to the (τ, σ) world-sheet, where the Ψ fields are fermions. We will later discuss if the above states are fermions or bosons in spacetime.

14.5 Ramond sector

With Ramond boundary conditions (14.26) the field Ψ^I is periodic and can be expanded using integrally moded oscillators:

$$\Psi^I(\tau, \sigma) \sim \sum_{n\in\mathbb{Z}} d_n^I \, e^{-in(\tau-\sigma)}. \tag{14.42}$$

Since Ψ^I is anticommuting, the oscillators d_n^I are anticommuting operators. Again, the negatively moded oscillators $d_{-1}^I, d_{-2}^I, d_{-3}^I, \ldots$, are creation operators, while the positively moded ones $d_1^I, d_2^I, d_3^I, \ldots$ are annihilation operators. The Ramond oscillators satisfy the anticommutation relation

$$\{d_m^I, d_n^J\} = \delta_{m+n,0}\eta^{IJ}. \tag{14.43}$$

As in the NS sector, the Ramond creation operators are all anticommuting and, as a result, can appear at most only once on any given state.

Ramond fermions are more complicated than NS fermions because the eight fermionic zero modes d_0^I must be treated with care. It turns out that these eight operators can be organized by simple linear combinations into four creation operators and four annihilation operators. Let us call the four creation operators

$$\xi_1, \ \xi_2, \ \xi_3, \ \xi_4. \tag{14.44}$$

Being zero modes, these creation operators do not contribute to the mass-squared of the states. Postulating a unique vacuum $|0\rangle$, the creation operators allow us to construct $16 = 2^4$ degenerate Ramond ground states. In fact, eight of these states have an even number of ξs acting on $|0\rangle$ and the other eight have an odd number of ξs acting on $|0\rangle$. Explicitly, the eight states $|R_a\rangle$, $a = 1, 2, \ldots, 8$, with an even number of creation operators are

$$\text{states } |R_a\rangle : \begin{cases} |0\rangle, \\ \xi_1\xi_2|0\rangle, \ \xi_1\xi_3|0\rangle, \ \xi_1\xi_4|0\rangle, \ \xi_2\xi_3|0\rangle, \ \xi_2\xi_4|0\rangle, \ \xi_3\xi_4|0\rangle, \\ \xi_1\xi_2\xi_3\xi_4|0\rangle. \end{cases} \tag{14.45}$$

The eight states $|R_{\bar{a}}\rangle$, $\bar{a} = \bar{1}, \bar{2}, \ldots, \bar{8}$, with an odd number of creation operators are

$$
\text{states } |R_{\bar{a}}\rangle : \begin{cases} \xi_1|0\rangle, \; \xi_2|0\rangle, \; \xi_3|0\rangle, \; \xi_4|0\rangle, \\ \xi_1\xi_2\xi_3|0\rangle, \; \xi_1\xi_2\xi_4|0\rangle, \; \xi_1\xi_3\xi_4|0\rangle, \; \xi_2\xi_3\xi_4|0\rangle. \end{cases} \tag{14.46}
$$

The states $|R_a\rangle$ and $|R_{\bar{a}}\rangle$ comprise the full set of degenerate Ramond ground states, denoted as $|R_A\rangle$ with $A = 1, \ldots, 16$. The Ramond sector of the state space contains the states

$$
\text{R sector:} \quad |\lambda\rangle = \prod_{I=2}^{9} \prod_{n=1}^{\infty} (\alpha_{-n}^I)^{\lambda_{n,I}} \prod_{J=2}^{9} \prod_{m=1}^{\infty} (d_{-m}^J)^{\rho_{m,J}} |R_A\rangle \otimes |p^+, \vec{p}_T\rangle. \tag{14.47}
$$

Here the $\rho_{m,J}$ are either zero or one.

Just like in the NS sector, the Ramond sector has an $(-1)^F$ operator. This operator anticommutes with all the fermion oscillators, including the zero modes:

$$
\{(-1)^F, d_n^I\} = 0, \tag{14.48}
$$

and, additionally, we conventionally declare $|0\rangle$ to be fermionic

$$
(-1)^F |0\rangle = -|0\rangle. \tag{14.49}
$$

It thus follows that all eight $|R_a\rangle$ states are fermionic and all $|R_{\bar{a}}\rangle$ states are bosonic.

The mass-squared operator in the R sector, before normal ordering, is given by

$$
M^2 = \frac{1}{\alpha'} \left(\frac{1}{2} \sum_{p \neq 0} \alpha_{-p}^I \alpha_p^I + \frac{1}{2} \sum_{n \in \mathbb{Z}} n d_{-n}^I d_n^I \right). \tag{14.50}
$$

For the R fermions the terms that require ordering are

$$
\begin{aligned}
\frac{1}{2} \sum_{n=-1,-2,\ldots} n d_{-n}^I d_n^I &= -\frac{1}{2} \sum_{n=1,2,\ldots} n d_n^I d_{-n}^I \\
&= \frac{1}{2} \sum_{n=1,2,\ldots} n d_{-n}^I d_n^I - \frac{1}{2}(D-2)(1+2+3+\cdots) \\
&= \frac{1}{2} \sum_{n=1,2,\ldots} n d_{-n}^I d_n^I + \frac{1}{24}(D-2).
\end{aligned} \tag{14.51}
$$

We thus learn that the ordering contribution from a (periodic) Ramond fermion is

$$
a_{\text{R}} = \frac{1}{24}. \tag{14.52}
$$

This number is precisely the opposite of a_B. Since there are equal numbers of bosonic coordinates X^I and Ramond fermions, their respective normal-ordering contributions to the mass-squared cancel out and (14.50) gives:

$$
M^2 = \frac{1}{\alpha'} \sum_{n \geq 1} \left(\alpha_{-n}^I \alpha_n^I + n d_{-n}^I d_n^I \right). \tag{14.53}
$$

This formula implies that all the Ramond ground states are massless.

Let us now list the states at various mass levels

$$\alpha' M^2 = 0 : \qquad\qquad\qquad\qquad |R_a\rangle \;\|\; |R_{\bar{a}}\rangle$$

$$\alpha' M^2 = 1 : \qquad\qquad \alpha^I_{-1}|R_a\rangle,\, d^I_{-1}|R_{\bar{a}}\rangle \;\|\; \alpha^I_{-1}|R_{\bar{a}}\rangle,\, d^I_{-1}|R_a\rangle,$$

$$\alpha' M^2 = 2 : \quad \{\alpha^I_{-2},\, \alpha^I_{-1}\alpha^J_{-1},\, d^I_{-1}d^J_{-1}\}|R_a\rangle,\; \|\; \{\alpha^I_{-2},\, \alpha^I_{-1}\alpha^J_{-1},\, d^I_{-1}d^J_{-1}\}|R_{\bar{a}}\rangle,$$

$$\{\alpha^I_{-1}d^J_{-1},\, d^I_{-2}\}|R_{\bar{a}}\rangle \;\|\; \{\alpha^I_{-1}d^J_{-1},\, d^I_{-2}\}|R_a\rangle. \qquad (14.54)$$

We have separated the states in two groups with identical number of states: to the left of the bars we find the states with $(-1)^F = -1$ (fermionic states) and to the right of the bars the states with $(-1)^F = +1$ (bosonic states). Note that for each state to the left there is a corresponding state to the right built on a ground state of opposite fermion number. The appearance of an equal number of bosonic and fermionic states at every mass level is a signal of supersymmetry. This is, however, supersymmetry on the world-sheet. As we will soon see, spacetime supersymmetry arises only after we combine states from the Ramond and Neveu–Schwarz sectors.

14.6 Counting states

Before assembling a supersymmetric theory we digress to learn how to count the number of states that a string theory has at any given mass level. Our goal is to obtain generating functions that encode these numbers for the NS and R sectors.

The generating functions contain the information about numbers of states in their power series expansions. Typically we want a function $f(x)$ such that

$$f(x) = \sum_{n=0}^{\infty} a(n)\, x^n, \qquad\qquad (14.55)$$

where $a(n)$ is the number of states with, say, $N^{\perp} = n$. Suppose we have just one oscillator $a_1^{\dagger}$. There is then just one state $|0\rangle$ with $N^{\perp} = 0$, one state $a_1^{\dagger}|0\rangle$ with $N^{\perp} = 1$ and, in fact, one state $(a_1^{\dagger})^k|0\rangle$ with $N^{\perp} = k$. It follows that the function $f_1(x)$ corrresponding to this system is

$$f_1(x) = 1 + x + x^2 + x^3 + \cdots = \frac{1}{1-x}. \qquad\qquad (14.56)$$

Suppose we had just one oscillator $a_2^{\dagger}$ with mode number two. Then we get one vacuum and one state for each even value of $N^{\perp}$, resulting in a function $f_2(x)$ that takes the form:

$$f_2(x) = 1 + x^2 + x^4 + x^6 + \cdots = \frac{1}{1-x^2}. \qquad\qquad (14.57)$$

We now ask: what is the function $f_{12}(x)$ corresponding to the states built using both $a_1^{\dagger}$ and $a_2^{\dagger}$? To obtain these states we form all the products where the first factor is a state built with $a_1^{\dagger}$s and the second factor is a state built with $a_2^{\dagger}$s. In forming these products one removes the ground states, multiplies the oscillators, and restores the ground state. It follows that f_{12} is given by the product of f_1 and f_2:

$$(1 + x + x^2 + x^3 + \cdots)(1 + x^2 + x^4 + x^6 + \cdots). \tag{14.58}$$

To aid the imagination one can place the oscillator content of the states on the generating functions:

$$(1 + a_1^\dagger x + (a_1^\dagger)^2 x^2 + (a_1^\dagger)^3 x^3 + \cdots)(1 + a_2^\dagger x^2 + (a_2^\dagger)^2 x^4 + (a_2^\dagger)^3 x^6 + \cdots).$$

It is clear that the product forms all possible states, and states with $N^\perp = k$ appear together with x^k. This makes it clear that (14.58) is the correct answer:

$$f_{12}(x) = f_1(x) f_2(x) = \frac{1}{1-x} \frac{1}{1-x^2}. \tag{14.59}$$

If we use oscillators $a_1^\dagger, a_2^\dagger, a_3^\dagger, \ldots$ of all mode numbers, the generating function is

$$f(x) = \prod_{n=1}^{\infty} \frac{1}{1-x^n} = \frac{1}{1-x} \frac{1}{1-x^2} \frac{1}{1-x^3} \cdots . \tag{14.60}$$

The above use of multiplication is quite general. Consider a state space with oscillators of type A and generating function $f_A(x)$ as well as a state space with oscillators of type B and generating function $f_B(x)$. Then, the generating function for the state space built with oscillators of types A and B is $f_{AB}(x) = f_A(x) f_B(x)$.

As an application we compute the generating function for bosonic open string theory. In this theory we have oscillators of all mode numbers, but they come in 24 species, one for each transverse light-cone direction. Since each species gives a generating function (14.60), the full generating function for bosonic open string theory is obtained by raising the right-hand side of (14.60) to the 24th power:

$$\prod_{n=1}^{\infty} \frac{1}{(1-x^n)^{24}}. \tag{14.61}$$

Since mass-squared is more physical than $N^\perp$ it is useful to use $\alpha' M^2$ instead of $N^\perp$ to define the generating function. In a generating function based on $\alpha' M^2$ the coefficient of x^k counts the number of states with $\alpha' M^2 = k$. For open bosonic strings $\alpha' M^2 = N^\perp - 1$, so the $\alpha' M^2$ generating function is obtained by dividing the $N^\perp$ generating function by one power of x. As a result, the generating function $f_{os}(x)$ for bosonic open string theory is

$$f_{os}(x) = \frac{1}{x} \prod_{n=1}^{\infty} \frac{1}{(1-x^n)^{24}}. \tag{14.62}$$

With the help of a symbolic manipulator we readily find that

$$f_{os}(x) = \frac{1}{x} + 24 + 324\, x + 3200\, x^2 + 25\,650\, x^3 + 176\,256\, x^4 + \cdots . \tag{14.63}$$

This equation reminds us that we have a tachyon with $\alpha' M^2 = -1$, 24 massless states of a Maxwell field, and 324 states with $\alpha' M^2 = +1$.

Quick calculation 14.5 Use the binomial formula to calculate, by hand, the terms in $f_{os}(x)$ up to and including $\mathcal{O}(x^2)$.

Quick calculation 14.6 Construct explicitly all the states with $\alpha'M^2 = 2$ and count them, verifying that there are indeed a total of 3200 states. You may find the counting formula in Problem 12.11 useful.

To obtain the generating functions for the NS and R sectors we must learn how to count states built with fermionic oscillators. Happily, this is easy to do. Again, we begin by using $N^\perp$ and assuming that we have a single fermionic creation operator f_{-r} that contributes r to $N^\perp$. For this oscillator we can only build two states $|0\rangle$ and $f_{-r}|0\rangle$. The generating function $f_r(x)$ is therefore

$$f_r(x) = 1 + x^r. \tag{14.64}$$

Since the NS sector contains oscillators $b^I_{-1/2}, b^I_{-3/2}, \ldots$ coming in eight species, the generating function associated with these is

$$\left[(1+x^{1/2})(1+x^{3/2})(1+x^{5/2})\cdots\right]^8 = \prod_{n=1}^{\infty}(1+x^{n-\frac{1}{2}})^8. \tag{14.65}$$

Recalling that $\alpha'M^2 = N^\perp - \frac{1}{2}$ and that we have eight bosonic coordinates as well, the $\alpha'M^2$ based generating function $f_{NS}(x)$ for the NS sector is

$$f_{NS}(x) = \frac{1}{\sqrt{x}}\prod_{n=1}^{\infty}\left(\frac{1+x^{n-\frac{1}{2}}}{1-x^n}\right)^8. \tag{14.66}$$

Expanding for the first few orders we get

$$f_{NS}(x) = \frac{1}{\sqrt{x}} + 8 + 36\sqrt{x} + 128x + 402x\sqrt{x} + 1152x^2 + \cdots. \tag{14.67}$$

This expansion shows the tachyon at $\alpha'M^2 = -1/2$, the eight massless states, and the 36 states at $\alpha'M^2 = 1/2$. The corresponding states were listed in (14.38).

For the Ramond sector we have $\alpha'M^2 = N^\perp$, with no offset. Since the fermionic oscillators $d^I_{-1}, d^I_{-2}, \ldots$ are integrally moded we get

$$f_R(x) = 16\prod_{n=1}^{\infty}\left(\frac{1+x^n}{1-x^n}\right)^8. \tag{14.68}$$

The overall multiplicative factor appears because each combination of oscillators gives rise to 16 states by acting on each of the available ground states. The power series expansion of (14.68) gives

$$f_R(x) = 16 + 256x + 2304x^2 + 15360x^3 + \cdots. \tag{14.69}$$

The NS generating function contains both integer and half-integer powers of x, while the R generating function only has integer powers of x. We note that the R coefficients are actually double the corresponding NS coefficients. This is not a coincidence, as we will see in the following section.

14.7 Open superstrings

We have seen that the Ramond sector has world-sheet supersymmetry: there are equal numbers of fermionic and bosonic states at each mass level. Consider, for example, the ground states. There are sixteen of them, obtained by acting on $|0\rangle$ with four linear combinations built from the zero modes d_0^I. The states break into two groups $|R_a\rangle$ and $|R_{\bar{a}}\rangle$ of eight states each, with opposite values of $(-1)^F$.

Since the d_0^I carry a Lorentz index they form a Lorentz vector and transform as such under Lorentz transformations. The ground states, however, are built in an intricate way using the zero modes (see (14.45) and (14.46)), so they do *not* transform as vectors. In fact, under Lorentz transformations the $|R_a\rangle$ transform among themselves and so do the $|R_{\bar{a}}\rangle$. Both transform as *spinors*, the kind of transformation that is appropriate for states that are *spacetime* fermions. Both the index a and the index $\bar{a}$ are spinor indices, but they label somewhat different spinors. This reflects the fact that there are two different kinds of fermions in a ten-dimensional spacetime. Curiously, if you have just one fermion there is no way of telling which kind it is, but once you have two of them, you can tell if they are of the same type or of different type.

So, do we get two spacetime fermions from the R sector ground states? There are two reasons to believe that the answer should be no. First, there is something strange about getting spacetime fermions from *both* $|R_a\rangle$ and $|R_{\bar{a}}\rangle$, given that these states have opposite values of $(-1)^F$ and thus rather different commuting character. Second, with two spacetime fermions we would not get spacetime supersymmetry. Identifying $|R_a\rangle$ as spacetime fermions and $|R_{\bar{a}}\rangle$ as spacetime bosons is not an alternative either, since spacetime bosons cannot carry a spinor index.

A strategy then emerges. Since all states in the R sector have a spinor index, we will only attempt to get spacetime fermions from this sector. We also recognize that all fermions must arise from states with the same value of $(-1)^F$. Following Gliozzi, Scherk, and Olive (GSO) we proceed to truncate the Ramond sector down to the set of states with $(-1)^F = -1$. These are the states to the left of the bars in (14.54). With our conventions, these are world-sheet fermionic states that are now recognized to be states of spacetime fermions. The resulting, truncated sector is called the R− sector. The R+ sector is defined as the set of R states with $(-1)^F = +1$. At each mass level it contains the same number of states as the R− sector.

After this truncation, the generating function (14.68) for the Ramond sector reduces to

$$f_{\text{R}-}(x) = 8 \prod_{n=1}^{\infty} \left(\frac{1 + x^n}{1 - x^n} \right)^8, \tag{14.70}$$

since each combination of oscillators now acts on eight ground states only, the ones of the type required to get $(-1)^F = -1$. Expanding the above power series we have

$$f_{R-}(x) = 8 + 128\,x + 1152\,x^2 + 7680\,x^3 + 42112\,x^4 + \cdots. \qquad (14.71)$$

There are eight massless fermionic states.

Let us now reconsider the states of the NS sector. No states here carry spinor indices, so we attempt to get spacetime bosons from this sector. The ground states $|NS\rangle \otimes |p^+, \vec{p}_T\rangle$ are tachyonic and have $(-1)^F = -1$. The massless states $b_{-1/2}^I|NS\rangle \otimes |p^+, \vec{p}_T\rangle$ carry a Lorentz vector index, so they are naturally identified as the eight photon states that arise from a ten-dimensional Maxwell gauge field. We wish to keep these eight states to match the eight massless fermionic states of the R sector. Since all bosons should arise from states with the same value of $(-1)^F$, we truncate the NS sector to the set of states with $(-1)^F = +1$. The resulting states comprise the so-called NS+ sector. The NS+ sector contains the massless states and throws away the tachyonic states. Moreover, as explained earlier, the states with $(-1)^F = +1$ all have an odd number of fermionic oscillators and integer $\alpha' M^2$. The set of mass-squared levels in the NS+ sector coincides with the set of mass-squared levels in the R− sector. The NS− sector is defined to contain all the states of the NS sector with $(-1)^F = -1$. The NS− sector contains a tachyon.

Our results strongly hint that the full open string theory, defined by combining additively the set of states from the R− and NS+ sectors, has a supersymmetric spectrum. Indeed, the integer mass-squared levels in the NS generating function (14.67) have degeneracies that match those of (14.71) for the R− sector.

In order to see if the number of fermionic and bosonic states match for all levels we need a generating function $f_{NS+}(x)$ for the NS+ sector. If we take $f_{NS}(x)$ in (14.66) and change the sign inside each factor in the numerator

$$\frac{1}{\sqrt{x}} \prod_{n=1}^{\infty} \left(\frac{1 - x^{n-\frac{1}{2}}}{1 - x^n}\right)^8, \qquad (14.72)$$

the *only* effect is changing the sign of each term in the generating function whose states arise with an odd number of fermions. Since these are precisely the states we want to keep, we can obtain the desired generating function by subtracting (14.72) from (14.66) and dividing by two

$$f_{NS+}(x) = \frac{1}{2\sqrt{x}} \left[\prod_{n=1}^{\infty} \left(\frac{1 + x^{n-\frac{1}{2}}}{1 - x^n}\right)^8 - \prod_{n=1}^{\infty} \left(\frac{1 - x^{n-\frac{1}{2}}}{1 - x^n}\right)^8\right]. \qquad (14.73)$$

In order to have spacetime supersymmetry we need $f_{NS+}(x) = f_{R-}(x)$, or explicitly:

$$\frac{1}{2\sqrt{x}} \left[\prod_{n=1}^{\infty} \left(\frac{1 + x^{n-\frac{1}{2}}}{1 - x^n}\right)^8 - \prod_{n=1}^{\infty} \left(\frac{1 - x^{n-\frac{1}{2}}}{1 - x^n}\right)^8\right] = 8 \prod_{n=1}^{\infty} \left(\frac{1 + x^n}{1 - x^n}\right)^8. \qquad (14.74)$$

This intricate identity was proven by Carl Gustav Jacob Jacobi in his treatise on elliptic functions, published in 1829. Jacobi called (14.74) an obscure identity: "*aequatio identica satis abstrusa*". Presently, we recognize it as a key equation at the basis of supersymmetric string theory. The critical dimension of ten is also visible from the powers of eight on both sides of the equation.

The constructed theory of open superstrings is the theory of a single space-filling stable D9-brane. The D-brane is stable since the theory has no tachyon.

14.8 Closed string theories

We saw in Chapter 13 that closed strings are roughly obtained by combining multiplicatively left-moving and right-moving copies of an open string theory. The same is true for closed superstring theory. Since an open superstring has two sectors (NS and R), closed string sectors can be formed in four ways by combining a left-moving sector (NS or R) with a right-moving sector (NS or R). The result is four closed string sectors:

$$\text{closed string sectors:} \quad (\text{NS}, \text{NS}), \ (\text{NS}, \text{R}), \ (\text{R}, \text{NS}), \ (\text{R,R}) \,. \tag{14.75}$$

Conventionally, the first input in $(\cdot, \cdot)$ is the left-moving sector and the second input is the right-moving sector. We also have operators $(-1)^{F_L}$ and $(-1)^{F_R}$ that count fermions in the L and R sectors, respectively. In open superstrings spacetime bosons arise from the NS sector and spacetime fermions arise from the R sector. In closed superstring theories spacetime bosons arise from the (NS,NS) sector and also from the (R,R) sector, since this sector is "doubly" fermionic. The spacetime fermions arise from the (NS,R) and (R, NS) sectors.

In order to get a closed string theory with supersymmetry we must truncate the four sectors above. A consistent truncation arises if we use truncated left and right sectors to begin with. Suppose we take, for example,

$$\text{left sector:} \quad \left\{ \begin{array}{c} \text{NS+} \\ \text{R}- \end{array} \right\}, \quad \text{right sector:} \quad \left\{ \begin{array}{c} \text{NS+} \\ \text{R}+ \end{array} \right\}. \tag{14.76}$$

Combining these sectors multiplicatively we find the four sectors of the type IIA superstring:

$$\text{type IIA:} \quad (\text{NS+}, \text{NS+}), \ (\text{NS+}, \text{R+}), \ (\text{R}-, \text{NS+}), \ (\text{R}-, \text{R+}). \tag{14.77}$$

In a closed string theory the value of the mass-squared is given by

$$\tfrac{1}{2}\alpha' M^2 = \alpha' M_L^2 + \alpha' M_R^2, \tag{14.78}$$

where M_L^2 and M_R^2 denote the mass-squared operators for the open string theories that are used to build the left and right sectors, respectively. As befits closed strings there is also the level-matching condition $\alpha_0^- = \bar{\alpha}_0^-$ on the states. This condition guarantees that the left and right sectors give identical contributions to the mass-squared: $\alpha' M_L^2 = \alpha' M_R^2$. No closed string states can be formed if the left and right mass levels do not match.

The type IIA superstring has no tachyons and its massless states are obtained by combining the massless states of the various sectors:

$$(\text{NS+}, \text{NS+}) : \quad \bar{b}^I_{-1/2}|\text{NS}\rangle_L \; \otimes \; b^J_{-1/2}|\text{NS}\rangle_R \; \otimes |p^+, \vec{p}_T\rangle, \quad (14.79)$$

$$(\text{NS+}, \text{R+}) : \quad \bar{b}^I_{-1/2}|\text{NS}\rangle_L \; \otimes \; |\text{R}_{\bar{b}}\rangle_R \; \otimes |p^+, \vec{p}_T\rangle, \quad (14.80)$$

$$(\text{R−}, \text{NS+}) : \quad |\text{R}_a\rangle_L \; \otimes \; b^J_{-1/2}|\text{NS}\rangle_R \; \otimes |p^+, \vec{p}_T\rangle, \quad (14.81)$$

$$(\text{R−}, \text{R+}) : \quad |\text{R}_a\rangle_L \; \otimes \; |\text{R}_{\bar{b}}\rangle_R \; \otimes |p^+, \vec{p}_T\rangle. \quad (14.82)$$

The states in (14.79) carry two independent vector indices I, J that run over eight values. There are therefore 64 bosonic states. Just like the massless states in bosonic closed string theory they carry two vector indices. We therefore get a graviton, a Kalb–Ramond field, and a dilaton:

$$(\text{NS+}, \text{NS+}) \text{ massless fields}: \; g_{\mu\nu}, \; B_{\mu\nu}, \; \phi. \quad (14.83)$$

Quick calculation 14.7 Count the number of graviton, Kalb–Ramond, and dilaton states in ten dimensions. Add these numbers up and confirm that you get 64.

The states in (14.80) and (14.81) include only one Ramond vacuum and are therefore spacetime fermions. With $a = 1, \ldots, 8$ and $\bar{b} = \bar{1}, \ldots, \bar{8}$, these give a total of $2 \times 8 \times 8 = 128$ fermionic states. Finally, the states in (14.82) include the product of two R ground states, they are "doubly" fermionic, and thus spacetime *bosons*. There are $8 \times 8 = 64$ massless (R−, R+) bosonic states. Together with the NS–NS states in (14.79), they add up to the 128 massless bosonic states of the closed superstring. As required by supersymmetry, these match with the 128 massless fermionic states of the R–NS and NS–R sectors.

It should be said that the same type IIA string theory arises if the R+ and R− sectors in (14.76) where interchanged. Plainly, the type IIA superstring arises when the left and right truncated R sectors are of different types.

A different theory, a type IIB superstring, arises if the chosen Ramond sectors are of the same type:

$$\text{left:} \begin{Bmatrix} \text{NS+} \\ \text{R−} \end{Bmatrix}, \quad \text{right:} \begin{Bmatrix} \text{NS+} \\ \text{R−} \end{Bmatrix}. \quad (14.84)$$

We then get

$$\text{type IIB:} \; (\text{NS+}, \text{NS+}), \; (\text{NS+}, \text{R−}), \; (\text{R−}, \text{NS+}), \; (\text{R−}, \text{R−}). \quad (14.85)$$

The massless states of this theory are

$$(\text{NS+}, \text{NS+}) : \quad \bar{b}^I_{-1/2}|\text{NS}\rangle_L \; \otimes \; b^J_{-1/2}|\text{NS}\rangle_R \; \otimes |p^+, \vec{p}_T\rangle, \quad (14.86)$$

$$(\text{NS+}, \text{R−}) : \quad \bar{b}^I_{-1/2}|\text{NS}\rangle_L \; \otimes \; |\text{R}_b\rangle_R \; \otimes |p^+, \vec{p}_T\rangle, \quad (14.87)$$

$$(\text{R−}, \text{NS+}) : \quad |\text{R}_a\rangle_L \; \otimes \; b^J_{-1/2}|\text{NS}\rangle_R \; \otimes |p^+, \vec{p}_T\rangle, \quad (14.88)$$

$$(\text{R−}, \text{R−}) : \quad |\text{R}_a\rangle_L \; \otimes \; |\text{R}_b\rangle_R \; \otimes |p^+, \vec{p}_T\rangle. \quad (14.89)$$

The same type IIB theory would arise had we replaced both R− sectors in (14.84) by R+ sectors.

While the (NS+, NS+) bosons of type IIA and type IIB theories are the same, the R−R bosons are rather different. In the type IIA theory the massless R–R bosons include a Maxwell field A_μ and a three-index antisymmetric gauge field $A_{\mu\nu\rho}$. In the type IIB theory the massless R–R bosons include a scalar field A, a Kalb–Ramond field $A_{\mu\nu}$, and a totally antisymmetric gauge field $A_{\mu\nu\rho\sigma}$ with four indices. Summarizing

$$\text{R–R massless fields in type IIA}: \quad A_\mu, \quad A_{\mu\nu\rho}, \tag{14.90}$$

$$\text{R–R massless fields in type IIB}: \quad A, \quad A_{\mu\nu}, A_{\mu\nu\rho\sigma}. \tag{14.91}$$

The above R–R fields are deeply related to the existence of *stable* D-branes in type II superstring theories. We discuss this in Section 16.4, where we explain that stable D-branes are actually charged. In bosonic string theory all Dp-branes are unstable and none of them is charged.

The two truncations of (14.75) discussed above led to supersymmetric closed string theories. Other truncations of (14.75) are actually consistent, but the resulting theories are not supersymmetric. These truncations use the NS− sector, leading to a spectrum with tachyons.

Quick calculation 14.8 What sector(s) can be combined with a left-moving NS− to form a consistent closed string sector?

In addition to the type II theories, there are also two *heterotic* superstring theories. These are remarkable closed string theories. While a type II closed superstring arises by combining together left-moving and right-moving copies of open superstrings, in the heterotic string we combine a left-moving open *bosonic* string with a right-moving open *superstring*! Out of the 26 left-moving bosonic coordinates of the bosonic factor only ten of them are matched by the right-moving bosonic coordinates of the superstring factor. As a result, this theory effectively lives in ten-dimensional spacetime. Heterotic strings come in two versions: $E_8 \times E_8$ type and $SO(32)$ type. These labels characterize the groups of symmetries that exist in the theories. E_8 is a group, in fact, it is the largest exceptional group (the E is for exceptional). The group $SO(32)$ is the group generated by 32-by-32 matrices that are orthogonal and have unit determinant. A discussion of the heterotic $SO(32)$ theory can be found in Problem 14.5.

Finally, in addition to both type II and heterotic theories, there is the type I theory. This is a supersymmetric theory of open and closed *unoriented* strings, also with gauge group SO(32). A string theory is unoriented (see Problems 12.12 and 13.5) if the states of the theory are invariant under an operation that reverses the orientation of the strings. Both the type II theories and the heterotic theories are theories of oriented closed strings.

The complete list of ten-dimensional supersymmetric string theories is therefore

- type IIA,
- type IIB,
- $E_8 \times E_8$ heterotic,

- $SO(32)$ heterotic,
- type I.

These five theories have all been known since the middle of the 1980s. Some relationships between them were found soon after their discovery, but a clearer picture emerged only in the late 1990s. The limit of type IIA theory as the string coupling is taken to infinity was shown to give a theory in eleven dimensions. This theory is called M-theory, with the meaning of M to be decided when the nature of the theory becomes clear. It is known, however, that M-theory is *not* a string theory. M-theory contains membranes (2-branes) and 5-branes, and these branes are not D-branes. M-theory may end up playing a prominent role in understanding string theory. The discovery of many other relationships between the five string theories listed above and M-theory has made it clear that we really have just *one theory*. This is a fundamental result: there is a unique theory, and the five superstrings and M-theory are different limits of this unique theory.

It is not clear if bosonic strings are part of this interrelated set of theories. They certainly seem quite different from the superstrings. It would be very interesting, however, if *all* string theories were one single theory. There have been suggestions that bosonic string theories and superstrings are related via cosmological solutions. We have certainly not heard the last word on this subject.

Problems

Problem 14.1 Counting bosonic states.

(a) Consider k ordinary commuting oscillators a^i, with $i = 1, \ldots, k$. How many products of the form $a^{i_1} a^{i_2}$ can be built? How many $a^{i_1} a^{i_2} a^{i_3}$? How many $a^{i_1} a^{i_2} a^{i_3} a^{i_4}$? [Hint: use the result in Problem 12.11.]

(b) List and count the states in the $\alpha' M^2 = 3$ level of the open bosonic string. Confirm that you get the same number of states predicted by the generating function $f_{os}(x)$ in (14.63).

Problem 14.2 Generating function for the unoriented bosonic open string theory.

Write a generating function for the unoriented bosonic open string theory by starting with the generating function $f_{os}(x)$ for the full oriented theory and adding a term that implements the projection to unoriented states.

Problem 14.3 Massive level in the open superstring.

(a) Consider eight anticommuting variables b^i, with $i = 1, \ldots, 8$. Ignoring signs, how many inequivalent products of the form $b^{i_1} b^{i_2}$ can be built? How many $b^{i_1} b^{i_2} b^{i_3}$? How many $b^{i_1} b^{i_2} b^{i_3} b^{i_4}$?

(b) Consider the first and second excited levels of the open superstring ($\alpha' M^2 = 1$ and $\alpha' M^2 = 2$). List the states in the NS sector and the states in the R sector. Confirm that you get the same number of states.

Problem 14.4 Closed string degeneracies.

For closed string states the left-moving and right-moving excitations are each described like states of open strings with identical values of $\alpha' M^2$. The value of $\alpha' M^2$ for the closed string state is four times that value.

(a) State the values of $\alpha' M^2$ and give the degeneracies for the first five mass levels of the closed bosonic string theory.
(b) State the values of $\alpha' M^2$ and give the separate degeneracies of bosons and fermions for the first five mass levels of the type IIA closed superstrings. Would the answer have been different for type IIB?

Problem 14.5 Counting states in heterotic $SO(32)$ string theory.

In heterotic (closed) string theory the right-moving part of the theory is that of an open superstring. It has an NS sector whose states are built with oscillators α^I_{-n} and b^I_{-r} acting on the NS vacuum. It also has an R sector whose states are built with oscillators α^I_{-n} and d^I_{-n} acting on the R ground states. The index I runs over 8 values. The standard GSO projection down to NS+ and R− applies.

The left-moving part of the theory is that of a peculiar bosonic open string. The 24 transverse coordinates split into eight bosonic coordinates X^I with oscillators $\bar\alpha^I_{-n}$ and 16 peculiar bosonic coordinates. A surprising fact of two-dimensional physics allows us to replace these 16 coordinates by 32 two-dimensional left-moving *fermion* fields λ^A, with $A = 1, 2, \ldots, 32$. The (anticommuting) fermion fields λ^A imply that the left-moving part of the theory also has NS′ and R′ sectors, denoted with primes to differentiate them from the standard NS and R sectors of the open superstring.

The left NS′ sector is built with oscillators $\bar\alpha^I_{-n}$ and λ^A_{-r} acting on the vacuum $|NS'\rangle_L$, declared to have $(-1)^{F_L} = +1$:

$$(-1)^{F_L}|NS'\rangle_L = +|NS'\rangle_L.$$

The *naive* mass formula in this sector is

$$\alpha' M_L^2 = \frac{1}{2}\sum_{n\neq 0}\bar\alpha^I_{-n}\bar\alpha^I_n + \frac{1}{2}\sum_{r\in\mathbb{Z}+\frac{1}{2}} r\,\lambda^A_{-r}\lambda^A_r.$$

The left R′ sector is built with oscillators $\bar\alpha^I_{-n}$ and λ^A_{-n} acting on a set of R′ ground states. The naive mass formula in this sector is

$$\alpha' M_L^2 = \frac{1}{2}\sum_{n\neq 0}\left(\bar\alpha^I_{-n}\bar\alpha^I_n + n\,\lambda^A_{-n}\lambda^A_n\right).$$

Momentum labels are not needed in this problem so they are omitted throughout.

(a) Consider the left NS′ sector. Write the precise mass-squared formula with normal-ordered oscillators and the appropriate normal-ordering constant. The GSO projection here keeps the states with $(-1)^{F_L} = +1$; this defines the left NS′+ sector. Write explicitly and count the states we keep for the three lowest mass levels, indicating the corresponding values of $\alpha' M_L^2$. [This is a long list.]

(b) Consider the left R' sector. Write the precise mass-squared formula with normal-ordered oscillators and the appropriate normal-ordering constant. We have 32 zero modes λ_0^A and 16 linear combinations behave as creation operators. As usual, half of the ground states have $(-1)^{F_L} = +1$ and the other half have $(-1)^{F_L} = -1$. Let $|R_\alpha\rangle_L$ denote ground states with $(-1)^{F_L} = +1$. How many ground states $|R_\alpha\rangle_L$ are there? Keep only states with $(-1)^{F_L} = +1$; this defines the left R'+ sector. Write explicitly and count the states we keep for the two lowest mass levels, indicating the corresponding values of $\alpha' M_L^2$. [This is a shorter list.]

At any mass level $\alpha' M^2 = 4k$ of the *heterotic* string, the spacetime bosons are obtained by "tensoring" *all* the left states (NS'+ and R'+) with $\alpha' M_L^2 = k$ with the right-moving NS+ states with $\alpha' M_R^2 = k$. Similarly, the spacetime fermions are obtained by tensoring all the left states (NS'+ and R'+) with $\alpha' M_L^2 = k$ with the right-moving R− states with $\alpha' M_R^2 = k$. At any mass level where either left states or right states are missing, one cannot form heterotic string states.

(c) Are there tachyonic states in heterotic string theory? Write out the massless states of the theory (bosons and fermions) and describe the fields associated with the bosons. Calculate the total number of states in heterotic string theory (bosons plus fermions) at $\alpha' M^2 = 4$. *Answer:* 18 883 584 states.

(d) Write a generating function $f_L(x) = \sum_r a(r) x^r$ for the full set of GSO-truncated states in the left-moving sector (include both NS'+ and R'+ states). Use the convention where $a(r)$ counts the number of states with $\alpha' M_L^2 = r$. Use $f_L(x)$ and an algebraic manipulator to find the total number of states in heterotic string theory at $\alpha' M^2 = 8$. *Answer:* 6 209 372 160.

PART II

DEVELOPMENTS

15

D-branes and gauge fields

The open strings we have studied so far were described by coordinates all of which satisfy Neumann boundary conditions. These open strings move on the world-volume of a space-filling D25-brane. Here we quantize open strings attached to more general D-branes. We begin with the case of a single Dp-brane, with $1 \leq p < 25$. We then turn to the case of multiple parallel Dp-branes, where we see the appearance of interacting gauge fields and the possibility of massive gauge fields. We continue with the case of parallel D-branes of different dimensionalities.

15.1 Dp-branes and boundary conditions

A Dp-brane is an extended object with p *spatial* dimensions. In bosonic string theory, where the number of spatial dimensions is 25, a D25-brane is a space-filling brane. The letter D in Dp-brane stands for Dirichlet. In the presence of a D-brane, the endpoints of open strings must lie on the brane. As we will see in more detail below, this requirement imposes a number of Dirichlet boundary conditions on the motion of the open string endpoints.

Not all extended objects in string theory are D-branes. Strings, for example, are 1-branes because they are extended objects with one spatial dimension, but they are not D1-branes. Branes with p spatial dimensions are generically called p-branes. A 0-brane is some kind of particle. Just as the world-line of a particle is one-dimensional, the world-volume of a p-brane is $(p + 1)$-dimensional. Of these $p + 1$ dimensions, one is the time dimension and the other p are spatial dimensions. We first discussed the concept of D-branes in Section 6.5. In addition, Problem 6.11 examined the classical motion of open strings ending on D-branes of various dimensionalities. Our main subject in the present chapter is the quantization of open strings in the presence of various kinds of D-branes. This is a rich subject with important implications for the problem of constructing realistic physical models using strings. Furthermore, the study of D-branes and the gravitational fields they produce has led to surprising new insights in the study of strongly interacting gauge theories.

In this section, we set up the notation needed to describe D-branes, and then we state the appropriate boundary conditions. We let d denote the total number of spatial dimensions in the theory; in the present case, $d = 25$. The total number of spacetime dimensions is $D = d + 1 = 26$. A Dp-brane with $p < 25$ extends over a p-dimensional subspace of the 25-dimensional space. We will focus on simple Dp-branes: those that are p-dimensional *hyperplanes* inside the d-dimensional space. How can we specify such hyperplanes? We need $(d - p)$ linear conditions. In three spatial dimensions $(d = 3)$, a 2-brane $(p = 2)$ is a plane, and it is specified by one linear condition $(d - p = 3 - 2 = 1)$. For example, $z = 0$, specifies the (x, y) plane. Similarly, a string along the z axis $(p = 1)$ is specified by two linear conditions $(d - p = 3 - 1 = 2)$: $x = 0$ and $y = 0$. We need as many conditions as there are spatial coordinates normal to the brane.

Consider now a Dp-brane. We introduce spacetime coordinates x^μ, with $\mu = 0, 1, \ldots, 25$, that are split into two groups. The first group includes the coordinates tangential to the brane *world-volume*. These are the time coordinate and p spatial coordinates. The second group includes the $(d - p)$ coordinates normal to the brane world-volume. We write

$$\underbrace{x^0, x^1, \ldots, x^p}_{\text{D}p \text{ tangential coordinates}} \quad \underbrace{x^{p+1}, x^{p+2}, \ldots, x^d}_{\text{D}p \text{ normal coordinates}}. \tag{15.1}$$

The location of the Dp-brane is specified by fixing the values of the coordinates normal to the brane. With this split in mind we write

$$x^a = \bar{x}^a, \quad a = p + 1, \ldots, d. \tag{15.2}$$

Here the $\bar{x}^a$ are a set of $(d - p)$ constants. In a completely analogous fashion, the string coordinates $X^\mu(\tau, \sigma)$ are split as

$$\underbrace{X^0, X^1, \ldots, X^p}_{\text{D}p \text{ tangential coordinates}} \quad \underbrace{X^{p+1}, X^{p+2}, \ldots, X^d}_{\text{D}p \text{ normal coordinates}}. \tag{15.3}$$

Since the endpoints of the open string must lie on the Dp-brane, the string coordinates normal to the brane must satisfy Dirichlet boundary conditions

$$X^a(\tau, \sigma)\Big|_{\sigma=0} = X^a(\tau, \sigma)\Big|_{\sigma=\pi} = \bar{x}^a, \quad a = p + 1, \ldots, d. \tag{15.4}$$

The string coordinates X^a are called DD coordinates, because both endpoints satisfy a Dirichlet boundary condition. The open string endpoints can move freely along the directions tangential to the D-brane. As a result, the string coordinates tangential to the D-brane satisfy Neumann boundary conditions:

$$X^{m\prime}(\tau, \sigma)\Big|_{\sigma=0} = X^{m\prime}(\tau, \sigma)\Big|_{\sigma=\pi} = 0, \quad m = 0, 1, \ldots, p. \tag{15.5}$$

These string coordinates are called NN coordinates because both endpoints satisfy a Neumann boundary condition. We see that the split (15.3) into tangential and normal coordinates is also a split into coordinates which satisfy Neumann and Dirichlet boundary conditions, respectively:

$$\underbrace{X^0, X^1, \ldots, X^p}_{\text{NN coordinates}} \underbrace{X^{p+1}, X^{p+2}, \ldots, X^d}_{\text{DD coordinates}}. \tag{15.6}$$

In order to use the light-cone gauge we need at least one spatial NN coordinate that can be used together with X^0 to define the coordinates $X^\pm$. We therefore need to assume $p \geq 1$, and our analysis does not apply to strings attached to a D0-brane. D0-branes are perfectly consistent objects, but in order to study them we need a Lorentz covariant quantization (see Chapter 24 and Problem 24.4). We will label the light-cone coordinates as

$$\underbrace{X^+, X^-, \{X^i\}}_{\text{NN}} \; \underbrace{\{X^a\}}_{\text{DD}} \quad i = 2, \ldots, p \quad \text{and} \quad a = p+1, \ldots, d. \tag{15.7}$$

15.2 Quantizing open strings on D*p*-branes

Having specified the boundary conditions on the various string coordinates, we can proceed to the quantization of open strings in the presence of a D*p*-brane. The purpose of the analysis that follows is to determine the spectrum of open string states and to use this result to understand more deeply what goes on in the world-volume of a D*p*-brane.

Our earlier work in Chapter 12 is quite useful here. The NN coordinates $X^i(\tau, \sigma)$ satisfy exactly the same conditions that are satisfied by the light-cone coordinates $X^I(\tau, \sigma)$ of open strings attached to a D25-brane. All expansions and commutation relations for the X^i coordinates can be obtained from those of X^I by replacing $I \to i$ in the relevant equations.

We recall that the X^- coordinate was determined in terms of the transverse light-cone coordinates in equation (9.65):

$$\dot{X}^- \pm X^{-\prime} = \frac{1}{2\alpha'} \frac{1}{2p^+} (\dot{X}^I \pm X^{I\prime})^2. \tag{15.8}$$

Moreover, the mode expansion of $\dot{X}^I \pm X^{I\prime}$ was given in (9.74):

$$\dot{X}^I \pm X^{I\prime} = \sqrt{2\alpha'} \sum_{n \in \mathbb{Z}} \alpha_n^I \, e^{-in(\tau \pm \sigma)}. \tag{15.9}$$

A completely analogous mode expansion held for the coordinate X^-; this expansion continues to hold without change since X^- remains an NN coordinate. The above equations, together with the X^- expansion, led to (12.105) and (12.106), summarized here as

$$2p^+ p^- \equiv \frac{1}{\alpha'} \left(\frac{1}{2} \alpha_0^I \alpha_0^I + \sum_{n=1}^{\infty} \alpha_{-n}^I \alpha_n^I + a \right). \tag{15.10}$$

The ordering constant a was determined to be equal to minus one for the quantization of strings on a D25-brane. The light-cone index $I = 2, \ldots, 25$, takes values that, for a D*p*-brane, run over NN coordinates labeled by i and DD coordinates labeled by a. As a result, (15.8) now becomes

$$\dot{X}^- \pm X^{-\prime} = \frac{1}{2\alpha'} \frac{1}{2p^+} \left\{ (\dot{X}^i \pm X^{i\prime})^2 + (\dot{X}^a \pm X^{a\prime})^2 \right\}. \tag{15.11}$$

As explained before, the X^i coordinates are expanded as

$$\dot{X}^i \pm X^{i\prime} = \sqrt{2\alpha'} \sum_{n \in \mathbb{Z}} \alpha_n^i \, e^{-in(\tau \pm \sigma)}. \tag{15.12}$$

The X^a coordinates are the ones we must investigate. If an expansion analogous to (15.12) holds for X^a, we will be able to find p^- by letting $I \to (i, a)$ in (15.10), just as we did in order to obtain (15.11).

We are now in a position to address the novel part of the quantization of open strings attached to a Dp-brane. The coordinates X^a normal to the brane satisfy the wave equation, so the general solution is a superposition of two waves:

$$X^a(\tau, \sigma) = \frac{1}{2}\Big(f^a(\tau + \sigma) + g^a(\tau - \sigma)\Big). \tag{15.13}$$

Let us examine the boundary conditions (15.4). At $\sigma = 0$ we obtain

$$X^a(\tau, 0) = \frac{1}{2}\Big(f^a(\tau) + g^a(\tau)\Big) = \bar{x}^a, \tag{15.14}$$

so that $g^a(\tau) = -f^a(\tau) + 2\bar{x}^a$, and as a result,

$$X^a(\tau, \sigma) = \bar{x}^a + \frac{1}{2}\Big(f^a(\tau + \sigma) - f^a(\tau - \sigma)\Big). \tag{15.15}$$

The boundary condition at $\sigma = \pi$ then gives us

$$f^a(\tau + \pi) = f^a(\tau - \pi). \tag{15.16}$$

This simply means that $f^a(u)$ is a periodic function with period 2π. This information is incorporated into the following expansion:

$$f^a(u) = \tilde{f}_0^a + \sum_{n=1}^{\infty}\Big(\tilde{f}_n^a \cos nu + \tilde{g}_n^a \sin nu\Big). \tag{15.17}$$

It is interesting to note that there is no term linear in u. Such a term was present when the coordinate satisfied a Neumann boundary condition because in that case it was the derivative $f'(u)$ which was periodic. Replacing (15.17) in (15.15) and performing some trigonometric simplification, we find

$$X^a(\tau, \sigma) = \bar{x}^a + \sum_{n=1}^{\infty}\Big(-\tilde{f}_n^a \sin n\tau \sin n\sigma + \tilde{g}_n^a \cos n\tau \sin n\sigma\Big). \tag{15.18}$$

Redefining the expansion coefficients that are arbitrary anyway, we can write

$$X^a(\tau, \sigma) = \bar{x}^a + \sum_{n=1}^{\infty}\Big(f_n^a \cos n\tau + \tilde{f}_n^a \sin n\tau\Big) \sin n\sigma. \tag{15.19}$$

Since there is no term linear in τ, the string has no net time-averaged momentum in the x^a direction. This is reasonable since strings must remain attached to the brane. If there were a $p^a \tau$ term present, the endpoints $\sigma = 0, \pi$ would not remain at $x^a = \bar{x}^a$ when $\tau \neq 0$.

In order to define the quantum theory associated with X^a, we focus on the classical parameters that describe the motion of the open string in equation (15.19). Since we are trying to quantize strings attached to a *fixed* Dp-brane, the values $\bar{x}^a$ are not parameters that can be adjusted to describe various open string motions. The $(f^a, \tilde{f}^a)$, on the other hand, *are* parameters of the open string motion. Therefore, in quantizing the open string, the $\bar{x}^a$ remain numbers and *do not* become operators, while the $(f^a, \tilde{f}^a)$ turn into operators.

We now rewrite (15.19) in terms of oscillators, defined in order to simplify the following analysis:

$$X^a(\tau, \sigma) = \bar{x}^a + \sqrt{2\alpha'} \sum_{n \neq 0} \frac{1}{n} \alpha_n^a e^{-in\tau} \sin n\sigma. \tag{15.20}$$

The string coordinate X^a is Hermitian if $(\alpha_n^a)^\dagger = \alpha_{-n}^a$, which is the usual Hermiticity property of oscillators. Note that the zero mode α_0^a does not exist. Additionally,

$$\dot{X}^a = -i\sqrt{2\alpha'} \sum_{n \neq 0} \alpha_n^a e^{-in\tau} \sin n\sigma, \quad X^{a\prime} = \sqrt{2\alpha'} \sum_{n \neq 0} \alpha_n^a e^{-in\tau} \cos n\sigma, \tag{15.21}$$

and therefore

$$X^{a\prime} \pm \dot{X}^a = \sqrt{2\alpha'} \sum_{n \neq 0} \alpha_n^a e^{-in(\tau \pm \sigma)}. \tag{15.22}$$

The analogy with (15.12) is quite close, but there are two differences. First, when the lower sign applies, the combinations of derivatives differ by an overall minus sign. Second, the zero mode is absent in (15.22).

The quantization is now straightforward. With $\mathcal{P}^{\tau a}(\tau, \sigma) = \dot{X}^a/2\pi\alpha'$, the nonvanishing commutators are postulated to be

$$\left[X^a(\tau, \sigma), \dot{X}^b(\tau, \sigma') \right] = 2\pi\alpha' i \, \delta^{ab} \delta(\sigma - \sigma'). \tag{15.23}$$

Following the analysis of Section 12.2, this commutator can be rewritten in the form (12.30), with (I, J) replaced by (a, b). Since the mode expansions (15.22) take the standard form, the earlier analysis applies. The overall sign difference alluded to above is of no import since $(X^{a\prime} - \dot{X}^a)$ appears twice in the relevant commutators. We thus find

$$[\alpha_m^a, \alpha_n^b] = m \, \delta^{ab} \delta_{m+n,0}, \quad m, n \neq 0. \tag{15.24}$$

The zero modes work out consistently: $\bar{x}^a$ is a constant, and there is no conjugate momentum since $\alpha_0^a \equiv 0$. The sign difference is also immaterial for the evaluation of (15.11) since $(X^{a\prime} - \dot{X}^a)$ appears squared. Therefore, equation (15.10) can be split as

$$2p^+ p^- \equiv \frac{1}{\alpha'} \left(\alpha' p^i p^i + \sum_{n=1}^{\infty} \left[\alpha_{-n}^i \alpha_n^i + \alpha_{-n}^a \alpha_n^a \right] - 1 \right). \tag{15.25}$$

A few comments are needed here. Since $p^a \sim \alpha_0^a \equiv 0$, the term $\frac{1}{2}\alpha_0^I \alpha_0^I$ simply became $\alpha' p^i p^i$ (recall that $\alpha_0^\mu = \sqrt{2\alpha'} p^\mu$). The ordering constant has been set to minus one, as for the D25-brane. The critical dimension has not been changed either. This is reasonable since only the zero mode structure is different in X^a and X^i. In particular, the naive contributions needed to normal order $L_0^\perp$ are the same for X^a and for X^i. It follows from (15.25) that

$$M^2 = -p^2 = 2p^+p^- - p^ip^i = \frac{1}{\alpha'}\Big(\sum_{n=1}^{\infty}\big[\alpha_{-n}^i\alpha_n^i + \alpha_{-n}^a\alpha_n^a\big] - 1\Big). \tag{15.26}$$

Using creation and annihilation operators, we get

$$M^2 = \frac{1}{\alpha'}\Big(-1 + \sum_{n=1}^{\infty}\sum_{i=2}^{p} n\, a_n^{i\,\dagger}a_n^i + \sum_{m=1}^{\infty}\sum_{a=p+1}^{d} m\, a_m^{a\,\dagger}a_m^a\Big). \tag{15.27}$$

Let us now consider the state space of the quantum string. The ground states in the D25-brane background were $|p^+, \vec{p}_T\rangle$, where $\vec{p}_T = (p^2, \ldots, p^{25})$ is the vector with components p^I. The I index now runs over both i and a values, but there are no p^a operators, so the ground states of the theory are labeled by p^+ and p^i only:

$$|p^+, \vec{p}\rangle \quad \text{with} \quad \vec{p} = (p^2, \ldots, p^p). \tag{15.28}$$

We build additional states by acting with oscillators on the ground states. We have oscillators along the brane:

$$a_n^{i\,\dagger}, n \geq 1, \quad i = 2, \ldots, p, \tag{15.29}$$

and oscillators normal to the brane:

$$a_n^{a\dagger}, n \geq 1, \quad a = p+1, \ldots, d. \tag{15.30}$$

So the states take the form

$$\Big[\prod_{n=1}^{\infty}\prod_{i=2}^{p}\big(a_n^{i\,\dagger}\big)^{\lambda_{n,i}}\Big]\Big[\prod_{m=1}^{\infty}\prod_{a=p+1}^{d}\big(a_m^{a\,\dagger}\big)^{\lambda_{m,a}}\Big]|p^+, \vec{p}\rangle. \tag{15.31}$$

Schrödinger wavefunctions take the schematic form

$$\psi_{i_1\ldots i_p\, a_1\ldots a_q}(\tau, p^+, \vec{p}). \tag{15.32}$$

Just like the indices on the oscillators, the indices on the wavefunctions are of two types: indices along the directions tangent to the brane (i-type) and indices along the directions normal to the brane (a-type).

In the field theories that describe the states of the string, the fields take the same form as the string Schrödinger wavefunctions. We can therefore ask: where do the fields corresponding to (15.32) live? Are these fields defined over all of spacetime, or only in some subspace of spacetime?

Since the fields are functions of the momenta p^i, by Fourier transformation they can be viewed as fields that depend on the coordinates x^i. The τ dependence is in fact an x^+ dependence, and the p^+ dependence can be Fourier transformed into an x^- dependence. All together, we have fields that depend on x^+, x^-, and x^i, with $i = 2, \ldots, p$. These are precisely the $(p + 1)$ coordinates that span the world-volume of the Dp-brane. It is reasonable to conclude that the fields actually live on the Dp-brane. Indeed, this world-volume is the only natural candidate for a $(p + 1)$-dimensional subspace of spacetime.

Our analysis suggests, but does not prove, that the fields live on the Dp-brane; the positions $\bar{x}^a$ of the Dp-brane did not appear in the states nor in the wavefunctions. How could we *prove* that the fields live on the Dp-brane? We would have to study interactions. Since closed strings have no endpoints, they are not fixed by D-branes and can exist over all of spacetime. By scattering closed strings off the Dp-brane we can investigate whether the interactions between fields from the closed string sector and fields from the open string sector take place on the D-brane world-volume. The answer appears to be yes. Statements about where open string fields live, however, are likely to be ambiguous or even gauge dependent. Different answers could be completely consistent.

We conclude our analysis of the Dp-brane by giving a list and detailed description of the fields that satisfy $M^2 \leq 0$. All these fields live on the Dp-brane, so we must state how they transform under the Lorentz transformations that preserve the Dp-brane. These are Lorentz transformations in $p + 1$ dimensions, and the fields may be scalars or vectors, for example. Let us begin with the simplest states, the ground states:

$$|p^+, \vec{p}\,\rangle, \quad M^2 = -\frac{1}{\alpha}. \tag{15.33}$$

These states are tachyon states on the brane, and they have exactly the same mass as the tachyon states we found on the D25-brane. The corresponding tachyon field, of course, is just a Lorentz scalar on the brane.

The next states have one oscillator acting on them. Consider first the case in which the oscillator arises from a coordinate tangent to the brane:

$$a_1^{i\,\dagger}|p^+, \vec{p}\,\rangle, \quad i = 2, \ldots, p, \quad M^2 = 0. \tag{15.34}$$

For any momenta, these are $(p + 1) - 2$ massless states. Moreover, the index they carry lives on the brane. They are therefore states that transform as a Lorentz vector on the brane. Since the number of states equals the spacetime dimensionality of the brane minus two, these are clearly photon states. The associated field is a Maxwell gauge field living on the brane. This is a fundamental result:

a Dp-brane has a Maxwell field living on its world-volume.   (15.35)

Finally, let us consider the case in which the oscillator acting on the ground state arises from a coordinate normal to the brane:

$$a_1^{a\dagger}|p^+, \vec{p}\,\rangle, \quad a = p + 1, \ldots, d, \quad M^2 = 0. \tag{15.36}$$

For any momenta, these are $(d - p)$ states living on the brane. Since the index a is not a Lorentz index for the brane, this index is merely a counting label. These states transform as Lorentz *scalars* on the brane. Therefore, we get a massless scalar field for each direction normal to the Dp-brane:

a Dp-brane has a massless scalar for each normal direction.   (15.37)

These massless scalars have a physical interpretation. In Section 12.8 we indicated that open string states represent D-brane excitations. Our Dp-brane and a slightly displaced parallel Dp-brane are actually states of the same energy. The displaced Dp-brane can be viewed as a zero-energy excitation of the original brane. It is also an excitation with zero momentum, since it represents a displacement that is constant over all of the Dp-brane. A zero-momentum excitation with zero energy is a massless excitation because it obeys the energy-momentum relation $E = p$ of a massless particle. The massless scalars identified in (15.36) give rise to these excitations. This interpretation is supported by the fact that we have as many massless fields as there are directions normal to the Dp-brane. Those are the independent directions in which the Dp-brane can be moved. Note that a space-filling D25-brane has no massless scalars on its world-volume, consistent with the fact that it cannot be displaced.

All in all, the massless states on the Dp-brane are $(p-1)$ photon states and $(d-p)$ scalar field states. Apart from the momentum labels which are different, we have the same number of massless states as on the D25-brane. The $(d-1)$ states on the D25-brane are accounted for on the Dp-brane by $(p-1)$ photon states and $(d-p)$ scalar states.

15.3 Open strings between parallel Dp-branes

We will now consider the quantization of open strings that extend between two parallel Dp-branes. In describing such branes we will continue to use the notation of the previous sections. Two parallel branes of the same dimensionality have the same set of longitudinal coordinates and the same set of normal coordinates. Recall that the values $\bar{x}^a$ of the normal coordinates specify the position of a Dp-brane. This time the first Dp-brane is located at $x^a = \bar{x}_1^a$ and the second at $x^a = \bar{x}_2^a$. If we happen to have $\bar{x}_1^a = \bar{x}_2^a$ for all a, the two Dp-branes coincide in space – they are on top of each other. Otherwise, they are separated. In Figure 15.1 we show two parallel, separated D2-branes.

What kinds of open strings does this configuration of parallel Dp-branes support? There are actually four different classes of strings, each of which must be analyzed separately. The first two classes are made up of open strings that begin and end on the same D-brane, either brane one or brane two. These strings we already studied and quantized in the previous section. The other two classes consist of strings that start on one brane and end on the other. These are *stretched strings*. The strings that begin on brane one and end on brane

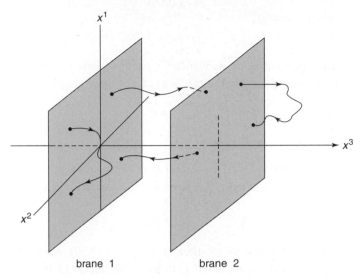

Fig. 15.1 Two parallel D2-branes. Here x^1 and x^2 are longitudinal coordinates, and x^3 is a normal coordinate. The positions of brane one and brane two are specified by the coordinates $\bar{x}_1^3$ and $\bar{x}_2^3$, respectively. We show the four types of strings that this configuration supports.

two are different from the strings that begin on brane two and end on brane one. These strings are oppositely oriented, and the orientation of a string (the direction of increasing σ) matters. As we will show in Chapter 16, the charge of a string changes sign when we reverse its orientation. The classes of open strings that are supported on a particular configuration of D-branes are called *sectors*. The quantum theory of open strings in the presence of two parallel Dp-branes has four sectors. In Figure 15.1 we show a string for each of the four sectors.

Let us consider the sector consisting of open strings that begin on brane one and end on brane two. The NN string coordinates X^+, X^-, and X^i are quantized just as before, since the corresponding boundary conditions are still given by (15.5). On the other hand, for the DD string coordinates the boundary conditions (formerly given by (15.4)) are now

$$X^a(\tau, \sigma)\Big|_{\sigma=0} = \bar{x}_1^a, \quad X^a(\tau, \sigma)\Big|_{\sigma=\pi} = \bar{x}_2^a, \quad a = p+1, \ldots, d. \tag{15.38}$$

The solution of the wave equation subject to these boundary conditions can be studied starting from (15.15), which already incorporates the boundary condition at $\sigma = 0$. In the present case we just change $\bar{x}^a$ to $\bar{x}_1^a$:

$$X^a(\tau, \sigma) = \bar{x}_1^a + \frac{1}{2}\Big(f^a(\tau+\sigma) - f^a(\tau-\sigma)\Big). \tag{15.39}$$

The boundary condition at $\sigma = \pi$ now gives us

$$f^a(\tau+\pi) - f^a(\tau-\pi) = 2\,(\bar{x}_2^a - \bar{x}_1^a), \tag{15.40}$$

or, equivalently,

$$f^a(u + 2\pi) - f^a(u) = 2(\bar{x}_2^a - \bar{x}_1^a). \tag{15.41}$$

This means that the derivative $f^{a\prime}(u)$ is a periodic function with period 2π and has an expansion of the type indicated in (15.17). Integrating, the function $f^a(u)$ has an expansion of the form

$$f^a(u) = f_0^a u + \sum_{n=1}^{\infty} (h_n^a \cos nu + g_n^a \sin nu). \tag{15.42}$$

We have not included a constant term because it would drop out of X^a, as can be seen in (15.39). The constant f_0^a is fixed by the boundary condition (15.41):

$$f_0^a = \frac{1}{\pi}(\bar{x}_2^a - \bar{x}_1^a). \tag{15.43}$$

It is now possible to substitute $f^a(u)$ into (15.39). Aside from handling the zero modes, the computations are identical to those that led to (15.19). This time we obtain

$$X^a(\tau, \sigma) = \bar{x}_1^a + (\bar{x}_2^a - \bar{x}_1^a)\frac{\sigma}{\pi} + \sum_{n=1}^{\infty}\left(f_n^a \cos n\tau + \tilde{f}_n^a \sin n\tau\right)\sin n\sigma. \tag{15.44}$$

Note that the boundary conditions are manifestly satisfied. To describe strings that extend from brane two to brane one we merely exchange $\bar{x}_1^a$ and $\bar{x}_2^a$ in the above equation. We can rewrite (15.44) in terms of oscillators, using (15.20) as a model:

$$X^a(\tau, \sigma) = \bar{x}_1^a + (\bar{x}_2^a - \bar{x}_1^a)\frac{\sigma}{\pi} + \sqrt{2\alpha'}\sum_{n\neq 0}\frac{1}{n}\alpha_n^a e^{-in\tau}\sin n\sigma. \tag{15.45}$$

The constants $\bar{x}_1^a$ and $\bar{x}_2^a$ do not become quantum operators because for fixed D-branes, just as before, they are not parameters of the open string fluctuations. Note the absence of terms linear in τ; the open strings have no time-averaged momentum in the x^a directions. Even though we are not giving new names to the oscillators above, they are different operators from those we obtained during the quantization of strings that begin and end on the same Dp-brane. The oscillators in different sectors must not be confused. This time the derivatives give

$$\dot{X}^a = -i\sqrt{2\alpha'}\sum_{n\in\mathbb{Z}}\alpha_n^a e^{-in\tau}\sin n\sigma, \quad X^{a\prime} = \sqrt{2\alpha'}\sum_{n\in\mathbb{Z}}\alpha_n^a e^{-in\tau}\cos n\sigma, \tag{15.46}$$

where

$$\sqrt{2\alpha'}\alpha_0^a = \frac{1}{\pi}(\bar{x}_2^a - \bar{x}_1^a). \tag{15.47}$$

Although the strings do not carry momentum in the x^a direction, there is still a nonvanishing α_0^a. There is no contradiction because the interpretation of α_0 as momentum requires that α_0 appear in $\dot{X}$. As you can see, α_0^a appears in $X^{a\prime}$, but not in $\dot{X}^a$. A nonvanishing α_0^a implies stretched strings: α_0^a vanishes precisely when the two D-branes coincide. Similar operators emerge in the expansion of closed strings that wrap around compact dimensions (see Chapter 17).

The two derivatives in (15.46) can be combined to form

$$X^{a\prime} \pm \dot{X}^a = \sqrt{2\alpha'} \sum_{n\in\mathbb{Z}} \alpha_n^a e^{-in(\tau\pm\sigma)}. \tag{15.48}$$

It follows from this result, and our comments in the previous section, that the oscillators satisfy the expected commutation relations. To calculate the mass-squared operator, we reconsider equation (15.10). As before, we let $I \to (i, a)$ and set the subtraction constant a equal to minus one, giving

$$2p^+ p^- = \frac{1}{\alpha'}\left(\alpha' p^i p^i + \frac{1}{2}\alpha_0^a \alpha_0^a + \sum_{n=1}^{\infty}\left[\alpha_{-n}^i \alpha_n^i + \alpha_{-n}^a \alpha_n^a\right] - 1\right). \tag{15.49}$$

We therefore have

$$M^2 = 2p^+ p^- - p^i p^i = \frac{1}{2\alpha'}\alpha_0^a \alpha_0^a + \frac{1}{\alpha'}\left(\sum_{n=1}^{\infty}\left[\alpha_{-n}^i \alpha_n^i + \alpha_{-n}^a \alpha_n^a\right] - 1\right). \tag{15.50}$$

Using the explicit value of α_0^a in (15.47), we finally obtain

$$M^2 = \left(\frac{\bar{x}_2^a - \bar{x}_1^a}{2\pi\alpha'}\right)^2 + \frac{1}{\alpha'}(N^\perp - 1), \tag{15.51}$$

where

$$N^\perp = \sum_{n=1}^{\infty}\sum_{i=2}^{p} n a_n^{i\,\dagger} a_n^i + \sum_{m=1}^{\infty}\sum_{a=p+1}^{d} m a_m^{a\dagger} a_m^a. \tag{15.52}$$

The first term on the right-hand side of (15.51) is a new contribution to the mass-squared of the states. Since the string tension is $T_0 = 1/(2\pi\alpha')$, this term is simply the square of the energy of a classical static string stretched between the two D-branes. It is reasonable to find that the mass-squared operator is altered by the addition of this constant. The constant vanishes precisely when the branes coincide.

Let us now consider the ground states. In fact, let us examine the ground states from each of the four open string sectors available in this D-brane configuration. The momentum labels of these states are the same for each sector: p^+ and $\vec{p}$. To distinguish the various sectors, we include as additional ground-state labels two integers $[ij]$, each of which can take the value one or two. The first integer denotes the brane on which the $\sigma = 0$ endpoint lies, and the second integer denotes the brane on which the $\sigma = \pi$ endpoint lies. In short, the open strings in the $[ij]$ sector extend from brane i to brane j. The ground states are written as $|p^+, \vec{p}; [ij]\rangle$, and they are of four types:

$$|p^+, \vec{p}; [11]\rangle, \quad |p^+, \vec{p}; [22]\rangle, \quad |p^+ \vec{p}; [12]\rangle, \quad |p^+, \vec{p}; [21]\rangle. \tag{15.53}$$

The states of open strings in the $[ij]$ sector are constructed from oscillators acting on $|p^+, \vec{p}; [ij]\rangle$. The states take the form indicated in (15.31), with the exception that the ground state is replaced by $|p^+, \vec{p}; [ij]\rangle$. The oscillators in the four sectors are the same in number and in type, but they are fundamentally different operators. We could label them with the $[ij]$ labels for clarity, but this is seldom necessary because the ground states carry the sector labels.

Where do the fields corresponding to the [12] string states live? This question is difficult to answer. They are clearly $(p + 1)$-dimensional fields, since the momentum structure of the states is the same as the one we had for the states of strings fully attached to a single brane. As far as the stretched strings are concerned, the two D-branes are on a similar footing, so we cannot say that the fields live on any single one of them. In some sense, the fields must live on *both* D-branes. Operationally, the fields are declared to live on some fixed $(p + 1)$-dimensional space (not necessarily identified with any of the two D-branes), and are seen to have nonlocal interactions that reflect the fact that the D-branes are separated. The spacetime interpretation of fields that arise from stretched strings appears to require a new way of thinking, the basis of which may be provided by a branch of mathematics called noncommutative geometry.

We continue our discussion of the state space by giving a list and a detailed description of the fields which comprise the two lowest levels of the stretched-string state space. Just as we did for the single brane, we will determine whether the states are scalars or vectors with respect to the $(p + 1)$-dimensional Lorentz symmetry. The simplest states are the ground states:

$$|p^+, \vec{p}\,; [12]\rangle, \quad M^2 = -\frac{1}{\alpha'} + \left(\frac{\bar{x}_2^a - \bar{x}_1^a}{2\pi\alpha'}\right)^2. \tag{15.54}$$

If the separation between the branes vanishes, these states are tachyon states of the usual mass-squared. If the branes are separated, the mass-squared gets a positive contribution. In fact, for the critical separation

$$|\bar{x}_2^a - \bar{x}_1^a| = 2\pi\sqrt{\alpha'}, \tag{15.55}$$

the ground states represent a massless scalar field. For larger separations, the ground states represent a massive scalar field.

The next states have one oscillator acting on them. Assume, until stated otherwise, that the separation between the branes is nonzero. If the oscillator acting on the ground states arises from a coordinate normal to the brane we have

$$a_1^{a\dagger}|p^+, \vec{p}\,; [12]\rangle, \quad a = p + 1, \ldots, d, \quad M^2 = \left(\frac{\bar{x}_2^a - \bar{x}_1^a}{2\pi\alpha'}\right)^2. \tag{15.56}$$

For any momenta, these are $(d - p)$ massive states. Since the index a is not a Lorentz index for the $(p + 1)$-dimensional spacetime, these states are Lorentz scalars. We therefore get $(d - p)$ massive scalar fields. If the oscillator arises from a coordinate tangent to the brane we have

$$a_1^{i\,\dagger}|p^+, \vec{p}\,; [12]\rangle, \quad i = 2, \ldots, p, \quad M^2 = \left(\frac{\bar{x}_2^a - \bar{x}_1^a}{2\pi\alpha'}\right)^2. \tag{15.57}$$

For any momenta, these are $(p + 1) - 2 = p - 1$ massive states. Moreover, they carry an index corresponding to the $(p + 1)$-dimensional spacetime. We might think that these states make up a massive Maxwell gauge field, but this is not exactly right.

A massive gauge field has more degrees of freedom than a massless gauge field. The results of Problem 10.7 indicate that a massive gauge field has *one more state* than a massless gauge field for each allowed value of the momentum. In a D-dimensional spacetime, a massless gauge field has $D - 2$ states for each p_μ which satisfies $p^2 = 0$, while a massive gauge field has $D - 1$ states for each p_μ which satisfies $p^2 + m^2 = 0$. Therefore, in the case at hand, one of the scalar states in (15.56) must join the $(p - 1)$ states in (15.57) to form the massive vector. At the end we have one massive vector and $(d - p - 1)$ massive scalars.

Can we guess which scalar state in (15.56) becomes part of the massive gauge field? If $p = d - 1$ the answer is simple. In that case, the D-branes are separated along one coordinate, and there is only one scalar in (15.56). The scalar uses the oscillator labeled with the direction along which the branes are separated. When $p < d - 1$, there is more than one state in (15.56). The scalar state that becomes part of the vector is the linear combination

$$\sum_a (\bar{x}_2^a - \bar{x}_1^a)\, a_1^{a\dagger}\, |p^+, \vec{p}\,; [12]\rangle. \tag{15.58}$$

Of all directions normal to the D-branes, the direction defined by the spatial vector with components $\bar{x}_2^a - \bar{x}_1^a$ is unique: it takes us from one brane to the other one. To visualize this, think of two parallel D1-branes in three-dimensional space. There are clearly many normal directions that do not take us from one brane to the other, and just one direction that does. The educated guess (15.58) can be proven to be the correct one.

We obtain a very interesting situation in the limit as the separation between the branes goes to zero. Even though the D-branes are then coincident, they are still distinguishable and we still have the four open string sectors. The massless open string states which represent strings extending from brane one to brane two include a massless gauge field and $(d - p)$ massless scalars. This is the same field content as that of a sector where strings begin and end on the same D-brane. When the two D-branes coincide we therefore get a total of four massless gauge fields. These gauge fields actually interact with one another – in the string picture they do so by the process of joining endpoints. Theories of interacting gauge fields are called Yang–Mills theories. They were discovered in the 1950s and later on used successfully to build the theories of electroweak and strong interactions. On the world-volume of two coincident D-branes we indeed get a $U(2)$ Yang–Mills theory. More precisely, we get a $U(2)$ Yang–Mills theory with some additional interactions that become negligible at low energies. The two in $U(2)$ is there precisely because we have two coincident D-branes. The meaning of $U(2)$ will be discussed below.

Suppose that we have N Dp-branes. This time the sectors will be labeled by pairs $[ij]$, where i and j are integers that run from 1 to N. The $[ij]$ sector consists of open strings that start on the ith brane and end on the jth brane. It is clear that there are N^2 sectors. In this setup, string interactions can be visualized neatly. In a typical process, a first open string joins with a second open string to form a new open string. To do so, the *end* of the first string ($\sigma = \pi$) joins with the *beginning* of the second string ($\sigma = 0$). The new string

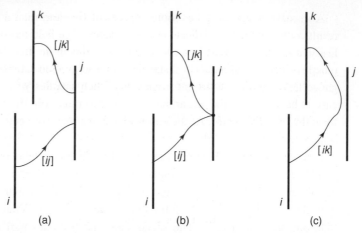

Fig. 15.2 (a) Three D-branes, labeled i, j, and k, and strings in the $[ij]$ and $[jk]$ sectors. (b) The end of the string in the $[ij]$ sector meets the beginning of the string in the $[jk]$ sector and the interaction takes place. (c) The resulting string in the $[ik]$ sector.

begins at the beginning of the first string and ends at the end of the second string. If the open strings are stretched between D-branes, a first string from the $[ij]$ sector can be joined by a second open string from the $[jk]$ sector to give a "product" open string in the $[ik]$ sector. This interaction is possible since both the end of the first string and the beginning of the second string lie on the same D-brane. The physical process can be imagined to take place as in Figure 15.2. The result is a single string, which does not remain attached to the j D-brane since the joining point is no longer an endpoint. The new string belongs to the $[ik]$ sector. We write this possible interaction as

$$[ij] * [jk] = [ik], \quad j \text{ not summed.} \tag{15.59}$$

If the N Dp-branes are coincident, the N^2 sectors result in N^2 interacting massless gauge fields. This defines a $U(N)$ Yang–Mills theory on the world-volume of the N coincident D-branes:

$$N \text{ coincident D-branes carry } U(N) \text{ massless gauge fields.} \tag{15.60}$$

In fact, the full spectrum of the open string theory consists of N^2 copies of the spectrum of a single Dp-brane.

If we have a single brane, (15.60) tells us that we get a $U(1)$ Yang–Mills theory. In fact, $U(1)$ Yang–Mills theory is Maxwell theory, so (15.60), for $N = 1$, is consistent with (15.35). Here $U(1)$ denotes a group: the elements of the group are complex numbers of unit length and group multiplication is just multiplication. The $U(1)$ group is relevant because, at any spacetime point, the gauge parameters in Maxwell theory are actually elements of $U(1)$. So far, our study of Maxwell gauge transformations has made no use of the $U(1)$

group structure, but this group structure is needed to understand the gauge symmetry in the presence of compact spatial dimensions (Chapter 18). For $U(N)$ Yang–Mills theory, $U(N)$ is also a group of symmetries: the elements of the group are $N \times N$ unitary matrices and group multiplication is matrix multiplication. At any spacetime point, the gauge parameters of $U(N)$ Yang–Mills theory are elements of the group $U(N)$.

Quick calculation 15.1 Recall that a group is a set which is closed under an associative multiplication; it contains an identity element, and each element has a multiplicative inverse. Verify that $U(1)$ and $U(N)$, as described above, are groups.

The discrete labels i, j used to label the branes and the various open string sectors are sometimes called *Chan–Paton* indices. These indices were introduced during the early stages of string theory, long before D-branes were known, as an algebraic device to obtain Yang–Mills theories from open strings. With the discovery of D-branes it became clear that the Chan–Paton indices are simply labels of D-branes in a multi-D-brane configuration.

The appearance of Yang–Mills theories on the world-volume of a D-brane configuration is of great relevance because Yang–Mills theories are used to describe the Standard Model of particle physics. The electroweak theory is described by a $U(2)$ Yang–Mills theory. The four gauge bosons of this theory include the photon γ, the W^+, the W^-, and the Z^0. The latter three are massive gauge bosons. The mechanism by which massless gauge fields become massive is known as the Higgs mechanism of field theory. A possible D-brane realization of the Higgs mechanism is obtained by separating D-branes which, when coincident, give the corresponding massless gauge particles. If we have two coincident D3-branes we obtain a $U(2)$ Yang–Mills theory, with four massless gauge fields living on the four-dimensional world-volume of the branes. Is this a good model for the electroweak gauge theory? Not quite. If we separate the D3-branes to give mass to some of the gauge bosons, two of them acquire a mass – the two arising from the stretched strings – and two remain massless. In the electroweak gauge theory only one gauge field remains massless. A more sophisticated D-brane configuration is needed to produce a model of the electroweak theory. A setup with intersecting D-branes that can be used to construct a particle physics model will be examined in Section 21.1. The construction of semi-realistic models is discussed in Section 21.4.

15.4 Strings between parallel Dp- and Dq-branes

In this section we examine the configuration of two parallel D-branes with different dimensionality. Let p and q be two integers which satisfy $1 \leq q < p \leq 25$, and consider a configuration consisting of a Dp-brane and a Dq-brane. We assume $p > q$, since the case $p = q$ was already considered. The branes are coincident if the Dq-brane world-volume is a subset of the Dp-brane world-volume. We take the branes to be parallel. This means the same as what we mean when we say that a line is parallel to a plane: there is a plane

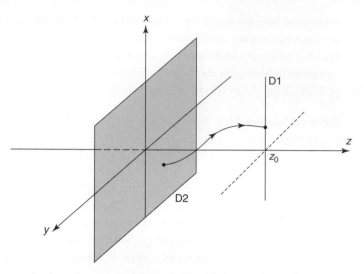

Fig. 15.3 A D2-brane stretched on the (x, y) plane and a parallel D1-brane stretched along the x axis and located at $y = 0$, $z = z_0$. Also shown is an open string going from the D2-brane to the D1-brane. For such a string, the string coordinate Y is of ND type.

parallel to the given plane which contains the line. Thus, if the Dp-brane and Dq-brane are separate, there is a p-dimensional hyperplane parallel to the Dp-brane that contains the Dq-brane.

A case that can be easily visualized is that of a D2-brane and a D1-brane, as illustrated in Figure 15.3. The D2-brane stretches along the x and y directions and is located at $z = 0$. The D1-brane stretches along the x direction and is located at $y = 0$ and at $z = z_0$. This D1-brane is parallel to the D2-brane, and they are coincident only when $z_0 = 0$. Table 15.1 summarizes the relevant spatial information about the D-branes. A dash (—) indicates a direction along the D-brane, and a bullet (•) indicates a direction normal to the D-brane. The coordinate x is a common tangential direction. The coordinate y is a mixed direction: one brane extends along this direction while the other does not. The z coordinate is a common normal direction. More generally, for the Dp-, Dq-brane configuration we have

$$\underbrace{x^0, x^1, \ldots, x^q}_{\text{common tangential coordinates}} \quad \underbrace{x^{q+1}, x^{q+2}, \ldots, x^p}_{\text{mixed coordinates}} \quad \underbrace{x^{p+1}, x^{p+2}, \ldots, x^d}_{\text{common normal coordinates}} . \tag{15.61}$$

There are $(q + 1)$ common tangential coordinates (all the world-volume coordinates of the Dq-brane, including time), $(p - q)$ directions that are tangential to the Dp-brane and normal to the Dq-brane, and $(d - p)$ common normal directions.

We have already studied strings that begin and end on the same D-brane, so we focus here on the strings that go from one D-brane to the other. For definiteness, consider the strings that stretch from the Dp-brane to the Dq-brane. We have partial knowledge about these strings: the common tangential coordinates are NN, and the common normal coordinates are DD. We have already studied these two types of coordinates. The mixed coordinates, which are tangential to one of the branes and normal to the other, are new. In

Table 15.1	D-brane configuration and boundary conditions		
Coordinate	x	y	z
D2	–	–	•
D1	–	•	•
[D2, D1]	NN	ND	DD

Note. In the second and third rows a dash – indicates coordinates along which the D-brane stretches and a bullet • indicates coordinates normal to the brane. In the last row we indicate the boundary conditions for open strings belonging to the sector [D2, D1] of strings that stretch from the D2-brane to the D1-brane.

our D2-, D1-brane example the y direction is mixed. For a string that stretches from the D2-brane to the D1-brane, the string coordinate Y associated with y is N at $\sigma = 0$, since y is tangential to the D2-brane, and D at $\sigma = \pi$, since y is normal to the D1-brane; in short, Y is an ND coordinate. For a string that stretches from the D1-brane to the D2-brane, Y is a DN coordinate.

In the D*p*-, D*q*-brane configuration, the mixed spatial coordinates were listed in (15.61). For open strings that stretch from the D*p*-brane to the D*q*-brane, the analogously labeled string coordinates satisfy a Neumann boundary condition on the starting D*p*-brane and a Dirichlet boundary condition on the ending D*q*-brane; they are ND coordinates. The full set of string coordinates splits into

$$\underbrace{X^0,\ X^1,\ \ldots,\ X^q}_{\text{NN coordinates}}\ \ \underbrace{X^{q+1},\ X^{q+2},\ \ldots,\ X^p}_{\text{ND coordinates}}\ \ \underbrace{X^{p+1},\ X^{p+2},\ \ldots,\ X^d}_{\text{DD coordinates}}. \qquad (15.62)$$

In the light-cone, we use three types of indices to label the string coordinates:

$$\underbrace{X^+,\ X^-,\ \{X^i\}}_{\text{NN}}\ \ \underbrace{\{X^r\}}_{\text{ND}}\ \ \underbrace{\{X^a\}}_{\text{DD}}, \qquad (15.63)$$

where

$$i = 2, \ldots, q, \quad r = q+1, \ldots, p, \quad \text{and} \quad a = p+1, \ldots, d. \qquad (15.64)$$

We think of the D*p*-brane as the first brane and the D*q*-brane as the second brane. The position of the D*p*-brane is specified by the coordinates $\bar{x}_1^a$, and the position of the D*q*-brane is specified by the coordinates $\bar{x}_2^r$ and $\bar{x}_2^a$. In our D2-, D1-brane example of Figure 15.3, the role of $\bar{x}_2^r$ is played by the y coordinate of the D1-brane. This coordinate can be set to zero by a suitable choice of axes.

Let us begin our analysis of the ND coordinates X^r. The boundary conditions are

$$\frac{\partial X^r}{\partial \sigma}(\tau, \sigma)\bigg|_{\sigma=0} = 0, \quad X^r(\tau, \sigma)\bigg|_{\sigma=\pi} = \bar{x}_2^r. \qquad (15.65)$$

The $\bar{x}_2^r$ coordinates can be set equal to zero by a suitable choice of axes, but we will not do so. As opposed to the coordinate differences $\bar{x}_1^a - \bar{x}_2^a$ that define the separation between the D-branes, the $\bar{x}_2^r$ will not play any significant role. Consider now the usual expansion

$$X^r(\tau, \sigma) = \frac{1}{2}\Big(f^r(\tau + \sigma) + g^r(\tau - \sigma)\Big). \tag{15.66}$$

The boundary condition at $\sigma = 0$ gives us

$$f^{r\prime}(u) = g^{r\prime}(u) \;\rightarrow\; g^r(u) = f^r(u) + c_0^r. \tag{15.67}$$

Bearing in mind that the second boundary condition will set X^r equal to $\bar{x}_2^r$ at $\sigma = \pi$, we choose $c_0^r = 2\bar{x}_2^r$ so that

$$X^r(\tau, \sigma) = \bar{x}_2^r + \frac{1}{2}\Big(f^r(\tau + \sigma) + f^r(\tau - \sigma)\Big). \tag{15.68}$$

The condition at $\sigma = \pi$ then gives

$$f^r(u + 2\pi) = -f^r(u). \tag{15.69}$$

The function $f^r(u)$ goes into *minus* itself when its argument increases by 2π. We encountered a similar situation for the twisted sector of the $\mathbb{R}^1/\mathbb{Z}_2$ orbifold (Section 13.6) and for Neveu–Schwarz fermions (Section 14.4). What are needed are trigonometric functions with half-integer moding:

$$f^r(u) = \sum_{n\in\mathbb{Z}_{\text{odd}}^+} \Big[f_n^r \cos\Big(\frac{nu}{2}\Big) + h_n^r \sin\Big(\frac{nu}{2}\Big) \Big]. \tag{15.70}$$

Substituting back into (15.68) and relabeling the expansion coefficients, we get

$$X^r(\tau, \sigma) = \bar{x}_2^r + \sum_{n\in\mathbb{Z}_{\text{odd}}^+} \Big[A_n^r \cos\Big(\frac{n\tau}{2}\Big) + B_n^r \sin\Big(\frac{n\tau}{2}\Big) \Big] \cos\Big(\frac{n\sigma}{2}\Big). \tag{15.71}$$

This is our expansion of the ND coordinates. To proceed with the quantization we define oscillators with half-integer moding. Another useful guide is the desired simplicity of $\dot{X}^r \pm X^{r\prime}$. We are thus led to write

$$X^r(\tau, \sigma) = \bar{x}_2^r + i\sqrt{2\alpha'} \sum_{n\in\mathbb{Z}_{\text{odd}}} \frac{2}{n}\, \alpha_{\frac{n}{2}}^r\, e^{-i\frac{n}{2}\tau} \cos\Big(\frac{n\sigma}{2}\Big), \tag{15.72}$$

where the sum runs over both positive and negative odd integers. The factor of i in front of the sum is necessary so that the Hermiticity of X^r imposes the standard Hermiticity property on the oscillators:

$$\big(\alpha_{\frac{n}{2}}^r\big)^\dagger = \alpha_{-\frac{n}{2}}^r. \tag{15.73}$$

The $\bar{x}_2^r$ are constants and do not become operators. There are no zero modes in the expansion of X^r, and therefore ND coordinates carry no average momentum. We also record the derivatives

$$\dot{X}^r \pm X^{r\prime} = \sqrt{2\alpha'} \sum_{n\in\mathbb{Z}_{\text{odd}}} \alpha_{\frac{n}{2}}^r\, e^{-i\frac{n}{2}(\tau\pm\sigma)}, \tag{15.74}$$

which are indeed of the expected form. This expansion reminds us of similar ones (13.106) for the twisted sector orbifold coordinate X.

The (nontrivial) commutation relations for the string coordinates take the form

$$[X^r(\tau, \sigma), \dot{X}^s(\tau, \sigma')] = i(2\pi\alpha')\delta(\sigma - \sigma')\delta^{rs}. \tag{15.75}$$

This equation implies that we can use the appropriate version of (12.30):

$$\left[(\dot{X}^r \pm X^{r\prime})(\tau, \sigma), (\dot{X}^s \pm X^{s\prime})(\tau, \sigma')\right] = \pm 4\pi\alpha' i\eta^{rs}\frac{d}{d\sigma}\delta(\sigma - \sigma'). \tag{15.76}$$

Since the expansions (15.74) take the standard form, equation (12.40) also holds, with minor modifications, for $\sigma, \sigma' \in [0, 2\pi]$:

$$\sum_{m',n'\in\mathbb{Z}_{\text{odd}}} e^{-i\frac{m'}{2}(\tau+\sigma)}e^{-i\frac{n'}{2}(\tau+\sigma')}\left[\alpha^r_{\frac{m'}{2}}, \alpha^s_{\frac{n'}{2}}\right] = 2\pi\, i\eta^{rs}\frac{d}{d\sigma}\delta(\sigma - \sigma'). \tag{15.77}$$

The commutators are extracted just like we did below (13.107) and the answer is

$$\left[\alpha^r_{\frac{m}{2}}, \alpha^s_{\frac{n}{2}}\right] = \frac{m}{2}\delta^{rs}\delta_{m+n,0}. \tag{15.78}$$

This is the expected form of the commutation relations.

Let us now calculate the mass-squared operator. This operator receives contributions from all the coordinates in this sector: the NN, the ND, and the DD coordinates. This is clear from (15.8) since the original light-cone index I now runs over i, r, and a labels. Given that the linear combination of derivatives (15.74) takes the standard form, the contribution from the ND coordinates takes a familiar form. The formula for $2p^+p^-$ can be obtained by a minor modification of equation (15.49):

$$2p^+p^- = \frac{1}{\alpha'}\left(\alpha' p^i p^i + \frac{1}{2}\alpha^a_0\alpha^a_0 + \sum_{n=1}^{\infty}\left[\alpha^i_{-n}\alpha^i_n + \alpha^a_{-n}\alpha^a_n\right] + \sum_{m\in\mathbb{Z}^+_{\text{odd}}}\alpha^r_{-\frac{m}{2}}\alpha^r_{\frac{m}{2}} + a\right). \tag{15.79}$$

In writing this equation we have restored the ordering constant a, which, as you recall, arises heuristically by ordering the oscillators in $L_0^\perp$ (see (12.102), (12.107), and (12.110)). Since all the oscillators for both the NN and DD directions are integrally moded, their normal-ordering constants are the same and equal to $\frac{1}{2}(\frac{-1}{12}) = -\frac{1}{24}$. With a total of 24 transverse light-cone coordinates, if we only have NN and DD coordinates we get $a = -1$. The ND coordinates, however, give a different contribution. The sum that must be rearranged in this case is

$$\frac{1}{2}\sum_{m\in\mathbb{Z}_{\text{odd}}}\alpha^r_{-\frac{m}{2}}\alpha^r_{\frac{m}{2}} = \sum_{m\in\mathbb{Z}^+_{\text{odd}}}\alpha^r_{-\frac{m}{2}}\alpha^r_{\frac{m}{2}} + \frac{1}{2}\sum_{m\in\mathbb{Z}^+_{\text{odd}}}\left[\alpha^r_{\frac{m}{2}}, \alpha^r_{-\frac{m}{2}}\right]. \tag{15.80}$$

The first term on the right-hand side is the one that appears in (15.79), and the second term is the ordering contribution. Since we have $(p - q)$ ND coordinates, the ordering constant is

$$\frac{1}{2}\sum_{m\in\mathbb{Z}^+_{\text{odd}}}\left[\alpha^r_{\frac{m}{2}}, \alpha^r_{-\frac{m}{2}}\right] = \frac{1}{4}(p - q)\sum_{m\in\mathbb{Z}^+_{\text{odd}}}m = \frac{1}{48}(p - q), \tag{15.81}$$

where we used (15.78) and recalled that the sum over positive odd integers is equal to $\frac{1}{12}$. This calculation shows that each ND coordinate contributes $+\frac{1}{48}$ to the ordering constant in $\alpha' M^2$. A DN string coordinate will contribute the same amount. In summary, the normal ordering contributions to $\alpha' M^2$ for the various string coordinates are

$$a_{\text{NN}} = a_{\text{DD}} = -\frac{1}{24}, \quad a_{\text{ND}} = a_{\text{DN}} = \frac{1}{48}. \tag{15.82}$$

Returning to the problem at hand, the total ordering constant a is given by (15.81) plus the contribution of the $(24 - (p - q))$ coordinates that are either NN or DD:

$$a = -\frac{1}{24}\big(24 - (p - q)\big) + \frac{1}{48}(p - q) = -1 + \frac{1}{16}(p - q). \tag{15.83}$$

With this information we can now find M^2. Following the same steps as in (15.50),

$$M^2 = \left(\frac{\bar{x}_2^a - \bar{x}_1^a}{2\pi\alpha'}\right)^2 + \frac{1}{\alpha'}\left(N^\perp - 1 + \frac{1}{16}(p - q)\right), \tag{15.84}$$

where

$$N^\perp = \sum_{n=1}^\infty \sum_{i=2}^q n\, a_n^{i\,\dagger} a_n^i + \sum_{k \in \mathbb{Z}_{\text{odd}}^+} \sum_{r=q+1}^p \frac{k}{2}\, a_{\frac{k}{2}}^{r\,\dagger} a_{\frac{k}{2}}^r + \sum_{m=1}^\infty \sum_{a=p+1}^d m\, a_m^{a\,\dagger} a_m^a. \tag{15.85}$$

This formula for M^2 incorporates all the effects we have discussed: a shifted ordering constant, a contribution from stretched strings, and a number operator that includes contributions from NN, DD, and ND coordinates.

Let us examine the state space and the fields associated with the two lowest mass levels. The ground states are labeled as

$$|p^+, \vec{p}\,; [12]\rangle, \quad \vec{p} = (p^2, \ldots, p^q). \tag{15.86}$$

The momentum labels on the states indicate that the corresponding fields live in a $(q + 1)$-dimensional spacetime. Roughly, they live on the world-volume of the Dq-brane, the brane of lower dimensionality. The general rule is clear: the spacetime dimensionality of the fields which arise in any given sector equals the number of NN string coordinates in the sector. The state space is built by letting the three types of oscillators – $a_p^{i\dagger}$, $a_{k/2}^{r\dagger}$, and $a_m^{a\dagger}$ – act on the ground states.

The ground states have $N^\perp = 0$ and correspond to a single scalar field on the Dq-brane. This scalar is in general massive, but it can be tachyonic or massless depending on the separation of the branes and the value of $p - q$. Assume, for simplicity, that the branes coincide. If, additionally, $p - q = 16$, then the scalar is massless. The next states are of the form

$$a_{\frac{1}{2}}^{r\dagger} |p^+, \vec{p}\,; [12]\rangle, \quad N^\perp = 1/2. \tag{15.87}$$

These states give rise to $(p - q)$ scalar fields, since the index r does not correspond to a world-volume direction on the Dq-brane. All other states are necessarily massive since they have $N^\perp \geq 1$, and this together with $p > q$ implies that $M^2 > 0$. In particular, we do not find any massless gauge fields.

Problems

Problem 15.1 A Dp-brane with orientifolds.

In this problem (a sequel to Problem 13.6), we study the effects of orientifolds on open strings.

The space-filling O25-plane truncates the spectrum down to the set of states that are invariant under the operation Ω which reverses the orientation of strings. When we have a Dp-brane, Ω acts on the open string coordinates as follows:

$$\Omega\, X^a(\tau, \sigma)\, \Omega^{-1} = X^a(\tau, \pi - \sigma), \tag{1}$$

$$\Omega\, X^i(\tau, \sigma)\, \Omega^{-1} = X^i(\tau, \pi - \sigma). \tag{2}$$

As usual, we demand that $\Omega\, x_0^-\, \Omega^{-1} = x_0^-$ and $\Omega\, p^+ \Omega^{-1} = p^+$.

(a) Give the Ω action on the oscillators α_n^a and α_n^i. What is the expected Ω action on α_n^-? Does it work out?

(b) Assume that the ground states $|p^+, \vec{p}\,\rangle$ are Ω invariant. Find the states of the theory for $N^\perp \leq 2$. As you will see, some massless states survive. Interpret these states along the lines of the discussion below (15.37).

Replace the O25-plane by an Op-plane coincident with the Dp-brane at $\bar{x}^a = 0$. Let Ω_p denote the operator for which this theory keeps only the states with $\Omega_p = +1$.

(c) How should equations (1) and (2) change when Ω is replaced by Ω_p? Give the Ω_p action on the oscillators α_n^a and α_n^i.

(d) Describe the full spectrum of the theory as a simple truncation of the Dp-brane spectrum. You will find no massless scalars in this case. What does this suggest regarding the possible motions of the Dp-brane?

Problem 15.2 String products and orientation reversing symmetries.

Equation (15.59) tells how open string *sectors* combine under interactions. The same product notation can be used for strings. By

$$|A\rangle * |B\rangle \tag{1}$$

we mean the string state that is obtained when a string in state $|A\rangle$ interacts with a string in state $|B\rangle$. The string product must obey the rule of sectors: the state in (1) must belong to the sector $[A] * [B]$, where $[A]$ and $[B]$ denote the sectors where string states $|A\rangle$ and $|B\rangle$ belong, respectively.

Use pictures of strings A and B to motivate the equations

$$\Omega\left(|A\rangle * |B\rangle\right) = \left(\Omega|B\rangle\right) * \left(\Omega|A\rangle\right), \tag{2}$$

$$\Omega_p\left(|A\rangle * |B\rangle\right) = \left(\Omega_p|B\rangle\right) * \left(\Omega_p|A\rangle\right). \tag{3}$$

Here Ω is string orientation reversal and Ω_p is orientifolding (orientation reversal plus reflection about a set of coordinates).

Problem 15.3 N coincident Dp-branes and orientifolds.

Let N coincident Dp-branes coincide with an Op-plane, all of them located at $\bar{x}^a = 0$. The orientifolding symmetry Ω_p, as usual, includes reflection of the coordinates normal to the orientifold and simultaneous orientation reversal of strings. Assume that the reflection of coordinates leaves each of the Dp-branes invariant (as opposed to mapping them into each other). The states of the theory are those for which $\Omega_p = +1$.

(a) Explain why it is reasonable to postulate that

$$\Omega_p|p^+, \vec{p}\,; [ij]\rangle = |p^+, \vec{p}\,; [ji]\rangle.$$

What are the ground states of the theory? How many are there?

(b) Describe the full open string spectrum of the theory in terms of the spectrum of a single Dp-brane. Check that for $N = 1$ you reproduce the result of Problem 15.1 (d).

Problem 15.4 Separated Dp-branes and an Op-plane.

We have learned that an orientifold acts as a kind of mirror. If we are to have D-branes that do not coincide with an orientifold, then there must be mirror D-branes at the reflected points. Therefore, to analyze the theory of Dp-branes off an orientifold Op-plane we begin with N Dp-branes and N mirror Dp-branes at the reflected positions. We must then define the orientifold action on all the states of the theory of $2N$ Dp-branes. Finally, we use this action to truncate down to the invariant states, obtaining in this way the states of the orientifold theory.

Consider the situation illustrated in Figure 15.4, where we show the configuration as seen in a plane spanned by two coordinates normal to the branes and the orientifold. The N Dp branes are labeled $1, 2, \ldots, N$, and the mirror images are labeled $\bar{1}, \bar{2}, \ldots, \bar{N}$. Two strings are exhibited: one in the [24] sector and the other in the [1$\bar{1}$] sector.

(a) Show the two strings obtained by the orientifold symmetry. Since the arguments $p^+, \vec{p}$ of the ground states are always present, let us omit them for brevity. The ground states are of four types:

$$|[ij]\rangle, \quad |[i\,\bar{j}]\rangle, \quad |[\bar{i}\,j]\rangle, \quad |[\bar{i}\,\bar{j}]\rangle. \tag{1}$$

Each class contains N^2 ground states since i and j run from 1 to N and $\bar{i}$ and $\bar{j}$ run from $\bar{1}$ to $\bar{N}$. Define an expected action of Ω_p on the ground states in (1). Show that your choice satisfies $\Omega_p^2 = 1$ acting on the ground states.

(b) What are the possible interactions between strings in the four types of sectors built on the states (1)? Write your answers using the notation of (15.59).

(c) It is a fact about string interactions that the string product of ground states gives states that have a component along a ground state. Thus, for example,

$$|[i\,\bar{j}]\rangle * |[\bar{j}\,k]\rangle = |[i\,k]\rangle + \cdots. \tag{2}$$

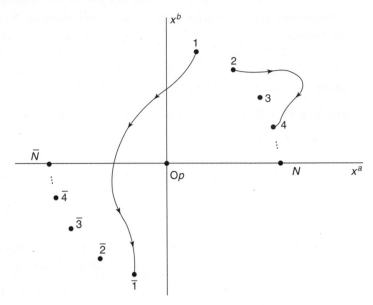

Fig. 15.4 Problem 15.4. A set of N Dp-branes together with a set of N image Dp-branes. The orientifold Op plane is at the origin.

Write the other possible ground state products. Test the consistency of your definition of Ω_p by acting with Ω_p on both sides of the equations giving ground state products. To act on products, use equation (3) of Problem 15.2.

(d) Find the Ω_p action on the α_n^i oscillators using $\Omega_p X^i(\tau, \sigma)\Omega_p^{-1} = X^i(\tau, \pi - \sigma)$. Since the strings are stretched along, or have nonzero values for, the x^a coordinates, an equation of the type $\Omega_p X^a(\tau, \sigma)\Omega_p^{-1} = -X^a(\tau, \pi - \sigma)$ cannot be fully implemented. Using (15.20) for X^a, for example, we would need $\Omega_p \bar{x}^a \Omega_p^{-1} = -\bar{x}^a$, which cannot hold since the $\bar{x}^a$ are numbers. A legal derivation of the Ω_p action on the α_n^a oscillators can be obtained by requiring $\Omega_p \dot{X}^a(\tau, \sigma)\Omega_p^{-1} = -\dot{X}^a(\tau, \pi - \sigma)$. Verify that for any arbitrary product R of oscillators of both types

$$\Omega_p R \Omega_p^{-1} = (-1)^{N^\perp} R, \tag{3}$$

where $N^\perp$ is the total number of R.

(e) Describe the orientifold spectrum in terms of the spectrum of a single Dp-brane. For this, consider an arbitrary product R of oscillators and build the general states

$$\sum_{ij}\left(r_{ij} R|[ij]\rangle + r_{i\bar{j}} R|[i\bar{j}]\rangle + r_{\bar{i}j} R|[\bar{i}j]\rangle + r_{\bar{i}\bar{j}} R|[\bar{i}\bar{j}]\rangle\right), \tag{4}$$

where $r_{ij}, r_{i\bar{j}}, r_{\bar{i}j}$, and $r_{\bar{i}\bar{j}}$ are four N by N matrices. Find the conditions that Ω_p invariance imposes on these matrices. There are two cases to consider, depending on the number $N^\perp$ of R. You should find that for $N^\perp$ odd there are $N(2N-1)$ linearly independent states in (4). For $N^\perp$ even there are $N(2N+1)$ linearly independent states.

Since the gauge fields arise from $N^\perp = 1$ states, there are $N(2N - 1)$ of them. This is the number of entries in a $2N$-by-$2N$ antisymmetric matrix. The interacting theory of such gauge fields is an $SO(2N)$ Yang–Mills gauge theory (SO stands for special orthogonal). The $SO(10)$ gauge theory, for example, can be used to build a grand unified theory of the strong and electroweak interactions.

Problem 15.5 Separated Dp-branes and a different Op-plane.

The brane setup here is that of Problem 15.4. In part (a) of that problem you defined a simple action of Ω_p on the ground states $|[ij]\rangle$, $|[i\,\bar{j}]\rangle$, $|[\bar{i}\,j]\rangle$, and $|[\bar{i}\,\bar{j}]\rangle$. Find an alternative $\Omega_{p'}$ action where some of the relations have a minus sign: $\Omega_p|[\dots]\rangle = \pm|[\dots]\rangle$. Test its consistency by verifying that $\Omega_p^2 = 1$ on ground states and that the products of ground states are compatible with the Ω_p action, as you tested in part (c) of Problem 15.4. Determine the spectrum of this variant orientifold theory.

The gauge fields arise from $N^\perp = 1$, and you will find that there are $N(2N + 1)$ of them. The interacting theory of such gauge fields is a $USp(2N)$ Yang–Mills gauge theory (USp stands for unitary symplectic).

Problem 15.6 DN string coordinates.

In Section 15.4 we considered strings that stretch from a Dp- to a Dq-brane, focusing on the coordinates X^r which satisfy ND boundary conditions. Consider now strings that stretch from the Dq- to the Dp-brane. For such strings the coordinates X^r are of DN type.

(a) Write the boundary conditions satisfied by the X^r coordinates and use them to derive a mode expansion along the lines of our result (15.72) for an ND coordinate.

(b) Find also the equations that replace (15.74). Explain briefly why the mass-squared formula (15.84) needs no modification.

(c) If we let $\sigma \to \pi - \sigma$ in (15.72) we automatically get a function with DN boundary conditions. Compare with the mode expansion you found in (a), and explain why the Hermiticity properties are consistent.

Problem 15.7 D1-branes at an angle.

Consider two infinitely long D1-branes stretched on the (x^2, x^3) plane. The *first* brane is defined by $x^3 = 0$, and the *second* brane is at an angle γ measured counterclockwise from the x^2 axis. Let the open string coordinates be $X^2(\tau, \sigma)$ and $X^3(\tau, \sigma)$, and consider only open strings which begin on the first brane and end on the second brane. Determine the boundary conditions satisfied by X^2 and X^3 at $\sigma = 0$ and $\sigma = \pi$.

Problem 15.8 Strings in a configuration with a Dp-brane and a D25-brane.

Consider the full state space of open string theory in a configuration with a Dp-brane and a D25-brane. Assume $1 \le p \le 24$. For each sector of the theory give the M^2 operator, and examine explicitly the states which arise in the two lowest levels, indicating the types of fields they correspond to and where these fields live.

Problem 15.9 A pair of intersecting D22-branes.

We study here a configuration of two D22-branes. One of them, henceforth called brane 1, is defined by $x^{25} = x^{23} = x^{22} = 0$. The other one, brane 2, is defined by $x^{24} = x^{23} = x^{22} = 0$.

Draw a picture of the brane configuration as it appears in the (x^{24}, x^{25}) plane. Construct a table, such as Table 15.1, adding more columns to include all coordinates and more rows to include all sectors. For each sector of the theory give the M^2 operator and examine explicitly the states which arise in the two lowest levels, indicating the types of fields they correspond to and where these fields live.

A basic point: if x^a is a common Dirichlet direction in a configuration of two intersecting D-branes, then the x^a coordinates of the two D-branes must be the same. Explain why.

16 String charge and electric charge

If a point particle couples to the Maxwell field then that particle carries electric charge. Strings couple to the Kalb–Ramond field; therefore, strings carry a new kind of charge – string charge. For a stretched string, string charge can be visualized as a current flowing along the string. Strings can end on D-branes without violating conservation of string charge because the string endpoints carry electric charge and the resulting electric field lines on a D-brane carry string charge. Certain D-branes in superstring theory carry electric charge for Ramond–Ramond fields. If a charged brane is fully wrapped on a compact space, it appears to a lower-dimensional observer as a point particle carrying the electric charge of a Maxwell field that arises from dimensional reduction.

16.1 Fundamental string charge

As we have seen before, a point particle can carry electric charge because there is an interaction which allows the particle to couple to a Maxwell field. The world-line of the point particle is *one*-dimensional and the Maxwell gauge field A_μ carries *one* index. This matching is important. The particle trajectory has a tangent vector $dx^\mu(\tau)/d\tau$, where τ parameterizes the world-line. Because it has one Lorentz index, the tangent vector can be multiplied by the gauge field A_μ to form a Lorentz scalar. Working with natural units ($\hbar = c = 1$), the interaction for a point particle of charge q is written as a term in the action taking the form

$$q \int A_\mu(x(\tau)) \frac{dx^\mu(\tau)}{d\tau} \, d\tau. \tag{16.1}$$

It is convenient to require that q be dimensionless in natural units. Since the action is also dimensionless in natural units, the gauge field A_μ must carry units of inverse length, or mass: $[A_\mu] = M$. The field strength has units $[F_{\mu\nu}] = M^2$ because it is obtained by differentiation of the gauge field with respect to the spacetime coordinates.

The complete interacting system of the charged particle and the Maxwell field is defined by the action considered in Problem 5.6:

$$S' = -m \int_{\mathcal{P}} ds + q \int_{\mathcal{P}} A_\mu(x)dx^\mu - \frac{1}{4\kappa_0^2} \int d^D x \, F_{\mu\nu} F^{\mu\nu}. \tag{16.2}$$

The first term on the right-hand side is the particle action, and the last term is the action for the Maxwell field. We have included the dimensionful constant κ_0, with units

$[\kappa_0^2] = M^{4-D}$, in order to make this term dimensionless. This constant is necessary whenever $D \neq 4$.

Can a relativistic string be charged? The above argument makes it clear that Maxwell charge is naturally carried by points. Closed strings do not have special points, but open strings have distinguished endpoints. It is therefore plausible that the endpoints of open strings carry electric Maxwell charge. We will show later that this is indeed the case. At the moment, however, we are looking for something fundamentally different. Since electric Maxwell charge is naturally associated with points, we may wonder whether there is some new kind of charge that is naturally associated with strings. For a new kind of charge, we need a new kind of gauge field. Thus, we may ask: is there a field in string theory that is related to the string in the same way that the Maxwell field is related to a particle? The answer is yes. The field is the Kalb–Ramond antisymmetric two-tensor $B_{\mu\nu}(= -B_{\nu\mu})$. This is a massless field that arises in closed string theory (see Section 13.3 and Problem 10.6).

Let us now mimic the logic that led to (16.1). At any point of the string trajectory we have two linearly independent tangent vectors. Indeed, with world-sheet coordinates τ and σ, the two tangent vectors can be chosen to be $\partial X^\mu / \partial \tau$ and $\partial X^\mu / \partial \sigma$. With these two tangent vectors and the two-index field $B_{\mu\nu}$ we can construct a Lorentz scalar:

$$ - \int d\tau d\sigma \, \frac{\partial X^\mu}{\partial \tau} \frac{\partial X^\nu}{\partial \sigma} \, B_{\mu\nu}\left(X(\tau,\sigma)\right). \tag{16.3} $$

This is how the string couples to the antisymmetric Kalb–Ramond field. It is called an *electric* coupling because it is the natural generalization of the electric coupling of a point particle to a Maxwell field. Thus we say that the string carries *electric Kalb–Ramond charge*. The coupling (16.3) must be dimensionless in natural units, so $B_{\mu\nu}$ carries units of inverse length-squared or mass-squared: $[B_{\mu\nu}] = M^2$. This coupling must also be invariant under (τ, σ) reparameterizations, just like the Nambu–Goto string action is. You will see in Problem 16.1 that the antisymmetry of $B_{\mu\nu}$ is necessary to ensure the reparameterization invariance of (16.3). An important point regarding the extent to which reparameterization invariance holds will be addressed later in this section.

Just as (16.2) represents the complete dynamics of a particle and a Maxwell field, the string coupling (16.3) must be supplemented by the string action S_{str} and a term giving dynamics to the $B_{\mu\nu}$ field:

$$ S = S_{\text{str}} - \frac{1}{2} \int d\tau d\sigma \, B_{\mu\nu}\left(X(\tau,\sigma)\right) \frac{\partial X^{[\mu}}{\partial \tau} \frac{\partial X^{\nu]}}{\partial \sigma} - \frac{1}{6\kappa^2} \int d^D x \, H_{\mu\nu\rho} H^{\mu\nu\rho}. \tag{16.4} $$

Here we have defined the antisymmetrization

$$ a^{[\mu} b^{\nu]} \equiv a^\mu b^\nu - a^\nu b^\mu \tag{16.5} $$

and the field strength $H_{\mu\nu\rho}$ associated with $B_{\mu\nu}$:

$$H_{\mu\nu\rho} \equiv \partial_\mu B_{\nu\rho} + \partial_\nu B_{\rho\mu} + \partial_\rho B_{\mu\nu}, \tag{16.6}$$

as in Problem 10.6. In the last term on the right-hand side of (16.4) we have introduced a dimensionful constant κ needed to make this term dimensionless ($[\kappa^2] = M^{6-D}$). In closed string theory the constant κ is a calculable function of the string coupling and α'. The antisymmetrization is responsible for the factor of $1/2$ which multiplies the second term on the right-hand side of (16.4). We have antisymmetrized the factor multiplying $B_{\mu\nu}$ because it is natural to do so: since $B_{\mu\nu}$ is antisymmetric, any symmetric part of the factor does not contribute to the product. Note the hybrid nature of the action (16.4): part of it is an integral over the string world-sheet, and part of it is an integral over all of spacetime.

In order to appreciate the nature of string charge, we reconsider the Maxwell equations (3.34), where the electric current appears as a source of the electromagnetic field:

$$\frac{\partial F^{\mu\nu}}{\partial x^\nu} = j^\mu. \tag{16.7}$$

Here the electric charge density is j^0. A static particle gives rise to electric charge, but zero electric current. The particle is a source of the Maxwell field, while the string is a source of the $B_{\mu\nu}$ field. There is an equation of motion for the $B_{\mu\nu}$ field, analogous to (16.7). We obtain this equation by calculating the variation of the action (16.4) under a variation $\delta B_{\mu\nu}(x)$. The variation of the last term in the action was calculated in Problem 10.6:

$$\delta \left[-\frac{1}{6\kappa^2} \int d^D x\, H_{\mu\nu\rho} H^{\mu\nu\rho} \right] = \frac{1}{\kappa^2} \int d^D x\, \delta B_{\mu\nu}(x)\, \frac{\partial H^{\mu\nu\rho}}{\partial x^\rho}. \tag{16.8}$$

To vary the second term in S we must vary $B_{\mu\nu}(x)$, but in this term the field is evaluated on the string world-sheet. The field $B_{\mu\nu}(X)$ can be rewritten as an integral over all of spacetime of $B_{\mu\nu}(x)$ times a delta function which localizes the field to the world-sheet:

$$B_{\mu\nu}\left(X(\tau,\sigma)\right) = \int d^D x\, \delta^D\left(x - X(\tau,\sigma)\right) B_{\mu\nu}(x). \tag{16.9}$$

With this identity, the second term in S is rewritten as

$$-\int d^D x\, B_{\mu\nu}(x) \frac{1}{2} \int d\tau d\sigma\, \delta^D\left(x - X(\tau,\sigma)\right) \frac{\partial X^{[\mu}}{\partial \tau} \frac{\partial X^{\nu]}}{\partial \sigma} \equiv -\int d^D x\, B_{\mu\nu}(x) j^{\mu\nu}(x), \tag{16.10}$$

where we have introduced the symbol $j^{\mu\nu}$ with value

$$j^{\mu\nu}(x) = \frac{1}{2} \int d\tau d\sigma\, \delta^D\left(x - X(\tau,\sigma)\right) \left(\frac{\partial X^\mu}{\partial \tau} \frac{\partial X^\nu}{\partial \sigma} - \frac{\partial X^\nu}{\partial \tau} \frac{\partial X^\mu}{\partial \sigma} \right). \tag{16.11}$$

It is noteworthy that $j^{\mu\nu}$ is only supported (i.e. it does not vanish) on spacetime points that belong to the string world-sheet. Indeed, if x is not on the world-sheet, then the argument of the delta function is never zero and the integral vanishes. The object $j^{\mu\nu}$ will play the role of a current. By construction, it is antisymmetric under the exchange of its indices:

$$j^{\mu\nu} = -j^{\nu\mu}. \tag{16.12}$$

We have now done all the work needed to find the equation of motion for $B_{\mu\nu}$. Combining equations (16.8) and (16.10), the total variation of the action S is

$$\delta S = \int d^D x \, \delta B_{\mu\nu}(x) \left(\frac{1}{\kappa^2} \frac{\partial H^{\mu\nu\rho}}{\partial x^\rho} - j^{\mu\nu} \right). \qquad (16.13)$$

If this variation is to vanish for arbitrary but antisymmetric $\delta B_{\mu\nu}$, the antisymmetric part of the factor multiplying $\delta B_{\mu\nu}$ must vanish (Problem 16.2). Since the entire term in parentheses is antisymmetric, it must vanish:

$$\frac{1}{\kappa^2} \frac{\partial H^{\mu\nu\rho}}{\partial x^\rho} = j^{\mu\nu}. \qquad (16.14)$$

Quick calculation 16.1 To test your understanding of antisymmetric variations, consider indices $i, j = 1, 2$ that run over two values and arbitrary antisymmetric variations δB_{ij} such that $\delta B_{ij} G^{ij} = 0$. Show explicitly that the only condition you get is $G^{ij} - G^{ji} = 0$.

The similarity between (16.14) and (16.7) is quite remarkable. It suggests that $j^{\mu\nu}$ is some kind of conserved current. The vector j^μ on the right-hand side of (16.7) is a conserved current because

$$\frac{\partial j^\mu}{\partial x^\mu} = \frac{\partial^2 F^{\mu\nu}}{\partial x^\mu \partial x^\nu} = 0, \qquad (16.15)$$

on account of the antisymmetry of $F^{\mu\nu}$ and the exchange symmetry of partial derivatives. In a similar fashion, equation (16.14) gives

$$\frac{\partial j^{\mu\nu}}{\partial x^\mu} = \frac{1}{\kappa^2} \frac{\partial^2 H^{\mu\nu\rho}}{\partial x^\mu \partial x^\rho} = 0. \qquad (16.16)$$

The μ index in $j^{\mu\nu}$ is tied to the conservation equation, but the ν index is free. The tensor $j^{\mu\nu}$ can thus be viewed as a set of currents labeled by the index ν. For each fixed ν, the current components are given by the various values of μ. Since the zeroth component of a current is a charge density, we have several charge densities $j^{0\nu}$. More precisely, since $j^{00} = 0$ (16.12), the nonvanishing charge densities are j^{0k}, with k running over spatial values. Therefore the charge densities of a string define a spatial vector:

Kalb–Ramond charge density is a vector $\vec{j}^0$ with components j^{0k}.   (16.17)

We will soon prove that the charge density vector is tangent to the string. Consider equation (16.16) for $\nu = 0$:

$$\frac{\partial j^{\mu 0}}{\partial x^\mu} = -\frac{\partial j^{0k}}{\partial x^k} = 0. \qquad (16.18)$$

This is the statement that the string charge density is a divergenceless vector:

$$\nabla \cdot \vec{j}^0 = 0. \qquad (16.19)$$

The vector string charge $\vec{Q}$ is naturally defined as the space integral of the string charge density:

$$\vec{Q} = \int d^d x \, \vec{j}^0. \qquad (16.20)$$

To understand string charge more concretely, let us evaluate $j^{\mu\nu}$ in the static gauge $X^0 = \tau$. With this condition, the delta function in equation (16.11) takes the form

$$\delta\left(x^0 - X^0(\tau, \sigma)\right) \delta\left(\vec{x} - \vec{X}(\tau, \sigma)\right) = \delta(t - \tau) \, \delta\left(\vec{x} - \vec{X}(\tau, \sigma)\right), \qquad (16.21)$$

and we can perform the τ integral to find

$$j^{\mu\nu}(\vec{x}, t) = \frac{1}{2} \int d\sigma \, \delta\left(\vec{x} - \vec{X}(t, \sigma)\right) \left[\frac{\partial X^\mu}{\partial t} \frac{\partial X^\nu}{\partial \sigma} - \frac{\partial X^\nu}{\partial t} \frac{\partial X^\mu}{\partial \sigma}\right](t, \sigma). \qquad (16.22)$$

Clearly, at any fixed time t_0, the current $j^{\mu\nu}$ is supported on the string – the set of points $\vec{X}(t_0, \sigma)$. For j^{0k} the second term in (16.22) does not contribute on account of $X^0 = t$, and we find

$$\vec{j}^0(\vec{x}, t) = \frac{1}{2} \int d\sigma \, \delta\left(\vec{x} - \vec{X}(t, \sigma)\right) \vec{X}'(t, \sigma). \qquad (16.23)$$

The $\vec{X}'$ factor on the right-hand side of this equation tells us that at every point on the string the string charge density $\vec{j}^0$ is tangent to the string. It points, in fact, along the tangent defined by increasing σ. Since the orientation of a string is said to be the direction of increasing σ, the charge density vector lies along the orientation of the string!

This might seem puzzling. We have emphasized that the reparameterization invariance of the string action means that a change of parameterization cannot change the physics. Changing the direction of increasing σ is a reparameterization, so how can this change the string charge density? While the Nambu–Goto action is invariant under any reparameterization, the coupling (16.3) of the string to the Kalb–Ramond field is *not*. If we let $\sigma \to \pi - \sigma$ while keeping τ invariant, the measure $d\tau d\sigma$ does not change sign but $X^{\nu\prime}$ does. As a result, (16.3) changes sign. In fact, any reparameterization that changes the orientation of the world-sheet will reverse the sign of this term (Problem 16.1).

Open strings are therefore *oriented* curves. At any fixed time they are fully specified by a curve in space together with an identification of the endpoint that corresponds to $\sigma = 0$ (or, equivalently, the endpoint $\sigma = \pi$). Although closed strings do not have endpoints, they still have an orientation, which is also defined by the direction of increasing σ. The open and closed string theories we examined in previous chapters were theories of oriented open strings and oriented closed strings, respectively. Theories of unoriented strings do exist. These are consistent theories obtained by truncating the state space of (oriented) string theories down to the subspace of states that are invariant under the operation of orientation reversal. We examined these theories in a series of problems beginning with Problems 12.12 and 13.5. The theory of unoriented closed strings has no Kalb–Ramond field in the spectrum. This fits in nicely with our discussion since states of unoriented strings do not carry string charge (where could it point to?).

The integral in (16.23) is easily evaluated for an infinitely long static string stretched along the x^1 axis (a similar configuration was studied in Section 6.7). This string is described by the equations

$$X^1(t, \sigma) = f(\sigma), \quad X^2 = X^3 = \cdots = X^d = 0, \tag{16.24}$$

where $f(\sigma)$ is a function of σ whose range is from $-\infty$ to $+\infty$. The function f must be either a strictly increasing or a strictly decreasing function of σ. We expect this distinction to be significant, as these two alternatives correspond to strings with opposite orientation. Only X^1 has σ dependence, so equation (16.23) implies that the only nonvanishing component of $j^{\mu\nu}$ is j^{01} $(= -j^{10})$:

$$
\begin{aligned}
j^{01}(\vec{x}, t) &= \frac{1}{2} \int d\sigma\, \delta\left(x^1 - X^1(\tau, \sigma)\right) \delta(x^2)\delta(x^3) \ldots \delta(x^d)\, f'(\sigma) \\
&= \frac{1}{2} \delta(x^2)\delta(x^3) \ldots \delta(x^d) \int_{-\infty}^{\infty} d\sigma\, \delta(x^1 - f(\sigma)) f'(\sigma).
\end{aligned}
\tag{16.25}
$$

Letting $\sigma(x^1)$ denote the unique solution of $x^1 - f(\sigma) = 0$, a familiar property of delta functions gives

$$\int_{-\infty}^{\infty} d\sigma\, \delta(x^1 - f(\sigma)) f'(\sigma) = \frac{f'(\sigma(x^1))}{|f'(\sigma(x^1))|} = \mathrm{sgn}(f'(\sigma(x^1))), \tag{16.26}$$

where $\mathrm{sgn}(a)$ denotes the sign of a. Since the function f is strictly increasing or strictly decreasing, this sign is either positive or negative for all x^1. Thus, back to $j^{01}(\vec{x}, t)$,

$$j^{01}(x^1, \ldots, x^d; t) = \tfrac{1}{2} \,\mathrm{sgn}(f')\, \delta(x^2) \ldots \delta(x^d) = \tfrac{1}{2}\, \mathrm{sgn}(f') \delta(\vec{x}_\perp), \tag{16.27}$$

where $\vec{x}_\perp$ is the vector whose components comprise the directions orthogonal to the string. The string charge density is localized on the string, and we see explicitly the orientation dependence in the sign of f'. For an arbitrary static string the spatial string coordinates X^k are time independent. As a result, equation (16.22) implies that

$$j^{ik} = 0, \quad \text{for a static string.} \tag{16.28}$$

For a static string only the string charge densities j^{0k} are nonvanishing.

Before concluding this section, let us briefly discuss the issue of background fields. We have argued here that the string action must be supplemented by the coupling (16.3). You may ask: were we wrong in our earlier quantization of the string, where we did not consider this extra term in the string action? No, our quantization was valid for zero *background* Kalb–Ramond field. The field $B_{\mu\nu}$ in (16.3) is a called a background Kalb–Ramond field. A Kalb–Ramond background is a $B_{\mu\nu}$ field that satisfies its classical equations of motion. The logical process that led us to consider backgrounds ran as follows. We quantized the closed string and discovered Kalb–Ramond particle states. From these quantum states, we deduced the existence of $B_{\mu\nu}$ fields and derived their (linearized) equations of motion. By postulating that backgrounds exist, we are implying that there are nontrivial $B_{\mu\nu}$ fields that satisfy their full equations of motion. One speaks of background electromagnetic fields, referring to $\vec{E}$ and $\vec{B}$ fields that satisfy Maxwell's equations. In a related vein, a gravitational background is a spacetime whose metric $g_{\mu\nu}$ satisfies Einstein's equations of general relativity.

Suppose we were handed a $B_{\mu\nu}$ field configuration. How could we decide whether it provides a background? If we had available the full nonlinear field equations, we could simply test whether the field configuration provides a solution. Since we do not have such field equations, a less direct procedure is necessary. We must re-quantize the string, this time using the coupling (16.3) to take into account the effects of the candidate background $B_{\mu\nu}$. If the quantization is successful then we may conclude that the field configuration provides a background. This procedure can be carried out in practice, and physicists have discovered several $B_{\mu\nu}$ backgrounds.

16.2 Visualizing string charge

In classical Maxwell electromagnetism there are many different configurations of charge: point charges, line charges, surface charges, and continuous charge distributions. Since we have seen that the string charge is localized on the string, you may perhaps think that string charge density can be imagined as some Maxwell line charge density on the string. Not really. String *charge density* can be visualized as a Maxwell *current* on the string. Indeed, we saw that string charge density is a spatial vector that points in the direction of the string – this is just what a Maxwell current on the string looks like.

The string charge density can be integrated over space to define the total string charge $\vec{Q}$ (16.20). This charge has some shortcomings: it is infinite for an infinitely long stretched string (see (16.27)), and it vanishes for a contractible closed string (Problem 16.3). It is possible, however, to use the string charge density to count strings. The string number $\mathcal{N}$, to be defined below, counts the number of strings linked by a given space.

The string charge density j^0 behaves like an electric Maxwell current because of (16.19). Electric charge conservation in electromagnetism requires

$$\frac{\partial \rho}{\partial t} + \nabla \cdot \vec{j} = 0. \tag{16.29}$$

In magnetostatics, the electric charge density ρ is time independent, and, as a result, the electric current density is divergenceless. This means that charge does not accumulate anywhere at any time. Divergenceless currents cannot stop. When they flow on wires either the wires form loops (closed strings for us) or they are infinitely long (infinite strings). We learned that $\nabla \cdot j^0 = 0$ vanishes even if we have time-dependent string configurations. Thus electric string charge density is always analogous to an electric current in magnetostatics. String charge conservation requires strings to form closed loops or to be infinitely long.

To elaborate further on the magnetostatic analogy, we examine the Kalb–Ramond fields created by static strings. We will see that the Kalb–Ramond field strength can be encoded in an effective magnetic field. To simplify matters we work in four-dimensional spacetime. There are two possibilities regarding equation (16.14): either both free indices are space indices or one is a time index and the other is a space index. In the first case we have

$$\frac{\partial H^{ik\rho}}{\partial x^\rho} = 0, \tag{16.30}$$

since j^{ik} vanishes for static strings. We satisfy this equation with the following ansatz: all components of H are *time independent* and

$$H^{ijk} = 0. \tag{16.31}$$

The other equation to consider is

$$\frac{\partial H^{0kl}}{\partial x^l} = \kappa^2 \, j^{0k}. \tag{16.32}$$

We cast this equation into the form of a Maxwell equation by introducing a vector $\vec{B}_H$ with components $B_{H\,m}$ defined by

$$H^{0kl} = \epsilon^{klm} B_{H\,m}. \tag{16.33}$$

Here ϵ^{ijk} is totally antisymmetric and satisfies $\epsilon^{123} = 1$. The vector $\vec{B}_H$ is called the field strength dual to H. Substituting back into (16.32), we find

$$\epsilon^{klm} \frac{\partial B_{H\,m}}{\partial x^l} = \kappa^2 \, j^{0k} \quad \longrightarrow \quad (\nabla \times \vec{B}_H)_k = \kappa^2 \, j^{0k}. \tag{16.34}$$

At this stage, the relevant components of H have been encoded in a dual "magnetic field," and equation (16.32) has been recast in the form

$$\nabla \times \vec{B}_H = \kappa^2 \, \vec{j}^0. \tag{16.35}$$

This is Ampère's equation for the magnetic field of a current $\kappa^2 \vec{j}^0$. Note that, given our ansatz, equation (16.35) is equivalent to the original equations for H; if we cannot solve it, there is no solution for H. The consistency condition for (16.35) is familiar. Since the divergence of a curl is zero, the existence of a solution requires (once again) that $\vec{j}^0$ be divergenceless. Alternatively, given a closed one-dimensional curve Γ that is the boundary of a two-dimensional surface S, the integral form of equation (16.35) is

$$\frac{1}{\kappa^2} \oint_\Gamma \vec{B}_H \cdot d\vec{\ell} = \int_S \vec{j}^0 \cdot d\vec{a}. \tag{16.36}$$

A curve Γ is said to link a string if the string pierces every surface whose boundary is Γ. If the string ended at some point, the current j^0 would end at that point, as well, leading to a nonvanishing $\nabla \cdot \vec{j}^0$ and, consequently, to an inconsistency in (16.35). If the string ended at some point, then for any fixed Γ the left-hand side in (16.36) would be well defined, but the right-hand side would depend on the choice of surface S. This is also an inconsistency.

Equation (16.36) naturally leads to the definition of the string number $\mathcal{N}$ announced at the beginning of this section. The string number $\mathcal{N}$ associated with a curve Γ is defined as

$$\tfrac{1}{2}\mathcal{N} \equiv \frac{1}{\kappa^2} \oint_\Gamma \vec{B}_H \cdot d\vec{\ell} = \int_S \vec{j}^0 \cdot d\vec{a}. \tag{16.37}$$

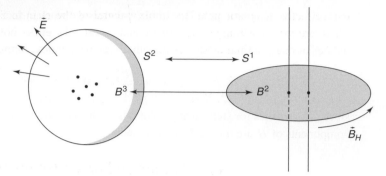

Fig. 16.1 Comparing the computations of Maxwell charge and of string number in a world with three spatial dimensions. The two-sphere which encloses the Maxwell charge is analogous to the circle which links the strings.

We expect $\mathcal{N}$ to give the number of strings linked by the curve Γ. Let us calculate, as an illustration, the value of $\mathcal{N}$ for the string stretched along the x^1 axis that we considered in the previous section. We assume, however, that there are only three spatial dimensions, so that the results of the present section apply. Choosing the orientation so that $f'(\sigma) > 0$, equation (16.27) gives

$$j^{01} = \tfrac{1}{2}\delta(y)\delta(z). \tag{16.38}$$

Consider now a closed curve Γ linking the string and lying in a plane of constant x. Assume that on this plane the curve encloses a surface S whose oriented normal points in the positive x direction. Since both the area vector and $\vec{j}^0$ point in the x direction, we find

$$\tfrac{1}{2}\mathcal{N} = \int_S \vec{j}^0 \cdot d\vec{a} = \int_S j^{01}\, dy dz = \tfrac{1}{2}\int_S \delta(y)\delta(z)\, dy dz = \tfrac{1}{2}. \tag{16.39}$$

As expected, we got $\mathcal{N} = 1$. In general, $\mathcal{N} = N$, where N is the number of strings linked by the chosen curve. The orientation matters: if the curve links two strings with opposite orientation, then their individual contributions to $\mathcal{N}$ will cancel. The $\vec{B}_H$ field in the above example can be calculated easily. It is also possible to write an explicit expression for the antisymmetric tensor field $B_{\mu\nu}$ (Problem 16.4).

Let us compare with electromagnetism. For a localized Maxwell charge distribution, the charge is calculated by integrating the electric charge density over a three-ball B^3 enclosed by a suitable two-sphere S^2 that encloses the charges (see Figure 16.1). A set of parallel, infinite strings is surrounded – but not enclosed – by a suitable circle S^1. This is the natural analog: in the same way as electric charges do not touch the surface S^2 that encloses them, strings do not touch the "surface" S^1 that links them. You cannot remove a Maxwell charge without puncturing the two-sphere, nor can you remove a string without breaking the circle. The computation analogous to the volume integral of Maxwell charge density, gives the number of strings linked by an S^1 as an integral of the local flux of string charge density over a two-ball B^2 (a disk) whose boundary is the S^1. Finally, in Maxwell theory the charge can also be computed as a flux integral of the electric field over the surface S^2 which encloses the charges. The string number is computed analogously as an integral of the Kalb–Ramond dual field strength $\vec{B}_H$ along the curve which links the strings.

Quick calculation 16.2 A string lies along the x^1 axis in a world with four spatial dimensions x^1, x^2, x^3, and x^4. Write a couple of equations that define a sphere that links the string.

16.3 Strings ending on D-branes

We learned in Section 15.2 that there is a Maxwell field living on the world-volume of every D-brane. Indeed, photon states arise from the quantization of open strings whose endpoints lie on the D-brane. The quantization of closed strings in Section 13.3 revealed states that arise from a Kalb–Ramond field $B_{\mu\nu}$ living over all spacetime. We have seen that the string couples electrically to the $B_{\mu\nu}$ field. There is therefore an obvious question: if D-branes have Maxwell fields, is there any object that carries electric charge for these fields? This puzzle is related to another one: what happens to the string charge density – which as we learned can be visualized as a current – when a string ends on a D-brane? Does string charge conservation fail to hold?

Puzzles with charge conservation have led to interesting insights in the past. It led, for example, to the recognition of the displacement current in time-dependent electromagnetic processes. In string theory, the solution to the puzzle involves the realization that the ends of the open string behave as electric point charges! They are charged under the Maxwell field that lives on the D-brane where the string ends. Moreover, the electric field lines of those point charges carry string charge. The interplay between string charge and electric charge, and between the associated Kalb–Ramond and Maxwell fields, results in string charge conservation.

Current conservation is intimately related to gauge invariance. In electromagnetism, the coupling of the gauge field to a current is a term in the action which takes the form

$$S_{\text{coup}} = \int d^D x \, A_\mu(x) \, j^\mu(x). \tag{16.40}$$

In equation (16.2), for example, the coupling term is the middle term on the right-hand side. The gauge transformations are

$$\delta A_\mu(x) = \partial_\mu \epsilon, \tag{16.41}$$

and the field strength $F_{\mu\nu} = \partial_\mu A_\nu - \partial_\nu A_\mu$ is gauge invariant: $\delta F_{\mu\nu} = 0$. The first and last terms on the right-hand side of (16.2) are manifestly gauge invariant. This is generic, terms in the action other than (16.40) are gauge invariant by themselves. The gauge invariance of the action then requires the gauge invariance $\delta S_{\text{coup}} = 0$ of the coupling (16.40). Assuming that the current j^μ is itself gauge invariant,

$$\delta S_{\text{coup}} = \int d^D x \ (\partial_\mu \epsilon) \, j^\mu(x) = - \int d^D x \ \epsilon \, \partial_\mu j^\mu(x), \qquad (16.42)$$

where we integrated by parts and set the boundary terms to zero by assuming that the parameter ϵ vanishes sufficiently rapidly at infinity. We now see that current conservation ($\partial_\mu j^\mu = 0$) implies gauge invariance ($\delta S_{\text{coup}} = 0$).

Similar ideas hold for the coupling of the Kalb–Ramond field $B_{\mu\nu}$. The gauge transformations of $B_{\mu\nu}$ were given in Problem 10.6:

$$\delta B_{\mu\nu} = \partial_\mu \Lambda_\nu - \partial_\nu \Lambda_\mu. \qquad (16.43)$$

The totally antisymmetric field strength $H_{\mu\nu\rho}$ (16.6) is invariant under these gauge transformations. As indicated on the right-hand side of (16.10), the coupling of $B_{\mu\nu}$ to a current $j^{\mu\nu}(= -j^{\nu\mu})$ is of the general form

$$- \int d^D x \, B_{\mu\nu}(x) \, j^{\mu\nu}(x). \qquad (16.44)$$

Quick calculation 16.3 Prove that the coupling term (16.44) is invariant under the gauge transformations (16.43) if $j^{\mu\nu}$ is a conserved current.

The above results indicate that we can investigate potential failures of current conservation by focusing on the gauge invariance properties of the actions. Let us therefore reconsider the term in the action (16.4) that couples the string to the $B_{\mu\nu}$ field:

$$S_B = -\frac{1}{2} \int d\tau d\sigma \, \epsilon^{\alpha\beta} \partial_\alpha X^\mu \partial_\beta X^\nu B_{\mu\nu}(X(\tau,\sigma)). \qquad (16.45)$$

Here we have introduced two-dimensional indices, $\alpha, \beta = 0, 1$, as well as $\partial_0 = \partial/\partial\tau$ and $\partial_1 = \partial/\partial\sigma$. Also, $\epsilon^{\alpha\beta}$ is totally antisymmetric with $\epsilon^{01} = 1$. Since the gauge invariance of S_B is a little subtle, we will study a simpler case first. We will check the gauge invariance of the term that couples a point particle to the Maxwell field:

$$q \int A_\mu(x) dx^\mu. \qquad (16.46)$$

Why is this invariant under (16.41)? Using a parameter τ that ranges from $-\infty$ to $+\infty$, we see that the variation is proportional to

$$\int_{-\infty}^{\infty} d\tau \, \delta A_\mu(x(\tau)) \frac{dx^\mu}{d\tau} = \int_{-\infty}^{\infty} d\tau \frac{\partial \epsilon(x(\tau))}{\partial x^\mu} \frac{dx^\mu}{d\tau} = \int_{-\infty}^{\infty} d\tau \frac{d\epsilon(x(\tau))}{d\tau} \qquad (16.47)$$

$$= \epsilon(x(\tau = \infty)) - \epsilon(x(\tau = -\infty)).$$

Since τ parameterizes time, $t(\tau \to \pm\infty) = \pm\infty$. Gauge invariance then follows if we assume that the gauge parameter vanishes in the infinite past and in the infinite future: $\epsilon(t = \pm\infty, \vec{x}) = 0$.

Let us now return to our problem, the gauge invariance of the action (16.45). Since the arguments of $B_{\mu\nu}$ are the string coordinates, the gauge transformations take the form

$$\delta B_{\mu\nu}(X) = \frac{\partial \Lambda_\nu}{\partial X^\mu} - \frac{\partial \Lambda_\mu}{\partial X^\nu}, \qquad (16.48)$$

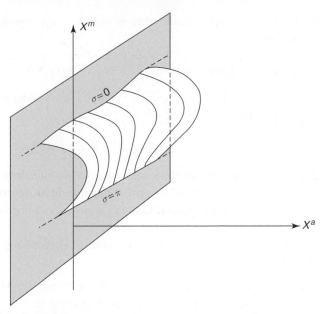

Fig. 16.2   A D-brane and the world-sheet of an open string. The world-sheet boundaries lie on the D-brane: they are the world-lines of the open string endpoints $\sigma = 0$ and $\sigma = \pi$. X^m and X^a are string coordinates along the brane and normal to the brane, respectively.

where the arguments of Λ are also the string coordinates $X(\tau, \sigma)$. The terms multiplying $B_{\mu\nu}$ in (16.45) are antisymmetric in μ and ν (check this!). As a result, each term in (16.48) gives the same contribution to the variation:

$$\delta S_B = -\int d\tau d\sigma \, \epsilon^{\alpha\beta} \, \frac{\partial \Lambda_\nu}{\partial X^\mu} \, \partial_\alpha X^\mu \partial_\beta X^\nu = -\int d\tau d\sigma \epsilon^{\alpha\beta} \, \partial_\alpha \Lambda_\nu \, \partial_\beta X^\nu. \qquad (16.49)$$

Writing out the various terms,

$$\begin{aligned}
\delta S_B &= -\int d\tau d\sigma \left(\partial_\tau \Lambda_\nu \, \partial_\sigma X^\nu - \partial_\sigma \Lambda_\nu \, \partial_\tau X^\nu \right) \\
&= -\int d\tau d\sigma \left(\partial_\tau (\Lambda_\nu \partial_\sigma X^\nu) - \partial_\sigma (\Lambda_\nu \partial_\tau X^\nu) \right).
\end{aligned} \qquad (16.50)$$

Note that we have two total derivatives. The ∂_τ term gives no contribution since we can assume that Λ vanishes at the endpoints of time. If the string under consideration is closed, then there is no boundary in σ and the ∂_σ term gives no contribution, either. This shows the gauge invariance of S_B for closed strings.

For an open string, however, the ∂_σ term in (16.50) gives rise to boundary contributions that do not vanish. The open string world-sheet has boundaries, which appear as lines on the world-volume of a D-brane (Figure 16.2). Let us now calculate δS_B for open strings. We will call the string coordinates along the brane X^m and the string coordinates normal to the brane X^a:

$$X^\mu = (X^m, X^a), \quad \mu = (m, a). \qquad (16.51)$$

If the D-brane is a Dp-brane, then $m = 0, 1, \ldots, p$. We drop the ∂_τ term in (16.50), as before, and focus on

$$\delta S_B = \int d\tau d\sigma \, \partial_\sigma (\Lambda_\nu \partial_\tau X^\nu) = \int d\tau \left[\Lambda_m \partial_\tau X^m + \Lambda_a \partial_\tau X^a \right]_{\sigma=0}^{\sigma=\pi}. \qquad (16.52)$$

Since the X^a are DD coordinates, $\partial_\tau X^a = 0$ at both endpoints, and the second term above gives no contribution. As a result,

$$\delta S_B = \int d\tau \, \Lambda_m \partial_\tau X^m \Big|_{\sigma=\pi} - \int d\tau \, \Lambda_m \partial_\tau X^m \Big|_{\sigma=0}. \qquad (16.53)$$

Gauge invariance has failed because of these two boundary terms. This demonstrates that string charge conservation fails at the endpoints of an open string. We must restore gauge invariance. As we have already said, this requires coupling the Maxwell fields on the brane to the ends of the string.

So let us add to the string action a couple of terms that give electric charge to the string endpoints:

$$S = S_B + \int d\tau \, A_m(X) \frac{dX^m}{d\tau} \Big|_{\sigma=\pi} - \int d\tau \, A_m(X) \frac{dX^m}{d\tau} \Big|_{\sigma=0}. \qquad (16.54)$$

Since the terms above have opposite signs, the string endpoints are oppositely charged. As a convention, we have chosen the string to begin at the negatively charged endpoint and to end at the positively charged endpoint. We have also set $q = \pm 1$ for the endpoint charges, and we will keep this convention throughout. The physical strength of charges can only be determined if we know the normalization of the F^2 terms on the D-brane. This normalization is fixed by the constant κ_0^2 in (16.2). The F^2 terms, together with the couplings in (16.54), determine how the string endpoints create electromagnetic fields. The normalization of the F^2 terms on the D-brane involve the string coupling and α'. They will be determined in Section 20.3.

More briefly, and in the notation of (16.53), we rewrite (16.54) as

$$S = S_B + \int d\tau \, A_m \, \partial_\tau X^m \Big|_{\sigma=\pi} - \int d\tau \, A_m \partial_\tau X^m \Big|_{\sigma=0}. \qquad (16.55)$$

How can we use these terms to restore gauge invariance? By letting the Maxwell field vary under the gauge transformation of the $B_{\mu\nu}$ field! This is a little strange and surprising, but without an interplay between the two types of fields we could not fix our problem of gauge invariance.

So we postulate that, whenever we vary $B_{\mu\nu}$ with a gauge parameter $\Lambda_\mu = (\Lambda_m, \Lambda_a)$, we must also vary the Maxwell field A_m on the D-brane:

$$\delta B_{\mu\nu} = \partial_\mu \Lambda_\nu - \partial_\nu \Lambda_\mu,$$

$$\delta A_m = -\Lambda_m. \qquad (16.56)$$

If we vary A_m in this way, then the variation of the last two terms in (16.55) cancels the variations found in (16.53), thus restoring gauge invariance.

Letting A vary as in (16.56) solves the problem at hand, but it raises some interesting questions. Besides the entire string action, we also want the Maxwell action to be gauge invariant. Since this action is proportional to F^2, we ask: is F_{mn} gauge invariant? It is not! Indeed,

$$\delta F_{mn} = \partial_m \delta A_n - \partial_n \delta A_m = -\partial_m \Lambda_n + \partial_n \Lambda_m = -\delta B_{mn}, \qquad (16.57)$$

where in the last step we recognized that the variation coincides with the gauge transformation of B_{mn}. This is significant, because it follows that the fully gauge invariant combination is

$$\delta(F_{mn} + B_{mn}) = 0. \qquad (16.58)$$

We call the new invariant quantity $\mathcal{F}_{mn}$:

$$\mathcal{F}_{mn} \equiv F_{mn} + B_{mn}, \quad \delta \mathcal{F}_{mn} = 0. \qquad (16.59)$$

On the D-brane, $\mathcal{F}_{mn}$ is the physically significant field strength. The familiar field strength F_{mn} is not fully physical because it is not gauge invariant. Maxwell's equations will be modified by replacing F by $\mathcal{F}$. In many circumstances this will be a small modification, and for zero B, $\mathcal{F}$ equals F. The interplay between these fields helps us to understand intuitively the fate of the string charge when a string ends on a D-brane. We turn to this issue now.

We have seen that string charge density can be thought of as a kind of current flowing down the string. Suppose we have a string ending on a D-brane, as shown in Figure 16.3. The current cannot stop flowing at the string endpoint, so it must flow out *into* the D-brane. How can it do so? We know that the string endpoint is charged, so electric field lines emerge from it spreading out inside the D-brane. The field lines cannot go into the ambient space since the Maxwell field only lives on the D-brane. We will see that, in fact, the electric field lines carry the string charge!

As equation (16.10) indicates, string charge density j^{0k} is, by definition, the quantity that couples to B_{0k}. Whatever couples to B_{0k} appears on the right-hand side of (16.14) as a contribution to j^{0k}. On the D-brane, a Lagrangian density proportional to $-\frac{1}{4} \mathcal{F}^{mn} \mathcal{F}_{mn}$, is the gauge invariant generalization of the Maxwell Lagrangian density. Expanding it out,

$$-\frac{1}{4} \mathcal{F}^{mn} \mathcal{F}_{mn} = -\frac{1}{4} B^{mn} B_{mn} - \frac{1}{4} F^{mn} F_{mn} - \frac{1}{2} F^{mn} B_{mn}. \qquad (16.60)$$

The last term above is particularly interesting. We can expand it further as

$$-\frac{1}{2} F^{mn} B_{mn} = -F^{0k} B_{0k} + \cdots. \qquad (16.61)$$

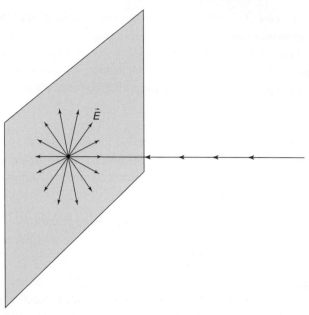

Fig. 16.3   A string ending on a D-brane. The string charge density carried by the string can be viewed as a current flowing down the string. The current is carried on the D-brane by the electric field lines.

The complete action for the D-brane and the string will include this $F^{0k} B_{0k}$ term. Anything that couples to B_{0k} carries string charge, so F^{0k} represents string charge on the brane. But $F^{0k} = E_k$ is the electric field. Therefore, electric field lines on the D-brane carry string charge.

16.4 D-brane charges

We have learned that a string carries electric charge for the Kalb–Ramond field of closed string theory. It is natural to wonder whether there are other extended objects in string theory that carry charge, as well. In addition to strings, the only extended objects we have encountered so far are Dp-branes, with various values of p. Do they carry charge?

Point particles have one-dimensional world-lines and carry electric charge if they couple to a one-index massless gauge field. Strings have two-dimensional world-sheets and carry electric charge if they couple to the Kalb–Ramond gauge field, a massless, two-index anti-symmetric tensor field. A Dp-brane has a $(p + 1)$-dimensional world-volume and is said to be electrically charged if it couples to a massless antisymmetric tensor field with $(p + 1)$ indices. The world-volume of the Dp-brane is parameterized by τ and the set of coordinates $\sigma^1, \sigma^2, \ldots, \sigma^p$. The spacetime coordinates that describe the position of the brane are $X^\mu(\tau, \sigma^1, \ldots, \sigma^p)$, with $\mu = 0, 1, \ldots, d$, and the antisymmetric tensor field is denoted by

$A_{\mu\mu_1\ldots\mu_p}(x)$. The required coupling is a generalization of (16.3):

$$S_p = -\int d\tau d\sigma_1 \ldots d\sigma_p \, \frac{\partial X^\mu}{\partial \tau} \frac{\partial X^{\mu_1}}{\partial \sigma^1} \cdots \frac{\partial X^{\mu_p}}{\partial \sigma^p} \, A_{\mu\mu_1\cdots\mu_p}\left(X(\tau, \sigma^1, \ldots, \sigma^p)\right).$$
(16.62)

In natural units S_p is dimensionless, so the antisymmetric tensor field has natural units $[A_{\mu\mu_1\ldots\mu_p}] = M^{p+1} = L^{-(p+1)}$.

In bosonic closed string theory, the Kalb–Ramond field is the only massless antisymmetric tensor field. This field is sourced by strings, as we have seen. In the absence of additional massless antisymmetric tensors, the bosonic string Dp-branes cannot be charged. On the other hand, type IIA and type IIB closed superstring theories have additional antisymmetric tensors in the Ramond–Ramond sector. These were listed in equations (14.90) and (14.91):

$$\text{IIA}: \; A_\mu, \; A_{\mu\nu\rho},$$
$$\text{IIB}: \; A, \; A_{\mu\nu}, \; A_{\mu\nu\rho\sigma}.$$
(16.63)

It turns out that the R–R gauge fields couple electrically to the appropriate D-branes. In type IIA superstring theory, A_μ couples to D0-branes and $A_{\mu\nu\rho}$ couples to D2-branes. In type IIB superstring theory, $A_{\mu\nu}$ couples to D1-branes and $A_{\mu\nu\rho\sigma}$ couples to D3-branes. The field A in IIB theory carries no index, so it does not couple electrically to any conventional D-brane (it couples electrically to an object called a D-instanton). Summarizing, the electrically charged D-branes are

$$\text{IIA}: \; \text{D0}, \; \text{D2},$$
$$\text{IIB}: \; \text{D1}, \; \text{D3}.$$
(16.64)

Charge and energy conservation together imply that a charged object cannot decay if there are no lighter mass candidate decay products that can carry the charge. The D-branes in (16.64) are in fact stable D-branes, and they cannot decay into open or closed string states. The bosonic D-branes carry no charge and are unstable, as demonstrated by the existence of a tachyon field on their world-volume. It is known that Dp-branes with even p are stable in type IIA theory but unstable in type IIB theory. The D6-branes of type IIA theory, for example, are stable. Additionally, Dp-branes with odd p are stable in type IIB theory but unstable in type IIA theory. The stable D3-branes of type IIB theory are particularly intriguing because their world-volume is a four-dimensional spacetime. All stable D-branes of type II string theory are charged. Nevertheless, the ones that do not appear in the list (16.64) – the D4, D6, and D8 of type IIA theory, and the D5, D7, and D9 of type IIB theory – turn out to carry *magnetic* charge for either the R–R gauge fields in (16.63) or for other subtle R–R states that we have not included in our discussion. We will not study magnetic charge in this book.

The (electric) charge of a Dp-brane has a simple description when p spatial dimensions are curled up into circles and the Dp-brane is wrapped around the resulting compact space. In this case, the p compact space directions lie along the D-brane (they are of type – in the notation of Table 15.1). The other spacetime directions, which define the effective lower-dimensional spacetime, are normal to the brane (they are of type •). The lower-dimensional observer, who only has access to these noncompact directions, sees the brane as a point particle. Our claim is that this particle is electrically charged under a Maxwell field that originates from the antisymmetric tensor field $A_{\mu\mu_1\dots\mu_p}$.

Let $x^1,\dots,x^p$ denote the p compact directions, and let $X^1,\dots,X^p$ denote the corresponding brane coordinates. If the compact directions are circles of radii $R^1,\dots,R^p$ and we use parameters $\sigma^k \in [0, 2\pi]$, then

$$X^k(\tau,\sigma^1,\dots,\sigma^P) = R^k\sigma^k, \quad k=1,\dots,p \text{ (not summed)}, \tag{16.65}$$

represents the wrapped Dp-brane. Indeed, when σ^k runs from zero to 2π, the coordinate X^k runs from zero to $2\pi R^k$, thus going once around the kth circle. Let X^m, with m an index for noncompact directions, take the form

$$X^m(\tau,\sigma^1,\dots,\sigma^P) = x^m(\tau). \tag{16.66}$$

This equation states that the Dp-brane appears as a point particle to the lower-dimensional observer: for any τ, the full Dp-brane is mapped to a single point in the lower-dimensional spacetime.

Since only X^k depends on σ^k, the nonvanishing contributions to (16.62) arise when $\mu_k = k$, for $k=1,\dots,p$:

$$S_p = -\int d\tau d\sigma_1 \dots d\sigma_p \frac{\partial X^\mu}{\partial\tau} R^1 R^2 \dots R^P A_{\mu\,12\dots p}\left(X(\tau,\sigma^1,\dots,\sigma^P)\right). \tag{16.67}$$

The tensor field $A_{\dots}$ is fully antisymmetric and all compact indices have been used, so the index μ can only take values over the noncompact directions: $\mu = m$. As a result,

$$S_p = -\int d\tau d\sigma_1 \dots d\sigma_p \frac{dX^m}{d\tau} R^1 R^2 \dots R^P A_{m\,12\dots p}\left(X^m(\tau),X^k(\sigma^k)\right). \tag{16.68}$$

Finally, let us restrict our attention to the part of the $A_{\dots}$ field that is independent of the compact coordinates: $A_{m\,12\dots p}(x^m(\tau))$. Equation (16.68) then becomes

$$S_p = -R^1 R^2 \dots R^P \int d\tau d\sigma_1 \dots d\sigma_p \frac{dx^m}{d\tau} A_{m\,12\dots p}(x(\tau)). \tag{16.69}$$

The σ integrals can now be done, giving a factor of $(2\pi)^P$, which, together with the product of the radii, yields the volume $V_p = (2\pi R^1)\dots(2\pi R^P)$ of the compact space. Noting that, for all intents and purposes, $A_{m\,12\dots p}$ is a one-index gauge field, we introduce the gauge field $\bar{A}_m$, defined by

$$\frac{1}{(\alpha')^{P/2}}\bar{A}_m(x(\tau)) \equiv A_{m\,12\dots p}(x(\tau)). \tag{16.70}$$

The factor of α' was introduced in order to give the gauge field $\bar{A}_m$ the expected dimension of mass, or inverse length. The field $\bar{A}_m$ is said to be the Maxwell field that arises from the tensor field $A_{\dots}$ by dimensional reduction. In this process of dimensional reduction we did two things: (1) all indices except one were taken to run over compact dimensions, and (2) we dropped the dependence of the field on the compact dimensions. We will examine dimensional reduction further in Section 17.6. Using (16.70), the value of S_p in (16.69) becomes

$$S_p = -\frac{V_p}{(\alpha')^{P/2}}\int d\tau \frac{dx^m}{d\tau}\bar{A}_m(x(\tau)) = -\frac{V_p}{(\ell_s)^p}\int \bar{A}_m dx^m, \tag{16.71}$$

which is recognized as the coupling of a point particle to a Maxwell field $\bar{A}_m$. The Dp-brane appears as a charged point particle. The Maxwell charge Q of the brane is

$$Q = \frac{V_p}{(\ell_s)^p}. \tag{16.72}$$

The charge Q is given by the volume of the brane, measured in units of string length to the pth power. Q is dimensionless, as it should be.

Problems

Problem 16.1 Reparameterization invariance of the string/Kalb–Ramond coupling.

Consider a world-sheet with coordinates (τ, σ) and a reparameterization that leads to coordinates $(\tau'(\tau, \sigma), \sigma'(\tau, \sigma))$. Show that the coupling (16.3) transforms as follows:

$$\int d\tau' d\sigma' \, \frac{\partial X^\mu}{\partial \tau'} \frac{\partial X^\nu}{\partial \sigma'} \, B_{\mu\nu}(X) = \mathrm{sgn}(\gamma) \int d\tau d\sigma \, \frac{\partial X^\mu}{\partial \tau} \frac{\partial X^\nu}{\partial \sigma} \, B_{\mu\nu}(X),$$

where $\mathrm{sgn}(\gamma)$ denotes the sign of γ and

$$\gamma = \frac{\partial \tau}{\partial \tau'} \frac{\partial \sigma}{\partial \sigma'} - \frac{\partial \tau}{\partial \sigma'} \frac{\partial \sigma}{\partial \tau'}.$$

Note that your proof requires the antisymmetry of $B_{\mu\nu}$. If $\mathrm{sgn}(\gamma) = +1$, the reparameterization is orientation preserving. If $\mathrm{sgn}(\gamma) = -1$, the reparameterization is orientation reversing. Give two examples of nontrivial orientation preserving reparameterizations and two examples of nontrivial orientation reversing reparameterizations.

Problem 16.2 Antisymmetric variations and equations of motion.

Let $\delta B_{\mu\nu} = -\delta B_{\nu\mu}$ be an arbitrary antisymmetric variation ($\mu, \nu = 0, 1, \ldots, d$). Show that

$$\delta B_{\mu\nu} G^{\mu\nu} = \sum_{\mu > \nu}^{d} \sum_{\nu=0}^{d} \delta B_{\mu\nu} (G^{\mu\nu} - G^{\nu\mu}).$$

Now show that if $\delta B_{\mu\nu} G^{\mu\nu} = 0$ for all antisymmetric variations $\delta B_{\mu\nu}$ then $G^{\mu\nu} - G^{\nu\mu} = 0$.

Problem 16.3 Properties of the string charge $\vec{Q}$.

(a) Consider a string at some fixed time t_0 and a region $\mathcal{R}$ of space that contains a portion of this string: the string enters the region $\mathcal{R}$ at a point $\vec{x}_i$ and leaves the region $\mathcal{R}$ at a point $\vec{x}_f$ (assume there is no compactification of space). Use (16.23) to calculate the string charge $\vec{Q} = \int_{\mathcal{R}} d^d x \, \vec{j}^0$ contained in $\mathcal{R}$ at time t_0. Use your result to show that the total string charge $\vec{Q}$ associated with a closed string is zero.

(b) A more abstract proof that $\vec{Q}$ is zero for any localized configuration of closed strings requires showing that $\nabla \cdot \vec{j}^0 = 0$ implies $\int d^d x \, \vec{j}^0 = 0$. If you have trouble showing this you may look in your favorite E&M book: the same proof is needed in magnetostatics to demonstrate that the multipole expansion for the magnetic field of a localized current has no monopole term.

(c) Assume now that one space coordinate x is curled up into a circle of radius R, and consider a closed string wrapped around this circle. Calculate the string charge $\vec{Q}$. Explain why the answer is not zero, and compare with the result obtained in (16.72).

Problem 16.4 Kalb–Ramond field of a string.

Following the discussion in Section 16.2, calculate the field $H^{\mu\nu\rho}$ created by a string stretched along the x axis. Find a simple $B^{\mu\nu}$ that gives rise to the field strength H. Interpret your answer in terms of a magnetostatics analog.

Problem 16.5 Explicit checks of current conservation.

In Problem 5.3 you constructed the current vector of a charged point particle:

$$j^{\mu}(\vec{x}, t) = qc \int d\tau\, \delta^{D}(x - x(\tau)) \frac{dx^{\mu}(\tau)}{d\tau}.$$

Verify directly that this is a conserved current ($\partial_{\mu} j^{\mu} = 0$). Now extend this result to the case of a closed string. Verify directly that the current in (16.11) is conserved.

Problem 16.6 Equation of motion for a string in a Kalb–Ramond background.

Consider the string action (6.39) supplemented by the coupling (16.3) to the Kalb–Ramond field. Perform a variation δX^{μ}, and prove that the equations of motion for the string are

$$\frac{\partial \mathcal{P}^{\tau}_{\mu}}{\partial \tau} + \frac{\partial \mathcal{P}^{\sigma}_{\mu}}{\partial \sigma} = -H_{\mu\nu\rho} \frac{\partial X^{\nu}}{\partial \tau} \frac{\partial X^{\rho}}{\partial \sigma}. \tag{1}$$

Problem 16.7 H-field and a circular closed string.

Assume we have a constant, uniform H field that takes the value $H_{012} = h$, with all other components equal to zero. Assume we also have a circular closed string lying in the (x^1, x^2) plane. The purpose of this problem is to show that the tension of the string and the force on the string due to H can give rise to an equilibrium radius. As we will also see, the equilibrium is unstable.

We will analyze the problem in two ways. First, we use the equation of motion (1) derived in Problem 16.6, working with $X^0 = \tau$ ($c = 1$):

(a) Find simplified forms for $\mathcal{P}^{\tau}_{\mu}$ and $\mathcal{P}^{\sigma}_{\mu}$ for a static string.
(b) Check that the $\mu = 0$ component of the equation of motion is trivially satisfied. Show that the $\mu = 1$ and $\mu = 2$ components give the same result: for a suitable orientation of the closed string, the radius R of the string is fixed at the value $R = T_0/|h|$.

Second, we evaluate the action using the simplified geometry of the problem. For this, assume that the radius $R(t)$ is time dependent.

(c) Find $B_{\mu\nu}$ fields that give rise to the H field. In fact, you can find a solution where only B_{01} or B_{02}, or both, are nonzero.
(d) Show that the coupling term (16.3) for the string in question is equal to

$$\int dt\, \pi h R^2(t)$$

if the string is oriented counterclockwise in the (x_1, x_2) plane. Explain why this term represents (minus) potential energy.

(e) Consider the full action for this circular string, and use this to compute the energy functional $E(R(t), \dot{R}(t))$ (your analysis of the circular string in Problem 6.7 may save you a little work).

(f) Assume $\dot{R}(t) = 0$, and plot the energy functional $E(R(t))$ for $h > 0$ and $h < 0$. Show that the equilibrium value of the radius coincides with the one obtained before, and explain why this equilibrium value is unstable.

17

T-duality of closed strings

If a spatial dimension is curled up into a circle then closed strings are affected in two ways: their momentum along the circle gets quantized, and new winding states that wrap around the circle arise. The complementary behavior of momentum and winding states, as a function of the radius of the circle, results in a surprising symmetry: in closed string theory, the physics when the circle has radius R is indistinguishable from the physics when the circle has radius α'/R. This equivalence is proven by exhibiting an operator map between the two theories that respects all commutation relations.

17.1 Duality symmetries and Hamiltonians

Duality symmetries are some of the most interesting symmetries in physics. The term "duality" is generally used by physicists to refer to the relationship between two systems that have very different descriptions but identical physics. The main subject of this chapter is one such situation that arises in closed string theory. You may think that a world where one dimension is curled up into a circle of radius R could easily be distinguished from a world in which the circle has radius α'/R (recall that α' has units of length-squared), but in closed string theory these two worlds are indistinguishable for any value of R. There is a duality symmetry that relates them to each other. This symmetry is called T-duality, where the T stands for toroidal. A compactification is called toroidal if the compact space is a torus. With this terminology, a one-dimensional torus is defined to be a circle.

The AdS/CFT correspondence, to be described in Chapter 23, is an example of a duality: a type IIB superstring background and a supersymmetric Yang–Mills theory are in fact physically equivalent systems. In this section we will discuss duality symmetries that can be found in two familiar situations. Our first example is from electromagnetism and the second is from mechanics.

Consider Maxwell's equations in the absence of sources:

$$\nabla \cdot \vec{E} = 0\,, \quad \nabla \times \vec{B} = \frac{1}{c}\frac{\partial \vec{E}}{\partial t}\,,$$

$$\nabla \cdot \vec{B} = 0\,, \quad \nabla \times \vec{E} = -\frac{1}{c}\frac{\partial \vec{B}}{\partial t}\,. \tag{17.1}$$

One immediately notices that these equations remain invariant under the following *duality transformation*:

$$(\vec{E}, \vec{B}) \rightarrow (-\vec{B}, \vec{E}). \tag{17.2}$$

This invariance of the equations of motion is called the duality symmetry of electromagnetism. Had we simply looked at the familiar Lagrangian for electromagnetism, we might have missed this symmetry. Indeed, expanding the Lagrangian density (see, for example, Problem 5.6) in terms of electric and magnetic fields, we have

$$
\begin{aligned}
\mathcal{L} &= -\frac{1}{4} F_{\mu\nu} F^{\mu\nu} = -\frac{1}{4}(2 F_{0k} F^{0k} + F_{ij} F^{ij}) \\
&= -\frac{1}{2}\left(-F_{0k} F_{0k} + \frac{1}{2} F_{ij} F_{ij}\right) \\
&= \frac{1}{2}(E^2 - B^2),
\end{aligned}
\tag{17.3}
$$

where we wrote $E^2 = \vec{E} \cdot \vec{E}$ and $B^2 = \vec{B} \cdot \vec{B}$. The Lagrangian density $\mathcal{L}$ is *not* invariant under the duality transformation (17.2) – it changes sign. Of course, $\mathcal{L}$ and $-\mathcal{L}$, both written in terms of potentials, lead to the same equations of motion.

Lagrangians treat kinetic and potential energies differently: kinetic energy enters with a positive sign while potential energy enters with a negative sign. Any symmetry that exchanges these types of energy fails to leave the Lagrangian invariant. Both potential and kinetic energies enter into the Hamiltonian with the same sign, so duality symmetries are often exhibited using the Hamiltonian. In electromagnetism, the square of the electric field accounts for the kinetic energy since the electric field involves the time derivative of the vector potential (see equation (3.8)). The square of the magnetic field accounts for the potential energy since the magnetic field involves spatial derivatives of the vector potential. The Hamiltonian, or energy functional, is proportional to the volume integral of $(E^2 + B^2)$. The duality transformation (17.2) leaves the Hamiltonian unchanged.

Since the dynamical variables in electromagnetism are the gauge potentials, one may ask: are there transformations of the potentials that induce (17.2)? Not exactly, but close. One can formulate electromagnetism (without sources) using dynamical variables $\vec{E}$ and $\vec{A}$, both divergenceless. Duality transformations are then written as spatially nonlocal transformations of $\vec{E}$ and $\vec{A}$. (A typical spatially nonlocal transformation expresses the output at a point $\vec{x}$ in terms of the values of the input at all points $\vec{x}'$.)

Consider now a second example. The system is a simple harmonic oscillator consisting of a mass m attached to a spring with spring constant k. The Hamiltonian is given by

$$H(m, k) = \frac{p^2}{2m} + \frac{1}{2} k x^2, \tag{17.4}$$

where we have included the parameters m and k as arguments of the Hamiltonian. This Hamiltonian leads to oscillatory motion with angular frequency $\omega = \sqrt{k/m}$. The form of ω suggests a symmetry under the following duality transformation:

$$(m, k) \longrightarrow \left(\frac{1}{k}, \frac{1}{m}\right). \tag{17.5}$$

The Lagrangian for the original oscillator

$$L = \frac{1}{2}m\dot{x}^2 - \frac{1}{2}kx^2 , \tag{17.6}$$

just as in the electromagnetic case, does not remain invariant under the duality transformation. On the other hand, the equation of motion

$$m\ddot{x} = -kx , \tag{17.7}$$

is invariant under the duality. The Hamiltonian associated with the dual parameters is

$$H\left(\frac{1}{k}, \frac{1}{m}\right) = \frac{1}{2}kp^2 + \frac{1}{2m}x^2 . \tag{17.8}$$

Unlike before, now the Hamiltonian is *not* invariant. To exhibit the connection to the original Hamiltonian $H(m, k)$ we must use canonical transformations. This means changing the canonical variables in such a way that all commutation relations are preserved and therefore the physics remains the same. Consider the canonical transformation K that acts on x and p as follows:

$$K: \quad x \longrightarrow p , \quad p \longrightarrow -x . \tag{17.9}$$

This transformation is canonical because all the relevant commutation relations, which in this case is only $[x, p] = i$, are preserved:

$$K: [x, p] \longrightarrow [p, -x] = -(-i) = i . \tag{17.10}$$

Under this canonical transformation the Hamiltonian in (17.8) becomes

$$K: H\left(\frac{1}{k}, \frac{1}{m}\right) \longrightarrow \frac{1}{2}k(-x)^2 + \frac{1}{2m}p^2 = H(m, k) . \tag{17.11}$$

The Hamiltonian of the system with dual parameters is canonically equivalent to the original Hamiltonian, so the physics is indeed unchanged by the transformation (17.5).

17.2 Winding closed strings

To explore T-duality of closed string theory, we must first understand what effects there are on closed strings when one spatial dimension has been made into a circle. The closed strings we considered in Chapter 13 were moving in Minkowski space, and they could all be shrunk to zero size continuously. If we have one compact dimension then not all closed strings can be reduced continuously to zero size.

The easiest way to visualize this phenomenon is to imagine a world with only two spatial dimensions, one of which is compact. Such a world can be thought of as the surface of an infinitely long cylinder. Let x be the coordinate that has been made compact via the identification

$$x \sim x + 2\pi R , \tag{17.12}$$

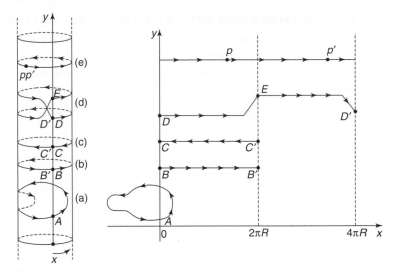

Fig. 17.1 Left: a collection of closed strings living on the surface of a two-dimensional cylinder. Right: the same set of strings represented on the covering space of the cylinder. Strings with nontrivial winding numbers appear in the covering space as open strings.

and let y denote the coordinate that extends along the length of the cylinder. This cylinder is shown on the left side of Figure 17.1. On the right side of the figure we show the (x, y) plane from which the cylinder was formed using the identification (17.12). This plane is known as the covering space of the cylinder. As usual, the x coordinate of a string will be denoted by X.

Let us consider what kinds of closed strings can live on the two-dimensional surface of the cylinder. We will also examine, on the right, how these strings appear in the covering space. While every string on the cylinder actually has infinitely many copies in the covering space (recall Section 2.7), to avoid confusion we will show only one copy.

The simplest strings are those that do not wrap around the cylinder. String (a) in Figure 17.1 is such a string. This string, without its copies, is shown in the covering space directly to the right, where it appears as a string that is actually closed. String (a) can be contracted down to a point in a continuous way. This is clear both on the cylinder and in the covering space. We say that string (a) has zero winding number because it does not "wind around" the compact dimension. The string coordinate in the covering space satisfies

$$\text{string (a):} \quad X(\tau, \sigma = 2\pi) - X(\tau, \sigma = 0) = 0. \tag{17.13}$$

This is the periodicity condition that we encountered in Chapter 13, where, more generally, we wrote $X(\tau, \sigma + 2\pi) = X(\tau, \sigma)$.

Consider now string (b) in the figure. This string is oriented in the direction of increasing x and wraps once around the cylinder. String (b) cannot be contracted to a point without first cutting it. We say that string (b) has winding number $+1$, since it winds once around the circle in the direction of positive x. The picture of this string in the covering space is interesting. If we mark point B on the cylinder as a reference starting point for the string, then in the covering space the string looks like an open curve that begins at B and ends at the point B'. We recognize that the string is actually closed because B and B' are identified by (17.12). The closed string is parameterized with $\sigma \in [0, 2\pi]$, where B corresponds to $\sigma = 0$, and B' corresponds to $\sigma = 2\pi$. The string coordinate in the covering space then satisfies

$$\text{string (b):} \quad X(\tau, \sigma = 2\pi) - X(\tau, \sigma = 0) = 2\pi R. \tag{17.14}$$

Analogous remarks hold for string (c). This string wraps around the cylinder once in the opposite direction, so it is said to have winding number -1. In the covering space we have

$$\text{string (c):} \quad X(\tau, \sigma = 2\pi) - X(\tau, \sigma = 0) = -2\pi R. \tag{17.15}$$

Strings (b) and (c) illustrate a general fact. Strings which wind around the cylinder appear in the covering space as open strings. But not every open string in the covering space represents a wound closed string. Some represent truly open strings on the cylinder. An open string in the covering space represents a closed string that is wound around the cylinder precisely when its endpoints are identified by (17.12). Since string (a) has zero winding, it is a closed string in the covering space.

Consider now a string, such as string (d), that wraps twice around the cylinder. The starting reference point is taken to be D, at which point both $x = 0$ and $\sigma = 0$. After the string has wrapped once it reaches point E on the cylinder. It then winds once more to finally reach point D' with $\sigma = 2\pi$. Points D and D' are the same on the cylinder. In the covering space they are separate, but they are identified by (17.12). This string is said to have winding number $+2$, and it satisfies

$$\text{String (d):} \quad X(\tau, \sigma = 2\pi) - X(\tau, \sigma = 0) = 2\,(2\pi R). \tag{17.16}$$

String (e) is a string with winding number $+2$, just like string (d). But the y coordinate of string (e) is a constant function of σ. Since it has the same winding number as string (d), it also satisfies condition (17.16). Back on the cylinder the windings overlap. In the covering space the string is represented as a straight horizontal line of length $4\pi R$.

A small confusion is possible here. Consider two distinct points p and p' *on* string (e) that are separated by a distance $2\pi R$ along the string. These two points happen to lie on top of the same point on the cylinder, but they are *not* the same point on the wrapped string. In the covering space the points p and p' of the string are separated by a horizontal distance of $2\pi R$. They lie on top of points that are identified, but p and p' are themselves not identified. You can imagine painting one half of the string yellow and the other half green. Point p would belong on the yellow part of the string and point p' would belong on the green part. The piece of string that stretches along the interval $[2\pi R, 4\pi R]$ in the covering space is not a copy of the piece of string that stretches along $[0, 2\pi R]$.

In Figure 17.1 we omitted all duplicate strings from the covering space. Each portion of string that is drawn represents a distinct portion of the string on the cylinder. This was done precisely in order to avoid the above confusion. If, for example, we had included the copies of string (b) in the covering space, then it would be impossible to distinguish between strings (b) and (e) without using colors to draw them differently in the covering space.

All closed strings, however long they may be, and however many times they may wrap, are parameterized with a σ range of 2π. We say that a string has *winding number m*, with m an integer, if it wraps m times around the cylinder in the direction of positive x. In this case, the string coordinate in the covering space satisfies

$$\text{winding number } m: \quad X(\tau, \sigma + 2\pi) = X(\tau, \sigma) + m(2\pi R). \qquad (17.17)$$

Strings with different winding numbers cannot be continuously deformed into each other, and therefore the winding number of a closed string is a topological property. Mathematically, winding numbers appear because we have *two* circles: one with coordinate σ and one with coordinate x. The closed strings are mappings from the σ-circle into the x-circle. A mapping of one circle into another is characterized by an integer known as the winding number of the map. For closed strings, this integer is the winding number m.

For purposes that will become clear later, we define the *winding w* in terms of the winding number m and the radius of the space:

$$w \equiv \frac{mR}{\alpha'}. \qquad (17.18)$$

The winding has units of inverse length, or momentum. In fact, winding will turn out to be a new kind of momentum. Using this definition, equation (17.17) becomes

$$X(\tau, \sigma + 2\pi) = X(\tau, \sigma) + 2\pi\alpha' w. \qquad (17.19)$$

We are now ready to discuss the mode expansion and quantization of closed strings in the presence of a compact dimension.

17.3 Left movers and right movers

Let us consider now strings that are propagating in a 26-dimensional spacetime that has coordinates $x^0, x^1, \ldots, x^{25}$. Assume that the coordinate x^{25} is curled up into a circle of radius R. Since we are going to use the light-cone gauge (this is possible because the string coordinates X^0 and X^1 are not associated with compact directions), it is convenient to organize the string coordinates as

$$X^+, X^-, \underbrace{X^2, X^3, \ldots, X^{24}}_{X^i}, X^{25}. \qquad (17.20)$$

X^i denotes transverse light-cone coordinates, but we do not permit the i index to take the value 25. We will always write the coordinate index explicitly, so we can delete the superscript from X^{25} without risk of confusion. With this understanding, the above set of coordinates will be represented by

$$X^+, X^-, \{X^i\}, \quad \text{and} \quad X \quad \text{with} \quad i = 2, 3, \ldots, 24. \tag{17.21}$$

The periodicity condition for X is given by (17.19). Since X satisfies the wave equation, the general solution is

$$X(\tau, \sigma) = X_L(\tau + \sigma) + X_R(\tau - \sigma) = X_L(u) + X_R(v), \tag{17.22}$$

where $u = \tau + \sigma$ and $v = \tau - \sigma$. Applying condition (17.19) we get

$$X_L(u + 2\pi) + X_R(v - 2\pi) = X_L(u) + X_R(v) + 2\pi\alpha' w, \tag{17.23}$$

which can be rewritten as

$$X_L(u + 2\pi) - X_L(u) = X_R(v) - X_R(v - 2\pi) + 2\pi\alpha' w. \tag{17.24}$$

The last term in the right-hand side was not present when we studied closed strings in Chapter 13 (see (13.13)). Some of the earlier analysis still applies, however. Just as before, the derivatives $X'_L(u)$ and $X'_R(v)$ are periodic functions of their arguments. So the expansions (13.14) hold in the present case, and equations (13.15) hold as well:

$$X_L(u) = \frac{1}{2}x_0^L + \sqrt{\frac{\alpha'}{2}}\bar{\alpha}_0 u + i\sqrt{\frac{\alpha'}{2}}\sum_{n \neq 0}\frac{\bar{\alpha}_n}{n}e^{-inu},$$

$$X_R(v) = \frac{1}{2}x_0^R + \sqrt{\frac{\alpha'}{2}}\alpha_0 v + i\sqrt{\frac{\alpha'}{2}}\sum_{n \neq 0}\frac{\alpha_n}{n}e^{-inv}. \tag{17.25}$$

The new feature is that $\bar{\alpha}_0$ does not necessarily equal α_0. Instead, equation (17.24) gives

$$2\pi\sqrt{\frac{\alpha'}{2}}\,\bar{\alpha}_0 = 2\pi\sqrt{\frac{\alpha'}{2}}\,\alpha_0 + 2\pi\alpha' w, \tag{17.26}$$

so, after cancelling constants,

$$\bar{\alpha}_0 - \alpha_0 = \sqrt{2\alpha'}\,w. \tag{17.27}$$

We see that now $\bar{\alpha}_0$ is equal to α_0 if and only if the winding vanishes. We can also calculate the momentum p of the string along the compact direction:

$$p = \frac{1}{2\pi\alpha'}\int_0^{2\pi}(\dot{X}_L + \dot{X}_R)d\sigma = \frac{1}{\sqrt{2\alpha'}}(\alpha_0 + \bar{\alpha}_0), \tag{17.28}$$

where we used equations (17.25) and noted that only the terms that are linear in u and v contribute to the integral. The momentum is proportional to the average of α_0 and $\bar{\alpha}_0$. In summary, we can now write

$$p = \frac{1}{\sqrt{2\alpha'}}(\bar{\alpha}_0 + \alpha_0),$$

$$w = \frac{1}{\sqrt{2\alpha'}}(\bar{\alpha}_0 - \alpha_0). \tag{17.29}$$

These equations suggest that the winding w is on the same footing as the momentum p. We think of both w and p as momentum operators. For reference, we record the values of the zero modes:

$$\alpha_0 = \sqrt{\frac{\alpha'}{2}}\,(p - w)\,,$$

$$\bar{\alpha}_0 = \sqrt{\frac{\alpha'}{2}}\,(p + w)\,. \tag{17.30}$$

When there was no compactification we had $\alpha_0 = \bar{\alpha}_0$, and this meant that there was only one momentum. Because of this, we were led to expect that only one coordinate zero mode was relevant. Now, however, we have $\alpha_0 \neq \bar{\alpha}_0$, and so there are two different kinds of momenta. There is therefore room for two distinct coordinate zero modes. We rewrite $x_0^L = x_0 + q_0$ and $x_0^R = x_0 - q_0$, thus introducing the average coordinate x_0 and the coordinate difference q_0. These, together with (17.30), allow us to rewrite (17.25) as

$$X_L(\tau + \sigma) = \frac{1}{2}(x_0 + q_0) + \frac{\alpha'}{2}(p + w)(\tau + \sigma) + i\sqrt{\frac{\alpha'}{2}}\sum_{n \neq 0} \frac{\bar{\alpha}_n}{n} e^{-in(\tau+\sigma)}\,,$$

$$X_R(\tau - \sigma) = \frac{1}{2}(x_0 - q_0) + \frac{\alpha'}{2}(p - w)(\tau - \sigma) + i\sqrt{\frac{\alpha'}{2}}\sum_{n \neq 0} \frac{\alpha_n}{n} e^{-in(\tau-\sigma)}\,. \tag{17.31}$$

The full coordinate $X(\tau, \sigma)$ is obtained by adding the above expressions for X_L and X_R:

$$X(\tau, \sigma) = x_0 + \alpha' p \tau + \alpha' w \sigma + i\sqrt{\frac{\alpha'}{2}}\sum_{n \neq 0} \frac{e^{-in\tau}}{n}(\bar{\alpha}_n e^{-in\sigma} + \alpha_n e^{in\sigma})\,. \tag{17.32}$$

In this expansion the only evidence of a compact dimension is the winding term $\alpha' w \sigma$. The zero mode q_0 is not present here. For the record, we give the standard linear combinations of derivatives of X:

$$\dot{X} + X' = 2X_L'(\tau + \sigma) = \sqrt{2\alpha'}\sum_{n \in \mathbb{Z}} \bar{\alpha}_n\, e^{-in(\tau+\sigma)}\,,$$

$$\dot{X} - X' = 2X_R'(\tau - \sigma) = \sqrt{2\alpha'}\sum_{n \in \mathbb{Z}} \alpha_n\, e^{-in(\tau-\sigma)}\,. \tag{17.33}$$

17.4 Quantization and commutation relations

In this section we will derive the commutation relations for the modes of the string coordinate X. We will then discuss the spectrum of the operators p and w.

The starting point is the familiar set of canonical commutators. The commutator between the string coordinate X and the string momentum $\mathcal{P}^\tau$ is taken to be

$$\left[X(\tau,\sigma),\, \mathcal{P}^\tau(\tau,\sigma') \right] = i\,\delta(\sigma - \sigma'). \tag{17.34}$$

In addition, the commutators that involve two coordinates or two momenta are presumed to vanish. Since the quantization of all the extended coordinates exactly follows the discussion in Chapter 13, we only need to consider the compact direction. But even for this coordinate X the situation is not so different.

First consider the commutators that do not involve x_0. When we computed these commutators in Chapter 13, the key to the calculation was (13.28). This equation was simple to deal with because the linear combinations of derivatives of X^I that appear in it took the compact form recorded in (13.26). These combinations of derivatives of X, recorded in (17.33), presently take the same form except for two minor differences: there is no superscript index, and α_0 and $\bar\alpha_0$ are not the same. It follows that equation (13.29) applies also to the modes of the string coordinate X, and therefore

$$[\bar\alpha_m,\bar\alpha_n] = [\alpha_m,\alpha_n] = m\,\delta_{m+n,0}\,, \quad [\alpha_m,\bar\alpha_n] = 0, \tag{17.35}$$

for all integers m and n. Of particular interest are the commutators involving α_0 and $\bar\alpha_0$. We have $[\alpha_0,\bar\alpha_0] = 0$, so on account of (17.29) we find

$$[p,w] = 0. \tag{17.36}$$

Since both α_0 and $\bar\alpha_0$ commute with all the oscillators α_n and $\bar\alpha_n$, so too do p and w:

$$[p,\bar\alpha_n] = [p,\alpha_n] = [w,\bar\alpha_n] = [w,\alpha_n] = 0. \tag{17.37}$$

All that remains is to determine the commutation relations of x_0 with the other operators. The strategy is to use

$$[X(\tau,\sigma),\, (\dot X \pm X')(\tau,\sigma')] = 2\pi\alpha'\, i\,\delta(\sigma-\sigma'), \tag{17.38}$$

which is derived by combining (17.34) and the σ' derivative of $[X(\tau,\sigma), X(\tau,\sigma')] = 0$. Glancing at (17.32), we see that the terms including p and w do not contribute, since p and w commute with all the α_n and $\bar\alpha_n$ operators. Integrating over $\sigma \in [0, 2\pi]$, we find

$$\left[x_0,\, (\dot X \pm X')(\tau,\sigma') \right] = \alpha'\, i\,. \tag{17.39}$$

It follows from this equation and the expansions (17.33) that x_0 commutes with all the α_n and $\bar\alpha_n$ operators of nonzero label, and that

$$[x_0,\alpha_0] = [x_0,\bar\alpha_0] = i\sqrt{\frac{\alpha'}{2}}\,. \tag{17.40}$$

Comparing with (17.30), we see that

$$[x_0,p] = i\,, \quad [x_0,w] = 0. \tag{17.41}$$

This completes our analysis of the commutation relations. There is one surprising point worth noticing: every operator that appears in X commutes with the winding w. The most conservative interpretation of this result is to say that w is a constant number. This would mean that the above quantization is only able to describe the set of closed strings that have some particular fixed winding. Different windings would correspond to different sectors of the full closed string theory. This interpretation treats p and w in quite different ways. A more intriguing interpretation is that w is an operator, just like p is, and that the eigenvalues of w correspond to the various possible windings. This will turn out to be the more natural interpretation, and, as we will see in Section 17.8, it is possible to identify the coordinate zero mode q_0 as the coordinate conjugate to the momentum w.

Because we have compactified the x dimension, the zero mode x_0 is a coordinate that lives on a circle. The momentum operator p along the x direction is the momentum conjugate to x_0, so by a familiar result in quantum mechanics, the possible values of the p-momentum carried by the states are quantized. To derive this quantization condition consider the operator e^{-iap} that translates states along the x direction by an amount a. Since x_0 lives on a circle of radius R, the translation operator that translates by $2\pi R$ has no effect on the states of the theory. Thus $e^{-i\,2\pi Rp}$ behaves like the unit operator. From this we conclude that the states of the theory have momentum along x that is quantized to take the values

$$p = \frac{n}{R}, \quad n \in \mathbb{Z}. \tag{17.42}$$

It should be noted that x_0 is not a well defined operator: its eigenvalues are ambiguous because of the identification $x_0 \sim x_0 + 2\pi R$. As a result, the commutation relation $[x_0, p] = i$ is in fact a formal statement without a precise meaning. The simplest form of the uncertainty principle, which states that for momentum eigenstates the position uncertainty is infinite, does not apply: in a circle the maximum position uncertainty is $2\pi R$. We can use x_0 to construct the well defined operators

$$e^{i\ell x_0/R}, \quad \ell \in \mathbb{Z}, \tag{17.43}$$

which are invariant under the shift of x_0 by any multiple of $2\pi R$. We then have

$$e^{-i\ell x_0/R}\, p\, e^{i\ell x_0/R} = p + \frac{\ell}{R}. \tag{17.44}$$

This operator equation is a precise statement. It is easily derived using $[x_0, p] = i$. It is in this sense that the naive relation $[x_0, p] = i$ can be used to derive unambiguous results. Similar remarks actually apply to (17.34) since X is not a well defined operator either. The operators in (17.43) have a well defined action on the state space, described in terms of momentum eigenstates $|p\rangle$.

Quick calculation 17.1 Show that $e^{i\ell x_0/R}|p\rangle$ is a state of momentum $p + (\ell/R)$. Note that this momentum is properly quantized.

There is another quantization condition in play. Recalling the periodicity condition (17.17), we demand that

$$X(\tau, \sigma + 2\pi) = X(\tau, \sigma) + m(2\pi R) \tag{17.45}$$

acting on the allowed states of the theory. On account of (17.32), this requires that the operator w has eigenvalues satisfying

$$\alpha' w(2\pi) = m(2\pi R) \longrightarrow w = \frac{mR}{\alpha'}, \quad m \in \mathbb{Z}. \tag{17.46}$$

Thus both p and w have discrete spectra.

There is an interesting double strike mechanism at work here. The identification $x \sim x + 2\pi R$ had two effects, or strikes. First, we lost some states. Without the circle, the momentum operator has a continuous spectrum. After the circle is created, the momentum is quantized, so we lost the states that do not satisfy the quantization condition. Second, we gained some new states. These are the winding states that wrap around the newly created circle. Thus we both lost some states and gained some states! It is interesting to note that for particle states, as opposed to string states, we only lose states when we turn a line into a circle. The particle cannot wrap around the circle to give us new states. Another example of the double strike mechanism is provided by the quantization of closed strings on an orbifold. The quantization of closed strings on the $\mathbb{R}^1/\mathbb{Z}_2$ orbifold was studied in Chapter 13. There we lost the states that are not invariant under $X \to -X$, and we gained a sector of twisted states.

17.5 Constraint and mass formula

In our earlier study of light-cone closed strings, the constraint $\alpha_0^- = \bar{\alpha}_0^-$ led to the requirement that $L_0^\perp - \bar{L}_0^\perp$ annihilates the states of the theory. This is still true in the current situation, because x^- is not compactified and, as a result, its left-moving and right-moving momenta remain equal. There is one difference, though: this time the fact that $L_0^\perp - \bar{L}_0^\perp$ vanishes does not imply that $N^\perp - \bar{N}^\perp$ vanishes. To see this, we need explicit expressions for $L_0^\perp$ and $\bar{L}_0^\perp$. These are readily obtained after considering equations (13.37) and (13.42). The modifications are minor. The sums over I split into a sum over i and another term which corresponds to the compact dimension. We find

$$\bar{L}_0^\perp = \frac{1}{2}\bar{\alpha}_0^I\bar{\alpha}_0^I + \bar{N}^\perp = \frac{\alpha'}{4}p^i p^i + \frac{1}{2}\bar{\alpha}_0\bar{\alpha}_0 + \bar{N}^\perp,$$

$$L_0^\perp = \frac{1}{2}\alpha_0^I\alpha_0^I + N^\perp = \frac{\alpha'}{4}p^i p^i + \frac{1}{2}\alpha_0\alpha_0 + N^\perp. \tag{17.47}$$

The operators $N^\perp$ and $\bar{N}^\perp$ include contributions from all the oscillators: those with superscript i and those corresponding to the string coordinate X. We can now calculate

$$L_0^\perp - \bar{L}_0^\perp = \frac{1}{2}(\alpha_0\alpha_0 - \bar{\alpha}_0\bar{\alpha}_0) + N^\perp - \bar{N}^\perp$$

$$= -\alpha' pw + N^\perp - \bar{N}^\perp, \tag{17.48}$$

where we made use of (17.30). It follows that the constraint on physical states takes the form

$$N^\perp - \bar{N}^\perp = \alpha' \, pw \,. \tag{17.49}$$

On states that have either zero momentum or zero winding, the number of left-moving and the number of right-moving excitations must agree. But on states that have both nonzero momentum and nonzero winding $N^\perp - \bar{N}^\perp$ cannot vanish. It must equal $\alpha' pw$. Since both $N^\perp$ and $\bar{N}^\perp$ are operators with integer eigenvalues, the left-hand side of (17.49) always takes integer values. Because of the quantization conditions (17.42) and (17.46),

$$\alpha' pw = \alpha' \cdot \frac{n}{R} \cdot \frac{mR}{\alpha'} = nm \,, \tag{17.50}$$

so the right-hand side of (17.49) also takes integer values. In terms of these integers, (17.49) takes the simple form

$$N^\perp - \bar{N}^\perp = nm \,. \tag{17.51}$$

   An important tool that we have used to study the string spectrum is the formula for the mass-squared of states. To obtain such a formula we take the viewpoint of an observer that lives in the 25-dimensional Minkowski spacetime that does not include the compact dimension. For this observer $M^2 = -p^2$, where p is the 25-dimensional momentum of the states. This momentum vector has components p^+, p^-, and p^i; it does not include the component of momentum that is along the compact dimension. We therefore have

$$M^2 = 2p^+p^- - p^i p^i = \frac{2}{\alpha'}(L_0^\perp + \bar{L}_0^\perp - 2) - p^i p^i \,, \tag{17.52}$$

where we used equation (13.46) to write p^- in terms of Virasoro operators. Replacing the values of $L_0^\perp$ and $\bar{L}_0^\perp$ from (17.47) we find

$$M^2 = \frac{1}{\alpha'}(\alpha_0 \alpha_0 + \bar{\alpha}_0 \bar{\alpha}_0) + \frac{2}{\alpha'}(N^\perp + \bar{N}^\perp - 2) \,. \tag{17.53}$$

Our last step makes use of (17.30) to give

$$M^2 = p^2 + w^2 + \frac{2}{\alpha'}(N^\perp + \bar{N}^\perp - 2) \,. \tag{17.54}$$

This is our result for the mass-squared operator. Here p and w are the quantized momentum and winding associated with the compact dimension. The last term is familiar from the case of closed strings in Minkowski space.

Consider the contribution of the momentum p to M^2, setting for the moment all other terms to zero. This momentum then gives a rest mass, or rest energy, of $M = |p|$. The internal momentum therefore contributes to the rest energy of the string in the same way as the momentum of a massless particle contributes to its energy. Consider now the contribution of the winding w to M^2, setting again all other terms to zero. In this case $M = |w|$, and we can provide a simple interpretation. Consider a string state that is wound $|m|$ times around the compact dimension. The length of such a string is $|m|2\pi R$, and since the string tension is $1/2\pi\alpha'$, the rest energy of the string is equal to

$$M = \frac{1}{2\pi\alpha'}|m|2\pi R = \frac{|m|R}{\alpha'} = |w|. \qquad (17.55)$$

Thus the contribution of the winding to the mass is naturally understood as the energy associated with the stretching that is required to wrap the string around the compactified dimension.

When a string has both nonzero momentum p and winding w, their contributions to the rest energy of the state do not simply add. They must be added in quadrature. The number operators contribute linearly to the square of the rest energy. In some sense a state with momentum, winding, and oscillator excitations can be viewed as a bound state that is composed of various building blocks which interact to give a total rest energy that is smaller than the sum of the energies of the separate constituents.

Quick calculation 17.2 Show that the Hamiltonian of the closed string theory under study is given by

$$H = \frac{\alpha'}{2}(p^i p^i + p^2 + w^2) + N^\perp + \bar{N}^\perp - 2. \qquad (17.56)$$

17.6 State space of compactified closed strings

We now construct explicitly the state space of the quantum closed string we have been studying. Let us begin with the states that have no oscillator excitations. These states have the familiar momentum labels that are associated with the 25-dimensional Minkowski space, but they also carry additional labels that specify the momentum and the winding along the compact dimension. Since the momentum is quantized as $p = n/R$, we can use the integer n as an alternative label for the momentum of the state. Similarly, with $w = mR/\alpha'$, we can use the integer m as a label for the winding of the state. Letting $\vec{p}_T$ denote the vector with components p^i, for $i = 2, \ldots, 24$, we have

$$\text{ground states:} \quad |p^+, \vec{p}_T; n, m\rangle, \quad n, m \in \mathbb{Z}. \qquad (17.57)$$

Although we refer to these as "ground states," this does not actually mean that they are all allowed states in our theory. Indeed, since $N^\perp = \bar{N}^\perp = 0$ for all of these states, the constraint (17.51) tells us that allowed ground states must have either n or m, or both, equal to zero. We call these "ground states" only to emphasize the fact that each of them is killed by all of the annihilation operators of the theory. Note also that although those states in (17.57) that have both n and m nonzero are not allowed, by themselves, to be states in our theory, they can be acted upon by appropriate combinations of creation operators in order to produce allowed states.

A basis for the state space is constructed by applying creation operators to the above states. The general *candidate* basis state of the theory is

$$\left[\prod_{r=1}^{\infty}\prod_{i=2}^{24}\left(a_r^{i\,\dagger}\right)^{\lambda_{i,r}}\right]\left[\prod_{s=1}^{\infty}\prod_{j=2}^{24}\left(\bar{a}_s^{j\,\dagger}\right)^{\bar{\lambda}_{j,s}}\right]\left[\prod_{k=1}^{\infty}\left(a_k^{\dagger}\right)^{\lambda_k}\right]\left[\prod_{l=1}^{\infty}\left(\bar{a}_l^{\dagger}\right)^{\bar{\lambda}_l}\right]|\,p^+,\vec{p}_T;\,n,m\rangle\,. \tag{17.58}$$

We separated out the oscillators that arise from the compact dimension because they carry no 25-dimensional Lorentz index. The number operators $N^{\perp}$ and $\bar{N}^{\perp}$ act on the above state to give

$$N^{\perp}=\sum_{r=1}^{\infty}\sum_{i=2}^{24}r\lambda_{i,r}+\sum_{k=1}^{\infty}k\lambda_k\,,\quad \bar{N}^{\perp}=\sum_{s=1}^{\infty}\sum_{j=2}^{24}s\bar{\lambda}_{j,s}+\sum_{l=1}^{\infty}l\bar{\lambda}_l\,. \tag{17.59}$$

The candidate state (17.58) is a member of the state space if and only if it satisfies the constraint (17.51):

$$N^{\perp}-\bar{N}^{\perp}=nm\,. \tag{17.60}$$

The mass-squared of the state is given by (17.54):

$$M^2=\left(\frac{n}{R}\right)^2+\left(\frac{mR}{\alpha'}\right)^2+\frac{2}{\alpha'}(N^{\perp}+\bar{N}^{\perp}-2)\,. \tag{17.61}$$

In order to familiarize ourselves with the spectrum we now examine some of the closed string states in more detail.

States with $m=n=0$ These states have neither momentum nor winding in the compact dimension. One such string might be string (a) in Figure 17.1 provided, of course, that it has no net momentum along the x axis. Equation (17.60) tells us that $N^{\perp}=\bar{N}^{\perp}$, so we must be sure to match the number of left-moving and right-moving oscillators that act on the state. The vacuum state is simply

$$|p^+,\vec{p}_T;0,0\rangle\,,\quad M^2=-\frac{4}{\alpha'}\,. \tag{17.62}$$

This is the closed string tachyon state. Next come the massless states that arise when $N^{\perp}=\bar{N}^{\perp}=1$. Since in both the left and right sectors we have two kinds of oscillators (those that belong to the compact direction and those that do not) there are four ways we can combine the oscillators to form massless states:

$$a_1^{\dagger}\bar{a}_1^{\dagger}\,|p^+,\vec{p}_T;0,0\rangle\,,$$
$$a_1^{\dagger}\bar{a}_1^{i\,\dagger}\,|p^+,\vec{p}_T;0,0\rangle\,,$$
$$a_1^{i\,\dagger}\bar{a}_1^{\dagger}\,|p^+,\vec{p}_T;0,0\rangle\,,$$
$$a_1^{i\,\dagger}\bar{a}_1^{j\,\dagger}|p^+,\vec{p}_T;0,0\rangle\,. \tag{17.63}$$

The first line contains only one state. Since it carries no 25-dimensional index, this is a state of a massless scalar field. The states in the second and third line each carry the index i, which is a complete light-cone index for the 25-dimensional spacetime. As a result, these are photon states, and each set of states corresponds to a Maxwell field. So we get a total of two Maxwell fields. This is interesting, since ordinary closed strings in Minkowski space did not even give rise to a single Maxwell field, let alone two! Finally, the states in the fourth line have exactly the same structure as the massless closed string states of Minkowski space, except that the dimensionality is now reduced to 25. Therefore these states comprise a gravity field, a Kalb–Ramond field, and a dilaton, all of them living in 25 spacetime dimensions.

If you look at the above states, you can see that all that has really happened is that the compactification reorganized the various massless states of the original 26-dimensional theory. The states now organize themselves into 25-dimensional Lorentz tensors. The number of oscillators has not changed, but since we now have one index that is inert under 25-dimensional Lorentz transformations, the types of fields have changed.

This reorganization of states upon compactification has been known in particle physics since the early work of Kaluza and Klein, who in the early 1920s attempted to build a four-dimensional unified theory of gravitation and electromagnetism by compactification of a purely gravitational theory in five dimensions. It is possible to understand heuristically what happens in this case. Let $g_{\mu\nu}$, where μ and ν are five-dimensional indices, represent the gravity field in five dimensions. Suppose that the dimension we are going to compactify is the fifth dimension, and let m and n be four-dimensional indices that run along the extended spacetime. As a matrix, $g_{\mu\nu}$ can be decomposed into a matrix g_{mn}, which represents the four-dimensional gravity field, a vector g_{m5}, which represents a four-dimensional Maxwell field (g_{5m} is not a new field since the original metric is symmetric), and a single component g_{55}, which represents a four-dimensional scalar. This is the main result of Kaluza and Klein: a five-dimensional gravity theory, after compactifying down to four dimensions, produces a gravity theory that is coupled to a Maxwell field and to a massless scalar.

How is it that Kaluza and Klein were only able to produce one Maxwell field from their compactified theory of gravity, while our list in (17.63) gave two? This is possible because string theory is more than simply a theory of gravity. The second Maxwell field arises from the higher-dimensional Kalb–Ramond field. So, even in string theory, only one Maxwell field arises from the gravitational field. It is just that now the compactification affects other fields in addition to gravity. If $B_{\mu\nu}$ denotes a five-dimensional Kalb–Ramond field, then, upon compactification, B_{m5} represents a four-dimensional Maxwell field. Indeed, forming linear combinations of the states in the second and third lines of (17.63) we have

$$\left(a_1^\dagger \bar{a}_1^{i\dagger} + a_1^{i\dagger} \bar{a}_1^\dagger\right)|p^+, \vec{p}_T; 0, 0\rangle \,,$$
$$\left(a_1^\dagger \bar{a}_1^{i\dagger} - a_1^{i\dagger} \bar{a}_1^\dagger\right)|p^+, \vec{p}_T; 0, 0\rangle \,. \tag{17.64}$$

The states in the first line correspond to photon states that arise from the 26-dimensional graviton states. The states in the second line correspond to photon states that arise from the 26-dimensional Kalb–Ramond states.

States with $n = 0$ or with $m = 0$ Since these states have either nonzero momentum or winding, but not both, they must still satisfy $N^\perp = \bar{N}^\perp$. The ground states are

$$|p^+, \vec{p}_T; n, 0\rangle, \quad M^2 = \frac{n^2}{R^2} - \frac{4}{\alpha'},$$

$$|p^+, \vec{p}_T; 0, m\rangle, \quad M^2 = \frac{m^2 R^2}{\alpha'^2} - \frac{4}{\alpha'}. \tag{17.65}$$

Because they have no 25-dimensional Lorentz indices, both sets of states correspond to scalar fields. For any fixed n or m, the states may be tachyonic, massless, or massive, depending on the value of the radius R. Acting with oscillators on these vacua produces heavier states. Such states have $N^\perp + \bar{N}^\perp \geq 2$ and, as a result, they are massive for all values of the radius R. It is therefore impossible to find massless vectors in this sector of the state space.

States with $n = m = \pm 1$ or $n = -m = \pm 1$ These states have both momentum and winding, so $N^\perp$ and $\bar{N}^\perp$ must be different. The two situations we consider are

$$n = \quad m = \pm 1 \longrightarrow N^\perp - \bar{N}^\perp = \quad 1,$$

$$n = -m = \pm 1 \longrightarrow N^\perp - \bar{N}^\perp = -1. \tag{17.66}$$

The lowest-mass solutions of $N^\perp - \bar{N}^\perp = 1$ arise when $N^\perp = 1$ and $\bar{N}^\perp = 0$. There are two kinds of states satisfying this condition:

$$a_1^\dagger |p^+, \vec{p}_T, \pm 1, \pm 1\rangle,$$

$$a_1^{i\dagger} |p^+, \vec{p}_T, \pm 1, \pm 1\rangle. \tag{17.67}$$

Similarly, the lowest-mass solutions of $N^\perp - \bar{N}^\perp = -1$ arise when $N^\perp = 0$ and $\bar{N}^\perp = 1$, so these are

$$\bar{a}_1^\dagger |p^+, \vec{p}_T; \pm 1, \mp 1\rangle,$$

$$\bar{a}_1^{i\dagger} |p^+, \vec{p}_T; \pm 1, \mp 1\rangle. \tag{17.68}$$

Both groups of states above have mass

$$M^2(R) = \frac{1}{R^2} + \frac{R^2}{\alpha'^2} - \frac{2}{\alpha'} = \left(\frac{1}{R} - \frac{R}{\alpha'}\right)^2. \tag{17.69}$$

It is interesting to note that there is a particular radius R^* at which $M^2(R^*) = 0$, so that all of the states are massless:

$$\frac{1}{R^*} = \frac{R^*}{\alpha'} \quad \rightarrow \quad R^* = \sqrt{\alpha'}. \tag{17.70}$$

R^* is precisely the string length. At this radius, which is called the self-dual radius (for reasons that will become clear later), the interpretation of the above states is simple. The states in the first rows of (17.67) and (17.68) comprise a total of four massless scalars. The states in the second rows of (17.67) and (17.68) comprise a total of four *massless gauge fields*.

Since all of these states have nonzero winding, they are truly "stringy" states that could not arise in a theory of particles.

Quick calculation 17.3 Prove that in the sector of states with $n \neq 0$ and $m \neq 0$ there are no additional states that can ever become massless.

Previously, we identified two Maxwell fields in the sector where momentum and winding are both zero (see (17.64)). These two $U(1)$ fields (using the language introduced in Section 15.3) arise from the gravity and Kalb–Ramond fields in the 26-dimensional theory via a mechanism that applies both in particle theory and in string theory. Now we have obtained four additional gauge fields, all of them massless at the self-dual radius and all of them truly "stringy." Something interesting now occurs, whose description uses the two physically equivalent $U(1)$ fields corresponding to the sum and differences of the states (17.64). Each of these latter $U(1)$ fields happens to combine with two of the stringy gauge fields to form a set of three gauge bosons that interact in a way described by an $SU(2)$ Yang–Mills theory. More explicitly, the combinations are

$$a_1^\dagger \bar{a}_1^{i\dagger} |p^+, \vec{p}_T; 0, 0\rangle \quad \text{with} \quad \bar{a}_1^{i\dagger} |p^+, \vec{p}_T; \pm 1, \mp 1\rangle,$$

$$a_1^{i\dagger} \bar{a}_1^\dagger |p^+, \vec{p}_T; 0, 0\rangle \quad \text{with} \quad a_1^{i\dagger} |p^+, \vec{p}_T, \pm 1, \pm 1\rangle. \tag{17.71}$$

In the compactification of a *particle* theory of gravity and Kalb–Ramond fields the resulting theory would have a $U(1) \times U(1)$ gauge group. In string theory, the $U(1) \times U(1)$ gauge group gets enhanced to an $SU(2) \times SU(2)$ symmetry at the self-dual radius. We have previously seen how Yang–Mills theories arise on the world-volume of coincident D-branes. Now we see one way in which Yang–Mills theories can appear in closed string theory.

17.7 A striking spectrum coincidence

We have already seen that the mass spectrum of the compactified string is very much dependent on the radius of compactification. At the self-dual radius R^*, for example, we get some additional massless gauge fields. Now we will discover a surprising property of the spectrum. To bring this property out into the open, we must look into equations (17.61) and (17.60), which read

$$M^2 = \frac{n^2}{R^2} + \frac{m^2 R^2}{\alpha'^2} + \frac{2}{\alpha'}(N^\perp + \bar{N}^\perp - 2), \quad N^\perp - \bar{N}^\perp = nm. \tag{17.72}$$

Now here is the remarkable property: the closed string spectrum for a compactification with radius R is identical to the closed string spectrum for a compactification with radius $\tilde{R} = \alpha'/R$. We will verify this property shortly, but it turns out that even more is true: the two compactifications are physically indistinguishable. This is the T-duality of closed string theory. Note that this means that, in closed string theory, a compactification with extremely large radius is equivalent to a compactification with extremely small radius! The radii R and α'/R are called dual radii:

$$R \longleftrightarrow \frac{\alpha'}{R} \equiv \tilde{R}. \tag{17.73}$$

To verify the coincidence of the spectrum, let us write an expression for the mass-squared for each of these radii:

$$M^2\left(R;\ n,m\right) = \frac{n^2}{R^2} + \frac{m^2 R^2}{\alpha'^2} + \frac{2}{\alpha'}(N^\perp + \bar{N}^\perp - 2),$$

$$M^2\left(\tilde{R};\ n,m\right) = \frac{n^2 R^2}{\alpha'^2} + \frac{m^2}{R^2} + \frac{2}{\alpha'}(N^\perp + \bar{N}^\perp - 2). \tag{17.74}$$

These do not look the same, but the difference is merely superficial. As n and m run over all possible integers, the *lists* of masses that result are the same. More explicitly, for all $n, m \in \mathbb{Z}$,

$$M^2\left(R;\ n,m\right) = M^2\left(\tilde{R};\ m,n\right). \tag{17.75}$$

In writing this equality, of course, we are comparing states with identical oscillator structure, otherwise the contributions from the number operators would not agree. Note also that the exchange of n and m does not affect the constraint in (17.72). This proves that the mass spectra of theories with dual radii are identical. The key to this equality is the opposite dependences on the radius of the contributions to the mass-squared from the momentum and the winding. The exchange of n and m in (17.75) is just the exchange of winding and momentum quantum numbers.

In the following section we will give evidence that dual radii are actually physically indistinguishable. This is a property of string theory that has been proven beyond doubt. The special radius R^* in (17.70) is the unique radius that is mapped to itself under the transformation in (17.73). The duality then implies that each radius smaller than R^* is equivalent to some radius larger than R^*. In this sense, the radius R^* represents the minimal radius that can be attained in toroidal compactification.

In the present analysis the value of the radius of the circle is adjustable. The theory did not select for us a particular radius the way it selected, for example, a particular space-time dimension. The radius of the circle must be viewed as a parameter of a particular class of compactified spacetimes that allows a consistent definition of string theory. The compactification radius is *not* a parameter of string theory itself, but rather a parameter of a spacetime allowed in string theory. It is an adjustable parameter. Such parameters are sometimes called *moduli*, and the set of values the parameters can take is called the *moduli space*. What we learn from T-duality is that the moduli space of compactifications into a circle can be taken to include only radii larger than or equal to R^*.

To appreciate how a very small circle can appear to be very large, consider the following heuristic argument. On a circle of radius R momentum is quantized in units of $1/R$. If R is very large the spacing between momentum eigenvalues is very small and the spectrum is almost continuous. When the radius is very small, the momentum eigenvalues are largely spaced. In standard particle theory, this is a signal of a small circle. In string theory the situation is different. The winding eigenvalues are quantized in units of R/α', and as the radius R becomes very small, the winding spectrum becomes almost continuous. An observer seeing this continuum might conclude that the compact dimension is very large.

17.8 Duality as a full quantum symmetry

We have seen that T-duality is a symmetry of the mass spectrum of compactified closed string theory. This result, by itself, does not imply that the physics is identical at dual radii. In this section we will prove that the full theory of *free* closed strings is T-duality invariant. It can be proven that T-duality holds even when interactions between strings are included, but we will not attempt to do so in this book.

How do we go about showing the equivalence of the dual theories? We will first bring out into the open some additional structure that exists in compactified string theory. We then explain the equivalence of dual theories in two related ways. In the first we show that T-duality arises as an interpretative ambiguity of a *single* theory: one possible choice of string coordinate suggests that the radius of the circle is R, while another, physically equivalent choice, suggests that the radius of the circle is α'/R. In the second, T-duality is exhibited as an equivalence between two distinct theories. We consider a theory at radius R and a theory at radius α'/R. We then find a one-to-one correspondence between the operators of these theories that preserves all the commutation relations and takes one Hamiltonian into the other. This is the same strategy that we used to demonstrate the duality symmetry of the harmonic oscillator back in Section 17.1.

The additional structure of compactified string theory is related to the coordinate zero mode q_0 that appeared both in X_L and in X_R but did not appear in the full coordinate $X = X_L + X_R$ (see (17.31)). Recall that the winding operator w had a vanishing commutator with all the operators that appear in X. It is natural to make (q_0, w) into a conjugate pair of variables. To this end, we now introduce a dual "coordinate" operator, defined by

$$\tilde{X}(\tau, \sigma) \equiv X_L(\tau + \sigma) - X_R(\tau - \sigma). \tag{17.76}$$

Making use of (17.31), we find that this coordinate takes the form

$$\tilde{X}(\tau, \sigma) = q_0 + \alpha' w \, \tau + \alpha' \, p \, \sigma + i \sqrt{\frac{\alpha'}{2}} \sum_{n \neq 0} \frac{e^{-in\tau}}{n} (\bar{\alpha}_n e^{-in\sigma} - \alpha_n e^{in\sigma}). \tag{17.77}$$

This coordinate is interesting because in it p multiplies σ and w multiplies τ, which reverses the familiar pairings in X. Moreover, q_0 appears in place of x_0. It follows that the momentum associated with the coordinate q_0 is w (since it appears with τ), and the winding associated with q_0 is p (since it appears with σ). The coordinate $\tilde{X}$ brings q_0 into play. The precise meaning of q_0 will be explained below.

We can now define a momentum $\tilde{\mathcal{P}}^\tau$ conjugate to the coordinate $\tilde{X}$:

$$\tilde{\mathcal{P}}^\tau \equiv \frac{1}{2\pi\alpha'} \, \partial_\tau \tilde{X} = \frac{1}{2\pi\alpha'} (\dot{X}_L - \dot{X}_R). \tag{17.78}$$

We postulate the commutator

$$[\tilde{X}(\tau, \sigma), \tilde{\mathcal{P}}^\tau(\tau, \sigma')] = i\delta(\sigma - \sigma') \tag{17.79}$$

and demand that the commutator between two coordinates or between two momenta vanishes. You may be concerned because these commutators involve the same oscillators that we encountered before, so that their commutation relations are already determined. This is true, but the same commutation relations emerge, because the only difference between the pairs $(X, \mathcal{P}^\tau)$ and $(\tilde{X}, \tilde{\mathcal{P}}^\tau)$ is that *the sign of X_R is reversed*. As you can see from (17.31), this sign reversal is implemented by changing the sign of all the α_n oscillators, exchanging x_0 and q_0, and exchanging p and w. These changes do not affect the commutators (17.35), which are therefore reproduced. The pair (q_0, w) appears in $\tilde{X}$ just as the pair (x_0, p) appears in X, so we find

$$[q_0, w] = i, \tag{17.80}$$

as well as

$$[q_0, p] = [q_0, \alpha_n] = [q_0, \bar{\alpha}_n] = 0, \quad n \neq 0. \tag{17.81}$$

We know that x_0 is a coordinate that lives on a circle of radius R, because the associated canonical momentum p is quantized with eigenvalues n/R. Similarly, from the quantization $w = mR/\alpha'$ of the winding momentum, we infer that the associated coordinate q_0 lives on a circle of radius $\tilde{R} = \alpha'/R$. Thus the string coordinate $\tilde{X}$ itself is a coordinate on a circle of radius α'/R. The relation $[q_0, w] = i$ is formal in the same way that $[x_0, p] = i$ is. Well defined operators analogous to (17.43) are given by

$$e^{i\ell q_0/\tilde{R}}, \quad \ell \in \mathbb{Z}. \tag{17.82}$$

We also have

$$e^{-i\ell q_0/\tilde{R}} \, w \, e^{i\ell q_0/\tilde{R}} = w + \frac{\ell}{\tilde{R}}. \tag{17.83}$$

The operators (17.82) have a well defined action on the spectrum.

Quick calculation 17.4 Verify that acting on a state with winding number m, the operator $e^{i\ell q_0/\tilde{R}}$ gives a state with winding number $m + \ell$.

The only commutator that is not fixed is that of x_0 with q_0. In terms of well defined operators we declare

$$\left[e^{-i\ell x_0/R}, \, e^{-imq_0/\tilde{R}} \right] = 0, \quad \ell, m \in \mathbb{Z}. \tag{17.84}$$

This is consistent with our understanding that the two kinds of operators involved here simply act on momentum or winding labels independently of each other. Naively, we write $[x_0, q_0] = 0$.

The Hamiltonian derived from $(\tilde{X}, \tilde{\mathcal{P}}^\tau)$ coincides with the one derived from $(X, \mathcal{P}^\tau)$. This is clear from equation (17.56): the exchange of p and w has no effect, and the minus sign in the α_n oscillators does not affect the number operators. T-duality emerges as an *interpretative ambiguity*. We began with a theory that used the coordinate X to describe a radius R compactification. When the operator content of the theory is examined, we find a different coordinate $\tilde{X}$ which gives an equally compelling interpretation of the theory as one for which the compactification radius is α'/R. In both interpretations we get the same Hamiltonian. In summary, duality arises because of the possibility of replacing X with $\tilde{X}$:

$$\text{T-duality:} \quad X = X_L + X_R \longrightarrow \tilde{X} = X_L - X_R\,. \tag{17.85}$$

This expression for the coordinate $\tilde{X}$ dual to X will be helpful when we discuss open strings in Chapter 18.

To describe duality as a map between two theories, we attempt to formulate the passage from $(X, \mathcal{P}^\tau)$ operators to $(\tilde{X}, \tilde{\mathcal{P}}^\tau)$ operators as a map:

$$(X, \mathcal{P}^\tau) \longrightarrow (\tilde{X}, \tilde{\mathcal{P}}^\tau)\,. \tag{17.86}$$

It is clear from (17.34) and (17.79) that the commutation relations are preserved. From our earlier comments, (17.86) is equivalent to the map

$$(X_L, X_R) \to (X_L, -X_R)\,. \tag{17.87}$$

Finally, by inspection of (17.31), we see that this map is implemented by the following map of oscillators and zero modes:

$$\left\{ \begin{array}{c} x_0 \to q_0 \\ q_0 \to x_0 \end{array} \right\} \quad \left\{ \begin{array}{c} p \to w \\ w \to p \end{array} \right\} \quad \left\{ \begin{array}{c} \alpha_n \to -\alpha_n \\ \bar{\alpha}_n \to \ \bar{\alpha}_n \end{array} \right\}\,. \tag{17.88}$$

This is not a map between two theories, but it will help us to construct one.

To find such a map, we now consider two distinct theories: one for which the familiar interpretation is that the compactification radius is R and another for which the familiar interpretation is that the compactification radius is α'/R. Our task is then to produce a map between the operators of the two theories that is consistent with the commutation relations and takes one Hamiltonian into the other. One important point to be aware of is that the map must respect the operator quantization conditions.

In Table 17.1 we have recorded the operators, quantization conditions, and commutation relations for a theory with radius R. Similarly, in Table 17.2 we have recorded the operators, quantization conditions, and commutation relations for a theory with radius α'/R. In this table we have placed a tilde over all of the operators in order to distinguish them from the operators of the other theory. The requisite operator map is suggested by (17.88) but now takes the form

$$\left\{ \begin{array}{c} x_0 \to \tilde{q}_0 \\ q_0 \to \tilde{x}_0 \end{array} \right\} \quad \left\{ \begin{array}{c} p \to \tilde{w} \\ w \to \tilde{p} \end{array} \right\} \quad \left\{ \begin{array}{c} \alpha_n \to -\tilde{\alpha}_n \\ \bar{\alpha}_n \to \ \tilde{\bar{\alpha}}_n \end{array} \right\}\,. \tag{17.89}$$

For all operators associated with the 25-dimensional spacetime the map is the identity map. The transformations (17.89) respect the commutation relations and map one Hamiltonian into the other. Moreover, we see explicitly that they map operators to others that live on

Table 17.1	A theory where the X coordinate lives on a circle of radius R; listed are the Hamiltonian, the zero modes, and a list of commutation relations

Theory with radius R

$H(R) = \frac{1}{2}\alpha' p^i p^i + \frac{1}{2}\alpha'(p^2 + w^2) + N^\perp + \bar{N}^\perp - 2$

x_0 lives on a circle of radius R

p has eigenvalues n/R

q_0 lives on a circle of radius α'/R

w has eigenvalues mR/α'

$[x_0, p] = [q_0, w] = i$

$[\bar{\alpha}_m, \bar{\alpha}_n] = [\alpha_m, \alpha_n] = m\delta_{m+n,0}$

$[\alpha_m, \bar{\alpha}_n] = 0$

Table 17.2	A theory where the X coordinate lives on a circle of radius α'/R; listed are the Hamiltonian, the zero modes, and a list of commutation relations

Theory with radius $\tilde{R} = \alpha'/R$

$\tilde{H}(\tilde{R}) = \frac{1}{2}\alpha' p^i p^i + \frac{1}{2}\alpha'(\tilde{p}^2 + \tilde{w}^2) + \tilde{N}^\perp + \bar{\tilde{N}}^\perp - 2$

$\tilde{x}_0$ lives on a circle of radius α'/R

$\tilde{p}$ has eigenvalues mR/α'

$\tilde{q}_0$ lives on a circle of radius R

$\tilde{w}$ has eigenvalues n/R

$[\tilde{x}_0, \tilde{p}] = [\tilde{q}_0, \tilde{w}] = i$

$[\bar{\tilde{\alpha}}_m, \bar{\tilde{\alpha}}_n] = [\tilde{\alpha}_m, \tilde{\alpha}_n] = m\delta_{m+n,0}$

$[\tilde{\alpha}_m, \bar{\tilde{\alpha}}_n] = 0$

similar spaces and have identical spectra. Both x_0 and $\tilde{q}_0$, for example, live on identical circles. Both p and $\tilde{w}$ have the same spectrum. The map of oscillators includes a sign factor that does not affect $N^\perp$. This map establishes the physical equivalence of the theories under consideration, and it proves that T-duality is an exact symmetry of free closed string theory compactified on a circle.

Problems

Problem 17.1 Zero mode Hamiltonian.

We can use an action like (12.81) to study the dynamics of the compact coordinate X:

$$S = \frac{1}{4\pi\alpha'} \int d\tau \int_0^{2\pi} d\sigma \left(\dot{X}\dot{X} - X'X' \right).$$

Consider now the zero mode expansion of the string coordinate in the sector of winding number m:

$$X(\tau, \sigma) = x(\tau) + m R \sigma .$$

Find the action for $x(\tau)$. Calculate the Hamiltonian, and show that you recover the contributions to (17.56) arising from the compact dimension.

Problem 17.2 Counting massless gauge fields.

Consider a string compactification where k coordinates are made into circles of critical radius. Describe the candidate ground states of this theory, and give the expression for the Hamiltonian. Find the number of massless vector fields that arise in the lower-dimensional spacetime. Give the explicit list of states corresponding to these fields.

Problem 17.3 Charge carried by winding strings.

The zero mode description of a string with winding number ℓ is given by

$$X^m(\tau, \sigma) = x^m(\tau), \quad X(\tau, \sigma) = \ell R \sigma , \tag{1}$$

where the index m is used to denote all string coordinates except for $X^{25} \equiv X$. Consider the coupling term (16.3) of the Kalb–Ramond field to the string:

$$S = - \int d\tau d\sigma \, \frac{\partial X^\mu}{\partial \tau} \frac{\partial X^\nu}{\partial \sigma} B_{\mu\nu}(X) .$$

Calculate the terms in S that couple $B_{m,25} = -B_{25,m}$ to the string trajectory $x^\alpha(\tau)$. The field $B_{m,25}$ plays the role of a Maxwell field in the 25-dimensional space, as explained in Section 17.6. Conclude that the winding string state described by (1) carries an electric charge proportional to ℓ.

Problem 17.4 Compactification on T^2 with a constant Kalb–Ramond field.

Assume that x^2 and x^3 are each compactified into a circle of radius R. The corresponding string coordinates are called X^r, with $r = 2, 3$. Moreover, there is a nonvanishing Kalb–Ramond field with expectation value

$$B_{23} \equiv \frac{1}{2\pi\alpha'} b , \tag{1}$$

where b is a dimensionless constant. All other components of $B_{\mu\nu}$ vanish.

(a) Build an action for the $X^r(\tau, \sigma)$ by adding an action of the type used in Problem 17.1 to the action in Problem 17.3.

(b) Consider the following expansion for the zero mode part of the coordinates:

$$X^r = x^r(\tau) + m_r R \sigma , \quad r = 2, 3. \tag{2}$$

Show that the Lagrangian for $x^r(\tau)$ is

$$L = \frac{1}{2\alpha'} \left((\dot{x}^2)^2 + (\dot{x}^3)^2 \right) - \frac{\alpha'}{2} \left((w_2)^2 + (w_3)^2 \right) - b \left(\dot{x}^2 w_3 - \dot{x}^3 w_2 \right) . \tag{3}$$

Here $w_r = m_r R/\alpha'$, as usual. The last term in (3) is a total derivative, but it is important in the quantum theory, as you will see.

(c) Define momenta canonical to x^r, compute the Hamiltonian, and show that it takes the form

$$H = \frac{\alpha'}{2}\left((p_2 + bw_3)^2 + (p_3 - bw_2)^2 + w_2^2 + w_3^2\right). \tag{4}$$

Verify that the Hamiltonian generates the correct equations of motion. Note that the quantization conditions on the momenta are $p_r = n_r/R$.

(d) While we have only looked explicitly at the zero modes, the oscillator expansion of the coordinates works just as before. Write the appropriate expansions for the coordinates $X^2(\tau,\sigma)$ and $X^3(\tau,\sigma)$ along the lines of equation (17.32). Explain why the mass-squared operator takes the form

$$M^2 = \left(\frac{n_2}{R} + b\frac{m_3 R}{\alpha'}\right)^2 + \left(\frac{n_3}{R} - b\frac{m_2 R}{\alpha'}\right)^2$$
$$+ \left(\frac{m_2 R}{\alpha'}\right)^2 + \left(\frac{m_3 R}{\alpha'}\right)^2 + \frac{2}{\alpha'}\left(N^\perp + \bar{N}^\perp - 2\right). \tag{5}$$

(e) Show that the constraint $L_0^\perp - \bar{L}_0^\perp = 0$ yields

$$N^\perp - \bar{N}^\perp = n_2 m_2 + n_3 m_3. \tag{6}$$

Problem 17.5 Dualities in the T^2 compactification with Kalb–Ramond field.

In Problem 17.4 you obtained equations (5) and (6), which define the spectrum of a compactification on a square torus T^2 with radius R and with Kalb–Ramond field $B_{23} = b/(2\pi\alpha')$. Show that the spectrum is unchanged under the following duality transformations on the background parameters R and b.

(a) The value of b is changed as

$$b \to b' = b + \ell\frac{\alpha'}{R^2}, \quad \ell \in \mathbb{Z}, \tag{1}$$

while R is left unchanged. Equation (1) states that b is a *periodic* variable. Let A denote the area of the torus. Use (1) to show that the "flux" parameter $f_B = B_{23}A$ is an *angle* variable (i.e., $f_B \sim f_B + 2\pi$). To prove that the spectrum is unchanged, you must find an appropriate compensating change in the quantum numbers, as in (17.75), for example. [Hint: $n_2 \to n_2 - \ell m_3$ is one needed change.]

(b) The values of R and b are changed as

$$R \to R' = \frac{\alpha'}{R}\frac{1}{\sqrt{1+b^2}}, \quad b \to b' = -b. \tag{2}$$

When $b = 0$ this is the familiar T-duality transformation of the radius. The compensating change here is the expected $n_r \leftrightarrow m_r$. [Hint: the algebra is easier if you first expand (5).]

(c) $b \to -b$, with R unchanged. Use $m_r \to -m_r$ as the compensating change of quantum numbers. What additional change must be done regarding oscillators to make (6) work?

The identifications in (1) and (2) are given a geometrical interpretation in Problem 26.3.

18

T-duality of open strings

T-duality relates a world in which a spatial coordinate on a Dp-brane is stretched around a circle to a different looking, but equivalent, world in which a D$(p-1)$-brane has a fixed position on a circle of dual radius. In the first world open strings can have momentum along the circle but cannot wind around it, while in the second world they have no momentum along the dual circle but, as we will see, they can in fact wind around it. We use Maxwell gauge transformations to show that, on a circle, the values of the gauge field line integral $\oint A dx$ are periodically identified. The holonomy of the gauge field along a Dp-brane direction that is wrapped on a circle is related by T-duality to the angular position of a D$(p-1)$-brane on the dual circle.

18.1 T-duality and D-branes

Let us consider the propagation of open strings in a spacetime in which one spatial dimension has been curled up into a circle. Assume that we have a space-filling D25-brane, so that the open string endpoints are free to move all over space. As before, we choose the x^{25} dimension to be compactified:

$$x^{25} \sim x^{25} + 2\pi R. \tag{18.1}$$

All open string coordinates, including X^{25}, satisfy Neumann boundary conditions at both endpoints, so they are all of NN type. In the presence of a compact dimension, closed strings exhibit fundamentally new states: closed strings can wrap around the compact dimension so that they cannot be shrunk to a point. Open strings on a space-filling D-brane have endpoints that are free to move anywhere, so open strings can always be shrunk away. Open strings exhibit no fundamentally new states in the presence of a compact dimension. The open string momentum in the x^{25} direction is quantized: $p^{25} = n/R$. This contributes an amount n^2/R^2 to the mass-squared of the string. There is no winding number w^{25}.

Now consider an open string in a related spacetime, which also has a space-filling D25-brane and a compactified x^{25}, but in which the radius of compactification is $\tilde{R} = \alpha'/R$. T-duality of closed strings tells us that the physics of these two spacetimes is indistinguishable as far as closed strings are concerned. In this new spacetime, however, open strings have their momentum quantized as $p^{25} = nR/\alpha'$, and therefore the mass-squared gets a contribution equal to $n^2 R^2/\alpha'^2$. It is clear that the spectrum of this open string theory does

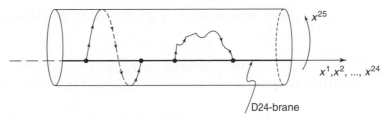

Fig. 18.1 A D24-brane, represented schematically, with x^{25} a normal direction that is curled up into a circle. The open string shown to the left wraps around the compact direction and is non-contractible since its endpoints must remain fixed on the D24-brane. The open string to the right can be contracted.

not coincide with that of the original open string theory on a circle of radius R. This implies that open strings on a D25-brane can tell the difference between a compactification with radius R and another with radius $\tilde{R} = \alpha'/R$. The inclusion of open strings seems to throw a monkey-wrench into the idea of T-duality.

But there is a solution that preserves T-duality, even in the presence of open strings. We will see that T-duality relates the spacetime with compactification radius R and a D25-brane to a spacetime with compactification radius $\tilde{R} = \alpha'/R$ and a D24-brane! The physics is equivalent for both open and closed strings if the D25-brane in the original world is replaced by a D24-brane in the dual world. In the dual world, x^{25} is the Dirichlet direction for the D24-brane, and the corresponding open string coordinate is of DD type. By convention, we set $x^{25} = 0$ to be the position of the brane along the compact direction.

In the dual world all open string endpoints must remain attached to points with $x^{25} = 0$. As a result, there are new open string configurations that cannot be contracted away. An open string that stretches from $x^{25} = 0$ up to $x^{25} = 2\pi \tilde{R}$, for example, winds around the compact direction once. This string cannot be contracted because the endpoints are not free to move along the compact dimension (see Figure 18.1). Open strings can wind any number of times, just as closed strings do. Winding open strings resemble closed strings, but they are not closed: the open string endpoints need not coincide. They typically lie on different points on the D24-brane.

When we had a D25-brane and an x^{25} circle of radius R, the open string had quantized momentum p^{25} but no winding. After the duality transformation we have a D24-brane and an x^{25} circle of radius $\tilde{R}$. The Dirichlet boundary condition imposes a zero momentum constraint, but the open string now has winding. The open string spectrum of the two theories coincide when $\tilde{R} = \alpha'/R$, because the momentum states contribute to M^2 in the first theory in the same way as the open string winding states contribute to M^2 in the second theory. By allowing the duality transformation to modify the D-brane, we can preserve T-duality in the presence of open strings. In summary,

$$\text{T-duality along } x^{25}: \quad \left(\text{D25}; \, R \right) \longrightarrow \left(\text{D24}; \, \tilde{R} = \alpha'/R \right). \tag{18.2}$$

Since the closed string spectrum is not affected by the D-branes, we have a complete physical equivalence.

To show how this works explicitly, we begin by recalling the expansion (12.32) of an NN-type open string coordinate. For $X^{25}(\tau, \sigma) \equiv X(\tau, \sigma)$, we write

$$X(\tau, \sigma) = x_0 + \sqrt{2\alpha'}\, \alpha_0 \tau + i\sqrt{2\alpha'} \sum_{n \neq 0} \frac{1}{n}\alpha_n \cos n\sigma\, e^{-in\tau}. \tag{18.3}$$

We also have

$$\alpha_0 = \sqrt{2\alpha'}\, p = \sqrt{2\alpha'}\, \frac{n}{R}, \tag{18.4}$$

since the momentum on the circle is quantized. The Hamiltonian for this open string is

$$H = L_0^\perp - 1 = \frac{1}{2}\alpha_0^I \alpha_0^I + N^\perp - 1 = \alpha' p^i p^i + \frac{1}{2}\alpha_0\alpha_0 + N^\perp - 1, \tag{18.5}$$

where $i = 2, \ldots, 24$, and $N^\perp$ includes the contribution from the oscillators α_n^i and α_n. We now separate the string coordinate X into left-moving and right-moving components:

$$X(\tau, \sigma) = X_L(\tau + \sigma) + X_R(\tau - \sigma), \tag{18.6}$$

where

$$X_L = \frac{1}{2}(x_0 + q_0) + \sqrt{\frac{\alpha'}{2}}\alpha_0(\tau + \sigma) + \frac{i}{2}\sqrt{2\alpha'}\sum_{n \neq 0}\frac{1}{n}\alpha_n e^{-in\tau}e^{-in\sigma},$$

$$X_R = \frac{1}{2}(x_0 - q_0) + \sqrt{\frac{\alpha'}{2}}\alpha_0(\tau - \sigma) + \frac{i}{2}\sqrt{2\alpha'}\sum_{n \neq 0}\frac{1}{n}\alpha_n e^{-in\tau}e^{+in\sigma}. \tag{18.7}$$

The constant q_0 is arbitrary. Inspired by closed string T-duality, where we reversed the sign of the right movers in (17.85), we now define

$$\tilde{X}(\tau, \sigma) \equiv X_L - X_R, \tag{18.8}$$

and then find

$$\tilde{X}(\tau, \sigma) = q_0 + \sqrt{2\alpha'}\, \alpha_0\, \sigma + \sqrt{2\alpha'}\sum_{n \neq 0}\frac{1}{n}\alpha_n e^{-in\tau}\sin n\sigma. \tag{18.9}$$

This is, in fact, the expansion for a string that stretches from one D-brane to another. We recall equations (15.45) and (15.47), which give:

$$X^a(\tau, \sigma) = \bar{x}_1^a + \sqrt{2\alpha'}\, \alpha_0^a\, \sigma + \sqrt{2\alpha'}\sum_{n \neq 0}\frac{1}{n}\alpha_n^a e^{-in\tau}\sin n\sigma, \tag{18.10}$$

together with

$$\sqrt{2\alpha'}\alpha_0^a = \frac{1}{\pi}(\bar{x}_2^a - \bar{x}_1^a). \tag{18.11}$$

The coordinate difference $\bar{x}_2^a - \bar{x}_1^a$ is the separation between the D-branes in the a-direction. Equations (18.9) and (18.10) are in full correspondence. If we delete the superscript a from (18.10) and identify the constants $\bar{x}_1$ and q_0, we recover the expansion in (18.9).

Before giving the physical interpretation of the new coordinate $\tilde{X}$, we explain why the duality $X \to \tilde{X}$ is a symmetry of the open string theory. Our work with X^a and $\mathcal{P}^a$ in Section 15.3 proves that $\tilde{X}$ and $\tilde{\mathcal{P}} = \partial_\tau \tilde{X}/(2\pi\alpha')$ satisfy the canonical commutation relations. Therefore the duality transformation does not alter the commutation relations. In addition, the Hamiltonian is unchanged. The Hamiltonian for the sector including X^a is obtained from (15.49):

$$H = 2\alpha' p^+ p^- = \alpha' p^i p^i + \frac{1}{2} \alpha_0^a \alpha_0^a + \sum_{n=1}^{\infty} \left[\alpha_{-n}^i \alpha_n^i + \alpha_{-n}^a \alpha_n^a \right] - 1. \tag{18.12}$$

If we delete the superscript a, the correspondence implies that this is the Hamiltonian that arises from $\tilde{X}$ and $\tilde{\mathcal{P}}$ together with the other string coordinates. It is then clear that the Hamiltonian coincides with the one given earlier in (18.5).

We can now turn to the physical interpretation. The string coordinate $\tilde{X}$ is of DD type, since the endpoints are fixed: $\partial_\tau \tilde{X} = 0$ for $\sigma = 0$ and $\sigma = \pi$. When σ goes from 0 to π, the open string stretches an interval

$$\tilde{X}(\tau, \pi) - \tilde{X}(\tau, 0) = \sqrt{2\alpha'}\, \alpha_0 \, (\pi - 0) = 2\pi\alpha' p = 2\pi \frac{\alpha'}{R} n = 2\pi \tilde{R} \, n. \tag{18.13}$$

Since n can take all possible integer values, the picture that emerges is that of an infinite collection of D24-branes with a uniform spacing $2\pi\tilde{R}$ along the x^{25} direction. Such a configuration is indeed physically equivalent to a single D24-brane at some fixed position on a circle of radius $\tilde{R}$.

It is interesting to note that duality interchanges boundary conditions. We have

$$\partial_\sigma X = X_L'(\tau + \sigma) - X_R'(\tau - \sigma) = \partial_\tau \tilde{X}, \tag{18.14}$$

and similarly

$$\partial_\tau X = X_L'(\tau + \sigma) + X_R'(\tau - \sigma) = \partial_\sigma \tilde{X}. \tag{18.15}$$

This makes it clear that N and D boundary conditions are exchanged by T-duality. Indeed, X is of NN type, and $\tilde{X}$ is of DD type. We summarize our facts on open string T-duality by

$$\begin{aligned} X &= X_L + X_R, & \tilde{X} &= X_L - X_R, \\ \partial_\sigma X &= \partial_\tau \tilde{X}, & \partial_\tau X &= \partial_\sigma \tilde{X}. \end{aligned} \tag{18.16}$$

The above differential relations can be used to prove again, more conceptually, that in the dual spacetime the open string winds around the compactified dimension:

$$\tilde{X}(\tau, \pi) - \tilde{X}(\tau, 0) = \int_0^\pi d\sigma\, \partial_\sigma \tilde{X} = \int_0^\pi d\sigma\, \partial_\tau X = 2\pi\alpha' \int_0^\pi d\sigma\, \mathcal{P}^\tau = 2\pi\alpha' p,$$

in agreement with (18.13).

The present considerations extend trivially to cases in which more than one dimension is compactified. Consider a D25-brane in a world where k spatial dimensions are curled up into circles. A simultaneous T-duality transformation on each circle gives a physically equivalent world where we have a D$(25 - k)$-brane and each circle is replaced by a circle with the dual radius. There is also no need to start with a D25-brane. If we curl up one circle only and wrap one direction of a D3-brane around it, a T-duality along the circle will give a D2-brane on a spacetime with a circle of dual radius. In general, if a Dp-brane stretches around a compact dimension, T-duality along this direction will give a D$(p - 1)$-brane at some fixed point on a circle of dual radius. All of these results hold trivially because a T-duality along a given direction does not affect the open string coordinates corresponding to other directions.

Since we use the light-cone gauge throughout, the T-duality that takes a Dp-brane into a D$(p - 1)$-brane has only been established for $p \geq 2$. Indeed, two or more spatial coordinates must be Neumann: X^1, to form together with X^0 the light-cone coordinates $X^\pm$, and X along the compact direction. It is nevertheless true that T-duality holds for $p = 1$. If we have a D1-brane wrapped around a circle, the equivalent T-dual configuration has a D0-brane at some point on a circle of dual radius. This result can be proven using covariant quantization of open strings.

18.2 $U(1)$ gauge transformations

In this section we will examine Maxwell gauge transformations in detail. Gauge field configurations related by gauge transformations are physically equivalent. It is therefore necessary to understand gauge transformations in order to find the possible inequivalent gauge field configurations. This is interesting because, just as D-branes change type under T-duality, gauge field configurations on a D-brane also change in a dramatic way under T-duality. Our previous understanding of gauge transformations does not suffice because of two complicating factors. First, we will deal with compact dimensions, where topological effects become important. Second, we include the effects of charges. Indeed, we learned in Section 16.3 that an open string endpoint lying on a D-brane is seen as a charged particle by the Maxwell field that lives on the D-brane.

The analysis of gauge transformations in the presence of charges is easily done by considering the Schrödinger equation for a nonrelativistic charged particle. In natural units ($\hbar = c = 1$), the Hamiltonian for a particle with mass m and charge q (Problem 5.4) is

$$H = \frac{1}{2m}(\vec{p} - q\vec{A})^2 + q\Phi. \tag{18.17}$$

Recall that the potentials have natural units of mass, or inverse length, and that the charge q is dimensionless. The Schrödinger equation is then

$$i\frac{\partial \psi}{\partial t} = \frac{1}{2m}\left(\frac{\nabla}{i} - q\vec{A}\right)^2 \psi + q\Phi\psi. \tag{18.18}$$

To test the gauge invariance of the classical mechanics of a charged point particle (Section 16.3), we only had to vary the electromagnetic potentials A_μ. The Schrödinger equation, however, is not invariant under just a change in the vector potential. A gauge transformation in quantum mechanics involves changes in both the potentials *and* the wavefunction. The content of the Schrödinger equation remains invariant when the following changes are made simultaneously:

$$\vec{A} \longrightarrow \vec{A}' = \vec{A} + \nabla\chi,$$
$$\Phi \longrightarrow \Phi' = \Phi - \frac{\partial\chi}{\partial t}, \tag{18.19}$$
$$\psi \longrightarrow \psi' = \exp(iq\chi)\,\psi.$$

Here $\chi(x)$ is a function of spacetime. One can readily show (Problem 18.1) that the Schrödinger equation for the primed variables,

$$i\frac{\partial\psi'}{\partial t} = \frac{1}{2m}\left(\frac{\nabla}{i} - q\vec{A}'\right)^2\psi' + q\Phi'\psi', \tag{18.20}$$

is equivalent to the original Schrödinger equation (18.18). For future reference, let $U(x)$ denote the phase factor in the new wavefunction:

$$U(x) \equiv \exp(iq\chi(x)), \quad \psi' = U\psi. \tag{18.21}$$

Now comes the key shift in viewpoint. In the past (and in the first two lines of (18.19), as well!), we have always written the vector potential gauge transformations as

$$A'_\mu = A_\mu + \partial_\mu\chi, \tag{18.22}$$

thinking of $\chi(x)$ as the gauge parameter. We will see that, in fact,

$$\text{the gauge parameter is } U(x). \tag{18.23}$$

While this change is sometimes inconsequential, it has effects when there are compact dimensions. If U is the gauge parameter, we must be able to write the gauge transformation of A_μ in terms of U. The object $(\partial_\mu U)U^{-1}$ is useful for this purpose, because

$$(\partial_\mu U)U^{-1} = \left(\partial_\mu \exp(iq\chi)\right)\exp(-iq\chi) = iq\,\partial_\mu\chi. \tag{18.24}$$

It then follows that (18.22) is reproduced by

$$A'_\mu = A_\mu - \frac{i}{q}(\partial_\mu U)U^{-1}. \tag{18.25}$$

Maxwell theory is called a $U(1)$ gauge theory, because the gauge parameter $U(x)$ can be viewed, for any fixed x, as an element of the group $U(1)$. To understand this we must explain what the $U(1)$ group is and why it is relevant to gauge transformations.

The group $U(1)$ is literally defined as the group of unitary matrices of size 1-by-1. A 1-by-1 matrix has just one entry u, and the unitarity condition, stating that the Hermitian conjugate matrix equals the inverse matrix, gives $u^*u = 1$. We conclude that u is a complex number of unit norm: a phase factor $u = \exp(i\theta)$. The set of complex numbers of unit norm forms a group under multiplication: the multiplication of two complex numbers of unit norm gives a complex number of unit norm, multiplication is associative, there is an identity element $u = 1$ in the group, and any number $\exp(i\theta)$ in the group has an inverse $\exp(-i\theta)$ in the group.

For any fixed spacetime point x, the gauge parameter $U(x)$ is a phase factor; the possible values of $U(x)$ are therefore in one-to-one correspondence with the elements of the group $U(1)$. Since U is spacetime dependent, we have, in fact, a $U(1)$-group worth of possible gauge parameters at *each* spacetime point. The gauge parameter U is best thought of as a function from spacetime to the $U(1)$ group; for each spacetime point we get a $U(1)$ group element.

The concept of a group is relevant to gauge theory because gauge transformations performed in sequence combine according to the rule of group multiplication. We claim that

> a gauge transformation with parameter U_2, followed by a gauge transformation with parameter U_1, is a gauge transformation with parameter $U_1 U_2$.

This composition law is easily verified for the wavefunction ψ. Equation (18.21) implies that the sequence of gauge transformations gives

$$\psi(x) \longrightarrow U_2(x)\psi(x) \longrightarrow U_1(x)(U_2(x)\psi(x)) = (U_1 U_2)(x)\,\psi(x). \qquad (18.26)$$

The gauge field A_μ transforms nontrivially under a gauge transformation only if U is not a constant. The composition law also holds for the gauge field:

$$A_\mu \longrightarrow A_\mu - \frac{i}{q}(\partial_\mu U_2)U_2^{-1} \longrightarrow A_\mu - \frac{i}{q}(\partial_\mu U_2)U_2^{-1} - \frac{i}{q}(\partial_\mu U_1)U_1^{-1}, \qquad (18.27)$$

since the last two terms on the right-hand side are equal to

$$-\frac{i}{q}\left(\partial_\mu (U_1 U_2)\right)(U_1 U_2)^{-1}. \qquad (18.28)$$

18.3 Wilson lines on circles

In this section we apply our discussion of gauge transformations to a world with a compactified dimension. We will find gauge field configurations and effects that are reminiscent of the classic Aharonov–Bohm effect. In the setup for the Aharonov–Bohm effect, there is a solenoid which produces a magnetic field that is confined to its interior. Although a charged particle that is moving outside the solenoid is moving in a region with $\vec{B} = 0$, the wavefunction of the particle is affected by the vector potential $\vec{A}$ outside the solenoid. Since the solenoid produces a nonzero magnetic field, the vector potential outside the solenoid cannot vanish. In a simple gauge, for example, the vector potential goes around the solenoid. The lesson of the Aharonov–Bohm effect is that in quantum mechanics there are magnetic effects in regions of space that have zero magnetic field. This happens because the vector potential is nontrivial. In the Aharonov–Bohm effect, the physical source of the effect is ultimately the magnetic field, since without it the interference effects would vanish. As we will soon see, if we have a compact dimension, there are physical effects from vector potentials even when the magnetic field vanishes *everywhere*.

Quick calculation 18.1 Show that, if a solenoid has magnetic flux $\Phi = \int \vec{B} \cdot d\vec{a}$, then $\oint \vec{A} \cdot d\vec{l} = \Phi$ for a closed curve that surrounds the solenoid. How is this curve oriented?

Suppose the spatial dimension x is compactified into a circle. In addition, assume that the vector potential $\vec{A}$ vanishes except for a component A_x along the circle. In this case there is no magnetic field anywhere! This may at first seem surprising, since a similar vector potential wraps around a circular solenoid's magnetic field. But because there is no space "inside" the circular dimension, there is no magnetic field. As we will see, this vector potential $\vec{A}$ is not a gauge artifact; it affects the physics of particles just as if it were the consequence of a magnetic field. In string theory, the analysis of such a setup is necessary to understand how T-duality works for configurations of D-branes that do not coincide.

Consider a nonzero A_x that is a constant, independent of all coordinates, including the compact dimension x. Can such a field exist? A field configuration can exist if it satisfies the field equations of the theory. A constant A_x, with all other components of A_μ equal to zero, gives zero $F_{\mu\nu}$, and this satisfies the source-free Maxwell equations. This solution can be understood intuitively in three spatial dimensions. Consider a circular solenoid of radius r_0, whose axis coincides with the z axis. If the solenoid flux is $\Phi > 0$, the vector potential can be chosen to be in the azimuthal direction and to have a magnitude $\Phi/(2\pi\rho)$, where ρ is the distance to the axis of the solenoid. This configuration satisfies all equations of motion outside the solenoid. Imagine now selecting a thin cylinder $R < \rho < R + \epsilon$, where $R > r_0$ and ϵ is infinitesimal. The equations of motion are satisfied on this thin cylinder. The solution that represents a constant $\vec{A}$ which wraps around a compact dimension is obtained by throwing away all the space inside and outside the cylinder and letting $\epsilon \to 0$.

Similar reasoning explains why a constant electric field can wrap around a compact dimension. It is possible to have

$$\oint_\Gamma \vec{E} \cdot d\vec{l} \neq 0 \qquad (18.29)$$

if Γ wraps around a compact dimension. Such an electrostatic field is not allowed in ordinary space. For any closed curve Γ in ordinary space there is a surface S whose boundary is Γ. Using Stokes' theorem and the Maxwell equation $\nabla \times \vec{E} = 0$, we then have

$$\oint_\Gamma \vec{E} \cdot d\vec{l} = \int_S (\nabla \times \vec{E}) \cdot d\vec{a} = 0. \qquad (18.30)$$

If Γ wraps around a compact dimension, however, there is no surface for which Γ is the boundary, and (18.30) then does not apply. Again, the existence of gauge potentials that satisfy all relevant equations is the trustworthy guide. Since $\vec{E} = -\partial\vec{A}/\partial t$, a constant E_x in the x direction can be obtained with $A_x = -E_x t$. This gauge field satisfies all the relevant field equations.

So let us consider a vector potential $A_x(x)$ along the compact dimension x. We will try to classify the physically inequivalent configurations. Under a gauge transformation A_x changes as (18.19)

$$A_x \longrightarrow A_x + \frac{\partial \chi}{\partial x}. \tag{18.31}$$

One might think that, since x and $x + 2\pi R$ represent the same point, the parameter χ should satisfy

$$\chi(x + 2\pi R) \overset{?}{=} \chi(x). \tag{18.32}$$

To understand the implications of this candidate condition, we examine the line integral of the vector potential around the circle:

$$w \equiv q \oint dx\, A_x. \tag{18.33}$$

Here we have defined the dimensionless quantity w associated with the gauge field along the circle. We also define the *holonomy* W of the gauge field:

$$W \equiv \exp(iw) = \exp\left(iq \oint dx\, A_x\right). \tag{18.34}$$

W is called a *Wilson line*. In our present context, the Wilson line associated with a closed curve is simply the calculable phase factor that depends on the values of the gauge field along the curve. After the gauge transformation (18.31) w is changed into w':

$$w' = q \oint dx\left(A_x + \frac{\partial \chi}{\partial x}\right) = w + q\big(\chi(x_0 + 2\pi R) - \chi(x_0)\big), \tag{18.35}$$

where we used some arbitrary reference point x_0 on the circle to do the integral. If we assume the periodicity condition (18.32) then $w' = w$. We will soon see that this is not the correct picture.

Our analysis is flawed because of a failure of imagination, rather than a technical mistake. Equation (18.32) is sensible if $\chi(x)$ is the fundamental gauge parameter. But we claimed in Section 18.2 that the gauge parameter is $U(x)$. If U is the gauge parameter, it must be periodic on the circle:

$$U(x + 2\pi R) = U(x). \tag{18.36}$$

Since $U = \exp(iq\chi)$, the above equation implies

$$q\,\chi(x + 2\pi R) = q\,\chi(x) + 2\pi m, \quad m \in \mathbb{Z}, \tag{18.37}$$

or, equivalently,

$$q\big(\chi(x + 2\pi R) - \chi(x)\big) = 2\pi m. \tag{18.38}$$

Note the dramatic change from (18.32). If U is the gauge parameter, $\chi(x)$ need only be quasi-periodic! Moreover, looking back at equation (18.35), we see that

$$w' = w + 2\pi m. \tag{18.39}$$

Therefore the physics does not change when we replace w by $w + 2\pi m$. We write $w \sim w + 2\pi m$, or

$$q \oint dx \, A_x \sim q \oint dx \, A_x + 2\pi m. \tag{18.40}$$

As before, we interpret the above equation to mean that w lives on a unit circle, or that all physically inequivalent values of w are represented on the fundamental domain

$$w = q \oint dx \, A_x \in [\, 0, 2\pi). \tag{18.41}$$

The most natural interpretation is that w is really an angle. Thus we will write

$$\theta \equiv w = q \oint dx \, A_x. \tag{18.42}$$

This is an important result. In the presence of compact dimensions, line integrals of the vector potential are angular variables. Gauge transformations act as $\theta \to \theta + 2\pi m$. We will see that in string theory the abstract angle θ has a concrete physical interpretation. It is worth noting that the Wilson line $W = \exp(i\theta)$ is gauge invariant. Gauge equivalent θ give the same holonomy W.

It is possible to write explicitly χ that are not single valued, but that, nevertheless, lead to a single-valued U because they obey (18.38). There are infinitely many physically equivalent choices, but the simplest one takes χ to be linear in the compact coordinate x:

$$q \chi = (2\pi m) \frac{x}{2\pi R} = \frac{mx}{R}. \tag{18.43}$$

As x changes from 0 to $2\pi R$, the quantity $q\chi$ changes by $2\pi m$, as required in (18.38). For this choice of χ, the gauge transformation changes the gauge field by a constant:

$$q A_x(x) \longrightarrow q A_x(x) + \frac{m}{R}. \tag{18.44}$$

This gauge transformation implies that we can identify

$$q A_x \sim q A_x + \frac{m}{R}. \tag{18.45}$$

The natural realization of a Wilson line is with *constant* A_x. In this case, (18.42) gives

$$q A_x = \frac{\theta}{2\pi R}. \tag{18.46}$$

The presence of a Wilson line on a circle changes significantly the physics of a charged particle. To see how this happens, we examine the Schrödinger equation

$$\frac{1}{2m} \left(\frac{1}{i} \frac{\partial}{\partial x} - q A_x \right)^2 \psi = i \frac{\partial \psi}{\partial t}. \tag{18.47}$$

With A_x given in (18.46), energy eigenstates $\psi(x, t) = e^{-iEt}\psi(x)$ satisfy

$$\frac{1}{2m}\left(\frac{1}{i}\frac{\partial}{\partial x} - \frac{\theta}{2\pi R}\right)^2 \psi(x) = E\psi(x). \tag{18.48}$$

The wavefunction $\psi(x)$ must be periodic: $\psi(x + 2\pi R) = \psi(x)$. Therefore the solutions take the form $\psi_\ell(x) \sim \exp(i\ell x/R)$, with $\ell \in \mathbb{Z}$. The corresponding energy levels are

$$E_\ell = \frac{1}{2m}\left(\frac{\ell}{R} - \frac{\theta}{2\pi R}\right)^2. \tag{18.49}$$

Note that the energy levels are shifted if $\theta \neq 0$. In particular, the $\theta = 0$ degeneracy between $\pm\ell$ energy levels is broken. Since θ is an angle variable the energy levels must remain unchanged when we let $\theta \to \theta + 2\pi$. Since ℓ varies over all $\mathbb{Z}$, we can easily see that the *set* of energy levels is unchanged under this transformation. The shift $\theta \to \theta + 2\pi$ is compensated by letting $\ell \to \ell + 1$. This confirmation that θ is naturally an angle variable gives credence to our claim that $U(x)$ is the fundamental gauge parameter. If χ had been the gauge parameter all values of θ would be gauge inequivalent.

Quick calculation 18.2 Consider a gauge transformation with χ linear in x that takes $\theta \to \theta + 2\pi$. What does it do to ψ_ℓ?

18.4 Open strings and Wilson lines

We are now finally in a position to study the physics of open strings on D-branes that have gauge fields characterized by holonomies. T-duality will provide a physical interpretation for the angle variable that represents the gauge-field holonomy.

Consider a Dp-brane that wraps around a compact dimension x. This simply means that one of the spatial directions along the Dp-brane is the x direction. A gauge field lives on the world-volume of the Dp-brane. Let us now assume that this gauge field is such that $q \oint A_x dx$ takes the value θ. What is the T-dual picture of this brane configuration? We learned before that the dual world has a D$(p-1)$-brane located at some position on a circle of dual radius. But what does the parameter θ correspond to? The most obvious guess is correct: θ parameterizes the position of the D$(p-1)$-brane on the dual circle! This gives a very concrete meaning to the formerly abstract angle variable. The situation is illustrated in Figure 18.2.

What is the evidence for this interpretation of θ? First, the periodicity makes physical sense since a D$(p-1)$-brane at θ is the same as a D$(p-1)$-brane at $\theta + 2\pi$. Second, just as the position of the D$(p-1)$-brane on the circle does not affect the spectrum of open strings that end on it, the Wilson line does not affect the spectrum of open strings with endpoints on the Dp-brane. This is not readily apparent, given that charged particles are in fact affected by Wilson lines. In the case of strings, however, the endpoints have opposite charges so the string as a whole is neutral and the Wilson line has no effect. If a D-brane wraps the compact dimension, the mass-squared for open string states is given as

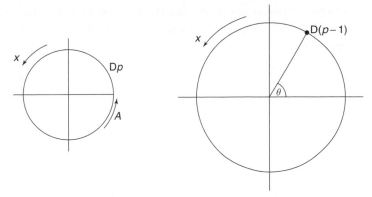

Fig. 18.2   A Dp-brane wrapped on a circle, with a vector potential along the circle. In the T-dual picture, the line integral of this gauge field becomes the angle that parameterizes the position of the D($p - 1$)-brane on the dual circle.

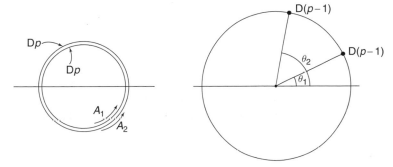

Fig. 18.3   Two Dp-branes that wrap on a circle, one with parameter θ_1 and the other with parameter θ_2. In the T-dual picture, we have two D($p - 1$)-branes separated by an angle $\theta_2 - \theta_1$.

$$M^2 = p^2 + \frac{1}{\alpha'}(N^{\perp} - 1), \quad p = \frac{\ell}{R}, \tag{18.50}$$

where p is the quantized momentum in the compact direction, and $N^{\perp}$ is the appropriate number operator. For a particle, the addition of a Wilson line resulted in p changing into $p - qA$ in the Hamiltonian. This implied letting

$$\frac{\ell}{R} \to \frac{\ell}{R} - \frac{\theta}{2\pi R}, \tag{18.51}$$

as we saw in (18.49). What happens with strings? Since strings have two endpoints with opposite charges, if both endpoints lie on the same Dp-brane the effect is cancelled: $p \to p - qA + qA = p$. To give direct evidence for the interpretation of θ we need more than one D($p - 1$)-brane on the circle. In this case, the relative positions of the branes matter.

Consider therefore a string stretching between the two Dp-branes. The situation is shown in Figure 18.3. Each D-brane has its own Maxwell field. Suppose the negatively charged endpoint lies on the first Dp-brane, with Wilson line parameter θ_1, and the positively charged endpoint lies on the second Dp-brane, with Wilson line parameter θ_2. The momentum is then shifted from p to $p + qA_1 - qA_2$, and therefore

$$\frac{\ell}{R} \rightarrow \frac{\ell}{R} - \frac{\theta_2}{2\pi R} + \frac{\theta_1}{2\pi R}. \tag{18.52}$$

As a result, the mass-squared formula reads

$$M^2 = \left(\frac{2\pi \ell - (\theta_2 - \theta_1)}{2\pi R}\right)^2 + \frac{1}{\alpha'}(N^\perp - 1), \quad \ell \in \mathbb{Z}. \tag{18.53}$$

If $\theta_1 = \theta_2$ the effects of the holonomies cancel out. As we argued before, the same would happen if both endpoints of the string were on the same brane.

The configuration T-dual to the two Dp-branes with different θ parameters consists of two D$(p-1)$-branes with different positions, corresponding to the two values of θ. We can easily verify that the mass formulae are consistent. For example, consider the $\ell = 0$ states of the string that stretches between the two Dp-branes:

$$M^2 = \left(\frac{\theta_2 - \theta_1}{2\pi R}\right)^2 + \frac{1}{\alpha'}(N^\perp - 1). \tag{18.54}$$

If, as we claim, θ_1 and θ_2 are physical angles, then in the dual world there must be two D$(p-1)$-branes such that the angular difference between them is $\theta_2 - \theta_1$. The mass-squared for a string that is stretched between these branes receives a contribution equal to the length of the string multiplied by its tension, squared:

$$\begin{aligned}
M^2 &= \left((\theta_2 - \theta_1)\tilde{R}T_0\right)^2 + \frac{1}{\alpha'}(N^\perp - 1) \\
&= \left((\theta_2 - \theta_1)\frac{\alpha'}{R}\frac{1}{2\pi\alpha'}\right)^2 + \frac{1}{\alpha'}(N^\perp - 1) \\
&= \left(\frac{\theta_2 - \theta_1}{2\pi R}\right)^2 + \frac{1}{\alpha'}(N^\perp - 1),
\end{aligned} \tag{18.55}$$

which exactly matches (18.54). This is strong evidence for our physical interpretation of the θ parameter (18.42) as the angular position of D-branes in the T-dual configuration.

Two important points about T-duality are examined in the problems. The first concerns the fact that the string coupling constant must change under a T-duality transformation along a circle. We learned in Section 3.9 that Newton's constant in the effective lower-dimensional world is related to the higher-dimensional Newton constant and the volume of the extra dimensions. This volume changes under T-duality since one circle changes its radius. Moreover, the higher-dimensional Newton constant is set by the value of the string coupling and α' (see Section 13.4). If T-duality is a symmetry of the theory, then the lower-dimensional Newton constant must be unchanged. After all, this constant is observable. If the string coupling is g and the radius of the circle to be dualized is R, after T-duality the value $\tilde{g}$ of the string coupling is (Problem 18.5)

$$\tilde{g} = \frac{\sqrt{\alpha'}}{R}g. \tag{18.56}$$

The second point deals with the tension of D-branes (Problem 18.6). For a static string, the tension multiplied by the length gives the mass. D-branes also have tension and mass. For a static Dp-brane, the tension T_p of the brane multiplied by its volume V_p gives the mass of the brane. Our work with D-branes has assumed that they are fixed hyperplanes, or very heavy objects that are hardly affected by the open strings that are attached to them. In fact, the tension T_p of a D-brane goes to infinity as the string coupling $g \to 0$. The fact that the mass of a wrapped D-brane must be unchanged under T-duality will allow you to prove that

$$T_p(g) = \frac{\tau_p}{g} \quad \text{and} \quad T_{p-1}(g) = T_p(g) \cdot 2\pi \sqrt{\alpha'}. \tag{18.57}$$

Here τ_p is a p-dependent, but g-independent constant that includes a suitable factor of α' to give the correct units to the tension. The second relation implies that the tensions of all D-branes are related. It is conventional to choose the precise definition of the string coupling (see (13.85)) in such a way that the tension of the D1-brane is given by

$$T_1(g) = \frac{1}{2\pi\alpha'} \frac{1}{g}. \tag{18.58}$$

This formula is reminiscent of the string tension $1/(2\pi\alpha')$, which lacks the factor $1/g$. With the help of (18.57) the tensions of all D-branes are now determined. This formula also gives the correct value for a D0-brane, whose tension is simply its mass.

Quick calculation 18.3 Find the explicit p-dependent expression of $T_p(g)$. Confirm that it has the correct units.

Problems

Problem 18.1 Gauge invariance of the Schrödinger equation.

To prove the gauge invariance of the Schrödinger equation it is useful to note that, for any function M, and for U defined as in (18.21), we have

$$\left(\frac{\nabla}{i} - q\vec{A} - q\nabla\chi \right) UM = U\left(\frac{\nabla}{i} - q\vec{A} \right) M.$$

Prove this result and use it to show that (18.20) is equivalent to (18.18).

Problem 18.2 Explicit T-duality of DN string coordinates.

Consider the expansion (15.72) for an ND string coordinate. Find the separate left-moving and right-moving pieces, construct the dual string coordinate, and verify that it is of DN type.

Problem 18.3 T-duality invariance of the Hamiltonian.

Find the relations between $\dot{X} \pm X'$ and $\dot{\tilde{X}} \pm \tilde{X}'$. Use these relations, together with (15.8), to explain why the string Hamiltonian is unchanged under T-duality.

Problem 18.4 T-duality of a Dp-brane with an electric field.

Assume we have a constant electric field E_x on the world-volume of a Dp-brane that is wrapped around a circle of radius R in the x direction. Give the time-dependent A_x that corresponds to this electric field. Using the T-dual of a Dp-brane with a holonomy, show that the T-dual of the Dp-brane with an electric field is a D$(p-1)$-brane on a circle of dual radius, moving along the circle with a velocity $v_x = 2\pi \alpha' E_x$. As an additional consistency check, use the original world of the Dp-brane with the electric field to calculate how long the D$(p-1)$-brane takes to go around the dual circle.

Problem 18.5 T-duality and the string coupling.

Consider a D-dimensional spacetime where p spatial dimensions have been curled up. One of them is a circle of radius R, and the rest form a space of volume V_{p-1}. Let $\hat{G}$ denote Newton's constant in the $(D-p)$-dimensional effective spacetime. Write an expression for $\hat{G}$ in terms of the string coupling g, α', R, and V_{p-1}. Imagine now performing a T-duality along the circle of radius R. Show that the invariance of $\hat{G}$ requires changing the string coupling g into $\tilde{g}$, as given in (18.56).

Problem 18.6 D-brane tension and descent relations.

The tension T_p of a D-brane can be written in the form $T_p = \tau_p \, h(g)$, where g is the string coupling, h is a function to be determined below, and τ_p is a p-dependent constant that includes some suitable power of α'. Consider the setup of Problem 18.5, and imagine wrapping a Dp-brane around the compact dimensions. The mass of this object, perceived as a point particle by the lower-dimensional observer, must be unchanged under T-duality. Use this condition to show that

$$\tau_p \, h(g) \, 2\pi R = \tau_{p-1} \, h(g \sqrt{\alpha'}/R).$$

Use this equation to prove (18.57). The first relation in (18.57) requires absorbing a p-independent constant into the definition of τ_p.

19

Electromagnetic fields on D-branes

We now begin a study of D-branes that carry electric or magnetic fields on their world-volume. Open strings couple to these electromagnetic fields at their endpoints. Using the tools of T-duality we show that a D-brane with an electric field is physically equivalent to a moving D-brane with no electric field. The constraint that a D-brane cannot move faster than light implies that the strength of an electric field cannot exceed a certain maximum value. We also show that a Dp-brane with a magnetic field is T-duality equivalent to a tilted $D(p-1)$-brane with no magnetic field. Alternatively, the magnetic field on the Dp-brane can be thought of as being created by a distribution of dissolved $D(p-2)$-branes.

19.1 Maxwell fields coupling to open strings

Among the quantum states of open strings attached to a D-brane we found photon states with polarizations and momentum along the D-brane directions. We thus deduced that a Maxwell field lives on the world-volume of a D-brane. The existence of this Maxwell field was in fact necessary to preserve the gauge invariance of the term that couples the Kalb–Ramond field to the string in the presence of a D-brane. We also learned that the endpoints of open strings carry Maxwell charge.

Since any D-brane has a Maxwell field, it is physically reasonable to expect that background electromagnetic fields can exist: there may be electric or magnetic fields that permeate the D-brane. If the universe was the world-volume of a D-brane, an intergalactic magnetic field would be an example of a background field. It is customary to think of backgrounds as solutions of classical field equations. Historically, electromagnetism was first studied using familiar backgrounds, such as the magnetic field of the earth, or the static electric field of a charge. The study of these and other backgrounds led to a set of classical equations, Maxwell's equations, which gave these backgrounds as classical solutions. Eventually, physicists developed a quantum theory of electromagnetism, and this theory predicts photon states. Note that our string theory discovery of electromagnetism ran completely in reverse. We found string quantum states that could be identified as photon states, and through an analysis of the Schrödinger equations satisfied by the quantum states, we recovered the classical equations of electromagnetism. Now, we want to study background electromagnetic fields!

Candidate background fields can be tested to see if they are consistent with string theory. As we first discussed at the end of Section 16.1, to do so, we must re-quantize the string, taking into account the effects of the candidate background fields. If the quantization is successful then we may conclude that the candidate background fields are permitted in string theory. After all, a successful quantization establishes that quantum string states can consistently propagate on the given background. Although the quantizations we performed previously only applied to strings in the absence of nontrivial background fields, we already studied the effects of at least one background field. Our quantization of closed strings gave rise to Kalb–Ramond states, and as closed strings are not confined to D-branes, we concluded that Kalb–Ramond fields live on the full spacetime. Our previous discussion was therefore an analysis of nontrivial Kalb–Ramond backgrounds. In that discussion we introduced into the string action a new term, which couples the string to the Kalb–Ramond background field (see (16.3)). We did not re-quantize the closed string with this new term, except in the simple case considered in Problem 17.4. Rather, we examined the implications of the new coupling for string motion and gauge invariance.

In this chapter we study the effects of electromagnetic backgrounds on open strings. We will not embark on the details of the quantization; instead, we will assume that the quantization works out properly (as it does, for the backgrounds we will consider). In the present section we derive the equations of motion for open strings in the presence of background electromagnetic fields, and later we will use the tools of T-duality to gain new physical insights. The discussion of electromagnetic fields on D-branes continues in Chapter 20, where we will show that their dynamics is governed by the Born–Infeld theory of nonlinear electrodynamics.

We learned in Chapter 16 how to describe the coupling of Maxwell fields to strings. The string endpoints couple to the Maxwell potential A_m in the same way as a charged particle does. The coupling terms were given in equation (16.54); adding them to the string action gives

$$S = \int d\tau d\sigma \, \mathcal{L}(\dot{X}, X') + \int d\tau \, A_m(X) \frac{dX^m}{d\tau}\Big|_{\sigma=\pi} - \int d\tau \, A_m(X) \frac{dX^m}{d\tau}\Big|_{\sigma=0}. \quad (19.1)$$

Here $\mathcal{L}$ denotes the Nambu–Goto Lagrangian density. In our index convention $\mu, \nu, \ldots$ are spacetime indices, which run from 0 to d, while $m, n, \ldots$ are brane world-indices, which run from 0 to p. The index on the gauge potential A_m, for example, is a brane world-index. The indices $i, j, \ldots$ are spatial indices on the brane, which run from 1 to p, and $a, b, \ldots$ are indices for directions normal to the brane, which run from $p+1$ to d. We will only consider backgrounds for which the electromagnetic field strength F_{mn} is a constant. If the only nonvanishing entries are the $F_{0i}(= -F_{i0})$ then the background is purely electric. If the only nonvanishing entries are the F_{ij} then the background is purely magnetic. For a constant F_{mn} the gauge potentials can be chosen to be

$$A_n(x) = \frac{1}{2} F_{mn} \, x^m. \quad (19.2)$$

Quick calculation 19.1 Confirm that $\partial_m A_n - \partial_n A_m$ is equal to F_{mn}.

Making use of (19.2) the action S becomes

$$S = \int d\tau d\sigma \, \mathcal{L}(\dot{X}, X') + \frac{1}{2} \int d\tau \, F_{mn}\left(X^m \partial_\tau X^n |_{\sigma=\pi} - X^m \partial_\tau X^n |_{\sigma=0}\right). \qquad (19.3)$$

To find the equations of motion we use our earlier notation where $\mathcal{P}_\mu^\tau = \partial \mathcal{L}/\partial \dot{X}^\mu$ and $\mathcal{P}_\mu^\sigma = \partial \mathcal{L}/\partial X^{\mu\prime}$. Note that $\mathcal{P}_\mu^\tau$ is not the complete momentum conjugate to X^μ, since the present $\mathcal{L}$ does not include the contributions from the endpoints and is therefore not the complete Lagrangian density. Since the two endpoints enter into the string action almost symmetrically, we calculate the variation of the action by focusing only on the $\sigma = \pi$ endpoint:

$$\delta S = \int d\tau d\sigma \, \left(\mathcal{P}_\mu^\tau \partial_\tau \delta X^\mu + \mathcal{P}_\mu^\sigma \partial_\sigma \delta X^\mu \right)$$

$$+ \frac{1}{2} \int d\tau \, F_{mn}\left(\delta X^m \partial_\tau X^n + X^m \partial_\tau \delta X^n\right)\Big|_{\sigma=\pi} + \cdots, \qquad (19.4)$$

where the dots indicate terms contributed by the $\sigma = 0$ endpoint. The wave equation $\partial_\tau \mathcal{P}_\mu^\tau + \partial_\sigma \mathcal{P}_\mu^\sigma = 0$ continues to apply, but at the string endpoints there is a new constraint. The only effect of electromagnetic fields is a change in the boundary conditions. We can determine the boundary conditions by paying attention to the terms that involve variations at the string endpoints. Since total τ derivatives are not relevant to the equations of motion, the boundary terms that we should examine are

$$\delta S = \int d\tau d\sigma \, \partial_\sigma (\mathcal{P}_\mu^\sigma \delta X^\mu) + \int d\tau \delta X^m \, F_{mn} \partial_\tau X^n |_{\sigma=\pi} + \cdots. \qquad (19.5)$$

For the coordinates normal to the brane, we have the usual Dirichlet boundary condition $\delta X^a = 0$. Focusing now on the variations δX^m along the brane, we find

$$\delta S = \int d\tau \, \delta X^m \left(\mathcal{P}_m^\sigma + F_{mn} \partial_\tau X^n\right)\Big|_{\sigma-\pi} + \cdots. \qquad (19.6)$$

Since the X^m are coordinates on the brane, they carry free boundary conditions: no constraint can be imposed on the variations δX^m. As a result,

$$\mathcal{P}_m^\sigma + F_{mn} \partial_\tau X^n = 0, \quad \text{for} \quad \sigma = 0, \pi. \qquad (19.7)$$

Quick calculation 19.2 Convince yourself that (19.7) applies for $\sigma = 0$.

To simplify the boundary conditions we must choose a particular gauge. We impose the orthonormality conditions

$$\dot{X} \cdot X' = 0, \quad \dot{X}^2 + X'^2 = 0, \qquad (19.8)$$

which hold for both the static and light-cone gauges. We then have

$$\mathcal{P}_\mu^\sigma = -\frac{1}{2\pi\alpha'} \partial_\sigma X_\mu, \qquad (19.9)$$

and the boundary condition becomes

$$\partial_\sigma X_m - 2\pi\alpha' F_{mn}\partial_\tau X^n = 0, \quad \sigma = 0, \pi. \tag{19.10}$$

Let us look briefly at the physics that is contained in this boundary condition in the case of a pure magnetic field. Since $F_{0i} = 0$, there is no change in the boundary condition for X^0 – it is still Neumann. On the other hand, for spatial directions along which there is a magnetic field we get

$$\partial_\sigma X_i - 2\pi\alpha' F_{ij}\partial_\tau X^j = 0, \quad \sigma = 0, \pi. \tag{19.11}$$

This boundary condition is of mixed type; it is neither Neumann nor Dirichlet. Suppose that the only nonvanishing component of the magnetic field is $F_{23} = -F_{32} \equiv B$. The coordinates X_1 and X_i with $i > 3$ then satisfy Neumann boundary conditions, but X_2 and X_3 satisfy

$$\begin{aligned} \partial_\sigma X_2 - 2\pi\alpha' B\,\partial_\tau X^3 &= 0, \\ \partial_\sigma X_3 + 2\pi\alpha' B\,\partial_\tau X^2 &= 0. \end{aligned} \tag{19.12}$$

If B is very large, then the first term in each equation is negligible when compared to the second term. In this case the boundary conditions become, approximately,

$$\partial_\tau X^2 = \partial_\tau X^3 = 0, \quad \sigma = 0, \pi. \tag{19.13}$$

As $F_{23} = B$ becomes infinitely large, the motion of the string endpoints along the directions x^2 and x^3 of the brane is frozen! The string coordinates X^2 and X^3 have become Dirichlet. It is as if the original Dp-brane is now filled with infinitely many D$(p - 2)$-branes, one at each possible value of (x^2, x^3). The strings end on those D$(p - 2)$-branes and cannot change their positions in the (x^2, x^3)-plane. If we begin with a D2-brane, the motion of the string endpoints along the brane will be completely frozen. Although this seems like a rather bizarre interpretation of the boundary condition, we will see later that there is quantitative truth behind it.

19.2 D-branes with electric fields

In this section we will consider a D-brane with an electric field on its world-volume. We will not examine the classical solutions that describe the motion of an open string in this background (see, however, Problem 19.2); instead, our discussion will use T-duality to learn about the physics of electric fields. We will assume that the electric field is constant and points along a compact direction on the world-volume of a Dp-brane, and we will use T-duality to relate this configuration to one in which a D$(p - 1)$-brane is moving along the dual circle. This result was anticipated in Problem 18.4, but our analysis here will be more general. The constraint that the velocity of the D$(p - 1)$-brane cannot exceed the velocity of light will imply that the value of the electric field in the Dp-brane cannot exceed a critical value – electric fields on D-branes are bounded.

Our analysis will proceed in three steps. Our first step will be to find a user-friendly way of writing the boundary conditions at the string endpoints. We will then use the same language to express the equations that relate T-dual coordinates. Finally, applying a boost and a T-duality transformation we will prove the above equivalence.

Consider a Dp-brane that wraps around a compact dimension x^{25} of radius R, and assume that the brane carries an electric field along this direction:

$$F_{25,0} = E_{25} \equiv E. \tag{19.14}$$

Let us look at the string boundary conditions. The only interesting directions are X^0 and X^{25}, and from (19.10) we see that

$$\partial_\sigma X_0 - 2\pi\alpha' F_{0,25}\partial_\tau X^{25} = 0,$$
$$\partial_\sigma X_{25} - 2\pi\alpha' F_{25,0}\partial_\tau X^0 = 0. \tag{19.15}$$

Using (19.14), $X_0 = -X^0$, and writing $X^{25} \equiv X$, we find

$$\partial_\sigma X^0 - \mathcal{E}\,\partial_\tau X = 0,$$
$$\partial_\sigma X - \mathcal{E}\,\partial_\tau X^0 = 0, \tag{19.16}$$

where we define the dimensionless electric field $\mathcal{E}$ as

$$\mathcal{E} \equiv 2\pi\alpha'\,E. \tag{19.17}$$

Our next objective is to write the above boundary conditions in a more manageable form. We fit the two dynamical variables X^0 and X into a column vector and look for boundary conditions that take the form of *invertible linear relations* between derivatives of the column vector. The derivatives ∂_τ and ∂_σ are not appropriate for this. A Neumann boundary condition, for example, simply requires that the σ derivative of the column vector vanishes, and this is not an invertible linear relation between σ and τ derivatives of the column vector. To obtain relations of the desired form we introduce new partial derivatives:

$$\partial_+ \equiv \tfrac{1}{2}\,(\partial_\tau + \partial_\sigma), \quad \partial_- \equiv \tfrac{1}{2}\,(\partial_\tau - \partial_\sigma). \tag{19.18}$$

Solving for ∂_τ and ∂_σ, we obtain

$$\partial_\tau = \partial_+ + \partial_-, \quad \partial_\sigma = \partial_+ - \partial_-. \tag{19.19}$$

Rewriting the boundary conditions (19.16) in terms of $\partial_\pm$ and collecting the ∂_+ and ∂_- derivatives on the left- and right-hand sides, respectively, we find

$$\partial_+ X^0 - \mathcal{E}\,\partial_+ X = \partial_- X^0 + \mathcal{E}\,\partial_- X,$$
$$-\mathcal{E}\,\partial_+ X^0 + \partial_+ X = \mathcal{E}\,\partial_- X^0 + \partial_- X. \tag{19.20}$$

Solving for $\partial_+ X^0$ and $\partial_+ X$ gives

$$\partial_+ \begin{pmatrix} X^0 \\ X \end{pmatrix} = \begin{pmatrix} \dfrac{1+\mathcal{E}^2}{1-\mathcal{E}^2} & \dfrac{2\mathcal{E}}{1-\mathcal{E}^2} \\ \dfrac{2\mathcal{E}}{1-\mathcal{E}^2} & \dfrac{1+\mathcal{E}^2}{1-\mathcal{E}^2} \end{pmatrix} \partial_- \begin{pmatrix} X^0 \\ X \end{pmatrix}. \tag{19.21}$$

This is the desired form of the boundary conditions. These equations hold at the endpoints $\sigma = 0, \pi$. Note that the matrix above has unit determinant. Moreover, the entries of the matrix become singular when $\mathcal{E} = \pm 1$.

Quick calculation 19.3 Prove equation (19.21).

We also need to express in this new language the Neumann and Dirichlet boundary conditions, and the T-duality relations, as well. If a pair of coordinates X^0 and X satisfy Neumann boundary conditions $\partial_\sigma X^0 = \partial_\sigma X = 0$, then in terms of $\partial_\pm$ these conditions read

$$\partial_+ X^0 = \partial_- X^0,$$
$$\partial_+ X = \partial_- X, \tag{19.22}$$

or, in matrix form,

$$\partial_+ \begin{pmatrix} X^0 \\ X \end{pmatrix} = \begin{pmatrix} 1 & 0 \\ 0 & 1 \end{pmatrix} \partial_- \begin{pmatrix} X^0 \\ X \end{pmatrix} \quad \text{for} \quad \{X^0, X\} = \{\text{N}, \text{N}\}. \tag{19.23}$$

Thus, in terms of our linear relations, Neumann boundary conditions lead to an identity matrix. If X^0 is of Neumann type and $\tilde{X}$ is of Dirichlet type, then $\partial_\sigma X^0 = \partial_\tau \tilde{X} = 0$. In terms of ∂_+ and ∂_-, then,

$$\partial_+ X^0 = \partial_- X^0,$$
$$\partial_+ \tilde{X} = -\partial_- \tilde{X}, \tag{19.24}$$

or, in matrix form,

$$\partial_+ \begin{pmatrix} X^0 \\ \tilde{X} \end{pmatrix} = \begin{pmatrix} 1 & 0 \\ 0 & -1 \end{pmatrix} \partial_- \begin{pmatrix} X^0 \\ \tilde{X} \end{pmatrix} \quad \text{for} \quad \{X^0, \tilde{X}\} = \{\text{N}, \text{D}\}. \tag{19.25}$$

As we studied before (see (18.16)), the T-dual coordinate $\tilde{X}$ is obtained by changing the sign of the right movers in X:

$$X = X_L(\tau + \sigma) + X_R(\tau - \sigma),$$
$$\tilde{X} = X_L(\tau + \sigma) - X_R(\tau - \sigma). \tag{19.26}$$

Simple relations then follow for the $\partial_\pm$ derivatives:

$$\text{Duality relations:} \quad \partial_+ X = \partial_+ \tilde{X}, \quad \partial_- X = -\partial_- \tilde{X}. \tag{19.27}$$

Since T-duality exchanges Neumann and Dirichlet boundary conditions, (19.27) can be used to obtain equations (19.23) and (19.25) from one another. Equations (19.27) hold for all values of σ, including $\sigma = 0$ and $\sigma = \pi$.

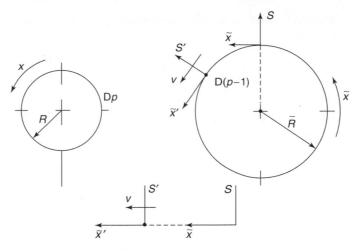

Fig. 19.1 On the left is a Dp-brane wrapped on a circle of radius R. On the right is a D(p − 1)-brane on the dual circle of radius $\tilde{R}$. The D(p − 1)-brane is at rest in the frame S', which is boosted with respect to the frame S.

We can now return to our present problem. We showed in Section 18.1 that a Dp-brane with one direction wrapped around a circle of radius R is T-dual to a world in which a D(p − 1)-brane is localized at some point on a circle of radius $\tilde{R} = \alpha'/R$. Our problem is to find the dual description of a Dp-brane which, in addition to being wrapped around a compact dimension, carries an electric field that points in this direction. The boundary conditions for such a Dp-brane are given by (19.21). We claim that the dual description of this configuration is a D(p − 1)-brane moving with constant velocity around the dual compact dimension. Our strategy will be to show that the T-dual of this moving brane configuration is described by boundary conditions that coincide with those in (19.21).

Let S be the frame which is at rest on the circle along which the D(p − 1)-brane is moving, and let S' be the rest frame of the D(p − 1)-brane (see Figure 19.1). The S' frame is boosted relative to the S frame by the boost parameter $\beta = v/c$, where v is the speed of the brane. In the S' frame the D(p − 1)-brane is at rest, so we can express the boundary conditions for strings ending on the D-brane in terms of the S' string coordinates. Let X'^0 and $\tilde{X}'$ denote the string coordinates in the S' frame (primes are not σ derivatives!). Since X'^0 is Neumann and $\tilde{X}'$ is Dirichlet, we can use equation (19.25) to write

$$\partial_+ \begin{pmatrix} X'^0 \\ \tilde{X}' \end{pmatrix} = \begin{pmatrix} 1 & 0 \\ 0 & -1 \end{pmatrix} \partial_- \begin{pmatrix} X'^0 \\ \tilde{X}' \end{pmatrix}. \tag{19.28}$$

We need the boundary conditions in the S frame, since this is the frame in which we know how to perform the T-duality transformation. To find them, we use the Lorentz boost

$$\begin{aligned} X'^0 &= \gamma(X^0 - \beta\tilde{X}), \\ \tilde{X}' &= \gamma(-\beta X^0 + \tilde{X}), \end{aligned} \tag{19.29}$$

where X^0 and $\tilde{X}$ are the string coordinates in the S frame and $\gamma = (1 - \beta^2)^{-1/2}$. In matrix language:

$$\begin{pmatrix} X'^0 \\ \tilde{X}' \end{pmatrix} = \gamma \begin{pmatrix} 1 & -\beta \\ -\beta & 1 \end{pmatrix} \begin{pmatrix} X^0 \\ \tilde{X} \end{pmatrix} \equiv M \begin{pmatrix} X^0 \\ \tilde{X} \end{pmatrix}, \qquad (19.30)$$

where we defined the constant matrix M. Substituting (19.30) into (19.28), noting that M commutes with the partial derivatives, and multiplying both sides of the equation by M^{-1}, we get

$$\partial_+ \begin{pmatrix} X^0 \\ \tilde{X} \end{pmatrix} = M^{-1} \begin{pmatrix} 1 & 0 \\ 0 & -1 \end{pmatrix} M \, \partial_- \begin{pmatrix} X^0 \\ \tilde{X} \end{pmatrix}. \qquad (19.31)$$

This equation gives the string boundary conditions in the S frame.

Now we can perform a T-duality transformation on the $\tilde{X}$ coordinate. Using the duality relations (19.27), we see that the $\tilde{X}$ on the left-hand side can be changed into X, and the $\tilde{X}$ on the right-hand side can be changed into $(-X)$, if we include another matrix:

$$\partial_+ \begin{pmatrix} X^0 \\ X \end{pmatrix} = M^{-1} \begin{pmatrix} 1 & 0 \\ 0 & -1 \end{pmatrix} M \begin{pmatrix} 1 & 0 \\ 0 & -1 \end{pmatrix} \partial_- \begin{pmatrix} X^0 \\ X \end{pmatrix}. \qquad (19.32)$$

A small calculation now gives

$$\partial_+ \begin{pmatrix} X^0 \\ X \end{pmatrix} = \begin{pmatrix} \dfrac{1 + \beta^2}{1 - \beta^2} & \dfrac{2\beta}{1 - \beta^2} \\[2mm] \dfrac{2\beta}{1 - \beta^2} & \dfrac{1 + \beta^2}{1 - \beta^2} \end{pmatrix} \partial_- \begin{pmatrix} X^0 \\ X \end{pmatrix}. \qquad (19.33)$$

These are the open string boundary conditions for the theory dual to the moving $D(p-1)$-brane. As promised, they coincide with those in (19.21), which were written for a Dp-brane carrying an electric field, if we set

$$\mathcal{E} = 2\pi\alpha' E = \beta. \qquad (19.34)$$

Our main result can thus be summarized:

a $D(p-1)$-brane that is moving with velocity parameter β on a circle is T-dual to a Dp-brane that is wrapped on the dual circle and that carries an electric field $\mathcal{E} = \beta$ along the direction of the circle.

Furthermore, since no object with a rest mass can reach the speed of light, $\beta < 1$, and we discover that the electric field is bounded:

$$|E| = \frac{|\beta|}{2\pi\alpha'} < \frac{1}{2\pi\alpha'} = E_{\text{crit}}. \qquad (19.35)$$

E_{crit} denotes the critical electric field, the maximum value of an electric field on a D-brane (in the absence of magnetic fields). Interestingly, the critical electric field coincides with the string tension:

$$E_{\text{crit}} = T_0. \qquad (19.36)$$

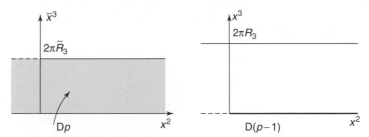

Fig. 19.2   A Dp-brane stretched on a cylinder of circumference $2\pi\tilde{R}_3$. By performing a T-duality transformation on $\tilde{X}^3$, we obtain a D$(p-1)$-brane stretched along the axis of a dual cylinder.

We can give an intuitive explanation for this equality by considering the motion of an open string in an electric field. As we have seen, an open string has charges of value ± 1 at the endpoints. Two forces that must cancel out act on each (zero mass) endpoint: the electric force of magnitude E and the effective tension $T_0(1 - v_\perp^2)^{1/2}$ (see (7.15)). For $E < T_0$ the two forces can be balanced for a suitable endpoint velocity $v_\perp$. When $E = T_0$ the endpoints must stop moving. Finally, for $E > T_0$ the forces at the endpoints cannot be balanced. This discussion is nicely illustrated in the detailed analysis of Problem 19.2.

19.3 D-branes with magnetic fields

We will now explore the properties of D-branes that carry magnetic fields on their world-volume. Again, our main tool will be T-duality, and we will gain considerable insight by constructing the T-dual version of a D-brane that is carrying a background magnetic field. The motion of open strings in a background magnetic field is of interest, but this subject is relegated to Problem 19.5. There you will show that an open string develops an electric dipole moment in a direction orthogonal to the motion; the magnitude of the electric dipole is proportional to both the magnetic field and the momentum of the string.

Consider a Dp-brane for which two directions of its world-volume lie on the $(x^2, \tilde{x}^3)$ plane. Assume that the dimension $\tilde{x}^3$ is compactified into a circle of radius $\tilde{R}_3$, so that x^2 and $\tilde{x}^3$ together define a cylinder of circumference $2\pi\tilde{R}_3$ (see Figure 19.2). The open string coordinates will be denoted by X^2 and $\tilde{X}^3$, and both of them are Neumann. If we perform a T-duality transformation on the string coordinate $\tilde{X}^3$, then the dual coordinate X^3 will live on a circle of radius $R_3 = \alpha'/\tilde{R}_3$. The coordinate X^3 is Dirichlet, and we now have a D$(p-1)$-brane stretching along x^2 at a fixed position $x^3 = 0$. This dual picture is shown on the right side of Figure 19.2. On the dual cylinder, the D$(p-1)$-brane appears as a line that runs parallel to the axis of the cylinder.

Now suppose that there is a nonvanishing magnetic field $F_{23} = B$ on the Dp-brane. What happens in the dual world? It turns out that the D$(p-1)$-brane tilts by an angle. Electric fields appear as boosts in the dual world; magnetic fields appear as rotations! To

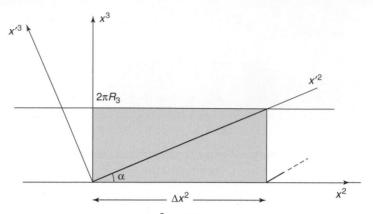

Fig. 19.3   The D-brane lies along the rotated axis x'^2. Δx^2 is the distance that one must move along
the length of the cylinder for the D-brane to have wrapped once along the compact vertical
direction. The rotation angle α is related to the magnetic field on the T-dual D-brane.

demonstrate this, we begin by writing the boundary conditions for open strings that end on
the Dp-brane that carries the magnetic field. From (19.12) we have

$$\partial_\sigma X^2 - \mathcal{B}\partial_\tau \tilde{X}^3 = 0,$$
$$\partial_\sigma \tilde{X}^3 + \mathcal{B}\partial_\tau X^2 = 0, \tag{19.37}$$

where we define the dimensionless magnetic field $\mathcal{B}$ as

$$\mathcal{B} \equiv 2\pi\alpha' B. \tag{19.38}$$

In terms of ∂_+ and ∂_- the boundary conditions take the form

$$\partial_+ \begin{pmatrix} X^2 \\ \tilde{X}^3 \end{pmatrix} = \begin{pmatrix} \dfrac{1-\mathcal{B}^2}{1+\mathcal{B}^2} & \dfrac{2\mathcal{B}}{1+\mathcal{B}^2} \\ -\dfrac{2\mathcal{B}}{1+\mathcal{B}^2} & \dfrac{1-\mathcal{B}^2}{1+\mathcal{B}^2} \end{pmatrix} \partial_- \begin{pmatrix} X^2 \\ \tilde{X}^3 \end{pmatrix}. \tag{19.39}$$

These boundary conditions encode all the effects of the magnetic field. The equations of
motion for the string coordinates are the usual wave equations, which are unchanged by
the presence of the magnetic field.

Quick calculation 19.4 Show that (19.39) follows from (19.37).

Consider now a D$(p-1)$-brane that is tilted on the cylinder, as illustrated in Figure 19.3.
This brane wraps around the cylinder as it advances along its length. If x^3 were not compactified, then a tilted brane would be physically equivalent to a horizontal brane. But with
the compactification of x^3 the tilting of the D-brane has consequences. Assume that the
tilting angle is α when the magnetic field is B. Our goal is to calculate α as a function of
B. The boundary conditions for the tilted D-brane are easily expressed in terms of a coordinate system x'^2, x'^3 that is rotated by the angle α. In this frame the D-brane coincides

with the x'^2 axis. Since X'^2 and X'^3 are Neumann and Dirichlet, respectively, equation (19.25) applies:

$$\partial_+ \begin{pmatrix} X'^2 \\ X'^3 \end{pmatrix} = \begin{pmatrix} 1 & 0 \\ 0 & -1 \end{pmatrix} \partial_- \begin{pmatrix} X'^2 \\ X'^3 \end{pmatrix}. \tag{19.40}$$

Now we perform a rotation in order to change coordinates to the unprimed frame. It is easy to figure out the proper rotation matrix from the geometry of the problem:

$$\begin{pmatrix} X'^2 \\ X'^3 \end{pmatrix} = \begin{pmatrix} \cos\alpha & \sin\alpha \\ -\sin\alpha & \cos\alpha \end{pmatrix} \begin{pmatrix} X^2 \\ X^3 \end{pmatrix} \equiv R \begin{pmatrix} X^2 \\ X^3 \end{pmatrix}, \tag{19.41}$$

where we defined the rotation matrix R. Back in (19.40) we then find

$$\partial_+ \begin{pmatrix} X^2 \\ X^3 \end{pmatrix} = R^{-1} \begin{pmatrix} 1 & 0 \\ 0 & -1 \end{pmatrix} R\, \partial_- \begin{pmatrix} X^2 \\ X^3 \end{pmatrix}. \tag{19.42}$$

We now perform the duality transformation that takes X^3 to $\tilde{X}^3$. This is done by including an additional matrix in the right-hand side above:

$$\partial_+ \begin{pmatrix} X^2 \\ \tilde{X}^3 \end{pmatrix} = R^{-1} \begin{pmatrix} 1 & 0 \\ 0 & -1 \end{pmatrix} R \begin{pmatrix} 1 & 0 \\ 0 & -1 \end{pmatrix} \partial_- \begin{pmatrix} X^2 \\ \tilde{X}^3 \end{pmatrix}. \tag{19.43}$$

Multiplying the matrices together, we finally get

$$\partial_+ \begin{pmatrix} X^2 \\ \tilde{X}^3 \end{pmatrix} = \begin{pmatrix} \cos 2\alpha & -\sin 2\alpha \\ \sin 2\alpha & \cos 2\alpha \end{pmatrix} \partial_- \begin{pmatrix} X^2 \\ \tilde{X}^3 \end{pmatrix}. \tag{19.44}$$

Now we can compare this result with (19.39). There is a clear similarity between the two matrices: in both cases the two terms on the diagonal are the same, and the off-diagonal terms differ by only a sign. Let us solve for B using the diagonal terms, and then we will check that for this solution the off-diagonal terms are correct. We have

$$\frac{1 - B^2}{1 + B^2} = \cos 2\alpha, \tag{19.45}$$

which gives

$$B^2 = \frac{1 - \cos 2\alpha}{1 + \cos 2\alpha} = \frac{1 - (1 - 2\sin^2\alpha)}{1 + (2\cos^2\alpha - 1)} = \tan^2\alpha, \tag{19.46}$$

and thus we have $B = \pm\tan\alpha$. We will check below that the negative sign is the correct one, so

$$\mathcal{B} = 2\pi\alpha' B = -\tan\alpha. \tag{19.47}$$

A zero magnetic field produces no rotation, and it requires an infinite magnetic field to rotate the D-brane by an angle of ninety degrees. Finally, we can use the value of B in (19.47) to confirm that the off-diagonal entries in the boundary conditions work out:

$$\frac{2B}{1 + B^2} = -\frac{2\tan\alpha}{\sec^2\alpha} = -\sin 2\alpha, \tag{19.48}$$

as expected, and confirming that we chose the correct sign in (19.47). Since the boundary conditions match perfectly, we have proven that the tilted D-brane is the dual version of the D-brane with the magnetic field. Moreover, we have found the precise relation between the angle α and the magnetic field B.

To learn more from this problem, we now examine the vector potentials in the Dp-brane picture. We must find potentials A_2 and A_3 that give $F_{23} = \partial_2 A_3 - \partial_3 A_2 = B$. Let us choose

$$A_2 = 0, \quad A_3 = B\, x^2. \tag{19.49}$$

We would have found the same B had we chosen instead $A_2 = -B\, \tilde{x}^3$ and $A_3 = 0$, but we would then have had some complications because $\tilde{x}^3$ is not a well defined coordinate.

The potential A_3 lies along a compact dimension and is therefore a periodic quantity (see Section 18.3). Using (18.45) with $q = 1$, we see that

$$A_3 \sim A_3 + \frac{n}{\tilde{R}_3}, \quad n \in \mathbb{Z}. \tag{19.50}$$

This identification implies that a change in A_3 will have no physical effect if this change is quantized in units of the inverse radius $\tilde{R}_3$. It follows that the linear growth of A_3 along x^2 encodes some periodicity along the x^2 direction. The configuration must be invariant under any displacement Δx^2 that satisfies

$$\Delta A_3 = B \Delta x^2 = \frac{n}{\tilde{R}_3}. \tag{19.51}$$

The interpretation of this periodicity is quite striking in the dual world. Using dual variables, recalling (19.47), and letting $n \to -n$ for convenience, we see that

$$\Delta x^2 = -\frac{n R_3}{\alpha' B} = -\frac{2\pi n R_3}{2\pi \alpha' B} = \frac{2\pi n R_3}{\tan \alpha}. \tag{19.52}$$

It is convenient to rewrite this equation as

$$\tan \alpha = \frac{2\pi n R_3}{\Delta x^2}, \quad n \in \mathbb{Z}. \tag{19.53}$$

The smallest displacement Δx^2 which leads to repetition corresponds to $n = 1$:

$$\tan \alpha = \frac{2\pi R_3}{\Delta x^2}. \tag{19.54}$$

As you can see in Figure 19.3, since the D-brane is at an angle α, it completes a full wrapping of the x^3 direction precisely after moving a distance Δx^2. In this dual picture things repeat explicitly each time the D-brane completes a wrapping of the compact dimension. This provides a concrete realization of the periodicity property of the gauge potential in the original picture.

Since the relevant physics repeats along the x^2 axis, we can compactify this direction into a circle of radius R_2, where $2\pi R_2$ is the smallest repetition length:

$$2\pi R_2 = \Delta x^2 = \frac{2\pi R_3}{\tan \alpha}. \tag{19.55}$$

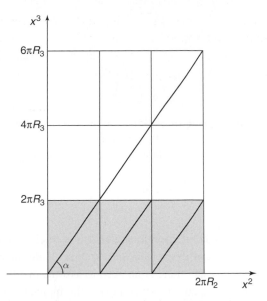

A D-brane wrapped around a torus of radii R_2 and R_3. The fundamental domain of the torus is shaded. The D-brane is tilted at an angle α such that, as it goes around the x^2 circle, it wraps three times around the x^3 circle.

This turns the cylinder around which the D-brane wraps into a torus with radii R_2 and R_3. The D-brane, once infinitely long, is now of finite length. It wraps diagonally across the fundamental domain of the torus, which is shown as shaded in Figure 19.3. Equivalently, as it wraps once around x^2, it also wraps once around x^3. The Dp-brane world also has a torus, with radii R_2 – since we do not T-dualize in this direction – and $\tilde{R}_3$.

Suppose now that on the torus where the D-brane wraps diagonally we change the angle α in such a way that

$$\tan \alpha = n \, \frac{2\pi R_3}{2\pi R_2} = n \, \frac{R_3}{R_2}, \tag{19.56}$$

with $n > 1$. In Figure 19.4 we show the case $n = 3$. Equation (19.56) represents the situation in which the D-brane wraps n times around the x^3 direction as it wraps once around the x^2 direction. What is the significance of the integer n in the dual world with the magnetic field? To find out, we calculate the total magnetic flux Φ on the torus. The flux is simply the magnetic field multiplied by the area of the fundamental domain:

$$\Phi = B \, (2\pi \tilde{R}_3) \, (2\pi R_2). \tag{19.57}$$

Using the value of the magnetic field (19.47), writing $\tilde{R}_3$ in terms of R_3, and using (19.56), we can simplify the expression for the flux:

$$\Phi = -\frac{\tan \alpha}{2\pi \alpha'} \, \frac{2\pi \alpha'}{R_3} \, 2\pi R_2 = -2\pi \, \frac{R_2}{R_3} \, \tan \alpha = -2\pi n. \tag{19.58}$$

This means that the magnetic flux is quantized! Since B is uniform, it must be quantized as well. Geometrically, this quantization emerges because, on a given torus, a D-brane

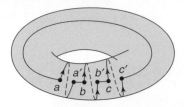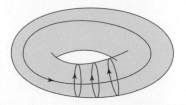

Fig. 19.5 A D-brane wrapping three times around the short circle of a torus as it travels along the long circle of the torus. In the limit that $a \to a'$, $b \to b'$, and $c \to c'$, the figure becomes one in which there are three D-branes wrapped around the short circle and one D-brane wrapped around the long circle.

can only be tilted at specific angles if it is to close after one winding in the x^2 direction. Actually, one can derive the quantization of the magnetic flux on a torus directly (see Problem 19.3), and there is no need to assume that the magnetic field is constant over the torus. What happens in our context if the magnetic field is not constant? A nonconstant, but static, magnetic field does not satisfy the relevant equations of electromagnetism, so this makes the question about the T-dual configuration somewhat ambiguous. The T-dual D-brane on the dual torus is not expected to appear as a straight line but rather as a curve. Because of the brane tension, this configuration cannot be static either.

There is a sense in which a tilted $D(p-1)$-brane that wraps n times around x^3 as it wraps once around x^2 is in the same class as a configuration composed of one $D(p-1)$-brane that only wraps in the x^2 direction and n $D(p-1)$-branes that only wrap in the x^3 direction:

$$D(p-1)\text{-brane (along } x^2) \quad \text{and} \quad n\, D(p-1)\text{-branes (along } x^3). \qquad (19.59)$$

Consider Figure 19.5, which shows the torus when $n = 3$. We have deformed the D-brane so that it is no longer a straight line. Most of the D-brane trajectory is horizontal except for the wrappings around x^3, which now occur rapidly. Since we can achieve this configuration by a continuous deformation of the original straight D-brane, this must preserve the quantized flux on the dual brane, although the B field would not be expected to remain constant. In the limit that $a \to a'$, $b \to b'$, and $c \to c'$, the D-brane turns into a single horizontal D-brane and three vertical D-branes. In general, we have the set of branes indicated in (19.59). The configuration in (19.59) may be static since the branes are once again straight lines on the torus. The original tilted D-brane configuration and the final configuration (19.59) are related by a deformation process. They are not physically equivalent configurations; instead, they are deformation equivalent, meaning that they can be deformed into each other. The precise mathematical statement is that the curves defining the two brane configurations are homologically equivalent.

Now consider performing, once again, a T-duality transformation in the x^3 direction. Previously, T-duality turned the tilted $D(p-1)$-brane into a Dp-brane with a magnetic field. This time the result of T-duality on (19.59) is different. The $D(p-1)$-brane along x^2 becomes a Dp-brane in the dual world. The n $D(p-1)$-branes in the x^3 direction, however, become n D($p-2$)-branes, located at certain fixed points on the torus. So in the dual world we have one Dp-brane and n D($p-2$)-branes:

$$\text{D}p\text{-brane} \quad \text{and} \quad n\,\text{D}(p-2)\text{-branes}. \tag{19.60}$$

This shows that the original Dp-brane with a constant magnetic field is deformation equivalent to a Dp-brane with a number n of D($p-2$)-branes on its world-volume. A D2-brane with a constant magnetic field of flux $2\pi n$, for example, is deformation equivalent to a D2-brane with n D0-branes on its world-volume. The deformation that takes the latter into the former is the process where the D0-branes *dissolve* into the torus. The physical picture is clear: D0-branes on a D2-brane represent a magnetic field that vanishes everywhere except at the positions of the D0-branes, where it has infinite magnitude but finite flux. A constant magnetic field with the same flux is possible, and the spread-out magnetic field represents the dissolved D0-branes. The technical implication of the word dissolved is that the resulting configuration is a bound state: it has less energy than the total energy of the constituent D2- and D0-branes taken separately. The energies actually add in quadrature, as you will prove in Problem 19.4. An extremely strong magnetic field can be viewed as a D2-brane with infinitely many dissolved D0-branes. This was the picture suggested by a direct analysis of the boundary conditions at the end of Section 19.1.

Problems

Problem 19.1 Constant electromagnetic fields on D-branes from constant $B_{\mu\nu}$ in spacetime.

We would like to show that a Dp-brane with a constant electromagnetic background field $\bar{F}_{mn}$ is equivalent to a Dp-brane with $F_{mn} = 0$ but in a spacetime with a constant Kalb–Ramond field $B_{mn} = \bar{F}_{mn}$. The indices m, n are brane indices.

(a) Refer to Section 16.3, and consider a situation where $B_{mn} = \bar{F}_{mn}$ is a nonzero constant and $F_{mn} = 0$. Find an explicit gauge parameter for which the transformed fields are $B_{mn} = 0$ and $F_{mn} = \bar{F}_{mn}$. Note that, as expected, the gauge invariant field strength $\mathcal{F}_{mn}$ is unchanged.

(b) Let B_{mn} be a constant Kalb–Ramond field. Consider the action (16.45), and show that the integrand is a total derivative. Drop the total τ derivatives, and show that for open strings the total σ derivatives contribute

$$S_B = \frac{1}{2} \int d\tau\, B_{mn} \Big(X^m \partial_\tau X^n |_{\sigma=\pi} - X^m \partial_\tau X^n |_{\sigma=0} \Big).$$

Compare with (19.3) and comment.

Problem 19.2 Motion of an open string in a constant electric field.

Consider the motion of an open string in the background of an electric field of magnitude $E = \mathcal{E}/2\pi\alpha'$ that points in the x^1 direction. When $\mathcal{E} \to 0$ the motion is rigid rotation in the (x^2, x^3) plane. We use string coordinates $X^\mu = (X^0, X^1, \vec{X})$ where $\vec{X} = (X^2, X^3)$.

All other coordinates are set equal to zero. The classical solution must satisfy the boundary conditions

$$\partial_\sigma X^0 - \mathcal{E} \, \partial_\tau X^1 = 0, \quad \partial_\sigma X^1 - \mathcal{E} \, \partial_\tau X^0 = 0, \quad \partial_\sigma \vec{X} = 0, \tag{1}$$

as well as the wave equations and constraints:

$$(\partial_\tau^2 - \partial_\sigma^2) X^\mu = 0, \quad \dot{X} \cdot X' = 0, \quad \dot{X}^2 + X'^2 = 0. \tag{2}$$

We look for a solution in which the string rotates around the x^1 axis with angular frequency ω but is not restricted to lie on the (x^2, x^3) plane. Construct the solution beginning with the ansatz

$$(X^0, X^1, X^2, X^3) = \left(\alpha\,\tau, \; \beta(\sigma - \tfrac{\pi}{2}), \; \gamma \cos\sigma \cos\tau, \; \gamma \cos\sigma \sin\tau \right). \tag{3}$$

(a) Examine equations (1) and (2) and fix the constants α, β, and γ in terms of ω and $\mathcal{E}$.

(b) Give $\vec{X}(t, \sigma)$, and show that the velocity of the endpoints is $v = \sqrt{1 - \mathcal{E}^2}$. Show also that the string is stretched along the x^1 axis a total distance $\Delta X^1 = \pi \mathcal{E}/\omega$.

(c) Plot the string in the (x^1, x^2) plane at $t = 0$. Show that at the endpoints, the string points in the direction of the electric field. Interpret this result by verifying that the effective string tension cancels the electric force at the endpoints. Show that at the origin

$$\frac{dX^2}{dX^1}\bigg|_{X^1=0} = -\sqrt{\frac{1}{\mathcal{E}^2} - 1}. \tag{4}$$

Does this make sense for $0 \le \mathcal{E} \le 1$? Describe the solution in the limit as $\mathcal{E} \to 1$ with ω finite.

(d) Prove that the string energy $U = \int T_0 ds (1 - v_\perp^2)^{-1/2}$ is equal to $\pi T_0/\omega$, for all values of the electric field.

The front cover of the book shows three surfaces generated by the motion of open strings. The surfaces are obtained for $\mathcal{E} = 0.3, 0.6$, and 0.95. Construct the associated plot of the three strings at $t = 0$ on the (x^1, x^2) plane, setting $\omega = 1$.

Problem 19.3 Quantization of magnetic flux on two-tori.

In this problem you will show that the flux Φ of the magnetic field on a two-torus is quantized: $\Phi = 2\pi n$, with $n \in \mathbb{Z}$. This quantization emerges when we try to construct a consistent vector potential. We assume, of course, that the Maxwell gauge group is $U(1)$ and, as explained in Section 18.2, the gauge parameter is $U = \exp(i\chi)$ (with $q = c = \hbar = 1$).

Consider a two-torus of length L_x along the x axis and length L_y along the y axis. Let $F_{12} = \partial_x A_y - \partial_y A_x = B$, where B is the value of the magnetic field.

(a) Assume the magnetic field $B = B_0$ is constant. Take $A_y(x, y) = B_0 x$ and $A_x = 0$. Note that $A_y(x + L_x, y) \ne A_y(x, y)$. Allow for $A_y(x + L_x, y)$ to be gauge equivalent to $A_y(x, y)$. Show that the condition of having a well defined gauge parameter leads to the desired quantization.

(b) Consider the flux integral of the magnetic field (not necessarily a constant one), and use Stokes' theorem to relate it to a line integral of the vector potential around the four sides of the rectangle $0 \leq x \leq L_x, 0 \leq y \leq L_y$. Use the gauge transformation properties of gauge-field line integrals to argue that the magnetic flux is properly quantized.

Problem 19.4 Dissolved $D(p-2)$-branes.

Consider a Dp-brane that is wrapped on a torus with radii R_2 and $\tilde{R}_3$ and carries a magnetic field with flux $|\Phi| = 2\pi n$ (as in Section 19.3). The other $p-2$ dimensions on the brane are wrapped around a compact space of volume V_{p-2}. Assume that the string coupling constant takes the value g. The energy E of this configuration is equal to the energy of the T-dual configuration (dualized along $\tilde{R}_3$) where a $D(p-1)$ brane wraps along the dual torus. Show that

$$E = T_{p-1}(\tilde{g}) \, V_{p-2} \sqrt{(2\pi R_2)^2 + (2\pi R_3 n)^2}, \tag{1}$$

where $\tilde{g}$ is the string coupling in the dual configuration. Now rewrite E in terms of variables in the original picture with the Dp-brane and the magnetic field, showing that

$$E = \sqrt{M_p^2 + (nM_{p-2})^2}, \tag{2}$$

where M_p is the mass of the Dp-brane and M_{p-2} is the mass of a $D(p-2)$ brane. Explain why this result shows that the energy E is smaller than the sum of the energies of the Dp-brane and n $D(p-2)$-branes. This justifies the description of the configuration as one where n $D(p-2)$-branes have *dissolved* inside the Dp-brane.

Problem 19.5 Motion of an open string in a constant magnetic field.

Consider the following expansion for a spatial open string coordinate on the world-volume of a D-brane with a constant magnetic field:

$$X_i(\tau, \sigma) = x_i + 2\alpha'(p_i \tau + 2\pi\alpha' F_{ij} \, p^j \sigma). \tag{1}$$

The coordinates X_i clearly solve the wave equations. Moreover, assume that

$$X^0 = 2\alpha' p^0 \tau. \tag{2}$$

(a) Prove that, for a calculable value of p^0 that you should determine, the ansatz in (1) and (2) satisfies: the boundary conditions (19.11), the constraint $\dot{X} \cdot X' = 0$, and the constraint $\dot{X}^2 + X'^2 = 0$.

(b) Let $\Delta X_i = X_i(\tau, \pi) - X_i(\tau, 0)$ denote the "span" of the string. Verify that

$$\Delta X_i = (2\pi\alpha')^2 F_{ij} \, p^j \quad \text{and} \quad p^i \Delta X_i = 0.$$

Since the electric dipole of an open string is a vector parallel to its span (why?), we conclude that the dipole is perpendicular to the momentum.

(c) For motion in three space dimensions, we have $F_{ij} = \epsilon_{ijk} B_k$. Show that

$$\Delta \vec{X} = (2\pi\alpha')^2 \, \vec{p} \times \vec{B}.$$

The dipole moment is orthogonal to both $\vec{B}$ and $\vec{p}$.

(d) Let $\mathcal{B}_k = 2\pi\alpha' B_k$. Show that the velocity of the string is given by

$$\vec{v} = \frac{\vec{n}}{\sqrt{1 + \vec{\mathcal{B}} \cdot \vec{\mathcal{B}} - (\vec{\mathcal{B}} \cdot \vec{n})^2}},$$

where $\vec{n}$ is a unit vector in the direction of the momentum. Note that when $B_k \to 0$ we have $|\vec{v}| \to 1$, so that the string approaches the speed of light.

(e) In flat space the dot product of two vectors $\vec{a}$ and $\vec{b}$ is defined by $\vec{a} \cdot \vec{b} = a^i b^j \eta_{ij} = a^i b^i$. Now introduce a new product $*$ defined by

$$\vec{a} * \vec{b} = a^i b^j \, \bar{\eta}_{ij}, \text{ where } \bar{\eta}_{ij} \equiv \eta_{ij} (1 + \vec{\mathcal{B}} \cdot \vec{\mathcal{B}}) - \mathcal{B}_i \mathcal{B}_j.$$

Show that the mass-shell condition for the string can be written as

$$-(p^0)^2 + \vec{p} * \vec{p} = 0.$$

Moreover, show that the velocity of the string satisfies

$$\vec{v} * \vec{v} = 1.$$

The metric $\bar{\eta}_{ij}$ is called the open string metric. It is a natural metric for the physics of open strings in the presence of a magnetic field. In the open string metric, the motion studied here shares the properties of free open string motion.

20 Nonlinear and Born–Infeld electrodynamics

We introduce nonlinear theories of electrodynamics, which generalize the linear Maxwell theory. Born–Infeld electrodynamics is a specific theory of nonlinear electrodynamics with particularly nice properties. It incorporates maximal electric fields, and in it point charges have finite electrostatic self-energy. We use T-duality arguments to explain why electromagnetic fields on the world-volumes of D-branes are governed by Born–Infeld theory.

20.1 The framework of nonlinear electrodynamics

Maxwell's equations are both the basis for classical electromagnetism and the starting point for the formulation of quantum electrodynamics, a theory that has been tested to a high degree of accuracy. Maxwell's equations are written in terms of electric and magnetic fields, which in turn arise from gauge potentials. Charges and currents are the sources in Maxwell's equations.

A somewhat different version of Maxwell's equations is used in the study of electromagnetic phenomena in the presence of materials. In this case, the materials contribute polarization charges to the charge density and magnetization currents to the current density. The original Maxwell equations hold, but one must include these contributions to the charges and currents. This is done efficiently by introducing, in addition to $\vec{E}$ and $\vec{B}$, the fields $\vec{D}$ and $\vec{H}$. The equations of electromagnetism in the presence of materials then take the form

$$\nabla \times \vec{E} = -\frac{1}{c}\frac{\partial \vec{B}}{\partial t} \,,$$
$$\nabla \cdot \vec{B} = 0 \,, \tag{20.1}$$

for the equations without sources, and

$$\nabla \cdot \vec{D} = \rho \,,$$
$$\nabla \times \vec{H} = \frac{\vec{j}}{c} + \frac{1}{c}\frac{\partial \vec{D}}{\partial t} \,, \tag{20.2}$$

for the equations with sources. Here ρ and $\vec{j}$ are called *free* sources. This means that they *do not* take into account polarization charges or magnetization currents. The free sources are charges and currents that are not bound to the materials. The contributions of polarization and magnetization have been incorporated through the fields $\vec{D}$ and $\vec{H}$. In fact, given a material, there are phenomenological relations that express $\vec{D}$ and $\vec{H}$ in terms of $\vec{E}$ and $\vec{B}$:

$$\vec{D} = \vec{D}(\vec{E}, \vec{B}), \quad \vec{H} = \vec{H}(\vec{E}, \vec{B}).\qquad(20.3)$$

Without these relations the two sets of equations in (20.1) and (20.2) would be unrelated. Linear dielectrics, for example, obey $\vec{D} = \epsilon\,\vec{E}$ (and $\vec{H} = \vec{B}$), where the constant ϵ is the electric permittivity. Linear magnetic materials obey $\vec{B} = \mu\vec{H}$ (and $\vec{D} = \vec{E}$), where the constant μ is the magnetic permeability. For more complicated materials the relations need not be linear. In such cases, equations (20.1), (20.2), and (20.3) define a nonlinear theory of electrodynamics.

In introductory courses on electromagnetism the point is often made that the "fundamental" Maxwell equations are those in terms of $\vec{E}$ and $\vec{B}$, together with the full charge and current densities. The equations above are regarded to be of limited validity, as they deal with materials, most of which do not generally lend themselves to exact analysis. Born–Infeld and related nonlinear theories of electrodynamics in fact suggest that the above equations are as fundamental as the original Maxwell equations, if not more so. These theories do not aim to describe electromagnetism in the presence of materials but rather electromagnetism in the vacuum. The point is that in nonlinear electrodynamics the vacuum itself behaves as some kind of material. As we will show, general Lagrangians lead directly to equations (20.2), together with nontrivial relationships between $(\vec{D}, \vec{H})$ and $(\vec{E}, \vec{B})$.

In nonlinear electrodynamics the electromagnetic fields $\vec{E}$ and $\vec{B}$ are still encoded in the field strength $F_{\mu\nu} = \partial_\mu A_\nu - \partial_\nu A_\mu$, as shown in equation (3.20). Using spatial indices i, j, we write

$$F_{i0} = E_i, \quad F_{ij} = \epsilon_{ijk} B_k.\qquad(20.4)$$

Equations (20.1) are automatically satisfied because $\vec{E}$ and $\vec{B}$ arise from potentials. How do we write equations (20.2)? Recall that the similar looking equations in Maxwell theory were written in (3.34): $\partial_\nu F^{\mu\nu} = (1/c)\,j^\mu$. It follows that (20.2) are given by

$$\frac{\partial G^{\mu\nu}}{\partial x^\nu} = \frac{1}{c}\,j^\mu,\qquad(20.5)$$

where $G^{\mu\nu} = -G^{\nu\mu}$ is obtained from $F^{\mu\nu}$ by replacing $\vec{E}$ by $\vec{D}$ and $\vec{B}$ by $\vec{H}$:

$$G^{\mu\nu} = \begin{pmatrix} 0 & D_x & D_y & D_z \\ -D_x & 0 & H_z & -H_y \\ -D_y & -H_z & 0 & H_x \\ -D_z & H_y & -H_x & 0 \end{pmatrix}.\qquad(20.6)$$

Although the matrix $G^{\mu\nu}$ written here applies only in four-dimensional spacetime, equations (20.5) make sense in any number of dimensions. Together with the definition of $F_{\mu\nu}$

in terms of potentials, they form the equations of nonlinear electrodynamics in an arbitrary number of dimensions. Of course, the relation between $G^{\mu\nu}$ and $F^{\mu\nu}$ must also be given.

We now show that equation (20.5) follows from the variation of an action. Moreover, the definition of $G^{\mu\nu}$ in terms of the field strength $F^{\mu\nu}$ also arises. Consider the action S, written for arbitrary spacetime dimensionality:

$$S = \int d^D x\, \mathcal{L}\,(F_{\mu\nu}) + \frac{1}{c} \int d^D x\, A_\mu\, j^\mu \,. \qquad (20.7)$$

For simplicity, the Lagrangian density $\mathcal{L}(F_{\mu\nu})$ is assumed to depend only on the field strength and not, for example, on its derivatives. Since the field strength is gauge invariant, $\mathcal{L}$ is also gauge invariant. The Lagrangian density $\mathcal{L}$ is otherwise arbitrary; it need not coincide with the Maxwell Lagrangian density. In order to perform the variation of the action efficiently, we need to *define* partial derivatives with respect to field strengths.

For this purpose we first note that the variations $\delta F_{\mu\nu}$ are constrained by antisymmetry: $\delta F_{\mu\nu} = -\delta F_{\nu\mu}$. For any function M of the field strengths we then write

$$\delta M = \frac{1}{2} \frac{\partial M}{\partial F_{\mu\nu}} \delta F_{\mu\nu} \,. \qquad (20.8)$$

As usual, repeated indices are summed over. Since the variations $\delta F_{\mu\nu}$ are antisymmetric, we can require that

$$\frac{\partial M}{\partial F_{\mu\nu}} = -\frac{\partial M}{\partial F_{\nu\mu}} \,. \qquad (20.9)$$

Equations (20.8) and (20.9) together define the partial derivatives $\partial M / \partial F_{\mu\nu}$.

Quick calculation 20.1 Use $M = F_{12}$ and the above equations to prove that $\partial F_{12}/\partial F_{12} = 1$ and that $\partial F_{12}/\partial F_{\mu\nu} = 0$ if $(\mu, \nu) \neq (1, 2)$ and $(\mu, \nu) \neq (2, 1)$.

It is also useful to learn how to use the chain rule. In order to calculate derivatives of M with respect to some variable U we write

$$\delta M = \frac{1}{2} \frac{\partial M}{\partial F_{\mu\nu}} \frac{\partial F_{\mu\nu}}{\partial U} \delta U \equiv \frac{\partial M}{\partial U} \delta U \,, \qquad (20.10)$$

thus learning that

$$\frac{\partial M}{\partial U} = \frac{1}{2} \frac{\partial M}{\partial F_{\mu\nu}} \frac{\partial F_{\mu\nu}}{\partial U} \,. \qquad (20.11)$$

We are now in a position to vary the action S. Using (20.8) with M set equal to $\mathcal{L}$, the variation of the first term in S gives

$$\delta \int d^D x\, \mathcal{L} = \int d^D x\, \frac{1}{2} \frac{\partial \mathcal{L}}{\partial F_{\mu\nu}} \delta F_{\mu\nu} = \int d^D x\, \frac{1}{2} \frac{\partial \mathcal{L}}{\partial F_{\mu\nu}} (\partial_\mu \delta A_\nu - \partial_\nu \delta A_\mu) \,. \qquad (20.12)$$

Integrating by parts, relabeling indices, and using (20.9) we find

$$\delta \int d^D x\, \mathcal{L} = \int d^D x\, \delta A_\mu\, \partial_\nu \left(\frac{\partial \mathcal{L}}{\partial F_{\mu\nu}} \right) . \qquad (20.13)$$

The variation of the whole action (20.7) then gives

$$\delta S = \int d^D x \, \delta A_\mu \left[\partial_\nu \left(\frac{\partial \mathcal{L}}{\partial F_{\mu\nu}} \right) + \frac{1}{c} j^\mu \right]. \tag{20.14}$$

Equating this variation to zero, we find (20.5) if we identify

$$G^{\mu\nu} \equiv -\frac{\partial \mathcal{L}}{\partial F_{\mu\nu}}. \tag{20.15}$$

The tensor $G^{\mu\nu}$ is antisymmetric by construction. If the nonlinear Lagrangian is known, (20.15) expresses $G^{\mu\nu}$ as a function of the field strength $F_{\mu\nu}$. To illustrate this, let us calculate $\vec{D}$. Equation (20.6) tells us that $G^{0i} = D_i$, and, for arbitrary spacetime dimensionality, this is taken to be the definition of the vector $\vec{D}$. We begin by calculating the derivative $\partial \mathcal{L}/\partial E_i$. Using the chain rule (20.11) and recalling that $E_i = F_{i0} = -F_{0i}$, we find

$$\frac{\partial \mathcal{L}}{\partial E_i} = \frac{1}{2} \frac{\partial \mathcal{L}}{\partial F_{0i}} \frac{\partial F_{0i}}{\partial E_i} + \frac{1}{2} \frac{\partial \mathcal{L}}{\partial F_{i0}} \frac{\partial F_{i0}}{\partial E_i} = -\frac{\partial \mathcal{L}}{\partial F_{0i}} = G^{0i} = D_i. \tag{20.16}$$

Therefore, we have shown that

$$\vec{D} = \frac{\partial \mathcal{L}}{\partial \vec{E}}. \tag{20.17}$$

In four-dimensional spacetime we can also write $\vec{H}$ in terms of derivatives of $\mathcal{L}$. For example, since $B_1 = F_{23} = -F_{32}$, we have

$$\frac{\partial \mathcal{L}}{\partial B_1} = -\frac{1}{2} \frac{\partial \mathcal{L}}{\partial F_{32}} + \frac{1}{2} \frac{\partial \mathcal{L}}{\partial F_{23}} = G^{32} = -H_1. \tag{20.18}$$

Working out the other two components, we find that the end result is

$$\vec{H} = -\frac{\partial \mathcal{L}}{\partial \vec{B}}. \tag{20.19}$$

If $\mathcal{L}(\vec{E}, \vec{B})$ is known, equations (20.17) and (20.19) give us $\vec{D}$ and $\vec{H}$. For Maxwell electrodynamics, for example, the Lagrangian density is

$$\mathcal{L}_M = -\frac{1}{4} F^{\mu\nu} F_{\mu\nu} = \frac{1}{2} (E^2 - B^2). \tag{20.20}$$

Here we defined $E^2 = \vec{E} \cdot \vec{E}$ and $E = |\vec{E}|$, with similar definitions for $\vec{B}$, $\vec{H}$, and $\vec{D}$. It follows from (20.20) that $\vec{D} = \vec{E}$ and $\vec{H} = \vec{B}$.

$\vec{E}$ is related to the time derivative of $\vec{A}$, so it may be viewed as a velocity. Equation (20.17) then implies that $\vec{D}$ is the canonical momentum associated with the velocity $\vec{E}$. This suggests that the Hamiltonian, or energy functional, is given as

$$\mathcal{H} = \vec{D} \cdot \vec{E} - \mathcal{L}. \qquad (20.21)$$

This formula is clearly correct for Maxwell theory: since $\vec{D} = \vec{E}$, we find $\mathcal{H} = \frac{1}{2}(E^2 + B^2)$, which is the familiar energy density of the electromagnetic field. In Problem 20.1 you will give a proof that $\mathcal{H}$ is indeed a conserved energy for arbitrary $\mathcal{L}$. While some of our results in this section are written in the language of four dimensions, the important ideas work in all dimensions. In fact, all of the shaded equations are valid in any spacetime dimension.

Let us conclude this section by considering some candidate Lagrangian densities in four-dimensional spacetime. The density $\mathcal{L}$ in (20.7) must be both gauge invariant and Lorentz invariant. Since it is built from field strengths it is clearly gauge invariant. To be Lorentz invariant the field strengths must form objects with no free indices. There are two independent nontrivial Lorentz invariant objects that we can build using $F_{\mu\nu}$ and none of its derivatives:

$$s \equiv -\frac{1}{4} F^{\mu\nu} F_{\mu\nu} = \frac{1}{2}(E^2 - B^2),$$

$$p \equiv -\frac{1}{4} \tilde{F}^{\mu\nu} F_{\mu\nu} = \vec{E} \cdot \vec{B}. \qquad (20.22)$$

In fact, it can be proven that s and p are the *only* independent invariants that can be built from $F_{\mu\nu}$. The invariant p makes use of the *dual* field strength $\tilde{F}^{\mu\nu}$, which is defined as

$$\tilde{F}^{\mu\nu} \equiv \frac{1}{2} \epsilon^{\mu\nu\rho\sigma} F_{\rho\sigma} . \qquad (20.23)$$

Here $\epsilon^{\mu\nu\rho\sigma}$ is a totally antisymmetric Lorentz tensor (more precisely, it is a pseudo-tensor). This implies that $\tilde{F}^{\mu\nu}$ is antisymmetric. Just like $\eta^{\mu\nu}$, the tensor $\epsilon^{\mu\nu\rho\sigma}$ takes the same values in all Lorentz frames. In any frame $\epsilon^{0123} = 1$, and $\epsilon^{\mu\nu\rho\sigma}$ vanishes if any index is repeated. For example,

$$\tilde{F}^{01} = \frac{1}{2} \epsilon^{01\rho\sigma} F_{\rho\sigma} = \frac{1}{2}(F_{23} - F_{32}) = F_{23} = B_x . \qquad (20.24)$$

Calculating all the other entries, we have

$$\tilde{F}^{\mu\nu} = \begin{pmatrix} 0 & B_x & B_y & B_z \\ -B_x & 0 & -E_z & E_y \\ -B_y & E_z & 0 & -E_x \\ -B_z & -E_y & E_x & 0 \end{pmatrix} . \qquad (20.25)$$

Quick calculation 20.2 Calculate the other entries in the matrix $\tilde{F}^{\mu\nu}$.

Quick calculation 20.3 Show that p takes the value quoted in (20.22).

In four dimensions, the most general Lorentz invariant Lagrangian density built out of $F_{\mu\nu}$ is an arbitrary function of s and p. The Maxwell Lagrangian density is simply s.

20.2 Born–Infeld electrodynamics

In this section we will write the Lagrangian that turns out to describe the electromagnetic fields that live on the world-volumes of D-branes. This is the Born–Infeld electromagnetic Lagrangian. In Born–Infeld theory, as we shall see, the electrostatic self-energy of a point charge is finite. This is an improvement over Maxwell theory, where the self-energy of a point charge is infinite. The endpoints of open strings are point charges, so it is reassuring to see that in string theory these do not carry infinite energy. In the following section we will explain how T-duality gives direct evidence that the Born–Infeld Lagrangian governs the dynamics of electromagnetic fields on D-branes.

We begin our work in four-dimensional spacetime. In addition to the natural requirements of gauge and Lorentz invariance, we impose two constraints on the nonlinear Lagrangian. First, it must reduce to the Maxwell Lagrangian for small $\vec{E}$ and $\vec{B}$. Second, there should be a maximal electric field when $\vec{B} = 0$. We showed in Section 19.2 that string theory has a critical electric field $E_{\text{crit}} = 1/(2\pi\alpha') \equiv b$. For the originators of nonlinear electrodynamics, a maximal electric field was needed to obtain point charges with finite self-energy.

How can we impose the existence of a maximal electric field on our Lagrangian? In special relativity the maximal velocity is clearly evident in the point particle Lagrangian (5.8): the argument of the square root must be positive, and this ensures that the particle velocity cannot exceed the velocity of light. To impose $E \leq b$ we also use a square root. There is a simple expression which uses only the Lorentz invariant s:

$$\mathcal{L} = -b^2 \sqrt{1 - \frac{(E^2 - B^2)}{b^2}} + b^2 = -b^2 \sqrt{1 - \frac{2s}{b^2}} + b^2 . \tag{20.26}$$

As required, $E \leq b$ for $B = 0$. Moreover, for small fields we have $s \ll b^2$, so

$$\mathcal{L} = -b^2 \left(1 - \frac{s}{b^2}\right) + b^2 + \mathcal{O}(s^2) = s + \mathcal{O}(s^2) . \tag{20.27}$$

For small fields we recover the Maxwell Lagrangian. While (20.26) satisfies the two conditions we requested, a somewhat more complicated Lagrangian has even nicer features. Consider the Born–Infeld Lagrangian density

$$\mathcal{L} = -b^2 \sqrt{1 - \frac{E^2 - B^2}{b^2} - \frac{(\vec{E} \cdot \vec{B})^2}{b^4}} + b^2 = -b^2 \sqrt{1 - \frac{2s}{b^2} - \frac{p^2}{b^4}} + b^2 . \tag{20.28}$$

Since it is built from s and p, this density is also Lorentz invariant. For small fields where s and p are comparable and both are much smaller than b^2, the weak field approximation is not changed: $\mathcal{L} \sim s$. In a generic theory of nonlinear electrodynamics, waves with different polarizations propagate with different velocities through a background electromagnetic field. In Born–Infeld theory the velocity is independent of the polarization. In all theories of nonlinear electrodynamics, Born–Infeld included, there are nontrivial dispersion relations; that is, waves of different frequencies travel with different velocities.

The Born–Infeld Lagrangian density is special for yet another reason. It can be written elegantly in terms of the square root of a determinant:

$$\mathcal{L} = -b^2 \sqrt{-\det\left(\eta_{\mu\nu} + \frac{1}{b} F_{\mu\nu}\right)} + b^2 \,. \tag{20.29}$$

As opposed to (20.28), this formula allows an obvious generalization to any number of dimensions. A calculation is needed, however, to show that the two densities are the same in four dimensions. The calculation is straightforward but a little long: one must simply write the 4-by-4 matrix and calculate its determinant. It is more instructive to understand why the equality is reasonable. To simplify the equations, we temporarily set $b = 1$ in (20.29):

$$\mathcal{L} = -\sqrt{-\det(\eta_{\mu\nu} + F_{\mu\nu})} + 1 \,. \tag{20.30}$$

The original Lagrangian density is recovered by letting $F_{\mu\nu} \to F_{\mu\nu}/b$ and $\mathcal{L} \to b^2\mathcal{L}$. The Lagrangian (20.28) is explicitly Lorentz invariant. How do we see the Lorentz invariance of (20.30)? To show the Lorentz invariance of $\det(\eta_{\mu\nu} + F_{\mu\nu})$ we first note that the determinant of an arbitrary matrix M with components $M_{\mu\nu}$ is the same as the determinant of the matrix $\bar{M}$ with components $M^{\mu\nu}$. To see this, we first write

$$M_{\mu\nu} = \eta_{\mu\rho}\eta_{\nu\sigma} M^{\rho\sigma} = \eta_{\mu\rho} M^{\rho\sigma} \eta_{\sigma\nu} \longrightarrow M = \eta\bar{M}\eta \,, \tag{20.31}$$

where η is the matrix with components $\eta_{\mu\nu}$. Taking determinants of both sides of the equation, we confirm that

$$\det M = \det(\eta M \eta) = (\det \bar{M})(\det \eta)^2 = \det \bar{M} \,. \tag{20.32}$$

Let $\bar{\eta}$, F, and $\bar{F}$ denote the matrices with entries $\eta^{\mu\nu}$, $F_{\mu\nu}$, and $F^{\mu\nu}$, respectively. On account of the result just established, we have

$$\det(\eta + F) = \det(\bar{\eta} + \bar{F}) \,. \tag{20.33}$$

It therefore suffices to prove the Lorentz invariance of $\det(\bar{\eta} + \bar{F})$.

Consider a Lorentz transformation $x'^{\mu} = L^{\mu}{}_{\nu} x^{\nu}$, as written in (2.38). The matrix L, with entries $L^{\mu}{}_{\nu}$ (μ is the row index and ν is the column index), satisfies $(\det L)^2 = 1$. Since both $\eta^{\mu\nu}$ and $F^{\mu\nu}$ are Lorentz tensors of the same type, they transform under Lorentz transformations in the same way. The tensors carry two indices, so the matrix L must be applied to them twice:

$$\eta'^{\mu\nu} + F'^{\mu\nu} = L^{\mu}{}_{\rho} L^{\nu}{}_{\sigma} (\eta^{\rho\sigma} + F^{\rho\sigma}) = L^{\mu}{}_{\rho} (\eta^{\rho\sigma} + F^{\rho\sigma}) L^{\nu}{}_{\sigma} \,. \tag{20.34}$$

In matrix notation,

$$\bar{\eta}' + \bar{F}' = L\,(\bar{\eta} + \bar{F})\,L^{\mathrm{T}} \,. \tag{20.35}$$

Since $\det L = \det L^{\mathrm{T}}$ and $(\det L)^2 = 1$, we can take determinants to conclude immediately that

$$\det(\bar{\eta}' + \bar{F}') = \det(\bar{\eta} + \bar{F}) \,. \tag{20.36}$$

This establishes the Lorentz invariance of the Born–Infeld Lagrangian density.

One additional fact is easily derived. Since the determinant of a matrix does not change under transposition, we have

$$\det(\eta + F) = \det(\eta^{\mathrm{T}} + F^{\mathrm{T}}) = \det(\eta - F). \qquad (20.37)$$

We conclude that the Born–Infeld Lagrangian is an even function of F. The expression for the argument of the square root in (20.30) can be simplified, using $\eta = \eta^{-1}$ and $\det \eta = -1$:

$$-\det(\eta + F) = -\det(\eta(1 + \eta F)) = \det(1 + \eta F). \qquad (20.38)$$

Explicitly, the matrix $1 + \eta F$ is given by

$$1 + \eta F = \begin{pmatrix} 1 & E_x & E_y & E_z \\ E_x & 1 & B_z & -B_y \\ E_y & -B_z & 1 & B_x \\ E_z & B_y & -B_x & 1 \end{pmatrix}. \qquad (20.39)$$

Each term of the determinant expansion is a product of terms, each of which contains precisely one element from each row and one element from each column. So the terms which are quadratic in the fields contain two elements from the diagonal. There are six ways of picking these two, and the corresponding terms give $-E^2 + B^2$, which is the expected answer. The quartic terms take a little more work to write out completely. In this case, one must not pick any term from the diagonal.

Our final topic in this section is the computation of the self-energy of a point charge in Born–Infeld theory. For this problem we can set $\vec{B} = 0$ in (20.28) and use the simplified Lagrangian density

$$\mathcal{L} = -b^2 \sqrt{1 - \frac{E^2}{b^2}} + b^2. \qquad (20.40)$$

As a first step, we calculate $\vec{D}$:

$$\vec{D} = \frac{\partial \mathcal{L}}{\partial \vec{E}} = \frac{\vec{E}}{\sqrt{1 - E^2/b^2}}. \qquad (20.41)$$

We see that $\vec{E}$ and $\vec{D}$ point in the same direction. To solve for $\vec{E}$ in terms of $\vec{D}$, we first square the above equation and solve for E^2:

$$D^2 = \frac{E^2}{1 - E^2/b^2} \rightarrow E^2 = \frac{D^2}{1 + D^2/b^2}. \qquad (20.42)$$

At this stage, by writing

$$E^2 = b^2 \left(\frac{D^2}{D^2 + b^2} \right) = D^2 \left(\frac{b^2}{b^2 + D^2} \right), \qquad (20.43)$$

we immediately see that

$$E \leq b, \quad E \leq D. \tag{20.44}$$

As expected, the electric field is bounded by b. Moreover, the magnitude E of the electric field is everywhere bounded by the magnitude D of $\vec{D}$. Note, however, that D can be arbitrarily large. In fact, it requires infinitely large D in order to get $E = b$.

To obtain $\vec{E}$, we take the "square root" of (20.42) by writing

$$\vec{E} \cdot \vec{E} = \frac{\vec{D}}{\sqrt{1 + D^2/b^2}} \cdot \frac{\vec{D}}{\sqrt{1 + D^2/b^2}}. \tag{20.45}$$

Since $\vec{E}$ points in the same direction as $\vec{D}$, the unique solution is

$$\vec{E} = \frac{\vec{D}}{\sqrt{1 + D^2/b^2}}. \tag{20.46}$$

With $\vec{E}(\vec{D})$ at hand we are close to being able to calculate the energy density $\mathcal{H}(\vec{D})$. As a final ingredient, we need $\mathcal{L}$ written in terms of $\vec{D}$:

$$\mathcal{L} = -b^2 \sqrt{1 - \frac{D^2}{b^2(1 + D^2/b^2)}} + b^2 = \frac{-b^2}{\sqrt{1 + D^2/b^2}} + b^2. \tag{20.47}$$

The Hamiltonian is then

$$\mathcal{H} = \vec{E} \cdot \vec{D} - \mathcal{L} = \frac{D^2}{\sqrt{1 + D^2/b^2}} + \frac{b^2}{\sqrt{1 + D^2/b^2}} - b^2, \tag{20.48}$$

and the final result is

$$\mathcal{H} = b^2 \sqrt{1 + \frac{D^2}{b^2}} - b^2. \tag{20.49}$$

This is the Born–Infeld energy density when $\vec{B} = \vec{H} = 0$. You will confirm in Problem 20.4 that this result holds for Born–Infeld theory in an arbitrary number of dimensions.

Let us calculate the self-energy of a point charge in four-dimensional spacetime. In Maxwell theory the infinite self-energy comes about because the energy density is proportional to $ED = E^2$, and $E \sim r^{-2}$, where r is the distance to the charge. As a result, $d^3x E^2 \sim dr/r^2$, and the energy integral diverges for small r. In Born–Infeld theory we also find $D \sim r^{-2}$, but for large D the energy density (20.49) becomes

$$\mathcal{H} \simeq bD = E_{\text{crit}} D, \quad \text{as } D \to \infty. \tag{20.50}$$

For large fields, the Maxwell energy density $u = \frac{1}{2}ED$ is replaced in Born–Infeld theory by $u = E_{\text{crit}} D$. For large fields, the Born–Infeld energy grows linearly with D. As a result, $d^3x E_{\text{crit}} D \sim d^3x \, r^{-2} \sim dr$, and the integral will converge. Let us now examine the details.

Suppose that we have a point charge Q. Because of spherical symmetry the field $\vec{D}$ is radial, and the equation $\nabla \cdot \vec{D} = \rho$ can be integrated over the volume bounded by a sphere of radius r to give

$$\oint_S \vec{D} \cdot d\vec{a} = Q \longrightarrow \vec{D}(\vec{r}) = \frac{Q}{4\pi r^2} \vec{e}_r , \tag{20.51}$$

where $\vec{e}_r$ is the unit vector in the radial direction. We must check that our solution satisfies $\nabla \times \vec{E} = 0$. Because of (20.46), we know that $\vec{E}$ is of the form $\vec{E} = f(r)\vec{r}$. Such an electric field has zero curl:

Quick calculation 20.4 Show that $\nabla \times (f(r)\vec{r}) = 0$.

The energy U_Q of the point charge can now be calculated. Making use of (20.49) and (20.51), we write:

$$U_Q = \int d^3x \, \mathcal{H} = b^2 \int_0^\infty 4\pi r^2 dr \left(\sqrt{1 + \left(\frac{Q}{4\pi b r^2} \right)^2} - 1 \right)$$

$$= 4\pi b^2 \int_0^\infty dr \left(\sqrt{r^4 + \left(\frac{Q}{4\pi b} \right)^2} - r^2 \right). \tag{20.52}$$

Letting $r = x\sqrt{Q/4\pi b}$, we obtain

$$U_Q = \sqrt{\frac{b}{4\pi}} \, Q\sqrt{Q} \int_0^\infty dx (\sqrt{1 + x^4} - x^2). \tag{20.53}$$

This integral converges. The short distance problem has disappeared: the integrand is regular around $x = 0$. We do not expect a problem at large x: the fields are small and the Born–Infeld energy is approximately equal to the Maxwell energy, for which no problem arises at large distances. Indeed,

$$\sqrt{1 + x^4} - x^2 = \frac{1}{\sqrt{1 + x^4} + x^2} < \frac{1}{2x^2}, \tag{20.54}$$

which is integrable around $x = \infty$. Doing the integral explicitly, one finds

$$\int_0^\infty dx (\sqrt{1 + x^4} - x^2) = \frac{(\Gamma(1/4))^2}{6\sqrt{\pi}} \simeq \frac{(3.6256)^2}{6(1.7725)} \simeq 1.236. \tag{20.55}$$

Here $\Gamma(x)$ is the gamma function (see (3.50)). Back in (20.53), we find our final answer for the energy of a point charge in Born–Infeld theory:

$$U_Q = \frac{1}{4\pi} \frac{1}{3} (\Gamma(1/4))^2 b^{1/2} Q^{3/2} \simeq \frac{1}{4\pi} \cdot 4.382 \, b^{1/2} Q^{3/2}. \tag{20.56}$$

If b corresponds to the critical electric field in string theory, then $b = 1/(2\pi\alpha')$. Moreover, $\sqrt{\alpha'} = \ell_s$, where ℓ_s is the string length. The self-energy of a point charge becomes

$$U_Q = \frac{1}{4\pi} \frac{1}{\sqrt{2\pi}} \frac{1}{3} (\Gamma(1/4))^2 \frac{Q^2}{\ell_s \sqrt{Q}} \simeq \frac{1}{4\pi} \cdot 1.748 \frac{Q^2}{\ell_s \sqrt{Q}}. \tag{20.57}$$

To appreciate this answer better, we compare it with the electrostatic energy of a charge distribution in Maxwell theory. If we assume that the charge is distributed uniformly over the volume of a ball of radius a, then the energy $U(Q)$ can be shown to be

$$U(Q) = \frac{1}{4\pi} \frac{3}{5} \frac{Q^2}{a}. \tag{20.58}$$

As $a \to 0$ the charge becomes point-like, and we obtain the expected infinite self-energy. We can think of a as a small smearing parameter that is introduced to make the self-energy finite. The classical electron radius, for example, is roughly the value of a for which $U(e)$ equals the rest energy of the electron. If we compare the Born–Infeld energy (20.57) with the Maxwell energy, we see that $\ell_s \sqrt{Q}$ plays the role of the smearing parameter. This quantity has the correct units because Q is dimensionless. It is interesting that the smearing parameter grows with the value of Q. The $Q^{3/2}$ dependence of the Born–Infeld energy U_Q is a sign of a nonlinear theory.

The above calculation of the self-energy of a static point charge does not seem to be directly applicable in string theory. On a space-filling D-brane open string endpoints are charged, but they also move. A tractable situation arises, however, when a semi-infinite open string ends on a Dp-brane, as you may examine in Problems 20.6 and 20.7. The string can then be static, and the Born–Infeld energy of the solution gives the energy associated with the semi-infinite string!

20.3 Born–Infeld theory and T-duality

In this chapter we have studied Born–Infeld electrodynamics in some detail. We were motivated by the existence of a critical electric field in string theory. In the present section we will show that our work on the T-duality properties of electric and magnetic fields on D-branes gives direct evidence that the dynamics of electromagnetic fields on D-branes is described by Born–Infeld theory.

For a static Dp-brane the product of its tension $T_p(g)$ and its volume gives the mass of the brane. Consider a world with one dimension curled up into a circle of radius R and $(p-1)$ dimensions curled up into some compact space of volume V_{p-1}. Now imagine a Dp brane that wraps around the full set of p compact dimensions. The mass of this Dp-brane is

$$T_p(g)\,(2\pi R)\,V_{p-1}\,. \tag{20.59}$$

Under a T-duality transformation along the circle of radius R we obtain a D$(p-1)$-brane placed at some point on the dual circle, while the other $(p-1)$ directions along the brane world-volume still wrap around the compact space of volume V_{p-1}. The mass of this D-brane is given by

$$T_{p-1}(\tilde{g})\,V_{p-1}\,, \tag{20.60}$$

where $\tilde{g}$ is the string coupling in the dual picture (see (18.56)). Since T-duality is a physical equivalence, the two masses obtained above must be equal. Indeed, each D-brane is seen by a lower-dimensional observer as a point mass, and unless these masses are the same, the observer can tell that the physics has changed. Equating the values of the two masses, we find

$$T_{p-1}(\tilde{g}) = 2\pi R\, T_p(g)\,. \tag{20.61}$$

This is a relation between tensions of D-branes in *dual* pictures. $T_p(g)$ is the tension of a brane in the world with coupling constant g (the world with the circle of radius R), while $T_{p-1}(\tilde{g})$ is the tension of a brane in the world with coupling constant $\tilde{g}$ (the world with the circle of dual radius $\tilde{R}$). This relation between tensions holds generally, regardless of which coordinates the branes wrap around. The different looking relation between D-brane tensions in (18.57) compares tensions in the *same* picture, that is, with the same string coupling.

Let us now reconsider the configuration discussed in Section 19.3. We had two compact dimensions of radii R_2 and R_3 which formed a two-torus. One direction on the world-volume of a D$(p-1)$-brane was stretched along the diagonal of the torus. Assume, for simplicity, that the other $(p-2)$ directions wrap around a compact space of volume V_{p-2}. Let $\tilde{g}$ denote the string coupling in this picture. A T-duality transformation along the circle of radius R_3 gave us a Dp-brane with two directions wrapped around a torus with radii R_2, $\tilde{R}_3$ and a magnetic field. The other $(p-2)$ directions on the brane also wrap the space of volume V_{p-2}. Let g denote the string coupling in this picture. Since the circle that is dualized in the Dp-brane world is of radius $\tilde{R}_3$, equation (20.61) gives

$$T_{p-1}(\tilde{g}) = 2\pi \tilde{R}_3 \, T_p(g) \,. \tag{20.62}$$

For the static D$(p-1)$-brane that stretches along the torus diagonal, the Lagrangian is the negative of the brane rest energy. This energy is just the brane tension $T_{p-1}(\tilde{g})$ times the brane volume. The volume, in turn, is V_{p-2} times the length L_{diag} along the diagonal, so

$$L = -V_{p-2} L_{\text{diag}} \, T_{p-1}(\tilde{g}) = -V_{p-2} \, \sqrt{(2\pi R_2)^2 + (2\pi R_3)^2} \;\, T_{p-1}(\tilde{g}) \,. \tag{20.63}$$

We showed in (19.47) that $2\pi \alpha' B = -\tan\alpha$, where α is the angle that the diagonal in the torus makes with the horizontal direction. In fact, $\tan\alpha = R_3/R_2$, as indicated in (19.55). Therefore, the magnetic field is related to the ratio of the radii by

$$2\pi \alpha' B = -\frac{R_3}{R_2} \,. \tag{20.64}$$

Equations (20.62) and (20.64) allow us to rewrite (20.63) as follows:

$$L = -V_{p-2} \, (2\pi R_2) \sqrt{1 + (2\pi \alpha' B)^2} \;\, (2\pi \tilde{R}_3) \, T_p(g) \,. \tag{20.65}$$

Since $(2\pi R_2)(2\pi \tilde{R}_3)$ is the volume of the torus wrapped by the Dp-brane, we finally have

$$L = -V_p \, T_p(g) \sqrt{1 + (2\pi \alpha' B)^2} \;\,. \tag{20.66}$$

If $B = 0$, we recover the Lagrangian for a static Dp-brane.

To compare with the Born–Infeld Lagrangian, we consider (20.28), which for $\vec{E} = 0$ reduces to

$$\mathcal{L} = -b^2 \sqrt{1 + \frac{B^2}{b^2}} + b^2 \,. \tag{20.67}$$

Both the b^2 multiplying the square root and the b^2 added to $\mathcal{L}$ were introduced for reasons that are no longer relevant. The additive contribution, for example, was originally included to cancel the constant term in $\mathcal{L}$. Now the constant term is needed to represent the rest energy of the D-brane in the absence of electromagnetic fields. The b^2 multiplying the square root was originally included to give a standard normalization to the Maxwell action. Now the overall normalization of the action is fixed in (20.66). We cannot even rescale the gauge potentials. The normalization of the gauge field was fixed when we wrote the coupling (16.54) to the open string endpoints.

As a result, the value $b = 1/(2\pi\alpha')$, that we read by comparing the terms inside the square roots in (20.66) and (20.67), is only to be used inside the square root. Therefore, on account of (20.29), the D-brane Lagrangian (20.66) is consistent with the Born–Infeld Lagrangian density

$$\mathcal{L} = -\, T_p(g) \, \sqrt{-\det(\eta_{mn} + 2\pi\alpha' F_{mn})} \; . \tag{20.68}$$

Here m, n are indices on the world-volume of the Dp-brane. The volume factor V_p was removed in order to write the density $\mathcal{L}$. This Lagrangian density describes the behavior of electromagnetic fields on D-branes.

Quick calculation 20.5 Assume that $F_{23} = B$ is the only nonvanishing magnetic field component. Evaluate the Lagrangian density (20.68) for an arbitrary Dp-brane ($p \geq 3$), and verify that your result is consistent with (20.66).

The Lagrangian density (20.68) allows us to calculate the electromagnetic fields that arise from open string endpoints, which, in our conventions, carry charges of unit magnitude. For a static oriented open string that ends at $\vec{x}_0$, for example, the charge density is given by $\rho(\vec{x}) = \delta(\vec{x} - \vec{x}_0)$, where $\vec{x}$ represents the spatial coordinate on the D-brane. The equation for $\vec{D}$ then reads

$$\nabla \cdot \vec{D} = \delta(\vec{x} - \vec{x}_0) \,, \quad \vec{D} = \frac{\partial \mathcal{L}}{\partial \vec{E}} \,. \tag{20.69}$$

A particular version of this equation is solved in Problem 20.7. The solution describes a string ending on a D-brane.

The T-duality analysis of electric fields in Section 19.2 supports the identification of (20.68) as the Lagrangian density. There we had a circle of radius $\tilde{R}$ and a D$(p-1)$-brane moving with velocity v along this circle. How do we write a Lagrangian for such a D-brane? For a point particle of mass m, the Lagrangian is $(-m)$ times the relativistic factor $\sqrt{1 - v^2/c^2}$, as shown in (5.8). For a string, the Lagrangian density is (minus) the rest energy $(-T_0 ds)$ of a piece of string times the analogous relativistic factor (see (6.89)). For a moving D$(p-1)$-brane we write

$$L = -V_{p-1} \, T_{p-1}(\tilde{g}) \, \sqrt{1 - \frac{v^2}{c^2}} \; . \tag{20.70}$$

We now re-express this Lagrangian in terms of the variables of the T-dual picture, where we have a Dp-brane with an electric field along a circle of radius R and the string coupling is g. The value of the electric field is related to the brane velocity by $v/c = \beta = 2\pi\alpha'E$ (see (19.34)). We thus find

$$L = -V_{p-1}\,(2\pi R)\,T_p(g)\,\sqrt{1 - (2\pi\alpha'E)^2}\ , \tag{20.71}$$

where we also made use of (20.61). Finally, the product of $2\pi R$ and V_{p-1} gives the total volume V_p of the Dp-brane:

$$L = -V_p\,T_p(g)\,\sqrt{1 - (2\pi\alpha'E)^2}\ . \tag{20.72}$$

This Lagrangian is correctly reproduced by (20.68) when we have a constant electric field and zero magnetic field. This can be easily checked by letting $F_{01} = -E_x$ be the only nonvanishing electric field component. We have thus obtained further evidence for the correctness of (20.68).

Problems

Problem 20.1 Energy functional in nonlinear electrodynamics.

(a) According to equation (20.21), the total energy U in a fixed volume V is

$$U = \int_V \left(\vec{D}\cdot\vec{E} - \mathcal{L}\right)d^3x\ . \tag{1}$$

Prove that, in four-dimensional spacetime,

$$\frac{dU}{dt} = \int_V \left(\vec{E}\cdot\frac{\partial\vec{D}}{\partial t} + \vec{H}\cdot\frac{\partial\vec{B}}{\partial t}\right)d^3x\ . \tag{2}$$

(b) Use equations (20.1) and (20.2) to show that, in the absence of sources,

$$\frac{dU}{dt} = -\int_S (\vec{E}\times\vec{H})\cdot d\vec{a}\ , \tag{3}$$

where S is the surface that bounds the volume V. This shows that U is constant if the electromagnetic fields vanish at the boundary and thus carry no energy out of V. U is a conserved energy. Indeed, you may recognize $\vec{E}\times\vec{H}$ as the Poynting vector, which represents local energy flow per unit area per unit time.

Problem 20.2 Capacitance in Born–Infeld theory.

The capacitance C of a two-conductor configuration is defined as $Q = CV$, where the conductors have charges Q and $-Q$, respectively, and V is the potential difference between them.

(a) Let C_M denote the capacitance in Maxwell theory. Explain why C_M is a constant, independent of Q and V.

(b) Consider a two-conductor configuration, with charges Q and $-Q$, and let $\vec{E}_M(\vec{x})$ be the electric field between the conductors in Maxwell theory. Prove that whenever ∇E_M^2 points in the direction of $\vec{E}_M$, the field $\vec{D}(\vec{x})$ in Born–Infeld theory, for the same charged configuration, is given by $\vec{D}(\vec{x}) = \vec{E}_M(\vec{x})$. Prove that in such cases the Born–Infeld capacitance is always greater than the Maxwell capacitance.

(c) Consider a parallel plate capacitor of area A and plate separation d. Show that the Born–Infeld capacitance $C(V)$ is

$$C(V) = \frac{C_M}{\sqrt{1 - (V/V_c)^2}},$$

where $C_M = A/d$ is the Maxwell capacitance and $V_c = bd$. What is the interpretation of V_c?

Problem 20.3 Dual field strength and the Lorentz invariant p.

(a) Show that the replacement $\tilde{F}^{\mu\nu} \to F^{\mu\nu}$ corresponds to $\vec{E} \to -\vec{B}$ and $\vec{B} \to \vec{E}$. Verify that these are the duality transformations of electromagnetism (Section 17.1).

(b) Show that the Lorentz invariant $p = -\frac{1}{4}\tilde{F}^{\mu\nu}F_{\mu\nu}$ can be written as a total derivative: $p = -\frac{1}{4}\partial_\mu(\epsilon^{\mu\nu\rho\sigma}F_{\nu\rho}A_\sigma)$.

Problem 20.4 Electric fields and point charges in higher dimensions.

(a) Examine the Lagrangian density (20.29) when there is only an electric field, $F_{0i} = -E_i$, in a world with $D = d + 1$ spacetime dimensions. Calculate explicitly the determinant, and show that it takes the form given in (20.40), with $E^2 = \vec{E} \cdot \vec{E}$.

(b) The calculation of the determinant in (a) can be simplified. We have proven the Lorentz invariance of the Born–Infeld Lagrangian density and, therefore, its rotational invariance. Imagine calculating the Lagrangian density at the origin, using a set of axes for which $\vec{E}$ is aligned along the first spatial coordinate. Show how the almost trivial computation in this frame can be used to anticipate the answer obtained in (a).

(c) The energy U_Q of a point charge Q in a D-dimensional spacetime is proportional to Q^δ, where δ is a constant. Calculate δ.

Problem 20.5 Calculating the Born–Infeld Hamiltonian.

Consider the full Born–Infeld Lagrangian density $\mathcal{L}$, with $b = 1$:

$$\mathcal{L} = -\sqrt{1 - E^2 + B^2 - (\vec{E} \cdot \vec{B})^2} + 1. \tag{1}$$

(a) Show that

$$\vec{D} = \frac{\vec{E} + (\vec{E} \cdot \vec{B})\vec{B}}{\sqrt{1 + B^2 - E^2 - (\vec{E} \cdot \vec{B})^2}}. \tag{2}$$

(b) Now the challenge is to solve for $\vec{E}$ in terms of $\vec{D}$. This requires quite a bit of trickery in vector algebra. As a first step note that equation (2) is of the form $f(\vec{E}, \vec{B})\,\vec{D} = \vec{E} + (\vec{E} \cdot \vec{B})\vec{B}$, where f is a scalar function. Show that

$$\vec{E} = \frac{f(\vec{E}, \vec{B})}{1 + B^2} \left(\vec{D} - (\vec{D} \times \vec{B}) \times \vec{B} \right). \tag{3}$$

The function f contains the combination $E^2 + (\vec{E} \cdot \vec{B})^2$. To write this combination in terms of $\vec{D}$ and $\vec{B}$, examine the values of D^2 and $(\vec{D} \times \vec{B})^2$. Now prove that

$$\vec{E} = \frac{\vec{D} - (\vec{D} \times \vec{B}) \times \vec{B}}{\sqrt{1 + B^2 + D^2 + (\vec{D} \times \vec{B})^2}}. \tag{4}$$

(c) As you are now equipped with the value of $\vec{E}$, the calculation of the Hamiltonian is relatively simple. Use (20.21) to show that

$$\mathcal{H} = \sqrt{1 + B^2 + D^2 + (\vec{D} \times \vec{B})^2} \; - 1. \tag{4}$$

Problem 20.6 String ending on a D-brane in Born–Infeld theory: Part 1.

The dynamics of electromagnetic fields on a Dp-brane is governed by (20.68). But we know that there are also $(d - p)$ massless scalar fields living on the world-volume of a Dp-brane, whose excitations represent transverse displacements of the brane (recall the discussion below (15.37)). Consider the scalar field X associated with displacements along an x axis normal to the brane. The Born–Infeld action can be generalized to describe both the electromagnetic and X fields:

$$\mathcal{L} = -T_p \sqrt{-\det \left(\eta_{mn} + 2\pi \alpha' F_{mn} + \partial_m X \partial_n X \right)} \; . \tag{1}$$

The field X here is a function of the coordinates x^m on the brane (X is *not* a string coordinate). The value $X(x^m)$ is the x coordinate of the point on the brane with coordinates x^m.

(a) Suppose there are no electromagnetic fields on the Dp-brane, but we let $X = vt$, attempting to represent the motion of the D-brane along the normal direction x with velocity v. Show that $\mathcal{L} = -T_p \sqrt{1 - v^2/c^2}$, which is the expected Lagrangian density for a moving D-brane. This result supports the interpretation of X given above.

We want to evaluate the Lagrangian density (1) in the case where there is only an electric field $\vec{E}$. For notational convenience introduce $\vec{\mathcal{E}} = 2\pi \alpha' \vec{E}$.

(b) To simplify the evaluation of the determinant we use symmetry arguments, as in part (b) of Problem 20.4. Choose a set of axes so that the electric field points along the first direction: only $\mathcal{E}_1$ is nonzero. There is another vector in this problem: the gradient ∇X of the field X. We choose the axes in such a way that this vector lies in the plane formed by the first and second directions, so that only $(\nabla X)_1 \equiv X_1$ and $(\nabla X)_2 \equiv X_2$ are nonzero. Show that under these conditions the determinant in (1) gives

$$\det(\eta_{mn} + \cdots) = -1 + \dot{X}^2 - X_1^2 - X_2^2 + \mathcal{E}_1^2 + \mathcal{E}_1^2 X_2^2. \tag{2}$$

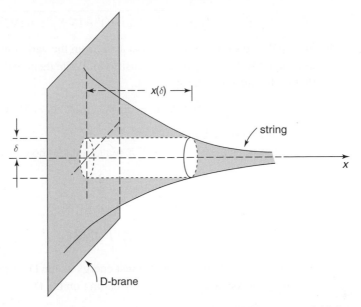

Fig. 20.1 **Problem 20.7: a string that ends on a D-brane is represented in Born–Infeld theory as a solution in which the brane itself is deformed.**

(c) The rotational invariance of the determinant implies that it can be written in terms of dot products of the vectors $\vec{\mathcal{E}}$ and ∇X and in terms of the rotation scalar $\dot{X}$. Use this requirement to deduce that

$$\mathcal{L} = -T_p\sqrt{(1-\mathcal{E}^2)(1+(\nabla X)^2)+(\vec{\mathcal{E}}\cdot\nabla X)^2-\dot{X}^2}\ . \tag{3}$$

Problem 20.7 String ending on a D-brane in Born–Infeld theory: Part 2.

The description of a string ending on a Dp-brane in Born–Infeld theory involves an electric field on the brane, due to the charged endpoint, and an excitation of the scalar field X that represents brane displacements in the direction along the stretched string (see Figure 20.1). This direction is assumed to be normal to the brane. We can therefore use the Lagrangian density (3), obtained in Problem 20.6:

$$\mathcal{L} = -T_p\sqrt{(1-\mathcal{E}^2)(1+(\nabla X)^2)+(\vec{\mathcal{E}}\cdot\nabla X)^2-\dot{X}^2}\ . \tag{1}$$

Here the electric field is: $\vec{\mathcal{E}} = 2\pi\alpha'\vec{E} = \nabla\mathcal{A}_0 - \partial_t\vec{\mathcal{A}}$, where $\mathcal{A}_\mu \equiv 2\pi\alpha' A_\mu$.

(a) Explain why the equations of motion that follow from the variation of $\vec{\mathcal{A}}$ are satisfied if all fields are time independent.

We will therefore assume that all fields are indeed time independent, and, moreover, we will set $\vec{\mathcal{A}} = 0$, which is consistent with our assumption that there are no magnetic fields F_{ij}. In this case, the Lagrangian density becomes

$$\mathcal{L} = -T_p\sqrt{(1-(\nabla\mathcal{A}_0)^2)(1+(\nabla X)^2) + (\nabla\mathcal{A}_0 \cdot \nabla X)^2} \,. \tag{2}$$

(b) Derive the equations of motion that arise from the variations δX and $\delta\mathcal{A}_0$ of the action associated with the Lagrangian density (2). Write them in the form $\nabla \cdot [\dots] = 0$. Show that both equations are satisfied if

$$\vec{\mathcal{E}} = \nabla\mathcal{A}_0 = \pm\nabla X \,, \quad \nabla^2\mathcal{A}_0 = \nabla^2 X = 0 \,, \tag{3}$$

for either choice of sign.

(c) Show that when $\vec{\mathcal{E}} = \nabla\mathcal{A}_0 = \pm\nabla X$ (equation (3)) holds we have

$$\vec{D} = 2\pi\alpha' T_p \vec{\mathcal{E}} \,, \tag{4}$$

and the energy $U = \int d^p x (\vec{D} \cdot \vec{E} - \mathcal{L})$ is given by

$$U = T_p \int d^p x \left(1 + \vec{\mathcal{E}} \cdot \vec{\mathcal{E}}\right). \tag{5}$$

The first term in U represents the rest energy of the Dp-brane. The second term gives the energy associated with the string ending on the D-brane, as we discuss next.

Let r denote radial distance on the D-brane. The solution for a string *ending* at $r = 0$ is obtained solving $\nabla \cdot \vec{D} = +\delta(r)$ (see (20.69)). The solution for $\vec{D}$ (and for $\vec{\mathcal{E}}$) is spherically symmetric; so is the solution for X, which is of the form $X(r)$. Additionally, we assume that $p \geq 3$, in which case we can require that $X(\infty) = 0$ (explain why!). The sign in $\vec{\mathcal{E}} = \pm\nabla X$ determines whether the string stretches along the positive x axis (as in the figure) or along the negative x axis. Which sign do you need to get the option in the figure?

(d) Show that the energy $U_s = T_p \int d^p x \, \vec{\mathcal{E}} \cdot \vec{\mathcal{E}}$ is infinite.

(e) To interpret the infinite value of U_s, consider the region on the D-brane with $r > \delta$, and let $U_s(\delta)$ denote the energy contained in this region. Show that

$$U_s(\delta) = T_p \int_{r>\delta} d^p x \, \nabla X \cdot \nabla X = T_p |X(\delta)| \cdot \text{Flux of } \vec{\mathcal{E}} \text{ across } S^{p-1}(\delta) \,, \tag{6}$$

where $S^{p-1}(\delta)$ denotes the $(p-1)$-dimensional sphere of radius δ. Conclude finally that

$$U_s(\delta) = \frac{1}{2\pi\alpha'} |X(\delta)| \,. \tag{7}$$

Since $1/(2\pi\alpha')$ is the string tension, this confirms that $U_s(\delta)$ is the energy of the piece of string that stretches from $x = 0$ up to $x = |X(\delta)|$. As $\delta \to 0$ the energy diverges because this is the energy of an infinitely long string.

Configurations of intersecting D6-branes in type IIA superstring theory define string models of particle physics. Open string states supported at the intersection of the branes naturally give chiral fermions, a key ingredient of the Standard Model. If orientifold planes are included, the models can display the massless spectrum of gauge bosons and chiral fermions of the Standard Model. Compactification moduli are adjustable parameters that give rise to undesirable massless scalars and must be stabilized. Flux compactifications achieve moduli stabilization and give rise to an extremely large landscape of string vacua. The existence of vacua in which the vacuum energy matches the presently observed value becomes statistically plausible.

21.1 Intersecting D6-branes

In this section we consider a D-brane configuration that has a set of features that make it a good starting point for the construction of a string model of particle physics. Since we need fermions, we use a ten-dimensional superstring theory. In this theory, six of the ten dimensions, $x^4, \ldots, x^9$, are taken to form a small compact space of finite volume. This is necessary in order to have an effectively four-dimensional spacetime (with coordinates x^0, x^1, x^2, and x^3). The compact space is as simple as possible: each dimension is turned into a circle, so that the resulting space is a six-dimensional torus T^6. We will assume that all circles have the same radius R, so we are taking $x^i \sim x^i + 2\pi R$ for $i = 4, \ldots, 9$.

In order to obtain an effective four-dimensional Yang–Mills theory, we need D-branes that have at least three spatial directions to stretch along the spatial coordinates x^1, x^2, and x^3 of the effective spacetime. Therefore, we will use Dp-branes with $p \geq 3$. In fact, in this section we will work with D6-branes in type IIA superstring theory. This is, of course, just one choice among many that we could have made in order to construct a model. In addition, we will let the D6-branes intersect. When two D-branes intersect, one discovers a sector of open strings that stretch from one brane to the other and are localized near the intersection. Under certain circumstances, such strings give rise to matter fields with the properties of Standard Model fermions. We will examine this point in detail in Sections 21.3 and 21.4. Our main goal in the present section is to understand the geometry of intersecting D-branes.

Let us describe two D6-branes that intersect on the torus T^6 introduced above. For a D6-brane, of the nine spatial directions in the ten-dimensional spacetime, three directions are Dirichlet and six are Neumann. Three of the Neumann directions must be x^1, x^2, and x^3, as mentioned above, and the other three lie on T^6. The three Dirichlet directions are also on T^6. The position of the D6-brane can be specified by giving the values of the coordinates along the three Dirichlet directions. Thus, let the first D6-brane be defined by

$$\text{D6-brane \#1:} \quad x^5 = x^7 = x^9 = 0, \tag{21.1}$$

and the second D6-brane be defined by

$$\text{D6-brane \#2:} \quad x^4 = x^6 = x^8 = 0. \tag{21.2}$$

A point belongs to the intersection if it belongs to both D6-branes. The conditions (21.1) must hold for a point to be on the first D6-brane, and the conditions (21.2) must hold for a point to be on the second D6-brane. As a result, a point belongs to the intersection if

$$\text{intersection conditions:} \; x^4 = x^5 = x^6 = x^7 = x^8 = x^9 = 0. \tag{21.3}$$

Since the torus T^6 is spanned by the above coordinates, the D6-branes intersect on a *point* on the torus, the point $(0, 0, 0, 0, 0, 0)$. On the spacetime, however, the intersection of the D6-branes is the set of points

$$\text{intersection set:} \; (x^0, x^1, x^2, x^3, 0, 0, 0, 0, 0, 0), \quad \text{with} \quad x^0, x^1, x^2, x^3 \in \mathbb{R}. \tag{21.4}$$

The intersection of the D6-branes fills the effective four-dimensional spacetime.

A simple way to visualize the intersection of the D6-branes is to describe the six-torus T^6 in terms of three square two-tori T^2. One writes $T^6 = T^2 \times T^2 \times T^2$, in the same sense that one writes $\mathbb{R}^3 = \mathbb{R} \times \mathbb{R} \times \mathbb{R}$. Indeed, a six-torus is equivalent to a two-torus in the x^4 and x^5 directions, a two-torus in the x^6 and x^7 directions, and a two-torus in the x^8 and x^9 directions. These three two-tori are shown in Figure 21.1. On each T^2, each brane appears as a line; in fact, brane #1 appears as a horizontal line on account of (21.1), and brane #2 appears as a vertical line on account of (21.2). On each T^2, these straight lines are in fact circles because their endpoints are identified. The two D6-branes intersect at a point on T^6 because these lines intersect at a point on *each* T^2. Note that the full intersection is characterized by three intersection angles, one for each T^2. In the present case, these angles are all equal to $\pi/2$. The figure also shows a string in the sector [12]. On each of the T^2 we see a projection of the string as it stretches from the first brane to the second brane.

In the same way as we did for other D-brane configurations, we can use a table to describe the two D-branes and the boundary conditions for the various open string sectors. The result is given in Table 21.1. It is noteworthy that in the [12] and [21] sectors, the string coordinates along the torus are of DN or ND types. This happens because the branes intersect orthogonally on each T^2. We have chosen axes such that any coordinate on T^6 lies along one brane and is orthogonal to the other one. For arbitrary intersection angles, the boundary conditions for the string coordinates are slightly more complicated (Problem 15.7).

We are interested in D6-brane configurations obtained by allowing more general intersection angles between the lines that represent the D-branes on each T^2. For this purpose,

Table 21.1	A configuration of two intersecting D6-branes						
Coordinate	$x^{1,2,3}$	x^4	x^5	x^6	x^7	x^8	x^9
D6 #1	–	–	•	–	•	–	•
D6 #2	–	•	–	•	–	•	–
Sector [1,1]	NN	NN	DD	NN	DD	NN	DD
Sector [2,2]	NN	DD	NN	DD	NN	DD	NN
Sector [1,2]	NN	ND	DN	ND	DN	ND	DN
Sector [2,1]	NN	DN	ND	DN	ND	DN	ND

Note: We denote by – coordinates along which the D-branes are stretched and by • coordinates normal to the branes. The last four rows give the boundary conditions for string coordinates in the four possible open string sectors.

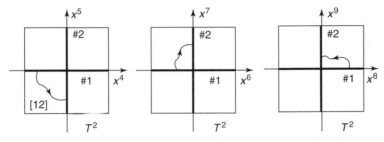

Fig. 21.1 Two D6-branes on a x^6, which is presented as the product of three two-tori x^2. Brane #1 stretches along x^4, x^6, and x^8 and appears as a thick horizontal line. Brane #2 stretches along x^5, x^7, and x^9 and appears as a thick vertical line. A string stretching from the first to the second brane is also shown.

we must first understand how two lines intersect on a two-torus. The lines that we will examine are in fact closed straight lines: the projections of the branes into the torus form circles, or line segments with identified endpoints. This is physically reasonable, the tension of a brane forces them to be straight. Moreover, unless they are closed, they will have infinite length, which requires infinite rest energy.

We can work with a torus defined on the (x, y) plane by the identifications $x \sim x + 1$ and $y \sim y + 1$, where by convention we have chosen the length between identified points to equal unity. The torus can be viewed as the unit square $0 \leq x, y \leq 1$ with boundary identifications. We can use two relatively prime integers (m, n) to describe an *oriented* closed line on this torus: the line is constructed on the (x, y) plane as an oriented straight segment from the origin $(0, 0)$ up to the point (m, n). The identifications can then be used to exhibit this segment fully on the square $0 \leq x, y, \leq 1$. Note that $(m, n) \sim (0, 0)$ under the identifications, so the segment is a closed line on the torus. Since m and n are relatively prime, the segment on the plane does not encounter a point with integer coordinates between its endpoints. This means that the line on the torus does not close before the segment ends. If m and n are not relatively prime, then their greatest common divisor represents the number

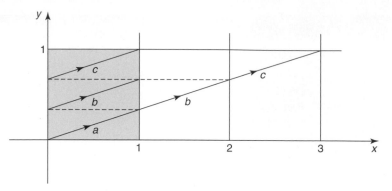

The segment joining (0, 0) to (3, 1) and the corresponding closed line $\ell = (3, 1)$ on the (shaded) torus. The segment goes through three copies of the fundamental domain, so it appears on the fundamental domain as three segments, denoted by a, b, and c.

of times that the line is wrapped on the torus. We will focus on lines for which m and n are relatively prime. If one of the two integers is zero, then the other must be either plus or minus one, otherwise the curve will be multiply wrapped. An example showing the line $(3, 1)$ is given in Figure 21.2.

Now consider a line $\ell_1 = (m_1, n_1)$ and a different line $\ell_2 = (m_2, n_2)$. The question is: how many times do these lines intersect on the torus? The answer is simple to state (and we will discuss its origins a little later on): the intersection number $\#(\ell_1, \ell_2)$ is

$$\#(\ell_1, \ell_2) = m_1 n_2 - m_2 n_1 = \det \begin{pmatrix} m_1 & n_1 \\ m_2 & n_2 \end{pmatrix}. \tag{21.5}$$

This intersection number can be interpreted as the z component of the vector cross product of $(m_1, n_1, 0)$ and $(m_2, n_2, 0)$. This means that the magnitude of the intersection number coincides with the area of the parallelogram defined by the vectors ℓ_1 and ℓ_2. As defined, the intersection number is antisymmetric under the exchange of the two lines: $\#(\ell_1, \ell_2) = -\#(\ell_2, \ell_1)$. The sign associated with the intersection has a meaning because we are dealing with oriented lines. When $\#(\ell_1, \ell_2) > 0$, the oriented line ℓ_1 aligns with the oriented line ℓ_2 after a *counterclockwise* rotation with angle less than π. When $\#(\ell_1, \ell_2) < 0$, the alignment occurs after a *clockwise* rotation with angle less than π. For example, the intersection number of $\ell_1 = (1, 0)$ with $\ell_2 = (0, 1)$ is equal to

$$\#(\ell_1, \ell_2) = 1 \times 1 - 0 \times 0 = +1. \tag{21.6}$$

Consistent with the plus sign, a counterclockwise rotation of $\ell_1 = (1, 0)$, with angle $\pi/2$, gives $\ell_2 = (0, 1)$. Since we are dealing with straight lines, the intersection angle at each intersection point is the same and the rotation needed to align the oriented lines is of the same type.

Quick calculation 21.1 Consider $\ell_1 = (-1, 1)$ and $\ell_2 = (1, 1)$. Calculate $\#(\ell_1, \ell_2)$, identify the intersection points on the torus, and confirm that the sign of the intersection number is in accord with expectations.

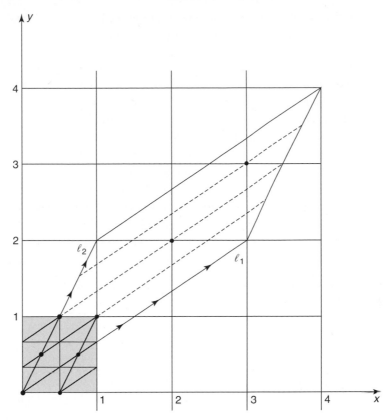

Fig. 21.3 The lines $\ell_1 = (3, 2)$ and $\ell_2 = (1, 2)$ intersect four times on the (shaded) two-torus. The cell $\mathcal{C}$ spanned by ℓ_1 and ℓ_2 is a parallelogram with area four.

In Figure 21.3 we show the intersection of the lines $\ell_1 = (3, 2)$ and $\ell_2 = (1, 2)$. When drawn on the square $0 \leq x, y \leq 1$, the line ℓ_1 appears as four segments, and the line ℓ_2 appears as two segments. We find two intersection points in the interior of the square. There is one intersection point on the horizontal sides (it is identified on the two sides). And there is one intersection point at the corners, all of which are identified. This gives a total of four intersection points. The intersection number can be calculated directly using (21.5), and we get the expected answer: $\#(\ell_1, \ell_2) = 3 \times 2 - 1 \times 2 = 4$.

The formula (21.5) for the intersection number gives zero for the intersection of a line with itself. A line coincides fully with itself, so what is the meaning of the zero? The meaning is topological: given two coincident lines on a torus, we can displace one of them a little and obtain a situation in which the lines no longer intersect. Consider the line $(1, 0)$, for example. A slightly displaced version of this closed line is $0 \leq x \leq 1$, $y = \epsilon$, with ϵ a small positive number. This new closed line has no point in common with the closed line $(1, 0)$. The intersection number is a topological quantity; roughly, the intersection number of two lines does not change under small deformations of the lines. That is why the intersection number of coincident lines is zero.

Quick calculation 21.2 Display on a torus the line $\ell = (1, 1)$ and a second line, parallel to ℓ, which is obtained by a small displacement of ℓ.

We now briefly explain the origin of (21.5). Consider the (x, y) plane with the earlier unit identifications $x \sim x + 1$ and $y \sim y + 1$, and think of the lattice of points with integer coordinates. The unit cell in this lattice is spanned by the vectors $(1, 0)$ and $(0, 1)$. The identifications on the unit cell give the unit torus. The cell $\mathcal{C}$ spanned by the vectors ℓ_1 and ℓ_2 is a parallelogram with some integer area I. We can use $\mathcal{C}$ to construct a big torus $\hat{\mathcal{C}}$ by gluing together the parallel sides of $\mathcal{C}$. If we denote by $\vec{x} = (x, y)$ an arbitrary point on the plane, the big torus $\hat{\mathcal{C}}$ arises from the identifications $\vec{x} \sim \vec{x} + \ell_1$ and $\vec{x} \sim \vec{x} + \ell_2$. We want to show that the area I of $\hat{\mathcal{C}}$ is the number of times that the closed lines ℓ_1 and ℓ_2 intersect on the unit torus. It is a known fact about lattices that the torus $\hat{\mathcal{C}}$ contains I copies of each point on the unit torus. Let us now count the total number of copies that arise from the set of intersection points on the unit torus. For this we must find all the intersections between all the copies of ℓ_1 and all the copies of ℓ_2. The torus $\hat{\mathcal{C}}$ contains I copies of the line ℓ_1: one copy is provided by the edge ℓ_1 of $\mathcal{C}$ (and the other parallel edge), and the other $I - 1$ copies are segments parallel to ℓ_1 that go through the interior of $\mathcal{C}$. Similarly, $\hat{\mathcal{C}}$ contains I copies of the line ℓ_2. On $\hat{\mathcal{C}}$, each copy of ℓ_1 intersects once with each copy of ℓ_2, so we have a total of $I \times I$ distinct intersection points on $\hat{\mathcal{C}}$. It follows that there are $I^2/I = I$ distinct intersection points on the unit torus. This is what we wanted to prove.

Quick calculation 21.3 Explain why the interior of $\mathcal{C}$ contains $I - 1$ points with integer coordinates. Explain also why the copies of ℓ_1 and ℓ_2 on the interior of $\mathcal{C}$ go through the points with integer coordinates.

Quick calculation 21.4 Find the coordinates of the intersection points between the lines $\ell_1 = (3, 2)$ and $\ell_2 = (1, 2)$ (Figure 21.3). Express your answers using points on the unit cell $0 \le x, y \le 1$.

We can now return to our D6-branes, which have three directions wrapped on T^6. Each wrapping is specified by three lines (ℓ_1, ℓ_2, ℓ_3). The line ℓ_i represents the direction of the D-brane on the ith T^2. The first D6-brane in our earlier example is now described by $\ell_1 = \ell_2 = \ell_3 = (1, 0)$. The second brane is described by $\ell_1 = \ell_2 = \ell_3 = (0, 1)$. When we specify a D6-brane using three oriented lines, we are actually giving an orientation to the three-dimensional subspace of the D6-brane that lies on the T^6. The orientation of a space of dimension k is defined by a choice of an ordered set of k linearly independent tangent vectors. The two-dimensional (x, y) plane, for example, can be oriented using the ordered pair of tangent vectors $((1, 0), (0, 1))$. Intuitively, this defines a circulation on the plane; the circulation is the direction in which one must rotate the first vector (with an angle smaller than π) to align it with the second vector. In more elementary descriptions, the orientation of a surface inside $\mathbb{R}^3$ is defined by an oriented direction normal to the surface. The two descriptions are related: the cross product of the ordered tangent vectors on the surface points in the direction of the oriented normal. For the three-dimensional subspace of a D6-brane lying on T^6, the set of ordered vectors is (ℓ_1, ℓ_2, ℓ_3).

Consider now a general situation where we have two D6-branes, a and b, each of which has three directions wrapped around T^6. Such a configuration is specified by

$$\text{D6-brane } a: \ (\ell_1^{(a)}, \ell_2^{(a)}, \ell_3^{(a)}),$$
$$\text{D6-brane } b: \ (\ell_1^{(b)}, \ell_2^{(b)}, \ell_3^{(b)}).$$

(21.7)

How many times do these D6-branes intersect on T^6? They intersect a number of times I_{ab} equal to the product of the intersection numbers of the corresponding lines on each of the three two-tori. Indeed, the general intersection point on T^6 is obtained by choosing one intersection point from each of the three two-tori. We thus have

$$I_{ab} = \#(\ell_1^{(a)}, \ell_1^{(b)}) \cdot \#(\ell_2^{(a)}, \ell_2^{(b)}) \cdot \#(\ell_3^{(a)}, \ell_3^{(b)}) = \prod_{i=1}^{3} \#(\ell_i^{(a)}, \ell_i^{(b)}).$$

(21.8)

If we let $\ell_i^{(a)} = (m_i^a, n_i^a)$ and $\ell_i^{(b)} = (m_i^b, n_i^b)$, then

$$I_{ab} = \prod_{i=1}^{3} (m_i^a \, n_i^b - m_i^b n_i^a).$$

(21.9)

The number I_{ab} carries a sign, the interpretation of which is straightforward. Since we have three intersections, there are three rotations that are needed to align brane a with brane b. The sign of I_{ab} is negative if the number of clockwise rotations is odd, and it is positive if the number of clockwise rotations is even. I_{ab} is the intersection number of the two oriented D6-branes. In particle physics models constructed with intersecting D6-branes, four-dimensional chiral fermions arise from open strings localized near each intersection. Even more, the sign of the intersection number determines the orientation of the open string which represents the chiral fermions. We will consider these matters in some detail in Sections 21.3 and 21.4.

If $I_{ab} = 0$, then at least one of the three intersection numbers $\#(\ell_i^{(a)}, \ell_i^{(b)})$, $i = 1, 2, 3$, must be equal to zero. Suppose that the intersection number vanishes only for $i = 1$. Then $\ell_1^{(a)}$ and $\ell_1^{(b)}$ must be either parallel or antiparallel. Since the lines we are considering are defined by relatively prime integers, we must have $\ell_1^{(a)} = \pm \ell_2^{(b)}$. The two D-branes then coincide over a set of n circles: the circle is the common closed line on the first torus, and n is the absolute value of the product of the intersection numbers on the second and third tori. If $\ell_1^{(a)}$ is displaced a little so that it does not coincide with $\ell_1^{(b)}$, then the two D-branes will also have no common points. With this understanding, we say that two branes with zero intersection number do not intersect.

21.2 D-branes and the Standard Model gauge group

We have seen that on the world-volume of N coincident D-branes there are $U(N)$ gauge fields, or gauge bosons, whose low energy dynamics is governed by a Yang–Mills theory

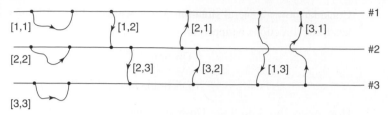

Fig. 21.4 Three D-branes that give rise to the gauge bosons of $U(3)$. The branes are shown separate, for convenience, but are supposed to be coincident. We show nine strings, each belonging to one of the nine open string sectors.

with gauge group $U(N)$. The gauge bosons of the Standard Model – the gluons, the $W^{\pm}$, the Z, and the photon – are all described by Yang–Mills theories. In this section we will examine the gauge group of the Standard Model, paying particular attention to the kind of D-brane configurations from which they could arise.

Let us start with the gluons, the gauge bosons that transmit the strong color force. Gluons are described by a four-dimensional $SU(3)$ Yang–Mills theory. This theory is closely related to the $U(3)$ Yang–Mills theory which arises at low energies on the world-volume of three coincident D3-branes (see Section 15.3). We will use the brane picture to understand what $SU(3)$ is. The D-brane configuration for $U(3)$ is shown in Figure 21.4, where the D-branes have been separated in order to be able to exhibit the various strings. There are nine open string sectors; they are labeled by $[ij]$, with $i, j = 1, 2, 3$. The figure shows one string from each sector. Every sector contains a string state that represents a gauge field.

Each of the three D-branes carries its own Maxwell field. These three Maxwell fields $A_{(i)\mu}$ ($i = 1, 2, 3$) are associated with the states $\alpha_{-1}|[ii]\rangle$, which represent open strings that begin and end on the same D-brane. In writing $\alpha_{-1}|[ii]\rangle$, we have suppressed both the momentum labels of the states and the spacetime indices of the oscillators. With the help of these gauge fields, we can build a general class of states:

$$\sum_{i=1}^{3} A_{(i)} \alpha_{-1}|[ii]\rangle, \qquad (21.10)$$

where we have suppressed the spacetime indices of the gauge fields, as well.

Let us now consider interactions, but restrict ourselves to low energies, where the string interactions become those of the Yang–Mills theory. The Maxwell fields $A_{(i)}$ do not interact with each other because the rule of combination of sectors (15.59) does not allow it: the endpoints of the different strings are never on the same brane. They do not have self-interactions, either, since Maxwell fields are free fields. The Maxwell fields $A_{(i)}$ do, however, interact with any state that carries their charge. Of the nine gauge fields in the D-brane configuration, the three Maxwell fields $A_{(i)}$ are on a special footing. Together, they comprise the largest possible set of gauge fields that have no interactions among its members.

Any string state is characterized by the values q_1, q_2, and q_3 of the charges that it carries with respect to the Maxwell fields $A_{(1)}$, $A_{(2)}$, and $A_{(3)}$, respectively. The charges of a state are collectively denoted by (q_1, q_2, q_3). In our convention, an oriented open string carries a unit negative charge at the $\sigma = 0$ endpoint and a unit positive charge at the $\sigma = \pi$ endpoint. A string in the [12] sector, for example, has charges $(-1, 1, 0)$. Here $q_1 = -1$ since the string begins on the first brane, $q_2 = +1$ since the string ends on the second brane, and $q_3 = 0$ since no endpoint lies on the third brane. For any string that is stretched from one brane to a different one, one charge is plus one, one charge is minus one, and one charge is zero. The sum of the three charges is zero. In fact, for a string beginning and ending on the same D-brane all three charges are zero.

If we think of a string in the limit in which it becomes a point with spacetime coordinates x^μ, the existence of the three charges implies the following coupling of the trajectory to the Maxwell fields:

$$\sum_{i=1}^{3} q_i \int A_{(i)} dx = \int \left(\sum_{i=1}^{3} A_{(i)} q_i \right) dx. \tag{21.11}$$

Because of (21.10), a change of basis in the space of states spanned by the $\alpha_{-1}|[ii]\rangle$ implies a linear redefinition of the Maxwell fields. On the other hand, (21.11) implies that a linear redefinition of the Maxwell fields results in a linear redefinition of the charges.

Since the idea of redefining fields and charges may be unfamiliar, let us give an example. Consider two Maxwell fields A_1 and A_2, and let (q_1, q_2) denote the charges of a given particle with respect to these Maxwell fields. The claim is that this physical system can be described using a different set of Maxwell fields, for which the charges are also different. To show this, we introduce new fields $A_\pm = (A_1 \pm A_2)/\sqrt{2}$ built from the original fields A_1 and A_2. In this setup, the sum on the right-hand side of (21.11) involves two terms only, and it can be rewritten as

$$A_1 q_1 + A_2 q_2 = \frac{A_1 + A_2}{\sqrt{2}} \frac{q_1 + q_2}{\sqrt{2}} + \frac{A_1 - A_2}{\sqrt{2}} \frac{q_1 - q_2}{\sqrt{2}}$$
$$= A_+ q_+ + A_- q_-, \tag{21.12}$$

where we defined $q_\pm = (q_1 \pm q_2)/\sqrt{2}$. The physics, we claim, can be described using Maxwell fields A_+ and A_- and particles that carry charges $[q_+, q_-]$. Imagine, for example, two charged particles. In the first description (using A_1 and A_2), the first particle has charges $(1, 0)$, and the second particle has charges $(0, 1)$. It is clear that these particles do not experience an electrostatic force: the second particle carries no A_1 charge, and the first particle carries no A_2 charge. Now we describe the same two particles using fields A_+ and A_-. The new charges are $[1/\sqrt{2}, 1/\sqrt{2}]$ for the first particle, and $[1/\sqrt{2}, -1/\sqrt{2}]$ for the second particle. The force between the particles now has two contributions that cancel: the two particles carry the same A_+ charges, but they also carry A_- charges of the same magnitude but opposite signs. The net force is still zero. More generally, you can verify that the electrostatic force between arbitrarily charged particles does not change.

Quick calculation 21.5 Consider two particles with charges (q_1, q_2) and (q'_1, q'_2), respectively. The electrostatic force between them is proportional to $q_1 q'_1 + q_2 q'_2$. Let the redefined charges of the two particles be $[q_+, q_-]$ and $[q'_+, q'_-]$, respectively. Show that $q_+ q'_+ + q_- q'_- = q_1 q'_1 + q_2 q'_2$.

Returning to our brane configuration, we now claim that the nine gauge fields can be split into two sets, such that the gauge fields in one set have no interactions with the gauge fields in the other set. One set has eight gauge fields, and the other set has one gauge field. That special gauge field is the Maxwell field $\bar{A}_{(3)}$ associated with the state

$$|s_3\rangle \equiv \frac{1}{\sqrt{3}} \left(\alpha_{-1}|[11]\rangle + \alpha_{-1}|[22]\rangle + \alpha_{-1}|[33]\rangle \right). \qquad (21.13)$$

Since Maxwell fields only interact with charged objects, it suffices to show that no gauge field carries its charge. In fact, we assert that the charge $\bar{q}_3$ associated with $\bar{A}_{(3)}$ is given by

$$\bar{q}_3 = \frac{1}{\sqrt{3}} \left(q_1 + q_2 + q_3 \right). \qquad (21.14)$$

If this is true, our claim holds; we have seen that the sum of the three charges is equal to zero for any string fully contained in the D-brane configuration.

To prove these claims, consider a change of basis states implemented by a general invertible linear transformation:

$$|s_i\rangle = \sum_{j=1}^{3} M_{ij}\, \alpha_{-1}|[jj]\rangle, \quad i = 1, 2, 3. \qquad (21.15)$$

We can use matrix notation to write this equation as

$$\mathbf{s} = M\,\mathbf{v}. \qquad (21.16)$$

Here $\mathbf{s}$ denotes the column vector with entries $|s_i\rangle$, $\mathbf{v}$ denotes the column vector with entries $\alpha_{-1}|[ii]\rangle$, and M is the invertible matrix with components M_{ij}. Letting $\mathbf{A}$ denote the row vector with entries $A_{(i)}$, we see that the general state in (21.10) can be rewritten as a dot product:

$$\mathbf{A} \cdot \mathbf{v} = \mathbf{A} \cdot M^{-1} M \mathbf{v} = \mathbf{A} M^{-1} \cdot \mathbf{s} \equiv \bar{\mathbf{A}} \cdot \mathbf{s}, \qquad (21.17)$$

where $\bar{\mathbf{A}} = \mathbf{A} M^{-1}$ denotes the Maxwell fields associated with the new basis states. Equation (21.11) can now be used to find the new charges, which are the linear combinations of the old charges that multiply the new fields. Focusing on the pairing between fields and charges and letting $\mathbf{q}$ denote the column vector of charges q_i, we have

$$\sum_{i=1}^{3} A_{(i)} q_i = \mathbf{A} \cdot \mathbf{q} = \mathbf{A} M^{-1} \cdot M \mathbf{q} = \bar{\mathbf{A}} \cdot M \mathbf{q} \equiv \bar{\mathbf{A}} \cdot \bar{\mathbf{q}}. \qquad (21.18)$$

Here the new charges $\bar{\mathbf{q}}$ are given in terms of the old charges $\mathbf{q}$ by

$$\bar{\mathbf{q}} = M \mathbf{q}. \qquad (21.19)$$

We see that the matrix M that defines the new states (21.16) also defines the new charges. This is reasonable, since equations (21.10) and (21.11) show that states and charges couple to the fields in the same way. Our result proves that (21.13) implies (21.14). This is what we had to establish in order to conclude that $\bar{A}_{(3)}$ decouples from the other eight gauge fields in the D-brane configuration. It can be proven that the other eight gauge fields cannot be split further into sets of fields that are mutually noninteracting.

There is one minor point that should also be addressed. In order to be able to read out charges correctly, the normalization of the F^2 terms in the action for the Maxwell fields must be preserved. Let $\mathbf{F}$ and $\bar{\mathbf{F}}$ denote the row vectors of field strengths obtained from the row vector gauge fields $\mathbf{A}$ and $\bar{\mathbf{A}}$, respectively. The relation $\mathbf{A} = \bar{\mathbf{A}}M$ then implies that $\mathbf{F} = \bar{\mathbf{F}}M$. The sum of F^2 terms is simply $\mathbf{F} \cdot \mathbf{F}$, where the entry to the right of the dot must be understood as a column vector, obtained by transposition from the row vector which is denoted with the same symbol. As a result, we have

$$\mathbf{F} \cdot \mathbf{F} = \bar{\mathbf{F}}M \cdot M^{\mathrm{T}}\bar{\mathbf{F}} = \bar{\mathbf{F}} \cdot MM^{\mathrm{T}}\bar{\mathbf{F}}. \qquad (21.20)$$

In order to obtain $\bar{F}^2$ terms with the same normalization as that of the F^2 terms, we need $MM^{\mathrm{T}} = \mathbf{1}$ – the matrix M must be orthogonal. This is not a severe constraint; an orthogonal M is obtained if we supplement (21.13) with the states

$$|s_1\rangle \equiv \frac{1}{\sqrt{6}}\left(\alpha_{-1}|[11]\rangle + \alpha_{-1}|[22]\rangle - 2\alpha_{-1}|[33]\rangle\right),$$

$$|s_2\rangle \equiv \frac{1}{\sqrt{2}}\left(-\alpha_{-1}|[11]\rangle + \alpha_{-1}|[22]\rangle\right). \qquad (21.21)$$

Quick calculation 21.6 Write out the matrix M explicitly and verify that it is orthogonal.

It is interesting to note that the decoupled gauge field is obtained by adding up similar states from each of the three D-branes. We can also consider the state

$$\alpha^a_{-1}|[11]\rangle + \alpha^a_{-1}|[22]\rangle + \alpha^a_{-1}|[33]\rangle, \qquad (21.22)$$

where the index a represents a direction normal to the D-branes. This state is associated with a displacement of the *full* collection of D-branes in the x^a direction (see the discussion below (15.37)). Just as the gauge field on a D-branc has massless scalar partners representing motion, the states above are partners of the decoupled gauge field.

We have mentioned before that the theory of a Maxwell gauge field is a $U(1)$ Yang–Mills theory. We have shown above that the $U(3)$ Yang–Mills theory of nine interacting gauge fields on three coincident D-branes contains a decoupled $U(1)$ theory, which we have identified explicitly. The remaining eight interacting gauge fields define the so-called $SU(3)$ gauge theory. This is the theory that governs the dynamics of the eight massless gluons of quantum chromodynamics (QCD). As groups, $U(3)$ is the group of 3-by-3 unitary matrices and $SU(3)$ is a subgroup of $U(3)$ obtained by considering only unitary matrices that have unit determinant. The relationship between the group $U(3)$ and the groups $SU(3)$ and $U(1)$ is written as

$$U(3) = SU(3) \times U(1), \qquad (21.23)$$

where the product notation is used for groups that act independently. Equation (21.23) describes a local relation between groups which are also manifolds; it ignores topological issues. Our physical analysis does not fully justify this notation, but it does show that the two factors correspond to gauge theories whose gauge fields do not interact with each other. The color force is described by $SU(3)$, so in order to obtain this theory from a configuration with three coincident D-branes we must understand the role of the additional Maxwell field. We will discuss this matter further in the following section. In general,

$$U(N) = SU(N) \times U(1), \tag{21.24}$$

where $U(N)$ is a theory with N^2 gauge fields and $SU(N)$ is a theory with $N^2 - 1$ gauge fields. As we have noted before, $U(N)$ gauge fields arise on the world-volume of N coincident D-branes.

The full set of Standard Model gauge bosons is described by the Yang–Mills theory with gauge group

$$SU(3)_c \times SU(2)_w \times U(1)_Y. \tag{21.25}$$

The subscript c is for color, w is for weak, and Y is for hypercharge. The $SU(2)_w \times U(1)_Y$ factors define the electroweak Yang–Mills theory. In $SU(2)_w$, there are $2^2 - 1 = 3$ gauge bosons. To realize $SU(2)_w$ with D-branes we need two additional coincident D-branes. This pair of D-branes should not coincide with the three color D-branes needed to obtain $SU(3)_c$, otherwise we would get a $U(5)$ Yang–Mills theory. If the two sets of branes are kept separate, the gauge group that emerges is

$$U(3) \times U(2) = SU(3) \times SU(2) \times U(1) \times U(1), \tag{21.26}$$

where we applied (21.24) twice and noted that the order of group factors has no significance. It is natural to ask if one of the $U(1)$ factors, or a combination of the two, can be identified with the hypercharge factor in (21.25). If we only cared about gauge bosons, the answer would be yes. In the Standard Model, however, we must also give the correct hypercharges to the fermions. This cannot be done with the $U(1)$s in (21.26). At least two additional D-branes appear to be necessary. We will discuss this matter further in the following section.

A process of symmetry breaking is necessary to reduce the gauge group in (21.25) to the one observed at low energies. When (21.25) applies, we get twelve massless gauge fields. At low energies, however, some of the Standard Model gauge fields are known to be massive. Indeed, three of the four gauge fields in $SU(2)_w \times U(1)_Y$ acquire mass through symmetry breaking, giving us the W^+, the W^-, and the Z^0. The fourth gauge field remains massless; it is the photon, which arises as a linear combination of the hypercharge field and one gauge field in the $SU(2)_w$ factor. After symmetry breaking, the gauge group is

$$SU(3)_c \times U(1)_{em}. \tag{21.27}$$

Symmetry breaking is triggered when certain charged scalar fields, the *Higgs* fields, acquire expectation values. In this process not only gauge fields, but also fermions, acquire mass. This brings us to the topic of fermions, which we discuss next.

21.3 Open strings and the Standard Model fermions

We have discussed the gauge bosons of the Standard Model, and we have begun to examine the D-brane configurations that are needed to obtain them. In order to use string theory to describe the full Standard Model, we must also learn about the matter particles and the charges that they carry. This is the main subject of the present section. We will also make preliminary observations regarding the representation of fermions as strings ending on the D-brane configurations that carry the gauge bosons. A detailed discussion of the embedding of the full Standard Model in a D-brane configuration is given in the following section.

Let us begin by examining some of the basic properties of fermions in four-dimensional spacetime. Consider for this purpose a prototype spin-1/2 fermion field; such a field represents a massive particle together with its (different) antiparticle. The classic example is the Dirac electron field, which describes the electron e^- and its antiparticle, the positron e^+. Imagine now a world where the mass of the electron and the mass of the positron are zero. The particle states would then be characterized by their *helicity*, the spin angular momentum along the direction of the motion. If the helicity takes the value $+1/2$ then the fermion state is said to be right handed. If the helicity takes the value $-1/2$ then the fermion state is said to be left handed. This is not altogether different from the way that we characterize photon states using a basis of right and left circularly polarized states; these are indeed photons of definite helicity. Massive fermion states can also be characterized by their helicity, but then the characterization is not Lorentz invariant. Imagine a massive fermion that moves with velocity $0 < v_x < c$ along the x axis of a Lorentz frame S and has spin angular momentum that points along the positive x axis. This is a fermion with positive helicity. In a Lorentz frame S' boosted along the x axis with velocity $v > v_x$, the particle moves in the negative x' direction, but its angular momentum does not reverse direction. In the S' frame the particle has negative helicity. If the particle is massless then it moves with the speed of light, and we cannot reverse its helicity by changing the Lorentz frame. We will be working with massless fermions.

Let f denote a fermionic particle and $\bar{f}$ denote the antiparticle. In this notation, e^- and e^+ correspond to f and $\bar{f}$, respectively. The quantum field theory which describes these particles will then include creation and annihilation operators for both the left- and right-handed states of the particle *and* for both the left- and right-handed states of the antiparticle. The creation operators, for example, would be written as:

$$(f_L^\dagger, \ f_R^\dagger), \ \ (\bar{f}_L^\dagger, \ \bar{f}_R^\dagger), \tag{21.28}$$

where L and R stand for left-handed and right-handed, respectively. The operator $f_L^\dagger$, for example, creates a left-handed particle state when it acts on the vacuum, while $\bar{f}_R^\dagger$ creates a right-handed antiparticle state when it acts on the vacuum. The creation operators contain other labels (such as momentum) that we are suppressing for simplicity.

Quantum field theory automatically incorporates an important property of fermions: if we specify the charges of the *left*-handed *particles*, then the charges of the *right*-handed *antiparticles* are determined; they are in fact opposite. If the charges are specified as electric charges for a set of noninteracting Maxwell fields, then opposite charges are simply charges of opposite sign. Similarly, the charges of the right-handed particles and those of the left-handed antiparticles are also opposite. To display these relationships, we write

$$f_L^\dagger \; \leftarrow \text{ opposite charge } \rightarrow \; \bar{f}_R^\dagger,$$
$$f_R^\dagger \; \leftarrow \text{ opposite charge } \rightarrow \; \bar{f}_L^\dagger, \tag{21.29}$$

where we have omitted, for brevity, the vacuum states. All charges are determined if we specify the charges of two states that do not appear on the same line. These charges can be specified independently. It suffices, for example, to fix the charges of the particles $f_L^\dagger$ and $f_R^\dagger$ to determine the charges of the antiparticles $\bar{f}_L^\dagger$ and $\bar{f}_R^\dagger$, or vice versa. Alternatively, we can fix the charges of the left-handed states $f_L^\dagger$ and $\bar{f}_L^\dagger$, or the charges of the right-handed states $f_R^\dagger$ and $\bar{f}_R^\dagger$. We will generally specify fermion charges by listing those of the left-handed states.

A central property of the Standard Model is that the spectrum of fermions is *chiral*. To understand this property we first consider a simpler case. Assume we have a theory whose entire fermion spectrum consists of the states in (21.29). The fermion is said to be chiral if the left- and right-handed particle states $f_L^\dagger$ and $f_R^\dagger$ do not have the same charges:

$$\text{if } f_L^\dagger \text{ and } f_R^\dagger \text{ do not have the same charges, the fermion is chiral.} \tag{21.30}$$

In fact, quantum field theory does not require both lines of (21.29) to exist in a theory. A theory can be consistent with only the states on the first line, or only the states on the second line. In such cases, unless the states are neutral, the fermion is automatically said to be chiral.

Charge describes the response of a particle to gauge bosons, so the left- and right-handed particle states of a chiral fermion respond differently to the same set of gauge bosons. Since the particle and antiparticle charges are correlated (see (21.29)), the left- and right-handed antiparticle states also respond differently to the same set of gauge bosons. The electron is chiral: left-handed electrons and right-handed electrons respond differently to the weak interactions, and so do left- and right-handed positrons. In fact, all the fermions in the standard model are chiral. Chirality is a very powerful property: in a gauge theory with chiral fermions, the fermions cannot acquire mass so long as the gauge symmetry that acts chirally remains unbroken. Thus the term chiral fermions is synonymous with massless fermions. In the Standard Model, the electroweak interactions $SU(2)_w \times U(1)_Y$ act chirally. The Standard Model fermions remain massless until symmetry breaking reduces the gauge group (21.25) down to $SU(3)_c \times U(1)_{\text{em}}$. Since neither the color force nor the electromagnetic force act chirally, the fermions can then acquire mass. The mass scale is set by the mass parameter that appears in the symmetry breaking sector of the theory. This sector is called the Higgs sector. As we mentioned earlier, the Higgs bosons are charged scalar fields that trigger symmetry breaking when they acquire expectation values.

Quick calculation 21.7 Convince yourself that a fermion is chiral if the charges of the left-handed particles and those of the left-handed antiparticles are *not* opposite.

Let us now describe the matter content of the Standard Model. The fermions of the Standard Model fall into three *generations*, or sets, all of which contain the same number of particles with exactly the same charges. Once the fermions acquire masses, the three generations are no longer identical. There is a hierarchy of masses; the first generation contains the lightest fermions and the third generation contains the heaviest fermions. Within each generation we have quarks and leptons. Quarks feel both the strong interactions and the electroweak interactions, while leptons feel only the electroweak interactions. Our discussion will focus on charges, so it suffices to consider a single generation. We will give the charges of all the left-handed states in one generation of quarks and leptons. In other words, we will list the charges of all the left-handed particles and left-handed antiparticles.

Consider first the quarks. The quarks feel the $SU(3)$ color force because they carry color. Quarks come in three colors, which we will call red (r), blue (b), and green (g). The left-handed states q_L of a quark are therefore of three types:

$$q_{Lr}, \ q_{Lb}, \quad \text{and} \quad q_{Lg}. \tag{21.31}$$

The collection of these three states is said to form the representation **3** of the group $SU(3)$. One writes $q_L \sim \mathbf{3}$. Under an $SU(3)$ gauge transformation, the three states are rotated by an $SU(3)$ matrix. Since the color force does not act chirally, the left-handed antiquarks $\bar{q}_L$ carry the opposite color charges. These are called anti-red (a-r), anti-blue (a-b), and anti-green (a-g):

$$\bar{q}_{La\text{-}r}, \ \bar{q}_{La\text{-}b}, \quad \text{and} \quad \bar{q}_{La\text{-}g}. \tag{21.32}$$

One writes $\bar{q}_L \sim \bar{\mathbf{3}}$, the bar included to mean opposite. In group theory, the **3** and the $\bar{\mathbf{3}}$ are said to be conjugate representations. When the states (21.31) are acted upon by an $SU(3)$ matrix M, the states (21.32) are acted upon by the complex conjugate matrix M^*.

This discussion about quarks and $SU(3)$ representations can be made more intuitive using D-branes and the open strings that can represent the quarks. We have seen that $SU(3)_c$ requires three coincident D-branes. The key insight is that quarks are simply open strings that have *one* endpoint on one of these three branes (recall that gluons have *both* endpoints on the collection of branes). We can use the labels red, blue, and green to refer to the first, second, and third brane, respectively. An open string that ends on the red brane is a left-handed red quark, a string that ends on the blue brane is a left-handed blue quark, and a

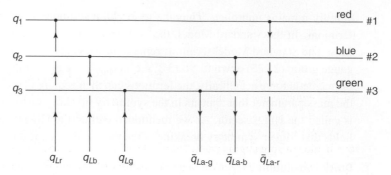

Fig. 21.5 The left-handed red, blue, and green quarks are open strings that *end* on the color D-branes. The left-handed anti-red, anti-blue, and anti-green antiquarks are open strings that *begin* on the color D-branes.

string that ends on the green brane is a left-handed green quark. The orientation of these strings points into the branes, or, equivalently, the endpoints on the brane are the $\sigma = \pi$ endpoints. How do the left-handed antiquarks arise? They are simply oppositely oriented open strings. An open string starting on the red brane, for example, would be a left-handed anti-red antiquark. All these strings are shown in Figure 21.5. We will refer to the three branes of $SU(3)_c$ as *color* branes, or *baryonic* branes. The name baryonic comes from baryon, a particle that is composed of three quarks.

We can also describe the $SU(3)$ charges of the quarks as follows. Recall that we characterize string states by their charges (q_1, q_2, q_3) with respect to the Maxwell fields that live on the branes. These three charges in fact define the $U(3)$ charge of a state. The charge for the decoupled $U(1)$ is proportional to $(q_1 + q_2 + q_3)$. We can label the $SU(3)$ charges using any *two* linear combinations of the charges that, together with $(q_1 + q_2 + q_3)$, define a linearly independent set. It is convenient to employ the pair (a_1, a_2) defined by

$$(a_1, a_2) \equiv (q_1 - q_2, \ q_2 - q_3). \tag{21.33}$$

The three left-handed quarks in (21.31), which comprise the representation **3** of $SU(3)$, are therefore characterized by

$$\mathbf{3}: \quad (1, 0), \quad (-1, 1), \quad (0, -1). \tag{21.34}$$

For the three left-handed antiquarks in (21.32) we have

$$\bar{\mathbf{3}}: \quad (-1, 0), \quad (1, -1), \quad (0, 1). \tag{21.35}$$

In the language of representation theory, a pair (a_1, a_2) is called a weight vector, and the entries a_1 and a_2 are the Dynkin labels of the weight vector. The **3** and the $\bar{\mathbf{3}}$ representations have three weight vectors.

Quick calculation 21.8 Give the eight weight vectors corresponding to the gauge field states in $SU(3)$. These states define the representation **8**, and the weights are the gluon charges. Each of the three gluons that begin and end on the same brane has weight $(0, 0)$ (why?). Only two of these belong to **8**, since one of them corresponds to the decoupled $U(1)$.

Only one endpoint of any open string that represents a quark lies on a color D-brane, so it is natural to ask: where does the other endpoint lie? For the left-handed quarks, we can find the answer by looking into their $SU(2)_w$ charges. Quark states fall into $SU(2)$ representations. The representations of $SU(2)$ are familiar from their role in the quantum mechanics of spin angular momentum. The $SU(2)_w$ of weak interactions has no relation to angular momentum, so one speaks of *isospin* instead of spin. The representation of isospin $I = 1/2$, for example, has two states, one with $I_3 = 1/2$ and the other with $I_3 = -1/2$, where I_3 denotes a third component of isospin.

Left-handed quark states fall into $I = 1/2$ representations of the weak interactions. New labels, called flavor labels, are needed to characterize the various types of quarks. For any fixed color, the left-handed u-quark is a state with $I_3 = 1/2$, and the (same color) left-handed d-quark is a state with $I_3 = -1/2$. Appropriately, the u and d stand for up and down, respectively. Thus u_L and d_L are members of an $SU(2)_w$ doublet, which we denote by **2**. Since quarks come in three colors, there are three $SU(2)$ doublets. Let us now think in terms of D-branes. We can build a $U(2) = SU(2)_w \times U(1)$ theory with two coincident D-branes. The left-handed u-quarks are strings that begin on one of these D-branes, call it the first brane, and end on one of the color branes. The left-handed d-quarks are strings that begin on the other D-brane, call it the second brane, and end on one of the color branes. The two D-branes we have introduced are called *left* branes. Left-handed quarks are open strings that begin on a left brane and end on a color brane.

The states of $SU(2)$ representations which arise from strings that have an endpoint on the left branes can also be characterized by charges. This time we have two charges $\bar{q}_1$ and $\bar{q}_2$ associated with the Maxwell fields living on brane one and brane two, respectively. As usual, in our convention $\bar{q}_i = +1$ for a string that ends on the brane i, and $\bar{q}_i = -1$ for a string that begins on the brane i. The charges $\bar{q}_1$ and $\bar{q}_2$ define the $U(2)$ charge of a state. The charge for the decoupled $U(1)$ is proportional to $(\bar{q}_1 + \bar{q}_2)$. We can label the $SU(2)$ charge using any combination of the charges that is independent from $(\bar{q}_1 + \bar{q}_2)$. It is convenient to use a Dynkin label

$$a_1 \equiv \bar{q}_1 - \bar{q}_2. \tag{21.36}$$

A string that begins on brane one has charges $(-1, 0)$, so $a_1 = -1$. Since this is a u-quark, which has $I_3 = 1/2$, we deduce a linear relation between the third component of isospin and the Dynkin label a_1:

$$I_3 = -a_1/2. \tag{21.37}$$

This relation also works for the d-quark: a string that begins on brane two has charges $(0, -1)$, which gives $a_1 = 1$ and $I_3 = -1/2$. Equation (21.37) illustrates how the quantum numbers of the states in a representation are related to the charges of the states.

Let us now try to visualize the full configuration of three coincident color branes and two coincident left branes. Whether or not the two sets of D-branes coincide, as long as they are parallel, the quark states stretching from one group of branes into the other fail to correspond to the Standard Model quarks in a dramatic way: they are not chiral. It can be shown that the spectrum contains left-handed quarks *and* right-handed quarks, with exactly the same charges. In a configuration with parallel D-branes, all states in (21.29) are in fact produced, and the left-handed and right-handed particles carry the same charges. The lack of chirality is manifest when the sets of D-branes are parallel and separate: the stretched strings are massive, and masses are not allowed for chiral fermions.

A physical situation is obtained if the coincident color D-branes *intersect* the coincident left D-branes. What we have in mind here is a configuration of intersecting D-branes of the type considered in Section 21.1. If the two sets of D-branes intersect, the fermion fields represented by strings that stretch from one set to the other will be localized near the intersection. Imagine beginning with coincident and parallel branes, where open strings can move in all the spatial directions along the branes. As the intersection angle grows away from zero, it is as if compactification had occurred: some directions of motion are lost. Some states that are massless at zero angle become massive. In fact, under suitable conditions, only the states in *one* line of (21.29) remain massless. We get a chiral fermion precisely because *only one* of the pairs is produced at the intersection: either a left-handed particle and its partner, or a left-handed antiparticle and its partner. If one pair is produced, the other pair may or may not be produced elsewhere on the D-brane configuration. There will be no reason for their charges to be correlated, if it is produced. This is why brane-intersection models naturally give a chiral fermion spectrum. In order to decide if a fermion spectrum is actually chiral, we must find all the fermion states that arise from the complete brane configuration.

We can now use Figure 21.6 to give a partial description of the situation. The color branes are shown horizontally, and the left branes are shown vertically. The three left-handed u-quarks and the three left-handed d-quarks are shown as stretched strings. The oppositely oriented strings (not shown) correspond to the oppositely charged right-handed antiquarks. The I_3 labels on the left branes indicate the values of I_3 for a string that *ends* on the branes.

The specification of charges for the left-handed quarks is completed by stating the values of the hypercharge Y. All three u_L quarks and all three d_L quarks are states of $Y = 1/6$. The information about representations and charges of fermions in the Standard Model is usually summarized using the notation

$$(\text{color, isospin})_Y. \tag{21.38}$$

Here color and isospin stand for the representations of $SU(3)$ and $SU(2)$, respectively. Y is simply the value of the hypercharge. This notation describes representations of the full Standard Model gauge group $SU(3) \times SU(2) \times U(1)$, since it specifies the representations with respect to each of the group factors. To illustrate the use of this notation, consider the states

$$\begin{pmatrix} u_{Lr} & u_{Lb} & u_{Lg} \\ d_{Lr} & d_{Lb} & d_{Lg} \end{pmatrix}, \tag{21.39}$$

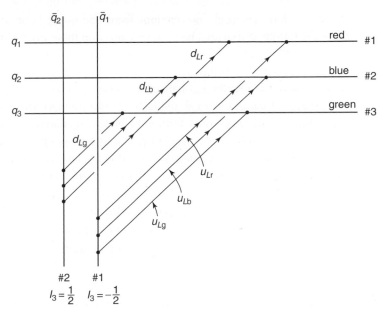

Fig. 21.6 The left-handed quarks are strings that stretch from the left branes to the baryonic branes. The three left-handed u-quarks are open strings that begin on the first left brane and end on a baryonic brane. The three left-handed d-quarks are open strings that begin on the second left brane and end on a baryonic brane.

which comprise the left-handed up and down quarks of all possible colors. The states can be viewed as a triplet of $SU(2)$ doublets (the three columns of the matrix) or, equivalently, as a doublet of $SU(3)$ triplets (the two rows of the matrix). In the notation of (21.38), the six states are denoted by

$$(\mathbf{3}, \mathbf{2})_{1/6} . \tag{21.40}$$

Since the labels for the color and isospin representations are actually equal to the number of states in the representations, the product $3 \times 2 = 6$ is the number of states in the full representation. All six states have the same hypercharge.

How do we calculate the hypercharge of states from the D-brane configuration? It turns out that the hypercharge receives contributions from both the decoupled $U(1)$ of the baryonic branes and the decoupled $U(1)$ of the left branes. Let us define conveniently normalized $U(1)$ charges Q_1 and Q_2, associated with the color and left branes, respectively:

$$Q_1 = q_1 + q_2 + q_3, \quad Q_2 = \bar{q}_1 + \bar{q}_2. \tag{21.41}$$

Any string that begins and ends on baryonic branes or that begins and ends on left branes has both Q_1 and Q_2 equal to zero. On the other hand, a left-handed quark is a string that begins on a left brane and ends on a color brane. Any such string has $Q_1 = 1$ and $Q_2 = -1$. The hypercharge of the left-handed quarks is obtained if we take

$$Y = -\tfrac{1}{3} Q_1 - \tfrac{1}{2} Q_2 - \cdots, \tag{21.42}$$

where the dots represent contributions from additional D-branes, to be included later. Since the left-handed quarks have no endpoints on these extra branes, $Y = -(1/3) \times 1 - (1/2) \times (-1) = 1/6$.

We must now consider the left-handed *antiquarks*, $\bar{u}_L$ and $\bar{d}_L$. At this point, we encounter the chirality of the electroweak interactions. Had the electroweak interactions been non-chiral, these antiquarks would form a doublet with charges opposite to those of the left-handed u and d quarks; they would be seen as strings ending on the left branes. It turns out, however, that the left-handed *antiquarks* $\bar{u}_L$ and $\bar{d}_L$ are each $SU(2)$ singlets (with representation denoted by **1**). The corresponding strings cannot then have endpoints on the left branes. Moreover, $\bar{u}_L$ has $Y = -2/3$ and $\bar{d}_L$ has $Y = 1/3$. A nonchiral coupling would have required $Y = -1/6$ for both of them. It follows from this and our earlier analysis of color that the left-handed antiquarks are in the representations

$$\bar{u}_L \sim \left(\bar{\mathbf{3}}, \mathbf{1}\right)_{-2/3} \quad \text{and} \quad \bar{d}_L \sim \left(\bar{\mathbf{3}}, \mathbf{1}\right)_{1/3}. \tag{21.43}$$

Here we have suppressed the color labels; thus $\bar{u}_L$, for example, stands for the three antiquarks of different colors. The representations of the left-handed quarks and antiquarks are then summarized as

$$\begin{pmatrix} u_L \\ d_L \end{pmatrix} \sim (\mathbf{3}, \mathbf{2})_{1/6}, \quad \bar{u}_L \sim \left(\bar{\mathbf{3}}, \mathbf{1}\right)_{-2/3} \quad \text{and} \quad \bar{d}_L \sim \left(\bar{\mathbf{3}}, \mathbf{1}\right)_{1/3}. \tag{21.44}$$

This is the set of left-handed quark and antiquark states in the first generation of the Standard Model. The generation also includes the corresponding right-handed states.

The $\bar{u}_L$ antiquarks are strings that begin on a color brane. Since they cannot end on a left brane (they are $SU(2)_w$ singlets), they must end on a new D-brane. If we let Q_3 denote the electric charge that couples to the Maxwell field on this D-brane, we find that (21.42) must be changed to

$$Y = -\tfrac{1}{3}Q_1 - \tfrac{1}{2}Q_2 - Q_3 - \cdots. \tag{21.45}$$

The hypercharge $Y = -2/3$ of the $\bar{u}_L$ states emerges because $Q_1 = -1$, $Q_2 = 0$, and $Q_3 = 1$. The $\bar{d}_L$ antiquarks are also strings that begin on a color brane and cannot end on a left brane. Their hypercharge $Y = 1/3$ is correctly reproduced by (21.45) with $Q_1 = -1$ and $Q_2 = 0$, if the strings end on a D-brane that does not contribute to Y. We will discuss such additional D-branes in the next section.

Let us now describe the left-handed leptons in the first generation. These are the left-handed electron-neutrino and the left-handed electron, together with the left-handed anti-neutrino and the left-handed positron. None of the leptons carries color; they are all color singlets, and their $SU(3)$ representation is denoted by **1**. The left-handed neutrinos and the left-handed electrons form an $SU(2)$ doublet with hypercharge $Y = -1/2$. Again, we have chirality. Both the left-handed positron and the left-handed anti-neutrino are $SU(2)$

singlets, and their hypercharges are one and zero, respectively. The lepton charges are summarized by

$$\begin{pmatrix} \nu_{eL} \\ e_L^- \end{pmatrix} \sim (\mathbf{1}, \mathbf{2})_{-1/2} \,, \quad e_L^+ \sim (\mathbf{1}, \mathbf{1})_1 \,, \quad \text{and} \quad \bar{\nu}_{eL} \sim (\mathbf{1}, \mathbf{1})_0 \,. \tag{21.46}$$

Note that the left-handed anti-neutrinos are color, weak, and hypercharge singlets. These states have not been detected directly, but they are likely to exist because neutrinos appear to have mass. The states in (21.44) and (21.46), together with the corresponding right-handed states, comprise the matter states in the first family of the Standard Model. As you have probably already realized, the open strings that represent leptons cannot end on color D-branes. The left-handed neutrino and the left-handed electron arise from strings that have one endpoint on a left brane and another endpoint on some other brane. Neither the left-handed positron nor the left-handed anti-neutrino have an endpoint on a color brane or a left brane.

The full set of left-handed states in a generation is obtained by listing the states in (21.44) together with those in (21.46). Suppressing the names of the states, and using the $+$ symbol to collect the representations together, we have

$$(\mathbf{3}, \mathbf{2})_{1/6} + \left(\bar{\mathbf{3}}, \mathbf{1} \right)_{-2/3} + \left(\bar{\mathbf{3}}, \mathbf{1} \right)_{1/3} + (\mathbf{1}, \mathbf{2})_{-1/2} + (\mathbf{1}, \mathbf{1})_1 + (\mathbf{1}, \mathbf{1})_0 \,. \tag{21.47}$$

As we have learned, charge reversal exchanges the $\mathbf{3}$ and $\bar{\mathbf{3}}$ of $SU(3)$ and reverses the sign of the hypercharge. Under charge reversal all singlets $\mathbf{1}$ remain unchanged, because they represent states with zero charge. For $SU(2)$, charge reversal exchanges the $\mathbf{2}$ and $\bar{\mathbf{2}}$, but it turns out that the $\bar{\mathbf{2}}$ is a representation that is equivalent to the $\mathbf{2}$. So charge reversal leaves the $\mathbf{2}$ of $SU(2)$ unchanged.

On account of the comments below (21.29), we can list the states in a generation in alternative ways.

Quick calculation 21.9 List the right-handed states in a generation, together with their corresponding charges.

Quick calculation 21.10 Describe the matter states in a generation by listing the left- and right-handed *particle* states, together with their corresponding charges.

In the Standard Model, the electric charge Q_{em} arises from a linear combination of the hypercharge Y and the third component I_3 of isospin:

$$Q_{\text{em}} = Y + I_3. \tag{21.48}$$

If we apply this formula to the u_L and d_L states, we find

$$Q_{\text{em}}(u_L) = Y(u_L) + I_3(u_L) = \tfrac{1}{6} + \tfrac{1}{2} = +\tfrac{2}{3},$$
$$Q_{\text{em}}(d_L) = Y(d_L) + I_3(d_L) = \tfrac{1}{6} - \tfrac{1}{2} = -\tfrac{1}{3}. \tag{21.49}$$

These are indeed the correct values. The proton is made up of two up quarks and one down quark. Its electric charge equals $2 \times \frac{2}{3} - \frac{1}{3} = 1$.

Quick calculation 21.11 Find the electric charges of the left-handed antiquarks, and verify that they are opposite to those of the left-handed quarks. Show that (21.48) gives zero electric charge to the left-handed neutrino and anti-neutrino states, electric charge minus one to the left-handed electron states, and electric charge plus one to the left-handed positron states. Conclude that electromagnetism does not couple chirally.

As far as states and charges are concerned, the other two generations of the Standard Model are copies of the first generation. The quark flavors in the second generation are called charm and strange, and denoted by c and s respectively. The leptons in the second generation are the muon-neutrino ν_μ and the muon μ^-. In the third generation we have a top quark t and a bottom quark b, as well as a tau-neutrino ν_τ and a tau τ^-. The full set of left-handed states in the Standard Model consists of three copies of the states in (21.47):

$$3 \times \left[(\mathbf{3}, \mathbf{2})_{1/6} + \left(\bar{\mathbf{3}}, \mathbf{1} \right)_{-2/3} + \left(\bar{\mathbf{3}}, \mathbf{1} \right)_{1/3} + (\mathbf{1}, \mathbf{2})_{-1/2} + (\mathbf{1}, \mathbf{1})_1 + (\mathbf{1}, \mathbf{1})_0 \right].$$

(21.50)

We can now state precisely the chirality property of the fermion spectrum in the Standard Model. A spectrum is said to be *non*chiral if the set of left-handed states can be split into pairs of left-handed states with opposite charges (see Quick calculation 21.7). Thus, given a list of charged left-handed states that defines a nonchiral spectrum, the operation of reversing the charges of all the states must leave the *list* of states invariant. The fermion spectrum of the standard model is chiral because the operation of reversing all charges in the list (21.50) changes the list.

The gauge group and the matter content of the Standard Model may seem to you rather intricate or perhaps even cumbersome. But this set of particles and interactions in fact provides a rather economical description of an extremely large number of experimental results obtained over the past few decades. The Standard Model of particle physics is indeed a magnificent achievement. It is not a final theory of particle physics, nor is it a complete one, but it seems certain that the Standard Model must appear in the low energy limit of any correct unified theory of all interactions. It is in this sense that the Standard Model of particle physics has become a permanent part of our knowledge about the physical world.

21.4 The Standard Model on intersecting D6-branes

In the two previous sections we acquainted ourselves with some of the ingredients needed to build a string theory model of elementary particles. In this section, we build a complete string model that has many of the features of the Standard Model. The model involves intersecting D6-branes wrapped on a T^6, in the framework of type IIA superstring theory.

Before symmetry breaking, it has all the massless particles of the Standard Model, but it also contains a few extra particles. It is in fact possible to construct a model with orientifold O6-planes and D6-branes that gives *precisely* the particle content of the Standard Model. An orientifold Op-plane is an extended object with p spatial dimensions. The basic properties of orientifold planes were studied earlier in a series of problems (13.6, 15.1, 15.3, 15.4, and 15.5). We will introduce some of the features of this second model, relegating most of the analysis to the problems at the end of the chapter.

One key property of the Standard Model matter content is its replication: there are three generations that contain fermionic matter states with identical charges. How do we explain such replication in terms of D-branes? We explain it in terms of multiple intersections. We found the left-handed u and d quarks at the intersection of the coincident baryonic branes with the coincident left branes. If these sets of branes wrap around the T^6 in such a way that they intersect three times, then the second intersection will give left-handed c and s quarks, and the third intersection will give left-handed t and b quarks. We learned in Section 21.1 that general classes of D6-branes wrapped on $T^6 = T^2 \times T^2 \times T^2$ are characterized by three lines ℓ_1, ℓ_2, and ℓ_3, each of which is, in turn, characterized by two integers. Moreover, we found the formula (21.9), which gives the number of times that two D6-branes intersect.

Let $N_1 = 3$ denote the number of baryonic branes, and let $N_2 = 2$ denote the number of left branes. Moreover, let them be wrapped on T^6 as follows:

$$N_1 = 3: \quad \ell_1^{(1)} = (1, 2), \quad \ell_2^{(1)} = (1, -1), \quad \ell_3^{(1)} = (1, -2),$$
$$N_2 = 2: \quad \ell_1^{(2)} = (1, 1), \quad \ell_2^{(2)} = (1, -2), \quad \ell_3^{(2)} = (-1, 5). \tag{21.51}$$

The intersection number I_{12} between a single baryonic brane and a single left brane is then

$$I_{12} = \left(1 \times 1 - 1 \times 2\right) \cdot \left(1 \times (-2) - 1 \times (-1)\right) \cdot \left(1 \times 5 - (-1) \times (-2)\right) = 3. \tag{21.52}$$

The intersection number is the desired one. Of course, the constraint $I_{12} = 3$ does not determine the wrappings in (21.51). These wrapping numbers are one of many possibilities that work.

We found that the left-handed quarks are obtained as strings that stretch from the N_2 branes to the N_1 branes. The orientation of these strings is correlated with the sign of I_{ab}. Let I_{ab} denote the intersection number for a D6-brane a and a D6-brane b. The precise rule that gives the number of fermions and specifies how they are represented by strings is given as follows.

> There are $|I_{ab}|$ left-handed fermions at the intersection set of brane a and brane b, one left-handed fermion at each intersection point. If $I_{ab} > 0$, the left-handed states are strings that stretch from brane b to brane a. If $I_{ab} < 0$, the left-handed states are strings that stretch from brane a to brane b.

The orientation of the strings determines the charges of the left-handed states. The states produced at an intersection are chiral in the sense discussed below (21.30): only the minimal set of states is produced. The oppositely oriented strings at the intersection represent the oppositely charged right-handed *antiparticles* that must necessarily accompany the left-handed particle states.

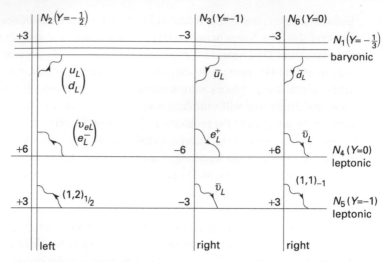

The brane configuration consisting of $N_1 = 3$ baryonic branes, $N_2 = 2$ left branes, $N_3 = N_6 = 1$ right branes, and $N_4 = N_5 = 1$ leptonic branes. The values of Y indicate the contribution to the hypercharge for a string that ends on the brane. At the intersections we give the intersection numbers I_{ab} with $a < b$. The name right branes is used because right-handed *particles* are attached to them (we actually show the left-handed antiparticles).

For the three generations of left-handed quarks obtained above, the value $Y = 1/6$ of the hypercharge is correctly given by (21.42). The picture of the branes is that of Figure 21.6, with the understanding that the displayed intersection actually occurs two additional times. To indicate this on the diagram, we insert a $+3$ at the intersection, as shown in Figure 21.7. We now try to obtain the three generations of left-handed antiquarks. Consider first the left-handed u-antiquark and its two copies (the left-handed c and t antiquarks). The open string that represents this quark begins on a baryonic brane, but it cannot end on a left brane because the state is an $SU(2)$ singlet. We thus need a new D-brane. Let $N_3 = 1$ represent another D-brane. According to the rule above, we need $I_{13} = -3$, since the open strings must start on the baryonic branes. Additionally, the contribution to the hypercharge from the charge Q_3 on this brane must be such that $Y(\bar{u}_L) = -2/3$. This requires that (21.42) be replaced by

$$Y = -\tfrac{1}{3} Q_1 - \tfrac{1}{2} Q_2 - Q_3 - \cdots. \tag{21.53}$$

We then find $Y(\bar{u}_L) = -(1/3)(-1) - (1/2)(0) - 1 = -2/3$, as desired. We fix the wrapping of the brane on T^6 by

$$N_3 = 1: \quad \ell_1^{(3)} = (1, 1), \quad \ell_2^{(3)} = (1, 0), \quad \ell_3^{(3)} = (-1, 5). \tag{21.54}$$

Including the result in (21.52), we then find

$$I_{12} = 3, \quad I_{13} = -3, \quad I_{23} = 0. \tag{21.55}$$

We see that the N_3 brane does not intersect the left branes. The N_3 brane is shown in Figure 21.7 as a vertical brane, placed to the right of the left branes. The equation $Y = -1$

is added to the label of the brane to indicate that a string that ends on this brane receives a contribution to the hypercharge equal to minus one (this number is equal to the coefficient of Q_3 in (21.53)). Similar labels are attached to the baryonic and to the left branes. The N_3 brane is called a right brane because right-handed quark states are attached to it. Indeed, instead of the $\bar{u}_L$ antiquark shown in the figure, we could have exhibited the oppositely oriented string which gives the right-handed quark state u_R.

In this model we still have to add three additional D-branes: another right brane to obtain the left-handed $\bar{d}_L$ antiquarks (and its copies) and two separate *leptonic* branes to obtain the leptons. Since we are not in a position to derive the wrappings of the D-branes, let us simply state them. Together with the

$$N_4 = 1: \quad \ell_1^{(4)} = (1, 2), \quad \ell_2^{(4)} = (-1, 1), \quad \ell_3^{(4)} = (1, 1),$$
$$N_5 = 1: \quad \ell_1^{(5)} = (1, 2), \quad \ell_2^{(5)} = (-1, 1), \quad \ell_3^{(5)} = (2, -7),$$
$$N_6 = 1: \quad \ell_1^{(6)} = (1, 1), \quad \ell_2^{(6)} = (3, -4), \quad \ell_3^{(6)} = (1, -5). \tag{21.56}$$

Both the N_4 and the N_6 branes are declared not to contribute to the hypercharge, but the N_5 brane is declared to contribute minus one unit. So the final formula for the hypercharge reads

$$Y = -\tfrac{1}{3} Q_1 - \tfrac{1}{2} Q_2 - Q_3 - Q_5. \tag{21.57}$$

The remaining intersection numbers are easily calculated and the complete list is

$$\begin{array}{lll}
I_{12} = 3, & I_{13} = -3, & I_{23} = 0, \\
I_{14} = 0, & I_{15} = 0, & I_{16} = -3, \\
I_{24} = 6, & I_{25} = 3, & I_{26} = 0, \\
I_{34} = -6, & I_{35} = -3, & I_{36} = 0, \\
I_{45} = 0, & I_{46} = 6, & I_{56} = 3.
\end{array} \tag{21.58}$$

Since $I_{ab} = -I_{ba}$, we have only listed intersection numbers I_{ab} with $a < b$. Those are the values exhibited at the intersections in Figure 21.7.

Quick calculation 21.12 Confirm the values of all the intersection numbers in (21.58), and check that they are correctly recorded in the figure. Check also that the strings shown in the figure have the orientation required by the rule for left-handed states.

We can organize the list of intersections using the horizontal branes on the figure, all of which intersect with the two coincident left branes and the two right branes. The intersections on the baryonic branes give

$$\underbrace{3\,(\mathbf{3}, \mathbf{2})_{1/6}}_{[12]} + \underbrace{3\left(\overline{\mathbf{3}}, \mathbf{1}\right)_{-2/3}}_{[13]} + \underbrace{3\left(\overline{\mathbf{3}}, \mathbf{1}\right)_{1/3}}_{[16]}, \tag{21.59}$$

where the $[ab]$ labels under the representations indicate that the corresponding fermions arise from the I_{ab} intersection. The above representations are precisely three copies of a single generation of quarks, as you can see by comparing with (21.44). The intersections on the leptonic brane N_4 give

$$\underbrace{6\,(\mathbf{1},\mathbf{2})_{-1/2}}_{[24]} + \underbrace{6\,(\mathbf{1},\mathbf{1})_{1}}_{[34]} + \underbrace{6\,(\mathbf{1},\mathbf{1})_{0}}_{[46]}. \qquad (21.60)$$

This result justifies the name leptonic. The above representations are six copies of a single generation of leptons, as you can see by comparing with (21.46). This is more than the three copies that we needed, but there is even more. The intersections on the leptonic N_5 brane give us

$$\underbrace{3\,(\mathbf{1},\mathbf{2})_{1/2}}_{[25]} + \underbrace{3\,(\mathbf{1},\mathbf{1})_{0}}_{[35]} + \underbrace{3\,(\mathbf{1},\mathbf{1})_{-1}}_{[56]}. \qquad (21.61)$$

Quick calculation 21.13 Use (21.57) and the picture of the strings in Figure 21.7 to confirm that the hypercharge assignments in (21.60) and in (21.61) are correct.

We found more leptons than we wanted, but this is unavoidable in consistent models that only have D6-branes. There is a simple rule that must be obeyed.

> Rule: the set of left-handed states that end on any collection of D-branes
>   must contain equal numbers of incoming and outgoing strings.   (21.62)

We will not derive this rule, but we can check that it holds for Figure 21.7. Consider the left branes, for example. There are nine outgoing quark doublets (three colors and three families). But there are also six incoming doublets from N_4 and three incoming doublets from N_5. This shows that we cannot obtain only three leptonic doublets.

Quick calculation 21.14 Verify that the numbers of ingoing and outgoing left-handed states are equal on each of the two right branes and on each of the two leptonic branes.

We can also state a general consistency condition that must be satisfied by the model. The condition has a topological character. Consider a point charge q at the origin of the (x, y) plane. In this two-dimensional world, the electric field lines extend radially from the charge all the way to infinity. On the other hand, if the plane is turned into a two-sphere of finite size, we have a problem with our charge: the field lines have nowhere to go. If the charge is at the north pole of the sphere, then the field lines will tend to go to the south pole. The solution is clear. Consistency requires a charge $(-q)$ at the south pole. The general conclusion is that the total charge in a compact space without boundary must be zero. This is easily derived.

Quick calculation 21.15 Consider the Maxwell equation $\nabla \cdot \vec{E} = \rho$, and integrate both sides of the equation over a compact space that has no boundary. Use the divergence theorem to show that the total charge must vanish.

In our present configuration we have D6-branes, which are objects that carry Ramond–Ramond (R–R) charge. Moreover, they wrap a boundaryless compact space T^6. So there is a requirement of zero total charge. The formulation of the condition goes as follows. Introduce three pairs of formal variables: (x_1, y_1), (x_2, y_2), and (x_3, y_3). For any D6-brane a specified by

$$\ell_1^{(a)} = (m_1^a, n_1^a), \quad \ell_2^{(a)} = (m_2^a, n_2^a), \quad \ell_3^{(a)} = (m_3^a, n_3^a), \tag{21.63}$$

we construct the following polynomial of the six formal variables:

$$\Pi_a(x_i, y_i) \equiv (m_1^a x_1 + n_1^a y_1)(m_2^a x_2 + n_2^a y_2)(m_3^a x_3 + n_3^a y_3). \tag{21.64}$$

The polynomial Π_a encodes the R–R charges of the D6-brane of type a. If we expand the product, the polynomial consists of eight independent monomials. The consistency condition simply states that the result of adding the polynomials associated with each of the D6-branes in the configuration must be zero. If N_a denotes the number of D6-branes of type a, the consistency condition takes the form

$$\sum_a N_a \Pi_a(x_i, y_i) = 0. \tag{21.65}$$

This gives a total of eight consistency conditions on the numbers N_a and on the windings of the various D6-branes. For example, the vanishing of the coefficient of $x_1 x_2 x_3$ gives

$$\sum_a N_a m_1^a m_2^a m_3^a = 0. \tag{21.66}$$

This equation is easily tested using (21.51), (21.54), and (21.56). We find that it requires the vanishing of $N_1 - N_2 - N_3 - N_4 - 2N_5 + 3N_6$. Indeed, this quantity vanishes: $3 - 2 - 1 - 1 - 2 + 3 = 0$. In fact, the rule (21.62) is a consequence of (21.65), as you will show in Problem 21.2.

In order to build a string model with the particle content of the Standard Model, one needs to introduce orientifold planes, which effectively introduce image D6-branes. Remarkably, it is possible to build a configuration of branes and orientifolds so that (1) the gauge group is the Standard Model gauge group (21.25) and all additional $U(1)$ factors disappear, and (2) the set of chiral fermions is precisely that of the Standard Model (21.50). Intersecting D6-brane models are the first string theory models that have given precisely the particle content of the Standard Model. A qualification is necessary: the closed string sector of the theory may contain additional unobserved particles and interactions.

We will not discuss these models in full detail, but in order to satisfy your curiosity we will show the brane configuration and make a few remarks (a number of consistency checks have been relegated to Problem 21.3). The set of intersecting branes is shown in Figure 21.8. Comparing with the model in Figure 21.7, we see that there are still $N_1 = 3$ baryonic branes and $N_2 = 2$ left branes, but there is only one right brane $N_3 = 1$ and only one leptonic brane $N_4 = 1$. The orientifolds introduce image D-branes, which are denoted by affixing an asterisk to the brane labels and are shown as dashed lines. The configuration is chosen such that no brane intersects with its own image.

The main purpose of this construction is to avoid the extra leptonic doublets that we obtained above. In this model the left-handed quark doublets arise in a new way. The doublet in the first generation appears at the intersection $I_{12} = 1$ of the baryonic branes and

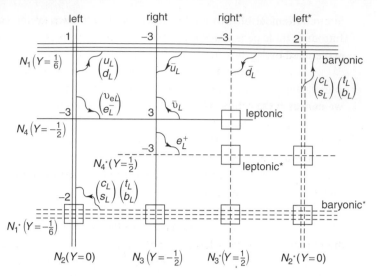

Fig. 21.8 The brane configuration that leads to a Standard Model gauge group and matter content. There are $N_1 = 3$ baryonic branes, $N_2 = 2$ left branes, $N_3 = 1$ right branes and $N_4 = 1$ leptonic branes. The image D-branes created by the orientifolds are shown in dashed lines. Intersections framed by a square are mirrors of previously accounted intersections and do not give new particles.

the left branes. The other two doublets appear at the intersection $I_{12*} = 2$ of the baryonic branes and the image of the left branes (top right of the figure). Equivalently, the rules of orientifolds allow us to view these two doublets as appearing at the intersection $I_{1*2} = -2$ of the left branes with the images of the baryonic branes (bottom left of the figure). Rule (21.62) continues to hold for the left branes: three quark doublets and three lepton doublets are outgoing, and six quark doublets are incoming.

We have implicitly been working with supersymmetric type IIA superstring theory; we are using stable D6-branes in a ten-dimensional spacetime. The branes and their intersections break all the supersymmetry. This had to be so, since we obtained the Standard Model spectrum, which does not have supersymmetry (the bosons and fermions do not appear in pairs of equal mass). In models without supersymmetry it is typically assumed that the string scale is low, perhaps on the order of a few TeV. This requires some large extra dimensions, otherwise the four-dimensional Planck length cannot be reproduced. In models with a low string scale, the lifetime of the proton can turn out to be too short. No proton decay has yet been observed, and the proton lifetime certainly exceeds 4×10^{33} years. In intersecting brane models the proton cannot decay by conventional processes where open strings join or split (see Problem 25.8), so proton decay appears to be safely suppressed.

You may ask: can we declare victory and state that the Standard Model has emerged from string theory? Not quite. The model must still do a lot more to be realistic. In particular, electroweak symmetry breaking must work out correctly. Recall that symmetry breaking is the process by which the Standard Model gauge group is reduced to (21.27), and the fermions acquire mass. Moreover, the brane configuration is not fixed uniquely by the requirement that the Standard Model spectrum emerges; there are discrete choices to be made in selecting the brane wrapping numbers. In addition, there are other choices that must be made. For example, the positions of some D-branes on T^6 can be adjusted without altering the massless spectrum. The sizes of the tori are not automatically fixed either. If no choices had been possible, we would have a single candidate for the Standard Model within this class of brane configurations.

In intersecting D-brane models, electroweak symmetry breaking happens through a process of brane recombination. For certain values of the intersection angles between two branes, tachyon states appear in the spectrum of open strings that stretch between them. These tachyons indicate an instability that can cause the intersecting branes to combine at the intersection and produce a single brane. With fewer branes present, the gauge group is reduced. Moreover, the total number of intersections will be reduced and, consequently, the number of massless fermions will be reduced, as well (Problem 21.4). It is not clear whether any of these models will be able to produce the expected spectrum *after* symmetry breaking. This is not easy: a large number of mass parameters and other couplings must work out correctly. If this were possible, it would be a truly exciting result. There is a lot more that one can say about intersecting brane models and the more recent work that focuses on models that incorporate supersymmetry. It is now time, however, to discuss briefly other ways in which the Standard Model may appear in string theory.

21.5 String theory models of particle physics

Intersecting brane models are particularly attractive, but they are by no means the only avenues that have been investigated in order to construct string models of particle physics. We had a detailed look at the intersecting brane models because they are simple enough and can be understood very concretely. Past and present-day attempts to obtain the Standard Model follow several lines of attack, which can be organized by the starting points of the constructions. These possible starting points are the five supersymmetric string theories and M-theory (see Section 14.1). Each of these theories can be used to investigate how the Standard Model could emerge. The intersecting brane models that we discussed, for example, use type IIA superstrings as the starting point. Since the various supersymmetric strings and M-theory are different limits of a single theory, the various approaches are no doubt related at some level. Nevertheless, each starting point gives somewhat different insights into the various phenomenological questions that arise in the process of constructing a model.

The early attempts to do string phenomenology were based on the heterotic $E_8 \times E_8$ superstring theory. In this theory, six of the nine spatial dimensions are curled up into a small six-dimensional compact space with rather special properties: a *Calabi–Yau* space. Calabi–Yau spaces have both discrete and continuous parameters, which determine the details of the four-dimensional theory that arises upon compactification. For all Calabi–Yau spaces the minimal amount of supersymmetry survives the compactification; the resulting four-dimensional theory is said to be $N = 1$ supersymmetric. If supersymmetry is present, it must be minimal for chiral fermions to exist. The compactification also allows one to break the original gauge symmetry $E_8 \times E_8$ down to $E_6 \times E_8$. One then views the E_8 factor as part of a hidden sector that has only indirect effects on our visible sector. The gauge group E_6 appears in the visible sector as the gauge group of a grand unified theory. The number of generations of chiral fermions depends on the topology of the Calabi–Yau space, and models with three generations can be obtained. The group E_6 contains $SU(3) \times SU(2) \times U(1)$ as a subgroup, so the Standard Model gauge group can arise upon further symmetry breaking. Calabi–Yau compactification of heterotic strings gave the first string models with semi-realistic particle physics. Since Calabi–Yau spaces are quite complicated, it turned out to be difficult to make progress with the questions of symmetry breaking and supersymmetry breaking.

The technical complications that must be faced with Calabi–Yau spaces prompted physicists to search for alternative six-dimensional spaces to compactify the heterotic string. If one simply uses a six-torus T^6, more than minimal supersymmetry survives and it becomes impossible to construct realistic models. Orbifolds provide a nice middle ground between Calabi–Yau spaces and tori. Orbifolds built from tori, for example, are much easier to analyze than general Calabi–Yau spaces, but they can still give $N = 1$ supersymmetric theories in four dimensions. It is possible to obtain semi-realistic models from orbifold compactifications of the heterotic string. Orbifolds also play an important role in the compactifications of other string theories.

More recently, physicists have been vigorously investigating the phenomenological possibilities of type II and type I superstring theories. One important class of models uses intersecting D-branes and orientifolds, as we considered in the previous section. Variations exist: one can use branes of different dimensionalities, or one can attempt to preserve some supersymmetry. A popular approach breaks the problem of constructing a fully consistent model into problems that can be analyzed separately.

One considers, for example, a type II string theory where the six extra dimensions form an infinite volume space that is everywhere flat except for a singularity at the origin. As you can imagine, flat extra dimensions with infinite volume cannot possibly give a realistic model: the extra dimensions would be visible. This complication, however, is ignored in the first part of the analysis. An example of such space is provided by the orbifold $\mathbb{C}^3/\mathbb{Z}_3$. Here $\mathbb{C}^3$ is six-dimensional flat space, viewed as the product of three copies of the complex plane, and $\mathbb{Z}_3$ describes the character of the identification that creates a singularity at the origin. Additionally, D3-branes are placed at the singularity. All the spatial directions on the branes lie along the four-dimensional space. It turns out that D-branes at orbifold singularities can give rise to gauge fields and chiral fermions. Models with a Standard Model-like spectrum do exist.

The second part of the analysis is a study of how to modify the non-compact six-dimensional space away from the singularity in order to turn it into a compact space of finite volume. If this is done in a way that leaves the region near the singularity unchanged, then the earlier results are preserved. Closing off the non-compact space may typically require the addition of other D-branes and/or orientifolds. A type II theory with D-branes and orientifolds can be analyzed alternatively as a type I superstring with D-branes. A type I superstring is a theory of unoriented open and closed strings. It can be viewed as the result of introducing a space-filling orientifold into type IIB string theory, which truncates the spectrum down to the subspace of states that are invariant under string orientation reversal. While this truncation does not give a consistent string theory, the inconsistencies can be cured by adding open string degrees of freedom. The result is type I superstrings. The perspective of type I theory can be quite valuable in model building.

Finally, there are models based on M-theory. These are sometimes closely related to models built with type IIA superstrings, because M-theory compactified on a circle is type IIA superstring theory with some finite value for the string coupling. In fact, intersecting D6-brane models are closely related to M-theory models. From the viewpoint of M-theory, which is a theory in eleven dimensions, realistic physics requires compactification on a seven-dimensional manifold. In order to obtain four-dimensional theories with $N = 1$ supersymmetry, the seven-dimensional manifold must have G_2 holonomy, a geometrical property that is effectively a constraint on the curvature of the space. Chiral fermions and a reasonable gauge group can appear if the seven-dimensional space has singularities. The M-theory approach may give valuable insight into the effects of finite string coupling. Another popular approach that gives semi-realistic models uses a Calabi–Yau space to reduce the spacetime dimensionality to five. The fifth dimension is then turned into a finite segment, a space that can be viewed as an orbifold (Problem 2.5). This is called the heterotic M-theory approach, because the compactification of M-theory on a segment has been shown to give the heterotic $E_8 \times E_8$ superstring.

In summary, while a fully realistic model of particle physics has not yet been built in string theory, consistent progress towards this goal has been made. As we have seen, there are string models on D-branes whose open strings yield the particle content of the Standard Model. The significance of this development will depend on the ultimate success or failure of the models and what we learn from them. The intersecting brane models are not fully realistic. Symmetry breaking, for example, remains to be worked out. It would be a major accomplishment to achieve correct electroweak symmetry breaking in any string model.

If symmetry breaking works out in detail in some consistent string model, we would have shown that the Standard Model in its full glory can occur as a solution of string theory. Such a string model (or models?) might make interesting predictions that can be tested by new experiments. We would not be done, however. String theory is a theory of all the interactions, and it includes gravity. Quite a few other features have to work out. One notorious problem is that of stabilization of closed string moduli. Another is that of the cosmological constant. Finally, we need to make sure that the Standard Model is embedded into a consistent cosmology, which presumably includes inflation. We turn now our attention to these issues, where there has been some interesting progress.

21.6 Moduli stabilization and the landscape

In our study of T-duality of closed strings we encountered the simplest example of a moduli space of compactifications. One spatial coordinate was curled up into a circle whose radius R was arbitrary. The moduli space is simply the space of possible choices of the modulus R. The fact that the parameter R is a modulus means that the potential $V(R)$ for R vanishes and, consequently, no particular value of R is selected. We can certainly choose a value of R and work with it but there is a complication. The theory in the reduced spacetime contains a massless scalar associated with fluctuations in the value of R. If we let x denote the coordinates of the reduced spacetime, the radius $R(x)$ of the circle at x is a massless field because the potential $V(R(x))$, constrained to vanish for constant $R(x)$, cannot contain a mass term $\frac{1}{2}m^2 R^2(x)$. Unless its interactions are unnaturally small, a massless scalar field is inconsistent with observation. It follows that realistic compactifications of string theory must have no moduli.

By choosing a physical setup in which nontrivial potentials arise for would-be moduli, we can force them to acquire specific values. Fluctuations around those values now cost energy and thus represent allowed massive scalars. In a typical setup, would-be moduli include parameters of the compact space, positions of D-branes, and the value of the closed string dilaton. The goal of moduli stabilization is to provide potentials that fix the values of all moduli so that we have an admissible string background or, as is usually called, a string vacuum. While this seemed extremely difficult to achieve for a long time, developments in the period 2002–2005 led to the construction of *flux* compactifications in which all moduli are stabilized. Fluxes, to be described below, allow the stabilization of moduli but lead to an extremely large number of vacua. This *landscape* of vacua is so vast that it becomes plausible that in some of them the cosmological constant takes the value that we observe in nature – an extraordinarily small value in Planckian units. Flux compactification has also allowed some exploration of cosmology, in particular, the identification of mechanisms that would implement an inflationary period.

To develop insight into the issue of moduli stabilization let us discuss a six-dimensional theory of Einstein's gravity and electromagnetism. In this theory we curl up two spatial dimensions into a compact space so that we get a reduced four-dimensional spacetime M_4. We can consider metrics of the form

$$-ds^2 = g_{\mu\nu}(x)\, dx^\mu dx^\nu + R^2(x)\, \bar{g}_{ab}(y) dy^a dy^b. \tag{21.67}$$

The first term on the right-hand side is the metric on M_4, with $\mu, \nu = 0, \ldots, 3$. The second term represents the metric of the compact two-dimensional space with coordinates y^1 and y^2 – the indices a, b can take values one and two. In this metric we separated out a scale factor $R(x)$ that depends only on M_4 and a fixed metric $\bar{g}_{ab}(y)$ that depends only on the coordinates of the two-dimensional space. The volume of the two dimensional space is $R^2(x)V_2$ where V_2 is the volume computed with the metric $\bar{g}_{ab}(y)$. We thus see that the ansatz (21.67) represents correctly a compact space whose volume can fluctuate in the reduced spacetime. Deriving the potential $V(R)$ associated with R is a straightforward but technical calculation in Einstein's general relativity. We will only motivate the result. The potential is proportional to a topological invariant of the two-dimensional manifold: the Euler number $\chi = 2 - 2g$, where g is a non-negative integer called the genus. The two-sphere S^2, for example, has genus zero and Euler number two. A two-dimensional torus has genus one and Euler number zero. The genus of a compact two-dimensional surface is g if the surface is topologically a sphere with g holes. The potential $V(R)$ is not just a constant but it also has some R dependence. This dependence arises in the process of disentangling the dynamics of gravity from that of R. As we have seen before, the dimensionally reduced gravitational constant G is equal to the higher-dimensional gravitational constant $G^{(6)}$ divided by the volume of the compact space. Since this volume can fluctuate, we seem to reach the paradoxical situation that G is not constant. Restoring a constant G requires a redefinition of the four-dimensional metric that introduces a factor of $1/R^4$ into the potential we are after. All in all, one obtains:

$$V(R) = -a_g \, \frac{\chi}{R^4}, \quad \chi = 2 - 2g, \tag{21.68}$$

where $a_g > 0$ is a constant. For a two-sphere $g = 0$ and the potential is negative definite $V \sim -1/R^4$. This means that left to its own devices a sphere will shrink to zero size. For a two-torus $g = 1$, the potential vanishes and R is a modulus. This is consistent with our experience with string theory T-duality where it is possible to curl up coordinates into circles of arbitrary radii; the space formed by two circles is a torus. For a space of $g > 1$, we have $V(R) \sim 1/R^4$ and left to its own devices the space will expand without limit. A compact space of genus zero has net positive curvature, a compact space of genus one has zero net curvature, and a compact space of genus greater than one has net negative curvature (this will be discussed further in §23.8). We thus see that positive net curvature induces collapse, negative net curvature induces blow up and zero net curvature gives us a modulus. None of these three situations is satisfactory, so we must consider extra elements that can help us stabilize the radius R.

If gravity is coupled to electromagnetism we can use magnetic flux to stabilize the radius. The magnetic flux on a torus is in fact quantized, as we learned in Section 19.3 and Problem 19.3. Assuming flux $\Phi = 2\pi n$, with n integer, we can easily estimate the contribution to the potential due to the magnetic field. If the characteristic area of the two dimensional space with the flux is R^2 the magnetic field magnitude goes like $B \sim n/R^2$. Since magnetic energy density is proportional to B^2 we get a total potential energy that goes like $R^2 B^2 \sim n^2/R^2$. With fixed flux the energy is reduced by making the space larger thus giving the space a tendency to expand. Due to the gravitational subtlety discussed above (21.68), the factor of $1/R^4$ also affects this contribution and the potential energy from the flux contributes to $V(R)$ a term that goes like n^2/R^6. We thus have

$$V(R) = -a_g \, \frac{\chi}{R^4} + a_f \, \frac{n^2}{R^6}, \tag{21.69}$$

where $a_f > 0$ is a constant. It is now clear that magnetic flux can stabilize a sphere: for $\chi = 2$ the potential above has a stable minimum for some $R > 0$. For different values n of the magnetic flux, we get different values of the critical radius. We have thus obtained an infinite family of stable solutions.

Quick calculation 21.16 Sketch the potential (21.69) for $\chi = 2$, determine the critical value of R, and confirm that it yields the minimum of the potential.

If we had wanted to fix R when it is a modulus ($\chi = 0$) and the first term in the potential (21.69) vanishes, in addition to magnetic flux we would need a negative contribution to $V(R)$. In string theory such negative contributions can arise from orientifolds. In a more complete superstring compactification the extra dimensions form a six-dimensional space and we would have contributions to the potential from orientifolds, D-branes, and fluxes. If we take, for example, type IIB superstrings, the NS-NS field $B_{\mu\nu}$ and the RR field $A_{\mu\nu}$ have three-index field strengths that can provide fluxes. Just like magnetic fluxes are integrals of two-index field strengths F_{ij} over two-dimensional manifolds, NS-NS and RR fluxes are integrals of three-index field strengths over three-dimensional manifolds. A typical six-dimensional Calabi–Yau space used for compactification can easily have hundreds of independent, non-contractible, three-dimensional submanifolds. Each one can support a flux characterized by an integer. As a result the fluxes on the Calabi–Yau can generate a potential that depends on hundreds of integers. A compactification of six dimensions into a Calabi–Yau with fluxes is an example of a *flux* compactification.

Faced with potentials arising from many fluxes, statistical considerations can help understand the space of possible vacua. As a very simplified but still illustrative example, we consider a potential for a single field ϕ characterized by two integers m and n:

$$V_{m,n}(\phi) = n\phi + \frac{1}{2}m\phi^2, \quad m, n \in \mathbb{Z}. \tag{21.70}$$

As m and n vary we get a set of potentials. In order to analyze this set concretely we limit the possible integers by the conditions

$$n^2 + m^2 \leq L, \quad m > 0. \tag{21.71}$$

The first constraint, with L a large integer, makes the set of potentials finite. The second constraint, $m > 0$, guarantees that the potentials have a stable minimum by making them bounded from below. Indeed, the critical point ϕ_* of (21.70) occurs at

$$0 = \frac{dV_{m,n}}{d\phi} = n + m\phi_* \quad \rightarrow \quad \phi_* = -\frac{n}{m}. \tag{21.72}$$

The value of ϕ_* represents the vacuum state in the potential $V_{m,n}$. We can also check that in this vacuum state the vacuum energy $\Lambda_{m,n}$ is

$$\Lambda_{m,n} = V_{m,n}(\phi_*) = -\frac{1}{2}\frac{n^2}{m}. \tag{21.73}$$

If we consider the set of all integers m, n allowed by (21.71), we can ask what is the approximate distribution of vacua ϕ_* in the set. We write

$$d\mathcal{N} = \rho(\phi_*)\, d\phi_*, \tag{21.74}$$

where $d\mathcal{N}$ is the number of critical points in the interval $(\phi_*, \phi_* + d\phi_*)$ and $\rho(\phi_*)$ is the distribution function we are after. We can represent the set of potentials as a set of points with integer coordinates on the (x, y) plane; each point (m, n) represents the potential $V_{m,n}(\phi)$ and the associated vacuum state ϕ_* (Figure 21.9). Because of (21.71) the allowed points lie to the right of the vertical axis and within a circle of radius $\sqrt{L}$ centered at the origin. For large L the total number of points, or total number $\mathcal{N}$ of vacua, is approximately equal to the area of the allowed region: $\mathcal{N} \simeq \frac{1}{2}\pi L$. We also note that $|\phi_*| = \frac{|n|}{m} < \sqrt{L}$ since $m \geq 1$, and $n < \sqrt{L}$.

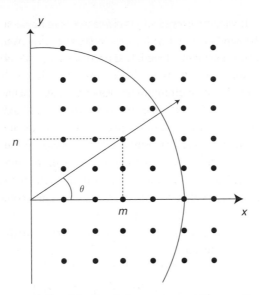

Fig. 21.9   A set of potentials $V_{m,n}(\phi)$ defined by two integers m, n, with $m > 0$.

To determine the distribution of vacua, we note that $\phi_* = -\frac{n}{m}$ implies that points on a fixed radial line give the same vacuum value. In fact

$$\phi_* = -\tan\theta \quad \text{or} \quad \theta = -\tan^{-1}\phi_*, \tag{21.75}$$

where θ is the angle that the radial line makes with respect to the positive x-axis. The number of vacua $d\mathcal{N}$ in the small sector $(\theta, \theta + d\theta)$ is equal to the area

$$d\mathcal{N} = \frac{L}{2}|d\theta| = \frac{L}{2}\frac{1}{1+\phi_*^2}|d\phi_*| \quad \rightarrow \quad \rho(\phi_*) = \frac{L}{2}\frac{1}{1+\phi_*^2}, \tag{21.76}$$

where we used (21.75) to relate $d\theta$ to $d\phi_*$ and (21.74) to read the value of $\rho(\phi_*)$. The distribution has the expected property $\rho(-\phi_*) = \rho(\phi_*)$, which arises because with $\phi_* = -n/m$, for each critical point with a given $n = n_0$ there is another with $n = -n_0$. On the given set, the distribution $\rho(\phi_*)$ applies for $|\phi_*| < \sqrt{L}$. The shape of ρ indicates that the density of vacua peaks at $\phi_* = 0$. In the limit of very large L we confirm that the number $\mathcal{N}$ of vacua is

$$\mathcal{N} = \frac{L}{2}\int_{-\sqrt{L}}^{\sqrt{L}} \frac{d\phi_*}{1+\phi_*^2} \simeq \frac{L}{2}\int_{-\infty}^{\infty} \frac{d\phi_*}{1+\phi_*^2} = \frac{1}{2}\pi L. \tag{21.77}$$

Another interesting distribution is that of vacuum energies (21.73). Letting $d\mathcal{N}$ denote the number of vacua with energy in the interval $(\Lambda, \Lambda + d\Lambda)$ we write

$$d\mathcal{N} = \rho(\Lambda)\,|d\Lambda|, \tag{21.78}$$

where $\rho(\Lambda)$ is the associated distribution function. A calculation (Problem 21.5) gives

$$\rho(\Lambda) = \frac{4}{3}\frac{1}{\sqrt{2|\Lambda|}}\left[\sqrt{\Lambda^2 + L} - |\Lambda|\right]^{3/2}. \tag{21.79}$$

Since all vacuum energies in this example are negative (see (21.73)), the above distribution must be supplemented with $\rho(\Lambda) = 0$ for $\Lambda > 0$. We note that $\rho(\Lambda)$ grows as $\Lambda \to 0$. For further discussion see Problem 21.5.

Statistical considerations are relevant when we discuss how string theory compactifications could reproduce the observed value of the so-called dark energy of the universe. This dark energy is the present value of the vacuum energy and, most likely, can be identified with a cosmological constant. While vacuum energy presently dominates the energy density of the universe, it is extremely small in Planckian units. To see this explicitly we first note that assuming that the Hubble expansion parameter takes the value $H_0 = 73.5$ km/s/Mpc, the present critical energy density ρ_c of the universe is

$$\rho_c = \frac{3H_0^2}{8\pi G} = 1.015 \times 10^{-29} \text{ g/cm}^3. \tag{21.80}$$

This density is equivalent to that of about six protons per cubic meter.

Quick calculation 21.17 Consider the natural Planckian mass density $\rho_\text{P} = m_\text{P}/\ell_\text{P}^3$. Show that $\rho_\text{P} = 5.20 \times 10^{93}$ g/cm^3.

Current experimental evidence indicates that the energy density of the universe is very close to the critical density and, apparently, vacuum energy contributes as much as 76% of it. We therefore find that the vacuum energy density is

$$\rho_\text{vac} \simeq 0.76 \, \rho_c = 0.771 \times 10^{-29} \frac{\text{g}}{\text{cm}^3} = 1.48 \times 10^{-123} \, \rho_\text{P}. \tag{21.81}$$

The present vacuum energy is about 123 orders of magnitude smaller than the natural value in Planckian units! Following conventional string theory notation, we use the symbol Λ to denote vacuum energy density. We do not follow the notation in general relativity where Λ is the "cosmological constant," a quantity obtained by multiplying the vacuum energy by $8\pi G/c^4$. We thus write (21.81) as

$$\Lambda_\text{obs} \simeq 1.48 \times 10^{-123} \, \Lambda_\text{P}. \tag{21.82}$$

Since string units are usually assumed to be only a couple of orders of magnitude different from Planckian units, the emergence of an extraordinarily small vacuum energy is a puzzle. Even in scenarios with large extra dimensions, Λ_obs would still be many orders of magnitude smaller than the natural string value. In string theory compactifications one expects vacuum energies of order Λ_P. If only a small number of string vacua existed, it would seem very unlikely that any of them would have an energy equal to $\sim 10^{-123}\Lambda_\text{P}$. As we will argue now, in a flux compactification with many fluxes the number of vacua is so large that it becomes possible to find a sizable number of vacua with energy in the experimentally allowed window. Consider then the potential for a set of moduli represented by a vector $\vec{\phi}$ in a theory with J fluxes characterized by integers $n_1, n_2, \ldots, n_J$:

$$V(\vec{\phi}) = V_0(\vec{\phi}) + \sum_{i=1}^{J} m_i(\vec{\phi}) \, n_i^2. \tag{21.83}$$

Here $V_0(\vec{\phi})$ is the part of the potential created by effects other than fluxes. It is the analog of the first term on the right-hand side of (21.69). The second term in (21.83) is the contribution from the fluxes and involves calculable functions $m_i(\vec{\phi})$. This term is the analog of the second term on the right-hand side of (21.69). When the number of moduli and the number J of fluxes are both large it becomes difficult to find the critical points of the potential. Following a strategy that was vindicated by more detailed analysis, we simply freeze the moduli $\vec{\phi}$ at some typical value $\vec{\phi}_0$ and skip the minimization process. Both V_0 and the functions m_i in (21.83) become just constants. Writing $\Lambda = V$ and $\Lambda_0 = V_0$ we then find,

$$\Lambda = \Lambda_0 + \sum_{i=1}^{J} q_i^2 \, n_i^2. \tag{21.84}$$

We have written $m_i = q_i^2$ to make it manifest that fluxes contribute positively to the energy density. Moreover, as in typical compactifications, Λ_0 will be taken to be a negative, Planckian size energy density:

$$\Lambda_0 \sim -\Lambda_{\mathrm{P}}. \tag{21.85}$$

We want to understand if there is a reasonable number of vacua for which the total vacuum energy is in the observational window, say between zero and $\Delta\Lambda$ where

$$\Delta\Lambda \simeq \Lambda_{\mathrm{obs}} \simeq 10^{-123}\Lambda_{\mathrm{P}}. \tag{21.86}$$

For the vacua in this interval we have

$$0 \leq \Lambda_0 + \sum_{i=1}^{J} q_i^2 \, n_i^2 \leq \Delta\Lambda. \tag{21.87}$$

Since Λ_0 is negative we write this as

$$|\Lambda_0| \leq \sum_{i=1}^{J} q_i^2 \, n_i^2 \leq |\Lambda_0| + \Delta\Lambda. \tag{21.88}$$

We want to know how many vacua satisfy this inequality. The number would be large if the q_i^2 were small compared to $\Delta\Lambda$. The problem is that the q_i^2 are *many* orders of magnitude larger than $\Delta\Lambda$. They are, in fact, Planckian, so at best a few orders of magnitude smaller than $|\Lambda_0|$.

If a vacuum characterized by nonvanishing integers $(n_1, n_2, \ldots, n_J)$ satisfies (21.88), we have at least 2^J degenerate vacua $(\pm n_1, \pm n_2, \ldots, \pm n_J)$ that do. If the q_i satisfy special relations further degeneracies can occur: if $q_i = q_j$, for example, the exchange of n_i with n_j yields additional degenerate vacua. We will call $d(J) \simeq 2^J$, the degeneracy of vacua.

To analyze the constraint (21.88) efficiently we introduce Cartesian coordinates

$$x_i = q_i n_i, \tag{21.89}$$

so that vacua become lattice points in x-space. Since lattice points are separated by distances q_i along x_i, the cells that contain a single lattice point have a volume

$$\text{vol}_{\text{cell}} = \prod_{i=1}^{J} q_i. \tag{21.90}$$

In x-space the constraint (21.88) becomes

$$|\Lambda_0| \le \sum_{i=1}^{J} x_i^2 \le |\Lambda_0| + \Delta\Lambda. \tag{21.91}$$

This is a spherical shell in x-space. To find the thickness dR of the shell we write $|\Lambda_0| = R_0^2$ and $|\Lambda_0| + \Delta\Lambda = (R_0 + dR)^2$ from which we find

$$dR = \frac{\Delta\Lambda}{2\sqrt{|\Lambda_0|}}. \tag{21.92}$$

The x-volume of the shell is equal to the volume of the sphere S^{J-1} of radius $R_0 = \sqrt{|\Lambda_0|}$ times the thickness dR:

$$\text{vol}_{\text{shell}} = \text{vol}(S^{J-1}) \sqrt{|\Lambda_0|}^{J-1} \frac{\Delta\Lambda}{2\sqrt{|\Lambda_0|}}, \quad \text{vol}(S^{J-1}) = \frac{2\pi^{J/2}}{\Gamma(J/2)}, \tag{21.93}$$

where we recalled (3.52). Since vacua are $d(J)$-fold degenerate, we have vacua in the shell if the number of lattice points in the shell, divided by $d(J)$, is greater than or equal to one:

$$\frac{1}{d(J)} \frac{\text{vol}_{\text{shell}}}{\text{vol}_{\text{cell}}} = \frac{\text{vol}(S^{J-1})}{2d(J)} \sqrt{|\Lambda_0|}^{J-1} \frac{\Delta\Lambda}{\sqrt{|\Lambda_0|}} \frac{1}{\prod_{i=1}^{J} q_i} \ge 1. \tag{21.94}$$

A rearrangement gives

$$\frac{\Delta\Lambda}{2|\Lambda_0|} \ge \frac{d(J)}{\text{vol}(S^{J-1})} \prod_{i=1}^{J} \frac{q_i}{\sqrt{|\Lambda_0|}}. \tag{21.95}$$

We want the ratio to the left of the inequality to be about 10^{-124}. For $J = 1$, ignoring the order one prefactor gives the unnaturally small $q \sim 10^{-124}\sqrt{|\Lambda_0|}$. For large J we need to pay some attention to the prefactor, which gets large. Indeed, $d(J) \simeq 2^J$ grows with J and $\text{vol}(S^{J-1})$ becomes smaller with J.

Quick calculation 21.18 Verify that the volume $\text{vol}(S^k)$ reaches a maximum for $k = 6$ and decreases monotonically for larger k.

In order to perform the large J analysis of (21.95) we use the asymptotic expansion

$$\text{vol}(S^{J-1}) \simeq \left(\frac{2\pi e}{J}\right)^{J/2}, \quad J \to \infty. \tag{21.96}$$

Moreover, we set the left-hand side of (21.95) equal to 10^{-124}, write $\prod_{i=1}^{J} q_i = q^J$, introducing an effective q, and set $d(J) = 2^J$. This gives

$$10^{-124} \ge \left(\frac{2\pi e}{J}\right)^{-J/2} \left(\frac{2q}{\sqrt{|\Lambda_0|}}\right)^{J}. \tag{21.97}$$

and leads to

$$\frac{2q}{\sqrt{|\Lambda_0|}} \simeq 10^{-\frac{124}{J}}\sqrt{\frac{2\pi e}{J}} \equiv f(J).\qquad(21.98)$$

For $J = 124$, for example, one gets $f(J) = 0.037$, a number that shows that qs just a little smaller than the natural Planckian value suffice to get string vacua in the experimentally allowed window.

Quick calculation 21.19 Use $\Gamma(x) \sim x^{x-1}\sqrt{2\pi x}\, e^{-x}(1 + \mathcal{O}(1/x))$ to prove (21.96).

Quick calculation 21.20 The function $f(J)$ attains a maximum for J near six-hundred. Determine the critical J and the corresponding value of $f(J)$.

We have produced, via flux compactification, a sizable number of vacua with energies in the observable window. Of course, there are many more vacua with unacceptably large energies. The argument does not tell us why the vacuum energy is so small, it just shows that it *can* be that small. If string theory incorporates inflation an unusual kind of explanation has been offered. Inflation, as presently understood, is eternal into the future. With eternal inflation, bubble universes nucleate forever and, most likely, all vacua in the landscape are eventually realized as physical bubble universes. This includes universes like ours, in which the vacuum energy is small. We find ourselves in such universe, it is said, because had the vacuum energy been substantially larger, galaxies and stars could not have formed and life could not exist. However outrageous this argument seems, it has had some success. Before its experimental discovery in 1999, most theoreticians believed that the vacuum energy was exactly zero. Every attempt to use theories to prove it had to be zero, however, had failed. Weinberg showed that structure formation in the universe gave an upper bound on the vacuum energy. He also argued that if no principle guaranteed a vanishing vacuum energy it would probably turn out to have a value smaller than the bound, but not dramatically smaller. This turned out to be correct.

Since the experimental evidence for a period of inflation has become much stronger in the last few years, one can ask if inflation occurs in string theory. Most investigations have been done using flux compactifications since moduli must be stabilized. Typically, one imagines a Calabi–Yau space with an elongation that has an anti-D-brane at its tip. A D-brane slowly moving along the elongation towards the anti-D-brane may produce inflation. It is too early to tell if these efforts will lead to complete models.

As one contemplates the current efforts to show that string theory describes the real world one cannot fail to experience mixed feelings. On the one hand, string models and flux compactifications are evidence that string theory *is* promising, after all, central features of the real world can emerge from the theory. On the other hand, the constructions seem contrived, at least in the sense that they are engineered to give the physics that we observe, rather than obtained naturally as the simplest solutions of string theory. The Standard Model of particle physics and the Standard Cosmological Model are intricate constructs, and present-day attempts to describe them within string theory do not result in simplification. Despite this complexity, you need not react as Alfonso X (the Wise, 1221–1284) appears to have done when faced with Ptolemy's astronomy:[1]

> Had I been present at the creation, I would have given some useful hints for the better ordering of the universe.

It is conceivable that the landscape idea is correct and our universe is just one out of a gigantic number of possibilities, one that is not especially natural. This would explain why the string constructions do not seem natural. At this stage, however, it may be more constructive to continue to explore the facts and the models, looking for naturalness to emerge as our understanding improves. This is based on the hope that, as stated by Maimonides (1135–1204) in *The Guide for the Perplexed* (1190),

> In the realm of Nature there is nothing purposeless, trivial, or unnecessary.

Problems

Problem 21.1 Oblique tori and orientifold action.

Consider the (x, y) plane with an orientifold line along the x axis. The orientifolding symmetry operation reverses string orientation (this will not play a role here) and acts on the plane by sending every point (x, y) into its image $(x, -y)$. The action is well defined because it takes points on the plane to points on the plane. The orientifold line is a line which is fixed under the orientifold action.

A two-torus is obtained by implementing two identifications on the (x, y) plane. If the identifications are those in (2.101) and (2.102) the result is a square torus. To get a rectangular torus we simply use

$$(x, y) \sim (x + a, y), \quad (x, y) \sim (x, y + b), \quad a, b > 0. \tag{1}$$

The fundamental domain of the identifications together with its boundary can be taken to be the region whose points (x, y) satisfy $0 \le x \le a$ and $0 \le y \le b$. This is a rectangle with sides a and b. Draw a picture illustrating the situation.

(a) Convince yourself that the orientifold action $(x, y) \to (x, -y)$ is well defined on the torus (1). On the torus, the line $0 \le x \le a$, $y = 0$ is a fixed line. There is another fixed line on the torus. Find it. Show that the orientifold action on the torus can be viewed as a reflection about either one of the two fixed lines. So, there are really two orientifold lines on the torus!

Consider now the class of oblique two-tori that are obtained by implementing the identifications

$$(x, y) \sim (x + a_1, y + a_2), \quad (x, y) \sim (x, y + b), \quad a_1, a_2, b > 0. \tag{2}$$

---

[1] I am grateful to J. Goldstone for providing this quote from Bartlett's *Familiar Quotations* (1919). Alfonso X was Spanish king of Castile and Leon. Moses Maimonides was born in Cordova, Spain. The intersecting D6-brane models we examined in Section 21.4 were developed by the "Spanish group."

The fundamental domain of the identifications together with its boundary can be taken to be the parallelogram $\mathcal{P}$ with a vertex at the origin and sides defined by the vectors $\vec{a} = (a_1, a_2)$ and $\vec{b} = (0, b)$. Let $\vec{x} = (x, y)$, so that the identifications can be written as $\vec{x} \sim \vec{x} + \vec{a}$ and $\vec{x} \sim \vec{x} + \vec{b}$. Draw a picture showing the parallelogram $\mathcal{P}$.

(b) Consider now the torus (2) when

$$a_2 = b/2. \tag{3}$$

Use an appropriate figure to convince yourself that the orientifold action $(x, y) \to (x, -y)$ is well defined on the torus. Find the two fixed lines on $\mathcal{P}$. Note that these lines represent a *single* closed orientifold line on the torus.

(c) Let $\vec{a}^* \equiv (a_1, -a_2)$. Show that (3) implies that $\vec{a}^* = \vec{a} - \vec{b}$. Note the significance of this relation on your sketch. Let points be denoted by vectors $t_1 \vec{a} + t_2 \vec{b}$, with constants t_1 and t_2. Show that the orientifold action takes

$$t_1 \vec{a} + t_2 \vec{b} \longrightarrow t_1 \vec{a} - (t_1 + t_2) \vec{b}. \tag{4}$$

(d) A closed line ℓ on the torus can be denoted by a pair of relatively prime integers $\ell = (m, n)$ with the understanding that the line is obtained from the straight segment between the origin and the point $m\vec{a} + n\vec{b}$. Examine the argument that established (21.5) for the intersection number on a square torus. Convince yourself that this formula also holds for lines on oblique tori. It follows from (4) that after orientifold action

$$\ell = (m, n) \longrightarrow \ell^* = (m, -m - n). \tag{5}$$

It is convenient to encode lines with another pair of numbers, one of which can be half-integral:

$$\ell = (m, n) \text{ is represented by } \ell = [m, n + \tfrac{m}{2}]. \tag{6}$$

(e) Show that the orientifold action takes

$$\ell = [r, s] \longrightarrow \ell^* = [r, -s]. \tag{7}$$

(f) Show that the intersection number for two lines $\ell_1 = [r_1, s_1]$ and $\ell_2 = [r_2, s_2]$ is

$$\#(\ell_1, \ell_2) = r_1 s_2 - r_2 s_1. \tag{8}$$

Show that

$$\#(\ell_1^*, \ell_2^*) = -\#(\ell_1, \ell_2). \tag{9}$$

(g) Orient the orientifold closed line on the torus from left to right. Show that it can be described as a vector of the form $t_1 \vec{a} + t_2 \vec{b}$ that is encoded as $[2, 0]$.

Problem 21.2 Intersection numbers and formal variables.

In Section 21.4 we introduced formal variables x_i, y_i, $i = 1, 2, 3$ to describe a polynomial Π that encodes the charge of a D6-brane wrapped on a T^6. Consider now the following rules for products of the formal variables:

$$x_i \, y_j = y_j \, x_i, \quad x_i \, x_j = x_j \, x_i, \quad \text{and} \quad y_i \, y_j = y_j \, y_i, \quad \text{for} \quad i \neq j$$
$$x_i \, x_i = y_i \, y_i = 0, \quad \text{and} \quad x_i \, y_i = -y_i \, x_i = 1, \quad i \text{ not summed.} \tag{1}$$

Moreover, assume that the product obeys the distributive law. Note that the product is not commutative (the order of the factors matters) and that the above rules cannot be used to simplify the polynomial Π associated with a brane.

(a) Show that the product of the polynomials associated with two wrapped D6-branes gives their intersection number: $\Pi_a \Pi_b = I_{ab}$.

(b) Use (21.65) to derive the consistency condition $\sum_a N_a I_{ab} = 0$, for any fixed b. Explain in detail why this condition gives the rule formulated in (21.62).

Problem 21.3 An intersecting brane model with the particle content of the Standard Model.

To solve this problem you must first solve Problem 21.1, which deals with orientifolds on oblique tori. The T^6 in the present brane model is composed of two rectangular tori (the first and the second tori) and one oblique torus (the third torus). On the oblique torus, we denote lines as $[p, q]$.

On each rectangular torus, the wrapping of the two fixed orientifold lines is represented by $(2, 0)$. On the oblique torus, the single orientifold line is represented by $[2, 0]$ (Problem 21.1, part (g)). As a result, the T^6 carries three of the spatial directions along the O6-planes. The other three directions along the O6-planes coincide with the spatial directions of the effective four-dimensional spacetime.

The four sets of D6-branes of the model are described by the lines

$$\begin{aligned}
N_1 &= 3: & \ell_1^{(1)} &= (1, 0), & \ell_2^{(1)} &= (5, 1), & \ell_3^{(1)} &= [1, 1/2], \\
N_2 &= 2: & \ell_1^{(2)} &= (0, -1), & \ell_2^{(2)} &= (1, 0), & \ell_3^{(2)} &= [1, 3/2], \\
N_3 &= 1: & \ell_1^{(3)} &= (4, 3), & \ell_2^{(3)} &= (1, 0), & \ell_3^{(3)} &= [0, 1], \\
N_4 &= 1: & \ell_1^{(4)} &= (1, 0), & \ell_2^{(4)} &= (1, 1), & \ell_3^{(4)} &= [1, -3/2].
\end{aligned} \tag{1}$$

The mirror image i^* of a D6-brane i is obtained by changing the sign of the second entry on each of the three lines that define the brane.

(a) Give the (m, n) values corresponding to the standard description of each of the four lines on the third torus.

(b) Verify that $I_{ii^*} = 0$ for $i = 1, 2, 3$, and 4. This ensures that no D6-brane intersects its own mirror image.

(c) Calculate all other intersection numbers I_{ij} and I_{ij^*}, with $i < j$. Verify that

$$I_{12} = 1, \quad I_{12^*} = 2, \quad I_{13} = -3, \quad I_{13^*} = -3, \quad I_{14} = I_{14^*} = 0,$$
$$I_{23} = I_{23^*} = 0, \quad I_{24} = -3, \quad I_{24^*} = 0, \quad I_{34} = 3, \quad I_{34^*} = -3. \tag{2}$$

Confirm that Figure 21.8 correctly represents the information encoded by the above intersection numbers.

(d) Since the orientifold wrappings are described by $(2,0)$, $(2,0)$, $[2,0]$, the polynomial (21.64) takes the form $\Pi_{O6} = 8x_1x_2x_3$. Since the Ramond–Ramond charge of an orientifold O6-plane is opposite to that of a D6-brane and four times larger, the condition of vanishing total R–R charge on T^6 is

$$\sum_{i=1}^{4} N_i\,(\Pi_i + \Pi_{i*}) - 4\,\Pi_{O6} = 0. \tag{3}$$

Verify that the wrapping numbers in (1) satisfy this constraint. Note: the last factor in the polynomial Π is built using the numbers in the $[p,q]$ representation.

(e) In this model, one of the four $U(1)$ factors remains massless. Its charge is proportional to $Q_1 - 3(Q_3 + Q_4)$. Here $Q_i = +1$ for a string ending on an i-brane or beginning on an i^*-brane. Consider the states in Figure 21.8. Verify that, with suitable normalization, the charge is precisely the hypercharge.

Problem 21.4 Symmetry breaking by recombination of intersecting branes.

In order to simplify our work, consider a pair of intersecting D4-branes. Each brane has one spatial direction wrapped as a closed line on a square torus T^2 with coordinates x^4 and x^5. The other three spatial directions on the world-volume of the branes stretch along the spatial directions of the four-dimensional spacetime. The details of the compactification of $x^6, \ldots, x^9$ are not relevant.

Since we have a single T^2, the wrapping of each D4-brane can be specified by a single line $\ell = (m, n)$, where m and n are relatively prime. Given two D4-branes with lines $\ell^{(1)} = (m_1, n_1)$ and $\ell^{(2)} = (m_2, n_2)$, the intersection number I_{12} which determines the number of left-handed fermions at the intersection set is simply given by

$$I_{12} = \#(\ell^{(1)}, \ell^{(2)}) = m_1 n_2 - m_2 n_1.$$

The process of brane recombination can be described using the corresponding lines on the torus. Take one of the branes, say the first one, and cut out an infinitesimal piece near the *endpoint* of its line. The result is an open line on the torus beginning at $(0,0)$ and ending near $(0,0)$. Take the second brane, and cut an infinitesimal piece near its *beginning*. The result is an open line beginning near $(0,0)$ and ending at $(0,0)$. Now glue the end of the first line with the beginning of the second. The result is a single closed line representing a recombined D-brane.

(a) Use the representation of the defining lines on the plane (where the torus arises as the unit square) to argue that the recombination of two D-branes $\ell^{(1)} = (m_1, n_1)$ and $\ell^{(2)} = (m_2, n_2)$ gives a D-brane represented by a line that begins at $(0,0)$, ends at $(m_1 + m_2, n_1 + n_2)$, and has one corner (where?). The line can be deformed continuously into a straight line joining the endpoints. This line, which represents the recombined D-brane Σ, is $\ell_\Sigma = (m_1 + m_2, n_1 + n_2)$. Explain why the final result of recombination does not depend on the order in which the D-branes are glued together.

(b) Carry out explicitly the recombination procedure on the torus for the branes $(1, 0)$ and $(0, 1)$. Show in a sequence of figures how the recombined branes can be continuously deformed into the brane $(1, 1)$.

(c) When two branes recombine, the total number of branes decreases by one unit and, as a result, the number of Maxwell fields decreases by one unit. Consider a D-brane i which intersects each of the two branes, brane 1 and brane 2, that recombine into Σ. Prove that

$$|I_{i\Sigma}| \leq |I_{i1}| + |I_{i2}|.$$

This result shows that the number of (massless) chiral fermions obtained after brane recombination is less than or equal to the total number of chiral fermions that were present before recombination. We thus have the expected signs of symmetry breaking: a number of Maxwell fields and a number of chiral fermions acquire mass.

Problem 21.5 Distribution of vacuum energies in toy model.

Derive the distribution $\rho(\Lambda)$ given in (21.79). One possible approach is to consider the continuous approximation in which $(m, n) = (x, y)$ and thus $\Lambda = -\frac{1}{2}\frac{y^2}{x}$ on the allowed space. Then calculate the area in between the curves associated with Λ and $\Lambda + d\Lambda$.

What is the upper bound for $|\Lambda|$ in the set of allowed vacua? Test the consistency of (21.79) by using it to calculate the total number of vacua. Take $L = 100$ and estimate the fraction of vacua that have $|\Lambda| < 1$.

Problem 21.6 Distribution of vacuum energies with large number of fluxes.

Consider equation (21.84) with $\prod_{i=1}^{J} q_i = q^J$ and J is large so that (21.96) applies.

(a) Determine the number $\mathcal{N}(\Lambda)$ of vacua with vacuum energy less than or equal to Λ. The answer depends on Λ and the parameters $|\Lambda_0|, q$, and J.
(b) Use differentiation to write an equation of the form

$$d\mathcal{N} = \rho(\Lambda)\frac{d\Lambda}{|\Lambda_0|},$$

and determine the distribution $\rho(\Lambda)$. The distribution depends parametrically on $|\Lambda_0|, q$, and J.
(c) Assume $J = 100$ and $q = 0.01\sqrt{|\Lambda_0|}$. Determine the value of $\rho(\Lambda = 0)$. Find the value of Λ_* (in terms of $|\Lambda_0|$) for which $\rho(\Lambda_*)$ differs from $\rho(0)$ by 1%. Calculate the ratio $\Lambda_*/\Lambda_{\text{obs}}$. Your result must demonstrate that the distribution $\rho(\Lambda)$ is accurately constant over ranges of conceivable experimental relevance. How many string vacua are there with Λ values within 1% of Λ_{obs}?

22

String thermodynamics and black holes

The thermodynamics of strings is governed largely by the exponential growth of the number of quantum states accessible to a string, as a function of its energy. We estimate such growth rates by counting the number of partitions of large integers. The behavior of the entropy indicates that at high energies the temperature approaches a finite constant, the Hagedorn temperature. The finite temperature single-string partition function for open bosonic strings is calculated. We explain how the counting of string states can be used to give a statistical mechanics derivation of the entropy of black holes. The calculations give results in qualitative agreement with the entropy of Schwarzschild black holes and in quantitative agreement with the entropy of certain charged black holes.

22.1 A review of statistical mechanics

Our study of string thermodynamics will make use of both the microcanonical and canonical ensembles. Recall that the *microcanonical ensemble* consists of a collection of copies of a particular system A, one for each state accessible to A at a particular fixed energy E. In the *canonical ensemble* we consider the system A in thermal contact with a reservoir at a temperature T. This ensemble contains copies of the system A together with the reservoir, one copy for each allowed state of the combined system. In the canonical ensemble the energy of system A varies among members of the ensemble.

Let us begin with the microcanonical ensemble. The system A is imagined to be in isolation and to have a fixed energy. We let $\Omega(E)$ denote the number of possible states of the system A when it has energy E. The entropy S of the system is defined in terms of the number of states as

$$S(E) = k \ln \Omega(E), \tag{22.1}$$

where k is Boltzmann's constant. The temperature T of the system is defined in terms of the derivative of the entropy with respect to the energy:

$$\frac{1}{T} = \frac{\partial S}{\partial E}. \tag{22.2}$$

It is sometimes easier to work with the canonical ensemble. Imagine a system A which has a fixed volume and which is in thermal contact with a reservoir at temperature T. This system could be a box full of strings, or it could be a box containing a single string. It is not necessary to specify what the reservoir is. Suppose we know the quantum states $\{\alpha\}$ of the system and their associated energies $\{E_\alpha\}$. Then the partition function Z for system A is defined as

$$Z \equiv \sum_\alpha e^{-\beta E_\alpha}, \quad \beta = \frac{1}{kT}. \tag{22.3}$$

The partition function is useful because it can be used to calculate interesting quantities. For instance, if system A is known to have temperature T, then, using Z, we can calculate the probability that A is in a particular quantum state. By definition, the partition function depends both on the temperature T and on the external parameters of the system. These are the parameters that determine the energy levels of the system. The systems we will consider have only one external parameter: the volume V occupied by the system. Thus we will think of Z as $Z(T, V)$, or

$$Z = Z(\beta, V). \tag{22.4}$$

The probability P_α that the system, in contact with the reservoir at temperature T, is in the state α is

$$P_\alpha = \frac{e^{-\beta E_\alpha}}{Z}. \tag{22.5}$$

Clearly $\sum_\alpha P_\alpha = 1$, as required by the interpretation of P_α as a probability. We can calculate the average energy E of the system A in the ensemble by differentiation of the partition function:

$$E = \sum_\alpha P_\alpha E_\alpha = -\frac{\partial \ln Z}{\partial \beta}. \tag{22.6}$$

The pressure p of the system can also be calculated from the partition function (Problem 22.1). It is given by

$$p = \frac{1}{\beta} \frac{\partial \ln Z}{\partial V}. \tag{22.7}$$

Another useful quantity is the Helmholtz free energy F. Its basic properties can be obtained in a few steps starting from the first law of thermodynamics. This law states that the energy change dE in a system whose only external parameter is the volume V can be written as

$$dE = T dS - p dV. \tag{22.8}$$

Here T is the temperature of the system and p is the pressure. Moreover, $T dS$ is the heat transferred into the system, and $(-p dV)$ is the mechanical work done *on* the system. Equation (22.8) implies that E should be viewed as a function $E(S, V)$ of S and V, and,

$$T = \left(\frac{\partial E}{\partial S}\right)_V, \quad p = -\left(\frac{\partial E}{\partial V}\right)_S. \tag{22.9}$$

We can also write the change in energy in (22.8) as

$$dE = d(TS) - S dT - p dV, \tag{22.10}$$

which means that

$$d(E - TS) = -S\,dT - p\,dV. \tag{22.11}$$

The free energy F is defined as

$$F \equiv E - TS, \tag{22.12}$$

and therefore we have

$$dF = -S\,dT - p\,dV. \tag{22.13}$$

We see that for processes at constant temperature the free energy represents the amount of energy that can go into mechanical work. For a chemical reaction that releases energy, for example, the entropy of the system typically decreases. Not all of the energy released can then be used for work; only the free energy can. Since the total entropy cannot decrease, the rest of the energy goes into heat, which increases the entropy of the world. It follows from (22.13) that F should be viewed as a function $F(T, V)$ of T and V, and,

$$S = -\left(\frac{\partial F}{\partial T}\right)_V, \quad p = -\left(\frac{\partial F}{\partial V}\right)_T. \tag{22.14}$$

The free energy can be calculated from the partition function (Problem 22.1). It is given by

$$F = -kT \ln Z. \tag{22.15}$$

Our aim is to use the basic thermodynamic relations reviewed above to compute interesting properties of the string. One central computation is that of the partition function for a string. This problem is a bit complex, so we first consider simpler problems that will help us build the necessary tools.

The first result we need is a formula for the number of partitions of large integers. We will obtain this mathematical result using a physical method: the analysis of the high-temperature behavior of a quantum *nonrelativistic* string, call it a "quantum violin string." With this result, we calculate the entropy/energy relation for an *idealized* quantum *relativistic* string, a string for which we ignore the momentum labels of the quantum states. The Hagedorn temperature already emerges in this context. After a discussion of the partition function for the relativistic point particle, we assemble all of our results to compute the partition function of the relativistic string.

In the latter part of this chapter we discuss a significant success of string theory: a statistical mechanics derivation of the entropy of black holes. This entropy, first arrived at via thermodynamical considerations, arises from the degeneracy of string states that have the macroscopic properties of black holes. The agreement between the string calculations and the thermodynamical expectation is only qualitative for the case of Schwarzschild black holes, but it is quantitative for certain types of extremal black holes.

22.2 Partitions and the quantum violin string

Consider a quantum mechanical nonrelativistic string with fixed endpoints: a quantum violin string. This string, studied classically in Chapter 4, has an infinite set of vibrating frequencies, all multiples of a basic frequency ω_0. Its idealization as a quantum string is a collection of simple harmonic oscillators with frequencies $\omega_0, 2\omega_0, 3\omega_0$, and so on. Each simple harmonic oscillator (SHO) has its own creation and annihilation operators, as well as its own Hamiltonian:

$$
\begin{aligned}
\text{SHO}_{\omega_0}: &\quad (a_1, a_1^\dagger), &H_{\omega_0} &= \hbar\omega_0\, a_1^\dagger a_1 , \\
\text{SHO}_{2\omega_0}: &\quad (a_2, a_2^\dagger), &H_{2\omega_0} &= 2\hbar\omega_0\, a_2^\dagger a_2 , \\
\text{SHO}_{3\omega_0}: &\quad (a_3, a_3^\dagger), &H_{3\omega_0} &= 3\hbar\omega_0\, a_3^\dagger a_3 , \\
\vdots &\qquad \vdots &\vdots \quad &\quad \vdots
\end{aligned}
\tag{22.16}
$$

Here we have discarded zero point energies, and all oscillators satisfy the conventional commutation relations

$$
[a_m, a_n^\dagger] = \delta_{mn}.
\tag{22.17}
$$

Since the quantum string is the union of all these oscillators, the Hamiltonian $\hat{H}$ is

$$
\hat{H} = \sum_{\ell=1}^{\infty} H_{\ell\omega_0} = \hbar\omega_0 \sum_{\ell=1}^{\infty} \ell\, a_\ell^\dagger a_\ell.
\tag{22.18}
$$

We recognize here the number operator $\hat{N}$:

$$
\hat{H} = \hbar\omega_0\, \hat{N}, \quad \hat{N} = \sum_{\ell=1}^{\infty} \ell\, a_\ell^\dagger a_\ell.
\tag{22.19}
$$

The vacuum state of the string is a state $|\Omega\rangle$ such that

$$
a_\ell|\Omega\rangle = 0, \quad \text{for all } \ell.
\tag{22.20}
$$

A quantum state $|\Psi\rangle$ of this string is obtained by letting creation operators act on the vacuum:

$$
|\Psi\rangle = (a_1^\dagger)^{n_1} (a_2^\dagger)^{n_2} \cdots (a_l^\dagger)^{n_l} \cdots |\Omega\rangle .
\tag{22.21}
$$

The state is therefore specified by the set $\{n_1, n_2, n_3, \ldots\}$ of occupation numbers. The number operator acting on the state $|\Psi\rangle$ gives us

$$
\hat{N}|\Psi\rangle = N|\Psi\rangle,
\tag{22.22}
$$

where

$$
N = n_1 + 2n_2 + 3n_3 + \cdots = \sum_{\ell=1}^{\infty} \ell n_\ell.
\tag{22.23}
$$

Table 22.1	Counting states of fixed total number eigenvalue N; $p(N)$ denotes the number of partitions of the integer N	
N	List of states	$p(N)$
1	$a_1^\dagger$	1
2	$a_2^\dagger$, $(a_1^\dagger)^2$	2
3	$a_3^\dagger$, $a_2^\dagger a_1^\dagger$, $(a_1^\dagger)^3$	3
4	$a_4^\dagger$, $a_3^\dagger a_1^\dagger$, $(a_2^\dagger)^2$, $a_2^\dagger(a_1^\dagger)^2$, $(a_1^\dagger)^4$	5

It then follows from (22.19) that the energy E of $|\Psi\rangle$ is given by

$$E = \hbar\omega_0\,N. \tag{22.24}$$

A counting question arises naturally here. For a fixed positive integer N, how many states are there with $\hat{N}$ eigenvalue equal to N? This number, denoted by $p(N)$, is so important that it has been given a name: the number of *partitions* of N. Before explaining the reason for this terminology, let us determine $p(N)$ for $N = 1, 2, 3$, and 4. Shown in Table 22.1 are the states with those values of N. For brevity, we show only the oscillators, omitting the vacuum state $|\Omega\rangle$ that they act on. The fourth line, for example, shows that there are five states with $\hat{N}$ eigenvalue equal to four. Thus $p(4) = 5$.

It is appropriate to name the quantity $p(N)$ the number of partitions of N. A partition of N is a set of positive integers that add up to N. The order of the elements in the set is immaterial. Thus, for example, $\{3, 2\}$ is a partition of 5, and so is $\{2, 1, 1, 1\}$. The partitions of 4 are

$$\{4\}, \quad \{3, 1\}, \quad \{2, 2\}, \quad \{2, 1, 1\}, \quad \{1, 1, 1, 1\}. \tag{22.25}$$

The number of states with $\hat{N}$ eigenvalue equal to N coincides with the number of partitions of N. Indeed, given a partition of N we can build a state by attaching each element of the partition as a subscript to an oscillator $a^\dagger$ and letting the resulting collection of oscillators act on the vacuum. Note that this is exactly how the states in the last line of Table 22.1 are built from the partitions of 4 given in (22.25). Conversely, given a state with number N, the set of subscripts of all the oscillators in the state gives a partition of N.

We would like to find a formula for $p(N)$, but our analysis will not give us that much. We will derive an expression that describes $\ln p(N)$ accurately for large N. A more refined calculation gives the famous approximation for $p(N)$ found by Hardy and Ramanujan.

Our strategy will be as follows. We know that the entropy S is given as a function of the energy E by (22.1). For a given E, $N = E/(\hbar\omega_0)$, and $\Omega(E)$ is simply $p(N)$. Therefore

$$S(E) = k \ln p(N) = k \ln p\left(\frac{E}{\hbar\omega_0}\right). \tag{22.26}$$

If we find $S(E)$ then we will have found the function $p(N)$. To find $S(E)$ we will calculate the partition function Z for the quantum violin string. From Z, we will find the free energy F. We will be able to evaluate the free energy explicitly only in the case of high temperature. It will then be easy to find the high energy behavior of the entropy $S(E)$. This we will use to find a large-N approximation for $p(N)$.

Let us now begin with the calculation of the partition function. We have

$$Z = \sum_{\alpha} \exp\left(-\frac{E_{\alpha}}{kT}\right) = \sum_{n_1,n_2,n_3,\dots} \exp\left[-\frac{\hbar\omega_0}{kT}(n_1 + 2n_2 + 3n_3 + \cdots)\right]. \qquad (22.27)$$

In writing this equation we have recognized that the set of all states is labeled by the set of occupation numbers. To sum over all states is to sum over all occupation numbers, each of which ranges from zero to infinity. Since the exponential of a sum can be written as a product of exponentials, the sums over different occupation numbers can be made independently,

$$Z = \sum_{n_1} \exp\left[-\frac{\hbar\omega_0}{kT} n_1\right] \cdot \sum_{n_2} \exp\left[-\frac{\hbar\omega_0}{kT} 2n_2\right] \cdots. \qquad (22.28)$$

Therefore we have

$$Z = \prod_{\ell=1}^{\infty} \sum_{n_\ell=0}^{\infty} \exp\left(-\frac{\hbar\omega_0 \ell n_\ell}{kT}\right). \qquad (22.29)$$

The sum over each n_ℓ is a geometric series, so we find

$$Z = \prod_{\ell=1}^{\infty} \left[1 - \exp\left(-\frac{\hbar\omega_0 \ell}{kT}\right)\right]^{-1}. \qquad (22.30)$$

Finally, the free energy F is found using (22.15):

$$F = -kT \ln Z = kT \sum_{\ell=1}^{\infty} \ln\left[1 - \exp\left(-\frac{\hbar\omega_0 \ell}{kT}\right)\right]. \qquad (22.31)$$

We cannot go any further unless we make some approximations. If the temperature T is high enough so that

$$\frac{\hbar\omega_0}{kT} \ll 1, \qquad (22.32)$$

then each term in the sum (22.31) differs very little from the previous one. This allows us to approximate the sum by an integral:

$$F \simeq kT \int_1^{\infty} d\ell \, \ln\left[1 - \exp\left(-\frac{\hbar\omega_0 \ell}{kT}\right)\right]. \qquad (22.33)$$

The choice $\ell = 1$ for the lower limit of integration, as opposed to any other finite, small number, plays no role. Indeed, changing the variable of integration to

$$x = \frac{\hbar\omega_0}{kT} \ell, \qquad (22.34)$$

we find that the lower limit of integration becomes $x = 0$ in the high temperature limit. As a result, we obtain

$$F \simeq \frac{(kT)^2}{\hbar\omega_0} \int_0^\infty dx \, \ln(1 - e^{-x}). \tag{22.35}$$

Using the expansion

$$\ln(1 - y) = -\left(y + \frac{1}{2}y^2 + \frac{1}{3}y^3 + \frac{1}{4}y^4 + \cdots\right), \tag{22.36}$$

which is valid for any $0 \le y < 1$, we have

$$F \simeq -\frac{(kT)^2}{\hbar\omega_0} \int_0^\infty dx \left(e^{-x} + \frac{1}{2}e^{-2x} + \frac{1}{3}e^{-3x} + \frac{1}{4}e^{-4x} \cdots\right)$$
$$\simeq -\frac{(kT)^2}{\hbar\omega_0}\left[1 + \frac{1}{2^2} + \frac{1}{3^2} + \frac{1}{4^2} + \cdots\right]. \tag{22.37}$$

The sum in brackets is a familiar one. It is, in fact, the zeta function (12.109) with an argument of two:

$$\zeta(2) = 1 + \frac{1}{2^2} + \frac{1}{3^2} + \frac{1}{4^2} + \cdots = \frac{\pi^2}{6}. \tag{22.38}$$

Thus we finally obtain the high temperature approximation for the free energy:

$$F \simeq -\frac{(kT)^2}{\hbar\omega_0}\frac{\pi^2}{6} = -\frac{1}{\hbar\omega_0}\frac{\pi^2}{6}\frac{1}{\beta^2}. \tag{22.39}$$

Evidently, for this string the free energy has no volume dependence.

We can now calculate the entropy as a function of temperature. Using (22.14) we find

$$S = -\frac{\partial F}{\partial T} = k\frac{\pi^2}{3}\left(\frac{kT}{\hbar\omega_0}\right). \tag{22.40}$$

Since we are interested in the entropy as a function of energy, we also compute the energy. Making use of (22.6) we have

$$E = -\frac{\partial \ln Z}{\partial \beta} = \frac{\partial}{\partial \beta}(\beta F) = -\frac{\pi^2}{6}\frac{1}{\hbar\omega_0}\frac{\partial}{\partial \beta}\left(\frac{1}{\beta}\right), \tag{22.41}$$

which gives

$$E = \frac{\pi^2}{6}\frac{1}{\hbar\omega_0}\frac{1}{\beta^2} = \frac{\pi^2}{6}\left(\frac{kT}{\hbar\omega_0}\right)^2\hbar\omega_0. \tag{22.42}$$

Quick calculation 22.1 Verify that the energy E can also be calculated from $F = E - TS$.

Combining (22.40) and (22.42) yields

$$S(E) = k\pi\sqrt{\frac{2}{3}\frac{E}{\hbar\omega_0}} = k\,2\pi\sqrt{\frac{N}{6}}. \tag{22.43}$$

Table 22.2	Comparing the exact values of $p(N)$ with the estimate $p(N)_{est}$ provided by the Hardy–Ramanujan formula		
N	$p(N)$	$p(N)_{est}$	$p(N)/p_{est}(N)$
5	7	8.94	0.7829
10	42	48.10	0.8731
100	190 569 292	199 281 893.25	0.9563
1000	2.406×10^{31}	2.440×10^{31}	0.9860
10 000	3.617×10^{106}	3.633×10^{106}	0.9956

Comparing with equation (22.26), we finally read

$$\ln p(N) \simeq 2\pi \sqrt{\frac{N}{6}}. \tag{22.44}$$

This was our goal, an estimate of $\ln p(N)$ for large N. Indeed, we must require large N since

$$N = \frac{E}{\hbar\omega_0} = \frac{\pi^2}{6}\left(\frac{kT}{\hbar\omega_0}\right)^2 \gg 1, \tag{22.45}$$

because of our high temperature assumption (22.32).

The result (22.44) is only the leading term of the celebrated Hardy–Ramanujan asymptotic expansion of $p(N)$:

$$p(N) \simeq \frac{1}{4N\sqrt{3}} \exp\left(2\pi\sqrt{\frac{N}{6}}\right). \tag{22.46}$$

This is not an exact formula either, but it is an accurate estimate of $p(N)$, as opposed to our accurate estimate of the logarithm of $p(N)$. We will not give here a derivation of the Hardy–Ramanujan result. It is fun, however, to test the accuracy of the Hardy–Ramanujan formula. In Table 22.2 we compare the values of $p(N)$, as calculated exactly, with the estimate $p_{est}(N)$ provided by (22.46). The estimate gives an error of about one-half of one percent for $N = 10\,000$.

We now need a minor generalization of (22.46). Assume the string can vibrate in b transverse directions. Then for each frequency $\ell\omega_0$ we have b harmonic oscillators that represent the possible polarizations of the motion. The associated occupation numbers need a superscript to label the b polarizations:

$$
\begin{array}{cccc}
n_1^{(1)} & n_1^{(2)} & \cdots & n_1^{(b)} \\
n_2^{(1)} & n_2^{(2)} & \cdots & n_2^{(b)} \\
\cdots & \cdots & \cdots & \cdots \\
n_\ell^{(1)} & n_\ell^{(2)} & \cdots & n_\ell^{(b)} \\
\cdots & \cdots & \cdots & \cdots
\end{array}
\tag{22.47}
$$

In order to sum over all possible states in the new partition function Z_b, we must sum over all possible values of the occupation numbers $n_k^{(q)}$, where $k = 1, 2, \ldots, \infty$, and $q = 1, 2, \ldots, b$. This gives

$$
Z_b = \sum_{n_k^{(1)}, \ldots, n_k^{(b)}} \exp\left[-\frac{\hbar\omega_0}{kT} \sum_{\ell=0}^{\infty} \sum_{q=1}^{b} \ell n_\ell^{(q)}\right].
\tag{22.48}
$$

The sums over the various $n^{(q)}$ factorize, so

$$
Z_b = \sum_{n_k^{(1)}} \exp\left[-\frac{\hbar\omega_0}{kT} \sum_{\ell=0}^{\infty} \ell n_\ell^{(1)}\right] \cdots \sum_{n_k^{(b)}} \exp\left[-\frac{\hbar\omega_0}{kT} \sum_{\ell=0}^{\infty} \ell n_\ell^{(b)}\right].
\tag{22.49}
$$

Each factor here is equal to the partition function Z calculated previously, so

$$
Z_b = (Z)^b.
\tag{22.50}
$$

The new free energy F_b is also easy to calculate:

$$
F_b = -kT \ln Z_b = -kTb \ln Z = bF.
\tag{22.51}
$$

The entropy, obtained by differentiation of the free energy, also acquires a multiplicative factor of b:

$$
S_b = bS.
\tag{22.52}
$$

For the energy E_b, the same multiplicative factor exists on account of (22.6):

$$
E_b = bE.
\tag{22.53}
$$

The four equations above are equalities of functions of temperature. For example, (22.52) is $S_b(T) = b\, S(T)$, where $S(T)$ is the entropy determined previously in (22.40). Since E is the previously determined energy (as a function of temperature), S and E are related by (22.43). Note that E_b is equal to $\hbar\omega_0 N$, where N is now the total occupation number

$$
E_b = bE = \hbar\omega_0 N, \quad N = \sum_{\ell,q} \ell n_\ell^{(q)}.
\tag{22.54}
$$

Using (22.52), our earlier result for $S(E)$ in (22.43), and (22.54) we find

$$
S_b = b\,(k\,2\pi)\sqrt{\frac{1}{6}\frac{E}{\hbar\omega_0}} = k\,2\pi\sqrt{\frac{b}{6}\frac{Eb}{\hbar\omega_0}} = k\,2\pi\sqrt{\frac{Nb}{6}}.
\tag{22.55}
$$

Let us call $p_b(N)$ the number of partitions of N into integers that carry any of b labels. This means, for example, that the partition $\{3, 2, 1\}$ of 6 now gives rise to many partitions written as $\{3_{p_1}, 2_{p_2}, 1_{p_3}\}$, where the subscripts p_1, p_2, and p_3 can take any value from one to b. A partition with different subscripts is considered a different partition. We now see that with b degenerate oscillators, the number of states with total number N is $p_b(N)$. Therefore $S_b = k \ln p_b(N)$, and comparing with (22.55) we conclude that for large N

$$\ln p_b(N) \simeq 2\pi \sqrt{\frac{Nb}{6}}. \tag{22.56}$$

The more accurate version of this result can be shown to be

$$p_b(N) \simeq \frac{1}{\sqrt{2}} \left(\frac{b}{24}\right)^{(b+1)/4} N^{-(b+3)/4} \exp\left(2\pi\sqrt{\frac{Nb}{6}}\right). \tag{22.57}$$

You can see that for $b = 1$ this reduces to $p(N)$, as given in (22.46). For $b = 24$, the number of transverse light-cone directions in the bosonic string, the expression simplifies a little:

$$p_{24}(N) \simeq \frac{1}{\sqrt{2}} N^{-27/4} \exp\left(4\pi\sqrt{N}\right). \tag{22.58}$$

Quick calculation 22.2 Show that for large N

$$\frac{p_{24}(N+1)}{p_{24}(N)} \simeq \exp\left(\frac{2\pi}{\sqrt{N}}\right). \tag{22.59}$$

This means that the fractional change in the number of partitions when the argument is increased by one unit goes down to zero as $N \to \infty$.

Quick calculation 22.3 Use direct counting to confirm that $p_{24}(1) = 24$, $p_{24}(2) = 324$, $p_{24}(3) = 3200$, and $p_{24}(4) = 25\,650$.

It is also interesting to count other types of partitions. Consider, for example, partitions of integers into unequal parts. The possible partitions of 6 into unequal integers are

$$\{6\}, \quad \{5, 1\}, \quad \{4, 2\}, \quad \{3, 2, 1\}. \tag{22.60}$$

We denote by $q(N)$ the number of partitions of N into unequal parts, so $q(6)$, for example, is equal to four. We can use a fermionic version of the violin string to determine the large-N behavior of $q(N)$. The frequencies of the oscillators are not changed, but this time we demand that each occupation number can only be equal to zero or to one. Since no creation operator can be used more than once, the total number N of any state is effectively split into contributions all of whose parts are unequal. Creation operators that cannot be used more than once create fermionic excitations. We call such oscillators fermionic. With a little abuse of language, the numbers that enter a partition into unequal parts are called fermionic numbers. You will show in Problem 22.2 that, for large N,

$$\ln q(N) \sim 2\pi \sqrt{\frac{N}{12}}. \tag{22.61}$$

We extended the earlier counting of $p(N)$ to the case in which the elements of a partition can carry any of b labels. If the elements of an unequal partition can carry any of f labels, then the number $q_f(N)$ of partitions is obtained from (22.61) by replacing $N \to Nf$. In such partitions a fermionic number can appear more than once if it carries a different label each time. This counting corresponds to a system with f species of fermionic oscillators.

A final generalization is also useful. We consider partitions of N into both ordinary and fermionic numbers, with b labels for the ordinary numbers and f labels for the fermionic numbers. In this case (Problem 22.4) we find that the large-N leading behavior of the number $P(N; b, f)$ of such partitions is

$$\ln P(N; b, f) \simeq 2\pi \sqrt{\frac{N}{6}\left(b + \frac{f}{2}\right)}. \qquad (22.62)$$

As an example, let us calculate $P(2; 1, 2)$, the number of partitions of 2 into ordinary and fermionic numbers, with the latter having two possible labels. The list of partitions is obtained by labeling the numbers in the ordinary partitions ({2} and {1, 1}) in all possible ways. We find

$$\{2\}, \ \{2_1\}, \ \{2_2\}, \ \{1, 1\}, \ \{1_1, 1\}, \ \{1_2, 1\}, \ \{1_1, 1_2\}. \qquad (22.63)$$

The labels on the fermionic numbers are shown as subscripts. We count $P(2; 1, 2) = 7$.

Equation (22.62) is useful for calculations in superstring theories, where the states are built with both bosonic and fermionic creation operators. An application to a supersymmetric black hole will be considered in Section 22.7.

22.3 Hagedorn temperature

Let us now return to the subject of relativistic strings. We will consider open strings that carry no spatial momentum. This will happen, for example, if the open string endpoints end on a D0-brane. With zero spatial momentum, the string has energy levels that are simply given by the rest masses of its quantum states. The mass-squared of a given state can be expressed in terms of the number operator $N^\perp$ (12.164):

$$M^2 = \frac{1}{\alpha'}(N^\perp - 1) \simeq \frac{N^\perp}{\alpha'}, \qquad (22.64)$$

in the approximation of large $N^\perp$. It follows that the energy $E = M$ is related to the number operator by the simple equality

$$\sqrt{N^\perp} = \sqrt{\alpha'}\, E. \qquad (22.65)$$

In the microcanonical ensemble, the number of states $\Omega(E)$ equals $p_{24}(N^\perp)$, because we have 24 transverse light-cone directions, and consequently 24 oscillator labels for each mode number. As a result, $S(E) = k \ln p_{24}(N^\perp)$. For large energy, $N^\perp$ is also large, and using equation (22.56) we find

$$S(E) = k \, 2\pi \sqrt{\frac{N^\perp \cdot 24}{6}} = k \, 4\pi \sqrt{N^\perp}. \tag{22.66}$$

Making use of the number–energy relation in (22.65) we find

$$S = k \, 4\pi \sqrt{\alpha'} \, E. \tag{22.67}$$

This is the entropy–energy relation at high energy. An entropy proportional to the energy is unusual because it leads to a constant temperature:

$$\frac{1}{kT} = \frac{1}{k} \frac{\partial S}{\partial E} = 4\pi \sqrt{\alpha'}. \tag{22.68}$$

This temperature is called the Hagedorn temperature T_{H}:

$$\frac{1}{\beta_{\mathrm{H}}} = k T_{\mathrm{H}} = \frac{1}{4\pi \sqrt{\alpha'}}. \tag{22.69}$$

Here $k T_{\mathrm{H}}$ is the thermal energy associated with the Hagedorn temperature. In the high energy approximation we are working with, we can arbitrarily increase the energy of the strings, and their temperature will remain fixed at the Hagedorn temperature. It is interesting to compare the energy $k T_{\mathrm{H}}$ to the rest mass of the particles found in the first massive level of the open string. This corresponds to $N^\perp = 2$ in (22.64) and gives $E = M = 1/\sqrt{\alpha'}$. The ratio of the Hagedorn thermal energy to this rest energy is

$$\frac{k T_{\mathrm{H}}}{(1/\sqrt{\alpha'})} = \frac{1}{4\pi} \simeq \frac{1}{12.6}. \tag{22.70}$$

This shows that the Hagedorn thermal energy is small compared with the rest energy of almost any particle state of the string. This is an important result that will play a role in our later work in this chapter.

The entropy–energy relation in (22.67) also holds for closed strings with no spatial momentum. Recalling (13.48), we find

$$M^2 = \frac{2}{\alpha'}(N^\perp + \bar{N}^\perp - 2) \simeq \frac{4}{\alpha'} N^\perp, \tag{22.71}$$

since closed string states satisfy $N^\perp = \bar{N}^\perp$. It follows that the energy $E = M$ is related to the number operator as

$$2\sqrt{N^\perp} = \sqrt{\alpha'} \, E. \tag{22.72}$$

This time, the number of states $\Omega(E)$ is equal to the product of available states in the left-moving and in the right-moving sectors:

$$\Omega(E) = p_{24}\left(N^{\perp}\right) p_{24}\left(\tilde{N}^{\perp}\right) = \left(p_{24}(N^{\perp})\right)^2. \tag{22.73}$$

As a result, the entropy S is precisely twice that indicated in (22.66):

$$S(E) = k\,4\pi\,(2\sqrt{N^{\perp}}) = k\,4\pi\,\sqrt{\alpha'}\,E. \tag{22.74}$$

This is the same entropy–energy relation we had for open strings. We conclude that the temperature T_{H} is also the approximate temperature of highly energetic closed strings.

22.4 Relativistic particle partition function

As a warmup for our computation of the partition function for a string, we compute here the partition function for a particle. We will work with a relativistic particle of mass m that lives in a D-dimensional spacetime, or equivalently, in $d = D - 1$ space dimensions. Moreover, we assume that this particle is confined to a box of volume V:

$$V = L_1 L_2 \ldots L_d. \tag{22.75}$$

This box is in thermal contact with a reservoir at temperature T. As usual, the energy and the momentum of the particle are related by

$$E(\vec{p}) = \sqrt{\vec{p}^2 + m^2}. \tag{22.76}$$

The quantum states of the particle in the box are labeled by the momenta $\vec{p}$, which are quantized. The partition function $Z(m^2)$ is given by

$$Z(m^2) = \sum_{\vec{p}} \exp(-\beta E(\vec{p})). \tag{22.77}$$

The volume dependence of this partition function arises because the quantized values of the momenta depend on the dimensions of the box. The quantum wavefunctions with momentum $\vec{p} = \hbar \vec{k}$ have a spatial dependence $\exp(i\vec{k} \cdot \vec{x})$. The periodicity of these wavefunctions in the box requires that for each spatial direction i

$$k_i L_i = 2\pi n_i, \quad i = 1, 2, \ldots, d, \quad n_i \in \mathbb{Z}. \tag{22.78}$$

Equivalently, in terms of momenta,

$$n_i = p_i \frac{L_i}{(2\pi\hbar)}. \tag{22.79}$$

It follows that summing over the various momenta is the same as summing over the various n_i. For an arbitrary smooth function $f[E]$ of the energy we can thus write

$$\sum_{\vec{p}} f[E(\vec{p})] = \sum_{\vec{n}} f[E(\vec{p}(\vec{n}))] \simeq \int dn_1 dn_2 \ldots dn_d\, f[E(\vec{p}\,(\vec{n}))], \tag{22.80}$$

where the approximation by an integral is allowed because, for large boxes, the momenta change very little when a counter n_i shifts by one unit. Using (22.79) and (22.75) we obtain

$$\sum_{\vec{p}} f[E(\vec{p})] \simeq V \int \frac{d^d\vec{p}}{(2\pi\hbar)^d} \, f[E(\vec{p})]. \tag{22.81}$$

This is the general prescription for dealing with sums over momenta. Applied to our case of interest (22.77) it gives

$$Z(m^2) = V \int \frac{d^d\vec{p}}{(2\pi\hbar)^d} \exp\left(-\beta\sqrt{\vec{p}^2 + m^2}\right). \tag{22.82}$$

This is the integral representation of the partition function for a relativistic point particle of rest mass m. The temperature and volume arguments of Z are implicit. Working with $\hbar = 1$, and changing variables of integration by letting $\vec{p} = m\vec{u}$, we find

$$Z(m^2) = Vm^d \int \frac{d^d\vec{u}}{(2\pi)^d} \exp\left(-\beta m \sqrt{1 + \vec{u}^2}\right). \tag{22.83}$$

This integral is not elementary, but it can be written in terms of derivatives of modified Bessel functions with argument βm (Problem 22.6). Rather than doing so, we will examine the integral in the domain of interest. For our string theory applications, the thermal energy is much smaller than the rest energy of the particle. Indeed, as we saw earlier, for temperatures below the Hagedorn temperature, almost all string states satisfy this condition. Thus we consider the situation where

$$\beta m \gg 1, \quad \text{low temperature.} \tag{22.84}$$

We now claim that the leading approximation to the integral can be found by expanding the square root in (22.83) for $\vec{u}^2$ small. This is explained as follows. Using spherical coordinates and letting $\vec{u}^2 = u^2$, we note that $d^d\vec{u} \sim u^{d-1}du$ (recall the familiar cases of $d = 2, 3$). As a plain one-dimensional integral, the integrand in (22.83) is thus of the form

$$\text{Integrand} \sim u^{d-1} e^{-\beta m\sqrt{1+u^2}}. \tag{22.85}$$

This integrand vanishes at $u = 0$ and $u = \infty$, and it peaks somewhere in between, giving the largest contribution to the integral. The maximum of the integrand can be found by setting the u derivative of (22.85) equal to zero. This gives the condition

$$\frac{d-1}{\beta m} = \frac{u^2}{\sqrt{1+u^2}}. \tag{22.86}$$

Since βm is large, the left-hand side is small, and u^2 must also be small. We can therefore neglect the u^2 in the square root, and we find that the integrand is largest for

$$u^2 \simeq \frac{d-1}{\beta m} \ll 1. \tag{22.87}$$

We are therefore allowed to expand the square root in (22.83) to write

$$Z(m^2) \simeq Vm^d e^{-\beta m} \int \frac{d^d\vec{u}}{(2\pi)^d} \exp\left(-\frac{1}{2}\beta m \, \vec{u}^2\right). \tag{22.88}$$

The integral is now Gaussian and is readily evaluated:

$$Z(m^2) \simeq V e^{-\beta m} \left(\frac{m}{2\pi\beta}\right)^{\frac{d}{2}}.$$

(22.89)

This is our final form for the partition function of a relativistic particle in the low temperature limit. One can verify that this partition function is dimensionless, as it should be. Except for the additional factor $e^{-\beta m}$, this partition function coincides with the exact partition function for a nonrelativistic particle. The exponential factor accounts for the contribution of the rest energy to the energy of the relativistic particle.

22.5 Single string partition function

We are now ready to evaluate the partition function for a single open string placed in a box of volume V. In order to calculate this, we must enumerate the quantum states of the string. The states are obtained by acting with the light-cone creation operators on the momentum eigenstates. A set of basis states is written as in (12.162):

$$|\lambda, p\rangle = \prod_{n=1}^{\infty} \prod_{I=2}^{25} (a_n^{I\dagger})^{\lambda_{n,I}} |p^+, \vec{p}_T\rangle,$$

(22.90)

where the notation $|\lambda, p\rangle$ emphasizes that the momentum components as well as the occupation numbers $\lambda_{n,I}$ are labels of the string states. The d components $(p^+, \vec{p}_T)$ listed in the momentum eigenstate specify the light-cone energy p^- via the on-shell condition:

$$M^2(\{\lambda_{n,I}\}) = -p^2 = 2p^+ p^- - p^I p^I,$$

(22.91)

where

$$M^2(\{\lambda_{n,I}\}) = \frac{1}{\alpha'}(N^\perp - 1), \quad N^\perp = \sum_{n,I} n\,\lambda_{n,I}.$$

(22.92)

Since both the spatial momentum and the energy are determined for the above states, we can label the string states with the set $\{\lambda_{n,I}\}$ of occupation numbers and the *spatial* momentum $\vec{p}$. We then write

$$E(\{\lambda_{n,I}\}, \vec{p}) = \sqrt{M^2(\{\lambda_{n,I}\}) + \vec{p}^2}.$$

(22.93)

To find the partition function Z_{str} of a single string, we must sum over all states $|\lambda, p\rangle$, or equivalently, over all spatial momenta $\vec{p}$ and all values of the occupation numbers $\lambda_{n,I}$:

$$Z_{\text{str}} = \sum_\alpha \exp(-\beta E_\alpha) = \sum_{\lambda_{n,I}} \sum_{\vec{p}} \exp\left[-\beta\sqrt{M^2(\{\lambda_{n,I}\}) + \vec{p}^2}\right].$$

(22.94)

The momentum sum simply gives the partition function Z for a relativistic particle of mass-squared $M^2(\{\lambda_{n,I}\})$. We thus write

$$Z_{\text{str}} = \sum_{\lambda_{n,I}} Z\left(M^2(\{\lambda_{n,I}\})\right).$$

(22.95)

Since the mass M^2 depends only on $N^\perp$, the sum over occupation numbers $\{\lambda_{n,I}\}$ can be traded for a sum over $N^\perp \equiv N$, as long as we keep in mind that there are $p_{24}(N)$ states with number eigenvalue N:

$$Z_{\text{str}} = \sum_{N=0}^{\infty} p_{24}(N) \, Z(M^2(N)). \tag{22.96}$$

So far, no approximations have been made, and the above result is exact.

Let N_0 denote an integer for which $p_{24}(N)$, with $N \geq N_0$, is reasonably approximated by (22.58). Moreover, let Z_0 denote the sum

$$Z_0 \equiv \sum_{N=0}^{N_0-1} p_{24}(N) \, Z(M^2(N)). \tag{22.97}$$

This definition allows us to rewrite Z_{str} in (22.96) as

$$Z_{\text{str}} = Z_0 + \sum_{N=N_0}^{\infty} p_{24}(N) \, Z(M^2(N)). \tag{22.98}$$

It is quite difficult to calculate Z_0 accurately. Our strategy will be to work in a regime where Z_0 is negligible compared to the second term on the right-hand side of (22.98). It will become clear below that this happens when the temperature T approaches the Hagedorn temperature. A few facts about Z_0 should be noted. The contribution from the tachyon states is problematic: equation (22.82) tells us that Z is a complex number if $m^2 < 0$. By neglecting Z_0 we are ignoring the tachyon instability in the theory. At any rate, Z_0 is a finite number for any value of the temperature, and its contribution to Z_{str} will be negligible when the second term on the right-hand side of (22.98) becomes very large.

To proceed further we approximate the sum in (22.98) by an integral. For $N \geq N_0$ we view $p_{24}(N)$ as the continuous function of N defined by the approximate relation (22.58). We then write

$$Z_{\text{str}} \simeq Z_0 + \int_{N_0}^{\infty} dN \, p_{24}(N) \, Z(M^2(N)). \tag{22.99}$$

It is customary to define a density of states $\rho(M)$ as a function of the mass M and to use the mass as the variable of integration. This is done using the relation

$$p_{24}(N)dN = \rho(M)dM. \tag{22.100}$$

We express the left-hand side in terms of mass by using $\alpha' M^2 \simeq N$:

$$dN = 2\alpha' M dM = 2(\sqrt{\alpha'}M) \, d(\sqrt{\alpha'}M). \tag{22.101}$$

Moreover, using (22.58) and (22.69) we find

$$p_{24}(N) \simeq \frac{1}{\sqrt{2}} (\sqrt{\alpha'}M)^{-27/2} \exp(\beta_H M). \tag{22.102}$$

Substituting these two equations back into (22.100) gives

$$\rho(M)dM = \sqrt{2} \, (\sqrt{\alpha'}M)^{-25/2} \exp(\beta_H M) \, d(\sqrt{\alpha'}M). \tag{22.103}$$

Note incidentally that

$$\rho(M) \sim M^{-25/2} \exp(\beta_H M), \qquad (22.104)$$

which shows that the exponential growth in the density of states is controlled by the Hagedorn temperature. As we will see shortly, the partition function does not converge for temperatures higher than the Hagedorn temperature.

With (22.103) and (22.100), the partition function in (22.99) becomes

$$Z_{str} \simeq Z_0 + \sqrt{2} \int_{M_0}^{\infty} (\sqrt{\alpha'}M)^{-25/2} \exp(\beta_H M) Z(M^2) \, d(\sqrt{\alpha'}M), \qquad (22.105)$$

where $\alpha' M_0^2 = N_0$. It only remains to write the particle partition function (22.89) in terms of M and kT_H. With the help of

$$\frac{M}{2\pi\beta} = 2(\sqrt{\alpha'}M) kT \, kT_H, \qquad \beta M = 4\pi(\sqrt{\alpha'}M)\frac{T_H}{T}, \qquad (22.106)$$

we find

$$Z(M^2) \simeq 2^{25/2} V \, (kT \, kT_H)^{25/2} (\sqrt{\alpha'}M)^{25/2} \exp\left(-4\pi \sqrt{\alpha'}M \frac{T_H}{T}\right). \qquad (22.107)$$

Substituting this result into (22.105), the string partition function becomes

$$Z_{str} \simeq Z_0 + 2^{13} V \, (kT \, kT_H)^{25/2} \int_{M_0}^{\infty} d(\sqrt{\alpha'}M) \exp\left(-4\pi \sqrt{\alpha'}M\left[\frac{T_H}{T} - 1\right]\right). \qquad (22.108)$$

Notice that the powers of M in the integrand cancelled out. Setting $x = \sqrt{\alpha'}M$, the above expression turns into

$$Z_{str} \simeq Z_0 + 2^{13} V \, (kT \, kT_H)^{25/2} \int_{\sqrt{N_0}}^{\infty} dx \, \exp\left(-4\pi x\left[\frac{T_H}{T} - 1\right]\right). \qquad (22.109)$$

The integral only converges for $T < T_H$, where we have

$$Z_{str} \simeq Z_0 + \frac{2^{11}}{\pi} V \, (kT \, kT_H)^{25/2} \left(\frac{T}{T_H - T}\right) \exp\left(-4\pi \sqrt{N_0}\left[\frac{T_H}{T} - 1\right]\right). \qquad (22.110)$$

In the limit when $T \to T_H$ from below, the argument of the exponential goes to zero and as a result, the exponential goes to one. In addition, the factor which multiplies the exponential grows without limit and becomes much larger than Z_0. It follows that for T sufficiently close to T_H, the partition function Z_{str} is well approximated by

$$Z_{str} \simeq \frac{2^{11}}{\pi} V \, (kT_H)^{25} \left(\frac{T_H}{T_H - T}\right), \qquad T \to T_H. \qquad (22.111)$$

In writing this formula we replaced T by T_H in all places where it is possible to do so. This is our final expression for the approximate partition function of a single open string that is enclosed in a box of volume V and is in thermal contact with a reservoir at a temperature T very close to T_H.

We can use this result to calculate the average energy of a string near the Hagedorn temperature. In view of (22.6), we need only the β dependence of $\ln Z_{\text{str}}$. Using $\beta \simeq \beta_{\text{H}}$, the result is

$$\ln Z_{\text{str}} \simeq -\ln(\beta - \beta_{\text{H}}) + \cdots, \tag{22.112}$$

where the dots represent terms without β dependence. It now follows that the average energy E_{str} of the string is

$$E_{\text{str}} = -\frac{\partial \ln Z_{\text{str}}}{\partial \beta} \simeq \frac{1}{\beta - \beta_{\text{H}}} \simeq k T_{\text{H}} \left(\frac{T_{\text{H}}}{T_{\text{H}} - T} \right). \tag{22.113}$$

The energy E_{str} grows without bound as the temperature approaches the Hagedorn temperature.

String thermodynamics is not yet well understood. Some of the complications arise because in string theory the relation between the canonical and microcanonical ensembles is not familiar. In general, if we let $\Omega(E)dE$ be the number of states in the energy interval dE, the canonical partition function Z is defined as

$$Z(\beta) = \int_0^\infty dE \, e^{-\beta E} \, \Omega(E). \tag{22.114}$$

The microcanonical distribution $\Omega(E)$ determines an energy–temperature relation $E(T)$ via the relation

$$\beta = \frac{\partial \ln \Omega}{\partial E}, \tag{22.115}$$

as you recall from (22.2). In the canonical ensemble, this energy–temperature relation is roughly reproduced whenever the integral (22.114) is dominated by a saddle point. The saddle point arises at the maximum of the integrand, and it is determined by the condition

$$\frac{d}{dE}(e^{-\beta E} \, \Omega(E)) = \left(-\beta \Omega(E) + \frac{\partial \Omega}{\partial E} \right) e^{-\beta E} = 0. \tag{22.116}$$

As you can see, the vanishing condition reproduces (22.115). If the integral is dominated by this saddle point, the average energy $\bar{E}(T)$ calculated from Z turns out to be approximately equal to $E(T)$. In familiar systems $\Omega(E) \sim E^\gamma$, with $\gamma > 0$. In this case a saddle point exists, and the two ensembles give approximately equal results. On the other hand, in string theory $\Omega(E) \sim \exp(\beta_{\text{H}} E)$, and the integrand is then proportional to $\exp([-\beta + \beta_{\text{H}}]E)$, which is a function of energy that has no critical point. As a consequence, there is no guarantee that the two ensembles give the same results. It becomes important to decide physically which ensemble is relevant for each specific problem. Some refined computations using the microcanonical ensemble are examined in Problem 22.7. The results are surprising.

22.6 Black holes and entropy

A black hole is a gravitationally collapsed object that is formed when the mass of an object is increased while its size remains fixed, or when the size of an object is reduced while its mass is kept constant. Black holes were initially predicted by theoreticians before there were any observational data to support their existence, but it has now become an experimental fact that they exist in our universe. The presence of a supermassive black hole at the center of our galaxy has been established convincingly. Most likely, there are millions of black holes in every galaxy. They are the remnants of ordinary stars that were a few times more massive than the sun.

Black holes pose very significant theoretical challenges. In Einstein's theory of general relativity they appear as classical solutions which represent matter that has collapsed down to a point with infinite density: a singularity. Although dealing with classical singularities is already a theoretical challenge, the real puzzles of black holes arise at the quantum level. Quantum mechanically, black holes radiate energy. They also have thermodynamical temperature and entropy, but these properties are difficult to understand at the fundamental level of statistical mechanics, where they should be derived from a counting of degrees of freedom. String theory has had some impressive success in understanding the entropy of black holes. In this section we review basic features of black holes and use string theory to discuss the entropy of four-dimensional Schwarzschild black holes. In the following section we will examine a particular five-dimensional black hole for which the entropy can be calculated exactly in string theory.

The simplest black holes are Schwarzschild black holes. These black holes are spherically symmetric static solutions of Einstein's equations that represent the gravitational field of a point mass M. For such a black hole, the point singularity is separated from the outside world by what is known as an *event horizon*. This is a mathematical two-sphere centered at the singularity, whose radius R is called the Schwarzschild radius, or simply the radius of the black hole. If any object ventures inside the event horizon it will fall irrevocably into the singularity. Classically, nothing can escape from the region enclosed by the event horizon. The value R of the Schwarzschild radius can be estimated by assuming that the total energy of any particle at the horizon is equal to zero. For a particle of mass m, this energy includes the rest energy mc^2 and the gravitational potential energy $-GMm/R$. Setting the sum of these two equal to zero, we find

$$mc^2 - \frac{GMm}{R} = 0 \longrightarrow R \simeq \frac{GM}{c^2}. \qquad (22.117)$$

The exact radius is calculated in general relativity and the answer is

$$R = \frac{2GM}{c^2}. \qquad (22.118)$$

Physicists often speak of the Schwarzschild radius associated with a mass, meaning by this the radius of a black hole carrying such mass. The Schwarzschild radius of the sun is about three kilometers. The Schwarzschild radius of the earth is about one centimeter. The

Schwarzschild radius of a billion-ton asteroid is of the order of 10^{-15} m. It is possible to use the Newtonian laws of gravitation to estimate the gravitational field at the horizon:

$$|\vec{g}| = \frac{GM}{R^2} = \frac{c^4}{4GM}. \tag{22.119}$$

This gravitational field becomes small for very massive black holes. A massive object is a black hole if it is enclosed by a sphere that has its Schwarzschild radius.

Quick calculation 22.4 Show that a spherical object of uniform mass density ρ is a black hole if its radius is *larger* than $c/\sqrt{8\pi G\rho/3}$.

If we believe the second law of thermodynamics, then the existence of black holes leads to some surprising conclusions. Assume that a certain amount of hot gas falls into a black hole, so that the black hole now has a slightly higher mass. Since the total entropy of the system consisting of the gas and the black hole cannot decrease, the black hole must have acquired at least as much entropy as was carried by the gas. We are thus led to believe that black holes have entropy. You know that a system has entropy when there are many microscopic states of the system that are consistent with its macroscopic properties. On the other hand, if the black hole is simply a point mass singularity, it is hard to see what are the microstates that give rise to the entropy.

Black holes emit thermal radiation with a well defined temperature: a Hawking temperature $\bar{T}_H$ that is proportional to the gravitational field at the horizon and, consequently, inversely proportional to the mass of the hole (do not confuse $\bar{T}_H$ with the Hagedorn temperature T_H). This proportionality is reasonable since the radiation from the black hole emerges from the near-horizon region, and is controlled by the intensity of gravity. In natural units the Hawking temperature of a black hole of mass M turns out to be $k\bar{T}_H = 1/(8\pi M)$, so by inserting back the factors of $\hbar, c$, and G (Problem 3.7) we find

$$k\bar{T}_H = \frac{\hbar c^3}{8\pi GM}. \tag{22.120}$$

This equation allows us to calculate the Bekenstein entropy S_B of the black hole. Using $E = Mc^2$ for the energy of the black hole and the first law of thermodynamics $dE = \bar{T}_H dS_B$, we write

$$dE = c^2 dM = \bar{T}_H dS_B = \frac{\hbar c^3}{8\pi GM} \frac{1}{k} dS_B. \tag{22.121}$$

A little rearrangement yields

$$\frac{1}{k} dS_B = \frac{4\pi G}{\hbar c} dM^2. \tag{22.122}$$

Integrating this equation, and assuming that the entropy of a zero-mass black hole is zero, we find

$$\frac{S_B}{k} = \frac{4\pi G}{\hbar c} M^2. \tag{22.123}$$

The entropy of the black hole is proportional to the *square* of its mass. A useful alternative expression for the entropy uses the area A of the event horizon. With $A = 4\pi R^2$ and R given by (22.118), one readily obtains

$$\frac{S_B}{k} = \frac{1}{4}\frac{c^3}{\hbar G}A = \frac{A}{4\ell_P^2}, \qquad (22.124)$$

where ℓ_P is the Planck length. The right-hand side in this equation has a simple interpretation: the Bekenstein entropy S_B is one-fourth of the area of the horizon expressed in units of Planck-length squared. Given that ℓ_P^2 is a remarkably small area, the entropy of any astrophysical-size black hole is extremely large. The entropy of a black hole is roughly reproduced if one imagines having a degree of freedom with a finite number of states for each horizon element of area ℓ_P^2. String theory provides candidate degrees of freedom for black holes, but they do not relate directly to the horizon area.

Quick calculation 22.5 Show that a photon of energy $k\bar{T}_H$ has a wavelength which is approximately 80 times the radius of the black hole.

In string theory, we attempt to relate a stationary Schwarzschild black hole to a string with a high degree of excitation but zero momentum. In the microcanonical ensemble, a string state with energy E has an entropy (22.67). This is true both for open and for closed strings (see (22.74)). Identifying $E = M$ and working henceforth with $\hbar = c = 1$, we have

$$\frac{S_{str}}{k} = 4\pi\sqrt{\alpha'}\,M, \qquad (22.125)$$

where we have added the subscript "str" to refer to the entropy of the string. This result should be compared to the black hole entropy (22.123):

$$\frac{S_B}{k} = 4\pi GM^2. \qquad (22.126)$$

The disagreement appears to be clear: the entropy of a black hole goes like the mass-squared, while the entropy of a string goes like the mass. We will soon show, however, that the apparent disagreement was to be expected. When they are properly understood, these equations display a surprising level of agreement. The linear dependence of the string entropy on the mass M of the string is not surprising. Entropy is an extensive quantity, and for a string the mass M is roughly proportional to its length L. The black hole entropy, on the other hand, exhibits a surprising feature: it is not proportional to the volume enclosed by the event horizon, but rather, to the area enclosed by the horizon. This failure of extensivity is a feature of gravitational physics.

Before considering the relation between equations (22.125) and (22.126), let us give a heuristic derivation of the string entropy. For this, we consider a string of mass M and estimate its length L to be roughly given by

$$M \sim T_0 L \sim \frac{1}{\alpha'}L, \qquad (22.127)$$

where $T_0 \sim 1/\alpha'$ is the string tension. We now imagine a string built by joining together bits of string, each of which is of length $\ell_s = \sqrt{\alpha'}$. Each string bit can point in any of n possible directions. The number n may be equal to the number of spatial dimensions, but since our arguments are rough, we will not be specific. Since the number of string bits is $L/\sqrt{\alpha'}$, the number of ways Ω that we can build this string is roughly

$$\Omega \sim n^{L/\sqrt{\alpha'}} \sim n^{M\sqrt{\alpha'}} \sim e^{M\sqrt{\alpha'}\ln n}. \qquad (22.128)$$

The entropy of the string is obtained by taking the logarithm of Ω:

$$\frac{S_{\text{str}}}{k} \sim M\sqrt{\alpha'} \sim M\,\ell_s, \qquad (22.129)$$

where we discarded the $\ln n$ factor, in keeping with the accuracy of the estimate. This result is consistent with the expression given in (22.125).

Equations (22.125) and (22.126) disagree because the black hole entropy S_{B} was calculated in a regime where interactions are necessary, while the string entropy S_{str} was calculated for free strings. We should not have expected agreement, unless for some reason interactions did not affect the calculation of the entropy of strings.

Interactions are necessary in the black hole entropy calculation because Newton's constant G vanishes if the string coupling constant g is set to zero. Indeed, we recall (13.83), which states that

$$G \sim g^2\alpha' = g^2\ell_s^2. \qquad (22.130)$$

The black hole entropy and the black hole radius are then given by

$$\frac{S_{\text{B}}}{k} \sim GM^2 \sim g^2\ell_s^2 M^2,$$
$$R \sim GM \sim g^2\ell_s^2 M. \qquad (22.131)$$

While they incorporate the string coupling dependence via Newton's constant, the above results use classical general relativity, where, for example, the concept of a horizon makes sense. We are allowed to neglect string theory corrections to general relativity as long as black holes are much larger than the string length.

Consider now a large black hole with Bekenstein entropy S_0, mass M_0, and radius $R_0 \gg \ell_s$. Fix also the string coupling to some finite value g_0. Equations (22.131) then give us

$$\frac{S_0}{k} \sim g_0^2\,\ell_s^2\,M_0^2,$$
$$R_0 \sim g_0^2\,\ell_s^2\,M_0. \qquad (22.132)$$

Since the calculation of the string entropy is valid for zero, and possibly small, string coupling, imagine now the process of dialing down the value of the string coupling. This is done by changing the expectation value of the dilaton, as explained in Section 13.4. It is reasonable to assume that this process can be carried out reversibly, so we can expect the black hole entropy to remain unchanged. On the other hand, as we dial down the string coupling g, the mass of the black hole must increase like $1/g$ to keep the entropy constant in (22.131). The mass is not increasing, however, if it is measured in units of Planck mass, since $G \sim 1/m_{\rm P}^2$. The radius R of the black hole decreases, as follows from the second relation in (22.131) bearing in mind that $M \sim 1/g$.

Let g_*, R_*, and M_* denote the final values of the string coupling, black hole radius, and black hole mass, respectively. The constancy of the entropy and the formula for the radius give

$$\frac{S_0}{k} \sim g_0^2 \ell_s^2 M_0^2 = g_*^2 \ell_s^2 M_*^2 \,,$$

$$R_* \sim g_*^2 \ell_s^2 M_*. \tag{22.133}$$

We do not expect these results to hold when the black hole becomes smaller than the string length, so let us fix $R_* = \ell_s$ as the minimum radius for which equations (22.133) can be trusted. The condition $R_* = \ell_s$ tells us that

$$g_*^2 \ell_s^2 M_* \sim \ell_s \longrightarrow M_* \sim \frac{1}{g_*^2 \ell_s}. \tag{22.134}$$

Back in the expression for the entropy S_0 we find

$$\frac{S_0}{k} \sim \frac{1}{g_*^2}. \tag{22.135}$$

The coupling g_* is clearly very small since S_0 was assumed to be very large. At such weak coupling we can reasonably trust the free string theory expression (22.129) for the entropy. Since the black hole we are comparing with has mass M_*, we consider a string of mass M_*. The entropy is then given by

$$\frac{S_{\rm str}}{k} \sim M_* \ell_s \sim \left(\frac{1}{g_*^2 \ell_s}\right) \ell_s \sim \frac{1}{g_*^2}, \tag{22.136}$$

where we made use of (22.134). Comparing with (22.135), we see that $S_{\rm str} \sim S_0$. This agreement is evidence for the hypothesis that a Schwarzschild black hole is the strong coupling version of a string with a very high degree of excitation. It is far from a proof, however. As you have seen, we have only written approximate relations, and we have made a series of assumptions about the ranges of validity of certain results. A proof remains to be found at this time. Nevertheless, there is additional circumstantial evidence that this picture is at least roughly correct. It is possible to estimate the "size" of a string using the picture of string bits and assuming that the string is a random walk. One can then show that for any fixed coupling g there is a mass beyond which any excited string state is smaller than its Schwarzschild radius (Problem 22.9). This suggests that very heavy string states will form black holes.

22.7 Counting states of a black hole

Our computation of the entropy of strings can be done in the limit when we neglect the effects of interactions. Since a black hole can only exist once interactions are turned on, an exact computation of the entropy of a black hole in string theory requires that the counting of states done with string coupling $g = 0$ remains valid when $g \neq 0$.

For the Schwarzschild black holes considered in the previous section this does not happen. As a result, we could only confirm qualitative agreement over a narrow range of couplings for which the gravity computation and the free string theory computation could both hold. In this section we wish to consider a particular five-dimensional black hole that appears in superstring theory. For this black hole, as we will explain below, the counting of states at zero string coupling will remain valid when the coupling becomes nonzero. It is the simplest known black hole with this property. Four-dimensional black holes with the same property are known, but they are are slightly more complicated. This is why we focus here on the five-dimensional black hole.

The remarkable property is due to supersymmetry. As long as supersymmetry is present, certain quantities can be calculated at zero coupling, and the results remain valid for all values of the coupling. Superstrings living in ten-dimensional Minkowski spacetime have supersymmetry. It is a challenge to compactify spacetime and preserve supersymmetry, but this happens if we curl up dimensions into circles. If we now include a black hole in the spacetime, supersymmetry can be lost. The black hole we are interested in is special: even when it is included some supersymmetry survives.

The starting point is ten-dimensional type IIB closed superstring theory. One can search for black hole solutions in the regime where the string theory is well approximated by a field theory of gravity, Kalb–Ramond fields, and other fields, including fermions. Such a theory is called type IIB supergravity. We curl up five of the spatial dimensions into circles. Let us denote these dimensions by x^5, x^6, x^7, x^8, and x^9. The black hole is a spherically symmetric configuration in the uncompactified effective spacetime M^5 defined by the coordinates x^0, x^1, x^2, x^3, and x^4. We cannot discuss here the full construction of the black hole, so we will simply summarize the results that are obtained.

(1) The black hole carries three different electric charges with respect to three Maxwell-like gauge fields that live on M^5. These charges are denoted by the integers

$$Q_1, \ Q_5, \ \text{and} \ N. \tag{22.137}$$

A specific black hole is obtained by choosing these three integers.

(2) The black hole is *extremal*: it has the minimal mass that is compatible with its charges. It does not radiate, since radiation would reduce its mass without the necessary change of charge. The black hole has zero temperature. In addition, its presence preserves a large part of the original supersymmetry of the IIB theory in ten-dimensional Minkowski spacetime.

(3) The black hole horizon is a three-sphere with finite volume A_H. The thermodynamical black hole entropy S_{bh} is calculated using the five-dimensional analog of (22.124):

$$\frac{S_{bh}}{k} = \frac{A_H}{4G^{(5)}} = 2\pi \sqrt{N Q_1 Q_5}. \tag{22.138}$$

Here $G^{(5)}$ is the five-dimensional Newton constant and we have set $\hbar = c = 1$. Interestingly, the entropy only depends on the charges carried by the black hole and not on other parameters, such as the string coupling or the size of the circles used for the compactification.

The goal is to use string theory to reproduce the entropy (22.138) by counting states. String theory must explain how this black hole can be constructed in many possible ways. We know how to count states in string theory when there are no interactions. This time, however, the black hole respects supersymmetry and this guarantees that the zero coupling counting continues to hold for nonzero coupling.

At zero coupling the black hole is constructed using type IIB superstring theory, with the five coordinates $x^5, \ldots, x^9$ curled up into circles. The charges Q_1 and Q_5 are generated by wrapping a number Q_1 of D1-branes around the circle x^5 and a number Q_5 of D5-branes around the five circles. These charges arise by the mechanism discussed in Section 16.4. Since a D5-brane has five spatial dimensions, the D5-branes wrap completely around the compact extra dimensions. How does this look to the five-dimensional observer in M^5? Since all spatial directions along M^5 are Dirichlet for the D5-branes, the D5-branes have fixed spatial coordinates on M^5. They appear, for all times, as a collection of static points. The same is true for the D1-branes. In the configuration we are trying to build, we require that all these points coincide. Thus all D-branes are coincident, and they are seen by the observer as a single point in space. This point is the center of the black hole that forms when the coupling is turned on. So far, this configuration of D-branes cannot be built in different ways that preserve supersymmetry. A few discrete choices are possible; we can choose, for example, another coordinate to wrap all of the D1-branes. But any number of choices that happens to be independent of the charges cannot give us the correct entropy. So, where does the entropy come from?

Recall that the macroscopic black hole had an additional charge N. What does this correspond to in the brane construction? It is a momentum quantum number. The momentum around the circle x^5 must equal

$$p^5 = \frac{N}{R}, \tag{22.139}$$

where R is the radius of the circle. This momentum cannot be carried by the D-branes since they are translationally invariant along the x^5 direction. The momentum is carried by open strings attached to the D-branes! We can now see how it is possible to get many states: there are many kinds of strings stretching between the Q_1 D1-branes and the Q_5 D5-branes. We have (1,1) strings going from D1-branes to D1-branes. We have (5,5) strings going from D5-branes to D5-branes. Finally, we have (1,5) and (5,1) strings, going from D1-branes to D5-branes and vice versa, respectively. Moreover, the total momentum quantum number N can be split between many open strings. Supersymmetry, however, makes one extra demand: all of the open strings must carry momentum in the same direction along x^5.

To proceed further, we need some known facts about the combined system of coincident D1- and D5-branes.

(1) The D1/D5-brane system is a bound state. Open strings of type (1,1) and (5,5) become massive and do not become excited in the configuration we are interested in. These strings can be dropped from the counting.

(2) The total number of ground states of a (1,5) string and the oppositely oriented (5,1) string is eight: four bosonic ground states and four fermionic ground states.

(3) The Q_1 D1-branes may join to form a single D1-brane wrapped Q_1 times around the circle. Similarly, the Q_5 D5-branes can join to form a single D5-brane wrapped Q_5 times around the full compact space. If this happens, the charges are not changed.

Bearing this information in mind, we see that the momentum number N must be split among open strings that stretch between D1-branes and D5-branes. We need a partition of N, but of which kind? Let us assume, for the time being, that $N \gg Q_1 Q_5$ and do a preliminary counting that will work but is not generally valid.

We have to partition N, and, for each element of each partition, we have to tell what kind of state is carrying the momentum quantum number. There are $Q_1 Q_5$ ways of picking a D1-brane and a D5-brane. Moreover, there are four additional ways to pick a bosonic excitation or, alternatively, four ways to pick a fermionic excitation (see (2) above). As a result, we have $b = 4Q_1 Q_5$ bosonic labels and $f = 4Q_1 Q_5$ fermionic labels. Making use of (22.62), the entropy is then

$$\frac{S_{\text{str}}}{k} = \ln P\left(N;\, 4Q_1 Q_5,\, 4Q_1 Q_5\right) \sim 2\pi \sqrt{\frac{N}{6}(4Q_1 Q_5)\frac{3}{2}} = 2\pi\sqrt{NQ_1 Q_5}, \quad (22.140)$$

in perfect agreement with (22.138). This is very nice, but it is not general enough. The restriction $N \gg Q_1 Q_5$ is needed because in (22.62) N must be much larger than both b and f. It can be shown that if N, Q_1, and Q_5 all grow large simultaneously then $\ln P$ fails to give the expected entropy. This means that we have not quite yet identified the general counting that gives the entropy.

The clue is given in item (3) of the list above. Imagine the D1-brane wrapped Q_1 times around the circle x^5. Consider then a (1,1) string moving along the D1-brane. How is the momentum of the string quantized? For such a string the circle has effectively become Q_1 times longer: $(2\pi R)Q_1$ is the distance the string must travel to return to its original starting point on the D1-brane. Accordingly, the string momentum is quantized in units of $1/(Q_1 R)$. This is true with one proviso. The individual open strings can have their momentum quantized with this finer unit, but the total momentum of all the open strings must still be quantized in units of $1/R$. This is because the system of the D1-brane and the attached open strings must be invariant under a translation by $2\pi R$ along the circle. As a result, the total momentum of the system must be quantized in units of $1/R$. Since the D1-brane has no momentum, the claim follows.

We must focus, however, on the strings stretching between D1-branes and D5-branes. Imagine now that the D5-branes are also wrapped. For simplicity, assume that Q_1 and Q_5 are relatively prime (we will relax this assumption shortly). Consider now a (1,5) string. How many times must it go around the x^5 circle so that *both* of its endpoints return to their original positions? After Q_1 turns the first endpoint does, but not the second. After Q_5 turns the second endpoint returns to its starting point, but the first does not. It takes $Q_1 Q_5$ turns to have both endpoints return to their original positions on the respective branes. As a result, the momentum of (1,5) and (5,1) strings is quantized with the even finer unit of $1/(Q_1 Q_5 R)$! This can be arranged to be approximately true even if Q_1 and Q_5 are not relatively prime. Take, for example $Q_1 = Q_5 = 100$. We can take the D1-brane and split off one turn to get a system with $Q'_1 = 99$ plus one extra D1-brane. Since Q'_1 and Q_5 are relatively prime, the momentum of most open strings is then quantized in units of $1/(Q'_1 Q_5 R)$, which is approximately equal to $1/(Q_1 Q_5 R)$. In general, for large Q_1 and Q_5 we can find relatively prime numbers $Q'_1 < Q_1$ and $Q'_5 < Q_5$ such that $Q'_1 \sim Q_1$ and $Q'_5 \sim Q_5$.

With this finer unit of quantization, the total momentum in (22.139) is suggestively written as

$$p^5 = \frac{N Q_1 Q_5}{Q_1 Q_5 R}. \tag{22.141}$$

This time we must partition the number $N Q_1 Q_5$. Since we just have one long D1-brane and one D5-brane, there is only one kind of string stretching between the branes. Therefore the labels on the elements of a partition are either four bosonic ones or four fermionic ones: $b = f = 4$. As a result, the entropy is given by

$$\frac{S_{\text{str}}}{k} = \ln P \, (N Q_1 Q_5; \, 4, \, 4) \sim 2\pi \sqrt{\frac{N Q_1 Q_5}{6} (4)\frac{3}{2}} = 2\pi \sqrt{N Q_1 Q_5}. \tag{22.142}$$

The agreement with the black hole entropy is now complete and holds generally.

The statistical mechanical derivation of black hole entropy is a significant accomplishment of string theory. Moreover, good progress has been made in extending the set of black holes for which the entropy is calculable. Still, much work remains to be done in string theory to understand black holes fully. Schwarzschild black holes are not under any precise control, and there are puzzles associated with the fate of the information that falls into a black hole.

String theory gives a clear picture of the zero-coupling degrees of freedom of a configuration that turns into a black hole for nonzero coupling. Moreover, we know that the counting continues to hold for nonzero coupling. We would like to know how these degrees of freedom look by the time the black hole is formed. Many mysteries remain.

Problems

Problem 22.1 Review of statistical mechanics.

(a) Prove equation (22.7). [Hint: the energy levels $E_\alpha(V)$ of the system depend on the volume. As the volume changes quasistatically, the change in mean energy is calculated using the equilibrium distribution of states. The change in mean energy can be interpreted as due to work against the pressure.]
(b) Prove equation (22.15). [Hint: consider the differential $d \ln Z(T, V)$.]

Problem 22.2 Fermionic violin string and counting unequal partitions.

Consider a system of simple harmonic oscillators with frequencies $\omega_0, 2\omega_0, \ldots$ identical to those of the bosonic violin string oscillators of Section 22.2. This time, however, each occupation number n_ℓ can only take the value 0 or 1. Oscillators with this property are said to be fermionic oscillators.

(a) Calculate the free energy of such a string in the high temperature limit. The answer involves the sum

$$1 - \frac{1}{2^2} + \frac{1}{3^2} - \frac{1}{4^2} + \frac{1}{5^2} - \cdots,$$

which can be calculated using (22.38).

(b) Let $q(N)$ denote the number of partitions of N into *unequal* pieces. Use your result in (a) above to show that the large-N behavior of $\ln q(N)$ is given by (22.61).

(c) Now assume this string is relativistic, with the energy related to the mode number as in (22.65): $\sqrt{N} = \sqrt{\alpha'} E$. What is the Hagedorn temperature for such a string?

Problem 22.3 Generating functions for partitions.

A particularly simple infinite product provides a generating function for the partitions $p(n)$:

$$\prod_{n=1}^{\infty} \frac{1}{(1 - x^n)} = \sum_{n=0}^{\infty} p(n) x^n \,.$$

Here $p(0) \equiv 1$. To evaluate the left-hand side, each factor is expanded as an infinite Taylor series around $x = 0$. Test this formula for $n \leq 4$, and explain (in words) why it works in general. Find a generating function for unequal partitions $q(n)$, and test it for low values of n.

Problem 22.4 Counting of generalized partitions.

Prove the formula (22.62) for the partitions $P(N; b, f)$ of N into ordinary integers with b labels and fermionic integers with f labels. Calling Z the partition function of ordinary oscillators and Z' the partition function of Problem 22.2, begin your derivation by explaining why the partition function Z_T for the composite system of bosonic and fermionic labeled oscillators is given by

$$Z_T = (Z)^b \, (Z')^f \,.$$

Problem 22.5 Open superstring Hagedorn temperature.

Consider the open superstring theory described in Section 14.1.

(a) Show that the total number of states (NS and R sectors) with number $N^\perp$ is $16 P(N^\perp; 8,8)$. [Hint: one of the two sectors is easier to count; then use supersymmetry.]

(b) Following the method of Section 22.3, calculate the Hagedorn temperature for an open superstring. Show that it is a factor of $\sqrt{2}$ larger than the Hagedorn temperature of the bosonic string.

Problem 22.6 Partition function of the relativistic particle.

Evaluate exactly the partition function (22.83) for the relativistic point particle in terms of modified Bessel functions (and derivatives thereof), making use of the integral representation

$$K_\nu(z) = \frac{\sqrt{\pi}\left(\frac{1}{2}z\right)^\nu}{\Gamma\left(\nu+\frac{1}{2}\right)} \int_0^\infty e^{-z\cosh t}\,\sinh^{2\nu} t\,dt\,.$$

Use the asymptotic expansion

$$K_\nu(z) \sim e^{-z}\sqrt{\frac{\pi}{2z}}\Big[1 + \frac{4\nu^2-1}{8z} + \cdots\Big],$$

which is valid for large z, to confirm our low temperature result in (22.89). Calculate the first nontrivial correction to this result.

Problem 22.7 Corrections to the temperature/energy relation in the microcanonical ensemble.

We found the Hagedorn temperature in the idealized string model by computing the entropy–energy relation in the high energy approximation where $\ln p_{24}(N) \sim 4\pi\sqrt{N}$. Use the more accurate expression for the partitions $p_{24}(N)$ as given in (22.58) to find the corrections to the temperature–energy relation. You will find the surprising result that as the energy goes to infinity the temperature goes to T_H from above! Plot $T(E)$, and calculate the (negative!) specific heat C in the high energy regime.

The above computations were done using $S = k\ln p_{24}(N)$ and the relation between E and N. In conventional systems with continuous energies one defines $S = k\ln\Omega(E)$, where $\Omega(E)dE$ is the number of states in the energy interval dE. Use the relation $\Omega(E)dE = p_{24}(N)dN$ to calculate $\Omega(E)$, and show that the resulting entropy $S(E)$ differs slightly from the one calculated before.

The negative specific heat obtained for the idealized string is not atypical. For open strings on a Dq-brane one can prove that

$$\Omega(E) \simeq E^{-\gamma}\exp(4\pi\sqrt{\alpha' E})\,, \quad \text{with} \quad \gamma = (25-q)/2\,.$$

Use the continuous energy formulation to calculate the specific heat C in the high energy regime. Give your answer in terms of γ, E, and kT_H.

Problem 22.8 Long strings are entropically favored.

The general approximate formula for partitions $p_b(N)$ takes the form

$$p_b(N) \sim \beta N^{-\gamma}\exp(\delta\sqrt{N})\,,$$

where β, γ, and δ are positive, b-dependent constants. Consider an open bosonic string with large excitation number $N^\perp = N_0$ and energy E_0 related by $\alpha' E_0^2 \simeq N_0$. Assume zero momentum for all strings.

(a) Find the ratio of the number of states available to the string and the number of states available when the string breaks into two (distinguishable) strings, each carrying half the energy. Express your answer in terms of N_0 and the constants in the problem. What is the change in entropy $\Delta S/k$ during the process in which the two half-strings join to make the original string? Show that this change is positive when N_0 is sufficiently large. Which constant out of β, γ, and δ is the one responsible for this effect?

(b) Show that, in general, the combination of two open strings with large excitation numbers into a single open string is a process that increases the entropy.

(c) Since the above results are valid for large N_0, it is of some interest to test them for smaller N_0. How many times are the number of available states increased when a string with $N_0 = 9$ is formed from two strings, each of which has the same energy? What is the change of entropy $\Delta S/k$? Use the exact formula $\alpha' E^2 = N - 1$ for all cases. (A little help: $p_{24}(9) = 143\,184\,000$.)

Problem 22.9 Estimating the size of a string state.

We used the heuristic picture of a string made out of string bits to estimate correctly the entropy (22.129) of a string. We now want to use this picture to estimate the size of an open string state. Assume that each string bit can point randomly in any of d orthogonal directions. The string can then be viewed as a random walk with a number of steps equal to the number of bits.

(a) Use the random walk formula for the average value of the square of the displacement to show that the "size" R_{str} of a string of mass M is

$$R_{\text{str}}(M) \sim M^{1/2} \ell_s^{3/2} \sim N^{1/4} \ell_s, \tag{1}$$

where N is the number eigenvalue associated with the mass M. Note that the size grows like the square root of the mass, while the length of the string grows like the mass. R_{str} is the size of the string at zero string coupling.

(b) Show that the (zero string coupling) size R_{str} of a string of mass M is smaller than the Schwarzschild radius of a mass M in a theory with coupling g, when $M > \bar{M}$ where

$$\bar{M} \sim \frac{1}{g^4 \ell_s} \sim \frac{m_P}{g^3}. \tag{2}$$

This result suggests that the sufficiently massive strings could form a black hole. Show that for $\bar{M}$ the radius of the hole is ℓ_s/g^2, so for couplings that are small, the size of the hole is too large to trust the string model. Show that the value of N for a zero coupling string of mass $\bar{M}$ is $N \sim 1/g^8$. Give a rough estimate of $\bar{M}$ in kg when $g \sim 0.01$.

(c) Consider a very large black hole of mass M and Schwarzschild radius R in a string theory at some finite coupling g. Assume now that the coupling is slowly dialed down to zero. Calculate the value of N that characterizes the zero coupling string that is obtained (write your answer in terms of M and m_P). Show that the size R_{str} of this string is equal to R/g. How much bigger is the length of the string?

The mass of the black hole at the center of our galaxy is approximately 2.6 million solar masses. Estimate the value of N for the corresponding zero coupling string. (Answer: $N \sim 10^{177}$.)

The random walk model of string states applies to strings with little or no angular momentum (recall that the size of a rigidly rotating open string is proportional to the mass). In this model the size of the string is much smaller than its length. The effect of string interactions appears to reduce further the size of the strings.

23

Strong interactions and AdS/CFT

String theory offers a number of insights into the theory of strong interactions. The quantum states of a rotating open string have key properties of hadronic excitations. The energy of a stretched string matches quite well the potential energy of a separated quark–antiquark pair. More surprisingly, certain strongly interacting gauge theories are physically equivalent to closed string theories. The closed strings propagate on a space whose boundary is roughly the space where the gauge theory lives. The prime example of this equivalence is the AdS/CFT correspondence, which states that supersymmetric four-dimensional $SU(N)$ gauge theory is fully described by type IIB closed superstrings in a spacetime that includes the five-dimensional anti-de Sitter space AdS_5. We motivate this correspondence and examine in detail the geometry of anti-de Sitter space and related hyperbolic spaces. The correspondence suggests that properties of the recently discovered quark–gluon plasma are related to properties of black holes in anti-de Sitter space.

23.1 Introduction

String theory was discovered in the attempts to understand the dynamics of strongly interacting hadrons. It had been noted that the plot of the angular momentum J of hadronic excitations against their energy-squared falls roughly into lines $J = \alpha' E^2$ called Regge trajectories. String theory seemed to be a reasonable candidate for a theory of strong interactions because this relationship between J and E^2 emerges naturally from a rotating classical open string, as we discussed in Section 8.6. Quantization picks discrete values for the angular momentum and modifies the linear relation between J and E^2 by the addition of a constant. Both changes are needed for agreement with the data.

Despite these encouraging indications, the early attempts to use relativistic strings to describe hadrons faced many problems. Among them was the presence of unwanted massless vectors and massless tensors, precisely the particles needed to make string theory a candidate for a unified theory of physics. The string theory approach to strong interactions was abandoned and quantum chromodynamics, or QCD, was adopted. QCD postulates that the basic constituents of hadrons are quarks and gluons. QCD is a quantum field theory, in fact, an $SU(3)$ Yang–Mills theory.

In QCD, a meson is viewed as a pair of quarks held together by gluons. The string picture, however, remains a useful approximate description. In this picture, a meson is a pair of quarks held at the ends of an open string that represents a thin tube of color flux

Fig. 23.1 A meson as a quark–antiquark pair held together by color field lines. When the quark and antiquark are separated the color field lines form a thin flux tube that can be viewed as a string.

lines (Figure 23.1). The confinement of quarks, that is, the fact that quarks are never seen in isolation, also has a simple explanation in the string picture. Since the tension of the string joining the quarks is independent of its length, it takes infinite energy to separate fully a pair of quarks. The potential energy between a quark and an antiquark has been the subject of much analytic and numerical work in QCD. As we will see, key features of this potential follow from a simple string model.

In the above string picture a meson is a string. When we discussed the Standard Model on a configuration of D6-branes, the quarks and leptons were the strings. There is no contradiction: a superstring may describe elementary particles and at the same time another string theory, perhaps an effective one, may describe the composite hadrons. In fact, it was a long-held belief that there should be some exact string theory description of QCD. While this string theory is not yet known, a surprising development has come tantalizingly close to achieving this objective.

As it turns out, certain strongly interacting gauge theories have an *exact* string picture, but the strings do not propagate in the spacetime where the strongly interacting theory lives. Roughly, the gauge theory lives on the boundary of the spacetime where the strings propagate. The equivalence of a gauge theory and a string theory was first exhibited in the AdS/CFT correspondence. In this correspondence, a maximally supersymmetric $SU(N)$ Yang–Mills theory in four-dimensional Minkowski spacetime is claimed to be equivalent to a type IIB closed superstring theory. The ten-dimensional spacetime in this superstring theory takes a particular form: five dimensions form a sphere S^5 and the other five dimensions form a noncompact *anti-de Sitter* spacetime, briefly denoted by AdS$_5$. One can view the Minkowski spacetime of the field theory as the boundary of the AdS$_5$ space. The maximally supersymmetric $SU(N)$ Yang–Mills theory has as much supersymmetry as a gauge theory can have. It is a conformal field theory (CFT), a type of theory that has no dimensionful parameters. In the AdS/CFT acronym, AdS stands for anti-de Sitter space and CFT stands for the gauge theory. After a discussion of the correspondence, we examine a variant that describes a "hot" gauge theory – the quark–gluon plasma – using strings moving on an AdS space that contains a black hole.

23.2 Mesons and quantum rotating strings

The classical linear relation $J = \alpha' M^2$ between angular momentum and mass-squared of a rotating open string suggests that the Regge trajectories of mesonic excitations could have

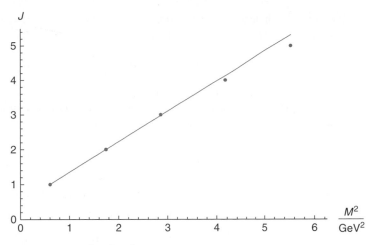

Fig. 23.2 Frautschi–Regge trajectory for the rho-mesons $\rho(776)$, $a_2(1320)$, $\rho_3(1690)$, $a_4(2040)$, and $\rho_5(2350)$. The line was fit to go through the $\rho(776)$ and the $a_2(1320)$.

a string theory explanation. Let us examine a particular meson and the Regge trajectory associated with it. We choose the rho-meson.

The rho-meson $\rho(776)$, of mass 776 MeV, is in fact a triplet of mesons (ρ^+, ρ^0, ρ^-), two of them charged and one uncharged. The rho-mesons are heavy analogs of the pions (π^+, π^0, π^-). Just like the pions, they are made of u and d quarks as well as antiquarks. As opposed to the pions, which are spin-zero combination of quarks, the rho-mesons have spin angular momentum S equal to one. With zero orbital angular momentum L, their total angular momentum is $J = 1$. The rho-mesons are unstable and decay mostly to a pair of pions, with lifetimes of about 10^{-23}s. The $\rho(776)$ belongs to a trajectory of mesons with $S = 1$ and higher values of L. Writing the masses (in MeV) inside parentheses, we have the $a_2(1320)$ with $J = 2$, the $\rho_3(1690)$ with $J = 3$, the $a_4(2040)$ with $J = 4$, and the $\rho_5(2350)$ with $J = 5$. These five mesons are plotted in Figure 23.2.

To get a sense of the accuracy of the linear trajectory we fit a line through the lowest two mesons. We write

$$J = \alpha' M^2 + \beta' , \tag{23.1}$$

and determine the constants α' and β' so that $M^2 = (0.776\,\mathrm{GeV})^2$ gives $J = 1$ and $M^2 = (1.320\,\mathrm{GeV})^2$ gives $J = 2$. This gives

$$J = 0.877\,02\,(\mathrm{GeV})^{-2}\,M^2 + 0.471\,88 . \tag{23.2}$$

This is the line shown in Figure 23.2. Using this fit one predicts a $J = 3$ meson of mass 1699 MeV, a $J = 4$ meson of mass 2006 MeV and a $J = 5$ meson of mass 2272 MeV. The errors in the masses are really small: 0.5%, 1.7%, and 3.3%, respectively.

Having seen that (23.1) is needed to fit the data, let us see how the classical relation $J = \alpha' M^2$ is modified in the quantum theory. Let us assume that the open string rotates in the (x^2, x^3) plane and has zero momentum in this plane. The relevant angular momentum operator $J = M^{23}$ is read from (12.147):

$$J = M^{23} = -i \sum_{n=1}^{\infty} \frac{1}{n} \left(\alpha_{-n}^{(2)} \alpha_n^{(3)} - \alpha_{-n}^{(3)} \alpha_n^{(2)} \right). \tag{23.3}$$

Naturally, the terms on the above right-hand side mix the oscillators from the two spatial coordinates. It is useful to define new oscillators α_n and $\bar{\alpha}_n$:

$$\alpha_n \equiv \frac{1}{\sqrt{2}} \left(\alpha_n^{(2)} + i\alpha_n^{(3)} \right), \qquad \bar{\alpha}_n \equiv \frac{1}{\sqrt{2}} \left(\alpha_n^{(2)} - i\alpha_n^{(3)} \right). \tag{23.4}$$

Note that $(\alpha_n)^\dagger = \bar{\alpha}_{-n}$. A short calculation shows that the commutation relations for the new oscillators are

$$[\alpha_m, \bar{\alpha}_n] = m\, \delta_{m+n,0}, \qquad [\alpha_m, \alpha_n] = [\bar{\alpha}_m, \bar{\alpha}_n] = 0. \tag{23.5}$$

Note that the first relation implies that $[\bar{\alpha}_m, \alpha_n] = m\delta_{m+n,0}$. In terms of the new oscillators the angular momentum operator is given by

$$J = \sum_{n=1}^{\infty} \frac{1}{n} \left\{ \alpha_{-n} \bar{\alpha}_n - \bar{\alpha}_{-n} \alpha_n \right\}. \tag{23.6}$$

The first term in the above sum counts the number of α oscillators in a state. The second term, up to a sign, counts the number of $\bar{\alpha}$ oscillators in a state. The convenience of the new basis is now apparent.

Quick calculation 23.1 Verify equation (23.6), and convince yourself of the claims that follow this equation.

Using the new oscillators we write states in the form

$$|\lambda\rangle = \dots \prod_{k=1}^{\infty} (\alpha_{-k})^{\lambda_k} (\bar{\alpha}_{-k})^{\bar{\lambda}_k} |p^+, \vec{p}_T\rangle, \tag{23.7}$$

where $\lambda_k \geq 0$ and $\bar{\lambda}_k \geq 0$ are integers and the dots represent products of oscillators in directions other than x^2 and x^3. Acting on $|\lambda\rangle$

$$J \text{ has eigenvalue } \mathcal{J} = \sum_{k=1}^{\infty} (\lambda_k - \bar{\lambda}_k). \tag{23.8}$$

Is it thus easy to build states with arbitrary values of the angular momentum.

Since we want to find the mass-squared values of states with definite angular momentum we consider the mass-squared relation (12.164) and write it as:

$$\alpha' M^2 + 1 = N^\perp = N_{23} + N', \tag{23.9}$$

where

$$N_{23} = \sum_{n=1}^{\infty} \left(\alpha_{-n}^{(2)} \alpha_n^{(2)} + \alpha_{-n}^{(3)} \alpha_n^{(3)} \right) \tag{23.10}$$

denotes the contribution to $N^\perp$ from the x^2 and x^3 directions and N' denotes the contribution to $N^\perp$ from the other transverse directions. One readily finds that in terms of the new oscillators

$$N_{23} = \sum_{n=1}^{\infty} (\alpha_{-n}\bar{\alpha}_n + \bar{\alpha}_{-n}\alpha_n) , \qquad (23.11)$$

and this implies that on the general states (23.7)

$$N_{23} \text{ has eigenvalue } \mathcal{N}_{23} = \sum_{k=1}^{\infty} k(\lambda_k + \bar{\lambda}_k) . \qquad (23.12)$$

Since $\lambda_k, \bar{\lambda}_k \geq 0$ and $k \geq 1$, we have

$$\mathcal{N}_{23} = \sum_{k=1}^{\infty} k(\lambda_k + \bar{\lambda}_k) \geq \sum_{k=1}^{\infty} \lambda_k + \sum_{k=1}^{\infty} \bar{\lambda}_k \geq \left| \sum_{k=1}^{\infty} \lambda_k - \sum_{k=1}^{\infty} \bar{\lambda}_k \right| = |\mathcal{J}| , \qquad (23.13)$$

where we used (23.8) and noted that for any two numbers $b_1 \geq 0$ and $b_2 \geq 0$, one has $b_1 + b_2 \geq |b_1 - b_2|$. The inequality we have obtained is

$$\mathcal{N}_{23} \geq |\mathcal{J}| . \qquad (23.14)$$

Noting that the eigenvalues of N' are greater than or equal to zero, equation (23.9) gives the inequality $1 + \alpha'\mathcal{M}^2 \geq \mathcal{N}_{23}$, where $\mathcal{M}^2$ is the eigenvalue of M^2. Finally, combining this inequality with (23.14) we obtain

$$|\mathcal{J}| \leq 1 + \alpha'\mathcal{M}^2 . \qquad (23.15)$$

This inequality holds for arbitrary states in (23.7). It is the quantum version of the classical equation $J = \alpha'M^2$. For states that saturate the inequality we have $\mathcal{J} = \alpha'\mathcal{M}^2 + 1$, an equation of the type (23.1). While α' must be determined by fitting the data, β' is predicted. The value $\beta' = 1$, from bosonic string theory, does not fit the rho-meson trajectory well. While most Regge trajectories give rather similar values of α', the values of β' vary more widely.

Let us examine states that saturate the inequality (23.15). For a state of the form $(\alpha_{-1})^N |p^+, \vec{p}_T\rangle$ we have $\lambda_1 = N$, all other λs and $\bar{\lambda}$s = 0, and $N' = 0$. So

$$\mathcal{J} = N , \qquad 1 + \alpha'\mathcal{M}^2 = \mathcal{N}_{23} = 1 \cdot \lambda_1 = N = \mathcal{J} , \qquad (23.16)$$

and the inequality is saturated. The state $(\alpha_{-1})^N |p^+, \vec{p}_T\rangle$ is said to belong to the Regge trajectory with maximal angular momentum per unit mass-squared.

The quantum states $|\psi_N\rangle \sim (\alpha_{-1})^N |p^+, \vec{p}_T\rangle$ resemble the classical rotating string: their angular momentum and mass-squared eigenvalues are related roughly like in the classical theory. In order to compute well defined expectation values the states must be normalized, and this requires delta functions for the momenta, as indicated in (12.171). In order to keep the notation simple and avoid working with superpositions of states we just write $\langle p^+, \vec{p}_T | p^+, \vec{p}_T \rangle = 1$ instead of delta functions of zero argument.

Perhaps surprisingly, the expectation values $\langle \psi_N | X^I(\tau, \sigma) | \psi_N \rangle$ do not behave at all like the classical coordinates $X^I(\tau, \sigma)$ of a rotating string. We find

$$\langle \psi_N | X^I(\tau, \sigma) | \psi_N \rangle = \langle x_0^I \rangle + 2\alpha' p^I \tau, \tag{23.17}$$

because the oscillators in the mode expansion of $X^I(\tau, \sigma)$ cannot contribute: adding a single oscillator to a nonvanishing expectation value makes it vanish. According to the expectation value (23.17) the string propagates as if it were a point. One can see evidence that the string is extended by computing the expectation values of *squares* of the coordinates (see Problem 23.1).

There are, however, coherent states for which the expectation values of the coordinates follow the classically expected trajectory. Take, for example,

$$|\Psi\rangle = e^{\mathcal{A}} |p^+, \vec{0}\rangle, \quad \text{with} \quad \mathcal{A} = v\left(\bar{\alpha}_1 - \alpha_{-1}\right), \tag{23.18}$$

where v is a real constant. Since $\mathcal{A}^\dagger = -\mathcal{A}$, we have $\langle \Psi | \Psi \rangle = 1$, up to the caveat discussed above. Let us first confirm that the expectation values of J and M^2 are the familiar ones. The expectation value of J is readily computed:

$$\langle J \rangle = \langle \Psi | J | \Psi \rangle = \langle \Psi | (\alpha_{-1} \bar{\alpha}_1 - \bar{\alpha}_{-1} \alpha_1) | \Psi \rangle, \tag{23.19}$$

since all other terms in (23.6) commute with $e^{\mathcal{A}}$ and annihilate the vacuum states to the right. In fact, α_1 also commutes with $\mathcal{A}$ and therefore the second term above does not contribute either. We then have

$$\langle J \rangle = \langle \Psi | \alpha_{-1} \bar{\alpha}_1 e^{\mathcal{A}} | p^+, \vec{0} \rangle = \langle \Psi | \alpha_{-1} [\bar{\alpha}_1, e^{\mathcal{A}}] | p^+, \vec{0} \rangle$$

$$= \langle \Psi | \alpha_{-1} [\bar{\alpha}_1, \mathcal{A}] e^{\mathcal{A}} | p^+, \vec{0} \rangle = -v \langle \Psi | \alpha_{-1} e^{\mathcal{A}} | p^+, \vec{0} \rangle. \tag{23.20}$$

Given that $\langle \Psi | e^{\mathcal{A}} \alpha_{-1} | p^+, \vec{0} \rangle = \langle p^+, \vec{0} | \alpha_{-1} | p^+, \vec{0} \rangle = 0$ we have

$$\langle J \rangle = -v \langle \Psi | [\alpha_{-1}, e^{\mathcal{A}}] | p^+, \vec{0} \rangle = -v \langle \Psi | [\alpha_{-1}, \mathcal{A}] e^{\mathcal{A}} | p^+, \vec{0} \rangle = v^2. \tag{23.21}$$

Since the computation of N_{23} involves exactly the same term $\alpha_{-1} \bar{\alpha}_1$ used for the computation of $\langle J \rangle$, we have, all in all:

$$\langle J \rangle = \langle N_{23} \rangle = v^2, \quad \langle \alpha' M^2 \rangle = v^2 - 1 \quad \to \quad \langle J \rangle = \langle (\alpha' M^2 + 1) \rangle, \tag{23.22}$$

as expected for a rotating string. Let us now consider the expectation values of the coordinates. A short computation using the oscillator expansion (12.66) gives

$$\langle (X^2 + iX^3)(\tau, \sigma) \rangle = \langle x_0^2 + ix_0^3 \rangle + 2i\sqrt{\alpha'} \langle (\alpha_1 e^{-i\tau} - \alpha_{-1} e^{i\tau}) \rangle \cos \sigma. \tag{23.23}$$

Noting that $\langle \alpha_1 \rangle = 0$ and $\langle \alpha_{-1} \rangle = -v$ and assuming $\langle x_0^2 \rangle = \langle x_0^3 \rangle = 0$, we find

$$\langle (X^2 + iX^3)(\tau, \sigma) \rangle = 2iv\sqrt{\alpha'} e^{i\tau} \cos \sigma. \tag{23.24}$$

Separating real and imaginary parts we indeed obtain the classical limit of a rotating string:

$$\langle X^2(\tau, \sigma) \rangle = -2v\sqrt{\alpha'}\, \sin\tau \, \cos\sigma \,,$$
$$\langle X^3(\tau, \sigma) \rangle = 2v\sqrt{\alpha'}\, \cos\tau \, \cos\sigma \,. \tag{23.25}$$

The parameter v can be related to familiar constants of the motion. The length ℓ of the above string is $\ell = 4v\sqrt{\alpha'}$. Moreover, since spatial momenta are zero, $M^2 = 2p^+ p^- = 2(p^+)^2$. From this and (23.22) we find

$$p^+ = \frac{1}{\sqrt{2\alpha'}}\sqrt{v^2 - 1}\,. \tag{23.26}$$

In the classical limit v is large and

$$p^+ \simeq \frac{v}{\sqrt{2\alpha'}} = \frac{\ell}{4\sqrt{2}\,\alpha'}\,. \tag{23.27}$$

This is the classical relation between p^+ and the length of a rotating string. Indeed, using (7.60), which states that the energy E of such a string is $\frac{\pi}{2} T_0 \ell$,

$$p^+ = \frac{1}{\sqrt{2}} E = \frac{1}{\sqrt{2}} \frac{\pi}{2} T_0 \ell = \frac{1}{\sqrt{2}} \frac{\pi}{2} \frac{1}{2\pi\alpha'} \ell = \frac{\ell}{4\sqrt{2}\,\alpha'}\,. \tag{23.28}$$

We have thus verified that the expectation values of the coordinates describe a rotating string of the correct size. The use of coherent states to describe quantum states with semiclassical limits is familiar from the simple harmonic oscillator, as reviewed in Problem 23.2.

23.3 The energy of a stretched effective string

The classical potential energy of a static stretched string is equal to the tension T_0 times its length ℓ. If we imagine a quark and an antiquark fixed at the ends of the string, this energy would represent the quark–antiquark potential $V(\ell)$ evaluated at a separation ℓ. The classical limit suggests that for large r one has $V(r) \sim T_0 r$.

How do we use quantum string theory to obtain information about $V(r)$? We look into the calculation of the mass-squared of a string stretched between two parallel D-branes. The D-branes are not good representations for the quarks, but they certainly hold the string stretched, and we are after the energy of the string, which we assume defines the potential $V(r)$. For D-branes separated a distance L we obtained (15.51), that written in terms of the string tension and an arbitrary spacetime dimension D reads:

$$M^2 = (T_0 L)^2 + \frac{1}{\alpha'}\left(N^\perp - \frac{1}{24}(D - 2)\right). \tag{23.29}$$

In order to represent a string without excitations we assume $N^\perp = 0$. Moreover, since the string is static we can identify M^2 with the potential energy squared. We thus write, using r for the length of the string,

$$(V(r))^2 = (T_0\, r)^2 - \frac{1}{\alpha'}\frac{(D-2)}{24}\,. \tag{23.30}$$

At this stage we reach a significant complication. We would like to determine the quark–antiquark potential in *four* spacetime dimensions. We are tempted to set $D = 4$ in the above equation, but the quantum consistency of the Nambu–Goto action requires $D = 26$. We cannot justify using this equation for other values of D.

There is evidence, however, that a variant of the Nambu–Goto action exists that can be formulated with both Lorentz and translational symmetry in spacetimes of arbitrary dimensionality D. The theory is complicated and the action contains an infinite number of additional terms. The quantization of a stretched string in this theory gives a potential that roughly agrees with (23.30): it reproduces the first few terms in the large r expansion of $V(r)$.

Thus reassured, we proceed to explore (23.30) for arbitrary D. It is customary to introduce the Luscher coefficient

$$\gamma_D \equiv -\frac{\pi}{24}(D-2)\,, \tag{23.31}$$

whose values in $D = 4$ and $D = 3$ are

$$\gamma_4 = -\frac{\pi}{12} = -0.262, \quad \gamma_3 = -\frac{\pi}{24} = -0.1309\,. \tag{23.32}$$

We can then write (23.30) as

$$V(r) = \sqrt{(T_0 r)^2 + 2T_0\gamma_D}\,. \tag{23.33}$$

Expanding the above right-hand side for large r,

$$V(r) = T_0\, r + \gamma_D \cdot \frac{1}{r} + \mathcal{O}(1/r^3)\,. \tag{23.34}$$

The above gives the leading quantum correction to the classical potential $V(r) = T_0 r$ for large r. The tension $T(r)$ on the string is the magnitude of the force associated with the potential:

$$T(r) = \frac{\partial V}{\partial r} = T_0 - \gamma_D \cdot \frac{1}{r^2} + \mathcal{O}(1/r^4)\,. \tag{23.35}$$

The tension $T(r)$ is equal to T_0, up to a correction linear in $1/r^2$. Since $\gamma_D < 0$ for $D > 2$, $T(r) > T_0$ in three or four spacetime dimensions. The Luscher coefficient can be extracted through derivatives of the force:

$$C(r) \equiv \frac{1}{2}r^3 \frac{\partial T}{\partial r} = \gamma_D + \mathcal{O}(1/r^2)\,. \tag{23.36}$$

Note that $C(r) \sim \gamma_D$, with small corrections for large r. The quantity $C(r)$ can be determined numerically in the lattice approximation. The computations evaluate the quark–antiquark potential in the prototypical $SU(3)$ gauge theory. The results give strong evidence that as r is increased $C(r)$ approaches the predicted values (23.32) of γ_D both for

$D = 4$ and for $D = 3$. This indicates that the string picture captures not only the leading term $T_0 r$ of the quark–antiquark potential but also the first nontrivial correction!

23.4 A large-*N* limit of a gauge theory

We now begin our study of the ideas relevant to the AdS/CFT correspondence. This correspondence of gauge theory with gravity is clearest when the gauge theory has a large number of degrees of freedom. We will thus examine an $SU(N)$ gauge theory with a large value of N. This theory has a dimensionless coupling constant g_{YM} which controls the strength of the interactions between gauge bosons. More concretely, each time a gauge boson turns into two gauge bosons, or two gauge bosons turn into one, the amplitude for the process must include a factor of g_{YM}. The key result we want to demonstrate is the following: there is a controllable $N \to \infty$ limit in which the physically relevant coupling is not g_{YM} but rather the 't Hooft coupling $\lambda = g_{YM}^2 N$ that is kept finite.

To show this we consider a system of N coincident branes. The strength of the interactions between the open strings is controlled by the open string coupling constant g_o: each time an open string splits into two strings, or two strings join to form a single one, the amplitude for the process must include a factor of g_o. In the low energy limit we find the $SU(N)$ gauge theory and a decoupled $U(1)$ theory that we can ignore. Since the gauge bosons are simply massless open strings, the coupling constant g_{YM} of the resulting gauge theory coincides with the open string coupling g_o. In the argument that follows we use open strings to derive a result in the gauge theory. Accordingly, we will simply write g_{YM} for the open string coupling constant.

We examine the propagation of an open string whose ends lie on the i and j branes, with $i \neq j$. To find the quantum-mechanical amplitude for propagation of this string from some initial to some final conditions we must sum over all possible intermediate states that are consistent with beginning and ending with the $[ij]$ string. This sum, of course, is very complicated to do explicitly, but we are only going to determine the g_{YM} and N dependence of the various contributions. The result will apply to the amplitude of propagation of a gauge boson in the $SU(N)$ gauge theory.

The various contributions can be organized with diagrams where we show the evolution of the string. The simplest diagram is one where the string does nothing – it just propagates freely from the initial to the final condition. Figure 23.3(a) shows the strip produced by the motion of the open string, with the edges labeled by the branes i and j. This diagram has no interactions and no N-dependence. We associate to it an amplitude

$$A_0 = c_0 \,, \tag{23.37}$$

where c_0 is some constant independent of g_{YM} and N.

The next diagram includes interactions. The simplest possibility is that the string splits and rejoins, as shown in Figure 23.3(b). The two interactions, indicated by the heavy dots,

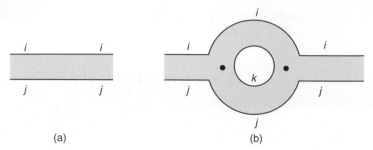

Fig. 23.3   (a) Strip created by the free evolution of an [ij] string. (b) Diagram produced when the [ij] string splits and then rejoins at the kth brane.

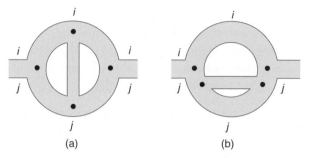

Fig. 23.4   Two diagrams, each with four interaction points, obtained by adding a strip to diagram (b) in Figure. 23.3.

give a factor $(g_{\mathrm{YM}})^2$. The inner boundary of the diagram has a label k, indicating that the string has split on the kth brane. Since that brane can be any of the N branes available and we must sum over all possibilities, the diagram also has a factor of N. Including a constant c_1, we write

$$A_1 = c_1 g_{\mathrm{YM}}^2 N \,. \tag{23.38}$$

A pattern emerges: each boundary that is a closed line contributes a factor of N, since it represents an open string endpoint that can be at any of the N branes. Suppose we add another strip to the diagram of Figure 23.3 (b). One simple way to do that will introduce two new interaction points and create one new boundary, as shown by the two diagrams of Figure 23.4. If this happens we get an extra factor of $g_{\mathrm{YM}}^2 N$ in the amplitude. We thus have

$$A_2 = c_2 (g_{\mathrm{YM}}^2 N)^2 \,. \tag{23.39}$$

As long as we keep adding strips that create a new boundary we get amplitudes of the form

$$A_p = c_p (g_{\mathrm{YM}}^2 N)^p \,. \tag{23.40}$$

The amplitude A obtained by adding all of the above contributions is

$$A = \sum_{n=0}^{\infty} A_n = \sum_{n=0}^{\infty} c_n (g_{\mathrm{YM}}^2 N)^n \,. \tag{23.41}$$

We see here the emergence of the 't Hooft coupling constant λ defined by

$$\lambda \equiv g_{\mathrm{YM}}^2 N \,. \tag{23.42}$$

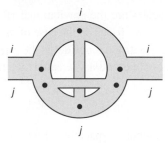

Fig. 23.5 A non-planar diagram obtained by adding a strip to any one of the two diagrams in Figure 23.4. Note that there is just one internal boundary.

The 't Hooft coupling controls the convergence of (23.41). We write

$$A = \sum_{n=0}^{\infty} c_n \lambda^n = f_0(\lambda) \,. \tag{23.43}$$

To find the complete amplitude we must add to A the contributions that arise when we add strips that do not create a new boundary. The simplest example involves adding a strip to any of the two diagrams of Figure 23.4 to obtain the diagram in Figure 23.5. We have introduced two new interactions, as usual, but the number of boundaries has been reduced by one. The two internal boundaries have become a single one, as you can check by moving the tip of a pencil along the boundary. This graph is actually non-planar: it cannot be drawn on the plane without making the strips cross. In fact a new strip will either increase the number of boundaries by one or decrease the number of boundaries by one. Each time we add a strip that decreases the number of boundaries we get an extra factor of g_{YM}^2/N or, equivalently, a factor of λ/N^2. Adding to A the contributions from diagrams that include all possible numbers of boundary-decreasing strips we find the full amplitude $\mathcal{A}$:

$$\mathcal{A} = f_0(\lambda) + f_2(\lambda) \cdot \frac{1}{N^2} + f_4(\lambda) \cdot \frac{1}{N^4} + \cdots \,. \tag{23.44}$$

When N is large and λ is fixed, the above is an expansion of the amplitude in powers of the small parameter $1/N^2$. The coefficient of N^{-2k}, with $k = 0, 1, 2, \ldots$, is controlled by the value of λ. If λ is small, each coefficient has an expansion in powers of λ. The limit $N \to \infty$ with $\lambda = g_{YM}^2 N$ fixed requires $g_{YM} \to 0$ in a controlled fashion. In this limit only the first term of the series contributes. The 't Hooft coupling λ fully controls the theory in the $N \to \infty$ limit.

23.5 Gravitational effects of massive sources

We examined a large-N gauge theory using open strings on N D-branes. We now ask: when are the gravitational effects of these D-branes important? After all, D-branes have energy (due to their tension) and must curve spacetime. More generally, we can ask: at what distance scales are the gravitational effects of massive objects important? After we discuss this point, we focus on the case of branes wrapped around extra dimensions.

The relevant quantity to consider is the gravitational potential energy per unit mass, that is, the gravitational potential V created by a mass M at a distance r from the mass:

$$V \simeq -\frac{G^{(D)}M}{r^{D-3}}. \tag{23.45}$$

Here D is the number of spacetime dimensions. Since V is dimensionless in natural units, the numerator defines a characteristic length scale R:

$$R^{D-3} \equiv G^{(D)}M. \tag{23.46}$$

The scale R coincides, up to factors of order one, with the Schwarzschild radius of a mass M black hole in arbitrary dimension D. Indeed, our derivation of the radius of a four-dimensional black hole in Section 22.6 follows a similar logic. With R so defined, the potential (23.45) can be written as

$$V \simeq -\left(\frac{R}{r}\right)^{D-3}. \tag{23.47}$$

We see that gravitational effects are negligible for $r \gg R$. Gravitational effects of point masses are important at scales of order R. If R, however, is much smaller than the size of the mass M, gravitational effects may be completely negligible. A billion ton asteroid, for example, has $R \sim 10^{-15}$m. There are no significant gravitational effects at this scale nor at larger scales.

Consider now a number N of Dp branes wrapped around a p-dimensional compact space of volume V_p. The tension T_p of a Dp-brane is given by (18.57):

$$T_p \simeq \frac{1}{g}\frac{1}{(\sqrt{\alpha'})^{p+1}}. \tag{23.48}$$

The mass of the branes is therefore

$$M = NT_p \cdot V_p \simeq N \cdot \frac{V_p}{g(\sqrt{\alpha'})^{p+1}}. \tag{23.49}$$

The dimensionally reduced spacetime is $D - p$ dimensional. Moreover, in this spacetime the branes appear as a point source of mass M. Using (23.46) the characteristic size R of the system is

$$R^{D-p-3} = G^{(D-p)}M = \frac{G^{(D)}}{V_p}M \simeq G^{(D)}N \cdot \frac{1}{g(\sqrt{\alpha'})^{p+1}}. \tag{23.50}$$

Note the cancellation of the volume factor V_p. We recall that $G^{(D)} \sim g^2(\sqrt{\alpha'})^{D-2}$ (Section 13.4), so we get

$$R^{D-p-3} \simeq gN(\sqrt{\alpha'})^{(D-p-3)}. \tag{23.51}$$

We write this result as

$$\left(\frac{R}{\sqrt{\alpha'}}\right)^{D-p-3} \simeq gN \,. \tag{23.52}$$

As $gN \to 0$ we have $R \to 0$ and the gravitational effects of the N D-branes go to zero. In fact, as R becomes smaller than the string length $\sqrt{\alpha'}$ there is no scale at which the gravitational effects are relevant. As $g \to 0$ the mass (23.49) of the D-branes diverges like $1/g$, but the gravitational coupling G goes to zero like g^2. The overall effect, proportional to GM, vanishes. Note also the independence of the estimate (23.52) on the volume V_p of the compact dimensions. This suggests that the characteristic scale R is relevant even for configurations of infinite D-branes. In that case, however, one need not get a black hole, but rather some other geometry where R plays a significant role.

23.6 Motivating the AdS/CFT correspondence

In this section we motivate the AdS/CFT correspondence, a surprising equivalence between a supersymmetric $SU(N)$ gauge theory that arises at low energies on a set of N coincident D3-branes and a type IIB superstring theory in a background spacetime closely related to the gravitational background created by the D3-branes.

Although gauge theories naturally arise as low-energy limits of open string theory, the correspondence involves a *closed* string theory. Moreover, the closed strings do not live on Minkowski space, the space where the gauge theory lives. In fact, Minkowski space is, in some sense, the boundary of the space where the closed strings live. The correspondence is sometimes called a duality because the same physics is described by two different looking systems (the closed strings and the gauge theory). Since the $SU(N)$ gauge theory has maximal supersymmetry, its physics is different from that of QCD, a gauge theory with no supersymmetry. We are still lacking a string theory description of QCD, but the discovery of the AdS/CFT correspondence has given strong evidence that such a description exists.

It should be emphasized that the correspondence has not yet been proven. Rather, it was originally motivated by some heuristic arguments and has since been tested extensively. There are no grounds to suspect that it fails to hold. The heuristic arguments fall into two groups. They are either based on ideas of symmetry or they are based on a low energy limit.

Let us first briefly consider the symmetry argument. A conformal field theory in four-dimensional Minkowski space has a set of conformal symmetries that are generated by fifteen operators. Ten of these generators are the familiar Lorentz generators (six of them) and the spacetime translation generators (four of them). The other five include four that generate the so-called special conformal transformations and one that generates scale transformations. The fifteen operators define the four-dimensional conformal Lie algebra, which includes the Lorentz algebra (11.81) as a subalgebra. The conformal symmetries act on the field theory as a set of field transformations. These symmetries must also appear on the string theory side of the correspondence. Indeed, the AdS_5 spacetime, which is part of the closed string background, has the property that the isometries of the space (the smooth, one-to-one maps of the space into itself that leave all distances invariant) are generated by fifteen operators that satisfy the same algebra as those that generate the conformal symmetry of the field theory. There is also a reason for the S^5 space: the supersymmetric Yang–Mills theory in question has a set of scalar fields and a set of fermions, the elements of which are rotated among each other by a set of symmetries that match the isometries of S^5. All in all, the $\text{AdS}_5 \times S^5$ spacetime carries as isometries the symmetries of the field theory, which provides some evidence for the correctness of the correspondence.

The argument based on a low energy limit is more direct. We will consider a set of N D3-branes, with N fixed and large, and we will dial up the string coupling constant g so that gN varies from very small $gN \ll 1$ to very large $gN \gg 1$. We will examine low energy limits in *both* extremes and extract some conclusions.

Let us start then with N coincident D3-branes at zero coupling g in flat ten-dimensional spacetime. We have free open strings on the branes and free type IIB closed strings on the spacetime. There are no interactions at all. Imagine now a minuscule increase of g to a fixed value such that $gN \ll 1$. Following our discussion in Section 23.5 we conclude that gravitational effects are negligible and we may continue to treat the D3-branes as if they were in flat space.

In a low energy limit we consider energies that are smaller than the string energy scale $1/\ell_s$:

$$E \ll \frac{1}{\sqrt{\alpha'}}. \tag{23.53}$$

Another way to think of the low energy limit is to imagine keeping all energies bounded while $\alpha' \to 0$:

$$E \leq E_0, \quad \alpha' \to 0. \tag{23.54}$$

For bounded energies, (23.53) holds for sufficiently small α'. In this limit, the massive states of the open strings on the D-branes are not accessible, so the physics on the branes is governed by the massless $U(N)$ Yang–Mills fields. As $\alpha' \to 0$, the closed string fields that propagate over the whole of the spacetime become free fields because the ten-dimensional Newton constant $G^{(10)} \sim g^2(\alpha')^4$ that governs their interactions (see (13.80)) goes to zero. Finally, the interactions between the spacetime fields and the $U(N)$ fields on the branes also go to zero because they too are controlled by $G^{(10)}$. The result is (1) a system of decoupled closed strings on the ten-dimensional Minkowski spacetime and (2) a supersymmetric four-dimensional $U(N)$ Yang–Mills theory. We learned in Section 21.2 that in a $U(N)$ Yang–Mills theory one gauge field actually decouples. The remaining, fully interacting gauge theory is $SU(N)$ Yang–Mills.

Consider now increasing the string coupling in the D3-brane system. As long as we consider low energy excitations, the excitations on the branes and the ones on the spacetime will continue to decouple. When $gN \gg 1$ gravitational effects are important. The branes carry energy and Ramond–Ramond charge. As a consequence, the N D3-branes are described by a nontrivial solution of the field equations for the massless fields of type IIB

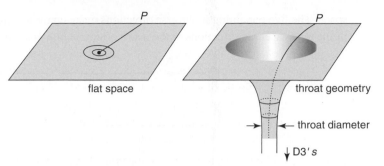

Fig. 23.6 Left: the circumference of the circle surrounding the origin goes to zero as we approach the origin. Right: in the throat geometry, the origin has moved an infinite distance down the throat, and the circumferences of the circles which surround the origin approach a constant value. For the D3-brane solution, the surface represents the six-dimensional space transverse to the branes, and the circles on the throat represent five-spheres which surround the branes. The radial direction in the transverse space is represented by the line starting at P and going into the throat.

string theory, much in the same way that a charged black hole is described. The solution which describes the D3-branes includes a horizon which lies at the end of an infinite throat.

To visualize such solution, consider first the origin of a flat world with two spatial dimensions (left side of Figure 23.6). The geometry of flat space is such that, at a distance r away from the origin, the circumference of the circle surrounding the origin is $2\pi r$. As we approach the origin, the circumference goes to zero. This changes dramatically if the geometry is that of a throat (right side of Figure 23.6). The origin is now an infinite distance down the throat. Moreover, down the throat, the circumference of a circle surrounding the origin approaches a constant called the circumference of the throat. Asymptotically, the throat becomes an infinite cylinder. The circle at the infinite "end" of the throat is called the horizon. The horizon is an infinite distance away from any point on the plane.

We can now return to our coincident D3-branes, which stretch along x^1, x^2, and x^3 but appear as a point along the six spatial coordinates $x^4, \ldots, x^9$ transverse to the brane. We need not focus on the longitudinal directions of the branes. In the transverse six-dimensional space the branes are surrounded by five-dimensional spheres. These are analogous to the circles that surround the origin in the two-dimensional example. A throat geometry with a horizon emerges in the transverse space. It takes an infinite distance to get to the horizon. The five-spheres that surround the horizon approach a constant volume as we travel down the throat. The radius R associated with this volume is called the radius of the horizon. It is worth noting that the geometry of the D3-brane solution is quite different from the geometry of the Schwarzschild black hole. In the latter, there is no infinite throat, and the horizon is a finite distance away from any point in the space.

The near-horizon geometry can be read directly from the metric which represents the gravitational solution: it turns out to be AdS$_5 \times S^5$. The D3-branes do not appear anymore in this geometry! The five-sphere was anticipated by our discussion of the throat geometry; its radius is the horizon radius R. The five-dimensional space AdS$_5$ arises from the four spacetime dimensions parallel to the branes plus the radial direction on the transverse space. The five remaining transverse directions make up the sphere S^5. This rearrangement of the ten spacetime directions at the near horizon region is summarized as follows.

$$
\begin{array}{cc}
\overbrace{}^{\text{AdS}_5} & \overbrace{}^{S^5} \\
x^0\ x^1\ x^2\ x^3\ r & y^1\ y^2\ y^3\ y^4\ y^5 \\
x^0\ x^1\ x^2\ x^3 & x^4\ x^5\ x^6\ x^7\ x^8\ x^9 \\
\underbrace{}_{\text{D3 tangential}} & \underbrace{}_{\text{D3 transverse}}
\end{array}
\tag{23.55}
$$

In here the six directions transverse to the brane became r – that went into the AdS space – and the five ys that make up the five-sphere.

A red-shift phenomenon that is familiar for black holes also occurs with the horizon. Finite energy excitations near the horizon are perceived as excitations of vanishingly small energy by an observer far away at infinity (the excitations are red shifted). As a result, low energy excitations for the observer at infinity can be of two types: *finite* energy excitations that emanate from the horizon or low energy (long wavelength) excitations far from the branes. There is evidence that these two types of excitations decouple. The ones that are far away are almost never captured by the horizon, which looks very small compared to the wavelength of the excitations. The excitations near the horizon cannot escape to infinity. As a result, the configuration is well approximated by two decoupled systems: (1) a system of low energy closed strings on flat space representing the far-away region and (2) a system of type IIB superstrings on the near-horizon $AdS_5 \times S^5$ geometry.

We have described N D3-branes in two regimes, a first one for $gN \ll 1$ and a second one for $gN \gg 1$. The two regimes are connected by dialing up g, since we imagine N fixed and large. In the low energy limit each regime gave rise to two decoupled systems. In fact, in the low energy limit decoupling occurs for *all values* of gN and one of the decoupled systems is free closed strings. When $gN \ll 1$ the other decoupled system is $SU(N)$ Yang–Mills. Since this theory makes sense for all values of gN it is reasonable to expect that $SU(N)$ Yang–Mills is the other decoupled system for all values of the coupling gN. Alternatively, for $gN \gg 1$ the other decoupled system is IIB superstrings on $AdS_5 \times S^5$. Since this theory makes sense for all values of gN it is reasonable to expect that IIB superstrings on $AdS_5 \times S^5$ is the other decoupled system for all values of gN. See the diagramatic representation of the situation in Figure 23.7.

Consider first the decoupled system of closed strings in flat spacetime. As gN goes from very small to very big, the closed string system from the $gN \ll 1$ regime goes into the closed string system of the $gN \gg 1$ regime. If gN goes from very large to very small, the closed string system of the $gN \gg 1$ regime goes into the closed string system of the $gN \ll 1$ regime going through the *same* set of theories. In fact, in the low energy limit the closed strings are non-interacting for all finite g, so physically all these theories are the same.

Having matched one of the decoupled systems, we now look at the second one. As we mentioned before, we have two candidates for the second decoupled system: $SU(N)$ Yang–Mills and IIB superstrings on $AdS_5 \times S^5$. The simplest possibility is that they are the same! Explicitly, as gN goes from very small to very big, the $SU(N)$ Yang–Mills theory

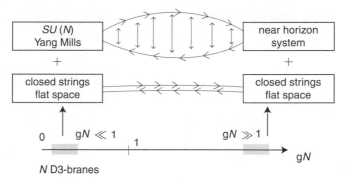

Fig. 23.7 A diagramatic representation of the decoupling limits of a system of N D3-branes. To the left we have the branes with $gN \ll 1$, and its two decoupled systems above. To the right we have the branes with $gN \gg 1$, and its two decoupled systems above. If the two lines connecting the left and right top boxes represent equivalent systems, as indicated by the vertical identifications, one has the AdS/CFT correspondence.

traces a line of theories that takes it to a complicated strong coupling regime that is, in fact, equivalent to the type IIB near horizon system that arises from the D3-branes at $gN \gg 1$. As gN goes from very large to very small, that near horizon system traces a line of theories that takes it to a complicated gravitational regime that is, in fact, equivalent to the weakly coupled $SU(N)$ Yang–Mills system that arises from the D3-branes at $gN \ll 1$. Moreover, the two lines of theories just discussed are in fact equivalent, as indicated by the vertical identifications in the figure. If this plausible scenario is realized, for any value of g and N there are two equivalent descriptions of the same physics, one using a gauge theory and another using the near horizon system, that is, closed superstrings on AdS$_5 \times S^5$. This is the AdS/CFT correspondence.

23.7 Parameters in the AdS/CFT correspondence

Let us discuss the parameters of the two theories in the AdS/CFT correspondence. In an $SU(N)$ Yang–Mills theory there are two dimensionless parameters: the coupling constant g_{YM} and the constant N. In IIB superstring theory on AdS$_5 \times S^5$ there are also two dimensionless parameters: the string coupling g and the radius $R/\sqrt{\alpha'}$ of the S^5 expressed in units of the string length. In summary:

$$\text{Yang–Mills: } g_{\mathrm{YM}}, \ N,$$
$$\text{IIB strings: } g, \ R/\sqrt{\alpha'}. \tag{23.56}$$

Out of these four parameters, two of them are well defined in the original string theory system of D3-branes: the number N of D3-branes and the string coupling g. We need two relations to connect g_{YM} and R to g and N.

One relation is suggested from the limit $gN \ll 1$. We have stated earlier ((13.84)) that the open string coupling g_o is related to the closed string coupling g by $g_o^2 \sim g$. Since open strings give rise to the Yang–Mills bosons of $SU(N)$, these bosons interact with $g_{YM} \sim g_o$. It follows that $g_{YM}^2 \sim g_o^2 \sim g$. With a precise definition of the various couplings one has

$$g_{YM}^2 = 4\pi\, g\,. \tag{23.57}$$

The second relation is suggested from the region $gN \gg 1$. How is R, the horizon size, determined by the number of branes N and the string coupling? We studied a closely related question in Section 23.5, reaching the conclusion that the characteristic gravitational size of a system of D-branes is given by (23.52). In fact, this relation applies, and with $D - p - 3 = 10 - 3 - 3 = 4$ we get $R^4/\alpha'^2 \simeq g\,N$. The precise relation is

$$\frac{R^4}{\alpha'^2} = 4\pi\, g\, N\,. \tag{23.58}$$

The two equations above determine the relations between the gravitational and gauge parameters needed for the equivalence:

$$g = \frac{1}{4\pi}\, g_{YM}^2 \quad \text{and} \quad \frac{R^4}{\alpha'^2} = g_{YM}^2\, N\,. \tag{23.59}$$

In terms of the 't Hooft coupling $\lambda = g_{YM}^2\, N$, the equations above can be rewritten as

$$g = \frac{\lambda}{4\pi N} \quad \text{and} \quad \frac{R}{\sqrt{\alpha'}} = \lambda^{1/4}\,. \tag{23.60}$$

The string coupling is smaller than the 't Hooft coupling by a factor of N. Moreover, in units of string length, the radius of the S^5 only depends on 't Hooft coupling.

The first relation in (23.59) shows that weak Yang–Mills coupling implies weak string coupling. Since theories are generally simpler to deal with at weak coupling, it may seem that the correspondence should be easy to test: both theories could be examined at weak coupling and the results compared. This argument is wrong on two accounts. First, the 't Hooft coupling is the relevant gauge coupling for N large, so we need small λ for simple gauge theory computations. Second, the ability to calculate and to make quantitative statements in the string theory side requires both weak coupling *and* large $R/\sqrt{\alpha'}$. If the sphere S^5 is large, its curvature is small and the superstring theory can be accurately approximated by a supergravity theory in which calculations are simpler. Since $R/\sqrt{\alpha'} = \lambda^{1/4}$ we have a clash: tractable IIB requires a large left-hand side while tractable gauge theory requires a small right-hand side.

Quick calculation 23.2 Let $\lambda' = \alpha'^2/R^4$ denote the string expansion parameter that must be small to have a tractable IIB theory. How are λ and λ' related? Consider a series like (23.44) in the Yang–Mills side of the correspondence for λ fixed and N large. What is the expansion parameter in the IIB side of the correspondence?

It is useful to bring in the precise value of the gravitational constant $G^{(10)} \sim g^2\alpha'^4$ in the superstring side of the correspondence:

$$16\pi\, G^{(10)} = (2\pi)^7\, g^2\alpha'^4\,. \tag{23.61}$$

We can now obtain a couple of interesting results:

Quick calculation 23.3 Use the matching relations to show that

$$G^{(10)} = \frac{\pi^4 R^8}{2N^2}.$$ (23.62)

Since $G^{(10)} = \ell_P^8$, where ℓ_P is the ten-dimensional Planck length, $R/\ell_P \sim N^{1/4}$. In Planckian units the radius of the sphere S^5 does not depend on the 't Hooft parameter.

Quick calculation 23.4 Show that the five-dimensional gravitational constant $G^{(5)}$ obtained after compactification on the S^5 of radius R is

$$G^{(5)} = \frac{\pi R^3}{2N^2}.$$ (23.63)

Despite the challenges, many nontrivial tests of the AdS/CFT correspondence have been carried out. The set of fields living on the gravitational side, for example, has been matched with the set of field operators that exist in the Yang–Mills theory. Supersymmetry is of help because it results in the existence of λ-independent quantities, known as protected observables. These can be calculated at zero coupling on the gauge theory side and compared with the gravitational predictions. Several such quantities have been compared successfully, and some new protected observables have been discovered. If we declare ourselves convinced that the AdS/CFT correspondence is correct, then the difficulty in testing it becomes a virtue: large-λ effects that are extremely difficult to compute in the gauge theory are calculable in the ten-dimensional supergravity theory!

Much of the recent work on the AdS/CFT correspondence has dealt with situations where the Yang–Mills theory has less, or no, supersymmetry. Such extensions are needed to obtain a correspondence that could apply to QCD. In addition, much work has also been devoted to develop other kinds of large-N limits, where the correspondence can be tested directly at weak coupling.

23.8 Hyperbolic spaces and conformal boundary

In this section we begin our preparation for the study of the geometry of anti-de Sitter spaces. We are already familiar with Minkowski space M_{n+1}, a flat space with one time dimension and n space dimensions. Anti-de Sitter space AdS_{n+1} also has one time dimension and n space dimensions, but it is curved. In fact, it is a negatively curved spacetime. De Sitter space dS_{n+1} has one time dimension and n space dimensions and is positively curved. Curved spacetimes are challenging to visualize, so we begin by discussing curved spaces that have no time direction. In such spaces the metric is positive definite and, consequently, all vectors have positive length squared (they are spacelike). We will focus on hyperbolic spaces, negatively curved spaces without a time direction. Hyperbolic spaces will help us understand anti-de Sitter spacetimes.

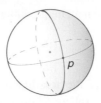

Fig. 23.8 Left: a sphere, positively curved at p, with two lines that are shown to bend towards the same side of the surface – the interior. Right: a manifold negatively curved at p, as implied by the existence of two lines through p that bend to opposite sides of the surface.

The unit sphere S^2 is the simplest example of a space of constant positive curvature. We can visualize it inside $\mathbb{R}^3$, with coordinates x^1, x^2, and x^3, as the surface $(x^1)^2 + (x^2)^2 + (x^3)^2 = 1$. This is an isometric embedding: the metric induced on the surface by the ambient metric (recall Section 6.2) is the familiar round metric on the sphere. It is generally easy to tell by inspection if a two-dimensional surface embedded in $\mathbb{R}^3$ is positively or negatively curved at a point p. Consider a vector $\vec{n}_p$ normal to the surface at p and a plane that contains $\vec{n}_p$. The intersection of the plane with the surface defines a line through p. By rotating the plane about the axis along $\vec{n}_p$ we get a family of lines that lie on the surface and go through p in all possible directions. A surface has positive curvature at p if all of the above lines bend towards the same side of the surface at p. A surface has negative curvature if one can find two lines that bend towards opposite sides of the surface at p. In Figure 23.8 we show a sphere and two lines through a point p. In fact, any line drawn in the manner explained above is a circle centered at the center of the sphere, and they all bend towards the interior of the sphere. We also show, to the right, a piece of a surface that is negatively curved at p. The horizontal circle bends in while the vertical line bends out.

The classic example of a two-dimensional space with constant negative curvature is hyperbolic space $\mathbb{H}_2$. It is actually a space with infinite volume and without a boundary. The first complication with this space is that, as opposed to the sphere S^2, it is not possible to fully embed it isometrically in $\mathbb{R}^3$. Certain portions of $\mathbb{H}_2$ can be embedded, and locally they look like the surface shown on the right side of Figure 23.8. Just like $\mathbb{R}^2$ and S^2, the space $\mathbb{H}_2$ is homogeneous: there are no special points, or equivalently, any point p can be moved to any point q by an isometry. For $\mathbb{R}^2$ the isometry is a translation and for S^2 the isometry is a rotation.

While $\mathbb{H}_2$ cannot be presented isometrically as a surface in $\mathbb{R}^3$ it can be presented isometrically as a surface in three-dimensional *Minkowski* space M$_3$. This presentation is very neat, but we lose perception of distances. Concretely, we have

$$\text{ambient metric:} \quad ds^2 = -(dz)^2 + (dx^1)^2 + (dx^2)^2\,,$$
$$\text{constraint:} \quad -z^2 + (x^1)^2 + (x^2)^2 = -R^2\,. \tag{23.64}$$

The first equation gives the metric of M$_3$, showing that z is the time coordinate and that x^1 and x^2 are the space coordinates. The second equation is the constraint that defines the surface. Since we have one constraint on points in a three-dimensional space, the surface

is two-dimensional. The constraint indicates that $z^2 \geq R^2$, so the surface has two disconnected leafs: it exists for $z \geq R$ and for $z \leq -R$. The space $\mathbb{H}_2$ includes only one leaf; we
can choose that to be the leaf $z \geq R$.

Since we claim that the surface has no time direction, no tangent to the surface can be
timelike. This can be verified as follows. First note that the constraint equation implies
that the position vector $v^\mu = (z, x^1, x^2)$ for a point on the surface is timelike. In fact,
$v^\mu v_\mu = v \cdot v = -R^2$. A tangent is an infinitesimal vector δv^μ such that $v^\mu + \delta v^\mu$ still
belongs to the surface: $(v + \delta v) \cdot (v + \delta v) = -R^2 + \mathcal{O}(\delta v^2)$. This means that $v \cdot \delta v = 0$.
Since two timelike vectors cannot be orthogonal, the tangent vector δv must be spacelike.
This is what we wanted to show.

Quick calculation 23.5 Why is the dot product of two timelike vectors always different
from zero?

Quick calculation 23.6 Consider the point $(z, x^1, 0)$ on the hyperboloid. How are z and
x^1 related? Exhibit two orthogonal (spacelike) tangents at the point. Note that this implies
that all tangent vectors are spacelike.

It is worth noting that for surfaces defined by vectors v with constant length, all tangents
δv at v satisfy $v \cdot \delta v = 0$. This means that the vector v is orthogonal to the surface. Moreover, since v^2 is nonvanishing, v cannot be a tangent at v. Since the normal to the surface
is timelike and the ambient space has no two orthogonal timelike directions, the surface
cannot have a timelike tangent vector.

Lorentz transformations preserve norm so acting on a vector that ends on the surface will
give another vector that ends on the surface. In fact, any two vectors v_1 and v_2 that end on
the surface can be mapped into each other by a Lorentz transformation: there are Lorentz
transformations L_1 and L_2 that take both v_1 and v_2 to the vector $(R, \vec{0})$, so $L_1 v_1 = L_2 v_2$
and $v_2 = (L_2)^{-1} L_1 v_1$. Since Lorentz transformations preserve the ambient metric they act
on the surface as isometries. All in all, this means that the surface is a homogeneous space,
the surface has no special points.

Our next task is to determine the induced metric on the surface. Since this is not more
complicated with a higher number of dimensions, let us consider the case of $\mathbb{H}_n$, described
as a surface in M_{n+1}. Using an index i that runs from 1 to n, we write

$$\text{ambient metric:} \quad ds^2 = -(dz)^2 + dx^i dx^i, \qquad i = 1, 2, \ldots, n;$$
$$\text{constraint:} \quad -z^2 + x^i x^i = -R^2. \tag{23.65}$$

As usual, repeated indices imply summation. Again, one selects the leaf $z \geq R$. Points in
the ambient space are denoted by coordinates $(z, x^1, x^2, \ldots, x^n) = (z, \vec{x})$. It is amusing to
note that the mass-shell condition $p^2 = -(p^0)^2 + (p^1)^2 + (p^2)^2 + (p^3)^2 = -m^2$ defines
the hyperbolic space $\mathbb{H}_3$ embedded in four-dimensional momentum space.

A nice form of the metric on $\mathbb{H}_n$ arises by a stereographic projection that takes the whole
leaf $z \geq R$ to the *interior* of the ball B^n of radius R. The new coordinates are denoted by

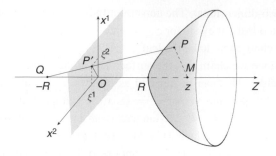

Fig. 23.9 The projection of a point P on the hyperboloid to a point P' on the screen $z = 0$. The ξ coordinates of P are equal to the x coordinates of P'.

ξ^i, with $i = 1, \ldots, n$. We use the vector notation $\vec{\xi} = (\xi^1, \xi^2, \ldots, \xi^n)$, and r is defined to be the radial coordinate in ξ variables:

$$r^2 \equiv \vec{\xi} \cdot \vec{\xi} = (\xi^1)^2 + (\xi^2)^2 + \cdots + (\xi^n)^2. \tag{23.66}$$

For a given point P on the hyperboloid consider the line from P to the point $Q = (-R, \vec{0})$ and the intersection P' of this line with the hyperplane $z = 0$, called the screen. We then set $\xi^i(P) = x^i(P')$, namely, the ξ coordinate of a point on the hyperboloid is declared equal to the x coordinate of its image on the screen. The three-dimensional representation in Figure 23.9 illustrates the construction for the case of $\mathbb{H}_2$. It follows from the projection that $\vec{\xi}$ is parallel to $\vec{x}$. Moreover, the ratio of their lengths is determined by the length r of $\vec{\xi}$: all points in a circle of radius r on the screen come from points with the same value of $|\vec{x}|$. We thus write

$$x^i = \rho(r)\,\xi^i \quad \rightarrow \quad |\vec{x}| = \rho\,r\,, \tag{23.67}$$

where $\rho(r)$ is a function to be determined. Using the similar triangles QOP' and QMP we have the relation

$$\frac{r}{R} = \frac{\rho r}{R + z} \quad \rightarrow \quad \frac{1}{R} = \frac{\rho}{R + \sqrt{R^2 + \rho^2 r^2}}\,, \tag{23.68}$$

where we have used

$$z^2 = R^2 + \vec{x} \cdot \vec{x} = R^2 + \rho^2 r^2\,. \tag{23.69}$$

Soving for ρ we find

$$\rho = \frac{2R^2}{R^2 - r^2} = \frac{2}{1 - \frac{r^2}{R^2}}\,. \tag{23.70}$$

Since ρ must be positive, the condition $r^2 < R^2$ guarantees that we get all possible values of ρ and no negative ones. Back in (23.67) we have that x^i and ξ^i coordinates are related by

$$x^i = \frac{2R^2}{R^2 - r^2}\,\xi^i\,. \tag{23.71}$$

To obtain the metric, we first calculate z as a function of r using (23.69) and (23.70). We find

$$z = R \frac{R^2 + r^2}{R^2 - r^2} . \tag{23.72}$$

Noting that $dr^2 = 2\xi^i d\xi^i$ a short computation gives

$$dz = \frac{4R^3}{(R^2 - r^2)^2} \xi^i d\xi^i . \tag{23.73}$$

Taking differentials of (23.71) we obtain:

$$dx^i = \frac{2R^2}{R^2 - r^2} d\xi^i + \frac{4R^2}{(R^2 - r^2)^2} \xi^i \xi^j d\xi^j . \tag{23.74}$$

Squaring dx^i and simplifying,

$$dx^i dx^i = \frac{4R^4}{(R^2 - r^2)^2} d\xi^i d\xi^i + \frac{16R^6}{(R^2 - r^2)^4} (\xi^i d\xi^i)^2 . \tag{23.75}$$

Finally, we combine the results in (23.73) and (23.75) into $ds^2 = -dz^2 + dx^i dx^i$, and immediately find

$$ds^2 = \frac{4 \, d\xi^i d\xi^i}{\left(1 - \frac{r^2}{R^2}\right)^2} , \tag{23.76}$$

where we recall that $r = \sqrt{\xi^i \xi^i} < R$. It is sometimes convenient to scale the disk to unit radius. For this we simply let $\xi^i \to R\xi^i$, which also gives $r \to R r$. The metric then becomes the standard

$$\mathbb{H}_n \text{ metric} : \ ds^2 = \frac{4R^2 \, d\xi^i d\xi^i}{(1 - r^2)^2}, \quad i = 1, \ldots, n, \quad r = \sqrt{\xi^i \xi^i} < 1 . \tag{23.77}$$

In this presentation the ξ coordinates have no units, the length scale is provided by the explicit factors of R, the radius of curvature of the space. Hyperbolic space has infinite volume. In fact a line from $r = 0$ to $r = 1$ has infinite length. To check this consider a line that goes from $\xi^1 = 0$ to $\xi^1 = \bar{r}$, with all other $\xi^i = 0$. Then the length $\ell(\bar{r})$ of this curve is

$$\ell(\bar{r}) = \int_0^{\bar{r}} \frac{2R \, d\xi^1}{1 - (\xi^1)^2} = R \ln\left(\frac{1 + \bar{r}}{1 - \bar{r}}\right) . \tag{23.78}$$

Indeed, as $\bar{r} \to 1$ the length $\ell(\bar{r})$ diverges.

It seems difficult to speak of the boundary $r \to 1$ of $\mathbb{H}_n$ since it lies an infinite distance away from any point. There is, however, the notion of a *conformal boundary* that can be made precise and carries very interesting information. Given a metric ds^2 we modify it by multiplication by an extra factor $\Omega^2 > 0$ to obtain a new metric $ds'^2 = \Omega^2 ds^2$. The factor Ω^2 is chosen so that in the new metric the distance from any point to all boundary points is finite. The conformal boundary is the boundary in the new metric ds'^2. This boundary can be examined clearly since it lies at a finite distance. The conformal boundary is only defined up to additional multiplication by factors Ω'^2 that keep all boundary points a finite distance away. While statements about the precise length or volume of the conformal boundary are meaningless, there is some invariant information. For some spaces the conformal boundary is always a point. For some spaces the conformal boundary can have an extent. If we are dealing with *spacetimes* with nontrivial conformal boundaries the character of tangent vectors to the boundary (spacelike, timelike, or null) is invariant information since the sign of the norm of a vector cannot be changed by multiplication by $\Omega^2 > 0$.

Let us consider, as a first example, two-dimensional flat space $\mathbb{R}^2$ with coordinates $-\infty \leq x, y \leq \infty$ and flat metric $ds^2 = (dx)^2 + (dy)^2 = (dr)^2 + r^2(d\theta)^2$, where r and θ are the standard polar coordinates. We produce a conformally related metric by multiplication by the square of $\mathcal{C}(r, \theta) > 0$:

$$ds'^2 = \mathcal{C}^2(r, \theta)ds^2 = \mathcal{C}^2(r, \theta)\left((dr)^2 + r^2(d\theta)^2\right). \qquad (23.79)$$

If lines from the origin to infinity along constant θ have finite length on the new metric we must have

$$\int_0^\infty \mathcal{C}(r, \theta)\, dr < \infty. \qquad (23.80)$$

Integrating over θ and exchanging the order of integration,

$$\int_0^\infty \frac{dr}{r} \int_0^{2\pi} \mathcal{C}(r, \theta)\, r d\theta < \infty. \qquad (23.81)$$

Noting that the integral over θ gives precisely the length $\ell(r)$ of the constant r circle, we write

$$\int_0^\infty \frac{dr}{r}\, \ell(r) < \infty. \qquad (23.82)$$

If $\ell(r)$ has a nonzero limit as $r \to \infty$ the integral would diverge. So we have learned that $\ell(r) \to 0$ as $r \to \infty$ in any suitable conformal metric. The length of circles approaching infinity goes to zero, and therefore the conformal boundary is just a point. The conformally related space is a sphere with a missing point. In Section 25.6 we will examine the construction of the "Riemann sphere" along related lines. One can also prove that the conformal boundary of $\mathbb{R}^n$ with $n > 2$ is always a point.

We can now turn to hyperbolic space (23.77). It is clear that we can get something quite simple by multiplication by the conformal factor $(1 - r^2)^2/(4R^2)$:

$$ds'^2 = \frac{1}{4R^2}(1 - r^2)^2\, ds^2 = d\xi^i d\xi^i. \qquad (23.83)$$

The conformally related space is just the interior $\xi^i \xi^i < 1$ of the unit ball B^n with the flat constant metric. The conformal boundary is the set $\xi^i \xi^i = 1$, that is, the unit sphere S^{n-1}. It is interesting that the conformal boundary of hyperbolic space is much more substantial than that of flat space.

23.9 Geometry of AdS and holography

In this section we examine in some detail the geometry of AdS spaces in order to appreciate the holographic aspects of the AdS/CFT correspondence. In optics, a hologram is a two-dimensional plate onto which an image of a three-dimensional object has been recorded. In photography, one directly focuses the image of an object into the film. In holography, one records an interference pattern between coherent light reflected from the object and coherent light from a reference beam. This recorded interference pattern is still two-dimensional, but it contains much more information than a photograph. It contains information about all three dimensions of the object. The pattern allows you to view a three-dimensional image that exhibits parallax: the image looks different depending on the direction of the observer.

In gravitational physics a system that extends over a macroscopic region of space is said to be holographic if all of its physics can be represented by a theory that lives on the boundary of this region. Moreover, one requires that the boundary theory should not contain more than one degree of freedom (with a finite number of states) per Planck area. Holography is motivated by the physics of black holes: the entropy of a black hole is not proportional to the volume enclosed by its horizon but rather to the area of the horizon. Moreover, as discussed previously (equation (22.124)), this entropy is reproduced by assuming that the horizon carries a degree of freedom per horizon element of Planck area. The AdS/CFT correspondence is a more concrete realization of holography. There is a precise sense in which the four-dimensional space where the Yang–Mills theory lives can be viewed as a boundary of AdS_5. The $SU(N)$ theory, which captures all the physics of the interior of the ten-dimensional spacetime, provides a holographic description of the gravitational world. Moreover, one can roughly argue that the holographic bound on the degrees of freedom at the boundary holds.

The space AdS_{n+1} is usually defined as a surface embedded in a flat space $\mathbb{R}^{2,n}$ with two time coordinates u and v and n space coordinates x^i:

$$\text{ambient metric:} \quad ds^2 = -(du)^2 - (dv)^2 + dx^i dx^i, \quad i = 1, \ldots, n;$$
$$\text{constraint:} \quad -u^2 - v^2 + x^i x^i = -R^2.$$
(23.84)

On the space $\mathbb{R}^{2,n}$ there are generalized Lorentz transformations, linear transformations that preserve the metric. The constraint states that a vector $V = (u, v, \vec{x})$ belongs to the surface if $V \cdot V = -R^2$, where the dot product uses the ambient metric. The vector V is timelike. The above equations are analogous to those that define hyperbolic space in (23.65), except that now we have two time directions. Since V is defined by a length condition, it is normal to the surface. Given that V is timelike and the ambient space has two orthogonal timelike directions, the surface must contain one timelike direction. In fact, this direction can be readily visualized. Fix a specific point $(u_0, v_0, \vec{x}_0)$ on the surface. Clearly,

$$u_0^2 + v_0^2 = R^2 + \vec{x}_0 \cdot \vec{x}_0.$$
(23.85)

Consider now the circle defined by all values of u and v that satisfy

$$u^2 + v^2 = R^2 + \vec{x}_0 \cdot \vec{x}_0.$$
(23.86)

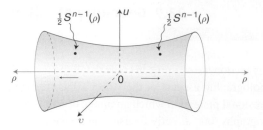

Fig. 23.10 AdS_{n+1} space described as a two-dimensional surface in which each point represents *half* of the $S^{n-1}(\rho)$ sphere. The horizontal axis has ρ increasing from zero to infinity *both* to the left and to the right. The half spheres at points with equal values of u and v (shown in figure) must be glued to form a single $S^{n-1}(\rho)$ sphere.

This circle lies on the surface and goes through the point $(u_0, v_0, \vec{x}_0)$. Moreover, the tangent vectors to the circle are timelike everywhere since they are vectors with components only along u and v.

Anti-de Sitter space is a homogeneous space: no point is special and any point can be mapped into any other point by a transformation that is an isometry. Indeed, two vectors V_1 and V_2 that have the same norm $V_1 \cdot V_1 = V_2 \cdot V_2 = -R^2$ can always be mapped into one another by a generalized Lorentz transformation (you may enjoy proving this statement). In order to convince yourself that the surface does not contain two independent timelike directions anywhere it suffices to verify that two orthogonal timelike directions do not exist at one point. This is readily done.

Quick calculation 23.7 Consider the point $(R, 0, \vec{0})$ on the surface. Show that the surface does not contain two orthogonal timelike directions at this point.

We can visualize AdS space using a two-dimensional surface on which each point represents a sphere. We write the constraint equation as

$$u^2 + v^2 = R^2 + \vec{x} \cdot \vec{x} \tag{23.87}$$

and plot the space using axes for u, v, and $\rho = \sqrt{\vec{x} \cdot \vec{x}}$. A point on the two-dimensional surface shown in Figure 23.10 is determined by u and v, since then the value of ρ is fixed. To see the whole AdS space we must include at each point on the surface the sphere S^{n-1} defined by the points $\vec{x}$ that satisfy: $\vec{x} \cdot \vec{x} = \rho^2 = u^2 + v^2 - R^2$. As shown in the figure, the surface extends both to the left and to the right, and in *both* regions ρ varies from 0 to ∞! Had we plotted the space as the region that only goes to the right, you may reach the incorrect conclusion that the space has a boundary at $\rho = 0$, or equivalently on the circle $u^2 + v^2 = R^2$. This conclusion is suspect because on this circle the S^{n-1} spheres have zero radius. Since we represent the range of ρ twice, there are two points associated with each value of the pair u, v. Consistency requires that on top of each of these two points we place *half* of the S^{n-1} sphere. We cut the sphere in two and place half of it on top of each point, with the understanding that the halves are to be glued.

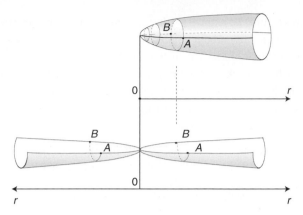

Fig. 23.11 Top: a cigar as a collection of circles over the half-line $r \geq 0$, with the radii of the circles going to zero as $r \to 0$. Bottom: the same space represented with an r that grows from zero to infinity both to the right and to the left. Above each point there is now half a circle. The two half circles at r are glued to form a full circle.

A lower-dimensional analog helps visualize the situation. Imagine a cigar represented as a circle S^1 over a semi-infinite line $r \geq 0$, as shown in the top part of Figure 23.11. The radius of the circle vanishes as $r \to 0$, where the cigar ends. The cigar, however, has no boundary at this point. A representation similar to the one we use for AdS, shown at the bottom, represents r in both directions and places half of each circle S^1 at symmetric points.

We identified a timelike direction on the AdS space, but in fact, we found something strange: closed timelike curves. Indeed, the (u, v) circles in (23.86) have timelike tangents everywhere and are closed. This is not the kind of space we want; in a universe with closed timelike curves one could travel for a while and get back before one's departure. What is sometimes called the "complete" AdS space, but we will simply call AdS space, is a space in which we unwrap the timelike circles in the space into open lines. The full AdS space can be imagined as a space rolled over the surface in Figure 23.10 an infinite number of times.

We now determine the metric on AdS_{n+1}. We begin by making the time coordinate explicit through the relations

$$ u = z \cos t, \quad v = z \sin t, \tag{23.88}$$

where we trade u and v for the time t and an additional coordinate z. It follows from these relations that

$$ u^2 + v^2 = z^2, \quad \text{and} \quad (du)^2 + (dv)^2 = (dz)^2 + z^2 (dt)^2. \tag{23.89}$$

We can now use these equations in (23.84) to find that the ambient metric and the constraint become

$$ \text{ambient metric:} \quad ds^2 = -z^2 (dt)^2 - (dz)^2 + dx^i dx^i, \quad i = 1, \ldots, n; $$
$$ \text{constraint:} \quad -z^2 + x^i x^i = -R^2. \tag{23.90}$$

Apart from the extra $-z^2(dt)^2$ appearing in the metric, the above are exactly the equations (23.65) for hyperbolic space. At any fixed time ($dt = 0$) the *spatial* geometry of the AdS space is that of $\mathbb{H}_n$. This is the way hyperbolic space appears in the anti-de Sitter geometry.

A little work remains now to find the full metric. As before we use ξ^i coordinates and write both x^i and z in terms of them. The value of z was determined in (23.72) and the portion $-(dz)^2 + dx^i dx^i$ of the metric was calculated in (23.76). We thus find that the metric in (23.90) becomes

$$ ds^2 = -\frac{(R^2 + r^2)^2}{(R^2 - r^2)^2} R^2 (dt)^2 + \frac{4R^4\, d\xi^i d\xi^i}{(R^2 - r^2)^2}\,. \tag{23.91}$$

As a last step we let $\xi^i \to R\xi^i$ and obtain the metric on AdS_{n+1}:

$$ \text{AdS}_{n+1} \text{ metric}: \ ds^2 = R^2\left[-\left(\frac{1+r^2}{1-r^2}\right)^2 (dt)^2 + \frac{4\, d\xi^i d\xi^i}{(1-r^2)^2}\right]. \tag{23.92}$$

Here, as before, $r^2 = \xi^i \xi^i$ and $i = 1, 2, \ldots, n$. We already know from our analysis of $\mathbb{H}_n$ that at any fixed time, the distance from any point to the boundary $r \to 1$ is infinite. What is quite interesting is that it does not take infinite time for a light ray to reach the boundary. Assuming the light-ray travels in the ξ^1 direction, the condition $ds^2 = 0$ gives

$$ \frac{1+r^2}{1-r^2} dt = \frac{2d\xi^1}{1-r^2} \ \to \ \frac{d\xi^1}{1+(\xi^1)^2} = \frac{1}{2} dt\,. \tag{23.93}$$

Integrating with $\xi^1 = 0$ for $t = 0$, we find $\xi^1 = \tan(t/2)$, which shows that the boundary point $\xi^1 = 1$ is reached at $t = \pi/2$.

To understand the conformal properties of the space and its boundary we use the metric (23.92) to define the conformally related

$$ ds'^2 = \frac{(1-r^2)^2}{4R^2} ds^2 = -\left(\frac{1+r^2}{2}\right)^2 (dt)^2 + d\xi^i d\xi^i\,. \tag{23.94}$$

The metric ds'^2 describes a spatial n-dimensional ball (the interior of $\xi^i \xi^i = 1$) and a time coordinate. Near $r = 1$ we have

$$ ds'^2 \simeq -(dt)^2 + d\xi^i d\xi^i\,, \quad \text{as} \quad r \to 1, \tag{23.95}$$

and the conformal boundary takes the form $\mathbb{R} \times S^{n-1}$, the $\mathbb{R}$ factor for the time and the S^{n-1} factor for the boundary $\xi^i \xi^i = 1$ of the hyperbolic spatial sections. For AdS_3 the boundary $\mathbb{R} \times S^1$ is the surface of a cylinder, and the full spacetime is inside of it (see Figure 23.12). It is a nontrivial fact that AdS spacetimes have a conformal boundary with both time and space directions. The conformal boundary of Minkowski space, for example, only contains null directions (see Problem 23.4).

For the AdS_5 spacetime the boundary is $\mathbb{R} \times S^3$. The gauge theory dual to the type IIB superstring background lives on this boundary. The scale ambiguity of the conformal

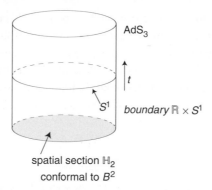

Fig. 23.12 The conformal view of AdS$_3$. The conformal boundary is the surface $\mathbb{R} \times S^1$ of the cylinder, where $\mathbb{R}$ is time and S^1 is the conformal boundary of the spatial sections $\mathbb{H}_2$. The full AdS$_3$ space is conformal to the interior of the cylinder.

boundary is consistent with the scale invariance of the field theory. The physics of the gauge theory does not care about the radius of the three-sphere. If we take the radius to infinity, the boundary is flat four-dimensional Minkowki spacetime.

We can obtain insight into AdS$_5$ holography by focusing on its fixed-time spatial geometry, which is that of $\mathbb{H}_4$. We calculate the volume and the boundary area of spherical subsets $\mathcal{S}(\bar{r})$ of $\mathbb{H}_4$ that include all points with $r \leq \bar{r}$. In the limit $\bar{r} \to 1$ $\mathcal{S}(\bar{r})$ approaches the full spatial section of AdS$_5$. The area of the boundary of $\mathcal{S}(\bar{r})$ is the "area" $A_3(\bar{r})$ of the three-sphere located at $r = \bar{r}$. The metric (23.92) tells us that physical length is obtained by multiplying coordinate length by $2R/(1 - r^2)$. It follows that

$$A_3(\bar{r}) = \left[\frac{2R}{1 - \bar{r}^2} \right]^3 \text{Vol}(S^3(\bar{r})) = \left[\frac{2\bar{r} R}{1 - \bar{r}^2} \right]^3 \cdot 2\pi^2 . \qquad (23.96)$$

The volume $V_4(\bar{r})$ of $\mathcal{S}(\bar{r})$ is given by

$$V_4(\bar{r}) = \int_0^{\bar{r}} \left[\frac{2R}{1 - r^2} \right]^4 \text{Vol}(S^3(r)) dr = 2\pi^2 (2R)^4 \int_0^{\bar{r}} \frac{r^3 dr}{(1 - r^2)^4} . \qquad (23.97)$$

The integral is readily evaluated and gives

$$V_4(\bar{r}) = 2\pi^2 (2R)^4 \frac{\bar{r}^4(3 - \bar{r}^2)}{12(1 - \bar{r}^2)^3} = \frac{R}{6} \bar{r} \left(3 - \bar{r}^2 \right) \left[\frac{2\bar{r} R}{1 - \bar{r}^2} \right]^3 \cdot 2\pi^2 . \qquad (23.98)$$

As expected, both the area $A_3(\bar{r})$ and the volume $V_4(\bar{r})$ diverge as $\bar{r} \to 1$. Their ratio, however, does not:

$$\frac{A_3(\bar{r})}{V_4(\bar{r})} = \frac{6}{R} \frac{1}{\bar{r}(3 - \bar{r}^2)} \quad \to \quad \lim_{\bar{r} \to 1} \frac{A_3(\bar{r})}{V_4(\bar{r})} = \frac{3}{R} . \qquad (23.99)$$

As the subsets grow without bound to cover $\mathbb{H}_4$, the boundary area and the bulk volume become proportional to one another! The constant of proportionality is the radius of curvature R of the AdS space. In flat space, a four-dimensional region of characteristic size L will have an area $A_3 \sim L^3$ and a volume $V_4 \sim L^4$. The ratio $A_3/V_4 \sim 1/L$ goes to zero as the size of the region grows without bound. This is why holography is hard to realize, the boundary is too small to capture the physics of the bulk. Less so in AdS space, where the area to volume ratio does not approach zero.

We can elaborate further if we introduce a regulator that makes both boundary areas and number of degrees of freedom finite. Consider a small number $\delta \ll 1$ and the sphere $\bar{r} = 1 - \delta$. It follows from (23.96) that the area A_3^δ of this sphere is

$$A_3^\delta \equiv A_3(1 - \delta) \simeq 2\pi^2 \frac{R^3}{\delta^3}. \tag{23.100}$$

Now let us consider the boundary theory at $\bar{r} = 1$. To estimate the number of degrees of freedom N_{dof} on this theory we take δ to define a short distance cutoff. We imagine having one degree of freedom per little cube of *coordinate* volume δ^3. Since the boundary has coordinate volume of order one, we have $1/\delta^3$ degrees of freedom. The boundary theory is made of $SU(N)$ fields, so we correct the estimate by multiplying by N^2; our final result being N^2/δ^3 degrees of freedom. We now verify that this number N_{dof} of degrees of freedom matches the holographic expectation of a degree of freedom per piece of surface of Planck size. We first note that

$$N_{\text{dof}} \sim \frac{N^2}{\delta^3} \sim \frac{N^2 A_3^\delta}{R^3}. \tag{23.101}$$

Recalling (23.63), which tells us that $R^3 \sim G^{(5)} N^2$, we find

$$N_{\text{dof}} \sim \frac{A_3^\delta}{G^{(5)}}. \tag{23.102}$$

This *is* the holographic expectation. If we take the total volume V of the space in the type II theory to be given by $V_4^\delta = V_4(1 - \delta)$ times the volume $V_5 \sim R^5$ of the S^5, the number of degrees of freedom per unit volume behaves as

$$\frac{N_{\text{dof}}}{V} \sim \frac{A_3^\delta}{V_4^\delta} \frac{1}{R^5 G^{(5)}} \sim \frac{1}{R} \frac{1}{G^{(10)}}. \tag{23.103}$$

For large R the number of degrees of freedom per unit volume in the bulk is very small. If we increase the number of branes, R increases and we can make the ratio N_{dof}/V arbitrarily small. This bulk theory, with such a small density of degrees of freedom, is string theory with a fixed string length $\sqrt{\alpha'}$.

23.10 AdS/CFT at finite temperature

The AdS$_5$ space supports black holes in the same sense that Minkowski space supports black holes: the geometry of the space is deformed by the presence of the hole but as we move away from the hole the spacetime metric approaches the original metric. A Schwarzschild black hole inside AdS$_5$ is described by the metric

$$ds^2 = -\left(1 + \frac{r^2}{R^2} - \frac{r_0^2}{r^2}\right)(dt)^2 + \left(1 + \frac{r^2}{R^2} - \frac{r_0^2}{r^2}\right)^{-1}(dr)^2 + r^2 d\Omega_3^2. \tag{23.104}$$

The above metric describes a space that is *asymptotically* AdS. That is, as $r \gg r_0$

$$r \gg r_0 : \quad ds^2 = -\left(1 + \frac{r^2}{R^2}\right)(dt)^2 + \left(1 + \frac{r^2}{R^2}\right)^{-1}(dr)^2 + r^2 d\Omega_3^2 , \qquad (23.105)$$

which is the metric of an AdS_5 space of radius R, written in a form different to the one we considered before (see Problem 23.5). Thus, in (23.104), R denotes the radius of the asymptotic AdS_5. The length parameter r_0 tells us that we have a black hole. One can write r_0 in terms of the mass M of the hole and the five-dimensional Newton's constant: $r_0^2 \sim G^{(5)} M$. In fact, as $R \to \infty$ the metric (23.104) becomes that of a black hole in five-dimensional Minkowski space with Schwarzschild radius r_0. Finally, $d\Omega_3^2$ denotes the metric on a three-sphere S^3 of unit radius.

The Schwarzschild radius r_+ of the AdS black hole is the value of r that makes the coefficient of $(dt)^2$ in the metric equal to zero:

$$1 + \frac{r_+^2}{R^2} - \frac{r_0^2}{r_+^2} = 0 . \qquad (23.106)$$

Solving for r_+^2 we find

$$r_+^2 = \frac{R^2}{2}\left(\sqrt{1 + \frac{4r_0^2}{R^2}} - 1\right). \qquad (23.107)$$

For a fixed AdS scale R, the Schwarzschild radius r_+ is a function of r_0 or, equivalently, a function of the black hole mass. For $r_0 \ll R$ one can see that $r_+ \sim r_0$ and for $r_0 \gg R$ one finds $r_+ \sim \sqrt{r_0 R}$. In fact, r_+ is always smaller than r_0, as one can see by rewriting (23.107) in the form

$$\frac{r_+^2}{r_0^2} = \frac{2}{1 + \sqrt{1 + \frac{4r_0^2}{R^2}}} \leq 1 . \qquad (23.108)$$

The scale R of the asymptotic AdS space is not a limit for the size of the black hole. A large black hole is one for which $r_+ \gg R$. A small black hole is one for which $r_+ \ll R$. A plot of r_+ as a function of r_0 is shown in Figure 23.13.

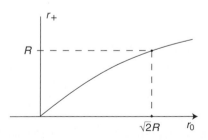

Fig. 23.13 The dependence of the Schwarzschild radius r_+ of the AdS black hole on the mass of the hole, encoded by r_0.

The black hole solution presented here has a Hawking temperature $\bar{T}_H$. A calculation of this temperature gives:

$$\bar{T}_H = \frac{R^2 + 2r_+^2}{2\pi r_+ R^2} . \tag{23.109}$$

While a justification of this result requires tools beyond those assumed in this text, it is worthwhile to know that this temperature emerges from a general formula as follows.

Quick calculation 23.8 Consider a metric $ds^2 = -f(r)(dt)^2 + (f(r))^{-1}(dr)^2 + \cdots$ with $f(r)$ a function of a radial coordinate that defines a horizon radius r_+ via $f(r_+) = 0$. The Hawking temperature $\bar{T}_H$ associated with this gravitational solution is given by $\bar{T}_H = f'(r_+)/(4\pi)$. Apply this result to (23.104) and derive (23.109).

The temperature (23.109) has interesting limits. For small black holes we find

$$\bar{T}_H \simeq \frac{1}{2\pi r_+} , \quad \text{for} \quad r_+ \ll R . \tag{23.110}$$

In Minkowski space the temperature of a black hole is inversely proportional to the Schwarzschild radius (see (22.120)); small black holes are hot. We have recovered this result in (23.110) because for small black holes the curvature of the AdS is a negligible effect. More surprising is the behavior for large black holes. We then have:

$$\bar{T}_H \simeq \frac{r_+}{\pi R^2} , \quad \text{for} \quad r_+ \gg R . \tag{23.111}$$

This is unusual: once the black hole is large enough, its temperature grows with its size. This is an important qualitative feature of black holes in AdS space. A sketch of the temperature $\bar{T}_H$ as a function of r_+/R is shown in Figure 23.14.

Quick calculation 23.9 Show that for fixed radius of curvature R, all black holes have temperature $\bar{T}_H$ that satisfies

$$\bar{T}_H \geq T_0 = \frac{\sqrt{2}}{\pi} \frac{1}{R} , \tag{23.112}$$

where the lower bound T_0 is realized for $r_+/R = 1/\sqrt{2}$. This information is displayed in Figure 23.14.

It is reasonable to expect that the AdS/CFT correspondence extends to the case of a black hole inside the now asymptotic AdS space. Indeed, there is an obvious candidate for the dual gauge theory: $SU(N)$ Yang–Mills theory at finite temperature $\bar{T}_H$, the temperature of the black hole. The metric (23.104) for fixed large r becomes

$$ds^2 \simeq \frac{r^2}{R^2}\Big[-(dt)^2 + R^2 d\Omega_3^2\Big]. \tag{23.113}$$

As in our discussion below (23.95) we see a boundary $\mathbb{R} \times S^3$, where the sphere is of radius R. While this radius is immaterial in the zero-temperature case, it is not immaterial now. The radius R is the radius of the sphere where the field theory at temperature $\bar{T}_H$ lives.

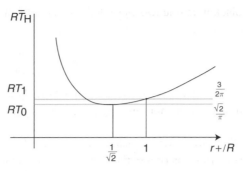

Fig. 23.14 The Hawking temperature $\bar{T}_H$ of the AdS black hole (multiplied by the AdS radius R) sketched as a function of the Schwarzschild radius r_+ scaled down by R. For each value of the Hawking temperature above T_0 there are two black holes. The temperature T_1 defines the Hawking–Page transition.

One cannot rescale with impunity the time coordinate t in (23.113) because at non-zero temperature the time coordinate carries information about temperature, which is an energy scale. The conformal ambiguity of the above metric still remains: if we multiply the factor in brackets by λ^2 this has the effect of changing the radius to λR and, by scaling time $t \rightarrow \lambda t$, changing the temperature to $\bar{T}_H/\lambda$. In the identification of the dual field theory the product $R\bar{T}_H$ of the radius and the temperature is the *only* invariant information.

As a test of the correspondence we calculate the entropy of the finite-temperature field theory and compare it with the entropy of the (large) black hole, hoping for an equality. Both of these numbers are finite (not infinite!). The field theory has a finite entropy density and lives on a space of finite volume, and the black hole has a horizon of finite area. Let us begin with the entropy S_{YM} of the field theory. We first recall that the entropy *density* s_γ of a photon gas in ordinary three-space at temperature T is given by

$$s_\gamma = \frac{2}{45}\pi^2 T^3 \cdot 2 \,, \tag{23.114}$$

where the last factor of two arises because the photon has two massless degrees of freedom. In the $SU(N)$ field theory the total entropy is given by

$$S_{YM} = \frac{2}{45}\pi^2 \bar{T}_H^3 \cdot \left(8 + 8 \cdot \frac{7}{8}\right) N^2 \cdot (2\pi^2 R^3) \,. \tag{23.115}$$

The last factor is the volume of the three-sphere of radius R where the field theory lives. In the limit of small 't Hooft coupling the massless particles in the $SU(N)$ theory are very weakly interacting and the total entropy is obtained by adding their separate contributions. The spectrum of the theory includes eight bosonic degrees of freedom and eight fermionic degrees of freedom, both repeated $N^2 - 1$ times, or just N^2 times, to the level of precision we need. Moreover, fermionic degrees of freedom contribute 7/8 of the entropy of bosonic degrees of freedom. All these considerations explain the result given in (23.115). Simplifying this expression, we get

$$S_{YM} = \frac{2}{3}\pi^2 N^2 \bar{T}_H^3 \cdot (2\pi^2 R^3) \,, \quad \lambda \ll 1 \,. \tag{23.116}$$

On the gravity side the entropy S_{BH} of the black hole is given by the "area" of the horizon divided by the five-dimensional Newton constant $G^{(5)}$:

$$S_{\text{BH}} = \frac{A_{\text{hor}}}{4G^{(5)}} . \tag{23.117}$$

The horizon is the locus of points with $r = r_+$. The last term in the metric (23.104) shows that the horizon is a three-sphere of radius r_+, so $A_{\text{hor}} = 2\pi^2 r_+^3$. This, together with the value of $G^{(5)}$ from (23.63) and the relation of r_+ to $\bar{T}_{\text{H}}$ in (23.111) gives

$$S_{\text{BH}} = \frac{2\pi^2 r_+^3}{\frac{2\pi R^3}{N^2}} = \frac{N^2 \, 2\pi^2 R^3 \, r_+^3}{2\pi R^6} = \frac{1}{2}\pi^2 N^2 \bar{T}_{\text{H}}^3 \cdot (2\pi^2 R^3) , \quad \lambda \gg 1 . \tag{23.118}$$

The assumption of large λ is implicit whenever we use the black hole picture. The result (23.118) is in qualitative agreement with the field theory estimate (23.116). The linear growth of the temperature with the radius of the black hole was essential to the match. Nevertheless, we got

$$S_{\text{BH}} = \frac{3}{4} S_{\text{YM}} , \tag{23.119}$$

so the agreement is not exact. It did not have to be so: in the field theory we assumed small 't Hooft parameter λ, while in the gravity side we assumed large λ. It is hoped that an exact calculation for arbitrary λ will show an entropy that decreases from the zero-coupling value obtained easily in the field theory to the large-coupling value obtained easily from the black hole.

There is an additional twist to the finite temperature correspondence. We have noted in (23.112) that, in a given AdS space with fixed radius of curvature R, all black holes have Hawking temperatures $\bar{T}_{\text{H}} \geq T_0$ (see Figure 23.14). Consider the $SU(N)$ gauge theory at some strong coupling that is dual to AdS with radius of curvature R. For temperatures smaller than T_0 the field theory cannot have a black hole dual since there is no black hole. The dual is then just thermal AdS, an AdS space filled with a gas of particles at the specified temperature. The field theory at these low temperatures is in the analog of a *confined* phase: the entropy of the theory is of $\mathcal{O}(N^0)$. For temperatures greater than T_0 we have two black holes: any horizontal line above T_0 in Figure 23.14 cuts the curve at two values of r_+/R, one smaller than $1/\sqrt{2}$ and one larger than $1/\sqrt{2}$. Thermal AdS also remains a possible gravitational background. It turns out that for temperatures T in the range

$$T_0 \leq T \leq T_1 , \quad \text{with} \quad T_1 = \frac{3}{2\pi R} , \tag{23.120}$$

thermal AdS remains the thermodynamically favored state – it has lower free energy than either black hole background. The field theory remains confined for $T \leq T_1$. Note that T_1 is only a bit larger than T_0 (see Figure 23.14).

Quick calculation 23.10 What are the values of r_+/R for the two black holes that exist for temperature T_1? Compare with Figure 23.14.

As the temperature exceeds T_1 the thermodynamically favored background ceases to be thermal AdS and becomes the larger black hole – this is referred to as the Hawking–Page transition. In fact, the smaller black hole is always least favored among the three options; like black holes in Minkowski space it is unstable to evaporation. At the temperature T_1 the large-N, strongly coupled gauge theory goes from confined to *deconfined*: the entropy becomes of $\mathcal{O}(N^2)$, indicating liberation of elementary degrees of freedom. The dual of the deconfined gauge theory is the (large) AdS black hole.

This finite temperature discussion, where the black hole spacetime is not exactly anti-de Sitter, points to a more general statement of the AdS/CFT correspondence. The correspondence can now be stated as the equivalence of four-dimensional, supersymmetric $SU(N)$ gauge theory with type IIB superstrings in a space whose geometry is *asymptotically* $AdS_5 \times S^5$. This statement clearly applies at zero and at finite temperatures. Further examples of asymptotically $AdS_5 \times S^5$ spaces that match with particular states of the field theory have been found.

23.11 The quark–gluon plasma

Experiments at the Relativistic Heavy Ion Collider (RHIC) in Brookhaven National Laboratory have created, for very brief instants, a deconfined state of QCD. Physicists accelerated and collided nuclei of gold against each other. In the center of mass frame each nucleon packed an energy of about 100 GeV. With 197 nucleons, each gold nucleus at collision carried an energy of about 20 TeV. The collision appears to create a quark–gluon plasma (QGP), a strongly interacting system of deconfined quarks and gluons.

We pause to note the scales of energies and distances relevant to these collisions. Nuclei are smaller by a factor of about 100 000 than atoms, which have sizes of about 10^{-10}m. The natural unit to describe a nucleus is therefore the femtometer (fm): 1 fm $= 10^{-15}$m. An approximate formula gives the radius r of a nucleus with A nucleons: $r \simeq r_0 A^{1/3}$, where $r_0 = 1.2$ fm. It is convenient to measure time in femtometers too, using the speed of light as a conversion unit. Thus, a time of 1fm is $\frac{1}{3} \times 10^{-23}$ s.

Quick calculation 23.11 Confirm that the estimate for the nuclear radius implies an energy density inside nuclei of about 0.13 GeV/fm^3. Confirm that the radius of the gold nucleus is about 7 fm.

Lattice calculations suggest that deconfinement in QCD occurs at a critical temperature $T_c \simeq 175$ MeV, with an uncertainty of about 10% due to systematic errors. This temperature corresponds to about two trillion degrees Kelvin, the temperature of the universe about 10^{-11} s after the Big Bang. At this temperature the energy density of the plasma is 0.7 GeV/fm^3, about five to six times the nuclear energy density.

Since the colliding nuclei at RHIC have Lorentz factors $\gamma \sim 100$, in the center of mass frame they look like thin pancakes with thickness below 14 fm/100 = 0.14 fm. The QGP is created when the pancakes go through each other and is believed to attain thermal equilibrium at times as early as 1 fm. The QGP lasts for no more than 15 fm, at which time the quarks and gluons have recombined into hadrons that are later captured by the detectors. By measuring the energies of these hadrons it is estimated that the energy density in the QGP at time 1 fm is at least 5 GeV/fm^3. For a head-on collision, at this time the plasma extends over a cylinder of length 2 fm and radius equal to the gold nucleus radius.

Quick calculation 23.12 Calculate the energy in the QGP for a head-on collision and show that it is about 4% of the energy available in the center-of-mass frame.

A temperature estimate is possible. Recalling that for a gas of massless particles the energy density $U \sim T^4$, we have

$$\left(\frac{T_{\mathrm{RHIC}}}{T_c}\right)^4 \simeq \frac{5 \text{ GeV/fm}^3}{0.7 \text{ GeV/fm}^3} \qquad \rightarrow \qquad T_{\mathrm{RHIC}} \simeq 1.6 \, T_c. \tag{23.121}$$

This estimate suggests that at time equal to 1 fm the temperature is still comfortably above the deconfinement temperature. At the Large Hadron Collider (LHC) in CERN one may reach temperatures of about $5 \, T_c$.

In general, collisions are non-central and the QGP is initially created in the approximate shape of an ellipse defined by the overlap of two "pancakes" with offset centers. The direction of the impact parameter is along the short axis of the ellipse. Consider the particles created by the collision with momenta orthogonal to the beam axis. One gets more particles along the direction of the short axis of the ellipse than along the direction of the long axis of the ellipse. Numerical simulations indicate that this anisotropy is consistent with the assumption that the QGP is a fluid with extraordinarily small viscosity. In a droplet of fluid one has maximal pressure at the center and zero pressure at the edges. The pressure gradient is larger along the shorter axis of the ellipse leading to a larger number of particles emitted in this direction.

It is natural to wonder if the QGP can be studied using the finite temperature version of the AdS/CFT correspondence. At first sight this seems unlikely since QCD, a non-supersymmetric $SU(3)$ gauge theory, is very different from supersymmetric $SU(N)$ gauge theory. Nevertheless there are some facts that make the proposition less outrageous. First, the QGP is strongly coupled: even with $N = 3$ one is likely to have a large 't Hooft coupling $\lambda \sim 20$, as is needed for a calculable gravity side. Moreover, at finite temperature supersymmetry is a broken symmetry and its effects are therefore somewhat hidden. It is plausible that finite temperature QCD and finite temperature supersymmetric gauge theory are not all that different.

It is striking that the general property of very small viscosity is a rather direct consequence of the AdS/CFT correspondence. The viscosity η of a fluid tells how the force F transmitted across fluid layers of area A depends on the velocity gradient $|\nabla v|$:

$$\frac{F}{A} = \eta |\nabla v|. \tag{23.122}$$

The units of viscosity are $[\eta] = M/(LT)$. The entropy density s is another important property of a fluid and has units $[s] = [k]/L^3$, where k is the Boltzmann constant. It follows that the η/s has units of $\hbar/k$. The AdS/CFT correspondence indicates that at strong coupling and in the large N limit, systems with a gravity dual have

$$\frac{\eta}{s} = \frac{\hbar}{4\pi k}. \tag{23.123}$$

In the simplest setup, the viscosity η is related to the absorption cross section of the black hole and the entropy density follows from (23.118). Almost all known gases and liquids in nature have much larger values of η/s. Exceptions are the QGP of RHIC and cold atomic gases at unitarity, both of which are considered strongly-coupled quantum liquids. Indeed, analysis of the data at RHIC is consistent with a value of η/s which is at most two or there times larger than the value indicated above. It is possible that the smallness of η/s is a benchmark for strongly-coupled quantum liquids.

Quick calculation 23.13 At $25\,°C$ and a pressure of 1 atm, liquid water has a viscosity of 0.9×10^{-3} kg/(m · s) and a molar entropy of $70\,$J/K. Calculate the ratio η/s and confirm that it exceeds the bound by a factor of nearly four hundred. (Useful constant: $k = 1.38 \times 10^{-23}$ J/K.)

Another interesting discovery at RHIC is that very high energy quarks (jets) moving through the QGP are stopped, or "quenched," after traveling only a few femtometers. This property of the QCD plasma is parameterized by a jet quenching parameter $\hat{q}$. The energy loss ΔE of the jet is proportional to $\hat{q}$ and, quite strikingly, to the *square* of the distance L travelled in the plasma: $\Delta E \sim \hat{q} L^2$. The above indicates that $\hat{q}$ has units of energy over length-squared. In natural units ($\hbar = c = 1$) this is equivalent to energy-squared over length, and this is the way that $\hat{q}$ is traditionally presented. The experimental value of $\hat{q}$ is fairly uncertain since the QGP expands and cools and one only sees the effects of a time-varying $\hat{q}$. At time 1 fm the value of $\hat{q}$ appears to lie in the range $5 - 15\,$GeV^2/fm. The parameter $\hat{q}$ of QCD can be given a natural gauge theory definition whose value in strongly coupled, finite temperature, supersymmetric Yang–Mills can be calculated using the dual black hole background. This result can be estimated at time 1 fm, using a 't Hooft coupling $\lambda \sim 20$, and the result is $\hat{q} \sim 4\,$GeV^2/fm. Although the analytic result for super-Yang–Mills gives a $\hat{q}$ lower than the one for hot QCD, given the various theoretical and experimental uncertainties this level of coincidence is encouraging. It seems likely that the AdS/CFT will be a valuable tool to understand and perhaps predict properties of the QGP to be measured at the higher energies of the LHC.

Problems

Problem 23.1 Length of a rotating string.

Consider the normalized string state that represents an open string rotating on the (x^2, x^3) plane:

$$|\psi_N\rangle = \frac{1}{\sqrt{N!}} (\alpha_{-1})^N |p^+, \vec{0}\rangle. \tag{1}$$

As discussed above (23.17), we write $\langle p^+, \vec{0} | p^+, \vec{0} \rangle = 1$, instead of the correct but cumbersome delta function of zero argument. To estimate the length of the string state we will evaluate the expectation value of a length-squared operator L^2 defined by

$$L^2(\tau) \equiv\, : (\Delta X^2(\tau) \Delta X^2(\tau) + \Delta X^3(\tau) \Delta X^3(\tau)) : . \tag{2}$$

Here $\Delta X^I(\tau) = X^I(\tau, \pi) - X^I(\tau, 0)$ is the difference of X^I coordinates at the string endpoints and the colons denote normal ordering. Given a product of oscillators that includes both creation and annihilation operators, normal ordering places all annihilators to the right of the creators. Thus, for example, $: \alpha_1^i \alpha_{-1}^j := \alpha_{-1}^j \alpha_1^i$ and normal ordering does not affect $\alpha_{-1}^i \alpha_{-1}^j$ nor $\alpha_1^i \alpha_1^j$. Normal ordering is needed in (2) because otherwise L^2 would have infinite expectation value even on the ground states. It is useful to define string coordinates

$$X \equiv \frac{1}{\sqrt{2}}(X^2 + iX^3), \qquad \bar{X} \equiv \frac{1}{\sqrt{2}}(X^2 - iX^3).$$

Verify that $L^2(\tau) =\, : 2\Delta X \Delta \bar{X} :$ and use this expression to calculate the expectation value $\langle L^2 \rangle = \langle \psi_N | L^2 | \psi_N \rangle$. Show that for states with large N, $\sqrt{\langle L^2 \rangle}$ gives the expected classical value of the length of the string. To test further your result, verify that a rotating string of $M^2 = 3/\alpha'$ has a length $\sqrt{\langle L^2 \rangle}$ equal to eight times the string length. [Hint: the expanded out $:2\Delta X \Delta \bar{X}:$ contains four kinds of terms, each with a different structure of α and $\bar{\alpha}$ oscillators. Only one contributes to the expectation value.]

Problem 23.2 Coherent states for quantum oscillator.

In order to understand better (23.18) we examine analogous coherent states

$$|z\rangle \equiv e^{za^\dagger - z^* a} |0\rangle,$$

of the simple harmonic oscillator. Here z^* is the complex conjugate of the complex number z. For the oscillator we use the Hamiltonian $H = a^\dagger a$ with $a = \frac{1}{\sqrt{2}}(x + ip)$.

(a) Explain why $\langle z | z \rangle = 1$. Show that $|z\rangle = e^{-\frac{1}{2}|z|^2} e^{za^\dagger} |0\rangle$.
(b) Show that the expectation values of x and p in the state $|z\rangle$ are encoded in the real and imaginary parts of z via the relation $\frac{1}{\sqrt{2}}(\langle x \rangle + i\langle p \rangle) = z$.
(c) Calculate the time-dependent physical state $e^{-iHt} |z_0\rangle$, where z_0 is an arbitrary complex number. Find the corresponding time-dependent expectations values of x and p in terms of the constants x_0 and p_0 defined by $z_0 \equiv \frac{1}{\sqrt{2}}(x_0 + ip_0)$.
(d) Confirm that the time-dependent expectation values of x and p satisfy the classical equations of motion of the oscillator.

Problem 23.3 Isometric embedding of a portion of $\mathbb{H}_2$ in $\mathbb{R}^3$ using the *tractrix*.

The metric (23.77) on $\mathbb{H}_2$ can be neatly written using a complex variable $w = \xi^1 + i\xi^2$:

$$ds^2 = R^2 \frac{4(d\xi^1 d\xi^1 + d\xi^2 d\xi^2)}{(1 - (\xi^1 \xi^1 + \xi^2 \xi^2))^2} = R^2 \frac{4\,dw\,d\bar{w}}{(1 - w\bar{w})^2}. \tag{1}$$

The complex variable w is constrained to lie within the unit disk: $|w| < 1$.

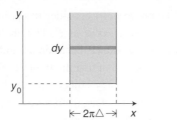 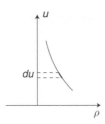

Fig. 23.15 Problem 23.3. Left: a portion of $\mathbb{H}_2$ to be isometrically embedded in $\mathbb{R}^3$. Right: The curve $u(\rho)$, rotated about the u axis, gives the embedded surface.

(a) It is more convenient to describe $\mathbb{H}_2$ using the upper-half plane (UHP), the part of the complex plane above the real line. If we use a complex coordinate $z = x + iy$ to describe the complex plane, the conformal map

$$z = \frac{1}{i}\frac{w-1}{w+1}, \tag{2}$$

takes the region $|w| < 1$ to the region $\Im(z) > 0$ of the complex z-plane. To convince yourself of this fact, show that the circle $w = e^{i\theta}$ is mapped by (2) to the full real line and give the resulting function $x(\theta)$. [Hint: the answer is a simple function of $\theta/2$.] Then verify that one point in the interior of the w-disk goes to a point above the real axis on the z-plane.

(b) Prove that in z coordinates the metric (1) takes the form

$$ds^2 = R^2 \frac{dz d\bar{z}}{(\Im(z))^2} = R^2 \frac{dx^2 + dy^2}{y^2}. \tag{3}$$

It is possible to embed isometrically in $\mathbb{R}^3$ a *piece* of the $\mathbb{H}_2$ surface. Consider Figure 23.15, where we show a semi-infinite vertical strip of coordinate width $\Delta x = 2\pi\,\Delta$ for some arbitrarily chosen positive number Δ. As we will see, the embedding of the strip works down to $y = y_0$, where $y_0 > 0$ depends on Δ. It is convenient to turn the strip into a cylinder by identifying the vertical edges. This identification is consistent with the metric, which is manifestly invariant under horizontal translations $x \to x + c, \; y \to y$. The embedded surface is the surface of revolution obtained by spinning the curve $u(\rho)$, shown to the right of the figure, about the u axis.

(c) Assume that the embedding is done with u increasing as y increases. It follows that ρ decreases with increasing u (why?). Consider the little circular strip of coordinate width dy shown to the left of the figure and the corresponding one of height du to the right of the figure, after spinning the curve. By equating the length and width of these pieces of surface derive the following differential equation for $u(\rho)$:

$$\frac{du}{d\rho} = -\frac{\sqrt{R^2 - \rho^2}}{\rho}. \tag{4}$$

Clearly, $\rho \leq R$. Find the relation between y_0 and Δ. The curve $u(\rho)$ is a *tractrix*.

(d) For any point P on the curve $u(\rho)$ consider the point P' on the u axis such that the segment PP' is tangent to the curve at P. Use the differential equation (4) (not its solution!) to show that the distance between P and P' is always equal to R. This property was used by Newton to define the tractrix. You can easily construct a tractrix as follows. Attach to a small heavy object a piece of string of length R and place the object on a table, a distance R away from a straight edge. If you drag the object by sliding the end of the string along the edge of the table, the object will move along the tractrix.

(e) Integrate (4) using $u(R) = 0$ and give the explicit form of $u(\rho)$. If you get the correct answer, $\rho = \frac{3}{5}R$ should give $u = R(-\frac{4}{5} + \ln 3)$. Make a plot of $u(\rho)$.

Problem 23.4 A conformal view of two-dimensional Minkowski space.

The metric of two-dimensional Minkowski space is $-ds^2 = -dt^2 + dx^2$, where the coordinates (x, t) take all possible real values: $-\infty < t, x < \infty$. Consider new coordinates (x', t') defined by the two relations

$$\tan(x' \pm t') = x \pm t.$$

(a) Show that in the new coordinates the metric takes the form

$$-ds^2 = \Omega^2(t', x')(-dt'^2 + dx'^2),$$

and determine the function Ω^2. Show that the full (x, t) plane is mapped to a finite region of the (x', t') plane. Describe this region and sketch it.

(b) Consider light-rays that depart from $x = 0$ at times $t = -1, 0$, and 1, and move towards x positive. Sketch the three light-rays in the (x', t') plane.

(c) Consider the timelike line $x = \alpha t$ with α a positive constant that satisfies $\alpha < 1$ and $t \in [0, \infty)$. Show that in the (x', t') plane the corresponding trajectory starts at the origin and eventually approaches the point $(x', t') = (0, \pi/2)$. Moreover, near that point, the trajectory follows the equation $t' = \frac{\pi}{2} - \frac{x'}{\alpha}$. Discuss the corresponding results for spacelike lines $x = \alpha t$ with $\alpha > 1$ and $t \in [0, \infty)$. You may find the following large u expansions useful:

$$\tan^{-1} u \simeq \pm \frac{\pi}{2} - \frac{1}{u}, \quad u \to \pm\infty.$$

Problem 23.5 Another view of the full AdS_{n+1} space.

(a) Consider the ambient metric and constraint (23.90) and now set $z = R \cosh \rho$, where $\rho \in [0, \infty)$ is a new coordinate. Moreover, solve the constraint by letting

$$x^i = R\,\Omega^i \sinh \rho \quad \text{with} \quad \Omega^i \Omega^i = 1. \tag{1}$$

The constrained variables Ω^i are coordinates on the unit sphere S^{n-1}. Note that $\Omega^i \Omega^i = 1$ requires the differentials to satisfy $\Omega^i d\Omega^i = 0$. Show that the metric becomes

$$ds^2 = R^2 \left[-(\cosh \rho)^2 (dt)^2 + (d\rho)^2 + (\sinh \rho)^2 d\Omega_{n-1}^2 \right], \tag{2}$$

where $d\Omega_{n-1}^2 \equiv d\Omega^i d\Omega^i$, given the constraint $\Omega^i\Omega^i = 1$, is the metric on the unit sphere S^{n-1}. Equation (2) is a useful form of the AdS_{n+1} metric.

(b) Consider the n-dimensional spatial region $\rho \leq \bar{\rho}$. Write an integral that gives its volume $V(\bar{\rho})$ and an expression for the area $A(\bar{\rho})$ of its boundary. Show that

$$\lim_{\rho\to\infty} \frac{A(\bar{\rho})}{V(\bar{\rho})} = \frac{c_n}{R}, \tag{3}$$

where the constant c_n that you will determine is consistent with (23.99). [Hint: the leading divergent term in the integral is easy to compute, the rest does not contribute in the limit.]

(c) In (2) introduce a new radial variable $r \in [0, \infty)$ by the relation $r = R \sinh \rho$. Also let $t \to t/R$. Show that the metric becomes

$$ds^2 = -\left(1 + \frac{r^2}{R^2}\right)(dt)^2 + \left(1 + \frac{r^2}{R^2}\right)^{-1}(dr)^2 + r^2 d\Omega_{n-1}^2. \tag{4}$$

This is another useful form of the AdS_{n+1} metric. The simple modification indicated in (23.104) gives the metric of the Schwarzschild black hole with asymptotically AdS geometry.

Problem 23.6 A partial but simple view of AdS_{n+1} space.

Consider the ambient metric and constraint (23.84) written as

$$\text{ambient metric:} \quad ds^2 = -(du)^2 + \sum_{i=1}^{n-1} dx^i dx^i + (dx^n)^2 \quad (dv)^2,$$

$$\text{constraint:} \quad -u^2 + \sum_{i=1}^{n-1} x^i x^i + (x^n)^2 - v^2 = -R^2.$$

(a) We now introduce coordinates z, t, and $\vec{y} = (y^1, \ldots, y^{n-1})$ by the relations:

$$v + x^n = \frac{R}{z}, \quad u = \frac{R}{z}t, \quad x^i = \frac{R}{z}y^i, \ i = 1, \ldots, n-1. \tag{1}$$

Here $t \in \mathbb{R}$, $\vec{y} \in \mathbb{R}^{n-1}$, and $z \in \mathbb{R}^+$. Since z is positive, so is $x^n + v$ and this presentation will not display the part of AdS space where $x^n + v$ is negative. Use the constraint to determine the value of $v - x^n$ in terms of the new coordinates.

(b) Show that the metric on the AdS_{n+1} space takes the remarkably simple form

$$ds^2 = \frac{R^2}{z^2}\left(dz^2 - (dt)^2 + (d\vec{y})^2\right). \tag{2}$$

[Hint: for the evaluation use $(dx^n)^2 - (dv)^2 = d(x^n + v)d(x^n - v)$.] The boundary of the space is at $z = 0$, where the metric is conformal to $-(dt)^2 + (d\vec{y})^2$, the n-dimensional Minkowski metric. A related form is obtained letting $z = R^2/r$:

$$ds^2 = R^2 \frac{dr^2}{r^2} + \frac{r^2}{R^2}\left[-(dt)^2 + (d\vec{y})^2\right]. \tag{3}$$

Problem 23.7 De Sitter spacetime.

The n-dimensional de Sitter spacetime dS_n has positive curvature. Its definition is a variant of that for hyperbolic spaces: one can obtain de Sitter spacetime as a surface in a Minkowski space of one higher dimension. This time, however, the normal to the surface is spacelike so that the surface contains a time direction. Moreover, just like hyperbolic spaces can describe spatial sections of AdS, spheres can be used to describe the spatial sections of dS.

With $i = 1, 2, \ldots, n$, the dS_n spacetime and its metric are defined by

$$\text{ambient metric:} \quad ds^2 = -(dz)^2 + dx^i dx^i \,,$$
$$\text{constraint:} \quad -z^2 + x^i x^i = R^2 \,. \tag{1}$$

(a) Explain in detail why the surface contains a timelike direction. Exhibit a timeline tangent at all points on the surface.
(b) To solve the constraint we set

$$z = R \sinh t \,, \quad x^i = R\,\Omega^i \cosh t \,, \quad \text{with} \quad \Omega^i \Omega^i = 1 \,. \tag{2}$$

See Problem 23.5 for comments on the constrained variables Ω^i. Show that the metric on the dS_n space takes the form

$$ds^2 = R^2 \big[-(dt)^2 + (\cosh t)^2 d\Omega_{n-1}^2 \big] \,, \tag{3}$$

where $d\Omega_{n-1}^2 \equiv d\Omega^i d\Omega^i$ is the metric on the unit sphere S^{n-1}. At any fixed time the spatial section is a sphere of radius $R \cosh t$. As time increases, the spatial section contracts while $t < 0$ and expands while $t > 0$.

(c) We noted in Section 23.8 that the mass-shell for a massive particle is hyperbolic space. What is the mass-shell for a tachyon?

Problem 23.8 Viscosity and entropy of a dilute gas.

Consider a container of volume V with N molecules of a gas. The gas molecules have mass m, an average velocity $\bar{v}$, and their mean free path is ℓ. Assuming that the gas is dilute, the viscosity of the gas is given roughly by

$$\eta = \frac{1}{3} \frac{N}{V} \bar{v} m \ell \,. \tag{1}$$

For a monoatomic ideal gas the entropy density s given by

$$\frac{s}{k} = \frac{1}{k} \frac{S}{V} = \frac{N}{V} \left[\ln\left(\frac{V/N}{\lambda_{\text{th}}^3} \right) + \frac{5}{2} \right] \,, \quad \lambda_{\text{th}} = \frac{h}{\sqrt{2\pi mkT}} \,. \tag{2}$$

Here λ_{th} is the thermal de Broglie wavelength. You can find the derivation of the above results for η and s in many textbooks on thermal physics.

(a) Use (2) to calculate the value of s/k per unit cm^3 for monoatomic hydrogen at $T = 25\,°C$ and 1 atm. Compare with the molar entropy of $114.7\,J/K$ quoted by chemical tables.

(b) Consult a textbook to find the Bose–Einstein condensation temperature T_B. Show that at this temperature the factor in brackets in (2) takes the approximate value 1.5397. So, as long as we have a gas, $s/k > (1.5397) \frac{N}{V}$. Show that for temperatures below one-thousand times T_B, we have $s/k < 11.901 \frac{N}{V}$.

(c) Estimate the mean free path ℓ in terms of N, V, and the molecular diameter d. Recalling that $\bar{v} \sim \sqrt{kT/m}$ conclude that $\eta \sim T^{1/2}$ and is independent of the pressure or the density. For a gas the viscosity increases with the temperature. For a liquid the viscosity decreases with temperature.

(d) In a gas $(\bar{v})^2 \simeq \langle v^2 \rangle$. Use this to show that $\eta = \frac{2}{3} \frac{N}{V} u \tau$, where u is the average energy of a molecule and τ is the mean free time. Conclude that in the range of temperatures considered in (b) one has

$$\frac{\eta}{s} \geq \frac{\eta}{s_{max}} \simeq 0.056 \frac{1}{k} u \tau . \tag{3}$$

The energy/time form of the uncertainty principle suggests that $u\tau \geq \hbar$, leading to η/s bounded below by some fraction of $\hbar/k$ (compare with (23.123)).

24 Covariant string quantization

In the Lorentz covariant quantization of string theory we treat all string coordinates $X^\mu(\tau, \sigma)$ on the same footing. To select physical states we use the constraints generated by a subset of the Virasoro operators. The states automatically carry time labels, so the Hamiltonian does not generate time evolution. We describe the Polyakov string action and show that it is classically equivalent to the Nambu–Goto action.

24.1 Introduction

In this book, the quantization of strings was carried out using light-cone coordinates and the light-cone gauge. String theory is a Lorentz invariant theory, but Lorentz symmetry is not manifest in the light-cone quantum theory. Indeed, the choice of a particular coordinate X^+ for special treatment hides from plain view the Lorentz symmetry of the theory. While hidden, the Lorentz symmetry is still a symmetry of the quantum theory, as we demonstrated by the construction of the Lorentz generator M^{-I}. This generator has the expected properties when the spacetime has the critical dimension.

Since Lorentz symmetry is of central importance, it is natural to ask if we can quantize strings preserving *manifest* Lorentz invariance. It is indeed possible to do so. The Lorentz covariant quantization has some advantages over the light-cone quantization. Our light-cone quantization of open strings did not apply to D0-branes because the light-cone gauge requires that at least one spatial open string coordinate has Neumann boundary conditions. Covariant quantization applies to D0-branes. The equations of motion for the fields that arise in string theory are better understood in Lorentz covariant notation. The calculation of tachyon potentials, alluded to in Section 12.8, appears to be possible only within the Lorentz covariant quantization of strings.

Why then, have we waited so long to discuss the Lorentz covariant quantization of strings? The covariant approach is very elegant but it is sometimes hard to extract the physical content from its equations. Moreover, covariant quantization has a series of features that are quite strange. We are accustomed to the idea that in quantum mechanics the position of a particle becomes an operator while time remains a parameter. In a Lorentz invariant quantization, all the coordinates x^μ of a particle, including x^0, become operators. Similar remarks apply to the string coordinates X^μ. We will also see that the string Hamiltonian annihilates the physical states of the theory. Finally, in covariant quantization it is necessary to discuss states whose norm is not positive; this takes us out of the usual Hilbert space postulates.

We chose to do the quantization of strings in the light-cone gauge because all of the above features would have distracted us from the task of extracting the physical content of the theory. The light-cone gauge has served us well, and it will continue to do so in the following chapters, where we use light-cone string diagrams to begin our discussion of string interactions. The proper treatment of covariant quantization requires tools that go beyond the level of this book. We will not be able to derive the critical dimension, for example. While our treatment will not be complete, we will still gain some important insights into the structure of the theory.

Let us begin our discussion by recalling some facts about parameterizations of the world-sheet. In Chapter 9 we described a large class of gauges characterized by a vector n^μ. By choosing n^μ appropriately we could produce the static or light-cone gauge. For open strings the choice of n^μ completely fixes the parameterization of the world-sheet. This is almost true for closed strings as well, except that we are still free to shift the coordinate σ rigidly along the strings. We showed that for any choice of n^μ the string coordinates satisfy the constraints

$$(\dot{X} \pm X')^2 = 0 \,. \tag{24.1}$$

Since *any* choice of n^μ results in these constraints, the constraints by themselves do not completely fix the parameterization of the world-sheet. In fact, as you may have seen in Problem 12.10, many reparameterizations preserve these constraints.

We were able to show that the constraints (24.1) cause the equations of motion to become simple wave equations:

$$\ddot{X}^\mu - X^{\mu\prime\prime} = 0 \,. \tag{24.2}$$

Furthermore, they imply that the momentum densities are given by

$$\mathcal{P}^{\sigma\mu} = -\frac{1}{2\pi\alpha'} X'^\mu \,, \quad \mathcal{P}^{\tau\mu} = \frac{1}{2\pi\alpha'} \dot{X}^\mu \,. \tag{24.3}$$

In the covariant formalism we use the constraints (24.1), and thus also (24.2) and (24.3), but we do *not* fix completely the parameterization of the world-sheet. The constraints (24.1) can be thought of as conditions for a *partial* gauge fixing. At the classical level, we solve the wave equations and check that the constraints are satisfied. At the quantum level, the constraints introduce subtle complications.

As you probably recall, we had to do plenty of work to get wave equations and simple momentum densities from the Nambu–Goto action. Is there an action for which these results follow quickly? The answer is yes. In fact, for transverse light-cone coordinates, such an action was given in (12.81). We used this action to give a more physical derivation of the oscillator commutation relations. In the present case, the desired action is

$$S = \int d\tau d\sigma \mathcal{L} = \frac{1}{4\pi\alpha'} \int d\tau d\sigma \left(\partial_\tau X^\mu \partial_\tau X_\mu - \partial_\sigma X^\mu \partial_\sigma X_\mu \right) \,. \tag{24.4}$$

Note that this action, as opposed to the Nambu–Goto action, contains no square root. It is quadratic in the dynamical variables X^μ. This action is useful because variation of X^μ

immediately produces the wave equations (24.2). Moreover, the simple expressions for the canonical momentum densities arise directly. For example,

$$\mathcal{P}^\tau_\mu = \frac{\partial \mathcal{L}}{\partial \dot{X}^\mu} = \frac{1}{2\pi\alpha'}\dot{X}_\mu .$$

(24.5)

Even the classical Hamiltonian is easy to calculate. In terms of coordinates and momenta, and dropping the τ superscript from $\mathcal{P}^\tau_\mu$, we find

$$
\begin{aligned}
H &= \int d\sigma\,(\mathcal{P}_\mu \dot{X}^\mu - \mathcal{L}) \\
&= \int d\sigma\left(\mathcal{P}_\mu \dot{X}^\mu - \frac{1}{4\pi\alpha'}\left((2\pi\alpha'\mathcal{P})^2 - X'^2\right)\right) \\
&= \pi\alpha' \int d\sigma\left(\mathcal{P}\cdot\mathcal{P} + \frac{X'\cdot X'}{(2\pi\alpha')^2}\right).
\end{aligned}
$$

(24.6)

This Hamiltonian is analogous to the light-cone Hamiltonian (12.15), but here the index contractions run over all spacetime dimensions. In the light-cone quantization we defined Heisenberg operators $X^I(\tau,\sigma)$ and $\mathcal{P}_I(\tau,\sigma)$ which had canonical commutation relations. Using these we verified that the Hamiltonian generates the correct operator equations of motion. In the covariant theory it is natural to introduce Heisenberg operators

$$X^\mu(\tau,\sigma) \quad \text{and} \quad \mathcal{P}_\mu(\tau,\sigma)$$

(24.7)

and to postulate the commutation relations

$$\left[X^\mu(\tau,\sigma),\,\mathcal{P}^\nu(\tau,\sigma')\right] = i\,\eta^{\mu\nu}\delta(\sigma-\sigma').$$

(24.8)

Note that even X^0 is a quantum operator. As usual, we set to zero the commutator of coordinates with coordinates and the commutator of momenta with momenta. Computations similar to those we performed using the light-cone gauge show that the quantum equations of motion take the form (24.2). This is strong evidence that H is the correct Hamiltonian.

24.2 Open string Virasoro operators

In light-cone quantization the constraints were used to solve for X^- in terms of the transverse coordinates X^I. The modes of the X^- coordinate were identified as transverse Virasoro operators. In the covariant approach the quantum constraints are not solved; rather, they are imposed on the states of the theory. We now examine the quantum constraints for the open string.

The quantization of the generic open string coordinate X^μ is similar to the quantization of the transverse light-cone coordinate X^I. Recall the oscillator expansion (9.56):

$$X^\mu(\tau,\sigma) = x_0^\mu + \sqrt{2\alpha'}\,\alpha_0^\mu\tau + i\sqrt{2\alpha'}\sum_{n\neq 0}\frac{1}{n}\alpha_n^\mu e^{-in\tau}\cos n\sigma .$$

(24.9)

This expansion led to particularly simple expressions for the linear combinations of derivatives in (9.59):

$$\dot{X}^{\mu} \pm X^{\mu\prime} = \sqrt{2\alpha'} \sum_{n \in \mathbb{Z}} \alpha_n^{\mu} \, e^{-in(\tau \pm \sigma)} \, . \tag{24.10}$$

This time there are no gauge conditions that we can use to simplify particular coordinates. The above expansions are to be used for all the string coordinates. The commutation relations (24.8), the expansions (24.10), and equation (24.5) together determine the oscillator commutation relations. Since the computations are analogous to the ones we performed in the light-cone gauge, the earlier results (12.45) and (12.64) are modified only by the replacement of η^{IJ} with $\eta^{\mu\nu}$:

$$[\, \alpha_m^{\mu} , \alpha_n^{\nu} \,] = m \, \eta^{\mu\nu} \delta_{m+n,0} \, , \quad [\, a_m^{\mu} , a_n^{\nu\dagger} \,] = \delta_{m,n} \, \eta^{\mu\nu} . \tag{24.11}$$

This time $\alpha_0^{\mu} = \sqrt{2\alpha'} \, p^{\mu}$, and the zero mode operators satisfy

$$[\, x_0^{\mu} , \, p^{\nu} \,] = i\eta^{\mu\nu} \, . \tag{24.12}$$

This equation should give us pause. We are familiar with the quantum mechanical procedure by which the spatial coordinates and the spatial momenta of a particle are turned into operators. In that scheme the spatial dependence of states is built-in, and the time dependence is determined by the Schrödinger equation. In covariant quantization even the time coordinate x^0 is made into an operator. So the states have a built-in time dependence. This means that the role of the Schrödinger equation must change. We will discuss this matter in detail in Section 24.4.

We can now explore the constraints (24.1) explicitly. Again, the computations are analogous to those of Chapter 9. By comparing with (9.79) we find

$$(\dot{X} \pm X')^2 = 4\alpha' \sum_{n \in \mathbb{Z}} L_n \, e^{-in(\tau \pm \sigma)} \, , \quad L_n = \frac{1}{2} \sum_{p \in \mathbb{Z}} \alpha_{n-p}^{\mu} \, \alpha_{p,\mu} \, . \tag{24.13}$$

The covariant Virasoro operators L_n differ from the transverse Virasoro operators $L_n^{\perp}$ because they include contributions from all the string coordinates. As before, the only Virasoro operator which has an ambiguous ordering is L_0. Again, we define L_0 to be the normal-ordered operator without any additional constant.

At the classical level, the constraint equations $(\dot{X} \pm X')^2 = 0$ require

$$\text{classically:} \quad L_n = 0 \, , \quad n \in \mathbb{Z} \, . \tag{24.14}$$

Classically, we have $L_n^* = L_{-n}$, so that the constraints in (24.14) need only be checked for $n \geq 0$. At the quantum level we have $L_n^{\dagger} = L_{-n}$. The analogous property for the transverse Virasoro operators was proven below (12.112).

Quick calculation 24.1 Verify that, classically, $L_n^* = L_{-n}$.

The quantum Virasoro operators have rather nontrivial commutation relations. We calculated these commutators for the transverse Virasoro operators, and the result was given in (12.133). The central term was proportional to $D - 2$, the number of transverse coordinates of string theory. Each transverse coordinate contributed the same amount to the central term. In the covariant treatment, where the Virasoro operators involve a sum over the full set of spacetime coordinates, that result becomes

$$[L_m, L_n] = (m - n)L_{m+n} + \frac{D}{12}(m^3 - m)\delta_{m+n,0}, \qquad (24.15)$$

where $D = 26$ is the full dimensionality of spacetime (we will not derive the critical dimension here). You may be wondering why the time coordinate, whose oscillators commute to give a minus sign (see (24.11)), contributes to the central term in the same way as the space coordinates. It does so because the central term arises from a commutator which involves four oscillators, so the basic commutator is used twice and the sign cancels out.

In (12.139) we established that the action of the transverse Virasoro operators on the string coordinates X^I describes the effects of a certain class of world-sheet reparameterizations. Exactly the same results hold for the action of the covariant Virasoro operators on the string coordinates X^μ. The reparameterizations have a special property: they preserve the constraints $(\dot{X} \pm X')^2 = 0$, as shown in Problem 12.10. This result gives a physical interpretation to the covariant Virasoro operators. These operators generate world-sheet reparameterizations that preserve the gauge conditions of the Lorentz-covariant formalism.

24.3 Selecting the quantum constraints

The intuition inspired by the classical theory suggests that the physical states of the quantum theory – the quantum analog of consistent classical string motions – should be annihilated by all the Virasoro operators. But we will see that no states would survive the imposition of all these Virasoro constraints.

It is absolutely clear, however, that some constraints must be imposed. In covariant quantization we have 26 sets of oscillators available, so if we did not impose any constraint the set of quantum states would be different from the set of quantum states we obtained earlier, when we had only 24 sets of oscillators. Different quantization procedures should not give different physical results. It is therefore reasonable to expect that at least some Virasoro operators must annihilate the physical states.

Let us start by exploring the constraint that the Virasoro operator L_0 would impose on states. We have *defined* the quantum L_0 as normal-ordered with no additional constant, but there is no reason why this L_0 should annihilate physical states. Rather, we expect that $(L_0 + a)$, where a is some constant, should annihilate physical states. We investigated a similar issue in the light-cone gauge. The light-cone energy p^- turned out to be proportional to $L_0^\perp - 1$ rather than proportional to $L_0^\perp$ (see (12.157)). The same constant is needed in the covariant formalism, so the quantum constraint is

$$(L_0 - 1)|\Phi\rangle = 0, \qquad (24.16)$$

for any physical state $|\Phi\rangle$. While we will not derive the ordering constant from first principles, we will see below that it is necessary for the covariant spectrum to agree with the one we obtained using the light-cone gauge. In the light-cone quantization, the relation between p^- and $L_0^\perp$ fixed the mass-squared of the states. The mass-squared is determined in the covariant treatment by the constraint (24.16). Writing the L_0 operator explicitly, we have

$$L_0 - 1 = \frac{1}{2}\alpha_0^\mu \alpha_{0,\mu} + \sum_{p=1}^{\infty} \alpha_{-p}^\mu \alpha_{p,\mu} - 1 = \alpha' p^2 - 1 + \sum_{n=1}^{\infty} n \, a_n^{\mu\dagger} a_{n,\mu} = 0 \,. \quad (24.17)$$

Since $M^2 = -p^2$, we find

$$M^2 = \frac{1}{\alpha'}\left(-1 + N\right), \quad N = \sum_{n=1}^{\infty} n \, a_n^{\mu\dagger} a_{n,\mu} \,. \quad (24.18)$$

This result takes the same form as the light-cone result, except that the number operator $N^\perp$ is replaced by its covariant counterpart N. The eigenvalues of the light-cone number operator $N^\perp$ were manifestly non-negative, and this allowed us to conclude that $M^2 \geq -1/\alpha'$. Are the eigenvalues of the number operator N also non-negative? To find out, we expand out the sum over μ in the formula for N. With due attention to the Minkowski metric, we find

$$N = \sum_{n=1}^{\infty} n \left(-a_{n,0}^\dagger a_{n,0} + \sum_{i=1}^{25} a_{n,i}^\dagger a_{n,i}\right) \,. \quad (24.19)$$

In the *classical* theory the dagger represents complex conjugation, so the number operator N could be negative with appropriate contributions from the time components. In the quantum theory the situation improves. Because the commutator of two timelike oscillators carries a negative sign ($[a_{n,0}, a_{n,0}^\dagger] = -1$), we have

$$[-n \, a_{n,0}^\dagger a_{n,0} \,, \, a_{n,0}^\dagger] = +n \, a_{n,0}^\dagger \,. \quad (24.20)$$

So even the timelike oscillators contribute positively to N. Thus N is non-negative, and $M^2 \geq -1/\alpha'$ continues to hold in covariant quantization. Even more, ground states in the covariant theory have $N = 0$, and therefore $M^2 = -1/\alpha'$. This value agrees with the mass-squared of the ground states that we obtained using the light-cone quantization. The agreement justifies our selection of the ordering constant in (24.16).

There is a problem, however, which indicates that additional constraints must be imposed: many states do not have positive norm! Consider an eigenstate $|\Phi\rangle$ of N with eigenvalue N_0 and positive norm $\langle\Phi|\Phi\rangle > 0$. We then have $a_{n,0}|\Phi\rangle = 0$ for $n > N_0$. Now consider the state $|\chi\rangle = a_{n,0}^\dagger|\Phi\rangle$, with Hermitian conjugate $\langle\chi| = \langle\Phi|a_{n,0}$. The norm of this state is

$$\langle\chi|\chi\rangle = \langle\Phi|a_{n,0} a_{n,0}^\dagger|\Phi\rangle = \langle\Phi|[a_{n,0}, a_{n,0}^\dagger]|\Phi\rangle = -\langle\Phi|\Phi\rangle < 0 \,. \quad (24.21)$$

Since $|\chi\rangle$ has negative norm, it is an unacceptable quantum state. We must impose extra conditions to remove such physically impossible states from the spectrum. Recall that in the light-cone gauge quantization all states have positive norm.

Let us therefore attempt to establish the full set of constraints. We cannot require that, in addition to $L_0 - 1$, all other Virasoro operators annihilate the physical states. We show this by considering a simple example. We will prove that there are no nontrivial states annihilated by the three operators $L_0 - 1$, L_2 and L_{-2}. Consider therefore a general state $|\Phi\rangle$, and impose the conditions

$$(L_0 - 1)|\Phi\rangle = 0\,, \quad L_2|\Phi\rangle = 0\,, \quad \text{and} \quad L_{-2}|\Phi\rangle = 0\,.$$

If these equations hold, then the commutator of L_2 with L_{-2} must annihilate the state:

$$[L_2, L_{-2}]|\Phi\rangle = L_2(L_{-2}|\Phi\rangle) - L_{-2}(L_2|\Phi\rangle) = 0\,. \tag{24.22}$$

On the other hand, equation (24.15) tells us that

$$[L_2, L_{-2}] = 4L_0 + \frac{D}{2}\,. \tag{24.23}$$

The left-hand side has been shown to annihilate $|\Phi\rangle$, so the right-hand side must annihilate $|\Phi\rangle$, as well,

$$\left(4(L_0 - 1) + 4 + \frac{D}{2}\right)|\Phi\rangle = \left(4 + \frac{D}{2}\right)|\Phi\rangle = 0\,. \tag{24.24}$$

Since D is positive, we find $|\Phi\rangle = 0$. This confirms that we cannot impose all the Virasoro conditions and expect to find nontrivial states.

Bearing this result in mind, we attempt to find a subset of the Virasoro constraints that can be imposed without setting to zero the states. The fact that a subset of the original set of constraints may suffice to define a consistent quantum theory is not obvious, but past experience suggests that it may. In the covariant quantization of electromagnetism, for example, the Lorentz gauge condition $\partial \cdot A = 0$ appears as a quantum constraint. This condition cannot be fully imposed on photon states. Instead, only "half" of the gauge condition needs to be imposed on states in order for the theory to work properly. We will follow a similar strategy here.

The complete set of operators that might annihilate physical states is given by

$$\{\ldots, L_{-3}, L_{-2}, L_{-1}\,, \; L_0 - 1\,, \; L_1, L_2, L_3, \ldots\}\,. \tag{24.25}$$

We have already seen that the constraint $L_0 - 1 = 0$ is necessary to fix the mass spectrum correctly. The simplest thing to try is to set to zero either all operators which have positive mode number or all operators which have negative mode number. In other words, we will impose the constraint

$$(L_0 - 1)|\Phi\rangle = 0 \tag{24.26}$$

together with *one* of the following:

$$L_n|\Phi\rangle \overset{?}{=} 0, \quad n > 0, \tag{24.27}$$

$$L_{-n}|\Phi\rangle \overset{?}{=} 0, \quad n > 0. \tag{24.28}$$

Only one of these choices works, even though they both have something in common. Both choices result in

$$\langle\Phi|L_n|\Phi\rangle = 0, \quad n \neq 0. \tag{24.29}$$

For instance, if we choose (24.27), then (24.29) holds for positive n because L_n kills the state on its right. Furthermore, the Hermitian conjugate of (24.27) yields

$$\langle\Phi|L_{-n} = 0, \quad \text{for } n > 0, \tag{24.30}$$

or, equivalently,

$$\langle\Phi|L_n = 0, \quad \text{for } n < 0, \tag{24.31}$$

showing that (24.29) holds for negative n as well. A similar argument can be used for the choice (24.28). So both choices result in physical states for which the *expectation values* of all the L_n with $n \neq 0$ are zero. The condition of vanishing expectation values is weaker than the condition that requires that the L_n annihilate the states. Happily, imposing only half of the constraints does provide a satisfactory theory. It has been proven that half of the constraints suffices to remove the states with negative norms.

We still have to decide which set of conditions to impose. When we were working in light-cone gauge we saw that the positively moded oscillators were the ones that functioned as annihilation operators. The negatively moded oscillators functioned as creation operators. It is natural to associate annihilation with the positively moded Virasoro operators. Therefore we declare the following.

> All positively moded Virasoro operators must annihilate physical states. (24.32)

States annihilated by all the positively moded Virasoro operators are called Virasoro *primaries*. We have learned that physical states are Virasoro primaries. Being a Virasoro primary is necessary but not sufficient to guarantee that a state is truly physical. Two additional conditions are necessary. The first is already familiar: the state must be annihilated by $L_0 - 1$. We will call the states that are primary and satisfy this first condition *admissible* states:

> $$|\Phi\rangle \text{ is admissible} \iff (L_n - \delta_{n,0})|\Phi\rangle = 0, \quad n \geq 0. \tag{24.33}$$

The second condition, as we shall see below, has to do with a class of states that are called Virasoro descendents. If a state is admissible and satisfies this additional condition it will be called *physical*. This terminology is not standard: our admissible states are usually called physical, and our physical states are usually called *real* physical states! (Here real is synonymous of truly.)

It is interesting to note that for any fixed state all but a finite number of positively moded Virasoro operators automatically annihilate the state without imposing any condition. Any state $|\Phi_0\rangle$ with number eigenvalue $N_0 \geq 0$ automatically satisfies

$$L_n|\Phi_0\rangle = 0, \quad \text{for } n > N_0. \tag{24.34}$$

To see this, first note that $L_0 = \alpha' p^2 + N$ implies

$$[N, L_n] = [L_0 - \alpha' p^2, L_n] = [L_0, L_n] = -nL_n. \tag{24.35}$$

The mode number of the state $L_n|\Phi_0\rangle$ is therefore

$$N L_n|\Phi_0\rangle = [N, L_n]|\Phi_0\rangle + L_n N|\Phi_0\rangle = (N_0 - n)L_n|\Phi_0\rangle. \tag{24.36}$$

If $n > N_0$, then $L_n|\Phi_0\rangle$ is a state of *negative* number eigenvalue. Since N only has non-negative eigenvalues, the state must vanish.

A Virasoro *descendent* of a given primary is a state that can be written as a finite linear combination of products of negatively moded Virasoro operators, acting on the primary state. For example, if $|p\rangle$ denotes a primary state, the state $L_{-1}|p\rangle$ is a descendent of $|p\rangle$, and so is $(L_{-2}L_{-1} + L_{-4}L_{-3})|p\rangle$. Since descendents play an important role, let us discuss them in some detail.

It follows from (24.36) that $N L_{-1}|p\rangle = (N_p + 1)L_{-1}|p\rangle$, where N_p is the number eigenvalue of the primary $|p\rangle$. The state $L_{-1}|p\rangle$ is, up to scale, the unique descendent of $|p\rangle$ with number $N_p + 1$. There are two basis descendents with number $N_p + 2$: $L_{-2}|p\rangle$ and $L_{-1}L_{-1}|p\rangle$. For descendents with number $N_p + 3$ the counting gets a bit more interesting. A list of candidate basis descendents is

$$L_{-3}|p\rangle, \quad L_{-2}L_{-1}|p\rangle, \quad L_{-1}L_{-2}|p\rangle, \quad (L_{-1})^3|p\rangle.$$

Because the Virasoro operators do not commute, the second and third states are not identical. There is, however, one linear relation among the above states, since

$$L_{-1}L_{-2} = [L_{-1}, L_{-2}] + L_{-2}L_{-1} = L_{-3} + L_{-2}L_{-1}.$$

This identity allows us to rewrite the third state in terms of the first two, so there are only three basis descendents at this level:

$$\text{descendents with number } N_p + 3: \quad L_{-3}|p\rangle, \quad L_{-2}L_{-1}|p\rangle, \quad (L_{-1})^3|p\rangle. \tag{24.37}$$

In general, for any fixed number $N_p + n$, one can choose a basis set of descendents such that each basis element is of the form:

$$L_{-n_1}L_{-n_2}\ldots L_{-n_k}|p\rangle, \quad \text{where} \quad n_1 \geq n_2 \geq \cdots \geq n_k \quad \text{and} \quad \sum_{i=1}^{k} n_i = n. \tag{24.38}$$

This is a useful conventional ordering of the Virasoro operators.

Quick calculation 24.2 Convince yourself that any descendent of $|p\rangle$ with number $N_p + n$ that is written as an arbitrary sequence of negatively moded Virasoro operators acting on $|p\rangle$ can be written as a linear superposition of states of the form (24.38). Note that the number of elements in the generating set (24.38) is equal to the number of partitions of n.

For any given primary $|p\rangle$, there may be linear relations between the basis descendents (24.38). These relations do not arise by manipulation of the Virasoro operators but rather from the specific properties of the state $|p\rangle$. There are simple examples of this phenomenon. The zero-momentum ground state $|0\rangle$, for example, is a primary whose descendent $L_{-1}|0\rangle$ vanishes identically (Problem 24.2). An important property of descendents is that they are all orthogonal to any primary. Indeed, any basis descendent $|d\rangle$ can be written as $|d\rangle = L_{-n_i}|\chi\rangle$ for some $n_i > 0$ and some state $|\chi\rangle$. It follows that for any primary $|p\rangle$,

$$\langle d \mid p \rangle = \langle \chi | L_{n_i} | p \rangle = 0, \tag{24.39}$$

since $|p\rangle$ is annihilated by all positively moded Virasoro operators.

We now suggest that a state that is both a primary and a descendent represents a state that is pure gauge. A state that is both a primary and a descendent is called a *null* state. It follows from (24.39) that a null state has zero inner product with itself, with any primary, and with any descendent. If we alter a primary state by the addition of a null state, then the new primary state has the same inner products with primary states as the original one. A null state behaves as a pure gauge if, in addition, physical operators in the theory map null states to null states; in this case the addition of null states to primary states cannot affect any physical expectation values. This motivates the following definition of a physical state.

> A nonvanishing state is said to represent a physical state if it is admissible and it is not a descendent, i.e., the state must be a primary, it must be annihilated by $(L_0 - 1)$, and it must not be a descendent. Two representatives of the same physical state must differ by a null state.

Note that we speak about representatives of physical states precisely because of the ambiguity that null states create. A physical state is *not* best thought of as one specific vector in the state space but rather as a class of vectors, all of which differ from one another by a null state. Any vector in this class is an equally valid representative of the *same* physical state. In order to give evidence that this definition is a good one, we will apply it in the following section to find the physical states for the two lowest-mass levels of the open string. We will recover the results previously derived in the light-cone gauge.

24.4 Lorentz covariant state space

Before constructing the state space of the covariantly quantized string, it is useful to recall some properties of the light-cone state space. The light-cone time-independent states are labeled by the light-cone momenta p^+ and $\vec{p}_T$. The ground states, for example, are written as $|p^+, \vec{p}_T\rangle$. Time-dependent states satisfy the Schrödinger equation which uses the light-cone Hamiltonian, as we saw in Section 11.4 and Section 12.7.

In the covariant formalism all components p^μ of the momentum are independent commuting operators, and we can label the ground states using the full momentum vector $|p\rangle = |p^0, p^1, \ldots, p^{25}\rangle$. The conjugate variables x^μ label the position states $|x\rangle = |x^0, x^1, \ldots, x^{25}\rangle$. There are as many position states as there are *spacetime* points, and the states carry time labels! Given a state $|\psi\rangle$, the associated wavefunction $\langle x|\psi\rangle$ will have time dependence even before we introduce a Schrödinger equation.

To understand the fate of the Schrödinger equation let us examine the Hamiltonian (24.6). Rewriting it in terms of $\dot{X}$, we have

$$
\begin{aligned}
H &= \frac{1}{4\pi\alpha'} \int_0^\pi d\sigma \left(\dot{X}^2 + X'^2 \right) \\
&= \frac{1}{4\pi\alpha'} \int_0^\pi d\sigma\, \frac{1}{2} \left((\dot{X} + X')^2 + (\dot{X} - X')^2 \right) \\
&= \frac{1}{2\pi} \sum_{n\in\mathbb{Z}} \int_0^\pi d\sigma\, L_n(e^{-in\sigma} + e^{in\sigma})\, e^{-in\tau},
\end{aligned} \tag{24.40}
$$

where we used (24.13). All integrals with $n \neq 0$ vanish, and for $n = 0$ the right-hand side gives L_0. This operator, however, is ambiguous because (24.13) does not yield the normal-ordered L_0 operator. Thus H equals L_0 up to a constant. The light-cone Hamiltonian is $L_0^\perp - 1$, suggesting that the covariant Hamiltonian must be chosen to be

$$
H = L_0 - 1 = \alpha' p^2 + N - 1. \tag{24.41}
$$

This ordering constant coincides with the one needed for the constraint associated with L_0. Since all physical states are annihilated by $L_0 - 1$, we reach the surprising conclusion that the Hamiltonian annihilates all physical states! This means that it is not possible to introduce a time variable and generate nontrivial time evolution through the Schrödinger equation. But then again, we do not need to introduce a time variable and generate time evolution as we did in the light-cone. The covariant states already have time labels. The Schrödinger equation has turned into the constraint $H|\Phi\rangle = 0$. This constraint – roughly speaking – fixes a relation between the time label and the position labels of the states. In momentum space, it imposes a mass-shell condition that fixes the energy p^0 in terms of the momentum $\vec{p}$.

The basis vectors of the covariant state space are constructed by acting on the ground states $|p\rangle$ with all possible creation operators:

$$
|r\rangle = \prod_{n=1}^{\infty} \prod_{\mu=0}^{25} \left(a_n^{\mu\dagger} \right)^{\lambda_{n,\mu}} |p\rangle. \tag{24.42}
$$

Here the $\lambda_{n,\mu}$ are non-negative integers, and, as usual, only finitely many of them are non-zero. All 26 values of μ can be used to build the basis states. In the light-cone gauge all the basis states we introduced were physical states. Not here: a vector $|\Phi\rangle$ represents a physical state if it satisfies the Virasoro constraints in (24.33) and it is not a descendent.

Let us examine concretely some of the physical states in the theory. Consider first the ground states $|p\rangle$. These are the states with number eigenvalue zero. On account of (24.34), they are automatically annihilated by the $L_{n\geq 1}$ operators, and the only nontrivial constraint is

$$0 = (L_0 - 1)|p\rangle = (\alpha' p^2 - 1)|p\rangle \longrightarrow p^2 = 1/\alpha'. \qquad (24.43)$$

This is the on-shell condition for a tachyon state with $M^2 = -1/\alpha'$. The states $|p\rangle$ are not descendents because if they were, they would have to have descended from a state with negative number eigenvalue, and there are no such states in the theory. We have therefore identified the physical tachyon states. In Chapter 12 we identified the same physical states using the light-cone quantization of the open string.

Quick calculation 24.3 Use (24.13) to verify explicitly that $L_1|p\rangle = 0$.

With a slight generalization we can also deal with tachyon *fields*, and the constraint will give us the classical field equation for the tachyon. A general tachyon state $|T\rangle$ can be constructed as a superposition of momentum ground states:

$$|T\rangle \equiv \int d^D p\, \phi(p)\, |p\rangle, \qquad (24.44)$$

where $\phi(p)$ is an arbitrary function of the spacetime momentum. For $|T\rangle$ to be physical it must be annihilated by $(L_0 - 1)$:

$$(L_0 - 1)|T\rangle = \int d^D p\, \phi(p)(L_0 - 1)|p\rangle = \int d^D p\, \phi(p) \left(\alpha' p^2 - 1\right) |p\rangle = 0. \quad (24.45)$$

Since the ground states $|p\rangle$ are linearly independent, the integrand must vanish for all p:

$$(\alpha' p^2 - 1)\phi(p) = 0. \qquad (24.46)$$

This is the field equation for the tachyon. The function $\phi(p)$, which was introduced to help construct the general tachyon state, turns out to be the tachyon field. Since the tachyon state $|T\rangle$ has number eigenvalue zero, there are no further Virasoro constraints, and the state cannot be a descendent.

Next we consider photon states, which we identified as $\xi_I a_1^{I\dagger}|p^+, \vec{p}_T\rangle$ in the light-cone quantization of the open string. Presently, we consider $N = 1$ states with fixed momentum p and polarization ξ_μ:

$$\xi_\mu \alpha_{-1}^\mu |p\rangle. \qquad (24.47)$$

The $L_0 - 1 = 0$ condition gives

$$0 = (\alpha' p^2 + N - 1)\, \xi_\mu \alpha_{-1}^\mu |p\rangle = \alpha' p^2 \xi_\mu \alpha_{-1}^\mu |p\rangle \longrightarrow p^2 = 0. \qquad (24.48)$$

With $L_1 = \alpha_0 \cdot \alpha_1 + (\alpha_{-1} \cdot \alpha_2 + \alpha_{-2} \cdot \alpha_3 + \cdots)$, the condition $L_1 \xi_\mu \alpha_{-1}^\mu |p\rangle = 0$ gives

$$0 = \alpha_0 \cdot \alpha_1\, \xi_\mu \alpha_{-1}^\mu |p\rangle = \sqrt{2\alpha'}\, p^\mu\, \xi_\mu\, |p\rangle \longrightarrow p \cdot \xi = 0. \qquad (24.49)$$

The $L_{n\geq 2}$ operators automatically annihilate the $\xi_\mu \alpha_{-1}^\mu |p\rangle$ states. So far, we have learned that physical photon states exist only for $p^2 = 0$ and must satisfy $p \cdot \xi = 0$. For any p_μ

which satisfies $p^2 = 0$, we can choose a Lorentz frame where $p_\mu = (p_0, p_0, 0, \ldots, 0)$, and then the constraint $p \cdot \xi = 0$ gives

$$\xi^0 + \xi^1 = 0. \tag{24.50}$$

This cannot be the final answer; (24.50) leaves $D - 1$ independent polarizations ξ_μ, and our light-cone analysis showed that photons only have $D - 2$ independent polarizations. We must impose the condition that physical states are defined up to null states. For this, consider the descendent

$$|d\rangle = L_{-1} \frac{1}{\sqrt{2\alpha'}} i\epsilon \, |p\rangle = \left(ip_\mu \epsilon\right) \alpha^\mu_{-1} |p\rangle, \quad \text{with} \quad p^2 = 0. \tag{24.51}$$

This state is of the form (24.47) with $\xi_\mu = ip_\mu\epsilon$. It follows from (24.49) that $|d\rangle$ is also primary since $p^\mu(i\epsilon p_\mu) = 0$. So $|d\rangle$ is null. We therefore have the equivalence of states

$$\xi_\mu \alpha^\mu_{-1} |p\rangle \; \sim \; \left(\xi_\mu + ip_\mu\epsilon\right) \alpha^\mu_{-1} |p\rangle, \tag{24.52}$$

and, consequently, the equivalence of polarizations

$$\xi^\mu \sim \xi^\mu + ip^\mu\epsilon, \quad \text{for} \quad p^2 = 0. \tag{24.53}$$

Again, using $p_\mu = (p_0, p_0, 0, \ldots, 0)$, this means that

$$\xi^0 \sim \xi^0 - ip_0\,\epsilon \quad \text{and} \quad \xi^1 \sim \xi^1 + ip_0\,\epsilon. \tag{24.54}$$

Since $\xi^0 - \xi^1 \sim \xi^0 - \xi^1 - 2ip_0\epsilon$, we can choose representative polarizations such that

$$\xi^0 - \xi^1 = 0. \tag{24.55}$$

Together with (24.50), we conclude that physical states have $D - 2$ independent polarizations. This is the result we had found in the light-cone gauge.

One can also work with general superpositions. We define a gauge field state $|A\rangle$ as

$$|A\rangle = \int d^D p \, A_\mu(p) \, \alpha^\mu_{-1} |p\rangle. \tag{24.56}$$

Quick calculation 24.4 Show that the $L_0 - 1$ and L_1 constraints give $p^2 A_\mu(p) = 0$ and $p \cdot A = 0$, respectively. The condition $p \cdot A = 0$ is the Lorentz gauge condition. The equation $p^2 A_\mu = 0$ is the familiar Maxwell field equation $p^2 A^\mu - p^\mu(p \cdot A) = 0$ in the Lorentz gauge.

For additional discussion of photon states, see Problem 24.3.

24.5 Closed string Virasoro operators

The covariant quantization of the closed string brings about no new complications. There are two sets of covariant Virasoro operators, and the operators with non-negative mode number annihilate the physical states. The vanishing of both Virasoro operators with mode number zero (with a proper subtraction constant) implies that the left and right number operators have the same eigenvalue when acting on physical states. We will only consider closed strings in the absence of compactification.

The mode expansion (13.24) for a generic closed string coordinate reads

$$X^\mu(\tau,\sigma) = x_0^\mu + \sqrt{2\alpha'}\,\alpha_0^\mu \tau + i\sqrt{\frac{\alpha'}{2}}\sum_{n\neq 0}\frac{e^{-in\tau}}{n}(\alpha_n^\mu e^{in\sigma} + \bar\alpha_n^\mu e^{-in\sigma})\,, \qquad (24.57)$$

where $\alpha_0^\mu = p^\mu \sqrt{\alpha'/2}$. The τ and σ derivatives of the coordinates were recorded in (13.26), with the understanding that $\bar\alpha_0^\mu = \alpha_0^\mu$. Just as in (13.36), we have

$$(\dot X^\mu + X^{\mu\prime})^2 = 4\alpha' \sum_{n\in\mathbb{Z}}\Big(\frac{1}{2}\sum_{p\in\mathbb{Z}}\bar\alpha_p^\mu \bar\alpha_{n-p,\mu}\Big)e^{-in(\tau+\sigma)} \equiv 4\alpha'\sum_{n\in\mathbb{Z}}\bar L_n e^{-in(\tau+\sigma)}\,,$$

$$(\dot X^\mu - X^{\mu\prime})^2 = 4\alpha' \sum_{n\in\mathbb{Z}}\Big(\frac{1}{2}\sum_{p\in\mathbb{Z}}\alpha_p^\mu \alpha_{n-p,\mu}\Big)e^{-in(\tau-\sigma)} \equiv 4\alpha'\sum_{n\in\mathbb{Z}} L_n e^{-in(\tau-\sigma)}\,,$$

$$\qquad (24.58)$$

where we defined

$$\bar L_n = \frac{1}{2}\sum_{p\in\mathbb{Z}}\bar\alpha_p^\mu \bar\alpha_{n-p,\mu}\,, \qquad L_n = \frac{1}{2}\sum_{p\in\mathbb{Z}}\alpha_p^\mu \alpha_{n-p,\mu}\,. \qquad (24.59)$$

These are the two sets of Virasoro operators of closed string theory. If we imposed the constraints (24.1), then all Virasoro operators would have to annihilate the physical states. Just like for open strings, this would leave no physical states. In analogy with the open string quantum constraints (24.33), we demand that for physical closed states $|\Psi\rangle$

$$(L_n - \delta_{n,0})|\Psi\rangle = 0\,, \qquad (\bar L_n - \delta_{n,0})|\Psi\rangle = 0\,, \qquad n \geq 0\,. \qquad (24.60)$$

A vector $|\Psi\rangle$ which satisfies these conditions is a representative for a physical closed string state if it is not a descendent. A descendent of a primary $|p\rangle$ is a state obtained by acting on the primary with a collection of negatively moded Virasoro operators. Both barred and unbarred operators can be used.

For $n = 0$, the conditions (24.60) give

$$(L_0 - 1)|\Psi\rangle = \Big(\frac{\alpha'}{4}p^2 + N - 1\Big)|\Psi\rangle = 0\,,$$

$$(\bar L_0 - 1)|\Psi\rangle = \Big(\frac{\alpha'}{4}p^2 + \bar N - 1\Big)|\Psi\rangle = 0\,, \qquad (24.61)$$

where N and $\bar N$ are the covariant number operators. Note that equations (24.61) imply that $L_0 - \bar L_0$ annihilates all physical states. As before, this condition is interpreted as an invariance of physical states under a constant shift of the σ coordinate on the world-sheet, or as the vanishing of the two-dimensional momentum along the string (Section 13.2). Equations (24.61) give *two* expressions for the mass-squared:

$$M^2 = -p^2 = \frac{4}{\alpha'}(N-1) = \frac{4}{\alpha'}(\bar{N}-1) , \qquad (24.62)$$

and therefore we have the constraint

$$N = \bar{N} . \qquad (24.63)$$

Making use of (24.63) we can write, more symmetrically,

$$M^2 = \frac{2}{\alpha'}(N + \bar{N} - 2) . \qquad (24.64)$$

The closed string state space is constructed by acting with arbitrary numbers of oscillators on the closed string ground states $|p\rangle$. To uncover the closed string tachyon states it suffices to consider the ground states and to impose the $L_0 - 1 = \bar{L}_0 - 1 = 0$ conditions. To uncover the massless closed string states – the Kalb–Ramond field, the graviton, and the dilaton – we must also examine the $L_1 = \bar{L}_1 = 0$ constraints. Those constraints suffice because the massless states have $N = \bar{N} = 1$, so Virasoro operators with higher mode number automatically annihilate these states. Of course, the equivalences generated by null states must also be considered (Problem 24.5).

24.6 The Polyakov string action

Despite the elegance of the action (24.4) and the ease with which we can obtain from it the equations of motion, the momenta, and the Hamiltonian, the constraint equations (24.1) still have to be imposed by hand. We will now develop another action, the Polyakov action, from which the constraint equations also emerge naturally. As a first step, we rewrite (24.4) in a more suggestive form:

$$S = -\frac{1}{4\pi\alpha'} \int d\tau d\sigma \, \eta^{\alpha\beta} \partial_\alpha X^\mu \partial_\beta X^\nu \, \eta_{\mu\nu} . \qquad (24.65)$$

The spacetime indices on the string coordinates are contracted with the Minkowski metric $\eta_{\mu\nu}$. The α and β indices run over two values, corresponding to the two world-sheet coordinates τ and σ:

$$\partial_\alpha = \frac{\partial}{\partial \xi^\alpha} , \quad \xi^\alpha = (\xi^1, \xi^2) = (\tau, \sigma). \qquad (24.66)$$

We have also introduced a *two-dimensional* Minkowski metric $\eta^{\alpha\beta}$. Just like the spacetime Minkowski metric, it is diagonal; the time–time entry is -1, and the space–space entry is $+1$:

$$\eta^{\alpha\beta} = \begin{pmatrix} -1 & 0 \\ 0 & 1 \end{pmatrix} . \qquad (24.67)$$

In (24.65) the repeated α and β indices are summed over the two values τ and σ. Writing out these sums we recover the action (24.4).

The above two-dimensional metric might remind you of the metric that emerged while we were making manifest the reparameterization invariance of the Nambu–Goto action. There we saw that the action takes the form indicated in equation (6.44):

$$S = -\frac{1}{2\pi\alpha'} \int d\tau d\sigma \sqrt{-\gamma} \,. \tag{24.68}$$

Here $\gamma = \det(\gamma_{\alpha\beta})$, and $\gamma_{\alpha\beta}$ is the world-sheet metric induced by the target-space Minkowski metric. It is explicitly given by (6.42):

$$\gamma_{\alpha\beta} = \frac{\partial X}{\partial \xi^\alpha} \cdot \frac{\partial X}{\partial \xi^\beta} = \partial_\alpha X \cdot \partial_\beta X \,. \tag{24.69}$$

The Polyakov action involves a new world-sheet metric, $h_{\alpha\beta}(\tau, \sigma)$. This is the kind of metric that one uses in two-dimensional general relativity. The metric $h_{\alpha\beta}(\tau, \sigma)$ is a dynamical variable in the action, so it leads to its own field equations. As it turns out, the equations of motion will relate the new metric $h_{\alpha\beta}$ to the induced metric $\gamma_{\alpha\beta}$. The metric $h_{\alpha\beta}$ enters the Polyakov action in a way that is analogous to the way in which $\eta^{\alpha\beta}$ enters the action (24.65). The Polyakov action is

$$S = -\frac{1}{4\pi\alpha'} \int d\tau d\sigma \sqrt{-h}\, h^{\alpha\beta} \partial_\alpha X^\mu \partial_\beta X^\nu \eta_{\mu\nu} \,. \tag{24.70}$$

Here $h \equiv \det(h_{\alpha\beta})$, and $h^{\alpha\beta}$ is the inverse of $h_{\alpha\beta}$:

$$h^{\alpha\beta} h_{\beta\gamma} = \delta^\alpha_\gamma \,, \quad h^{\alpha\beta} h_{\alpha\beta} = 2 \,. \tag{24.71}$$

The variation of the metric $h_{\alpha\beta}$ in the Polyakov action will give us the Virasoro constraints. Since $h_{\alpha\beta}$ is a symmetric two-by-two matrix, and such matrices have three independent entries, we should expect three constraints. But there are only two Virasoro conditions. How can these facts be reconciled?

First note that the metric $h^{\alpha\beta}$ enters the action through the specific combination $\sqrt{-h}\, h^{\alpha\beta}$. The square root factor is needed to make the measure $(d\tau d\sigma \sqrt{-h})$ reparameterization invariant, as you saw in Section 6.2. The $h^{\alpha\beta}$ factor is needed to contract against the indices carried by the derivatives of X. While $h^{\alpha\beta}$ is clearly determined by three real numbers at any fixed point on the world-sheet, the combination $\sqrt{-h}\, h^{\alpha\beta}$ is in fact determined by just two numbers! This is a peculiar property of two-dimensional metrics. Define this combination to be

$$M^{\alpha\beta} = \sqrt{-h}\, h^{\alpha\beta} \,. \tag{24.72}$$

We will show that the symmetric matrix M satisfies an additional constraint. To appreciate why two-dimensional metrics are special, let n denote the size of the matrix $h^{\alpha\beta}$. Then

$$\det(M^{\alpha\beta}) = (\sqrt{-h})^n \det(h^{\alpha\beta}) = \frac{(-h)^{\frac{n}{2}}}{\det(h_{\alpha\beta})} = \frac{(-h)^{\frac{n}{2}}}{h} = -(-h)^{\frac{n}{2}-1} \,. \tag{24.73}$$

The final simplification occurs only when $n = 2$. In that case we find

$$\det(M^{\alpha\beta}) = -1 \,. \tag{24.74}$$

A 2-by-2 symmetric matrix with determinant equal to a fixed constant is determined by two parameters. This is why the variation of the metric gives us only two independent constraints, which is the number we want.

Enough with preliminaries. Let us tackle the variation of the Polyakov action. First we vary the string coordinates X^μ. This variation gives

$$\delta S = -\frac{1}{2\pi\alpha'} \int d\tau d\sigma \sqrt{-h}\, h^{\alpha\beta} \partial_\alpha (\delta X^\mu) \partial_\beta X^\nu \eta_{\mu\nu}$$
$$= \frac{1}{2\pi\alpha'} \int d\tau d\sigma\, \delta X^\mu\, \partial_\alpha \left(\sqrt{-h}\, h^{\alpha\beta} \partial_\beta X^\nu \eta_{\mu\nu} \right), \qquad (24.75)$$

where we have discarded the complete derivatives which give the familiar boundary contributions in the case of open strings. The resulting equation of motion is therefore

$$\partial_\alpha \left(\sqrt{-h}\, h^{\alpha\beta} \partial_\beta X^\mu \right) = 0. \qquad (24.76)$$

This equation is not yet a wave equation for X^μ, but we will get there shortly.

In order to vary the metric $h^{\alpha\beta}$ we need a preliminary result. Let A be a 2-by-2 matrix, and let δA denote its variation:

$$A = \begin{pmatrix} a_{11} & a_{12} \\ a_{21} & a_{22} \end{pmatrix} \quad \text{and} \quad \delta A = \begin{pmatrix} \delta a_{11} & \delta a_{12} \\ \delta a_{21} & \delta a_{22} \end{pmatrix}. \qquad (24.77)$$

It is straightforward to check that the variation of $\det A$ can be written as

$$\delta \det A = (\det A)\, \mathrm{Tr}(A^{-1} \delta A), \qquad (24.78)$$

where A^{-1} denotes the inverse of A and Tr stands for trace. This identity actually holds for matrices of arbitrary dimension.

Quick calculation 24.5 Verify explicitly that equation (24.78) holds for 2-by-2 matrices.

Let $\delta h^{\alpha\beta}$ denote the variation of the metric. We can use (24.78) to calculate the variation of h:

$$\delta h = \delta \det(h_{\alpha\beta}) = h\, (h^{\alpha\beta} \delta h_{\beta\alpha}). \qquad (24.79)$$

We must write the variation $\delta h_{\alpha\beta}$ in terms of $\delta h^{\alpha\beta}$. For this we vary the second equation in (24.71) to find

$$\delta h^{\alpha\beta}\, h_{\alpha\beta} + h^{\alpha\beta}\, \delta h_{\alpha\beta} = 0 \longrightarrow h^{\alpha\beta}\, \delta h_{\beta\alpha} = -\delta h^{\alpha\beta}\, h_{\alpha\beta}, \qquad (24.80)$$

where we noted that $\delta h_{\alpha\beta} = \delta h_{\beta\alpha}$. We can then rewrite the variation of h as

$$\delta h = -h\, \delta h^{\alpha\beta}\, h_{\alpha\beta}. \qquad (24.81)$$

Of course, what we really need to vary is $\sqrt{-h}$. This can now be done as

$$\delta(\sqrt{-h}) = -\frac{1}{2} \frac{\delta h}{\sqrt{-h}} = -\frac{1}{2} \frac{(-h)\delta h^{\alpha\beta} h_{\alpha\beta}}{\sqrt{-h}} = -\frac{1}{2}\sqrt{-h}\, \delta h^{\alpha\beta}\, h_{\alpha\beta}. \qquad (24.82)$$

We can finally vary the metric in the action (24.70):

$$\delta S = -\frac{1}{4\pi\alpha'} \int d\tau d\sigma \sqrt{-h} \left(-\frac{1}{2}\delta h^{\alpha\beta} h_{\alpha\beta} (h^{\gamma\delta}\partial_\gamma X \cdot \partial_\delta X) + \delta h^{\alpha\beta}\partial_\alpha X \cdot \partial_\beta X \right)$$

$$= -\frac{1}{4\pi\alpha'} \int d\tau d\sigma \sqrt{-h}\, \delta h^{\alpha\beta} \left(\partial_\alpha X \cdot \partial_\beta X - \frac{1}{2}h_{\alpha\beta}(h^{\gamma\delta}\partial_\gamma X \cdot \partial_\delta X) \right). \quad (24.83)$$

The equation of motion which arises from the variation $\delta h^{\alpha\beta}$ is therefore

$$\partial_\alpha X \cdot \partial_\beta X - \frac{1}{2}h_{\alpha\beta}(h^{\gamma\delta}\partial_\gamma X \cdot \partial_\delta X) = 0. \quad (24.84)$$

Recalling that $\partial_\alpha X \cdot \partial_\beta X$ is just the induced metric $\gamma_{\alpha\beta}$ in (24.69), the above equation can be written as

$$\gamma_{\alpha\beta} - \frac{1}{2}h_{\alpha\beta}(h^{\gamma\delta}\gamma_{\gamma\delta}) = 0. \quad (24.85)$$

Since the factor $(h^{\gamma\delta}\gamma_{\gamma\delta})$ has no free indices, this equation sets the world-sheet metric $h_{\alpha\beta}$ proportional to the induced metric $\gamma_{\alpha\beta}$ at every point on the world-sheet. The factor of proportionality can be position dependent. This factor is *not* determined by (24.85): given a solution $h_{\alpha\beta}$ of this equation, the metric $h'_{\alpha\beta} = \Omega^2 h_{\alpha\beta}$, with Ω an arbitrary nonvanishing function, will also provide a solution. This happens because the second term on the left-hand side of (24.85) contains the product of the metric and its inverse. Equation (24.85) is therefore satisfied by

$$h_{\alpha\beta} = f^2(\xi)\, \gamma_{\alpha\beta}, \quad (24.86)$$

where $f(\xi)$ is some undetermined nonvanishing function on the world-sheet. By writing (24.86) with $f^2(\xi)$, we are making an additional statement: the proportionality factor that relates the metrics is positive. This has a physical implication: the notions of timelike and spacelike vectors defined by $h_{\alpha\beta}$ and $\gamma_{\alpha\beta}$ agree. Since the induced metric $\gamma_{\alpha\beta}$ is really the ambient metric referred to the world-sheet, if a world-sheet vector is spacelike or time-like, as determined by $h_{\alpha\beta}$, it will be spacelike or timelike, respectively, as determined by the Minkowski metric. Metrics related by a (positive) factor of proportionality are said to be conformal to each other. Thus the world-sheet metric is conformal to the induced metric.

Let us see what happens to the Polyakov action when we substitute into it the information we have gained. We first calculate $\sqrt{-h}h^{\alpha\beta} = M^{\alpha\beta}$ using (24.86). The good news is that the undetermined function $f(\xi)$ drops out of this calculation. To see this quickly, recall that the determinant of the matrix $M^{\alpha\beta}$ is (-1), independent of the metric $h^{\alpha\beta}$. If we scale the metric $h^{\alpha\beta}$, the most that could happen to $M^{\alpha\beta}$ is that it scales as well. But if $M^{\alpha\beta}$ were to scale then its determinant would change. Thus the scale of $h^{\alpha\beta}$ does not affect $M^{\alpha\beta}$. This argument can be confirmed explicitly. With $\gamma \equiv \det(\gamma_{\alpha\beta})$, we use (24.86) to determine the determinant of $h_{\alpha\beta}$ and the inverse metric $h^{\alpha\beta}$:

$$h = f^4 \gamma \quad \text{and} \quad h^{\alpha\beta} = \frac{1}{f^2}\gamma^{\alpha\beta}, \quad (24.87)$$

where $\gamma^{\alpha\beta}$ is the inverse of $\gamma_{\alpha\beta}$. It follows that

$$\sqrt{-h}\,h^{\alpha\beta} = f^2\,\sqrt{-\gamma}\,\frac{1}{f^2}\,\gamma^{\alpha\beta} = \sqrt{-\gamma}\,\gamma^{\alpha\beta}\,, \tag{24.88}$$

which confirms that the undetermined function f is irrelevant. If we now substitute back into the Polyakov action (24.70) we find

$$S = -\frac{1}{4\pi\alpha'}\int d\tau d\sigma\,\sqrt{-\gamma}\,\gamma^{\alpha\beta}\,\gamma_{\alpha\beta}\,. \tag{24.89}$$

The second equation in (24.71), which is valid for any two-dimensional metric, now gives

$$S = -\frac{1}{2\pi\alpha'}\int d\tau d\sigma\,\sqrt{-\gamma}\,. \tag{24.90}$$

This is exactly the Nambu–Goto action. We conclude that the Polyakov action is classically equivalent to the Nambu–Goto action.

Quick calculation 24.6 Verify that, had we chosen $h_{\alpha\beta} = -f^2\,(\xi)\,\gamma_{\alpha\beta}$ instead of (24.86), the Nambu–Goto action would have emerged with the wrong sign.

While we were showing that the factor $f^2(\xi)$ in (24.86) drops out during the evaluation of the string action, we also proved that the Polyakov action is invariant under the transformation

$$h_{\alpha\beta}(\tau,\sigma)\;\rightarrow\;\Omega^2(\tau,\sigma)\,h_{\alpha\beta}(\tau,\sigma)\,, \tag{24.91}$$

where Ω^2 is an arbitrary function on the world-sheet. This rescaling of the world-sheet metric is called a *Weyl transformation*. The invariance of the Polyakov action under a Weyl transformation indicates that, on the world-sheet, distances measured with the $h_{\alpha\beta}$ metric have no physical significance.

We have yet to show that we obtain the Virasoro constraints and wave equations for the string coordinates. The equation of motion (24.76) for X^μ looks complicated, and the equation of motion (24.84) for $h_{\alpha\beta}$ does not resemble the Virasoro constraints. In order to make progress we use the reparameterization invariance to choose a convenient form for $h_{\alpha\beta}$.

It is a well known result of two-dimensional geometry that a coordinate reparameterization allows an arbitrary metric $h_{\alpha\beta}$ on a surface to be cast locally in the form

$$h_{\alpha\beta} = \rho^2\,(\xi)\,\eta_{\alpha\beta}\,. \tag{24.92}$$

Here ρ is a *conformal factor* and $\eta_{\alpha\beta}$ is the two-dimensional Minkowski metric. The metric $h_{\alpha\beta}$ is said to be conformally flat. By restricting ourselves to the class of world-sheet coordinates that result in a conformally flat metric, we are making a partial gauge choice. This choice is called the conformal gauge. In the conformal gauge $\sqrt{-h}\,h^{\alpha\beta} = \eta^{\alpha\beta}$, so equation (24.76) becomes

$$\partial_\alpha\big(\eta^{\alpha\beta}\,\partial_\beta X^\mu\big) = \eta^{\alpha\beta}\,\partial_\alpha\partial_\beta X^\mu = 0\,. \tag{24.93}$$

Expanding out the sums over two-dimensional indices, we see that this is precisely the wave equation. For equations (24.84), the conformal gauge condition (24.92) results in

$$\partial_\alpha X \cdot \partial_\beta X - \frac{1}{2} \eta_{\alpha\beta} (\eta^{\gamma\delta} \partial_\gamma X \cdot \partial_\delta X) = 0 . \tag{24.94}$$

Expanding the expression in parentheses we have

$$\partial_\alpha X \cdot \partial_\beta X - \frac{1}{2} \eta_{\alpha\beta} (-\dot{X}^2 + X'^2) = 0 . \tag{24.95}$$

These are three equations, but only two of them are independent. Taking $\alpha = \beta = 1$ we get

$$\dot{X}^2 + \frac{1}{2}(-\dot{X}^2 + X'^2) = 0 \; \longrightarrow \; \dot{X}^2 + X'^2 = 0 , \tag{24.96}$$

which is one of the constraints. With $\alpha = 1, \beta = 2$ we get

$$\dot{X} \cdot X' = 0 , \tag{24.97}$$

which is the second constraint. Finally, when $\alpha = \beta = 2$ we get

$$X'^2 - \frac{1}{2}(-\dot{X}^2 + X'^2) = 0 \; \longrightarrow \; \dot{X}^2 + X'^2 = 0 , \tag{24.98}$$

which is redundant. We have thus shown that the equations of motion for the world-sheet metric reduce to the familiar Virasoro constraints in the conformal gauge.

Because of its elegance and convenience, the Polyakov string action is typically the starting point of any detailed analysis of covariant string quantization. In such an analysis, the various facts for which we have had to appeal to our previous light-cone results can be given independent justification.

Problems

Problem 24.1 Covariant quantization of the point particle.

The point particle Lagrangian is $L = -m\sqrt{-\dot{x}^2}$, where $\dot{x}^\mu = dx^\mu(\tau)/d\tau$.

(a) Calculate the momentum p_μ canonical to $x^\mu(\tau)$. Verify that the momentum satisfies the constraint $p^2 + m^2 = 0$. Prove that one cannot solve uniquely for the velocity in terms of the momentum by showing that, given one solution, one can easily construct a different one. In addition, only for momenta satisfying the constraint can one possibly find velocities.

(b) Show that $H \equiv p_\mu \dot{x}^\mu - L = 0$. Set up commutation relations and describe the state space.

(c) Physical states are those that satisfy the constraint. Build candidate states as general linear superpositions and compare with (11.48). Apply the constraint and show that the wavefunction for the physical states satisfies the Klein–Gordon equation for a scalar field of mass m.

Problem 24.2 States as Virasoro primaries or Virasoro descendents.

Consider the Virasoro operators $L_n = \frac{1}{2} \sum_p \alpha_{n-p} \alpha_p$ associated with a single open string coordinate X with oscillators that satisfy $[\alpha_m, \alpha_n] = m\delta_{m+n,0}$. The states that can be built from the zero-momentum vacuum $|0\rangle$ for $N \le 4$ are:

$$N = 0: \quad |0\rangle ,$$
$$N = 1: \quad \alpha_{-1}|0\rangle ,$$
$$N = 2: \quad \alpha_{-2}|0\rangle , \quad \alpha_{-1}\alpha_{-1}|0\rangle ,$$
$$N = 3: \quad \alpha_{-3}|0\rangle , \quad \alpha_{-2}\alpha_{-1}|0\rangle , \quad (\alpha_{-1})^3|0\rangle ,$$
$$N = 4: \quad \alpha_{-4}|0\rangle , \quad \alpha_{-2}(\alpha_{-1})^2|0\rangle , \quad \alpha_{-3}\alpha_{-1}|0\rangle , \quad \alpha_{-2}\alpha_{-2}|0\rangle , \quad (\alpha_{-1})^4|0\rangle . \tag{1}$$

We want to show that at each level N we can form an equivalent basis of states where each state is either a Virasoro primary or a Virasoro descendent from a primary in the list.

(a) Define the norm of $|0\rangle$ as $\langle 0|0\rangle \equiv 1$. Explain why any state $|\chi\rangle$ in (1) has positive norm $\langle \chi|\chi\rangle > 0$, and why any two different states $|\chi_1\rangle$, $|\chi_2\rangle$ are orthogonal: $\langle \chi_2|\chi_1\rangle = 0$. Argue that any nonzero linear combination of the above states is a state of positive norm, and that, as a result, no state in the state space is both a primary and a descendent.

(b) Explain why the state $|0\rangle$ is a primary. Now consider the state $\alpha_{-1}|0\rangle$ with $N = 1$. This state is either a descendent of $|0\rangle$ or a primary. Show that the only candidate descendent $L_{-1}|0\rangle$ vanishes. Prove that $\alpha_{-1}|0\rangle$ is a primary, and call it $|p_1\rangle$. Using a basis as in equation (24.38), explain why descendents of $|0\rangle$ with one or more L_{-1} operators vanish.

(c) Show that, for $N = 2$, one state is a descendent of $|0\rangle$, and one state is a descendent of $|p_1\rangle$.

(d) Show that, for $N = 3$, one state is a descendent of $|0\rangle$, and two states are descendents of $|p_1\rangle$.

(e) For $N = 4$ we have three candidate descendents of $|p_1\rangle$. From these states, show that one linear combination vanishes:

$$\left(L_{-3} - 2L_{-2}L_{-1} + \tfrac{1}{2}(L_{-1})^3 \right) |p_1\rangle = 0,$$

and the other two give states equivalent to the first two of the $N = 4$ list. Note that there are only two candidate descendents of $|0\rangle$. So the last three states on the list must break into two descendents and one new primary $|p_4\rangle$. Show that, up to an arbitrary normalization,

$$|p_4\rangle = \left(\alpha_{-3}\alpha_{-1} - \tfrac{3}{4}(\alpha_{-2})^2 - \tfrac{1}{2}(\alpha_{-1})^4 \right) |0\rangle ,$$

and write explicitly the three last states on the list as a linear superposition of this primary and the two descendents.

Problem 24.3 Photon states in the covariant description.

(a) Consider the field equation and gauge transformations for a Maxwell field:

$$p^2 A^\mu - p^\mu (p \cdot A) = 0, \quad \delta A^\mu = i p^\mu \epsilon.$$

Show that for $p^2 \neq 0$ it is possible to make the potentials satisfy the Lorentz gauge condition $p \cdot A = 0$. Show also that for $p^2 = 0$ the quantity $p \cdot A$ is gauge invariant and that it is set to zero by the field equation. This shows that the Lorentz gauge condition is generally valid for all p. It also shows, however, that for $p^2 = 0$ the Lorentz gauge does not fix the gauge symmetry. We have seen evidence of this: for $p^2 = 0$ the solutions of the equations of motion in the Lorentz gauge are characterized by $(D-1)$ independent degrees of freedom.

(b) Consider the coupling $\int A_\mu j^\mu d^D x$ of the gauge field to the conserved current j^μ. Show that for $p^2 = 0$ the part of the gauge field that can be gauged away drops out of this coupling.

(c) Examine the descendent $|D\rangle$ relevant to the discussion of the general superposition (24.56):

$$|D\rangle = L_{-1} \frac{1}{\sqrt{2\alpha'}} \int d^D p \; i\epsilon(p)|p\rangle.$$

When is $|D\rangle$ null? Show that for $p^2 = 0$ we have the identification $A^\mu(p) \sim A^\mu(p) + i p^\mu \epsilon(p)$. This is an on-shell gauge invariance.

Problem 24.4 D0-branes, open strings, and M-theory.

(a) We want to show that it is not possible to have only one string endpoint on a D0-brane. Intuitively, the string charge, visualized as a current on the string, has nowhere to go on the pointlike D0-brane. More quantitatively, as in any D-brane, on the D0-brane there is a gauge field that couples to the string endpoint as in (16.54). Since the D0-brane has no spatial coordinate the gauge field is just A_0. Show that the Maxwell action vanishes, and that, as a result, the variation of A_0 imposes an inconsistent equation of motion. Note that the inconsistency is removed if the two endpoints of an open string lie on the D0-brane.

(b) Consider the covariant quantization of an open string whose endpoints lie on a D0-brane. Describe the ground states, noting that the momentum has a single component. Construct the general states with $N = 0$ and discuss the equation of motion for the tachyon field $\phi(t)$. Construct the $N = 1$ physical states. Show that there are no relevant null states, and that we have D independent physical states of zero momentum.

(c) Give the mass m_0 of a D0-brane (see Section 18.4). In type IIA superstring theory the same formula applies, and a bound state of n D0-branes is known to have a mass exactly equal to nm_0. Such states can be identified with momentum states that arise from a compact eleventh dimension, assuming that the momentum p along this dimension contributes to the mass of the effective ten-dimensional particle as in $m = p$. What is the radius $\bar{R}$ of this eleventh dimension? How does it behave as a function of g? This result is one piece of evidence for the fact that eleven-dimensional M-theory compactified on a circle is type IIA superstring theory.

Problem 24.5 Graviton and dilaton states in covariant quantization.

Examine the closed string states $\xi_{\mu\nu}\, \alpha^{\mu}_{-1}\bar{\alpha}^{\nu}_{-1}|p\rangle$, with $\xi_{\mu\nu} = \xi_{\nu\mu}$.

(a) Show that the Virasoro constraints give the conditions $p^2 = 0$ and $p_\mu \xi^{\mu\nu} = 0$.
(b) Exhibit null states that generate the physical state equivalences $\xi^{\mu\nu} \sim \xi^{\mu\nu} + p^\mu \epsilon^\nu + p^\nu \epsilon^\mu$, which hold for $p^2 = 0$ and $p \cdot \epsilon = 0$.
(c) Show that there are $(D-2)(D-1)/2$ independent physical degrees of freedom in $\xi_{\mu\nu}\, \alpha^{\mu}_{-1}\bar{\alpha}^{\nu}_{-1}|p\rangle$ for each value of p_μ which satisfies $p^2 = 0$. These are the degrees of freedom of a graviton and a scalar dilaton.

25 String interactions and Riemann surfaces

The world-sheets of interacting open strings are recognized to be Riemann surfaces, and interaction processes are seen to construct the moduli spaces of these surfaces. Conformal mapping is used to provide canonical presentations for interacting light-cone world-sheets. The celebrated Veneziano amplitude for the interaction of open string tachyons is motivated and discussed.

25.1 Introduction

Interactions and the forces that mediate them make the world interesting. If the electron and the proton did not interact, there would be no hydrogen atom. The fine structure constant $\alpha = e^2/(4\pi\hbar c)$ quantifies the strength of electromagnetic interactions and determines the interaction potential between the electron and the proton (see Section 13.4). The hot filament of a light bulb emits photons, some of which are absorbed by your eye. Emission and absorption processes are also interactions. A neutron can turn into a proton, an electron, and an antineutrino. This process, called β-decay, is the result of a weak interaction.

In string theory the strength of interactions is parameterized by the string coupling g. The value of this dimensionless number is determined by the expectation value of the dilaton field, as we discussed in Section 13.4. The string coupling g, together with the slope parameter α', determines the value of Newton's constant. The constants g and α' also determine the tension of D-branes.

Interactions arise very elegantly in string theory because they are described by processes in which strings join and split. In these processes, the world-sheets of free strings combine to form a single world-sheet, which represents the interaction. Recall that world-sheets define the two-dimensional parameter space (τ, σ) for string propagation. For a free open string, this parameter space is an infinite strip; for a free closed string, it is an infinite cylinder.

Riemann surfaces are some of the most interesting two-dimensional surfaces. They are, roughly speaking, surfaces where the two coordinates make up a complex variable. Riemann surfaces are preserved by conformal maps, so they have no *a priori* concept of distance! Inequivalent Riemann surfaces can be distinguished by parameters called moduli. Finding a way to construct all Riemann surfaces with their associated moduli is a difficult mathematical problem. Our study of string interactions will give very strong evidence for a remarkable statement: the world-sheets of interacting strings give precisely this construction.

String world-sheets can be used to calculate quantum mechanical amplitudes for scattering. Our Riemann surface analysis will motivate the amplitude for the scattering of four tachyons, the famous Veneziano amplitude. String theory began with Veneziano's discovery of this amplitude. By reaching this point in our course, we have essentially traversed the historical path backwards!

One word about terminology before getting started. We defined the world-sheet as the spacetime surface $X^\mu(\tau, \sigma)$ which represents the history of a string. As we mentioned in the paragraph above equation (6.29), physicists sometimes use the term world-sheet to refer, in addition, to the parameter space (τ, σ). In this chapter we will use this alternative meaning of the term.

25.2 Interactions and observables

Before we begin our study of string interactions and Riemann surfaces, let us set the stage for the concept of interactions. The goal is to turn a picture of a set of interacting strings into a number which gives the probability for that event to occur. Doing so involves three steps.

(1) Drawing the string diagram and calculating the conformal map which gives a canonical representation of that diagram.
(2) Using conformal field theory to calculate the scattering amplitude from the canonical representation.
(3) Using formulae to turn the scattering amplitude into a cross section. The cross section is observable.

We will study step number (1) in some detail. This entails understanding the fascinating relation between string diagrams and Riemann surfaces. Step number (2), in all its generality, is the subject of more advanced string theory courses. Step (3) properly belongs to a quantum field theory course. Step (1) is in many ways the most nontrivial of the three.

Let us examine these three steps in the case of point particles in four-dimensional spacetime. We consider a specific theory with two types of scalar particles: ϕ particles with mass M and χ particles with mass m. Suppose that $M > 2m$, so that a ϕ particle can decay into two χ particles. Indeed, in the rest frame of the ϕ particle the total energy is Mc^2, and to create two χ particles we need at least $2mc^2$. Thus $M > 2m$ is required by energy conservation, and the difference $(M - 2m)c^2$ goes into the kinetic energy of the resulting χ particles.

The diagram of step (1) is simply a picture (shown in Figure 25.1). In this picture, called a Feynman diagram, three lines come together at a point. One of the lines, the dotted one, represents the ϕ particle, and the other two represent the χ particles. For step (2) we must calculate the decay amplitude. In order to do this, we introduce a constant λ which quantifies the strength of the interaction. The value of λ is a parameter in the theory. If we

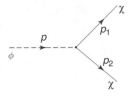

The diagram showing the interaction of a ϕ particle with two χ particles.

let p denote the momentum of the incoming ϕ particle and p_1, p_2, denote the momenta of the outgoing χ particles, then the amplitude for an initial state consisting of a ϕ particle to turn into a final state consisting of the two outgoing χ particles is

$$_{\text{out}}\langle p_1, p_2 \mid p\rangle_{\text{in}} = -i\,(2\pi)^4\delta(p - p_1 - p_2)\,T(p, p_1, p_2), \qquad (25.1)$$

where the decay amplitude T is given by

$$T(p, p_1, p_2) = \lambda. \qquad (25.2)$$

The four-dimensional delta function $\delta(p - p_1 - p_2)$ ensures that both energy and momentum are conserved. Step (3) is the construction of an observable. In this case we calculate the lifetime τ_ϕ of the ϕ particle. Converting the amplitude into a lifetime is a standard calculation. The answer is

$$\frac{1}{\tau_\phi} = \frac{\lambda^2}{32\pi}\frac{1}{M}\sqrt{1 - \frac{4m^2}{M^2}}. \qquad (25.3)$$

This is reasonable: as the interaction parameter λ goes to zero the ϕ particle lifetime becomes infinitely long. The factors multiplying λ^2 are kinematical; they take into account the state space available to the decay particles.

Things get more interesting when we consider the possibility of two χ particles scattering off each other. In that case both the initial and final states consist of two χ particles. We call p_1 and p_2 the momenta of the incoming particles and p'_1 and p'_2 the momenta of the outgoing particles. We now construct the simplest Feynman diagrams that can represent this scattering process. Since two χ can turn into a ϕ, and a ϕ can then turn into two χ, the scattering process will use the basic $\phi\chi\chi$ interaction twice. There are three possible Feynman diagrams for this scattering process, as shown in Figure 25.2. Feynman diagrams are a useful visual tool. In the path-integral formulation of quantum mechanics, the amplitude for a process is obtained by adding the amplitudes of all the "paths" consistent with the initial and final states. A Feynman diagram for a process is a representation of a specific class of allowed paths.

In diagram (a), the incoming χ with momenta p_1 and p_2 join to form a ϕ particle with momentum $p_1 + p_2$. This particle then decays into two χ with momenta p'_1 and p'_2. In this process, all external particles – all the χ – are on the mass-shell. That is, they all satisfy $p^2 = -m^2$ and thus represent physical particle states. How about the intermediate ϕ particle? The momentum of this particle is $p_1 + p_2$, and, generally, $(p_1 + p_2)^2 \neq -M^2$.

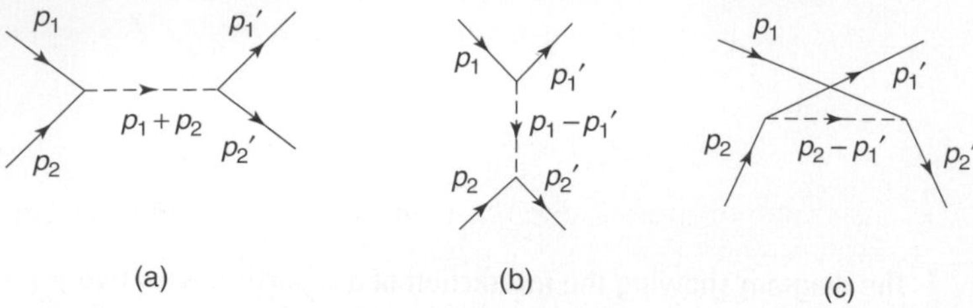

Fig. 25.2 The three possible Feynman diagrams for the scattering of two χ particles. In each diagram the intermediate particle is a ϕ particle.

The intermediate ϕ particle is typically *not* on the mass-shell; the intermediate states are not physical particle states. If the initial state is carefully prepared, ϕ may be on the mass-shell: the center-of-mass energy available would have to coincide precisely with Mc^2. Diagram (a) contributes to the scattering amplitude whether or not the ϕ particle is on the mass-shell.

Since all we measure are initial and final states, the interactions can occur in a more subtle way. In particular, there are diagrams where the intermediate ϕ particle could *never* be on the mass-shell! Consider diagram (b), where the incoming χ with momentum p_1 turns into the outgoing χ with momentum p_1' plus an intermediate ϕ. This intermediate ϕ could not possibly be physical – energy conservation rules out the production of physical χ and ϕ particles from a single physical χ particle (why?). The intermediate particle joins the incoming χ particle with momentum p_2 to give the outgoing χ with momentum p_2'. Diagram (c) shows the last possibility: the χ with momentum p_2 creates an intermediate ϕ and the outgoing χ with momentum p_1'. The intermediate particle joins the χ particle with momentum p_1 to give the outgoing χ with momentum p_2'. In these three diagrams, the intermediate momentum carried by the ϕ particle is given by

$$
\begin{aligned}
\text{first diagram} &: \quad p_1 + p_2, \\
\text{second diagram} &: \quad p_1 - p_1', \\
\text{third diagram} &: \quad p_2 - p_1'.
\end{aligned}
\tag{25.4}
$$

This completes step (1).

Given the three diagrams, we can now proceed to step (2), the calculation of the scattering amplitude. This is the analog of equations (25.1) and (25.2). The Feynman rules of field theory give

$$
_{\text{out}}\langle\, p_1', p_2' \,|\, p_1, p_2 \,\rangle_{\text{in}} = -i\,(2\pi)^4\delta(p_1 + p_2 - p_1' - p_2')\,T(p_1, p_2; p_1', p_2'),
\tag{25.5}
$$

where

$$
T(p_1, p_2; p_1', p_2') = -\lambda^2\Big(\frac{1}{(p_1 + p_2)^2 + M^2} + \frac{1}{(p_1 - p_1')^2 + M^2}
$$

$$
+ \frac{1}{(p_2 - p_1')^2 + M^2}\Big).
\tag{25.6}
$$

It is useful to define invariant quantities s, t, and u that represent the intermediate momenta squared. We let

$$s = -(p_1 + p_2)^2, \quad t = -(p_1 - p_1')^2, \quad u = -(p_2 - p_1')^2. \tag{25.7}$$

In the center-of-mass frame the total spatial momentum is zero, so $p_1 + p_2 = (E_1 + E_2, \vec{0})$, where E_1 and E_2 are the center-of-mass energies of the incoming particles. Since s is a Lorentz invariant, it can be evaluated in any frame. In the center-of-mass frame we find $s = (E_1 + E_2)^2$. The invariant s thus gives the square of the total energy available in the center-of-mass frame. Using the invariants s, t, and u the scattering amplitude is rewritten as

$$T(s, t, u) = \lambda^2 \left(\frac{1}{s - M^2} + \frac{1}{t - M^2} + \frac{1}{u - M^2} \right). \tag{25.8}$$

Each denominator is of the form $(p^2 + M^2)$, where p is the momentum carried by the intermediate particle. When the intermediate particle is on the mass-shell, the denominators vanish, and we get a pole in the amplitude. In this way, the poles of scattering amplitudes tell us about particles in the theory: the corresponding values of $-p^2$ give the particle masses. Thus, for example, if we only knew about the χ particles of mass m, but were given the scattering amplitude above, we could deduce that there is an intermediate particle with mass M. This completes step (2).

The scattering amplitude, while very interesting, is not an observable. It is used to calculate the scattering cross section σ, a quantity with the dimension of area. In the problem we considered above, the cross section is the effective area that a target χ particle presents to an incoming beam of χ particles. If the incoming beam has a cross sectional area A, the probability $\mathcal{P}$ that any of the beam particles will interact with the target is σ/A. Total cross sections can be refined into differential cross sections $d\sigma$, giving, for example, the effective area for scattering particles into a small solid angle $d\Omega$ in some specified direction. The differential cross section is the typical quantity that theorists must calculate and experimentalists must measure. The Standard Model has largely been tested by comparing predicted cross sections with measured cross sections.

Differential cross sections are calculated using field theory methods. They are integrals of $|T|^2$ (times suitable kinematic factors) over the available phase space. For the example we have considered, the result is

$$\frac{d\sigma}{d\Omega} = \frac{1}{4s} \left| \frac{T}{4\pi} \right|^2. \tag{25.9}$$

This completes step (3).

25.3 String interactions and global world-sheets

Strings can interact in a rather limited number of ways. Two possible interactions of open strings go as follows: an open string can split into two open strings, and two open strings

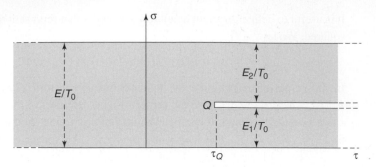

Fig. 25.3 The basic open string interaction, in which a string of energy E splits into strings of energies E_1 and E_2. The interaction occurs at $\tau = \tau_Q$.

can join to form a single open string. String diagrams are representations of the string interactions; they are analogous to the Feynman diagrams of particle physics. The focus of the string diagram is the (τ, σ) world-sheet. The string diagram does not represent the spacetime surface traced by the string. In a string diagram the world-sheets of free strings are put together to form the world-sheet of an interaction process.

Interactions must be analyzed using a specific gauge, since gauge choices affect the world-sheet description. Let us recall how world-sheets look in the static gauge. In this gauge

$$X^0(\tau, \sigma) = c\tau, \tag{25.10}$$

and strings are parameterized using the energy E that they carry:

$$\sigma \in \left[0, E/T_0\right] = \left[0, 2\pi\alpha' E\right], \tag{25.11}$$

where T_0 is the string tension. Since energy must be conserved and the energy of a string determines its σ length, the interactions must conserve the total σ length. In Figure 25.3 we show a string of energy E splitting into two strings with energies E_1 and E_2. Note that the incoming string with energy E is semi-infinite towards the past, while the outgoing strings are semi-infinite towards the future. In the far past $\tau \to -\infty$ there is only one string; in the far future $\tau \to +\infty$ there are two strings. The interaction point Q is also shown. It occurs at some value τ_Q of τ. At each $\tau < \tau_Q$ there is just one string, and at each $\tau > \tau_Q$ there are two strings.

Somewhat more complicated processes are also possible. Two strings, with energies E_1 and E_2, can come together to form a single open string, of energy $E_1 + E_2$. This intermediate string can then split into two open strings, of energies E_3 and E_4 with $E_3 + E_4 = E_1 + E_2$ (see Figure 25.4). This string diagram can be used to calculate a string amplitude regardless of the types of particles that are represented by the quantum states of the strings. For example, the initial quantum states of the two incoming strings could be two tachyons or two photons or a photon and a tachyon or any other two particle states in the quantum string spectrum. The same is true for the outgoing states. The string diagrams are universal; the information about the specific particles only enters in step (2) of the amplitude calculation.

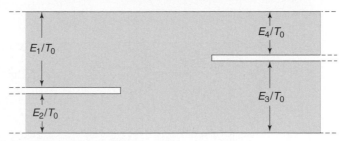

Incoming open strings one and two join to form an intermediate open string that then splits into strings three and four.

We now consider string diagrams in the light-cone gauge. In this gauge we have ((9.62) with $\beta = 2$)

$$X^+ = 2\alpha' p^+ \tau, \quad p^+ = \pi \, \mathcal{P}^{\tau+}. \tag{25.12}$$

The second equation, in particular, implies that $\sigma \in [0, \pi]$ for every string. The above equations are not suitable to deal with interactions. If every string has $\sigma \in [0, \pi]$, then σ length cannot be conserved in interactions, and it is not clear anymore how to draw string diagrams! There is an additional difficulty. Imagine that we have two open strings (strings one and two) which interact to form a single open string (string three). We select a Lorentz frame, use light-cone coordinates, and, with the above version of the light-cone gauge choice applied to each of the three strings, we have

$$X_1^+ = 2\alpha' p_1^+ \tau_1,$$
$$X_2^+ = 2\alpha' p_2^+ \tau_2,$$
$$X_3^+ = 2\alpha' p_3^+ \tau_3. \tag{25.13}$$

Here the subscripts label the strings. The interaction is a particular event in spacetime, and the three strings must agree on the value of the physical time X^+ at which it occurs. Moreover, since we are to combine the three world-sheets into a single world-sheet, with a single τ coordinate, the interaction point must have a unique τ. This value of τ is generically nonzero, so the above equations, with different values of p_i^+, will in general not give $X_1^+ = X_2^+ = X_3^+$.

How can we modify our light-cone setup in order to deal with interactions? Recall the more general version of equation (9.62), which applies for arbitrary β (see the remarks below (9.34)):

$$X^+ = \beta \alpha' p^+ \tau, \quad p^+ = \frac{2\pi}{\beta} \mathcal{P}^+_\mu. \tag{25.14}$$

Formerly, the choice $\beta = 2$ was taken in order to get open strings with $\sigma \in [0, \pi]$. The idea now is to choose β so as to eliminate the p^+ dependence from the relation between X^+ and τ. If we choose

$$\beta = \frac{1}{\alpha' p^+}, \tag{25.15}$$

then we will have the gauge

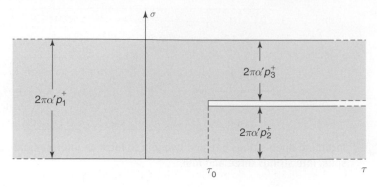

Fig. 25.5 The global world-sheet for a splitting interaction in the light-cone gauge. The conservation of p^+ implies that the total width of the strip is constant in τ.

$$X^+ = \tau, \tag{25.16}$$

which makes it possible to use a single τ for all interacting strings. For the case of the three strings just discussed, if we call τ_0 the interaction time, then we have

$$
\begin{aligned}
X_1^+ &= \tau, \quad \tau \leq \tau_0, \\
X_2^+ &= \tau, \quad \tau \geq \tau_0, \\
X_3^+ &= \tau, \quad \tau \geq \tau_0.
\end{aligned}
\tag{25.17}
$$

String one ceases to exist at τ_0, at which time strings two and three are born. Note that $X_2^+ = X_3^+$ when $\tau \geq \tau_0$. At $\tau = \tau_0$ the three X^+ agree. Along with (25.16), each string is parameterized by

$$\sigma \in \left[\, 0, \frac{2\pi}{\beta}\,\right] = [\, 0, 2\pi\alpha' p^+\,], \tag{25.18}$$

where p^+ is the momentum carried by the string. Had we used this convention to quantize the string in our earlier work, unwieldy factors would have appeared in many of the equations.

Now we can represent the splitting of string one into strings two and three using a single diagram in the (τ, σ) plane, as shown in Figure 25.5. The interaction looks like a constant width infinite strip with a cut. The figure is described by two parameters: the width of string one and the width of either string two or string three. Where we start the cut on an infinite strip is just a matter of convention, so the interaction time τ_0 is not really a parameter.

25.4 World-sheets as Riemann surfaces

We will now show that it is reasonable to view string world-sheets as Riemann surfaces. Riemann surfaces are two-dimensional surfaces. The simplest Riemann surface is the complex plane $\mathbb{C}$. This is the conventional (x, y) plane, but with the understanding that the complex coordinate $z \equiv x + iy$ is used to describe points on the plane. Subregions of the complex plane are also Riemann surfaces. For example, the upper half-plane $\mathbb{H}$, defined to include all points z with positive imaginary part, $\Im(z) > 0$, is an important Riemann surface. A closely related surface is the bordered upper half-plane $\bar{\mathbb{H}}$, which includes all points z with $\Im(z) \geq 0$ and the "point at infinity" (which is discussed in Section 25.6). An annulus is also a Riemann surface. More complicated Riemann surfaces include the Riemann sphere $\hat{\mathbb{C}}$ and the torus. We will have the chance to study all of these in some detail, either in this chapter or in the following one.

In this and in several of the following sections the level of mathematical rigor will be higher than before. Some familiarity with complex variables will be helpful. The definition of a Riemann surface (to be given below) is included for completeness and will not be strictly necessary for the material that follows.

The definition of a Riemann surface requires the concept of an analytic function. A function $f(z)$ of the complex variable z is analytic at z_0 if the derivative $f'(z)$ exists in some neighborhood of z_0. If we write $f(z) = u(x, y) + iv(x, y)$, then the functions u and v satisfy the Cauchy–Riemann equations

$$\frac{\partial u}{\partial x} = \frac{\partial v}{\partial y}, \quad \frac{\partial u}{\partial y} = -\frac{\partial v}{\partial x}, \tag{25.19}$$

over the domain of analyticity. A Riemann surface is a two-dimensional real manifold equipped with complex charts: homeomorphisms z_α which take open sets U_α into open subsets of the complex plane $\mathbb{C}$ (a homeomorphism is a one-to-one continuous map with a continuous inverse). Here α is a label for the open sets U_α that cover the manifold. Over the intersections $U_\alpha \cap U_\beta$, the transition functions $z_\alpha \circ z_\beta^{-1}$ must be analytic.

Two Riemann surfaces are considered equivalent if there is a continuous one-to-one and onto mapping relating them that is analytic. This means that the map, expressed in terms of the charts, is an analytic function from a subset of $\mathbb{C}$ to some subset of $\mathbb{C}$. An analytic map is *conformal*: angles between arcs meeting at a point are preserved under the map, except possibly at points where the derivative of the map vanishes (we will see examples of this soon). Even though angles are preserved locally, the overall shape of the surface can change quite dramatically under a conformal map. In this and in the next chapter we will take conformal to mean analytic and one-to-one. Accordingly, the existence of a conformal map taking one Riemann surface onto another Riemann surface implies the equivalence of the two Riemann surfaces.

Since conformally related surfaces are the same Riemann surface, the concept of distance cannot be defined intrinsically. Consider the complex z-plane and two points P_1 and P_2, with coordinates z_1 and z_2, respectively. The natural definition of distance between two points in $\mathbb{C}$ is given by the absolute value $|z_1 - z_2|$ of the difference of coordinates. Now map the z-plane into the w-plane via $w = 2z$. The two planes are the same Riemann surface. After the conformal map, the images of P_1 and P_2 (still called P_1 and P_2 in the w-plane), have coordinates $w_1 = 2z_1$, and $w_2 = 2z_2$. The distance between P_1 and P_2 in the w-plane is thus $|w_1 - w_2| = 2|z_1 - z_2|$. The distance is doubled. Since both planes are the

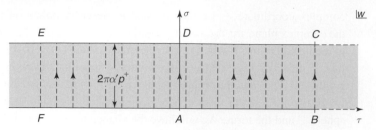

Fig. 25.6 An open string freely propagating from $\tau = -\infty$ to $\tau = +\infty$. The string diagram is a strip of constant width $2\pi\alpha'p^+$. The vertical dashed segments are the open strings.

same Riemann surface, we are at a loss when asked what is the distance between the points P_1 and P_2. Conformal maps locally scale distances. In fact, locally, lengths measured in all directions are scaled by the same factor. From $w = f(z)$ we find $dw = f'(z)dz$. It follows that at any point z_0 we have $|dw| = |f'(z_0)||dz|$. Since $|dw|$ is the length of the image of the vector dz, we see that, locally, all lengths are scaled by the same factor $|f'(z_0)|$.

Perhaps this reminds you of a metric whose scale turned out to be irrelevant. Indeed, a world-sheet metric $h_{\alpha\beta}$ enters in the Polyakov action (24.70), and Weyl transformations $h_{\alpha\beta} \to \Omega^2 h_{\alpha\beta}$ leave the string action invariant (see (24.91)). Weyl transformations scale all metric components by the same factor at each point on the world-sheet. As a result, locally, all lengths are scaled in the same way, just as a conformal map does for Riemann surfaces. The invariance of the string action under local scale transformations of the world-sheet metric indicates that the scale of lengths on the world-sheet is not physical. Since the scale of lengths is not well defined on Riemann surfaces, this suggests that viewing world-sheets as Riemann surfaces is natural.

While the scale of distances behaves similarly, there is one difference. The world-sheet metric can be set to be proportional to the two-dimensional Minkowski metric $\eta_{\alpha\beta}$, which defines distances as $-ds^2 = -d\tau^2 + d\sigma^2$. On the other hand, on the complex plane the natural metric is Euclidean: $ds^2 = |dz|^2 = dx^2 + dy^2$. Because of this sign difference, we cannot really prove that world-sheets can be treated as Riemann surfaces. But nothing stops us from trying to treat them as such, especially because it is known that much about a Minkowski theory can be learned from its Euclidean version. It turns out that thinking of world-sheets as Riemann surfaces leads to a consistent picture of string interactions.

In order to view world-sheets as Riemann surfaces, our first task is to find suitable complex coordinates. We assemble τ and σ into a complex coordinate w as follows:

$$w = \tau + i\sigma. \tag{25.20}$$

The simplest world-sheet is that of a freely propagating open string. As shown in Figure 25.6, the world-sheet for a string with light-cone momentum p^+ is the strip $0 \le \Im(w) \le 2\pi\alpha'p^+$ ($\Im(w)$ denotes the imaginary part of w and $\Re(w)$ denotes the real part of w). The strings are the vertical segments of constant τ. Since conformal maps do

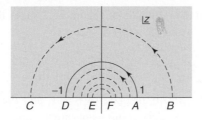
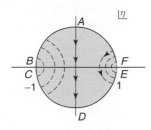

Fig. 25.7 Left: the infinite strip is mapped into $\bar{\mathbb{H}}$ via the exponential map. The infinite past is at $z = 0$, and the infinite future is at $z = \infty$. The open strings are the dashed lines. Right: $\bar{\mathbb{H}}$ is mapped onto the unit η disk via a linear fractional transformation. The infinite past is at $\eta = 1$, and the infinite future is at $\eta = -1$.

not change the physics, it is a good idea to familiarize ourselves with pictures that are conformally equivalent to the strip.

Let us see what happens to the strip when we map it conformally using the exponential function

$$z = \exp\left(\frac{w}{2\alpha' p^+}\right). \tag{25.21}$$

(Recall that $\exp(u + iv) = \exp(u)\,(\cos v + i \sin v\,)$.) The result of this mapping is shown in the left part of Figure 25.7. Note that the full string in the infinite past limit is mapped to a point! Indeed, for $w = -\tau_0 + i\sigma$, with $\tau_0 \to \infty$ we have

$$z = \exp\left(-\frac{\tau_0}{2\alpha' p^+}\right)\exp\left(i\frac{\sigma}{2\alpha' p^+}\right) \longrightarrow 0, \tag{25.22}$$

for all values of σ. As shown in the figure, the strings that approach the infinite past approach the point $z = 0$. Note also another curious fact. The two boundaries of the strip are mapped to the real line. The boundary $\sigma = 0$ is mapped to the positive half of the real line, while the boundary $\sigma = 2\pi\alpha' p^+$ is mapped to the negative half of the real line. The angle between the boundaries at the infinite past has been changed. Think of the boundaries as lines emerging from the past. While such boundaries make zero angle in the strip picture, they make an angle of 180° in the z-picture. The strip has been mapped into $\bar{\mathbb{H}}$, and the strings appear as semicircles centered at $z = 0$. The string at $\tau = 0$ is the unit semicircle. The strings in the far future are very large semicircles on the z-plane. They grow without bound as the strings approach the infinite future.

There is another useful presentation for the Riemann surface of a free open string. We can map $\bar{\mathbb{H}}$ onto the unit disk in the η plane via the transformation

$$\eta = \frac{1 + iz}{1 - iz}. \tag{25.23}$$

This mapping, illustrated in the right side of Figure 25.7, takes the real line $\Im(z) = 0$ into the unit circle. Indeed, for $z = x \in \mathbb{R}$ we have

$$|\eta|^2 = \eta\eta^* = \frac{1+ix}{1-ix} \cdot \frac{1-ix}{1+ix} = 1. \tag{25.24}$$

This time the infinite past $z = 0$ is mapped to $\eta = 1$, and the infinite future $z = \infty$ is mapped to $\eta = -1$. The two boundary components are the unit semicircles BAF and CDE.

We must draw one important lesson from the above maps. When the strip is mapped onto a bounded region, the complete strings in the infinite past and in the infinite future are mapped into two *points* on the boundary of the bounded region. More precisely, these points are limits that are not attained for any finite time string, so topologically these points are to be removed from the boundary of the bounded region: they are *punctures*. We say that strings are inserted at the punctures. Our conformal maps have shown that the infinite strip is conformally equivalent to a disk, twice-punctured on the boundary. The punctures break the boundary of the unit disk into two disjoint components. In the mapping of the strip to $\bar{\mathbb{H}}$, the point $z = 0$ is not attained for any string in the finite past. The far future on the strip is mapped to large z, and it will become clear in Section 25.6 that the "point at infinity" is not attained either. Therefore, the infinite strip maps onto $\bar{\mathbb{H}}$ with two points removed.

A free closed string can be represented by the string diagram in Figure 25.6, with the understanding that the top and bottom edges of the infinite strip are identified. More precisely, we identify points on the two edges that have the same value of $\Re(\tau)$. This identification results in an infinite cylinder of circumference $2\pi\alpha' p^+$. The vertical dashed segments become closed strings.

Quick calculation 25.1 Show that

$$z = \exp\left(\frac{w}{\alpha' p^+}\right) \tag{25.25}$$

maps the free closed string world-sheet to the full z-plane with the origin removed. Verify that the vertical dashed segments that represent closed strings in the w-plane become closed circles on the z-plane.

Note that under the map (25.25) the closed string in the infinite past is sent to $z = 0$, while the closed strings that approach the infinite future are mapped to larger and larger circles centered at the origin. The closed string world-sheet is the complex plane punctured at the origin.

25.5 Schwarz–Christoffel map and three-string interaction

Now that we have examined some conformal maps of the world-sheet of a freely propagating open string, we turn to the case of interacting strings. The simplest interaction is one where an open string breaks into two open strings, as shown in Figure 25.5. Following

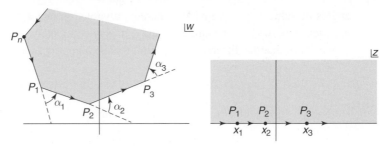

Fig. 25.8 The Schwarz–Christoffel map $w(z)$ takes $\bar{\mathbb{H}}$ onto the polygon. Shown are the vertices P_i, and the turning angles α_i. In the (bordered) upper half-plane, shown to the right, the x_i are mapped to the vertices of the polygon.

the intuition developed above, this surface should be conformally equivalent to a disk with three punctures on the boundary. In fact, the mapping theorem of Riemann guarantees that this interacting world-sheet can be conformally mapped into $\bar{\mathbb{H}}$, with the open string boundaries mapping into the real line.

Why do we want to map the world-sheet to $\bar{\mathbb{H}}$? $\bar{\mathbb{H}}$ gives us a canonical presentation that will allow us to compare different Riemann surfaces. This comparison will play a major role in the study of interactions that involve more than three strings. In mapping the string diagram to $\bar{\mathbb{H}}$ we fulfill step (1) of the list indicated at the beginning of Section 25.2.

It is not so easy to map the world-sheet in Figure 25.5 onto $\bar{\mathbb{H}}$. This world-sheet, however, can be viewed as the limit of a polygon, and there is a well established method for constructing the map that takes a polygon onto $\bar{\mathbb{H}}$. The map is conformal, except at the corners of the polygon. In general, one cannot write the conformal map explicitly, but one can always write a differential equation for the mapping function. The maps relating polygons and $\bar{\mathbb{H}}$ are called Schwarz–Christoffel maps.

Suppose we have a polygon in the complex w-plane. Let the polygon have n sides and therefore n vertices, denoted by $P_1, P_2, \ldots, P_n$. The Schwarz–Christoffel map $w(z)$ takes $z \in \bar{\mathbb{H}}$ into the polygon. The situation is illustrated in Figure 25.8. We choose the boundary of the polygon to be oriented so that the interior of the polygon lies to the left of the oriented boundary. The mapping takes the real line in the z-plane, oriented in the direction of increasing values, into the oriented polygon boundary.

The polygon has turning angles α_i at the vertices P_i. A turning angle is positive if the oriented side emerging from the vertex is obtained by counterclockwise rotation from the extension of the side that enters the vertex. By definition, we restrict turning angles to the interval $[-\pi, \pi]$

$$-\pi \leq \alpha_i \leq \pi. \tag{25.26}$$

A turning angle of value near to but smaller than π represents a corner where the polygon covers only a small sector. On the other hand, a turning angle of value near to but larger than $-\pi$ represents a corner where the polygon fails to cover a small sector. Turning angles of value π or $-\pi$ give degenerate polygons. In fact light-cone string diagrams are degenerate polygons. The angles π and $-\pi$ give completely different types of corners.

We claim that the differential equation for the mapping function $w(z)$ is

$$\frac{dw}{dz} = A(z - x_1)^{-\frac{\alpha_1}{\pi}}(z - x_2)^{-\frac{\alpha_2}{\pi}}\cdots(z - x_{n-1})^{-\frac{\alpha_{n-1}}{\pi}}. \tag{25.27}$$

Here the x_i are a collection of ordered real numbers:

$$x_1 < x_2 < \cdots < x_{n-1}, \tag{25.28}$$

and the map $z(w)$ takes the vertices to those points:

$$z(P_i) = x_i, \quad i = 1, 2, \ldots, n - 1. \tag{25.29}$$

Note that the turning angle α_n does not appear in the differential equation (25.27). In fact, the map constructed from (25.27) will take P_n to $z = \infty$:

$$z(P_n) = \infty. \tag{25.30}$$

Since a polygon is closed, the turning angles must sum to 2π:

$$\alpha_1 + \alpha_2 + \cdots + \alpha_n = 2\pi. \tag{25.31}$$

Thus α_n is determined by the other turning angles. This is why the map in (25.27) can properly handle the "missing" vertex (see Problem 25.6 for details).

Now we will explain why (25.27) is the correct differential equation for the Schwarz–Christoffel map. We do so by taking the argument of both sides of the equation. The argument of a complex number $z = r \exp(i\theta)$, with r real, is defined as $\arg(z) = \theta$. Since angles are multivalued, the arg function has an additive ambiguity, we can add or subtract any multiple of 2π. The function arg satisfies $\arg(z_1 z_2) = \arg(z_1) + \arg(z_2)$ and $\arg(z_1/z_2) = \arg(z_1) - \arg(z_2)$. Taking the argument of both sides of (25.27) and using these properties, we find

$$\arg(dw) - \arg(dz) = \arg A + \arg(z - x_1)^{-\frac{\alpha_1}{\pi}} + \cdots + \arg(z - x_{n-1})^{-\frac{\alpha_{n-1}}{\pi}}. \tag{25.32}$$

We fix our conventions by replacing the arg functions on the right-hand side with the principal value Arg, which satisfies

$$-\pi < \text{Arg}(z) \leq \pi. \tag{25.33}$$

Arg of a positive real number is zero, while Arg of a negative real number is π. If we define z^β as the principal branch $z^\beta \equiv \exp[\beta (\ln|z| + i\,\text{Arg}(z))]$, we then have

$$\text{Arg}(z^\beta) = \beta\,\text{Arg}(z), \quad \text{for} \quad |\beta| \leq 1. \tag{25.34}$$

The condition $|\beta| \leq 1$ guarantees that the right-hand side satisfies the inequalities appropriate for Arg of a complex number. Since $|\alpha_i/\pi| \leq 1$, we can use (25.34) to simplify (25.32):

$$\arg(dw) - \arg(dz) = \text{Arg}\,A - \frac{\alpha_1}{\pi}\text{Arg}(z - x_1) - \cdots - \frac{\alpha_{n-1}}{\pi}\text{Arg}(z - x_{n-1}). \tag{25.35}$$

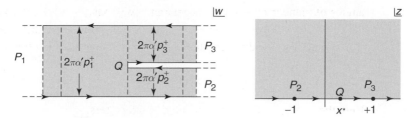

Fig. 25.9 The mapping of a three-string interaction into the upper half-plane. The string diagram is a degenerate polygon with vertices P_1, P_2, Q, and P_3. The interaction point Q is mapped to the point $z = x^*$. The vertex P_1 is mapped to infinity.

When $z = x$ is on the real line and $dz = dx > 0$, equation (25.35) gives

$$\arg(dw) = \mathrm{Arg}\, A - \frac{\alpha_1}{\pi}\mathrm{Arg}(x - x_1) - \cdots - \frac{\alpha_{n-1}}{\pi}\mathrm{Arg}(x - x_{n-1}). \qquad (25.36)$$

Since the real line in the z-plane maps to the boundary of the polygon, $\arg(dw)$ is the angle that an edge makes with respect to the horizontal axis. As long as x varies without crossing a turning point x_i, nothing changes on the right-hand side: the arguments of the positive quantities remain equal to zero, and the arguments of the negative quantities remain equal to π. This generates a straight edge in the w-plane. On the other hand, as x goes from the left side to the right side of a turning point x_i, $\mathrm{Arg}(x - x_i)$ changes from π to 0, a total change of $(-\pi)$. The right-hand side of the above equation then changes by $(-\alpha_i/\pi)(-\pi) = \alpha_i$. This means that dw along the new edge will have an argument larger by α_i, which is exactly what is supposed to happen. This confirms our claim that the differential equation does its job. Given a specific polygon, the turning points x_i as well as the constant A have to be calculated. These quantities depend on the length of the sides of the polygon. We will first apply the Schwarz–Christoffel transformation to the case of an open string splitting in two, as shown in Figure 25.9. This world-sheet can be viewed as a degenerate polygon with four vertices. If we orient the boundary of the polygon starting with the lower horizontal line, the first vertex we find is P_2, the second string in the infinite future. Here we have a turning angle of $+\pi$. The next edge travels back to the interaction point Q. This is the second vertex, and here the turning angle is $-\pi$. The next edge goes to the infinite future P_3 of the third string. Here we find our third vertex, where we have a turning angle of $+\pi$. Finally, the fourth vertex is P_1, the infinite past of string one. Here the turning angle is $+\pi$. In summary, we have

$$\begin{aligned}
&\text{turning angle at } P_2 = \alpha_2 = +\pi\,, \\
&\text{turning angle at } Q\; = \alpha_Q = -\pi, \\
&\text{turning angle at } P_3 = \alpha_3 = +\pi, \\
&\text{turning angle at } P_1 = \alpha_1 = +\pi.
\end{aligned} \qquad (25.37)$$

Note that the turning angles add up to 2π, as expected from (25.31). The turning angles are perhaps easier to appreciate in Figure 25.10, where the world-sheet is shown schematically

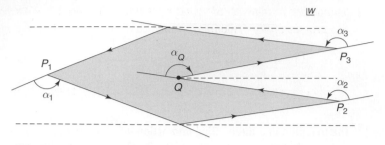

Fig. 25.10 A nondegenerate version of the degenerate polygon shown on the left side of Figure 25.9. In the degenerate limit, the small turning angles above and below Q disappear.

as an almost degenerate polygon by reducing the magnitude of the turning angles from the value of π and adding two small turning angles right above and right below Q.

We will map to $\bar{\mathbb{H}}$ by requiring that P_2 go to $z = -1$, P_3 go to $z = +1$, and P_1 go to $z = \infty$. Having one vertex at infinity is implicit in the differential equation we wrote, and it simplifies matters. You may ask: how do we know that we can map the other two points, P_2 and P_3, to the chosen values of z? The answer is based on a result obtained in the next section: three points on the real line can be mapped to any specific coordinates while preserving $\bar{\mathbb{H}}$ and the ordering of the points. These three points are P_2, P_3, and P_1, and we have chosen to map them to -1, $+1$, and ∞. We cannot, however, specify the coordinate x^* of the last vertex Q. As shown in the right part of Figure 25.9, x^* must lie between -1 and $+1$, since the vertex Q lies in between the vertices P_2 and P_3 of the oriented polygon.

We now proceed to write the differential equation for this map, following (25.27). With ordered turning points $z = -1, x^*, +1$ and turning angles $\alpha_2, \alpha_Q, \alpha_3$ we write

$$\frac{dw}{dz} = A(z - (-1))^{-\frac{\alpha_2}{\pi}} (z - x^*)^{-\frac{\alpha_Q}{\pi}} (z - 1)^{-\frac{\alpha_3}{\pi}}. \tag{25.38}$$

Using the values for the angles from (25.37), we obtain

$$\frac{dw}{dz} = A\frac{(z - x^*)}{(z + 1)(z - 1)}. \tag{25.39}$$

For z real and greater than one, dw/dz must be real and negative, since the image of $z > 1$ is the horizontal line between P_3 and P_1. Therefore, we conclude that A is real and negative. In order to integrate this equation, we decompose the right-hand side using partial fractions:

$$\frac{dw}{dz} = \frac{A}{2}\left(\frac{1 + x^*}{z + 1} + \frac{1 - x^*}{z - 1}\right). \tag{25.40}$$

Integrating, we obtain

$$w = \frac{A}{2}(1 + x^*)\ln(z + 1) + \frac{A}{2}(1 - x^*)\ln(z - 1). \tag{25.41}$$

Here we define the logarithms by $\ln(z) = \ln|z| + i\,\mathrm{Arg}(z)$, with the logarithm of a real number taken to be real. We still need to determine the constants A and x^* in terms of the parameters p_2^+ and p_3^+ of the string diagram. We need $\sigma = \Im(w)$ to increase by $2\pi\alpha' p_2^+$ as z crosses $z = -1$, the image of P_2. Since $\Im(\ln z) = \mathrm{Arg}(z)$, and A is real, we have

$$\sigma = \Im(w) = \frac{A}{2}(1 + x^*)\mathrm{Arg}(z + 1) + \frac{A}{2}(1 - x^*)\mathrm{Arg}(z - 1). \tag{25.42}$$

As z goes from the left to the right of -1, $\mathrm{Arg}(z + 1)$ goes from π to 0, and thus

$$\Delta\sigma = \frac{A}{2}(1 + x^*)(0 - (\pi)) = 2\pi\alpha' p_2^+. \tag{25.43}$$

This gives

$$\frac{A}{2}(1 + x^*) = -2\alpha' p_2^+. \tag{25.44}$$

Similarly, the behavior near $z = 1$ requires

$$\frac{A}{2}(1 - x^*) = -2\alpha' p_3^+. \tag{25.45}$$

With these two equations, back in (25.41) we have

$$w = -2\alpha' p_2^+ \ln(z + 1) - 2\alpha' p_3^+ \ln(z - 1). \tag{25.46}$$

This is the equation for the map, written in terms of known parameters.

Quick calculation 25.2 Convince yourself that the real axis of the w-plane coincides with the top edge of the strip on the left side of Figure 25.9.

In the future, we will be able to write equations like (25.46) by inspection. The rules are clear. We need a logarithm for each of the strings, except for the one whose turning point is mapped to infinity in the upper half-plane. The prefactor of each of these logarithms is related to the width of the corresponding string – it equals $(-2\alpha' p^+)$ when σ increases as we cross the turning point. We will use this understanding to write the map for a four-string interaction in Section 25.7.

The map (25.46) was written without calculating x^*. We can find x^* by dividing equations (25.44) and (25.45):

$$\frac{1 + x^*}{1 - x^*} = \frac{p_2^+}{p_3^+} \longrightarrow x^* = \frac{p_2^+ - p_3^+}{p_2^+ + p_3^+}. \tag{25.47}$$

If $p_2^+ = p_3^+$, then $x^* = 0$, as we would expect, since Q is then half-way between P_2 and P_3. If $p_2^+ \gg p_3^+$, the figure leads us to believe that, conformally speaking, Q is getting close to P_3. This is indeed the case, for then $x^* \to 1$. On the other hand, if $p_3^+ \gg p_2^+$ then $x^* \to -1$, and indeed Q is getting closer to P_2.

The value of x^* can also be calculated from (25.46). For this, note that $dw/dz = 0$ at $z = x^*$, because at Q, the derivative dw/dz goes from negative to positive (τ reaches a local minimum). Using (25.46) we find that

$$\frac{dw}{dz} = -\frac{2\alpha' p_2^+}{x^* + 1} - \frac{2\alpha' p_3^+}{x^* - 1} = 0, \tag{25.48}$$

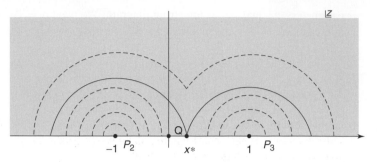

Fig. 25.11 Three-string interaction as seen in $\bar{\mathbb{H}}$. Shown as dashed lines are the images of the (vertical) strings in Figure 25.9. The string at the interaction time is shown as a solid line.

which leads to the same value for x^*. It is interesting to look at the strings in the z-plane, as shown in Figure 25.11. Observe the strings converging into $z = -1$ and $z = +1$. The strings arriving from P_1 at $z = \infty$ eventually reach the interaction point. The string at the interaction time is shown as a solid line. Its endpoints are to the left and to the right of P_2 and P_3, respectively, and one of its interior points touches the boundary at x^*.

Quick calculation 25.3 Consider the three-string interaction when $p_2^+ = p_3^+$, so that $x^* = 0$. Check that $\Re(w(x^*)) = 0$. Show that the endpoints of the string at the interaction time lie at $z = \pm\sqrt{2}$.

The amplitude for the three-string interaction considered here includes the open string coupling g_o as a multiplicative factor ($g_o^2 \sim g$, where g is the closed string coupling). This is analogous to the amplitude (25.2), which also includes the coupling constant.

25.6 Moduli spaces of Riemann surfaces

Since surfaces related by a conformal map are, by definition, the same Riemann surface, it is important to understand when we can map a pair of surfaces into each other. To make this question more concrete, consider a simple example. Take the upper half-plane $\bar{\mathbb{H}}$ with one puncture at a position x on the real axis. Can this configuration be mapped to the upper half-plane $\bar{\mathbb{H}}$ with a puncture $x' \neq x$? Are these two the same Riemann surface? They are. Use the coordinate z for the first $\bar{\mathbb{H}}$ and the coordinate z' for the second $\bar{\mathbb{H}}$. Then take

$$z' = z + (x' - x). \tag{25.49}$$

This conformal map (z' is an analytic function of z) takes the puncture at $z = x$ into the puncture at $z' = x'$. This map also preserves $\bar{\mathbb{H}}$ since $\Im(z') = \Im(z)$. The upshot of this example is that $\bar{\mathbb{H}}$ with one puncture on the boundary is a Riemann surface without a parameter. Whichever point is chosen, the Riemann surface is the same. We call the

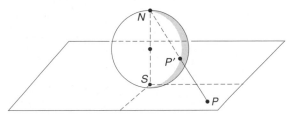

Fig. 25.12 A map from the plane to the sphere. The sphere is resting on the plane with the south pole S at the origin. For each point P in the plane we associate the point P' on the sphere. The point at infinity maps to the north pole N.

parameters of Riemann surfaces moduli, or modulus, if there is just one. Thus, $\bar{\mathbb{H}}$ with one puncture on the real axis has no moduli.

As we will see here, and in the exercises at the end of this chapter, the story is more interesting when we add additional punctures on the real line. $\bar{\mathbb{H}}$ with two punctures still has no moduli! The positions of the punctures can be changed at will. $\bar{\mathbb{H}}$ with three punctures still has no continuous modulus, but it does have one discrete modulus, which describes the ordering of the punctures on the boundary of $\bar{\mathbb{H}}$. Finally, $\bar{\mathbb{H}}$ with four punctures has one continuous modulus as well as a discrete one.

We begin our analysis with the complex plane $\mathbb{C}$, extended to define the Riemann sphere. We will then return to the case of $\bar{\mathbb{H}}$, which is immediately applicable to open string interactions. Our analysis of the Riemann sphere will be needed in the following chapter, where we discuss closed strings.

The complex plane $\mathbb{C}$ can be extended to construct the Riemann sphere $\hat{\mathbb{C}}$. Imagine a sphere sitting atop the complex plane, such that the south pole coincides with the origin and the north pole is directly above it (see Figure 25.12). Every point P on the complex plane can be matched to a point P' on the sphere using a stereographic projection: P' is the intersection of the sphere with the line that connects P to the north pole. Clearly, this is a one-to-one map. As we move out on the complex plane, the points on the sphere approach the north pole. The image of $\mathbb{C}$ under the map is the sphere without its north pole – the sphere minus one point. To build the complete sphere we must extend the complex plane by the inclusion of a "point at infinity," a point $\{\infty\}$ whose image under the map is declared to be the north pole. Via the stereographic map, the extended complex plane $\mathbb{C} \cup \{\infty\}$ is equivalent to the Riemann sphere $\hat{\mathbb{C}}$. This is why the Riemann sphere is sometimes called the extended complex plane. In practice, to work with a Riemann sphere, we use the complex plane with the understanding that $z = \infty$ is a point, to be treated just like every finite point.

The Riemann sphere $\hat{\mathbb{C}}$ is relevant to closed strings. Consider the map (25.25) from the infinite cylinder to the complex z-plane. There is some asymmetry here: in the cylinder picture the closed strings in the infinite past and infinite future appear symmetrically, but on the plane the first is seen as a puncture at $z = 0$, while the second is mapped to an infinite circle. To restore the symmetry the solution is clear: the infinite circle on the plane must be viewed as a single puncture. We therefore say that the world-sheet of a free closed string is the extended complex plane $\mathbb{C} \cup \{\infty\}$ with the point $z = 0$ removed and the point at infinity removed. Equivalently, the world-sheet of a free closed string is a Riemann sphere $\hat{\mathbb{C}}$ with two punctures. The punctures are $z = 0$ and the point at infinity; both are on the same footing in $\hat{\mathbb{C}}$.

What is the most general analytic one-to-one map from $\hat{\mathbb{C}}$ to $\hat{\mathbb{C}}$? Mathematicians have shown that it is a *linear fractional transformation*. Consider two copies of $\hat{\mathbb{C}}$, with coordinates z and w. A linear fractional transformation takes the form

$$w = \frac{az + b}{cz + d}, \tag{25.50}$$

where a, b, c, and d are complex numbers. Note that the point at infinity in the z-plane is mapped to $w = a/c$, which is typically a finite number. This is consistent with our view that in $\hat{\mathbb{C}}$ the point at infinity is a generic point. For the map to be one-to-one, we must, in addition, require

$$ad - bc \neq 0. \tag{25.51}$$

If $a = b = 0$, for example, this condition is violated, and (25.50) gives $w = 0$ for all z. To show that (25.51) guarantees a one-to-one map, we must show that $w(z_1) = w(z_2)$ implies $z_1 = z_2$. Using (25.50), the equality $w(z_1) = w(z_2)$ gives

$$\frac{az_1 + b}{cz_1 + d} = \frac{az_2 + b}{cz_2 + d}. \tag{25.52}$$

If $cz_1 + d = 0$ the left-hand side is infinite, so the right-hand side must be infinite as well. This requires $cz_2 + d = 0$, so then $z_1 = z_2$. If the denominators do not vanish, we can cross multiply, and, after cancelling common terms, we find:

$$(ad - bc)z_1 = (ad - bc)z_2, \tag{25.53}$$

which, on account of (25.51), implies $z_1 = z_2$. The linear fractional map is an invertible transformation. We can easily solve for z in terms of w to get

$$z = \frac{dw - b}{-cw + a}. \tag{25.54}$$

This is also a linear fractional map, and it is also one-to-one, as you can see by examining the condition in (25.51). The map in (25.50) can be rewritten, after cross multiplication, in the form

$$Awz + Bw + Cz + D = 0, \tag{25.55}$$

for complex constants A, B, C, and D. This is just another form that should be recognized as defining a linear fractional transformation. The symmetric role of w and z is manifest here.

Quick calculation 25.4 What is the condition on the coefficients A, B, C, and D for (25.55) to be a linear fractional transformation?

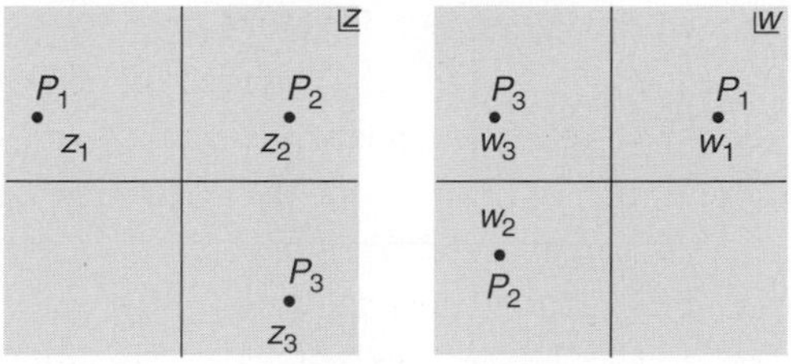

Fig. 25.13 Three punctures P_1, P_2, and P_3 with coordinates z_1, z_2, and z_3, respectively, can be conformally mapped to arbitrary positions w_1, w_2, and w_3 in the w-plane.

How many parameters does the map (25.50) have? Although the map is written using four complex numbers, a, b, c, and d, there are only three parameters. If we multiply a, b, c, and d by the same complex number $\alpha \neq 0$, the map in (25.50) will not change since the constant factor appears both in the numerator and in the denominator. Since $ad - bc \neq 0$, we can rescale the parameters a, b, c, and d by dividing them by $\sqrt{ad - bc}$. The new, rescaled parameters satisfy

$$ad - bc = 1. \tag{25.56}$$

It is now clear that we have only three adjustable complex parameters: given a, b, and c, for example, the equation fixes d.

Having three complex parameters, we might expect that a linear fractional transformation can map a z-sphere with three (different) punctures at z_1, z_2, and z_3, into a w-sphere where the three punctures go to *arbitrary* (different) coordinates w_1, w_2, and w_3 (see Figure 25.13). This is indeed the case, and it shows that $\hat{\mathbb{C}}$ with three punctures is a Riemann surface without moduli.

Let us now construct the map. We write

$$(z - z_1)\,(\cdots) = (w - w_1)\,(\cdots), \tag{25.57}$$

because we want that $w = w_1$ when $z = z_1$. The dots indicate additional multiplicative factors yet to be included. Clearly z_1 will go to w_1 since, regardless of the multiplicative factors, both sides of the equation are equal if $z = z_1$ and $w = w_1$. Next we want $w = w_2$ when $z = z_2$. To achieve this, we put the factors $w - w_2$ and $z - z_2$ in the denominators:

$$\frac{z - z_1}{z - z_2}\,(\cdots) = \frac{w - w_1}{w - w_2}\,(\cdots). \tag{25.58}$$

If $w = w_2$, the equality requires $z = z_2$. In order to incorporate the condition that $z = z_3$ when $w = w_3$, we simply include numerical factors that make each side equal to one when this happens:

$$\frac{z - z_1}{z - z_2}\,\frac{z_3 - z_2}{z_3 - z_1} = \frac{w - w_1}{w - w_2}\,\frac{w_3 - w_2}{w_3 - w_1}. \tag{25.59}$$

The above equation is our final result for the map that takes z_1, z_2, and z_3 to w_1, w_2, and w_3, respectively. It is a linear fractional transformation: by cross multiplying, we find that (25.59) is in the form of (25.55). Spheres with three punctures have no moduli because all

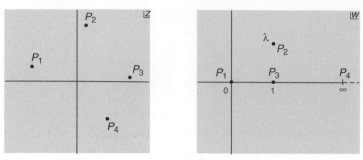

Fig. 25.14 Four punctures in the complex z-plane can be mapped uniquely into the w-plane by requiring that P_1, P_3, and P_4 go to $w = 0$, 1, and ∞, respectively. In this canonical presentation, the remaining puncture P_2 will go to some value $w = \lambda$.

such spheres are conformally equivalent; they can be mapped into each other using (25.59). In Problem 25.1 you will show that (25.59) is the *unique* map of $\hat{\mathbb{C}}$ taking an ordered set of three different points to another ordered set of three different points.

Consider now $\hat{\mathbb{C}}$ with four punctures at z_1, z_2, z_3, and z_4, and another copy of $\hat{\mathbb{C}}$ with *four* punctures at w_1, w_2, w_3, and w_4. Think of the punctures as labeled punctures, with labels 1, 2, 3, and 4. Can these copies of $\hat{\mathbb{C}}$ be mapped into each other with the punctures going into each other, that is, with $z_i \to w_i$, for $i = 1, 2, 3$, and 4? In general the answer is no. To map the four punctures into each other, we must certainly map z_1, z_2, and z_3 into w_1, w_2, and w_3. This requirement fixes the map (25.59), giving us the unique map capable of mapping the punctures. Being unique, we cannot incorporate an additional condition, and z_4 will go to some point that we cannot prescribe. Of course, the map may happen to send z_4 to the prescribed w_4, but that would be accidental, since z_4 and w_4 were not used to construct the map. Therefore, spheres with four labeled punctures are not necessarily conformally equivalent.

Suppose someone hands you a z-sphere and a z'-sphere, each with four punctures P_1, P_2, P_3, and P_4. How do you decide whether they are conformally equivalent? To find out, first take each sphere and map P_1, P_3, and P_4 to 0, 1, and ∞. This requires linear fractional maps $w = f(z)$ and $w' = f'(z')$. The question of conformal equivalence of the z- and z'-spheres is now the question of conformal equivalence of the w- and w'-spheres. Now look at P_2. Suppose P_2 is sent to $w = \lambda$ in the w sphere and to $w' = \lambda'$ in the w'-sphere. The four-punctured w- and w'-spheres are conformally equivalent *if and only if* $\lambda = \lambda'$. The only conformal map $w \to w'$ preserving the punctures P_1, P_3, and P_4 (set at 0, 1, and ∞ in both) is the identity map. If the P_2 punctures are to go into each other, their coordinates must be the same. This establishes the claim. In the canonical presentation, where three punctures are sent to 0, 1, and ∞, the position λ of the last puncture is therefore a modulus of the four-punctured sphere (see Figure 25.14). Since this position is a complex number, we have a complex modulus, or equivalently, two real moduli. These moduli are continuous parameters, since the position of the fourth puncture can be varied continuously.

The complex modulus λ which represents the position of a puncture can take all complex values except for 0, 1, and ∞, since in those cases two of the punctures would coincide. Shifting the viewpoint, consider a new Riemann sphere, with coordinate λ, and with the points 0, 1, and ∞ removed. To each point λ_0 of this sphere we associate a sphere with punctures at 0, 1, ∞ *and* λ_0. With this association, two different points in the λ-sphere represent conformally inequivalent four-punctured spheres. Finally, every four-punctured sphere is associated with a point on the λ-sphere. We call the λ-sphere, with 0, 1, and ∞ removed, the *moduli space* $\mathcal{M}_{0,4}$ of four-punctured spheres. The zero tells us that we are dealing with Riemann spheres (which are surfaces of genus zero), and the four indicates that the surfaces have four punctures, assumed to be labeled. Being itself a two-dimensional surface, the (real) dimensionality of $\mathcal{M}_{0,4}$ is two:

$$\dim(\mathcal{M}_{0,4}) = 2. \qquad (25.60)$$

Amusingly, the moduli space of four-punctured Riemann spheres is itself a Riemann surface, a sphere λ with three punctures. Understanding a moduli space means that we have control over all the inequivalent Riemann surfaces of a given type. The moduli space $\mathcal{M}_{0,4}$ will be important for our discussion of closed string amplitudes in Chapter 26. To understand open string scattering, we now consider $\bar{\mathbb{H}}$ with punctures on the real line and the moduli spaces of such Riemann surfaces.

Recall that $\bar{\mathbb{H}}$ is defined as the upper half-plane with the real line included and a "point at infinity" included, as well. This point at infinity is added in the same way as it was added for the Riemann sphere. In $\bar{\mathbb{H}}$, both plus and minus infinity on the real line are the same point, and so is the limit of $r \exp(i\theta)$, for $0 \leq \theta \leq \pi$, when $r \to \infty$. The point at infinity is part of the boundary of $\bar{\mathbb{H}}$; this boundary is a circle. Note that $\bar{\mathbb{H}}$ is conformally equivalent to the unit disk, as established by the map (25.23). In this map, the boundary of $\bar{\mathbb{H}}$ maps onto the boundary of the disk, and the point at infinity is mapped to $\eta = -1$. We can now clearly understand that the infinite strip is conformally equivalent to $\bar{\mathbb{H}}$ with the point $z = 0$ removed, and the point at infinity removed, as well. The point at infinity is removed because it is not attained for any finite time string. The infinite strip is conformally equivalent to $\bar{\mathbb{H}}$ twice punctured at the boundary

Let us consider the self-maps of $\bar{\mathbb{H}}$. A reasonable approach is to investigate the self-maps of $\hat{\mathbb{C}}$ that preserve the upper half-plane $\bar{\mathbb{H}}$. Following (25.50), we consider transformations of the form

$$w = \frac{az + b}{cz + d}. \qquad (25.61)$$

If a, b, c, and d are real, the real line in the z-plane maps to the real line in the w-plane. In fact, up to a common phase that cancels in (25.61), the condition that the real line is mapped to the real line implies that (Problem 25.3)

$$a, b, c, d \in \mathbb{R}. \qquad (25.62)$$

Finally, there is one additional condition. As before, the map is one-to-one if and only if

$$ad - bc \neq 0. \qquad (25.63)$$

Since we are now working with real coefficients, the left-hand side above is a number that is either positive or negative. Since we can only rescale the coefficients using real numbers, we can set $ad - bc = 1$ if the left-hand side is positive, and we can set $ad - bc = -1$ if the left-hand side is negative. It turns out that to preserve $\bar{\mathbb{H}}$ we need the former:

$$ad - bc = 1. \tag{25.64}$$

Indeed, taking the imaginary part of (25.61), we find

$$\Im(w) = \Im\left(\frac{az+b}{cz+d}\right) = \frac{1}{2i}\left(\frac{az+b}{cz+d} - \frac{a\bar{z}+b}{c\bar{z}+d}\right). \tag{25.65}$$

Simplifying this expression, we get

$$\Im(w) = \frac{1}{2i}\frac{(ad-bc)(z-\bar{z})}{|cz+d|^2} = \frac{ad-bc}{|cz+d|^2}\Im(z). \tag{25.66}$$

This equation makes it clear that $\Im(z) \geq 0$ implies $\Im(w) \geq 0$ if and only if $ad - bc > 0$. Therefore, (25.64) guarantees that (25.61) preserves the upper half-plane. All-in-all, the self-maps of $\bar{\mathbb{H}}$ are defined by equations (25.61), (25.62), and (25.64). They are characterized by three real parameters. In $\bar{\mathbb{H}}$ the point at infinity is treated like a regular point; its image under (25.61) is $w = a/c$.

Let us consider a few examples. Is the map

$$w = -z \tag{25.67}$$

a map of $\bar{\mathbb{H}}$ to $\bar{\mathbb{H}}$? It is not: a point z on the positive imaginary axis, for example, will be mapped to a point w on the negative imaginary axis. Therefore (25.67) cannot be written in the form (25.61) with the requisite conditions. Indeed, we can try to write

$$w = \frac{(-1)\cdot z + 0}{0\cdot z + 1} \quad\rightarrow\quad a = -1,\ b = 0,\ c = 0,\ d = 1, \tag{25.68}$$

but this gives $ad - bc = -1$, which violates condition (25.64). How about

$$w = z + 1\,? \tag{25.69}$$

This maps $\bar{\mathbb{H}}$ to $\bar{\mathbb{H}}$. In fact, it simply translates $\bar{\mathbb{H}}$ to the right by one unit. A more interesting map is

$$w = -\frac{1}{z}\,. \tag{25.70}$$

To see why this one works, first note that it can be written in the form (25.61) by choosing

$$w = \frac{0\cdot z + (-1)}{1\cdot z + 0} \quad\rightarrow\quad a = 0,\ b = -1,\ c = 1,\ d = 0. \tag{25.71}$$

Since $ad - bc = 1$, condition (25.64) is satisfied. More directly, write $z = re^{i\theta}$. Then $w = -\frac{1}{z} = -\frac{1}{r}e^{-i\theta}$. What does this transformation look like in the complex plane? Consider a complex number z with positive imaginary part, as shown in Figure 25.15. The vector $\frac{1}{r}e^{-i\theta}$ is obtained from z by reflection across the real axis and scaling. This vector is therefore in the lower half-plane. However, $w = -\frac{1}{z}$, and the negative sign reflects the

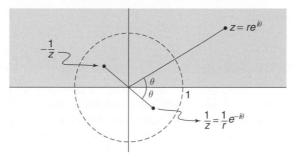

Fig. 25.15 Explaining why the map $z \to -1/z$ takes $\bar{\mathbb{H}}$ onto itself.

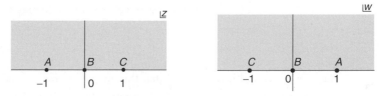

Fig. 25.16 There is no conformal map of the upper half-plane into itself that can reverse the order of three points on the real line.

vector across the origin. The final vector w is therefore in the upper half-plane. It is now clear that (25.70) maps the upper half-plane onto itself.

You might imagine that a real linear fractional transformation, having three real parameters, can be used to map $\bar{\mathbb{H}}$ with three punctures on the real line into $\bar{\mathbb{H}}$ with the three punctures at arbitrary real positions. This is almost true! You can indeed use (25.59) to construct a map, but one thing might go wrong. Condition (25.64) may fail to hold. It is interesting to see why this happens, as it has to do with an important feature of open strings.

The maps (25.61) that we are considering satisfy

$$\frac{dw}{dz} = \frac{(ad - bc)}{(cz + d)^2},\qquad(25.72)$$

and, since $ad - bc > 0$, this derivative is positive throughout the real line. This means that the map takes the *oriented* real line (say from left to right) into a real line with the same orientation! In fact, we can think of the real line in $\bar{\mathbb{H}}$ as a circle that includes the point at infinity. Three points on the real line can be mapped into three other points if and only if their *cyclic* orderings, as points on a circle, agree. Let $(P_1 P_2 P_3)$ denote the punctures 1, 2, and 3, if they appear in this order as we travel on the real line from left to right. Then this configuration can be mapped to any other configuration as long as the cyclic ordering is preserved. It can be mapped to the configuration $(P_2 P_3 P_1)$, for example, but not to the configuration $(P_2 P_1 P_3)$.

Consider an example. Let A, B, and C be points on the z-plane that lie at -1, 0, and 1, respectively (Figure 25.16). Let us try to build a map to the w-plane that takes the points

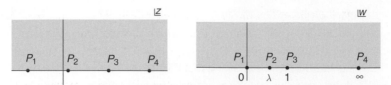

Fig. 25.17 The z upper half-plane with four punctures on the real line can be presented canonically by mapping it to the w upper half-plane and setting three of the punctures, say P_1, P_3, and P_4, to $w = 0, 1$, and ∞. The puncture P_2 will have $w = \lambda$ with $0 < \lambda < 1$.

to $1, 0$, and -1, respectively. In effect, we are trying to reverse the order of the points on the real axis. Using (25.59), we find that we need the following mapping:

$$\frac{z-1}{z+1} = \frac{w+1}{w-1}. \tag{25.73}$$

When we cross multiply, we find that this map is $w = -z$, and we showed before that this does not map $\bar{\mathbb{H}}$ to $\bar{\mathbb{H}}$. Since there are only two inequivalent cyclic orderings of three points on the boundary of a disk, the moduli space $\mathcal{N}_3$ of $\bar{\mathbb{H}}$ with three punctures on its boundary has two representatives. One represents $\bar{\mathbb{H}}$ with three punctures ordered as in $(P_1 P_2 P_3)$, and the other represents $\bar{\mathbb{H}}$ with three punctures ordered as in $(P_2 P_1 P_3)$. We can say that there is one discrete modulus taking two values, and that is all! Note that the moduli space $\mathcal{M}_{0,3}$ of three-punctured spheres had no discrete modulus – all such spheres were conformally equivalent. The moduli space $\mathcal{M}_{0,3}$ is just a point, while the moduli space $\mathcal{N}_3$ is two points. While there is an obvious way to define an ordering for points that lie on a circle (the boundary of $\bar{\mathbb{H}}$), there is no way to define an ordering for points on a sphere.

Consider $\bar{\mathbb{H}}$ with four labeled punctures, P_1, P_2, P_3, and P_4, on the real line, and fix their cyclic ordering to be $(P_1 P_2 P_3 P_4)$ (Figure 25.17). Can we map this $\bar{\mathbb{H}}$ to itself and take the four punctures to arbitrary positions that preserve the cyclic ordering? No. There are just three real parameters in the most general map of $\bar{\mathbb{H}}$ to $\bar{\mathbb{H}}$, and therefore only three positions can be specified. Following the same strategy as before, we can map three of the punctures to fixed w coordinates. For instance, let us map P_1 to $w = 0$, P_3 to $w = 1$, and P_4 to $w = \infty$. Under this map, P_2 will be taken to $w = \lambda$, where λ is some real number. Because the cyclic ordering is preserved, P_2 must still lie between P_1 and P_3, so $0 < \lambda < 1$.

We can view $0 < \lambda < 1$ as the space that represents all possible $\bar{\mathbb{H}}$ with four ordered punctures on the boundary. For any λ_0 in this interval there is an associated $\bar{\mathbb{H}}$ punctured at $0, \lambda_0, 1$ and ∞. Two points with different λ values represent conformally inequivalent $\bar{\mathbb{H}}$s with four boundary punctures. Finally, no $\bar{\mathbb{H}}$ with four boundary punctures is missed in this interval. We thus call $0 < \lambda < 1$ the moduli space $\mathcal{N}_4$ of *upper half-planes with four labeled boundary punctures of a given cyclic ordering*. This moduli space has one real parameter and is just an open interval. It is a one-dimensional space:

$$\dim \mathcal{N}_4 = 1. \tag{25.74}$$

The surfaces $\bar{\mathbb{H}}$ with four boundary punctures are important because they are the surfaces that arise for interactions which involve four open strings. As we will see below, the world-sheet for such a process is conformally equivalent to an $\bar{\mathbb{H}}$ with four boundary punctures. We will also see that the string diagram has one parameter, the time difference T between the interaction points. We will prove a remarkable result: as T varies over its natural range, the string diagrams produce all $\bar{\mathbb{H}}$ with four boundary punctures! That is, we generate the full moduli space $\mathcal{N}_4$. To prove this we must study the function $\lambda(T)$, which gives the modulus as a function of T. This, of course, requires the conformal map from the string diagram to $\bar{\mathbb{H}}$. We must show that, as T goes over its natural range, $\lambda(T)$ ranges from zero to one. We now turn to this analysis.

Quick calculation 25.5 How many inequivalent orderings of four points are possible on the boundary of $\bar{\mathbb{H}}$. What does this tell us about the moduli space of $\bar{\mathbb{H}}$ with four boundary punctures?

25.7 Four open string interaction

Let us consider the process where two open strings join to form an intermediate string, which then splits into two open strings. The incoming strings are strings three and four, and the outgoing strings are strings one and two. The string diagram is shown in Figure 25.18.

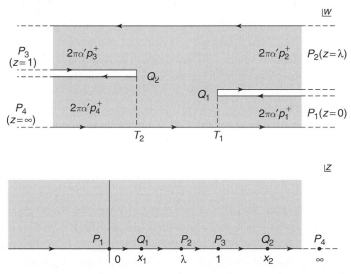

Fig. 25.18 Top: a light-cone string diagram showing strings three and four joining to form an intermediate string that then splits into strings one and two. Bottom: the light-cone diagram is mapped into $\bar{\mathbb{H}}$. P_1, P_3, and P_4 are go to 0, 1, and ∞. The interaction points Q_1 and Q_2 go to x_1 and x_2, and P_2 goes to λ.

Note that the interaction point Q_1 occurs at $\tau = T_1$, and Q_2 occurs at $\tau = T_2$. The time interval between interactions is

$$T = T_1 - T_2. \tag{25.75}$$

In the diagram shown in the figure $T > 0$. But the set of all possible interactions also includes the cases when $T < 0$. In those cases the slits move across each other. The complete range of T for this interaction is

$$-\infty < T < \infty. \tag{25.76}$$

There is an exception if $p_2^+ = p_3^+$. In this case the slits collide when $T = 0$, and T cannot be negative. You may examine such a situation in Problem 25.7.

We now use a Schwarz–Christoffel map to take this diagram into $\bar{\mathbb{H}}$. The special points on the boundary of the string diagram are ordered as $(P_1 Q_1 P_2 P_3 Q_2 P_4)$. Three points can go to arbitrary positions (consistent with the cyclic ordering), so we choose P_1, P_3, and P_4 to go to 0, 1, and ∞, respectively. Then P_2, whose position plays the role of modulus, must map to some $0 < \lambda < 1$. Since Q_1 lies between P_1 and P_2 and Q_2 lies between P_3 and P_4, their images x_1 and x_2 in $\bar{\mathbb{H}}$ satisfy

$$0 < x_1 < \lambda, \quad 1 < x_2 < \infty. \tag{25.77}$$

Our aim is to calculate the modulus λ of the $\bar{\mathbb{H}}$ with four boundary punctures as a function of the parameters T and p_i^+ of the string diagram:

$$\lambda = \lambda(T; \ p_1^+, p_2^+, p_3^+, p_4^+). \tag{25.78}$$

We want to prove a remarkable fact: the set of all possible string diagrams $T \in (-\infty, \infty)$ will give all $\lambda \in (0, 1)$, the full moduli space $\mathcal{N}_4$ of $\bar{\mathbb{H}}$ with four boundary punctures. λ is manifestly a well defined function of T: given T, we can calculate λ. Although we will not prove it, the function $\lambda(T)$ is in fact one-to-one. This shows that two string diagrams with different values of T give different Riemann surfaces. As a result, the set of string diagrams gives each surface in $\mathcal{N}_4$ once.

The differential equation for the conformal map is readily written using (25.27). At the points P_i the turning angles are $+\pi$, and at the interaction points Q_i the turning angles are $-\pi$. We therefore have

$$\frac{dw}{dz} = A \frac{1}{z} (z - x_1) \frac{1}{(z - \lambda)} \frac{1}{(z - 1)} (z - x_2). \tag{25.79}$$

We now need the partial fraction expansion of the right-hand side. We can use a trick valid when the factors in the denominator are all different, as they are in (25.79). The coefficient of $1/z$, for example, is obtained by deleting the $1/z$ factor and evaluating the rest of the right-hand side for $z = 0$. Similarly, to determine the coefficient of $1/(z - \lambda)$, you delete this factor, and evaluate the rest at $z = \lambda$. Removing the A factor, the right-hand side is therefore

$$\frac{1}{z} \frac{(-x_1)(-x_2)}{(-\lambda)(-1)} + \frac{1}{(z - \lambda)} \frac{(\lambda - x_1)(\lambda - x_2)}{\lambda(\lambda - 1)} + \frac{1}{(z - 1)} \frac{(1 - x_1)(1 - x_2)}{(1 - \lambda)}. \tag{25.80}$$

Simplifying, equation (25.79) becomes

$$\frac{dw}{dz} = A\left[\frac{1}{z}\frac{x_1 x_2}{\lambda} + \frac{1}{z-\lambda}\frac{(\lambda - x_1)(x_2 - \lambda)}{\lambda(1-\lambda)} - \frac{1}{z-1}\frac{(1-x_1)(x_2-1)}{1-\lambda}\right]. \qquad (25.81)$$

We also know a way to write the conformal map directly, as explained below equation (25.46). For the present case it gives

$$w(z) = -2\alpha' p_1^+ \ln z - 2\alpha' p_2^+ \ln(z-\lambda) + 2\alpha' p_3^+ \ln(z-1). \qquad (25.82)$$

Taking the derivative,

$$\frac{dw}{dz} = -2\alpha' p_1^+ \frac{1}{z} - 2\alpha' p_2^+ \frac{1}{z-\lambda} + 2\alpha' p_3^+ \frac{1}{z-1}. \qquad (25.83)$$

We can compare the two differential equations (25.81) and (25.83) to obtain three equations which relate the string diagram momentum parameters to the parameters in the $\overline{\mathbb{H}}$ presentation:

$$-2\alpha' p_1^+ = A\,\frac{x_1 x_2}{\lambda},$$
$$-2\alpha' p_2^+ = A\,\frac{(\lambda - x_1)(x_2 - \lambda)}{\lambda(1-\lambda)},$$
$$-2\alpha' p_3^+ = A\,\frac{(1-x_1)(x_2-1)}{1-\lambda}. \qquad (25.84)$$

So far we have three equations to solve for our four unknowns (A, λ, x_1, x_2). What piece of information is missing? We have not yet used the parameter $T = T_1 - T_2$ of the string diagram. Notice that T_1 is the real part of w when $z = x_1$, and T_2 is the real part of w when $z = x_2$:

$$T_1 = \Re(w(x_1)), \quad T_2 = \Re(w(x_2)). \qquad (25.85)$$

Since $\Re(\ln z) = \ln|z|$, we can use equation (25.82) to write

$$T_1 = -2\alpha' p_1^+ \ln x_1 - 2\alpha' p_2^+ \ln(\lambda - x_1) + 2\alpha' p_3^+ \ln(1 - x_1),$$
$$T_2 = -2\alpha' p_1^+ \ln x_2 - 2\alpha' p_2^+ \ln(x_2 - \lambda) + 2\alpha' p_3^+ \ln(x_2 - 1).$$

Here we have used $0 < x_1 < \lambda < 1 < x_2$. We can now evaluate $T = T_1 - T_2$:

$$\frac{1}{2\alpha'}T = p_1^+ \ln\frac{x_2}{x_1} + p_2^+ \ln\left(\frac{x_2 - \lambda}{\lambda - x_1}\right) + p_3^+ \ln\left(\frac{1-x_1}{x_2 - 1}\right). \qquad (25.86)$$

This last equation, together with those previous, fixes the values of the unknown parameters. It is difficult to solve this system of equations to obtain explicit values for the parameters, but, luckily, we can prove our claim without doing so. We will merely analyze them to derive the result.

Since A does not appear in (25.86), it is convenient to eliminate it from equations (25.7) by forming ratios:

$$\frac{p_2^+}{p_1^+} = \frac{(\lambda - x_1)(x_2 - \lambda)}{(1 - \lambda)x_1 x_2},$$

$$\frac{p_2^+}{p_3^+} = \frac{(\lambda - x_1)(x_2 - \lambda)}{(1 - x_1)(x_2 - 1)\lambda}. \tag{25.87}$$

As T varies, the values of x_1, x_2, and λ will vary continuously. T varies from $-\infty$ to $+\infty$. If we can show that λ approaches zero as $T \to \infty$ and that λ approaches one as $T \to -\infty$, then λ must take all values $0 < \lambda < 1$.

We have just suggested that $\lambda \to 0$ as $T \to \infty$. The intuition is that, when $T \to \infty$, the points P_1 and P_2 are separated from the points P_3 and P_4 by a long strip. Conformally, P_1 and P_2 are approaching each other. In our map to $\bar{\mathbb{H}}$ this requires $\lambda \to 0$ (if you want to develop your intuition further, solve Problem 25.5). We now prove that $T \to \infty$ as $\lambda \to 0$. Since $\lambda(T)$ is a well defined function of T, this means that $\lambda \to 0$ as $T \to \infty$.

Given that $0 < x_1 < \lambda$, we have $x_1 \to 0$ when $\lambda \to 0$. Let us see what happens to T in this case. Using (25.86), we find

$$\frac{1}{2\alpha'} T \simeq p_1^+ \ln \frac{x_2}{x_1} + p_2^+ \ln \frac{x_2}{\lambda - x_1} + p_3^+ \ln \frac{1}{x_2 - 1}. \tag{25.88}$$

To understand how T behaves, we must know how x_2 behaves. We use equations (25.7) to find out. With $\lambda \to 0$ and $x_1 \to 0$ they give

$$\frac{p_2^+}{p_1^+} \simeq \frac{(\lambda - x_1)}{x_1} = \frac{\lambda}{x_1} - 1,$$

$$\frac{p_2^+}{p_3^+} \simeq \frac{(\lambda - x_1)x_2}{(x_2 - 1)\lambda} = \frac{1 - \frac{x_1}{\lambda}}{1 - \frac{1}{x_2}}. \tag{25.89}$$

Since the ratios of light-cone momenta are fixed positive numbers, the top equation shows that, in the limit, λ/x_1 is finite and greater than one. The bottom equation then shows that x_2 is finite and greater than one. Using the finiteness of x_2, (25.88) becomes

$$\frac{1}{2\alpha'} T \simeq -p_1^+ \ln x_1 - p_2^+ \ln(\lambda - x_1) + \text{finite}. \tag{25.90}$$

Since both x_1 and $\lambda - x_1$ go to zero as $\lambda \to 0$, we find $T \to +\infty$. Therefore $\lambda \to 0$ when $T \to \infty$, as we wanted to show.

Now we would like to show that $\lambda \to 1$ as $T \to -\infty$. This is also reasonable from the viewpoint of the string diagram. In this case the slits have gone long past each other, so points P_2 and P_3 are separated from points P_1 and P_4 by a long strip. Conformally, P_2 and P_3 are approaching each other. Since P_3 sits at one, this requires $\lambda \to 1$.

Let us look at equation (25.89) to see what is happening with the parameters. As $\lambda \to 1$, the second ratio becomes

$$\frac{p_2^+}{p_3^+} = \frac{(\lambda - x_1)(x_2 - \lambda)}{(1 - x_1)(x_2 - 1)}. \tag{25.91}$$

If we naively take $\lambda \to 1$ in the above equation, we find $p_2^+/p_3^+ \to 1$, which is not correct. Therefore, as $\lambda \to 1$ we must have either $x_1 \to 1$ with x_2 finite or $x_2 \to 1$ with x_1 finite. If $x_1 \to 1$, we get

$$x_1 \to 1: \quad \frac{p_2^+}{p_3^+} \simeq \frac{(\lambda - x_1)}{(1 - x_1)} < 1, \tag{25.92}$$

where the last inequality follows because $x_1 < \lambda$. Having a look at Figure 25.18, we see that we chose conventionally $p_2^+ > p_3^+$. Therefore, we look at the possibility $x_2 \to 1$. In this case, we get

$$x_2 \to 1: \quad \frac{p_2^+}{p_3^+} = \frac{(x_2 - \lambda)}{(x_2 - 1)} > 1. \tag{25.93}$$

This is the correct regime:

$$p_2^+ > p_3^+, \quad \lambda \to 1, \quad x_2 \to 1, \quad \text{and} \quad x_1 \text{ finite.} \tag{25.94}$$

Now we can look at (25.86) to find what T does in this regime:

$$\frac{1}{2\alpha'} T \;\longrightarrow\; p_1^+ \ln \frac{1}{x_1} + p_2^+ \ln \left(\frac{x_2 - \lambda}{1 - x_1} \right) + p_3^+ \ln \left(\frac{1 - x_1}{x_2 - 1} \right). \tag{25.95}$$

Separating finite parts, we have

$$\frac{1}{2\alpha'} T \;\longrightarrow\; p_2^+ \ln(x_2 - \lambda) - p_3^+ \ln(x_2 - 1) + \text{finite.} \tag{25.96}$$

The first two terms on the above right-hand side can be rearranged to read

$$\frac{1}{2\alpha'} T \;\longrightarrow\; (p_2^+ - p_3^+) \ln(x_2 - \lambda) + p_3^+ \ln\left(\frac{x_2 - \lambda}{x_2 - 1}\right) + \text{finite.} \tag{25.97}$$

The ratio appearing on the right-hand side is finite (see (25.93)); therefore

$$\frac{1}{2\alpha'} T \;\longrightarrow\; (p_2^+ - p_3^+) \ln(x_2 - \lambda) + \text{finite}. \tag{25.98}$$

With $p_2^+ - p_3^+ > 0$, and $x_2 - \lambda \to 0$, we find $T \to -\infty$. Since $T \to -\infty$ as $\lambda \to 1$, we have that $\lambda \to 1$ as $T \to -\infty$, as we wanted to prove.

Since λ varies continuously and reaches both zero and one, it must attain all values in between. We have thus shown that the light-cone diagrams produce all relevant Riemann surfaces, that is, all $\bar{\mathbb{H}}$ with four boundary punctures. The string diagrams have produced the moduli space $\mathcal{N}_4$. As we mentioned before, λ varies monotonically with T, so each value of λ in the interval (0, 1) is taken only once.

The above is an illustration of an important and general result in string theory. For arbitrary string interactions an analogous result holds. Keeping the topological type of the string diagram fixed, as the parameters of the diagram are varied over the natural ranges, the complete set of inequivalent Riemann surfaces of the given topological type is produced. String interactions generate the moduli spaces of Riemann surfaces.

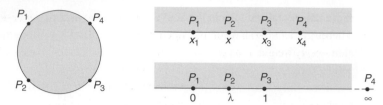

Fig. 25.19 The scattering amplitude for four open string tachyons uses a disk with four boundary punctures. Mapped to $\bar{\mathbb{H}}$, these four punctures P_i, $i = 1,\dots, 4$ have general positions x_1, x, x_3, x_4. They can also be mapped to 0, $\lambda, 1$, and ∞.

25.8 Veneziano amplitude

We will now use our understanding of the moduli space $\mathcal{N}_4$ to motivate the construction of the Veneziano amplitude, the amplitude for the scattering of four open string tachyons (two incoming and two outgoing). As you recall, open string tachyons are particles with $p^2 = -M^2 = 1/\alpha'$. In Section 25.2 we discussed the three steps necessary to get cross sections in string theory. We have done step (1), the drawing of string diagrams and their conformal mapping into a standard presentation. We now focus on step (2), where we must get the scattering amplitude. We will not consider step (3), which is the construction of a cross section.

We cannot derive here the scattering amplitude, but we can motivate it. For this, we note that it should be an integral over the *moduli* space of the surfaces that contribute to the process – the moduli space $\mathcal{N}_4$. Note that the integral is over the moduli space of surfaces, not over the surfaces themselves! Each surface represents a possible "path" connecting the initial and final states, so each surface contributes to the amplitude. To find the full amplitude, we must add up the contributions from all the surfaces; this is an integral over the parameter space of the surfaces.

Consider $\bar{\mathbb{H}}$ punctured at x_1, x, x_3, and x_4 with $x_1 < x < x_3 < x_4$, as shown in Figure 25.19. These are the points where the four tachyons are supposed to be inserted: the images of the far past and far future strips in the light-cone diagram. At x_1 we introduce the tachyon with spacetime momentum p_1, at x the tachyon with momentum p_2, and at x_3 and x_4 the tachyons with momenta p_3 and p_4, respectively. We could map x_1, x_3, and x_4 to 0, 1, and ∞, but we will not do so yet. Treating all punctures on the same footing makes for a clearer structure. Our strategy will be to use the conditions of conformal invariance to motivate an expression for the amplitude.

Using surfaces with punctures at x_1, x, x_3, and x_4, the moduli space $\mathcal{N}_4$ can be described as the set of surfaces obtained when x varies over the interval $x_1 < x < x_3$ (if you map x_1, x_3, and x_4 to 0, 1, and ∞, you recover the canonical presentation $x \in (0, 1)$). Writing the scattering amplitude $A(p_1, p_2, p_3, p_4)$ as an integral

$$A(p_1, p_2, p_3, p_4) = g_o^2 \int d\mu, \qquad (25.99)$$

the measure $d\mu$ must include a dx since we will integrate over x. The amplitude is proportional to the square of the open string coupling g_o, since the process involves two elementary interactions where two strings combine to form one string, and each such interaction carries one power of g_o. This amplitude A is an object similar to the amplitude T considered in (25.2). We will not attempt to fix the precise overall normalization of A. The measure $d\mu$ must also depend on the momenta of the tachyons and on the positions of the boundary punctures. Our momenta satisfy

$$p_1 + p_2 + p_3 + p_4 = 0, \quad p_1^2 = p_2^2 = p_3^2 = p_4^2 = \frac{1}{\alpha'}. \tag{25.100}$$

The first equation is the momentum conservation equation, with all momenta treated as incoming: if the outgoing tachyons are tachyons three and four, for example, then p_3 and p_4 above are actually *minus* the momenta of these two outgoing tachyons. The second equation implies that the four tachyons are on the mass-shell. We now propose a measure $d\mu$:

$$\begin{aligned} d\mu = dx\, |x_3 - x_1||x_4 - x_1||x_4 - x_3| \\ |x - x_1|^{2\alpha' p_2 \cdot p_1}|x_3 - x_1|^{2\alpha' p_3 \cdot p_1}|x_3 - x|^{2\alpha' p_3 \cdot p_2} \\ |x_4 - x_1|^{2\alpha' p_4 \cdot p_1}|x_4 - x|^{2\alpha' p_4 \cdot p_2}|x_4 - x_3|^{2\alpha' p_4 \cdot p_3}. \end{aligned} \tag{25.101}$$

This is a very symmetric expression. For each pair of punctures, there is a factor which is the distance between the points raised to the dot product of their full momenta. In this part of $d\mu$, occupying the second and third lines, all punctures are treated on the same footing. In the first line, however, the moving puncture x is treated differently from the other three fixed punctures, x_1, x_3, and x_4.

Now we consider the issue of conformal invariance. We have written an expression for $d\mu$ that arbitrarily selected three special punctures. We are integrating over the moduli space $\mathcal{N}_4$, presented as $x \in (x_1, x_3)$ for an arbitrary choice of x_1, x_3, and x_4. This choice must be irrelevant – the integral of $d\mu$ must give the same answer for any other choice. Since any two choices of three fixed punctures on the real line are related by a real linear fractional transformation, we can guarantee that the amplitude is unchanged *if the measure $d\mu$ is invariant under real linear fractional transformations.* The generic transformation of this type is

$$z \longrightarrow \frac{az + b}{cz + d}, \quad ad - bc = 1, \quad a, b, c, d \in \mathbb{R}. \tag{25.102}$$

It changes all of the quantities dx, x_1, x_2, x_3. Indeed, we will have

$$\begin{aligned} x &\longrightarrow \frac{ax + b}{cx + d}, \\ dx &\longrightarrow \frac{dx}{(cx + d)^2}, \\ x_i - x_j &\longrightarrow \frac{x_i - x_j}{(cx_i + d)(cx_j + d)}. \end{aligned} \tag{25.103}$$

The measure $d\mu$ must be invariant under all of these changes. This is a stringent condition. Now we can understand the logic behind the first line in (25.101). In including the extra factors along with dx, the first line transforms in a rather neat way. We have, with $x_2 \equiv x$,

$$dx\, |x_3 - x_1||x_4 - x_1||x_4 - x_3|$$

$$\longrightarrow dx\, |x_3 - x_1||x_4 - x_1||x_4 - x_3| \prod_{i=1}^{4} (cx_i + d)^{-2}. \qquad (25.104)$$

The transformation induced exactly the same factor for each of the punctures. Using the last equation in (25.103) to transform the second and third lines in the measure, we find that $d\mu$ transforms into itself times factors of $(cx_i + d)$ for each of the punctures. For $d\mu$ to be invariant, each factor must have the value one. Let us have a look at the factor $(cx_1 + d)$. We get a contribution from the first line of the measure, as calculated in (25.104), and three additional contributions, from the rest of the measure:

$$(cx_1 + d)^{-(2 + 2\alpha' p_2 \cdot p_1 + 2\alpha' p_3 \cdot p_1 + 2\alpha' p_4 \cdot p_1)}. \qquad (25.105)$$

This factor can only be equal to one if the exponent is equal to zero. Happily, this is the case. Looking at the exponent, we have

$$2 + 2\alpha' p_1 \cdot (p_2 + p_3 + p_4) = 2 - 2\alpha' p_1^2 = 2 - 2\alpha' \frac{1}{\alpha'} = 0. \qquad (25.106)$$

In obtaining this result, we used both momentum conservation and the on-shell condition (see (25.100)). Completely analogous results hold for the other punctures. This concludes our verification of the conformal invariance of the measure $d\mu$. The condition of conformal invariance is not strong enough to determine the measure uniquely, so we cannot prove that $d\mu$ is selected. It turns out that $d\mu$ is the correct measure for the scattering of open string tachyons.

We simplify the measure by choosing $x_1 = 0$, $x = \lambda$, $x_3 = 1$, and $x_4 = \infty$. It might seem that the placement of x_4 at infinity could give us a problem. If we trust our previous analysis, however, we know that no problem can arise. The measure $d\mu$ is finite for finite values of x_1, x_3, and x_4. Any linear fractional transformation, even if it takes one point to infinity, will not change the measure. Using $x_4 \gg x_1, x, x_2$, we can simplify (25.101):

$$d\mu = d\lambda\, |x_4|^2\, |\lambda|^{2\alpha' p_2 \cdot p_1}\, |1 - \lambda|^{2\alpha' p_3 \cdot p_2}\, |x_4|^{2\alpha' p_4 \cdot (p_1 + p_2 + p_3)}. \qquad (25.107)$$

The total exponent of $|x_4|$ adds up to zero, so $d\mu$ reduces to

$$d\mu = d\lambda\, |\lambda|^{2\alpha' p_2 \cdot p_1} |1 - \lambda|^{2\alpha' p_3 \cdot p_2}. \qquad (25.108)$$

Since $\lambda \in (0, 1)$, we find that the Veneziano amplitude is given by

$$A(p_1, p_2, p_3, p_4) = g_o^2 \int_0^1 d\lambda\, \lambda^{2\alpha' (p_1 \cdot p_2)} (1 - \lambda)^{2\alpha' (p_2 \cdot p_3)}. \qquad (25.109)$$

String theory began with this formula, written by Veneziano in the late 1960s. The formula was simply postulated, and physicists wondered what kind of theory would give rise to such an amplitude. It took a few years before it was demonstrated that this amplitude arose from a theory of strings.

Towards the end of Section 25.2, we realized that one can determine what kinds of particles exist in a theory by looking for poles in the scattering amplitudes. We shall do this for the Veneziano amplitude, just as physicists did shortly after it was discovered. As a first step, we express the momentum dot products in terms of the Lorentz invariants s and t. If strings one and two join to form an intermediate string, the string will carry momentum $p_1 + p_2$. On the other hand, if strings two and three join to form an intermediate string, the string will carry momentum $p_2 + p_3$. Since the punctures are cyclically ordered, there are no other possibilities: string one cannot join with string three to form an intermediate string (nor can string two join string four). Thus the relevant invariants are

$$s = -(p_1 + p_2)^2, \quad t = -(p_2 + p_3)^2. \tag{25.110}$$

Expanding s, and using the on-shell condition, we find

$$s = -p_1^2 - p_2^2 - 2p_1 \cdot p_2 = -\frac{2}{\alpha'} - 2p_1 \cdot p_2, \tag{25.111}$$

which leads to

$$2\alpha' p_1 \cdot p_2 = -\alpha' s - 2 = -(\alpha' s + 1) - 1. \tag{25.112}$$

For convenience, define

$$\alpha(s) \equiv \alpha' s + 1, \tag{25.113}$$

so that we may express (25.112) as

$$2\alpha' p_1 \cdot p_2 = -\alpha(s) - 1. \tag{25.114}$$

Using this relation, and the corresponding one for t, the Veneziano amplitude (25.109) is written as

$$A(p_1, p_2, p_3, p_4) = g_o^2 \int_0^1 d\lambda \, \lambda^{-\alpha(s)-1}(1 - \lambda)^{-\alpha(t)-1}. \tag{25.115}$$

This integral can be expressed in terms of gamma functions. Indeed, since

$$\int_0^1 dx \, x^{a-1}(1 - x)^{b-1} = \frac{\Gamma(a)\Gamma(b)}{\Gamma(a + b)}, \tag{25.116}$$

we find that

$$A(p_1, p_2, p_3, p_4) = g_o^2 \frac{\Gamma(-\alpha(s))\,\Gamma(-\alpha(t))}{\Gamma(-\alpha(s) - \alpha(t))}. \tag{25.117}$$

To find the poles of the amplitude A, we must know about the poles and zeros of the gamma function. We had a first look at the gamma function in Section 3.4, where we proved the recursion relation

$$\Gamma(z + 1) = z\,\Gamma(z), \quad \Re(z) > 0. \tag{25.118}$$

This equation shows that the gamma function has a pole at $z = 0$. Indeed, $\Gamma(1 + \epsilon) = \epsilon\Gamma(\epsilon)$ with ϵ small gives

$$\Gamma(\epsilon) = \frac{1}{\epsilon}\Gamma(1 + \epsilon) = \frac{1}{\epsilon}(\Gamma(1) + \mathcal{O}(\epsilon)) = \frac{1}{\epsilon} + \mathcal{O}(1). \tag{25.119}$$

We can extend the definition of the gamma function to negative real numbers using analytic continuation: equation (25.118) is taken to hold even when $\Re(z) \leq 0$. We then find, for example, a pole at $z = -1$:

$$\Gamma(\epsilon) = (-1 + \epsilon)\Gamma(-1 + \epsilon) \rightarrow \Gamma(-1 + \epsilon) \simeq -\frac{1}{\epsilon}. \tag{25.120}$$

Similar reasoning shows that the gamma function has poles at all of the negative integers (see also Problem 3.6). It is a famous (and nontrivial) result that the gamma function has no zeros.

We are now ready to investigate the pole structure of the Veneziano amplitude (25.117). Having no zeros, the gamma function in the denominator contributes no poles. All poles must arise from the gamma functions in the numerator. We have poles when

$$\alpha(s) = n, \quad n = 0, 1, 2, 3, \ldots, \tag{25.121}$$

and when

$$\alpha(t) = m, \quad m = 0, 1, 2, 3, \ldots. \tag{25.122}$$

Let us focus on the poles in the s channel. Recall that a pole for $s = M^2$ implies a particle of mass-squared equal to M^2 (see (25.8)). Therefore, in order to find which particles are present in the Veneziano amplitude, we need to find the values of s which give the poles in (25.121). Making use of (25.113), this condition is simply

$$\alpha's + 1 = n \rightarrow s = \frac{1}{\alpha'}(-1 + n) = M_n^2. \tag{25.123}$$

Since n runs from zero to infinity, there is an infinite number of poles, and an infinite number of particles. Note that the values of M_n^2 define precisely the levels of the relativistic string that we have been studying in this book. When $n = 0$ we find the tachyon ($M_0^2 = -1/\alpha'$). When $n = 1$ we find massless particles. For higher n, there are infinitely many massive particles. Exactly the same set of particles appears in the poles that occur in the t channel. It became clear to physicists early on that the model invented by Veneziano was somehow related to relativistic strings. This intuition was confirmed in the early 1970s when, using the light-cone gauge, the string was quantized and its spectrum was elucidated.

Problems

Problem 25.1 Self-mappings of the sphere $\hat{\mathbb{C}}$.

(a) Show that the composition of two linear fractional transformations (25.50) is a linear fractional transformation.

(b) A fixed point of a transformation $w = f(z)$ is a point z_0 such that $f(z_0) = z_0$. Prove that each linear fractional transformation, with the exception of the identity transformation $w = z$, has at most two fixed points.

(c) Show that (25.59) is the unique linear fractional transformation that maps the three different points z_1, z_2, and z_3 into the three different points w_1, w_2, and w_3, respectively. [Hint: let S and T be two distinct linear fractional transformations which satisfy this property, and consider the transformation obtained by composing S with the inverse of T.]

Problem 25.2 Conformal invariant of the sphere $\hat{\mathbb{C}}$ with four punctures.

Consider four points z_1, z_2, z_3, and z_4 in $\hat{\mathbb{C}}$. Write a map $w(z)$ such that $w(z_2) = 0$, $w(z_3) = 1$, and $w(z_4) = \infty$. Define $\lambda = w(z_1)$. Calculate λ, and show that it is given by

$$\lambda = \lambda(z_1, z_2, z_3, z_4) = \frac{(z_1 - z_2)(z_3 - z_4)}{(z_1 - z_4)(z_3 - z_2)}.$$

Verify explicitly that λ is a conformal invariant, that is, $\lambda(\phi(z_1), \phi(z_2), \phi(z_3), \phi(z_4)) = \lambda(z_1, z_2, z_3, z_4)$, where ϕ is any linear fractional transformation.

Problem 25.3 Self-mappings of $\bar{\mathbb{H}}$.

Prove that if a linear fractional transformation $w = (az + b)/(cz + d)$ maps the real line of the z plane into the real line of the w plane, then a, b, c, and d must all be real, except possibly for a common phase factor that can be removed without changing the map $z \to w$.

Problem 25.4 Upper half-plane $\bar{\mathbb{H}}$ with two boundary punctures.

Consider $\bar{\mathbb{H}}$ with two punctures P_1 and P_2 on the real line, with coordinates $z = x_1$ and $z = x_2$, respectively. Consider another copy of $\bar{\mathbb{H}}$ with two punctures P_1 and P_2 on the real line, with coordinates $z' = x_1'$ and $z' = x_2'$, respectively. Are these two surfaces the same Riemann surface? Prove that they are, by exhibiting the conformal map that takes the punctures into each other while preserving $\bar{\mathbb{H}}$. You may have to write two conformal maps, depending on the sign of $(x_2' - x_1')/(x_2 - x_1)$. What is the geometrical significance of this sign?

Problem 25.5 What does it mean for points to be close in a Riemann surface?

The concept of points approaching each other on a Riemann surface is subtle because we can always scale the coordinates. Consider $\bar{\mathbb{H}}$ with boundary punctures. If we only have two punctures, then they cannot be said to be close because we can always map them to arbitrary positions. The same is true for three punctures. For four punctures it is possible to have a notion of punctures coming together. The notion applies to a family of surfaces where punctures move.

Consider $\bar{\mathbb{H}}$ with four punctures P_i, $i = 1, 2, 3, 4$, with real coordinates $x_i(t)$, where $t \in [0, 1]$. This is a family of surfaces parameterized by the parameter t. As t varies, the coordinates of the punctures will vary. Of course, we can make (real) linear fractional transformations which change the values of the $x_i(t)$.

Definition We say that $P_i \to P_j$ as $t \to 0$ $(i \neq j)$ if the coordinate of P_i approaches the coordinate of P_j in a representation of $\bar{\mathbb{H}}$ where P_j and the other two punctures are held fixed.

To understand this definition properly, consider the following example:

$$x_1(t) = 0, \quad x_2(t) = t, \quad x_3(t) = 1, \quad x_4(t) = \infty.$$

It is clear from the definition that $P_2 \to P_1$ as $t \to 0$. Prove that the following statements hold for this family as $t \to 0$.

(a) $P_1 \to P_2$. This means that if one puncture approaches a second one, then the second also approaches the first.
(b) $P_3 \to P_4$. The other two punctures also approach each other!

Prove now, on general grounds, that

(c) The definition proposed above is conformal invariant. That is, if the relevant coordinates approach each other in a given representation of $\bar{\mathbb{H}}$ where P_j and the other two punctures are held fixed, then they will approach each other in any such representation.

Problem 25.6 Closing off the polygon in the Schwarz–Christoffel map.

The differential equation (25.27) does not show the turning angle α_n at P_n because this point has been mapped to $z = \infty$. We aim to understand how this point at infinity works out.

(a) To find the turning angle at $z = \infty$, consider the large-z limit of (25.27):

$$\frac{dw}{dz} \simeq A z^{-\frac{1}{\pi} \sum_{i=1}^{n-1} \alpha_i}.$$

Define $t = -1/z$, and calculate $\frac{dw}{dt}$ as a function of t. Explain why your result shows that the turning angle is α_n.

(b) The differential equation

$$\frac{dw}{dz} = A(z - x_1)^{-\frac{\alpha_1}{\pi}} (z - x_2)^{-\frac{\alpha_2}{\pi}} \cdots (z - x_n)^{-\frac{\alpha_n}{\pi}}, \tag{1}$$

with the last turning point P_n included as a finite point x_n, represents the situation where there is no corner at $z = \infty$. Prove that the polygon closes. For this, show that as we traverse the full real axis x, the change in w is zero:

$$w(x = \infty) - w(x = -\infty) = \int_{x=-\infty}^{x=\infty} dx \, \frac{dw}{dx} = 0.$$

[Hints: use (1) and contour deformation. Argue that there is no contribution from half-circles around the x_i and around ∞.]

Problem 25.7 Four open string interaction in a special configuration.

Consider the light-cone open string diagram of Figure 25.18 in the special configuration where all p^+ momenta are equal: $p_1^+ = p_2^+ = p_3^+ = p_4^+$. Let $T = T_1 - T_2$ denote the world-sheet time difference between the interaction points. Calculate the modulus $\lambda(T)$

explicitly. Show that the moduli space $\mathcal{N}_4$ is produced for $T \in [0, \infty)$. Confirm that λ is a monotonic function of T.

Problem 25.8 Proton decay in intersecting brane models.

We aim to show that in intersecting brane models there are no open string diagrams which represent proton decay into leptons and gauge bosons. We focus on the model of Figure 21.7 and restrict ourselves to string diagrams presented as a disk with boundary punctures, where the set of open string states relevant to the process (incoming and outgoing) are inserted. The punctures split the boundary of the disk into components, which are the loci of open string endpoints, and consequently, are labeled by D-branes.

Prove that no consistent assignment of labels to the boundary components is possible given that proton decay involves three quarks (and a number of leptons and gauge bosons), and a quark has one endpoint on a baryonic brane and the other endpoint elsewhere. [Hint: try drawing string diagrams. Your attempts should evolve into a clear and very brief argument.]

Loop amplitudes in string theory

To calculate scattering amplitudes with high accuracy, one must include the contribution from diagrams which contain loops that represent virtual processes. In Einstein's theory of gravity these diagrams give rise to ultraviolet divergences, which reveal intractable short-distance phenomena. String theory contains gravity, but there are no such ultraviolet divergences. The Riemann surfaces that are candidates for short-distance problems admit an interpretation where they clearly describe safe, long-distance phenomena. We illustrate this remarkable property for the case of annuli, which are the surfaces relevant to virtual open string processes, and for the case of tori, which are the surfaces relevant to virtual closed string processes.

26.1 Loop diagrams and ultraviolet divergences

When calculating scattering amplitudes in particle physics, one typically uses an approximation scheme in which the strength of the interactions is assumed to be small. The amplitude is then written in terms of a perturbative series expansion in powers of this small interaction parameter. The Feynman diagrams we considered in Chapter 25 and the similar looking string diagrams were all *tree diagrams*. This means that the graphs (see, for example, Figure 25.2) contain no nontrivial closed paths, or loops. Tree diagrams give the first term in the perturbative expansion of scattering amplitudes. To go beyond this lowest-order approximation, one must consider Feynman diagrams with loops.

Consider the Feynman diagram with a loop shown in Figure 26.1. This diagram represents an incoming particle which splits into two particles that rejoin to form an outgoing particle. The two particles with momentary existence are called virtual particles, and their appearance and subsequent disappearance is called a virtual process. When calculating the contribution of such graphs to amplitudes, one typically encounters divergent quantities. These divergences are called *ultraviolet* (UV) divergences if they arise from virtual processes involving very high energies or momenta or, alternatively, very short times or distances. Virtual processes that involve short distances can be roughly represented by graphs with small loops.

Ultraviolet divergences need not be fatal – in many cases they can be dealt with by the processes of regularization and renormalization. But for some theories regularization and renormalization do not work – the most important case being Einstein's theory of gravity. Faced with unrenormalizable ultraviolet divergences, one can sometimes work accurately at low energies using *effective* field theories which, at some loss of predictive power,

Fig. 26.1 A one-loop Feynman graph that represents a virtual process. In the limit in which the virtual process involves short distances or high momenta, one may find an ultraviolet divergence in the corresponding amplitude.

model in a controllable way the problematic high energy processes. At a fundamental level, however, this is far from being a fully satisfactory approach.

If string theory is a complete quantum theory, then there are two possibilities: either the ultraviolet divergences are manageable or there are simply no ultraviolet divergences. The wonderful news is that *there are no ultraviolet divergences in string theory*. String theory is the first quantum theory, ever, to include gravity without also including ultraviolet divergences. This is the reason why string theory has become the foremost candidate for a theory of quantum gravity. It is the purpose of this chapter to give you some understanding of this remarkable result.

The main property of string diagrams is that the relevant information is all encoded in the *Riemann surfaces* they define. Any conformal map of a string diagram gives an *equivalent* physical picture, since the Riemann surface is not changed. We will use conformal mapping to show that string diagrams that may appear to represent potentially dangerous short-distance physics are actually equivalent to diagrams that are manifestly free of ultraviolet problems. We learned in Chapter 25 that the string amplitude for a given process is obtained as an integral over the relevant moduli space of Riemann surfaces. In this integration, we are adding up the contributions to the amplitude from all the surfaces that are consistent with the initial and final states. The integration is not over the surfaces themselves.

For string loop diagrams, the simplest relevant Riemann surfaces are annuli and tori. We will look at both types of surfaces in some detail, and we will study their corresponding moduli spaces: the space of all possible annuli and the space of all possible tori. We will examine the regions of moduli space where Riemann surfaces seem to involve short-distance physics, and by using suitable conformal maps we will show that they represent safe, long-distance physics. We will take this as evidence that there is no room in string theory for ultraviolet divergences. While we will only examine one-loop string diagrams, this remarkable property of Riemann surfaces and their moduli spaces holds for string diagrams with an arbitrary number of loops.

26.2 Annuli and one-loop open strings

Let us begin by considering the one-loop open string diagram shown in Figure 26.2(a). This diagram involves one incoming open string, one outgoing open string, and two intermediate

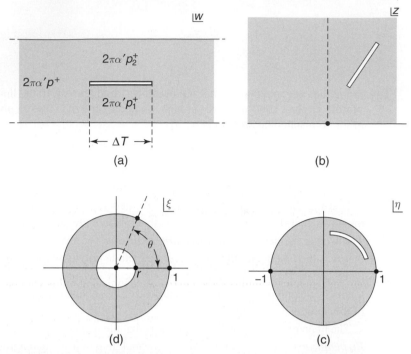

Fig. 26.2 (a) A string diagram that represents a one-loop open string process. This diagram has two parameters. (b) Mapping the diagram to $\bar{\mathbb{H}}$ with a slit. (c) Mapping the diagram into a disk with a slit. (d) Finally, mapping the diagram into a canonical annulus.

open strings. It is the string analog of the one-loop Feynman diagram in Figure 26.1. We will discuss a less obvious string analog in Section 26.4.

The light-cone diagram shows an open string with light-cone momentum p^+ that splits into two open strings. The two open strings propagate for a time ΔT and then recombine. By conservation of light-cone momentum we have $p^+ = p_1^+ + p_2^+$. For fixed external momentum p^+, this diagram has two parameters: $\Delta T \in (0, \infty)$, and the vertical position of the slit, parameterized by $p_1^+ \in (0, p^+)$. In the language developed in Chapter 25, we say that the class of Riemann surfaces corresponding to this process has two moduli. The amplitude is obtained by summing the contributions from all such string diagrams.

We now examine a sequence of conformal maps that turns this diagram into a canonical presentation. The first two steps focus on the full strip. As you saw in Section 25.4, the infinite strip can be mapped to $\bar{\mathbb{H}}$ using the exponential map in (25.21). This map, applied in the present case, gives the result indicated in Figure 26.2(b), where the new slit is the image of the original slit under the map. The incoming string has been sent to $z = 0$ and the outgoing one to $z = \infty$. With the help of (25.23) this surface can be mapped into a unit disk – again the slit goes along for the ride, and the incoming and outgoing strings end up at $\eta = 1$, and $\eta = -1$, respectively. At this stage, the string surface is topologically an annulus, a disk with a hole. The hole happens to be a slit.

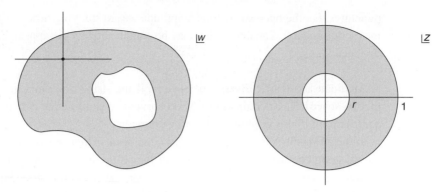

Fig. 26.3
A topological annulus – the region between the two closed curves in the w plane – can be mapped into a canonical annulus $r \leq |z| \leq 1$ for some value of the modulus r.

The last map is fairly nontrivial. The annulus shown in (c) can be mapped conformally into the canonical annulus shown in (d). A canonical annulus with parameter r in the range $0 < r < 1$ is the region $r \leq |\xi| \leq 1$ of the complex ξ plane. This map is a particular example of a general result: a region of the complex plane that is topologically an annulus can be mapped conformally to a canonical annulus (see Figure 26.3). The value r of the inner radius cannot be prescribed; the map fixes it uniquely. This quantity r is called the *modulus of the annulus*. In Figure 26.2(d) the incoming string can be placed at $\xi = 1$: if it originally lies elsewhere on the unit circle, the conformal map $\xi \to \exp(i\alpha)\xi$ will take it to $\xi = 1$ for suitable α. Once the ingoing string has been fixed, the position of the outgoing string cannot be adjusted. In the figure, the outgoing string appears as a puncture at an angle θ with respect to the positive horizontal axis. The Riemann surface is an annulus with two punctures. The boundary of an annulus has two components: the two circles which bound the annular region. In this string diagram, the two punctures lie on the same boundary component.

In the following section, we will use ideas from electrostatics to show explicitly that there exists a conformal map from any topological annulus to a canonical annulus. The modulus of an annulus is closely related to the capacitance of the cylindrical capacitor whose cross sectional region between the conductors is the annulus. We will also learn that two canonical annuli with inner radii r_1 and r_2 cannot be mapped conformally into each other unless $r_1 = r_2$. This is why the inner radius is properly called a modulus. Since the inner radius must be larger than zero and smaller than one, the moduli space of annuli is the space $0 < r < 1$.

Since the string diagram in Figure 26.2(a) has two parameters, the final diagram in part (d) also has two parameters. They are the modulus r of the annulus and the angle θ that defines the position of the outgoing string on the outer boundary. In fact, these two are the moduli of an annulus with two boundary punctures. To construct a string amplitude, we add up the contributions from all the Riemann surfaces that comprise the relevant moduli space. For the Veneziano amplitude we considered all disks with four boundary punctures. For the open string loop amplitude we are studying now, we must integrate over the moduli r and θ. The integration may be complicated by the appearance of an ultraviolet divergence.

A candidate ultraviolet divergence appears if the string diagrams in some region of the moduli space have a certain kind of short curves that suggest that they may give unbounded contributions to the amplitudes. Let us elaborate.

In the string diagram of Figure 26.2(a), the open strings are vertical segments, some of which stretch between the outer edges of the string diagram, and some of which stretch between an outer edge and the slit. Remember, the slit is one boundary component, and the outer edges of the string diagram, which are separated by the punctures, comprise the other boundary component. All the open strings are open curves that are nontrivial in the following sense: they cannot be contracted away if the endpoints remain on the boundary and are not allowed to slide past the punctures (the external states). Let us call *any* such nontrivial open curve a *possible* open string. There are many more possible open strings than there are open strings. Any curve that begins on the top edge and ends on the lower edge is a possible open string. So is any curve that begins on the left side of the top edge, goes under the slit, and ends on the right side of the top edge. As the name suggests, a possible open string is a curve that could be taken to be an open string with some suitable parameterization of the world-sheet. The concept of a possible open string on a Riemann surface is well defined: the nontrivial character of a possible open string is not changed by conformal mapping. Similarly, a possible closed string is a closed curve that cannot be shrunk away. In Figure 26.2(a), a closed curve looping around the slit is a possible closed string. In this string diagram there are no canonical closed strings, but there are many possible closed strings. The signal of short-distance physics that suggests ultraviolet divergences is the appearance of short possible open strings or short possible closed strings on a string diagram. The length is calculated in the obvious way: in complex coordinates z, a segment dz is assigned length $|dz|$. Since length is not conformal invariant, we can see that the appearance of short possible strings may have an alternative interpretation.

Let us now return to our example and search for the short possible strings that can arise as the modulus r of the annulus varies over its range. Interesting things happen near the ends of the moduli space $0 < r < 1$. As shown in Figure 26.4, an annulus with $r = 1 - \epsilon$ and $\epsilon \to 0$, is a vanishingly thin ribbon. The short radial lines going from one boundary to the other are short possible open strings of length ϵ. The diagram can be interpreted as a short open string traveling around the annulus. But by using a conformal map that scales the figure by the large factor $1/\epsilon$, we get a physically equivalent situation where an open string of unit length travels a very *long* distance $2\pi/\epsilon$. This region of moduli space has therefore the interpretation of long-distance open string physics and is not ultraviolet problematic. The existence of this "dual" interpretation is not the cure for a problem; rather, it simply reveals that the problem is not there. As you may imagine, the region $r \to 1$ corresponds to $\Delta T \to \infty$ in the original light-cone diagram. This is because the slit, which is the inner boundary of the annulus, is becoming very large.

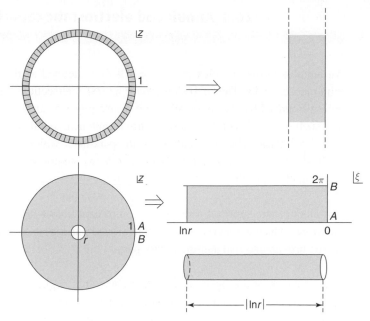

Top: an annulus with $r \to 1$. To the right we show a small region near $z = 1$ magnified by a large factor to give an open string of unit length that travels a long distance. Bottom: an annulus with $r \to 0$. Its physical interpretation is that of a closed string traveling a long distance.

The annuli in the region $r \to 0$ are particularly interesting. An $r \simeq 0$ annulus has a very small hole (see the bottom part of Figure 26.4), and we get short possible closed strings that go around the hole. This time we use the map $\xi = \ln z$ to give a nicer physical interpretation. This map turns the annulus into a long strip, with the top and bottom horizontal lines identified. This is simply a long cylinder of length $(-\ln r)$ and circumference 2π. Its physical interpretation is that of a closed string of circumference 2π propagating for a large distance $|\ln r|$. This shows that the region $r \to 0$ of the moduli space of annuli represents *long-distance* closed string physics. The short possible closed curves do not give rise to ultraviolet problems.

These results generalize to arbitrary string diagrams that involve any number of open and closed strings. Every time short possible open strings arise in a light-cone diagram, the diagram can be presented as one where a finite length open string propagates for a long time. Every time short possible closed strings arise in a light-cone diagram, the diagram can be presented as one where a finite length closed string propagates for a long time. A little more is actually required: the presentations in the various regions of the relevant moduli space must go continuously into each other as the moduli vary. This can be done. One can give a presentation that varies continuously with the moduli and that has two properties: (i) open and closed strings that propagate for long distances are represented by strips of constant width and by cylinders of constant circumference, respectively, and (ii) no short possible strings ever arise. This is the geometrical basis for the understanding of the absence of ultraviolet divergences in string theory.

26.3 Annuli and electrostatic capacitance

An annulus in the complex plane $z = x + iy$ is defined as the region between two disjoint closed curves. The curves must have no self-intersections, and one must lie inside the region bounded by the other. We want to use physical ideas from electrostatics to show that such an annulus can be mapped into a canonical annulus, a region $r \leq |\xi| \leq 1$ in the complex ξ plane. The inner radius r is the parameter which defines the annulus. We want to show that r is a modulus for annuli; that is, two annuli with different values of r are not conformally equivalent.

Our electrostatic approach uses the notion of capacitance and some properties of analytic functions, which we review now. As mentioned in Section 25.4, an analytic function $f(z)$ can be broken into real and imaginary parts,

$$f(z) = u(x, y) + iv(x, y), \tag{26.1}$$

that satisfy the Cauchy–Riemann equations

$$\frac{\partial u}{\partial x} = \frac{\partial v}{\partial y}, \quad \frac{\partial u}{\partial y} = -\frac{\partial v}{\partial x}. \tag{26.2}$$

By taking additional partial derivatives, one readily verifies that the functions u and v satisfy Laplace's equation $\nabla^2 u = \nabla^2 v = 0$. For example,

$$\left(\frac{\partial^2}{\partial x^2} + \frac{\partial^2}{\partial y^2} \right) v(x, y) = 0. \tag{26.3}$$

Additionally, the gradients of u and v are orthogonal vectors of equal magnitude. Indeed, the gradients are defined as

$$\nabla u = \left(\frac{\partial u}{\partial x}, \frac{\partial u}{\partial y} \right), \quad \nabla v = \left(\frac{\partial v}{\partial x}, \frac{\partial v}{\partial y} \right), \tag{26.4}$$

and the Cauchy–Riemann equations imply that we can write

$$\nabla u = \left(\frac{\partial v}{\partial y}, -\frac{\partial v}{\partial x} \right). \tag{26.5}$$

Using this version of ∇u, one readily sees that $\nabla u \cdot \nabla v = 0$ and $|\nabla u| = |\nabla v|$. Moreover, ∇v is obtained from ∇u by a counterclockwise rotation of $90°$ (the operation $(a, b) \to (-b, a)$ rotates the vector (a, b) by $90°$ in the counterclockwise direction). Since ∇u is orthogonal to the lines of constant u and ∇v is orthogonal to the lines of constant v, the orthogonality of ∇u and ∇v implies that the lines of constant u and the lines of constant v are orthogonal. These facts are illustrated in Figure 26.5. Given two points P_1 and P_2, the difference in u values $u(P_2) - u(P_1)$ is related to the flux of ∇v across any curve joining the two points:

$$u(P_2) - u(P_1) = \int_{P_1}^{P_2} \nabla v \cdot \vec{n} \, d\ell. \tag{26.6}$$

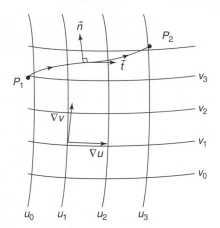

Fig. 26.5 Lines of constant u and lines of constant v. Here $u_0 < u_1 < u_2 < \cdots$ and $v_0 < v_1 < v_2 < \cdots$. The vectors ∇u and ∇v are orthogonal.

Here $d\ell$ is a length element, and $\vec{n}$ is the unit vector normal to the curve obtained by a 90° counterclockwise rotation of the unit tangent vector $\vec{t}$ of the curve, oriented from P_1 to P_2 (see Figure 26.5). The reason is simple: $\nabla v \cdot \vec{n} = \nabla u \cdot \vec{t}$ because the vectors on the left-hand side are obtained by a counterclockwise rotation of 90° from the vectors on the right-hand side, and $\nabla u \cdot \vec{t}$ is the change in u per unit length along the curve.

Having reviewed the basic facts needed, we can now discuss the electrostatic analogy. We will use the topological annulus to define a cylindrical capacitor. The cross section of the capacitor is such that the inner and outer boundaries of the annulus correspond to the inner and outer conductors of the capacitor. The annular region itself is the region between the two conductors. The annulus/capacitor is shown in Figure 26.6 as some region of the complex z plane ($z = x + iy$).

Now imagine setting the inner conductor at unit (electrostatic) potential and the outer conductor at zero potential. With $v(x, y)$ denoting the potential in the region between the conductors, we have the Dirichlet boundary conditions $v = 1$ and $v = 0$ at the inner and outer boundaries, respectively. The electric field in the annular region is given by

$$\vec{E} = -\nabla v. \tag{26.7}$$

As you learned in electrostatics, Dirichlet boundary conditions uniquely determine the potential $v(x, y)$ satisfying Laplace's equation in between the conductors. We use physics intuition as sufficient motivation for two facts, which we will not prove: (1) a solution for the potential $v(x, y)$ exists, and (2) the electric field lines begin on the inner conductor and end on the outer conductor.

Since the imaginary part of an analytic function satisfies Laplace's equation, we will identify $v(x, y)$ as the imaginary part of an analytic function $f(z) = u(x, y) + iv(x, y)$ and then construct $u(x, y)$. The lines of constant u will be orthogonal to the lines of constant v, the equipotentials. As a result, the lines of constant u must be the electric field lines.

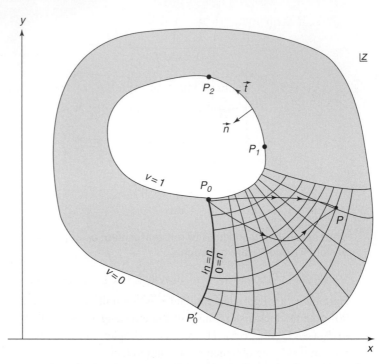

Fig. 26.6 The cross section of a cylindrical capacitor. The region in between the two conductors defines a topological annulus. We set the inner conductor at unit potential and the outer conductor at zero potential.

To construct the function $u(x, y)$, we make note of (26.6) and *define*

$$u(P) \equiv \int_{P_0}^{P} \nabla v \cdot \vec{n} \, d\ell, \qquad (26.8)$$

where P_0 is a fixed point chosen to lie on the inner conductor. The function $u(x, y)$ is well defined in the following sense: if we use two different paths γ_1 and γ_2 from P_0 to P, then the resulting values for $u(P)$ will be the same *if* the two paths can be deformed into each other (see Figure 26.6). In that case the paths γ_1 and γ_2, joined together by reversing the orientation of γ_2, form a closed path $\Gamma = \gamma_1 - \gamma_2$ which bounds a region R. The total integral around Γ then vanishes:

$$\int_{\gamma_1} \nabla v \cdot \vec{n} \, d\ell - \int_{\gamma_2} \nabla v \cdot \vec{n} \, d\ell = \oint_{\Gamma} \nabla v \cdot \vec{n} \, d\ell = \int_{R} da \, \nabla \cdot (\nabla v) = \int_{R} da \, \nabla^2 v = 0,$$
$$(26.9)$$

where we have used the divergence theorem to turn the flux of ∇v into the area integral of the divergence of ∇v. This proves that the two paths give the same value for $u(P)$. Note, however, that a path that wraps around the inner conductor need not give the same value for $u(P)$ as a path that does not wrap. This will play a role below.

We now prove that $u + iv$ is an analytic function. For this purpose it is useful to introduce a unit vector $\vec{k}$ which points out of the plane and satisfies $\vec{n} = \vec{k} \times \vec{t}$ (see Figure 26.5).

With the help of this vector, we write (26.8) as

$$u(P) = \int_{P_0}^{P} \nabla v \cdot (\vec{k} \times \vec{t}) \, d\ell = \int_{P_0}^{P} \nabla v \cdot (\vec{k} \times d\vec{\ell}) = \int_{P_0}^{P} d\vec{\ell} \cdot (\nabla v \times \vec{k}) \,, \qquad (26.10)$$

where we have used the cyclicity of the triple product. This equation implies

$$\nabla u = \nabla v \times \vec{k} \,. \qquad (26.11)$$

The component version of this equation gives precisely the Cauchy–Riemann equations, thus proving the desired result.

Consider now the field line that departs P_0 and reaches the outer boundary at P_0', as shown in Figure 26.6. On any field line, $\nabla v \cdot \vec{n}$ vanishes. It thus follows from (26.8) that the field line in question is a line with $u = 0$. The topology of the annulus, however, implies that u is not single valued. There are paths that cannot be deformed into each other. Using a path that wraps once around the annulus, we find that the field line $u = 0$ is also the line $u = u_f$, with u_f a constant whose value and physical interpretation we will determine next.

The constant u_f can be calculated using definition (26.8) and integrating around the inner conductor, starting at P_0 and moving counterclockwise until we return to P_0:

$$u_f = \oint \nabla v \cdot \vec{n} \, d\ell \,. \qquad (26.12)$$

Along the path, $\nabla v = -\vec{E}$ points into the inner conductor, and so does $\vec{n}$. Therefore $\nabla v \cdot \vec{n} > 0$ along the integration curve and, as a consequence, $u_f > 0$. Rewriting the above equation with $\nabla v = -\vec{E}$,

$$u_f = \oint \vec{E} \cdot (-\vec{n}) \, d\ell \,. \qquad (26.13)$$

Since $(-\vec{n})$ points out of the inner conductor, the above integral computes the flux of $\vec{E}$ out of the inner conductor. By Gauss' law, we find that

$$u_f = Q \,, \qquad (26.14)$$

where Q is the charge on the surface of the inner conductor per unit length, where length is measured along the axis of the cylindrical capacitor. The capacitor is neutral, so the outer conductor will have a charge of $(-Q)$ per unit length. For two conductors, with a potential difference V between them and charges Q and $(-Q)$ placed on them, the capacitance C per unit length is defined by the equation $Q = CV$. In our case $V = 1$, and therefore $C = Q$. Using (26.14), we find

$$C = u_f \,. \qquad (26.15)$$

The conformal map from the annulus to the canonical presentation is now readily obtained using $v(x, y)$ and $u(x, y)$ to define

$$w = f(z) = u(x, y) + iv(x, y) \,. \qquad (26.16)$$

The conformal map $z \to w$ takes the annular region into the configuration of a parallel plate capacitor! This is shown in Figure 26.7. The inner conductor, the surface of constant

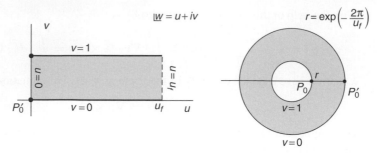

The capacitor mapped into a parallel plate capacitor and then into a round cylindrical capacitor.

$v = 1$, is mapped to $\Im(w) = 1$, while the outer conductor, the surface of constant $v = 0$, is mapped to $\Im(w) = 0$. Moreover, the full annulus lies between the vertical lines $u = 0$ and $u = u_f$. The two vertical segments $u = 0$ and $u = u_f$, with $0 \le v \le 1$, must be identified, as depicted in Figure 26.6. A final conformal map ξ takes the annular region in the w plane into canonical form:

$$\xi = \exp\left(2\pi i \, \frac{w}{u_f}\right). \tag{26.17}$$

This map takes the outer conductor ($0 \le u \le u_f, v = 0$) to the unit circle $|\xi| = 1$. It takes the inner conductor ($0 \le u \le u_f, v = 1$) to the circle $|\xi| = r$, where

$$r = \exp\left(-\frac{2\pi}{u_f}\right) = \exp\left(-\frac{2\pi}{C}\right). \tag{26.18}$$

This relates the inner radius r of the resulting annulus to the constant u_f and the capacitance C. With this final map, we have shown how to use the electrostatic problem to provide a map from the topological annulus to the canonical annulus. The region of the moduli space of annuli interpreted as long-time open string propagation ($r \to 1$) corresponds to large capacitance. The region interpreted as long-time closed string propagation ($r \to 0$) corresponds to small capacitance.

It remains to be shown that r is a modulus. To prove this, we first show that *capacitance is a conformal invariant*. This holds because the electrostatic solution for one capacitor can be used as a solution for any conformally related capacitor. Consider a capacitor in the $z = x + iy$ plane and a capacitor in the $\eta = \psi + i\phi$ plane, as shown in Figure 26.8. Assume that for the capacitor in the z plane we have found an analytic function $f(z) = u + iv$ such that v is the potential when the inner and outer conductors are kept at unit and zero potentials, respectively. Moreover, assume that the conformal map

$$z = h(\eta), \tag{26.19}$$

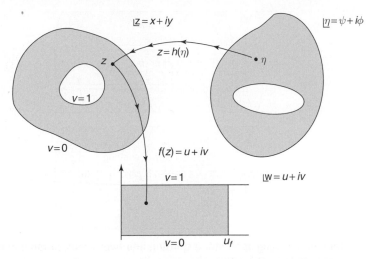

Fig. 26.8 If the annulus in the η plane can be mapped conformally to the annulus in the z plane, the capacitances of the associated cylindrical capacitors are the same. Here $z = h(\eta)$ denotes the conformal map between the annuli, and $w = f(z)$ maps the z plane capacitor into a parallel-plate capacitor.

takes the capacitor in the η plane to the capacitor in the z plane. This, of course, means that the inner conductors and the outer conductors are taken into each other by the map. We now claim that the function

$$f(h(\eta)) = u\Big(x(\psi, \phi), y(\psi, \phi)\Big) + i v\Big(x(\psi, \phi), y(\psi, \phi)\Big)$$
$$\equiv \tilde{u}(\psi, \phi) + i \tilde{v}(\psi, \phi), \qquad (26.20)$$

does for the capacitor in the η plane what $f(z)$ did for the capacitor in the z plane. This function assigns to the point η the same number that f assigns to the image of η under the map to the z plane. Since $f(h(\eta))$ is an analytic function, its real and imaginary parts together satisfy the Cauchy–Riemann equations and separately satisfy Laplace's equation. Additionally, the function $\tilde{v}(\psi, \phi)$ will assign unit potential to the inner conductor of the η capacitor and zero potential to the outer conductor. $\tilde{v}(\psi, \phi)$ is the unique solution for the potential in the η capacitor. Up to an additive constant, $\tilde{u}(\psi, \phi)$ is the unique function which can be combined with $\tilde{v}$ to form an analytic function. But the ambiguous constant is unimportant, since the capacitance is a function only of the total change in $\tilde{u}$ as we go around the inner conductor. It is clear that this change is the same as the change in u as we go around the inner conductor of the z capacitor. Thus the capacitance is a conformal invariant.

The conformal invariance of the capacitance implies that the inner radius of a canonical annulus is a modulus. Why? It is clear from (26.18) that two canonical annuli with different inner radii will have different capacitances. Since the capacitance is a conformal invariant, two canonical annuli with different radii cannot be mapped conformally into each other. This is what we wanted to show.

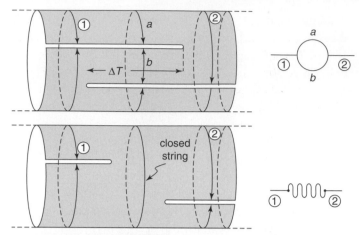

Fig. 26.9   Top: an incoming open string splitting into two strings, string *a* and string *b*, that then merge to form string two. The string diagram is nonplanar. Bottom: when the slits do not overlap, we find intermediate closed strings.

26.4 Non-planar open string diagrams

When we studied the one-loop open string diagram of Figure 26.2, we mentioned that there is another string diagram that corresponds to the Feynman diagram of Figure 26.1. That other string diagram is shown in the top part of Figure 26.9. The diagram may appear to be drawn in an unusual way: were it not for the slits, one extending far to the left and the other extending far to the right, we would have a closed string diagram. Both the incoming and the outgoing strings, called string one and string two, are open strings. When the two slits overlap we have two intermediate open strings, string *a* and string *b*. The process is that of an open string splitting into two open strings which, later on, recombine to form a single open string.

What is the difference between the one-loop diagram at the top of Figure 26.9 and the one-loop diagram of Figure 26.2? The difference is topological and very significant. In both diagrams, the boundary has two disjoint components. In Figure 26.2, both external open strings are attached to the same boundary component, and the other boundary component is provided by the intermediate slit. This is manifest in Figure 26.2(d). In Figure 26.9 the situation is completely different. The slits themselves are the boundary components, and each boundary component contains an external string. This string diagram is said to be *non-planar* because it cannot be flattened out without tearing it up or having one piece of the surface lie on top of some other piece of the surface. Another non-planar diagram is the infinite cylinder which represents free closed string propagation.

As we learned before, the set of string diagrams that contribute to an amplitude is obtained by varying the parameters that define the string diagram. The diagram at the top of Figure 26.9 has two parameters. The first specifies the light-cone momentum of one of the intermediate open strings. The second, more important for our present considerations, gives the length ΔT of the time interval during which the slits overlap. When ΔT is large, the slits go past each other for a long time. As $\Delta T \to 0$, the overlap time of the slits goes to zero. But $\Delta T = 0$ is not the end of the parameter space; ΔT can turn negative, in which case we get the string diagram shown in the bottom of Figure 26.9. Surprisingly, we get an intermediate closed string! The string interpretation of the string diagram changes completely. Now the process is described as an open string closing to form a closed string that propagates for some time and then breaks up to form an outgoing open string. To the right of the string diagram we see the corresponding graph. The straight line, representing the open string, turns into the wavy line, representing the closed string. The wavy line then turns back into a straight line. The graph has no loop!

There is an important lesson here. In a consistent theory of interacting quantum open strings, the inclusion of closed strings is generically required. This should not be too surprising. The process by which two open strings join at their endpoints is a *local* process; it is only dependent upon the regions of the strings near the endpoints. This means that it could just as well happen to the endpoints of a single string. Thus, if our theory includes processes in which two strings can join, then it will also, in general, contain processes in which a single open string closes to form a closed string.

A final comment on moduli. When ΔT is large and negative, the two boundaries of the annulus are far from each other. This is the region of moduli space where the inner radius of the canonical annulus is going to zero. This region is dominated by long-time closed string propagation, just as it was in our analysis at the end of Section 26.2. In the former case, the closed string propagates for a long distance and then stops. In the present case, after propagating for a long distance, the closed string turns again into an open string.

26.5 Four closed string interactions

We have seen before that the string diagram for a freely propagating closed string is a twice-punctured Riemann sphere $\hat{\mathbb{C}}$ (Section 25.6). The string diagrams for a closed string interaction which involves two incoming and two outgoing closed strings are Riemann spheres with four punctures. In Section 25.7 we saw that the string diagrams for the interaction of four open strings generate the moduli space of $\bar{\mathbb{H}}$ with four boundary punctures. It is reasonable to expect that the string diagrams for the interaction of four closed strings generate the moduli space $\mathcal{M}_{0,4}$ of four-punctured Riemann spheres. Recall from equation (25.60) that this moduli space has two real parameters. The moduli space is represented by the position of a puncture on $\hat{\mathbb{C}}$ when the other three punctures are fixed, conventionally, to be at 0, 1, and ∞. Our goal in this section is to give some evidence that the hypothesis is, in fact, true: the string diagrams generate the moduli space $\mathcal{M}_{0,4}$. Our arguments will only be qualitative.

Consider the interaction of four closed strings shown in the light-cone diagram of Figure 26.10. Note how closed strings interact: the two incoming closed strings meet at a point and form a longer closed string. After some propagation, two points on the intermediate closed string meet, and the closed string splits into the two outgoing closed strings. If

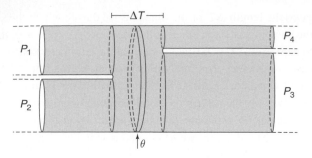

A light-cone string diagram for the interaction of four closed strings. The Riemann surface is a four-punctured sphere. This string diagram has two parameters, the time ΔT and the twist angle θ. String two approaches string one when ΔT is very large.

this diagram is to generate even only a portion of the two-dimensional moduli space $\mathcal{M}_{0,4}$, it must have two parameters. One parameter we have encountered already: the time ΔT between the interaction points. For simplicity, this diagram will be taken to include only the range $\Delta T \in [0, \infty)$. We will see that the region $\Delta T < 0$ actually comprises two diagrams! Where is the other parameter of the string diagram? Cut the middle cylinder with a pair of scissors along the double lines shown in the figure. Rigidly rotate the right side of the diagram by an angle θ, and then glue the diagram back together. The resulting diagram does not look like the original one. With a rotation of 90°, for example, string three would be in front of string four. Diagrams obtained from different values of θ can be shown *not* to be conformally equivalent. Therefore $0 \leq \theta < 2\pi$ is the second parameter in the string diagram.

With the parameters $0 \leq \Delta T < \infty$, and $0 \leq \theta < 2\pi$, what region of the moduli space $\mathcal{M}_{0,4}$ does the diagram in Figure 26.10 produce? To answer this question, we imagine mapping this diagram into $\hat{\mathbb{C}}$, letting P_1, P_3, and P_4 land at 0, 1, and ∞, respectively. The modular parameter is the position λ of P_2. How does λ vary as we change ΔT and θ? As $\Delta T \to \infty$, λ will approach P_1, which is at the origin. Moreover, for each fixed ΔT, as θ goes from zero to 2π, λ traces a closed curve around P_1. The string diagram generates a region of $\mathcal{M}_{0,4}$ with the topology of a disk. When $\Delta T = 0$, the curve that we get is the boundary of the disk domain generated by this string diagram.

In the string diagram of Figure 26.10 there is a single intermediate string. There are two diagrams in which the interaction points cross each other and there are three intermediate strings. In the first, shown in Figure 26.11, string one joins string four. In the second, shown in Figure 26.12, string two joins string four.

In Figure 26.11, the time between interactions is $\Delta \tilde{T}$. As $\Delta \tilde{T}$ becomes large, P_2 approaches P_3, which is located at $\lambda = 1$. The intermediate closed string with momentum $p_1^+ - p_4^+$ can be cut, rotated, and glued back, just as before, making the rotation angle θ a parameter. This string diagram will generate a disk domain around $\lambda = 1$. The curve arising from $\Delta \tilde{T} = 0$ represents the boundary of the domain generated by this string diagram.

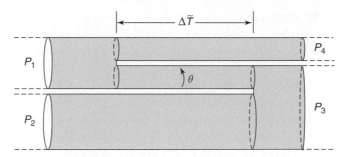

Fig. 26.11 In this string diagram P_2 approaches P_3 when $\Delta\tilde{T}$ becomes very large.

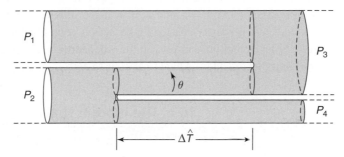

Fig. 26.12 In this string diagram P_2 approaches P_4 when $\Delta\hat{T}$ becomes very large.

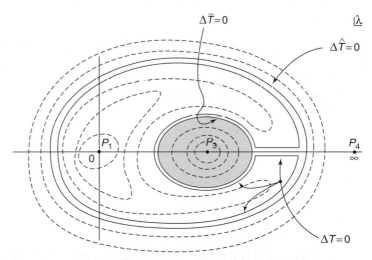

Fig. 26.13 The moduli space $\mathcal{M}_{0,4}$ of four-punctured spheres is generated by three string diagrams. The string diagram in Figure 26.10 generates the disk domain around $\lambda = 0$. The string diagrams in Figures 26.11 and 23.12 generate disk domains around $\lambda = 1$ and $\lambda = \infty$, respectively. The three disk domains fully cover $\mathcal{M}_{0,4}$, which is equivalent to $\hat{\mathbb{C}}$ with three points removed.

In Figure 26.12, the time between interactions is $\Delta\hat{T}$. As $\Delta\hat{T}$ becomes large, P_2 approaches P_4, which is located at $z = \infty$. The intermediate closed string, with momentum $p_2^+ - p_4^+$, also carries a θ parameter. This string diagram will generate a disk domain around $\lambda = \infty$. The curve arising from $\Delta\hat{T} = 0$ represents the boundary of the domain generated by this string diagram.

It turns out that the full moduli space $\mathcal{M}_{0,4}$ is precisely covered by the three string diagrams we have examined, each one producing a punctured-disk domain. Since no region in $\mathcal{M}_{0,4}$ is left unaccounted for, the domain boundaries must meet. So the images in $\mathcal{M}_{0,4}$ of the curves $\Delta T = 0$, $\Delta\tilde{T} = 0$, and $\Delta\hat{T} = 0$ must meet. The way this happens is actually quite intricate, as we show in Figure 26.13. The curves $\Delta\tilde{T} = 0$ and $\Delta\hat{T} = 0$ do not meet at all. Part of the $\Delta T = 0$ curve matches with the curve $\Delta\tilde{T} = 0$, and another part matches with the $\Delta\hat{T} = 0$ curve. Finally, some part of the $\Delta T = 0$ curve matches with itself! The disk domains around $\lambda = 1$ and $\lambda = \infty$ have ordinary shapes, but the one around $\lambda = 0$ has an unusual shape. You may take it as a good challenge to explain the features of Figure 26.13 by looking carefully at the θ dependence of each of the three string diagrams in the limit that the intermediate time goes to zero.

26.6 The moduli space of tori

We conclude this chapter with a look at the fascinating properties of loop amplitudes of closed strings. A light-cone string diagram which represents a one-loop process for a single closed string is shown in Figure 26.14. The slit in this diagram separates two cylinders – these are the intermediate closed strings. If there were no slit, the string diagram would be a two-punctured Riemann sphere. A torus can be viewed as a sphere with a hole. To do so, one can cut out a disk around the north pole of the sphere and a disk around the south pole of a sphere. The two resulting boundaries are then pushed towards each other and glued. The result is a torus. The slit in Figure 26.14 is the hole that turns the sphere into a torus. More precisely, we have a torus with two punctures.

To understand whether ultraviolet divergences can occur, the key questions we must answer are

(1) Why is the torus a Riemann surface?
(2) How many moduli does a torus have?
(3) What does the moduli space of tori look like?

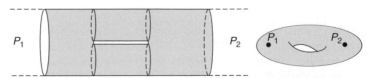

Fig. 26.14   A one-loop light-cone diagram for a closed string. The string diagram is a torus with two punctures.

647   26.6 The moduli space of tori

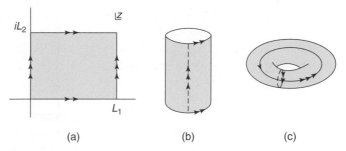

Fig. 26.15 (a) A rectangular torus is a rectangular region of the complex z-plane with identifications. (b) Gluing the vertical sides of the rectangular region gives a cylinder. (c) Gluing the horizontal sides as well gives a torus.

Since the answers to these questions are interesting even for tori without punctures, we will consider this case in detail. We will see that tori have two real moduli and that the moduli space is surprisingly intricate.

We begin with the simplest type of torus, the rectangular torus. As shown in Figure 26.15, the torus can be defined as a rectangular region of the complex plane $\mathbb{C}$ with some identifications. In particular, the horizontal sides must be identified as indicated by the double arrows, and the vertical sides must be identified as indicated by the triple arrows. To the right, we show the region after the vertical lines have been identified, and then, in the rightmost part of the figure, we have the final result.

The identifications that lead to the rectangular torus, starting from the full complex plane, are described by the equations

$$z \sim z + L_1 , \quad z \sim z + iL_2 . \tag{26.21}$$

The fundamental domain for these identifications is the rectangular region $0 \leq \Re(z) < L_1$, $0 \leq \Im(z) < L_2$. This torus is a Riemann surface because the complex plane $\mathbb{C}$ is one and because the identifications above are analytic identifications: $z \sim f(z)$, with f an analytic function.

It is important to note that neither L_1 nor L_2 is a parameter of the rectangular torus. We can scale the z coordinate by a constant factor. Letting $z' = z/L_1$, which is obviously a conformal map, the identifications in (26.21) become

$$z' \sim z' + 1 , \quad z' \sim z' + iT , \quad T \equiv \frac{L_2}{L_1} . \tag{26.22}$$

It follows that a rectangular torus has just one parameter, the value of T. The above equations define the canonical presentation of a rectangular torus. In the canonical presentation, the horizontal length is one.

Surprisingly, rectangular tori with different T parameters can sometimes be conformally equivalent. To prove this, consider the torus shown in the top left of Figure 26.16. This is a rectangular torus with $T < 1$, shown as a rectangular domain in the w plane. Now we

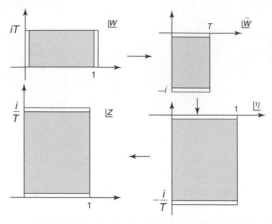

Fig. 26.16 Top left: a rectangular torus with parameter T in the w plane. Top right: the torus mapped to $\tilde{w} = -iw$. Bottom right: the torus mapped to $\eta = \tilde{w}/T$. Bottom left: the torus mapped to $z = \eta + \frac{i}{T}$. This final form is a rectangular torus with parameter $1/T$.

perform a sequence of three conformal maps. For visual aid, the vertical lines and their images under the map are shown as double lines. The first map is $\tilde{w} = -iw$. This is a clockwise rotation of the original rectangle by $90°$. Next we scale the figure by a factor of $1/T$: $\eta = \tilde{w}/T$. The rectangle now lies below the real line. In the final step, we lift it up: $z = \eta + i/T$. The result (bottom left of the figure) is the canonical presentation of a rectangular torus with parameter $1/T$! We have therefore shown that tori with parameters T and $1/T$ are conformally equivalent. This implies that the tori with $0 < T \le 1$ are conformally equivalent to the tori with $1 \le T < \infty$. As a result, the *moduli space* of rectangular tori can be chosen to be the interval $1 \le T < \infty$ or, alternatively, the interval $0 < T \le 1$. We are assuming that no conformal map exists that produces a further identification of rectangular tori, which is in fact true.

The above conclusion has some implications for ultraviolet divergences. The tori with $T \to 0$, give rise to short closed strings that appear as vertical segments in the top left of Figure 26.16. On the other hand, these tori are conformally equivalent to large tori with $T \to \infty$, so it is mathematically impossible for them to give rise to ultraviolet problems. In fact, if we include the tori $1 \le T < \infty$, then we should *not* include the tori $0 < T < 1$. String amplitudes are integrals over the space of inequivalent Riemann surfaces; if we included the short tori we would be double counting. Using the canonical presentation $1 \le T < \infty$ of the moduli allows us to work with long tori only.

The rectangular tori do not exhaust the set of all conformally inequivalent tori. Tori can be twisted as well. Referring to part (b) of Figure 26.15, one can glue the bottom and top edges of the cylinder with a twist. The way this is actually done is sometimes elusive, so we will explain it in detail. In particular, if you had a real-life rubber cylinder, then gluing its open edges after a twist by an angle θ and after a twist by an angle $\theta + 2\pi$ would be physically inequivalent processes. For a torus Riemann surface, a twist by θ and a twist by $\theta + 2\pi$ are equivalent.

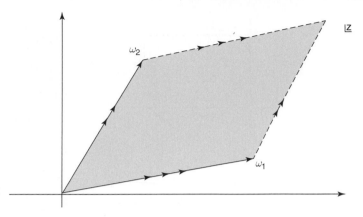

Fig. 26.17 A general torus is obtained from the complex z-plane by identifying $z \sim z + \omega_1$ and $z \sim z + \omega_2$.

To discuss twisted tori clearly, we begin by giving the general construction of a torus. To get a torus from the complex plane we need two identifications. We choose two complex numbers ω_1 and ω_2, both different from zero and satisfying

$$\Im\left(\omega_2/\omega_1\right) > 0 \,. \tag{26.23}$$

This condition is simpler than it looks. It just requires that the two complex numbers represent vectors that are not parallel. Then, by convention, we choose ω_2 to be the one which is obtained from ω_1 by a counterclockwise rotation through an angle smaller than $180°$. An example is shown in Figure 26.17. The torus is obtained by the identifications

$$z \sim z + \omega_1 \,, \quad z \sim z + \omega_2 \,. \tag{26.24}$$

The fundamental domain is the parallelogram shaded in the figure, and its edges are identified. This is the general construction of a torus Riemann surface, but it has extraneous parameters. We can define

$$\tau \equiv \omega_2/\omega_1 \,, \quad \Im(\tau) > 0 \,, \tag{26.25}$$

and, scaling z by a factor of $1/\omega_1$, we note that the identifications above are equivalent to the identifications

$$z \sim z + 1 \,, \quad z \sim z + \tau \,, \quad \Im(\tau) > 0 \,. \tag{26.26}$$

Observe that τ lives in the upper half-plane $\mathbb{H}$. The new picture of the parallelogram is shown in Figure 26.18. Note that the torus is rectangular precisely when $\Re(\tau) = 0$ (i.e., when τ is purely imaginary).

Could the moduli space of tori be just $\tau \in \mathbb{H}$? It cannot. Our previous work with rectangular tori indicates that tori with different τ parameters can be equivalent. Indeed, for a rectangular torus with parameter T, we have $\tau = iT$, as is clear by comparing (26.22) and (26.26). But this torus is equivalent to a torus with parameter $\tau' = i/T$. It follows that $\tau' = -1/\tau$, and, at least for rectangular tori, τ and $(-1/\tau)$ are conformally equivalent. In fact, we will see shortly that τ and $(-1/\tau)$ always define conformally equivalent tori.

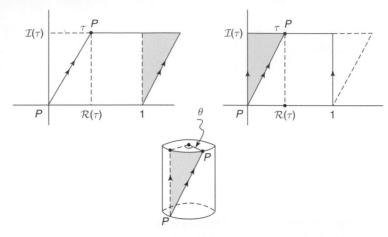

Fig. 26.18 The canonical presentation of a torus with parameter $\tau \in \mathbb{H}$. The shaded region in the top left figure can be moved, as shown in the top right figure. Using this rectangular fundamental domain, we form a cylinder and recognize the twist angle θ associated with the final identification.

Let us now explain the twisting of tori with $\Re(\tau) \neq 0$. Consider Figure 26.18. Let P denote the points $z = 0$ and $z = \tau$, which are equivalent after identification. Making use of the identification $z \sim z + 1$, we can move the shaded region, shown in the top left, to obtain the result shown in the top right. The fundamental domain has become a rectangle, even though the torus is not rectangular! While the vertical edges are identified, the identification of horizontal lines is shifted: the point P is not identified with a point that is directly above it, as would be the case for a rectangular torus. We can now roll up the rectangular domain to form a cylinder, whose ends remain to be joined together. This cylinder is shown in the bottom part of the figure. The identification of points on the two boundary components of the cylinder must be shifted, as P on the bottom end must be matched with P on the top end. We can use an angular variable running from zero to 2π to parameterize uniformly both boundary components of the cylinder. We take P on the lower boundary component to have zero angle. The upper boundary component is parameterized so that a given point takes the angular value assigned to the point on the lower boundary that is directly below it prior to identifications. Since the identifications are shifted by $\Re(\tau)$ along the boundary, and the boundaries have unit length, the angle θ associated with P on the top boundary is

$$\theta = 2\pi \, \Re(\tau) \,, \tag{26.27}$$

as shown in Figure 26.18. We call θ the twist angle.

We can now investigate the effects of letting the twist angle increase by 2π. This can be done by letting $\tau \to \tau + 1$, as can be seen in equation (26.27). If a torus with parameter τ and a torus with parameter $\tau + 1$ (Figure 26.19) are really the same, then increasing the twist angle by 2π changes nothing. We now use the arbitrariness in the choice of fundamental domain to demonstrate that two such tori are in fact the same. With the identification

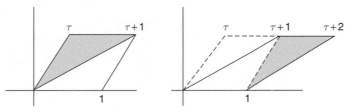

Fig. 26.19 A torus with parameter τ (left) and a torus with parameter $\tau + 1$ (right). By moving the shaded region on the left using $z \sim z + 1$, we see that the two tori are really the same.

$z \sim z + 1$, we can shift the shaded region on the left side of the figure to the position indicated on the right side of the figure. In doing so each point in the region has been moved horizontally by one unit. This operation does not change the torus; it is simply a different choice of fundamental domain. It follows that the torus on the left, with parameter τ, is the same torus as the torus on the right, with parameter $\tau + 1$:

$$\tau \sim \tau + 1. \tag{26.28}$$

Back in (26.27), this means that $\theta \sim \theta + 2\pi$. Thus, for example, twist angles beyond the interval $-\pi \leq \theta < \pi$ do not yield new tori. Now we can understand this concretely. Do *not* think of twisting as an application of physical torsion to wind up the cylinder before gluing the boundaries. Twisting is just a prescription for identifying points on the two boundary components of the cylinder: we arbitrarily pick a point P on the bottom boundary and then fix a point on the top boundary to identify it with. The identification is then extended uniformly over the boundaries. It is clear from this prescription that the twist identification parameter simply describes the position of a point on the top boundary, and thus the twist parameter lives on a circle.

Concerning the moduli space of tori, the identification (26.28) has an important implication. While τ certainly lives in $\mathbb{H}$, the space of inequivalent tori is much smaller. The identification shows that any infinite vertical strip of unit width contains all inequivalent tori. It is customary to choose the strip $\mathcal{S}_0$:

$$\mathcal{S}_0 \equiv \left\{ -\tfrac{1}{2} < \Re(\tau) \leq \tfrac{1}{2}, \quad \Im(\tau) > 0 \right\}. \tag{26.29}$$

The set of inequivalent tori is contained in $\mathcal{S}_0$. Note that the right boundary $\Re(\tau) = 1/2$ of the strip is included in the set. The left boundary $\Re(\tau) = -1/2$ is not included, because it is identified with the right boundary under $\tau \to \tau + 1$.

Our analysis of rectangular tori indicates that perhaps $\tau \sim -1/\tau$. Let us now prove that this is true. Consider a torus with parameter τ, as shown in the top left of Figure 26.20. Now define the conformally related $\tilde{z} = z/\tau$. This map is a rigid rotation plus a uniform scaling, and the original parallelogram turns into the parallelogram shown at the top right part of the figure. In a final step we do a rigid translation $z' = \tilde{z} - 1/\tau$. The final parallelogram in the z'-plane is in canonical form, and the parameter of the associated torus is $(-1/\tau)$. Thus $\tau \sim -1/\tau$.

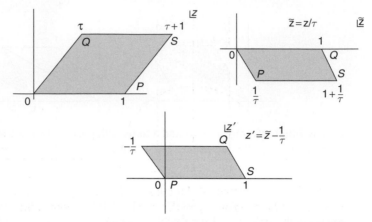

Fig. 26.20 A sequence of conformal maps that shows that a torus with parameter τ is conformally equivalent to a torus with parameter $-1/\tau$.

At this stage, we have found two equivalences for tori described by $\tau \in \mathbb{H}$:

$$\tau \sim \tau + 1, \quad \tau \sim -1/\tau. \tag{26.30}$$

It turns out that there are no additional independent identifications. It is therefore of great interest to find a fundamental domain for the identifications (26.30), for this will give us the set of all inequivalent tori. A fundamental domain for the first identification is the strip S_0. Since the second relation identifies points in the region $|\tau| < 1$ with points in the region $|\tau| > 1$, we can attempt to use the subset of S_0 which lies beyond the unit circle together with some points which lie on the unit circle. This is actually the answer, but it is by no means obvious. The same logic could have been used to argue that we can use the subset of S_0 which lies inside the unit circle, together with some points which lie on the unit circle, but this would have been wrong. The claim is that a fundamental domain is the region $\mathcal{F}_0$, defined by

$$\mathcal{F}_0 \equiv \Big\{ -\tfrac{1}{2} < \Re(\tau) \leq \tfrac{1}{2}, \Im(\tau) > 0, \ |\tau| \geq 1, \text{ with the}$$
$$\text{further restriction that } \Re(\tau) \geq 0 \text{ if } |\tau| = 1 \Big\}. \tag{26.31}$$

The fundamental domain $\mathcal{F}_0$ is shown as the shaded region in Figure 26.21. As stated in the above conditions, certain boundaries (shown by dashed lines in the figure) are not included in $\mathcal{F}_0$. These boundaries are composed of the points on the vertical line $\Re(\tau) = -1/2$ and by the points on the unit circle that lie to the left of $\tau = i$, which are removed by the further

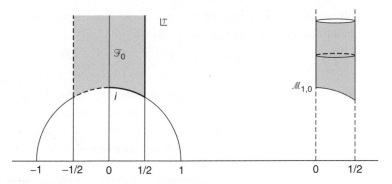

Fig. 26.21 The fundamental domain $\mathcal{F}_0$ for the identifications $\tau \sim \tau + 1$ and $\tau \sim -1/\tau$ of $\mathbb{H}$. The closure $\bar{\mathcal{F}}_0$ of the region $\mathcal{F}_0$, with suitable identifications is the moduli space $\mathcal{M}_{1,0}$ of tori shown to the right.

restriction in (26.31). We will give below part of the proof that $\mathcal{F}_0$ is a fundamental domain for the identifications (26.30), leaving the rest of the proof for Problem 26.6.

The moduli space of tori, denoted as $\mathcal{M}_{1,0}$, is the τ upper half-plane $\mathbb{H}$ subject to the identifications (26.30). We learned in Section 2.7 how to build a space which arises from identifications: we take the fundamental domain together with its boundary and implement the identifications on the boundary. In the present situation, the fundamental domain $\mathcal{F}_0$ together with its boundary defines the closure $\bar{\mathcal{F}}_0$:

$$\bar{\mathcal{F}}_0 = \left\{ -\tfrac{1}{2} \leq \Re(\tau) \leq \tfrac{1}{2}, \, \Im(\tau) > 0, \, |\tau| \geq 1 \right\}. \tag{26.32}$$

On $\bar{\mathcal{F}}_0$ we impose two identifications: $\Re(\tau) = -1/2$ and $\Re(\tau) = 1/2$ are identified by $\tau \to \tau + 1$, and the points with $|\tau| = 1$ are identified among themselves by $\tau \to -1/\tau$. These are, in fact, all the identifications in $\bar{\mathcal{F}}_0$ (see Problem 26.6). The resulting moduli space $\mathcal{M}_{1,0}$ of tori can be visualized by cutting out the region $\mathcal{F}_0$, folding it along the imaginary axis, and gluing the boundaries, as shown in the right side of Figure 26.21.

Quick calculation 26.1 Consider the points on the unit circle $|\tau| = 1$ that lie on $\mathbb{H}$ in between the vertical lines $\Re(\tau) = \pm 1/2$. Show that the points to the right of $\tau = i$ are identified with the points to the left of $\tau = i$ via $\tau \to -1/\tau$.

Let us consider the physical implications of this result. Since all inequivalent tori are contained in $\mathcal{F}_0$, one-loop closed string amplitudes must only include the contributions from tori with $\tau \in \mathcal{F}_0$. Since the tori in $\mathcal{F}_0$ are long tori, ultraviolet divergences are not an issue. If amplitudes used the full strip $\mathcal{S}_0$, the situation would have been quite problematic. This is not because we would have had some small tori to deal with, since, after all, they are conformally equivalent to (safe) long tori. The problem would have arisen because the complement of $\mathcal{F}_0$ in $\mathcal{S}_0$ contains an *infinite* number of copies of all the tori in $\mathcal{F}_0$! We can easily verify, for example, that it contains infinitely many copies of the torus $\tau = i$. Consider the set τ_n of values

$$\tau_n = i + n, \quad n \geq 1, \tag{26.33}$$

all of which, by virtue of $\tau \sim \tau + 1$, are just copies of the $\tau = i$ torus. Using the identification $\tau \sim -1/\tau$, we conclude that

$$-\frac{1}{\tau_n} = -\frac{n}{n^2+1} + \frac{i}{n^2+1} \tag{26.34}$$

are also copies of the $\tau = i$ torus. But for $n \geq 1$, these copies lie in $\mathcal{S}_0 - \mathcal{F}_0$. This is clear because the real part of $-1/\tau_n$ is in $[-1/2, 0]$ and $|1/\tau_n| = 1/\sqrt{n^2+1} < 1$. If we had to include all tori in $\mathcal{S}_0$, all one-loop amplitudes would be infinite. Instead, one-loop amplitudes use $\mathcal{F}_0$ because this space contains every inequivalent torus exactly once.

To establish that $\mathcal{F}_0$ is a fundamental domain for the identifications (26.30) requires some real work. Let us do some of it here to give you an idea about the necessary tools. To begin with, it is helpful to introduce some additional structure. Note that the identifications (26.30) are actually real linear fractional transformations, just like the ones we studied in Section 25.6. Let T and S denote the generators of linear fractional transformations that act as

$$T\tau = \tau + 1, \quad S\tau = -1/\tau. \tag{26.35}$$

Consider now transformations g of the form

$$g\tau = \frac{a\tau + b}{c\tau + d}, \quad a, b, c, d \in \mathbb{Z}, \quad ad - bc = 1. \tag{26.36}$$

Note that this is a linear fractional transformation with *integer* parameters. We can use matrix notation to describe these transformations. Associated with the transformation g we introduce the 2-by-2 matrix $[g]$ defined by

$$[g] = \begin{pmatrix} a & b \\ c & d \end{pmatrix}, \quad \det[g] = 1. \tag{26.37}$$

The matrix $[g]$ has an ambiguity. We can change the sign of all entries without changing the transformation. We must therefore consider a matrix and minus the matrix to be equivalent. Given two transformations g_1 and g_2 of the form indicated in (26.36), the composition $g_1 g_2$ is also a transformation of the same form, and

$$[g_1 g_2] = [g_1][g_2]. \tag{26.38}$$

Quick calculation 26.2 Prove equation (26.38).

The matrices associated with the transformations T and S are conventionally chosen to be

$$[T] = \begin{pmatrix} 1 & 1 \\ 0 & 1 \end{pmatrix}, \quad [S] = \begin{pmatrix} 0 & -1 \\ 1 & 0 \end{pmatrix}. \tag{26.39}$$

The composition of any number of S and T transformations, in any order, gives rise to a transformation of the type (26.36). A useful property of such transformations is that

$$\Im(g\tau) = \frac{\Im(\tau)}{|c\tau + d|^2}, \tag{26.40}$$

as we verified in a similar calculation earlier (see (25.66)).

The set of linear fractional transformations with integer coefficients (a, b, c, d) that satisfy $ad - bc = 1$ defines a group $\overset{\circ}{G}$ of transformations under function composition. The group is called the *modular group*, and it is the same group as the group of 2-by-2 matrices with integer entries and unit determinant – with the proviso that a matrix and minus the matrix are considered to be the same group element. Here group multiplication is matrix multiplication. The modular group is called $PSL(2, \mathbb{Z})$. Here L stands for linear transformations or matrices. The two is for 2-by-2, the size of the matrices. S is for special, since the matrices have unit determinant. $\mathbb{Z}$ indicates that the entries are integers. Finally, P is for projective, or the fact that the sign of all entries can be reversed.

Let G' denote the set of projective transformations that can be generated by composition of any number of S and T transformations and their inverses. G' is clearly a subgroup of the modular group G. We will now prove that, for any $\tau \in \mathbb{H}$, there is a $g \in G'$ such that $g\tau \in \mathcal{F}_0$. This states that $\mathcal{F}_0$ contains a copy of each torus – a necessary condition for $\mathcal{F}_0$ to be a fundamental domain. The proof is not a trivial matter, the problem being that it is not all that easy to see how the tori in $\mathcal{S}_0 - \mathcal{F}_0$ can be sent into $\mathcal{F}_0$. The strategy requires making sure that we can send any point to a new point with a sufficiently large imaginary part, so that it can then be brought into $\mathcal{F}_0$ by some number of T transformations.

First we show that, for each τ, there is a $g \in G'$ such that $\Im(g\tau)$ is largest. It follows from (26.40) that we need only search for a g such that $|c\tau + d|$ is smallest. Since c and d are integers, as we vary them, $(c\tau + d)$ forms the lattice of points generated by τ and 1 in the complex plane. For any number $\alpha > 0$, there are at most a finite number of points such that $|c\tau + d| < \alpha$. Thus we should be able to find a minimum for $|c\tau + d|$ among all g, which gives a g for which $\Im(g\tau)$ is largest. The transformation g is not unique.

Now act with T^n on $g\tau$, with $n \in \mathbb{Z}$ chosen so that $T^n g\tau \in \mathcal{S}_0$. We claim that, unless $T^n g\tau \in \bar{\mathcal{F}}_0$, we have a contradiction. Let

$$\tau' = T^n g \tau . \tag{26.41}$$

The imaginary part of τ' is equal to the imaginary part of $g\tau$, and, by our proof above, it should not be possible to find a $g' \in G'$ such that $g'\tau'$ has an imaginary part larger than that of τ'. If τ' is not in $\bar{\mathcal{F}}_0$ then $|\tau'| < 1$. But then, using (26.40), we find that

$$\Im(S\tau') = \frac{\Im(\tau')}{|\tau'|^2} > \Im(\tau') , \tag{26.42}$$

contradicting the fact that no transformation in G' can increase the imaginary part of τ'. This contradiction proves that any point $\tau \in \mathbb{H}$ can be brought into $\bar{\mathcal{F}}_0$ by transformations generated by T and S (and their inverses). Once it is in $\bar{\mathcal{F}}_0$, then, if it is in $\mathcal{F}_0$ we are done. If it is not in $\mathcal{F}_0$, then it must lie in the boundary of $\bar{\mathcal{F}}_0$. Applying S or T once will then send it to $\mathcal{F}_0$. This completes the proof that any point $\tau \in \mathbb{H}$ can be brought into $\mathcal{F}_0$ by transformations generated by T and S.

Of course, more remains to be shown in order to prove that $\mathcal{F}_0$ is a fundamental domain. It must be established that no two points in $\mathcal{F}_0$ are connected via a transformation in G'. In fact, it is possible to show a stronger result: no two points are connected via a transformation in the modular group G. Once this is proven, one can finally show that G' coincides with G. This means that S and T generate the full modular group (see Problem 26.6). It also follows that $\mathcal{F}_0$ is a fundamental domain for the modular group.

Problems

Problem 26.1 Open–closed string transition.

Examine the process by which an open string with light-cone momentum p^+ closes up to form a closed string.

(a) Draw the light-cone diagram as an infinite cylinder with a semi-infinite slit parallel to its axis. Show the incoming open string, the outgoing closed string, and the special closed string created at the interaction point.

(b) To draw this diagram as a region of the w plane, cut open the cylinder by following the direction of the slit. The result is the infinite strip $0 < \Im(w) \leq 2\pi\alpha' p^+$. The open strings exist for $\Re(w) < 0$, and the closed strings exist for $\Re(w) > 0$. The lines $\Im(w) = 0$ and $\Im(w) = 2\pi\alpha' p^+$ are glued when $\Re(w) > 0$. What is the boundary of the string diagram?

(c) Construct a map from this light-cone diagram into $z \in \bar{\mathbb{H}}$. Require that the incoming open string in the infinite past is mapped to $z = 0$ and that the outgoing closed string in the infinite future is mapped to $z = i$.
Hints: use $\xi = \exp(w/(2\alpha' p^+))$ to map the strip to $\bar{\mathbb{H}}$. Then let $\eta = \xi^2$. Why does this map remove the need for identifications? Show that the surface has gone to the exterior of a finite slit on the full η complex plane. To map this to $z \in \bar{\mathbb{H}}$ you will need a map that involves square roots.

(d) Plot in the z-plane the closed string emerging at the interaction time. Sketch also incoming open strings and outgoing closed strings.

Problem 26.2 Mapping of annuli into canonical form.

Find the error in the following argument which suggests that it is not possible to map conformally a topological annulus into a canonical annulus.

> The Riemann mapping theorem implies that any topological disk can be mapped to the unit disk. A topological annulus can therefore be mapped into the inside of a unit disk $|z| \leq 1$, with the outer boundary mapping into $|z| = 1$ and the inner boundary mapping into some closed curve inside the disk. To map this into a canonical annulus we must make the inner curve round and centered at the origin, while preserving the boundary $|z| = 1$. This requires a map of the disk to itself. The mappings of a disk to itself have three real parameters, just as the mappings of $\bar{\mathbb{H}}$ to itself. It is not possible to round a general curve using just three parameters, so it is impossible to map the annulus into canonical form.

Problem 26.3 Moduli space of T^2 compactifications with B field.

In Problem 17.5, the compactification parameters of the square torus T^2 are R and the Kalb–Ramond field b. Define the complex variable

$$\rho \equiv b\,\frac{R^2}{\alpha'} + i\,\frac{R^2}{\alpha'} = \frac{f_B}{2\pi} + i\,\frac{R^2}{\alpha'}\,.$$

Note that $\rho \in \mathbb{H}$. Prove that the transformations (1) and (2) in Problem 17.5 take the form

$$\rho \to \rho + 1\,, \quad \rho \to -1/\rho\,.$$

Describe the moduli space of compactifications on square tori T^2.

Problem 26.4 Shapiro–Virasoro amplitude.

The Shapiro–Virasoro amplitude is the scattering amplitude of four closed string tachyons. Recall that for a closed string tachyon we have

$$p^2 = -M^2 = 4/\alpha'\,. \tag{1}$$

For this amplitude we need the moduli space $\mathcal{M}_{0,4}$ of four-punctured spheres. When three of the punctures are placed at 0, 1, and ∞, the moduli space is parameterized by the position of the last puncture, a complex number $\lambda \in \hat{\mathbb{C}}$. The self-mappings of the z-sphere $z \in \hat{\mathbb{C}}$ are the linear fractional transformations

$$z \to \frac{az+b}{cz+d}\,, \quad ad - bc = 1\,. \tag{2}$$

(a) Suppose we want to integrate some quantity over the sphere. We then write an integral $\int d^2z \dots$. Here $d^2z \equiv dxdy$, with $z = x + iy$. How does d^2z transform under a linear fractional transformation? (Recall that you must use Jacobians!)

(b) Mimicking the open string Veneziano amplitude, consider an expression for the Shapiro–Virasoro amplitude A_{SV} of the form

$$A_{SV} = g^2 \int d^2z |z_1 - z_3|^\alpha \cdots |z - z_1|^{\beta p_2 \cdot p_1} \cdots \tag{3}$$

where the dots represent additional factors, which you must write, and α and β are constants you must determine by the condition that the integrand is invariant under linear fractional transformations.

(c) Simplify (3) now for the case $z_1 = 0$, $z_3 = 1$, $z_4 = \infty$, and $z = \lambda$. Show that you get an expression of the form

$$A_{SV} = g^2 \int d^2\lambda\, |\lambda|^{\beta p_2 \cdot p_1} |1 - \lambda|^{\beta p_2 \cdot p_3}\,. \tag{4}$$

(d) To do the integral, first prove the following identity:

$$|z|^{-a} = \frac{1}{\Gamma(a/2)} \int_0^\infty dt\, t^{\frac{a}{2}-1} \exp(-t|z|^2)\,. \tag{5}$$

Now use it twice in (4), once for each factor in the integrand. Call the needed parameters t and s. The λ integral now becomes Gaussian, and this is made clear by writing $\lambda = \lambda_1 + i\lambda_2$ and $d^2\lambda = d\lambda_1 d\lambda_2$. Do the Gaussian integral. To do now the t and s integrals let $t = xu$ and $s = (1-x)u$ with $0 < x < 1$ and $0 < u < \infty$. Your final answer should be the ratio of three gamma functions over three other gamma functions.

(e) Consider the invariants

$$s = -(p_1 + p_2)^2, \quad t = -(p_2 + p_3)^2, \quad u = -(p_1 + p_3)^2.$$

Show that your answer has the expected pole structure in each of the three possible channels.

Problem 26.5 Bringing τ into the fundamental domain.

Let $\tau = (9 + i)/10$. Find the transformation in the modular group that brings it into a point τ_0 of the fundamental domain $\mathcal{F}_0$. Show that the transformation is unique by proving that there is no transformation in the modular group (except the identity) that leaves τ_0 fixed. Partial answer: $\tau_0 = 5i$. Repeat the above computations for $\tau = 9/10 + i/100$. Partial answer: $\tau_0 = i + 1/10$. [Hint: use the procedure discussed in the proof that any τ can be brought into the fundamental domain.]

Problem 26.6 S and T generate the modular group.

(a) Let $z \in \bar{\mathcal{F}}_0$ and assume that $gz \in \bar{\mathcal{F}}_0$ is a different point, where g is an element of the modular group (as in (26.37)). Since we can replace (z, g) by (gz, g^{-1}), we can assume that $\Im(gz) \geq \Im(z)$. This requires $|cz + d| \leq 1$. Explain why $|c| \geq 2$ is impossible. Discuss the finitely many possible values of c and d and prove that either $\Re(z) = \pm 1/2$ and $gz = z \mp 1$ or $|z| = 1$ and $gz = -1/z$.
(b) Prove that the group G' generated by S and T (and its inverses) coincides with the modular group G. Since G' is a subgroup of G, it suffices to show that each element of G is an element of G'. Construct the proof using the following two facts: (i) any point outside $\mathcal{F}_0$ can be brought into $\mathcal{F}_0$ by a transformation in G', and (ii) for any point z_0 on $\mathcal{F}_0$ and any $g \in G$ different from the identity, the point gz_0 is outside $\mathcal{F}_0$ (this is a consequence of (a)).

References

Most works in string theory assume background knowledge in quantum field theory and general relativity. For this reason, the present bibliography consists mostly of review papers. Even reviews are not an easy read, but they are more pedagogical than research papers, and they give many references, some of which the reader may want to consult at some point. We have also included in this bibliography some readable articles that supplement the material that has been covered in this book.

Readers of this book who are looking for more advanced material in string theory are well advised to begin their search by consulting the textbook by Becker, Becker, and Schwarz (2007), the two-volume text by Polchinksi (1998), and the two-volume text by Green, Schwarz, and Witten (1987). A helpful reference text is that of Kiritsis (2007). A detailed discussion of matters related to D-branes can be found in the book by Johnson (2003). While all these books require a background in quantum field theory and general relativity, selected parts may be understood without such a background.

For a view of string theory at its earliest stage of development, see the book by Frampton (1974). For string field theory, see the book by Siegel (1988). A book with a broad scope is that by Kaku (2000). Useful lecture notes include those of Lüst and Theisen (1989), Ooguri and Zin (1997), and 't Hooft (2004). A number of popular-level books are good reads: Gell-Mann (2002) for string theory within physics, Greene (2000) for an introduction to string theory, Randall (2005) for extra dimensions, and Susskind (2006) for string theory and the landscape.

The main reference for Part I of this book is the article of Goddard *et al.* (1973), which developed the light-cone quantization of the bosonic open string. After studying Part I, the reader may enjoy having a close look at this paper. Related subjects were nicely reviewed early on by Scherk (1975). The discussion in Section 7.4 is based on unpublished work of Jeffrey Goldstone. Background material for Part I can be found in textbooks on special relativity, electrodynamics, classical mechanics, and quantum mechanics. Although not needed, some students may enjoy Hartle's introduction to general relativity (2003) and Mandl and Shaw's introduction to quantum field theory (1984).

We now refer to articles and reviews that touch on specific topics. Problem 2.9 is based on a calculation by Seiberg (1997). For experiments testing the existence of large extra dimensions, see Chiaverini *et al.* (2003), Hoyle (2003), Long and Price (2003), and Kapner *et al.* (2007). For lectures on extra dimensions see Sundrum (2005). Problem (4.6) was based on a suggestion by Alan Guth. Problem 6.10, on a circular string in de Sitter space, is based on a homework problem by Frey (2005). Cosmic strings are discussed in detail in Vilenkin and Shellard (2000). For a status report see Polchinski (2006), and for cosmological bounds see Jeong and Smoot (2004). The calculation of the central term in the

Virasoro algebra follows a method by 't Hooft (2004). On the subject of tachyons and D-brane decays (the Sen conjectures), two references are Gaiotto and Rastelli (2003) and Sen and Zwiebach (2000). See also the reviews by Ohmori (2001), Sen (1999), and Taylor and Zwiebach (2003). The analytic description of the tachyon vacuum was given by Schnabl (2006). To supplement the discussion of superstrings in Chapter 14, the reader may consult the lectures by Schwarz (2000) and many other long and short reviews. A selected sample is: Douglas (1996), Polchinski (1996), Sagnotti and Sevrin (2002), Schwarz (2003), Schwarz and Seiberg (1999), Sen (2001), Taylor (2001), and Witten (2002).

Many of the subjects discussed in Part II are covered (more technically) in the books cited above. For D-branes, the reader may also consult the lectures by Bachas (1998), Polchinski, Chaudhuri, and Johnson (1996), Taylor (1998), the review by Giveon and Kutasov (1999), and the article by Hanany and Witten (1997). For some perspective on QCD within the Standard Model, see Gross (1999).

Electromagnetic duality is discussed in Deser (1982) and the review by Harvey (1996). For reviews on T-duality of closed strings, see Giveon, Porrati, and Rabinovici (1994) and Lerche, Schellekens, and Warner (1989). Open string T-duality and electromagnetic fields on D-branes are interrelated subjects. Some reviews are Ambjorn *et al.* (2003) and Taylor (1998). For additional details on strings in electric fields, see Seiberg, Susskind, and Toumbas (2002). For strings in magnetic fields, see Seiberg and Witten (1999) and the readable discussion of Bigatti and Susskind (2000a). A clear and detailed discussion of nonlinear electrodynamics is given by Bialynicki-Birula (1983). For references and results on Born–Infeld theory, see Gibbons (2003). Problems 20.6 and 20.7 are based on a result by Callan and Maldacena (1998).

The discussion of intersecting brane models and string phenomenology is based on the articles by Ibanez, Marchesano, and Rabadan (2001) and Cremades, Ibanez, and Marchesano (2002a, 2003). For reviews, see Uranga (2003) and Cremades, Ibanez, and Marchesano (2002b). Earlier works on the subject include Cvetic, Shiu, and Uranga (2001) and Blumenhagen *et al.* (2000). For other aspects of string theory model building, see the reviews by Aldazabal *et al.* (2000), Angelantonj and Sagnotti (2002), Dienes (1997), Ovrut (2002), and Quevedo (1996, 2002). Our discussion of moduli stabilization and the landscape borrows from Denef, Douglas, and Kachru (2007). Key ideas on this subject appeared in Bousso and Polchinski (2000). For a recent discussion of the landscape and particle physics see Lüst (2007).

An insighful historical introduction to thermodynamics and statistical mechanics is found in the book by Atkins (1994). Since the subject of string thermodynamics is very broad, we cite Abel *et al.* (1999), a relatively recent paper with many references. Also recommended are the lectures by Deo, Jain, and Tan (1990). For a readable introduction to the physics of black holes see, for example, the book by Begelman and Rees (1998) and the book by Thorne (1994). See also the article by Bekenstein (2001). Das and Mathur (2001) review many aspects of black hole physics in string theory. Our discussion of the entropy of Schwarzschild black holes is based on Horowitz and Polchinski (1997). The five-dimensional black hole discussed in the text is that in Strominger and Vafa (1996). For a novel approach to black hole entropy see Mathur (2005). For a recent review on the subject see Sen (2007).

For Regge trajectories see Tang and Norbury (2000) and Selem and Wilczek (2006). Problem 23.1 is based on a similar problem in Frey (2005). The Poincaré-invariant modification of string theory with critical dimension four is in Polchinski and Strominger (1991). For a review of results on the quark-antiquark potential see Kuti (2005). For the AdS/CFT correspondence, a readable, brief review is Horowitz and Polchinski (2006). Other useful reviews are: Aharony *et al.* (2000), D'Hoker and Freedman (2002), Klebanov (2000), Maldacena (2003), Mateos (2007), and Petersen (1999). The AdS/CFT correspondence originated from the Maldacena conjecture (1998). For other reviews on related topics, see Banks (1999), Bigatti and Susskind (1997, 2000b), Taylor (2001), and Witten (1998). On the quark-gluon plasma, see the popular-level accounts by Riordan and Zajc (2006) and by Blau (2005). Recent technical references are Son and Starinets (2007), Liu, Rajagopal, and Wiedemann (2006), and Herzog *et al.* (2006).

Covariant quantization and string interactions are discussed in the string theory textbooks cited before. Readers who wish to go beyond our basic discussion of the Virasoro operators may look at the lecture notes by Ginsparg (1991), which give an efficient introduction to conformal field theory. A readable book on Riemann surfaces is that by Springer (1981). For a construction of all Riemann surfaces without short possible strings, see Zwiebach (1993). Our discussion of the modular group follows the book by Serre (1973).

References

Abel, S. A., Barbon, J. L., Kogan, I. I., and Rabinovici, E. (1999). String thermodynamics in D-brane backgrounds, *J. High Energy Phys.* **9904**, 015 [arXiv:hep-th/9902058].

Aharony, O., Gubser, S. S., Maldacena, J. M., Ooguri, H., and Oz, Y. (2000). Large N field theories, string theory and gravity, *Phys. Rep.* **323**, 183 [arXiv:hep-th/9905111].

Aldazabal, G., Ihanez, L. E., Quevedo, F., and Uranga, A. M. (2000). D-branes at singularities: a bottom-up approach to the string embedding of the standard model, *J. High Energy Phys.* **0008**, 002 [arXiv:hep-th/0005067].

Ambjorn, J., Makeenko, Y. M., Semenoff, G. W., and Szabo, R. J. (2003). String theory in electromagnetic fields, *J. High Energy Phys.* **0302**, 026 [arXiv:hep-th/0012092].

Angelantonj, C. and Sagnotti, A. (2002). Open strings, *Phys. Rep.* **371**, 1 [Erratum **376**, 339 (2003)] [arXiv:hep-th/0204089].

Atkins, P. W. (1994). *The Second Law*. New York: Scientific American Library.

Bachas, C. P. (1998). Lectures on D-branes, arXiv:hep-th/9806199.

Banks, T. (1999). TASI lectures on matrix theory, arXiv:hep-th/9911068.

Bartlett, J. (1919). *Familiar Quotations*. Boston: Little, Brown, and Company.

Becker, K., Becker, M., and Schwarz, J. H. (2007) *String Theory and M-Theory*. Cambridge: Cambridge University Press.

Begelman, B. and Rees, M. (1998). *Gravity's Fatal Attraction*. New York: Scientific American Library.

Bekenstein, J. D. (2001). The limits of information, *Stud. Hist. Philos. Mod. Phys.* **32**, 511 [arXiv:gr-qc/0009019].

Bialynicki-Birula, I. (1983). Non-linear electrodynamics: variations on a theme of Born and Infeld. In *Quantum Theory of Fields and Particles*, ed. B. Jancerwicz and J. Lukierski. Singapore: World Scientific.

Bigatti, D. and Susskind, L. (1997). Review of matrix theory, arXiv:hep-th/9712072.

(2000a). Magnetic fields, branes and noncommutative geometry, *Phys. Rev. D* **62**, 066004 [arXiv:hep-th/9908056].

(2000b). TASI lectures on the holographic principle, arXiv:hep-th/0002044.

Blau, S. K. (2005) A string-theory calculation of viscosity could have surprising applications. *Phys. Today* **58N5**:23.

Blumenhagen, R., Goerlich, L., Kors, B., and Lüst, D. (2000). Noncommutative compactifications of type I strings on tori with magnetic background flux, *J. High Energy Phys.* **0010**, 006 [arXiv:hep-th/0007024].

Bousso, R. and Polchinski, J. (2007). Quantization of four form fluxes and dynamical neutralization of the cosmological constant. *J. High Energy Phys.* **0006**:006 [arXiv: hep-th/0004134].

Callan, C. G. and Maldacena, J. M. (1998). Brane dynamics from the Born–Infeld action, *Nucl. Phys. B* **513**, 198 [arXiv:hep-th/9708147].

Chiaverini, J., Smullin, S. J., Geraci, A. A., Weld, D. M., and Kapitulnik, A. (2003). New experimental constraints on non-Newtonian forces below 100-mu-m, *Phys. Rev. Lett.* **90**, 151101 [arXiv:hep-ph/0209325].

Cremades, D., Ibanez, L. E., and Marchesano, F. (2002a). Intersecting brane models of particle physics and the Higgs mechanism, *J. High Energy Phys.* **0207**, 022 [arXiv:hep-th/0203160].

(2002b). More about the Standard Model at intersecting branes, arXiv:hep-ph/0212048.

(2003). Yukawa couplings in intersecting D-brane models, *J. High Energy Phys.* **0307**, 038 [arXiv: hep-th/0302105].

Cvetic, M., Shiu, G., and Uranga, A. M. (2001). Three-family supersymmetric standard like models from intersecting brane worlds, *Phys. Rev. Lett.* **87**, 201801 [arXiv:hep-th/0107143].

Das, S. R. and Mathur, S. D. (2001). The quantum physics of black holes: results from string theory, *Annu. Rev. Nucl. Sci.* **50**, 153 [arXiv:gr-qc/0105063].

Denef, F., Douglas, M. R., and Kachru, S. (2007). Physics of String Flux Compactifications, *Ann. Rev. Nucl. Part. Sci.* **57**, 119 [arXiv: hep-th/0701050].

Deo, N., Jain, S., and Tan, C. I. (1990). The ideal gas of strings, *Bombay Quant. Field Theory* **1990** 112–148 (scanned version: KEK library).

Deser, S. (1982). Off-shell electromagnetic duality invariance, *J. Phys. A: Math. Gen.* **15** 1053.

D'Hoker, E. and Freedman, D. Z. (2002). Supersymmetric gauge theories and the AdS/CFT correspondence, arXiv:hep-th/0201253.

Dienes, K. R. (1997). String theory and the path to unification: a review of recent developments, *Phys. Rep.* **287**, 447 [arXiv:hep-th/9602045].

Douglas, M. R. (1996). Superstring dualities, Dirichlet branes and the small scale structure of space, arXiv:hep-th/9610041.

Frampton, P. H. (1974). *Dual Resonance Models*. Frontiers in Physics Series. Benjamin.

Frey, A. (2005). http://theory.caltech.edu/~frey/ph135/.

Gaiotto, D. and Rastelli, L. (2003). Experimental string field theory, *J. High Energy Phys.* **0308**, 048 [arXiv:hep-th/0211012].

Gell-Mann, M. (2002). *The Quark and the Jaguar.* New York: Henry Holt and Company, LLC.

Gibbons, G. W. (2003). Aspects of Born–Infeld theory and string/M-theory, *Rev. Mex. Fis.* **49S1**, 19 [arXiv:hep-th/0106059].

Ginsparg, P. (1991). Applied conformal field theory, arXiv:hep-th/9108028.

Giveon, A. and Kutasov, D. (1999). Brane dynamics and gauge theory, *Rev. Mod. Phys.* **71**, 983 [arXiv:hep-th/9802067].

Giveon, A., Porrati, M., and Rabinovici, E. (1994). Target space duality in string theory, *Phys. Rep.* **244**, 77 [arXiv:hep-th/9401139].

Goddard, P., Goldstone, J., Rebbi, C., and Thorn, C. B. (1973). Quantum dynamics of a massless relativistic string, *Nucl. Phys. B* **56**, 109.

Green, M. B., Schwarz, J. H., and Witten, E. (1987). *Superstring Theory.* Vol. 1: *Introduction*, Vol. 2: *Loop Amplitudes, Anomalies and Phenomenology.* Cambridge: Cambridge University Press.

Greene, B. (2000). *The Elegant Universe.* New York: Vintage Books.

Gross, D. J. (1999). Twenty five years of asymptotic freedom, *Nucl. Phys. Proc. Suppl.* **74**, 426 [arXiv:hep-th/9809060].

Hanany, A. and Witten, E. (1997). Type IIB superstrings, BPS monopoles, and three-dimensional gauge dynamics, *Nucl. Phys. B* **492**, 152 [arXiv:hep-th/9611230].

Hartle, J. B. (2003). *Gravity.* San Francisco, CA: Addison-Wesley.

Harvey, J. A. (1996). Magnetic monopoles, duality, and supersymmetry, arXiv:hep-th/9603086.

Herzog, C. P., Karch, A., Kovtun, P., Kozcaz, C., and Yaffe, L. G. (2006). Energy loss of a heavy quark moving through N = 4 supersymmetric Yang-Mills plasma, *JHEP* **0607**, 013 [arXiv:hep-th/0605158].

Horowitz, G. T. and Polchinski, J. (1997). A correspondence principle for black holes and strings, *Phys. Rev. D* **55**, 6189 [arXiv:hep-th/9612146].

   (2006). Gauge/gravity duality, arXiv: gr-qc/0602037.

Hoyle, C. D. (2003). The weight of expectation, *Nature* **421**, 899.

Ibanez, L. E., Marchesano, F., and Rabadan, R. (2001). Getting just the Standard Model at intersecting branes, *J. High Energy Phys.* **0111** 002 [arXiv:hep-th/0105155].

Jeong, E. and Smoot, G. F. (2005). Search for cosmic strings in CMB anisotropies, *Astrophys. J.* **624**, 21 [astro-ph/0406432].

Johnson, C. V. (2003). *D-branes.* Cambridge: Cambridge University Press.

Kaku, M. (2000). *Strings, Conformal Fields and M-theory.* New York: Springer.

Kapner, D. J., Cook, T. S., Adelberger, E. G., Gundlach, J. H., Heckel, B. R., Hoyle, C. D., and Swanson, H. E. (2007). Tests of the gravitational inverse-square law below the dark-energy length scale, *Phys. Rev. Lett.* **98**, 021101 [arXiv:hep-ph/0611184].

Kiritsis, E. (2007). *String Theory in a Nutshell.* Princeton, New Jersey: Princeton University Press.

Klebanov, I. R. (2000). TASI lectures: introduction to the AdS/CFT correspondence, arXiv:hep-th/0009139.

Kuti, J. (2005). Lattice QCD and string theory, arXiv: hep-lat/0511023.

Lerche, W., Schellekens, A. N., and Warner, N. P. (1989). Lattices and strings, *Phys. Rep.* **177**, 1.

Liu, H., Rajagopal, K., and Wiedemann, U. A. (2006). Calculating the jet quenching parameter from AdS/CFT, *Phys. Rev. Lett.* **97**, 182301 [arXiv:hep-ph/0605178].

Long, J. C. and Price, J. C. (2003). Current short-range tests of the gravitational inverse square law, arXiv:hep-ph/0303057.

Lüst, D. (2007) String Landscape and the Standard Model of Particle Physics, arXiv: hep-th/0707.2305.

Lüst, D. and Theisen, S. (1989). *Lectures on String Theory*. Berlin: Springer Verlag.

Maldacena, J. M. (1998). The large N limit of superconformal field theories and supergravity, *Adv. Theor. Math. Phys.* **2**, 231 (*Int. J. Theor. Phys.* **38**, 1113 (1999)) [arXiv:hep-th/9711200].

  (2003). TASI 2003 lectures on AdS/CFT, arXiv:hep-th/0309246.

Mandl, F. and Shaw, G. (1984). *Quantum Field Theory*. New York: Wiley.

Mateos, D. (2007). String Theory and Quantum Chromodynamics, arXiv: hep-th/0709.1523.

Mathur, S. D. (2005). The Fuzzball proposal for black holes: An Elementary review. *Fortsch. Phys.* **53**, 793 [arXiv: hep-th/0502050].

Ohmori, K. (2001). A review on tachyon condensation in open string field theories, arXiv:hep-th/0102085.

Ooguri, H. and Zin, Y. (1997). Lectures on perturbative string theories. In *Fields, Strings, and Duality*, *TASI 1996*, ed. C. Efthimiou and B. Greene, pp. 5–82. Singapore: World Scientific [arXiv:hep-th/9612254].

Ovrut, B. A. (2002). Lectures on heterotic M-theory, arXiv:hep-th/0201032.

Petersen, J. L. (1999). Introduction to the Maldacena conjecture on AdS/CFT, *Int. J. Mod. Phys. A* **14**, 3597 [arXiv:hep-th/9902131].

Polchinski, J. (1996). String duality: a colloquium, *Rev. Mod. Phys.* **68**, 1245 [arXiv: hep-th/9607050].

  (1998). *String Theory*. Vol. 1: *An Introduction to the Bosonic String*, Vol. 2: *Superstring Theory and Beyond*. Cambridge: Cambridge University Press.

  (2006) Cosmic String Loops and Gravitational Radiation, arXiv:0707.0888.

Polchinski, J., Chaudhuri, S., and Johnson, C. V. (1996). Notes on D-branes, arXiv: hep-th/9602052.

Polchinski, J. and Strominger, A. (1991). Effective string theory, *Phys. Rev. Lett.* **67**, 1681.

Quevedo, F. (1996). Lectures on superstring phenomenology, arXiv: hep-th/9603074.

  (2002). Lectures on string/brane cosmology, *Class. Quant. Grav.* **19**, 5721 [arXiv: hep-th/0210292].

Riordan, M. and Zajc, W. (2006). The first few microseconds, *Sci. Am.* **294N5**, 24.

Randall, L. (2005). *Warped Passages*. New York: Harper Collins.

Sagnotti, A. and Sevrin, A. (2002). Strings, gravity and particle physics, arXiv: hep-ex/0209011.

Scherk, J. (1975). An introduction to the theory of dual models and strings, *Rev. Mod. Phys.* **47**, 123.

Schnabl, M. (2006). Analytic solution for tachyon condensation in open string field theory, *Adv. Theor. Math. Phys.* **10**, 433 [arXiv: hep-th/0511286].

Schwarz, J. H. (2000). Introduction to superstring theory, arXiv:hep-ex/0008017.

(2003). Update on string theory, arXiv:astro-ph/0304507.

Schwarz, J. H. and Seiberg, N. (1999). String theory, supersymmetry, unification, and all that, *Rev. Mod. Phys.* **71**, S112 [arXiv:hep-th/9803179].

Seiberg, N. (1997). Why is the matrix model correct?, *Phys. Rev. Lett.* **79**, 3577 [arXiv: hep-th/9710009].

Seiberg, N. and Witten, E. (1999). String theory and noncommutative geometry, *J. High Energy Phys.* **9909**, 032 [arXiv:hep-th/9908142].

Seiberg, N., Susskind, L., and Toumbas, N. (2002). Strings in background electric field, space/time noncommutativity and a new noncritical string theory, *J. High Energy Phys.* **0006**, 021 [arXiv:hep-th/0005040].

Selem, A. and Wilczek, F. (2006). Hadron systematics and emergent diquarks, arXiv: hep-ph/0602128.

Sen, A. (1999). Non-BPS states and branes in string theory, arXiv:hep-th/9904207.

(2001). Recent developments in superstring theory, *Nucl. Phys. Proc. Suppl.* **94**, 35 [arXiv:hep-lat/0011073].

(2005). Tachyon dynamics in open string theory, *Int. J. Mod. Phys.* **A20**:5513 [arXiv: hep-th/0410103].

(2007). Black Hole Entropy Function, Attractors and Precision Counting of Microstates, arXiv:0708.1270.

Sen, A. and Zwiebach, B. (2000). Tachyon condensation in string field theory, *J. High Energy Phys.* **0003**, 002 [arXiv:hep-th/9912249].

Serre, J. P. (1973). *A Course in Arithmetic.* Graduate Texts in Mathematics. Berlin: Springer-Verlag.

Siegel, W. (1988). Introduction to string field theory, arXiv:hep-th/0107094.

Son, D. T. and Starinets, A. O. (2007). Viscosity, Black Holes, and Quantum Field Theory, *Ann. Rev. Nucl. Part. Sci.* **57**, 95 [arXiv:0704.0240].

Springer, G. (1981). *Introduction to Riemann Surfaces.* New York: Chelsea Publishing Company.

Strominger, A. and Vafa, C. (1996). Microscopic origin of the Bekenstein–Hawking entropy, *Phys. Lett. B* **379**, 99 [arXiv:hep-th/9601029].

Sundrum, R. (2005). TASI 2004 lectures: To the fifth dimension and back, arXiv: hep-th/0508134.

Susskind, L. (2006). *The Cosmic Landscape.* New York: Little, Brown and Company.

Tang, A. and Norbury, J. W. (2000). Properties of Regge trajectories, *Phys. Rev.* **D62**, 016006 [arXiv: hep-ph/0004078].

Taylor, W. (1998). Lectures on D-branes, gauge theory and M(atrices), arXiv: hep-th/9801182.

(2001). M(atrix) theory: matrix quantum mechanics as a fundamental theory, *Rev. Mod. Phys.* **73**, 419 [arXiv:hep-th/0101126].

Taylor, W. and Zwiebach, B. (2003). D-branes, tachyons, and string field theory, arXiv:hep-th/0311017.

Thorne, K. S. (1994). *Black Holes and Time Warps*. New York: Norton & Company.

't Hooft, G. (2004). Introduction to string theory. http://www.phys.uu.nl/~thooft/lectures/stringnotes.pdf

Uranga, A. M. (2003). Chiral four-dimensional string compactifications with intersecting D-branes, *Class. Quant. Grav.* **20**, S373 [arXiv:hep-th/0301032].

Vilenkin, A. and Shellard, E. P. S. (2000). *Cosmic Strings and Other Topological Defects*. Cambridge: Cambridge University Press.

Witten, E. (1998). New perspectives in the quest for unification, arXiv:hep-ph/9812208.
	(2002). Comments on string theory, arXiv:hep-th/0212247.

Zwiebach, B. (1993). Closed string field theory: an introduction, arXiv:hep-th/9305026.

Index

Dp-brane: Oscillators: α_n^i, $\ i = 2, 3, \ldots, p$ and α_n^a, $\ a = p+1, \ldots, d$

Ground states: $|p^+, \vec{p}\rangle$, $\quad \vec{p} = (p^2, \ldots p^p)$

Mass-squared: $\alpha' M^2 = -1 + \sum_{n=1}^{\infty} \left(\alpha_{-n}^i \alpha_n^i + \alpha_{-n}^a \alpha_n^a \right).$

N Dp-branes: Ground states $|p^+, \vec{p}\,; [j\,k]\rangle$, $\ j, k = 1, 2, \ldots, N$, $\ \vec{p} = (p^2, \ldots p^p)$

$$a_{\mathrm{NN}} = a_{\mathrm{DD}} = -\frac{1}{24}, \quad a_{\mathrm{ND}} = a_{\mathrm{DN}} = \frac{1}{48}$$

Kalb-Ramond coupling to strings and open string endpoints on D-brane:

$$S = -\int d\tau d\sigma \, \frac{\partial X^\mu}{\partial \tau} \frac{\partial X^\nu}{\partial \sigma} \, B_{\mu\nu} \left(X(\tau, \sigma) \right) + \int d\tau \, A_m(X) \, \frac{dX^m}{d\tau} \Big|_{\sigma=\pi}$$
$$- \int d\tau \, A_m(X) \frac{dX^m}{d\tau} \Big|_{\sigma=0}$$

T-duality: Winding: $w = \dfrac{mR}{\alpha'}$, $\quad$ Momentum: $p = \dfrac{n}{R}$, $\quad$ Dual radius: $\tilde{R} = \dfrac{\alpha'}{R}$

$$M^2 = p^2 + w^2 + \frac{2}{\alpha'}(N^\perp + \bar{N}^\perp - 2), \quad N^\perp - \bar{N}^\perp = \alpha' p w = mn$$

$$X = X_L + X_R \longrightarrow \tilde{X} = X_L - X_R$$

Brane Tensions: $\dfrac{T_p}{T_{p-1}} = \dfrac{1}{2\pi\sqrt{\alpha'}}$, $\quad T_1 = \dfrac{1}{2\pi\alpha'}\dfrac{1}{g}$

Open strings coupling to EM fields: $\mathcal{P}_m^\sigma + F_{mn}\partial_\tau X^n = 0$, $\ $ for $\sigma = 0, \pi$

Born-Infeld Lagrangian density: $\mathcal{L} = -T_p \sqrt{-\det\left(\eta_{mn} + 2\pi\alpha' F_{mn}\right)}$

Hagedorn Temperature: $\dfrac{1}{\beta_{\mathrm{H}}} = kT_{\mathrm{H}} = \dfrac{1}{4\pi\sqrt{\alpha'}}$

Black hole temperature: $k\bar{T}_{\mathrm{H}} = \dfrac{\hbar c^3}{8\pi G M}$, $\quad$ Bekenstein Entropy: $\dfrac{S_{\mathrm{B}}}{k} = \dfrac{A}{4\ell_{\mathrm{P}}^2}$

AdS/CFT parameter matching: $g = \dfrac{1}{4\pi} g_{\mathrm{YM}}^2 = \dfrac{\lambda}{4\pi N}$, $\quad \dfrac{R^4}{\alpha'^2} = g_{\mathrm{YM}}^2 N = \lambda$

Polyakov string action: $S = -\dfrac{1}{4\pi\alpha'} \int d\tau d\sigma \, \sqrt{-h} \, h^{\alpha\beta} \partial_\alpha X^\mu \partial_\beta X^\nu \eta_{\mu\nu}$